A guide to the

Ordnance Survey

1:25,000 First Series

Dedicated to the memory of

Edward Christie Willatts O.B.E.

born 4 July 1908, died 8 January 2000

influential advocate and constructive critic

of postwar Ordnance Survey mapping

A guide to the Ordnance Survey 1:25,000 First Series

Roger Hellyer

with an introductory essay by
Richard Oliver

London
The Charles Close Society
2003

First published in 2003 by The Charles Close Society for the Study of Ordnance Survey Maps,
c/o The Map Library, British Library, 96 Euston Road, London NW1 2DB

Text © Roger Hellyer, 2003
Introductory essay © Richard Oliver, 2003

Quotations from unpublished Ordnance Survey documents are Crown copyright

Maps and diagrams are reproduced with the authority of the Ordnance Survey under permit number NC/00/1340

Colour plates 6 and 7 are reproduced by permission of the British Library (Maps GSGS 4627)

All rights reserved

No part of this work may be reproduced, stored in a retrieval system, or transmitted, in any form or by any means, electronic, photographic, electrostatic, recorded or otherwise without the prior written permission of the publisher and copyright owners

A catalogue record for this book is available from the British Library

ISBN 1-870598-20-2

Typeset by the author

Printed and bound by Bookcraft (CPI Group), Midsomer Norton

Colour printing by Oxford Cartographers, Oasis Park, Eynsham, Oxford OX8 1TP

Contents

List of illustrations   vi

Acknowledgements   vii

Introductory essay by Richard Oliver

    The antecedents and development of the Ordnance Survey 1:25,000 First Series map   1

Compiler's notes

    The layout described   54

    List of airfields   58

    The military list   60

    Specification changes   61

Index diagrams

    Regular Edition sheets   45

    England and Wales   64

    Scotland   66

Cartobibliography   74

Appendices

    Isle of Man   301

    GSGS 4627, later M821: Great Britain military edition   302

    GSGS 4627A: Great Britain military edition, coloured with layers   333

    Cross references to other issues based on the 1:25,000 First Series   334

Colour plates

Illustrations

Monochrome

1	Ordnance Survey leaflet describing the eleven Regular Edition sheets	45
2	Index diagram of England (west) and Wales	64
3	Index diagram of England (east)	65
4	Index diagram of Scotland and England (north)	66
5	The evolution of the frame of the three-colour map (34/52, later SD 52)	67
6	Legend types 1 and 2 (41/67 and 41/02)	68
7	Legend types 3 and 4 (34/95 and 41/55)	69
8	Legend types 5 and 6 (SX 45 & Part of SX 44)	70
9	Legend type 7 (*Isles of Scilly*)	71
10	Depiction of airfields (SU 07)	72
11	The application of stipple to indicate building development (SU 65)	73

Colour plates

1	Sheet 63/20, edition 30046, 1946, showing the city of Norwich
2	Sheet 4009, [1944]
3a,b	Sheet 40/09, in three colours, 1944, and four colours, 1945
4	Regular Edition sheet SX 45 & Part of SX 44, edition A, 1956
5	Regular Edition sheet SX 45 & part of SX 44, edition B/*/, 1967
6	GSGS 4627[A] sheet 41/33, edition A, 1951, with layer colouring
7a,b	Layer and legend boxes on GSGS 4627 sheet 41/55, 1949
8	Sheet NY 30, edition C, 1959, showing the Lake District countryside around Ambleside

Acknowledgements

I would like to extend especial thanks to Nick Millea and his staff in the Map Room of the Bodleian Library, Oxford, where the basic research for this cartobibliography was undertaken. But the Bodleian, as with other copyright libraries, for the most part only possesses those editions which appeared in the Ordnance Survey Publication Reports, as well as a generous supply of the military edition. The gaps were filled from other substantial collections, for which reason I would like also to thank the staff in the map rooms of the British Library, the Public Record Office and the Library of Congress, Washington D.C., also Anne Taylor (Cambridge University Library), Peter Milne (National Library of Scotland), Pauline Hull and Pete Clark (Ordnance Survey Record Map Library), Nicholas Hutchings (MOD Map Library), Francis Herbert and David McNeill (Royal Geographical Society (with the Institute of British Geographers)), and, in university geography departments, James Peart (Birmingham), Lilian Sherwood (Bristol), Richard Oliver (Exeter), Mike Shand (Glasgow), Chris Perkins (Manchester), Ann Rooke (Newcastle upon Tyne), Linda Atkinson (Oxford), Val Clinging (Sheffield). My thanks also go to several members of the Charles Close Society who granted me access to their private collections, to Ian Mumford for guidance on matters military, Yolande Hodson for her care and diligence in overseeing the illustrations into print, and David Archer and Richard Dean, both of whom generously permitted me to work through their sales stocks in my own home. I am further indebted to David for the time and trouble he took in proof reading, and weeding out errors in the lists. Any that remain are of course the responsibility of the compiler. I am grateful to Chris Higley for providing such admirable index diagrams, and finally to Richard Oliver, as ever unfailingly ready with assistance and advice whenever asked, and who has so ably set the history of this series into its wider context in his valuable introductory essay.

The antecedents and development of the Ordnance Survey 1:25,000 First Series map

Richard Oliver

'Respice, aspice, prospice': a summary and some preliminary considerations

This essay is not about a much-loved map series evoking warm memories of a time when the world seemed to get along better than it does now. Rather is it the story of a map series which at times only survived grudgingly, and which, reportedly, was described by a serving Ordnance Survey officer at a meeting of the British Cartographic Society some time in the late 1960s as 'this awful map'.[1]

Great Britain and Ireland were first fully covered by mapping at the scale of 1:25,000 (4 centimetres to 1 kilometre; 2.5344 inches to 1 mile; colloquially 'the two-and-a-half-inch map') in 1940-41, at the height of World War II. The mapping, series GSGS 3906, was produced by direct photo-reduction from the six-inch (1:10,560) scale, in response to a threat of invasion.[2] Some mapping at this scale for military use had been prepared earlier, but it only covered a limited part of Britain, and a proposal to extend it for civil use had not yet been realised. The improvised mapping proved to be of use in planning post-war reconstruction, and in 1943 work began on a successor for civil use, which by 1956 covered about 80 per cent of Britain: all but most of the Scottish highlands and islands. (It also excluded the Isle of Man, for which, since the 1870s, Britain has only taken responsibility for mapping at scales of 1:50,000 and smaller, except during World War II, when it was mapped in GSGS 3906.) This mapping, in 2,027 sheets, was known at first as the 'Provisional Edition', and from 1968 as the 'First Series'. Its particular distinction is that it is the only series at 1:25,000 or a comparable scale with national or near national cover to have been compiled and drawn wholly independently of larger scales. In 1948 work began on its successor, at first known as the 'Regular Edition', and then later as the 'Second Series' and the 'Pathfinder Series': in 1989 the last of the Provisional Edition sheets were replaced by Pathfinders.

The 1:25,000 came into being because there was believed to be a need for mapping at this scale which could not be met by either of the two nearest scales published by the OS: the six-inch (replaced from 1969 by the 1:10,000), and the one-inch (1:63,360, replaced in 1974-6 by the 1:50,000). That sounds like a platitude, and needs amplification. In British conditions, the 1:25,000 has the particular attribute that it is about the smallest scale at which it is possible to

---

[1] I owe this to Peter Clark: so far, neither the officer nor the occasion have been identified.
[2] GSGS: Geographical Section, General Staff.

show rural field boundaries with clarity, and in built-up areas it allows the general character of building - terraces, semi-detached, large detached, etc - to be indicated reliably. It also enables landforms to be shown in some detail (one millimetre on the map equals 25 metres on the ground), and over Britain as a whole is suitable for contouring at intervals of either 25 feet (7.62 metres) or 5 metres. As such, the scale has obvious attractions for users such as walkers who seek to cover a considerable tract of country, and who wish for a map which combines relative compactness with the depiction of details such as field boundaries, and to 'indoor' users such as geographers and planners who similarly need detailed information allied to ease of handling. This scale, or the similar one of 1:20,000, has also been found useful in the past for military purposes, more especially artillery firing by 'map shooting': the 1:20,000 scale was first used for this purpose by the British Army late in 1914, and during World War I both artillery techniques and mapping practices were refined. At a scale of 1:20,000, twenty metres on the ground is represented by one millimetre on the map, and this is roughly equal to paper distortion relative to the squaring or grid systems commonly used at this or similar scales; the ideal is a scale at which paper distortion does not affect the purpose to which the map is being put.[3] Although suitable for artillery use, the 1:20,000 was replaced in official favour by the 1:25,000 *circa* 1930, and it went largely out of military favour, at any rate in Britain, in the 1960s, and a map which had started with a primarily military justification perforce had to find a civil one were it to survive. That it did survive, and that more recently OS have been prepared to invest money in its latest form, the Explorer Series, is indicative that there is indeed a significant civil need for mapping at this scale.

This book is concerned with the Provisional Edition or First Series (for consistency the former title will be favoured in this essay), but it is in the nature of national topographic mapping that it does not suddenly arise, fully formed both in conception and cover: rather, it has predecessors and antecedents which inevitably have their effect on its origin and design, if only by reaction rather than imitation. For this reason, a substantial part of this essay is devoted to the antecedents of the Provisional Edition before 1939. Similarly, it is almost unknown for a national map series even at 1:50,000 or a similar scale to be replaced overnight: the recasting of the old 1:25,000 Second Series as the Explorers is expected to take about six years, even with the help of digital technology. The replacement of the Provisional Edition was a protracted and involved process, so much so that the first eleven 'Regular Edition' sheets themselves became part of the First Series, and a subject rather than an instrument of replacement. For this reason, another substantial part of this essay is concerned with the schemes for the replacement of the Provisional Edition.

The military predecessors: GSGS 3036, 2748 and 3906, 1910-39

Early precedents
Although the 1:25,000 was only adopted officially *circa* 1930-31, mapping at broadly related scales had been produced for military purposes for as long as parts of Britain had been subject to official mapping of extensive tracts of land. The military survey of Scotland of 1747-55 was at a scale of 1 inch to 1000 yards (1:36,000), and, after limited areas had been surveyed at the six-inch scale (1:10,560), the military survey of Great Britain begun in earnest in 1795 was made first at three inches to one mile (1:21,120) and then at two inches to one mile (1:31,680). These surveys were originally made for military use rather than for general publication, and often omitted precise field boundary information. After 1815, when the military justification ceased and they were intended purely to provide a basis for a one-inch (1:63,360) map, surveying at the two-inch scale was retained, and the two-inch drawings often contain minor names and other details which do not appear on the published one-inch maps. Surveying at the two-inch scale was abandoned in favour of the six-inch in 1840, but later in the century a number of maps at the two-inch and three-inch scales were produced of areas of particular military interest, for example a three-inch map of

---

[3] Peter Chasseaud, *Artillery's astrologers: a history of British survey and mapping on the Western Front 1914-1918*, Lewes, Mapbooks, 1999, 49.

the Medway of 1863, which was derived from 1:2500 survey.[4] However, the majority of these appear to have been obtained by direct photo-enlargement of one-inch material, with some additions.[5] For example, the two-inch map in two sheets of the Aldershot area of 1889 has several additional names and shows relief by close form-lines (in brown) as well as by contours and spot-heights. The two-inch Cannock Chase manoeuvre map of 1894, also enlarged from one-inch material (but this time derived from a drawn rather than an engraved original), has hill-shading and seems to be the first OS map on which appear the now familiar symbols for churches with towers and spires. Several other two-inch manoeuvre maps were produced at this time, but in the nature of things they were ephemeral, rather than contributions to what today would be called 'core mapping'. There were also occasional unofficial productions: for example, Messrs Gale and Polden of Aldershot published their three-inch scale *Military map of Aldershot and surrounding country*, 'From Special Survey … for Gale and Polden': the map omitted field boundaries and, whatever might have been gleaned specially for the publishers, the 100-foot (30.5 metre) contour intervals suggest that OS material made a more substantial contribution.[6]

The new church symbols were a result of a War Office Committee on the Military Map of the United Kingdom, which had sat in April 1892. The most tangible outcome of this committee's deliberations was the redesign of the New Series one-inch map, but the evidence given indicated some military desire for mapping at a scale or scales intermediate between one-inch and six-inch. Such mapping might be national, but it might possibly be of more limited areas: either the south-east of England or the vicinity of important military centres.[7] In the 1890s, when the authorisation of cyclic revision of OS mapping was offset by the withdrawal of state support for mapping at scales larger than 1:2500, any significant extension of official mapping was unlikely: the only new map to be published in Britain was the two-inch of Salisbury Plain of 1898. Indeed, when in 1902 national cover by a new scale was introduced, and at military behest, it was the half-inch (1:126,720). The South African War had proved a chastening experience for the British military, and maps were not excluded. The expectation was that future wars would be comparatively mobile, and thus would call for mapping which combined a suitably large area with compactness and portability.

Within a decade, there was returning military interest in a scale intermediate between one-inch and six-inch, at any rate for a limited area. According to Brigadier H.S.L. Winterbotham (Director-General of the Ordnance Survey (DGOS) 1930-35), 'The question began to be active in 1910'.[8] The result was the mapping of a large part of eastern England at a just such a scale: 1:25,344 or the true two and a half inches to one mile. Indeed, there were two series: one, planned in six sheets to cover London (GSGS 2525), seems to have been produced in 1911 as a map that was so 'secret' that no copies are known to survive: the other, called later 'Map of East Anglia' and numbered GSGS 3036, is one of the more obscure OS map series.[9]

The 'Map of East Anglia' (or Suffolk and adjoining parts), 1914 (or 1911?)

Earlier references in print to this map gave the date as 1914 or later, and say or imply that it was overtly defensive in its nature.[10] There is no reason to doubt this, but there is equally no doubt

---

[4] Copy in Public Record Office (PRO): WO 78/596.
[5] For a list, see Tim Nicholson, 'The Ordnance Survey and smaller scale military maps of Britain 1854-1914', *Cartographic Journal*, 25 (1988), 109-127. This is not complete: for example, it omits the Medway map cited above.
[6] The British Library Map Library has examples dated 1888 (at Maps 2560 (14)) and 1893 (at Maps 2560 (7)).
[7] *Report of committee on a military map of the United Kingdom*, London, War Office, 1892, pp 12, 13, 15, 17, 19, 20, 21, 24, 25; see also tabulated answers to circular.
[8] H.S.L. Winterbotham, 'Sidelights', (MS in Ordnance Survey library, Southampton), 177: he cites two files, in the War Office, 81/223 (M.O.4) 8/12/10, and in the OS, A/237/4/101c, both now lost.
[9] GSGS 2525 is listed in a handwritten register of GSGS numbers: there is a photocopy of the original in the Map Room in the Public Record Office.
[10] J.B. Harley, *Ordnance Survey maps: a descriptive manual*, Southampton, Ordnance Survey, 1975, 91 [this seems to be the first reference in print to the series]; Peter Clark, 'Maps for the army: the Ordnance Survey's

that this mapping was originally prepared in anticipation of war, rather than as a result of that which actually broke out.[11] Winterbotham wrote in 1934:

> ... parties from the Ordnance Survey were used.... East Anglia was considered a possible area for invasion. It became a matter of War Office policy to pay for the remapping of this area by the Ordnance Survey at the scale of 2½ inches to the mile. The map was to carry secret information and a recontouring which was also to be secret. There was no reason why the State Survey should shoulder the burden... The War Office paid, and the sheets were not 'published'. It is interesting to note however that the original area of resurvey was, afterwards, hurriedly enlarged by the rapid photographic reduction of the six inch showing no secret information and suitable (if required) for subsequent national use.[12]

Most of the first 36 sheets carry magnetic variation data for 1911. They appear to have been first printed in August and September 1914 from plates which had been prepared some time before.[13] As 'preparatory' rather than 'consequence' mapping they have much in common with the 1:100,000 and 1:80,000 mapping of Belgium and northern France prepared by the War Office and OS which was in hand by 1909, and of which large stocks were available in 1914, in anticipation of a war of rapid movement which failed to materialise. Otherwise, the mapping differs totally in that its evident purpose was for defensive operations in the event of a hostile invasion, although it would have been 'eminently suitable' as an artillery map.[14]

Although its sheet lines derived from the one-inch, being quarters of the Third Edition (Large Sheet Series), the *Map of East Anglia* (as it was called on the cover; there was no series title on the map face) was fundamentally an offshoot from the six-inch. Indeed, the linework was photo-reduced directly, though this operation was effectively disguised by rewriting the names, using the same fonts as those used on the six-inch. (In the 1880s the later sheets of the six-inch first edition had been produced by a similar process: 1:2500 linework was used for the smaller scale, and the names were typed separately. A similar procedure would be used in 1962-89 with the 1:25,000 Second Series maps.) The maps were printed in six colours: outline and names in black, water in blue, contours in brown, woodland in green, roads in burnt sienna, and some military information in red. With the exception of the red, the style has a particular affinity with the six-inch 'War Game' maps of the Reading-Wallingford-Marlow area produced in 1900-01, and a more general affinity with contemporary OS coloured smaller-scale mapping.[15] The maps measured 33.75 by 22.5 inches (85.76 by 57.17 cm) within the neat line, which was significantly larger than the 27 by 18 inches (68.61 by 45.74 cm) standard for one-inch and half-inch maps at this time. The map was covered by 2½-inch squares in black, providing a crude referencing system similar to that on contemporary one-inch and half-inch maps. The 'confidential' information not found on standard civil OS mapping included watering places for men, horses and traction engines, the routes of water mains, the construction of bridges and the suitability of church towers and hill-tops as observation points. Contours were at 10 feet (3.28 metres) intervals. As well as the military information, there was some limited piecemeal revision.[16] The apparent oddity of the lack of a GSGS number on these '1914' sheets can perhaps be explained by the mapping having been prepared by the OS, albeit at WO behest; in this it was similar to the

---

contribution', *Sheetlines* 7 (October 1983), 2-6; [Y. Hodson in] W.A. Seymour, ed., *A history of the Ordnance Survey*, Folkestone, Dawson, 1980, 263.

[11] I do not now wish to defend the suggestion made in *Sheetlines* 36 that it was otherwise (Richard Oliver, 'Episodes in the history of the Ordnance Survey 1:25,000 map family', *Sheetlines* 36 (April 1993), 1-27).

[12] Winterbotham, 'Sidelights', 168, 177.

[13] No copies were sent to the legal deposit ('copyright') libraries; there is an incomplete set formerly in the Ministry of Defence (Military Survey) (MoD) collection, now being transferred to the British Library Map Library (BLML), from the accession dates of which the date of printing has been inferred, and other copies survive in ones and twos in a few private collections. All information deriving from sets of maps formerly or still in the MoD collection is the work of Roger Hellyer.

[14] Chasseaud, *Artillery's astrologers*, 3, 6-10, 12, 15.

[15] See note by Bob McIntosh in *Sheetlines* 21 (April 1988), 22.

[16] For example, the addition of Radio Telegraph Station in square 7E on Sheet 85 N.E.

half-inch, which differed in that it lacked 'confidential' information and was offered for sale. The Geographical Section, General Staff only had a very small complement at this time, and thus much 'military mapping' was prepared by the OS, which also enjoyed the advantage of powers of entry to land under the Survey Act, 1841.

The 'Map of the Eastern Counties', 1914-1940, GSGS 3036

27 more 1:25,344 sheets were prepared in 1915-17, covering the London area and the country north of a line from Reigate to Dover: the covers retained the series title *Map of East Anglia*. In about March 1915 the series was allocated the series number GSGS 3036, and the first group of additional sheets was evidently prepared shortly afterwards.[17] Although photo-reduction from the six-inch was retained, the cartographic style was considerably modified. The most obvious change was in the writing of the names, now mostly in Egyptian, as on the better known contemporary Western Front maps. The contour interval was doubled to 20 feet, and the additional 'military' information characteristic of the 1911-14 *Map of East Anglia* was omitted; it was this later mapping which Winterbotham described in 1934 as suitable for 'national' (i.e. civil) purposes. The road classification was adapted, where possible, from that of the 'third revision' one-inch map which we now know as the Popular Edition, although it was less elaborate and probably more practical than that published at the smaller scale.[18] The 2½-inch squaring was now printed in red, and marginal instructions were provided so that more sophisticated references could be given, as on Western Front mapping. Although certainly six and probably ten more sheets were prepared in 1918-19, they were apparently only printed in 1925, when all 73 sheets were printed with the squaring replaced by the British System grid in purple. In 1931-3, the series was renamed the *Map of the Eastern Counties* and all the sheets were republished with the Modified British System grid, as part of the conversion operation which also gave birth to GSGS 3906, described later.[19] These post-1931 sheets omitted the green woods plate, thus coming into line both with GSGS 3906 and contemporary printings of the military edition of the one-inch (GSGS 3907, England and Wales; GSGS 3908, Scotland). GSGS 3906 superseded GSGS 3036 in 1940.

1:20,000 mapping, 1915-1933: GSGS 2748

Although GSGS 3036 had apparently made a good start in providing Britain with a topographic map intermediate in scale between one-inch and six-inch, further developments were rather different in nature, at any rate in the shorter term. Already by late September 1914 the war on the Western Front had become relatively static, and 1:20,000 mapping (GSGS 2742) was put in hand for artillery use and 'map shooting'. This was initially by enlargement from the French 1:80,000 or by copying the Belgian 1:20,000; subsequently much of the mapping was resurveyed and all of it was redrawn in the 'Egyptian' style of later GSGS 3036 sheets. A common drawing served for 1:10,000, 1:20,000 and 1:40,000 scales.[20] By late 1914 1:20,000 maps of military training districts in Britain were being prepared (initially as GSGS 2742; from early 1915 as GSGS 2748), by photographic reduction of the OS six-inch, and with interpolated metric contours and the Western Front squaring reference system added; imperial contours and spot heights were omitted.[21] The resulting map looked different both from GSGS 3036 and from the mature Western Front GSGS 2742, which was wholly redrawn, and printed in several colours. The question of sheet numbering and meridian was effectively side-stepped by identifying each sheet

---

[17] The date of allocation is by analogy with the dates of Western Front series cited in Chasseaud, *Artillery's astrologers*, 67, and the styles of mapping in the same author's *The topography of Armageddon*, Lewes, Mapbooks, 1991.

[18] Preparation of this mapping had begun in 1911-12, but the first sheets were only printed in 1918: see Yolande Hodson, *Popular maps,* London, Charles Close Society, 1999, esp. 134-7.

[19] Proof copies of six of the 1918-19 sheets are in the ex-MoD collection now transferred to BLML.

[20] Chasseaud, *Artillery's astrologers*, 27, 30, 31, 48, 49, 61.

[21] Chasseaud, *Artillery's astrologers*, 40; Peter Chasseaud, 'The development of artillery squares and artillery training maps of the U.K. 1914-1918', part I, *Sheetlines* 10 (August 1984), 2-8, and part II, *ibid*, 11 (December 1984), 12-14. A few of these 1:20,000 maps bore GSGS 2742.

by name only, and by making the squaring unique to each sheet or group of sheets. At first GSGS 2742 had a colour-scheme similar to the second group of GSGS 3036 sheets, with squaring in red and roads infilled in orange or 'burnt sienna'; this seems to reflect the thinking of a decade or more earlier, that one basic colour-scheme would serve for most colour-printed topographic mapping. From 1916 onwards the colouring was simplified, with the squaring now usually on the black plate (which itself was sometimes printed in grey), and the road and wood infills omitted. This three-colour map would need much less press-time for a given number of finished impressions, and this would have been an important consideration when stock was needed urgently, as it so often was. This reduction in colour made for a clearer general-purpose map (and at this scale road classification for transport purposes can hardly have been very important; one version of the 1:40,000 (GSGS 2743) retained the road and wood infills), and would have been more suitable for various sorts of overprint, notably trenches. By 1918 the military were fully convinced of the value of 1:20,000 mapping, and a notionally national series at this scale was inaugurated, retaining series number 2748. It was drawn by the Ordnance Survey but paid for by the War Office, and was laid out on a metric grid.[22] The grid was an adaptation of the French Lambert System, which British troops had been using in Italy since November 1917, and which was officially adopted by the Allies for the Western Front in June 1918; it was to have become operational in January 1919.[23]

The 'new' GSGS 2748 was a strong contrast to both the 'old' one and to the *Map of East Anglia*, and was much more akin to GSGS 2742. As with the latter series and with GSGS 3036, all lettering for cultural features was Egyptian, with italic reserved for water names and Roman for names of woods. Like GSGS 2742 in its later form, there was no colouring of roads or woods, and the map was printed in four colours: outline, grid lines and most names in black; water and water names in blue; contours in brown; and grid figures in red. (In this form, it was not quite so well suited to overprinting as had been GSGS 2742.) The sheets were laid out on the metric British System grid, on the Cassini projection and meridian of Dunnose, and were designed as a national series. Indeed, the only previous OS mapping of Great Britain to be laid out as a national map, with the sheets numbered in a single sequence, had been that at the ten-mile (1:633,600) and 1:1,000,000 scales. Sheets covered an area of 15 km by 10 km on the ground, and measured 75 by 50 cm (29.52 by 19.68 inches) within the neat line. This size may have been determined by the intended recasting of the Western Front mapping on Lambert grid sheet lines, based on 1:40,000 sheets measuring 75 by 50 cm within the neat line (i.e. equal to 30 by 20 km on the ground).[24] It was also fairly close to the 27 by 18 inch (68.61 by 45.74 cm) standard for the half-inch and one-inch, though GSGS 2748 had much wider margins.

The sheet layout and numbering had two peculiarities. One was that the sheet numbering ran from south to north, whereas the almost invariable OS practice was to number from north to south. The other was the method of numbering: each 1:20,000 sheet was one of 25 sub-sheets within a notional series of parent sheets, covering 75 by 50 km. The comparison with the arrangement of the 1:2500 by sub-sheets of the six-inch is irresistible, and suggests provision for 1:100,000 mapping to replace the half-inch for military use. Also peculiar, if not archaic, was the use of Roman numerals for the parent sheets and letters for the sub-sheets, so that 1:20,000 sheets were identified in the manner VI.G (rather than '6.7'), which suggests the less-than-critical adapting of six-inch practice, and strikes one as an unfortunate blemish on perhaps the most innovative and forward-looking topographical map ever to be introduced in Britain. Influence from the larger scales is also suggested by the use of the meridian of Dunnose, which was that employed for a large block of county six-inch and 1:2500 mapping; it would have been possible to plot the new grid in those counties (which included districts of military interest, including Aldershot and Salisbury Plain, which would be early candidates for 1:20,000 mapping) by adapting existing co-ordinates, and thereby reducing the computing work involved in extending

---

[22] Seymour, *A history*, 263.
[23] Chasseaud, *Artillery's astrologers*, 438-9.
[24] Chasseaud, *Artillery's astrologers*, 439.

the metric grid over the whole country. At a time of manpower shortages and general retrenchment such arguments would have been persuasive. Use of the Dunnose origin entailed retention of the Cassini projection, which, for an artillery map, was less than ideal as it distorted angles. The Transverse Mercator or Gauss Conformal projection, which would be adopted later, was free of angular distortion.

Between 1918 and 1930 at least 118 sheets were published in the 'British System' GSGS 2748, comprised of 12 redrawn numbered sheets designated 'Provisional Edition', 57 other redrawn numbered sheets, 21 Special Sheets (some varying from the standard specification), and 28 'Provisional Edition' which were direct photo-reductions from six-inch, with the addition of the grid in colour. (No sheet is known in both a six-inch photo-reduced and in a redrawn version.)[25] Most of the sheets covered military training areas. There was also a two-sheet special map of London (GSGS 3786), apparently first issued in 1926, which differed from the standard style in that building infill was omitted and the method of grid referencing was changed.[26] This map of London was almost immediately issued with a 'secret' overprint showing hospitals, police stations, etc, for military use during the General Strike and its aftermath.[27] The sheets were derived from the six-inch, with additional revision made for the one-inch Popular Edition. Larger blocks of building were hatched and smaller blocks were solid black. The more important streets in towns were named. Railways were shown by the same elongated chequer symbols as had been used on GSGS 2742 and other Western Front maps; the distinction between single and double-track lines was not very clear, but was nonetheless used for the civil 1:25,000 Provisional Edition introduced in the 1940s and finally replaced in 1989. (It was also used on the one-inch Fifth Edition, on which work began in 1928.) Contours were at 5 metre intervals, obtained by interpolation from contours and heights on the six-inch, and the six-inch and two-inch hill-sketches made for hachuring the one-inch. Symbols were used to show triangulation points on buildings (so that the depiction of churches differed fundamentally from that on the one-inch), and there were lists of trig. point co-ordinates in the right-hand margin. These last were to facilitate artillery operations ('map shooting' depended on precisely surveyed 'aiming points': church spires and factory chimneys were ideal), and were the direct outcome of Western Front experience. Also of military application was the inclusion of apparently minor detail. Mile posts and stones, with distances and sign-posts (by 'G.P.' for Guide Post, following long-standing OS practice) had their cartographic precedent in kilometre stones on Western Front mapping.[28] Pumps, wells and water taps ('W.T.') were shown when away from buildings. Boundary and similar stones were shown, except for War Department and Admiralty boundary stones. Minor detail shown on the six-inch which was omitted from the 1:20,000 included signal and mile posts on railways, and all boundaries except national ones.[29] The mile posts and stones on main roads and the wells continue to appear on the modern 1:25,000, presumably from habit rather than any reminiscence of military application. Although military in origin and bearing the note 'Not to be published', the War Office were quite prepared for the map to be put on sale, in order to offset the cost (and even printed a price on some sheets in *circa* 1920-23), but this was objected to by the OS; it has been suggested that this was possibly on the grounds that sales of the six-inch map would thereby be harmed.[30] This fear was probably groundless: being all-metric would hardly have helped sales at this time. Another objection would be that publishing such mapping for civil

---

[25] As with GSGS 3036, no copies were sent to the legal deposit libraries, and the only copies so far recorded are oddments in a few university or private collections, and the ex-MoD copies now in BLML.
[26] See Roger Hellyer, 'Some notes on the origin of the Modified British System of the War Office Cassini Grid', *Sheetlines* 55 (August 1999), 3-11.
[27] C. Board, 'The Secret Map of the County of London 1926, and its sequels', *London Topographical Record* 27 (1995), 257-80; Christopher Board, 'The three-inch Map of London and its predecessors of 1926', *Sheetlines* 43 (August 1995), 48-50.
[28] Chasseaud, *Artillery's astrologers*, 125.
[29] *Instructions for the preparation, drawing and typing of the G.S.G.S. 1:25,000 Series Great Britain*, Southampton, Ordnance Survey, 1933, 'For Departmental Use': copy in PRO OS 45/21.
[30] Seymour, *A history*, 263.

use would inspire demands for its extension to areas where the interest was preponderantly civil rather than military, at a time when there was insufficient money even for to maintain existing mapping in a proper state.

1:25,000: GSGS 3906

In 1931-33 GSGS 2748 was replaced by a 1:25,000 series, GSGS 3906. The change coincided with the metamorphosis of the Dunnose grid from the British System to the Modified British System. The reasons for the change of scale are unknown: it has been suggested that perhaps 1:25,000 was felt to be a better international standard than 1:20,000 to which to accustom troops.[31] The nearest that the writer has come to an explanation is the following tantalising reference by Major-General Geoffrey Cheetham in 1950:

> The scale of 1/25,000 was adopted for the Civil Map of [Great Britain] because the Military Edition at that scale was already in use. The scale had been adopted for the Military Map for reasons which, in the main, had no force if applied to the Civil Edition.[32]

One possible reason which has not been suggested hitherto is that the 1:25,000 could show a similar level of detail at a smaller scale and therefore be printed on smaller sheets of paper; more to the point, a map area of 60 by 40 cm - equivalent to 15 by 10 km at 1:25,000 - could just be fitted onto a sheet of paper suitable for a Demy press (maximum size 24.5 by 19.5 inches, about 62 by 49 cm), and therefore was therefore suitable for printing in the field by mobile units equipped with this size press. (It must be admitted that, as drawn, the printed area of a standard 1930s GSGS 3906 sheet was about 67 by 50 cm).[33] As will be seen later, such considerations were important when the sheet size of the 1:25,000 was under discussion between the OS and the War Office in 1956-7; in 1955 Brigadier J.C.T. Willis, then Director-General, claimed that the 10 by 10 km size of the Provisional Edition was 'originally geared to a Military requirement and the capacity of mobile printing machines'.[34]

Whether this is the true reason or not, it is indisputable that the first 54 sheets of GSGS 3906 were reissues of GSGS 2748 material. At least 174 sheets of GSGS 3906 were produced by the outbreak of war in 1939, and another 32 had been ordered by the War Office, some of which would have begun the replacement of GSGS 3036. There were also 29 'Provisional Edition' sheets covering eastern Yorkshire, published in 1934-9 and obtained by direct photo-reduction from the six-inch,; one was later redrawn as a standard sheet.[35] The change in scale was made the opportunity of minor changes in design. Red was replaced by purple for grid figures; purple had been standard since 1923 on military printings of the one-inch, which had supplanted the half-inch in military favour in 1922 because it was better suited to a 1 km grid.[36] Given the use of purple for grid lines as well as figures on GSGS 3036 and on military one-inch and smaller scale mapping, the retention of black grid lines on GSGS 3906 is odd; it is possible that this was on grounds of economy, rather than for design reasons. On new sheets, roads were drawn to standard conventional widths instead of nominally to scale. As a result, sheets drawn after 1930 for GSGS 3906 tend to have a more 'open' look compared with those previously drawn for GSGS 2748. The tables of trig. co-ordinates were now omitted.

Although the old 1:20,000 sheet lines were retained, the opportunity was taken to devise a new sheet numbering system, whereby each 15 by 10 km sheet was treated as a quarter of a parent 30 by 20 km sheet, numbered by an abbreviated grid reference to its south-west corner: thus sheet 32/10 NE denoted the north-east quarter of sheet 32/10, i.e. with a south-west corner 320 km east

---

[31] In conversation with Peter Clark and Ian Mumford.
[32] Cheetham to Brown, 28 June 1950, 158B in PRO OS 1/416.
[33] Paper size from Report, 26A in PRO OS 1/1192.
[34] Willis to Directorate of Military Survey (DMS), 4 October 1955: 25A in PRO OS 1/1123.
[35] Once again, no copies were sent to the legal deposit libraries, and the full extent of publication has had to be inferred from ex-MoD copies now in BLML and scattered holdings in university and private collections. Figures for sheets on order are derived from information in PRO OS 1/219.
[36] Hodson, *Popular maps*, 76.

and 100 km north of the false origin of the grid, from which it was possible to deduce that the south-west corner of sheet 32/10 NE was 335 km east and 110 km north. This was not as good as the grid-related system which would be devised for the civil 1:25,000 in the 1940s, but was certainly an improvement on the numbering system of GSGS 2748. The use of notional parent sheets divided into quarters was a convenient way of reconciling the existing 1:20,000 sheets with a grid-related sheet numbering system, and could have been the basis for a 1:50,000 series. Indeed, a few years later precisely this arrangement was adopted for mapping of Belgium and north-east France: the 1:25,000 series (GSGS 4041, in a style very similar to GSGS 3906, but with blue grid figures) was laid out as quarters of the 1:50,000 series (GSGS 4040), although a consecutive rather than a grid-related sheet numbering system was used.

No extant drawing instructions are known for the 1:20,000 series, GSGS 2748, but those for GSGS 3906 were printed in 1933, and several copies survive.[37] Sheets were drawn at 1:15,000 scale and were derived from the six-inch, with the addition of any later one-inch revision (which was surveyed at six-inch). The component six-inch sheets had the War Office Cassini (Dunnose) grid added to them in manuscript, together with any suitable one-inch revision, and were then fitted to a standard 1000-metre grid. (Although the instructions are silent on this point, analogy with the civil Provisional Edition begun in 1943 shows that 'shining-up' would have been necessary, i.e. making numerous small cuts in the six-inch mapping in order that grid intersections on it would be brought into sympathy with the master grid.)[38] The assembly was then photographed, producing a negative at 1:15,000, from which three prints were made - one each for outline, water and contours - on Whatman's paper in ferro-prussiate blue (which would be 'invisible' to further photography). These measured 39.370 inches (100.00 cm) by 26.247 inches (66.666 cm) within the neat line, and so were slightly larger (by a few centimetres) than a standard 1:2500 County Series sheet. The detail, water and contours were then drawn, in black, and the completed drawings photographed to produce three negatives at 1:25,000. A ferro-prussiate blue print of the 'detail' (black) plate at 1:25,000 was made, on which were typed the grid figures. Printing plates were produced from the negatives; it seems to have been usual practice for the OS to print the initial stock (often 1000 copies, though larger quantities were sometimes printed of areas of particular training interest), and then to pass the plates to the War Office for any reprinting.[39] A few special sheets were derived from air photography and were drawn in a rather different style, with uncased roads in solid black[40]

National cover and national use
In the early 1930s GSGS 3906 began, in a very small way, to serve civil wants. In 1933 there was published a four-sheet street map of London at the three-inch scale, using the drawings for GSGS 3786. The published map differed in a number of significant respects, including the use of a green plate and the civil Transverse Mercator yard grid. In 1933 there appeared the first of what was intended to be a set of six sheets of a series, *Celtic Earthworks of Salisbury Plain*. Only sheet 6, *Old Sarum*, went on sale: a second sheet, 4, *Amesbury*, was at proof stage in 1939 but went no further. The base-map was in grey and the archaeological information in red. Both omitted the military grid, and indications of their military origin. Old Sarum was printed in a run of 500 copies, and there was a reprint of 1,000 in 1937.[41]

---

[37] *Instructions for ... the G.S.G.S. 1:25,000...*, 1933: copy in PRO OS 45/21.

[38] 'Instructions for drawing the 1:25,000 map', mimeographed, September 1945, p.1: copy in PRO OS 45/34.

[39] Inferred from note on specimen of sheet 44/10 NW in PRO OS 1/355, that plates to WO on 2 May 1935, and from reprint codes on other sheets.

[40] That of Chatham was included in Lieut J.S.A. Salt, *A simple method of surveying from air photographs, Professional Papers of the Air Survey Committee* 8, London, HMSO, 1933 [another edition 1939]; others included the special sheets *West Down and Lark Hill* (1932) and *Imber* (1933). A 1:50,000 sheet of Aldershot in 'air survey style' was prepared in 1933 (GSGS 3949): copy in Bodleian Library, Oxford.

[41] Roger Hellyer, 'The archaeological and historical maps of the Ordnance Survey', *Sheetlines* 21 (April 1988), 6-15, esp. 10-11.

However, at the outbreak of war in September 1939 GSGS 3906 was still a training map, which, together with GSGS 3036, covered only about a seventh of Britain. A year later it was an operational map, produced in response to the threat of invasion, and covering the whole of Great Britain and Ireland by early 1941. The initiative for completing cover was taken by Major-General M.N.MacLeod, who had been Director-General of the Ordnance Survey since 1935 and who was a strong advocate of the 1:25,000 scale, as will be recounted later: the War Office was apparently not convinced of the necessity of this mapping for counter-invasion purposes. The new sheets were printed at the rate of 50 a day and the whole project was completed in about two months.[42] They were produced by adding the grid to the latest available six-inch mapping, compiling it into GSGS 3906 sheet lines, photo-reducing it to 1:25,000, and calling the result 'Provisional Edition'. Gaps between the component six-inch sheets are often visible on the printed maps. The result was a single colour map which was, to be charitable, not always very legible, but which had the merit that it was easy to reproduce by mobile field survey units. (Some pre-war sheets were reprinted in colour; some were converted to monochrome style; all were eventually replaced by Provisional Editions.) A '2nd Provisional Edition' of many sheets was produced by photo-enlarging the contours from the one-inch and overprinting them, usually in brown, but sometimes in red, orange or (experimentally) green; the clue to the overprinting is the '2nd' being on the contour plate. Four sheets used the three-inch London mapping which had been derived from the GSGS 3786 drawings. Sheets in areas which had been mapped in GSGS 3036 reused that series' 10 and 20 feet interval contours. A few sheets were subject to revision by Field Survey Companies and went to third or fourth editions (e.g. 53/24 SE, which included Luton): they are often distinguished by some relettering of more prominent names and redrawing of railways and main road casings but, unlike GSGS 3036, there was no question of disguising their six-inch origins.

Although produced for immediate military needs, GSGS 3906 was found to be useful for civil planning work. One such user was Professor Dudley Stamp, founder of the Land Utilisation Survey, who was working on behalf of the Ministry of Agriculture, and obtained three sets from the military[43]; another was the LUS's former Organising Secretary, Dr Edward Christie Willatts, who from 1941 was working for the Advisory Maps Committee of Lord Reith's Panel for Physical Reconstruction.[44] By his own account, Willatts learned of the existence of national GSGS 3906 cover in conversation with the then Director-General, Major-General M.N. MacLeod:

> I immediately asked if I could obtain copies for use in the Maps Office I had established... He referred me to the relevant GSGS colonel & I negotiated the transfer of, I think, 20 copies of each sheet.
>
> We used them for recording a variety of data relating to land use planning &, I believe, issued some to county planners. Steers used them as field record sheets in his coastal survey he made for us.[45]

By the autumn of 1943 stocks were held by the OS for issue to government departments and local authorities.[46] Some time in the later 1940s GSGS 3906 was placed on sale, and some sheets were reprinted, so that copies can be found in various local collections, though they occasionally reached civilian hands even before the end of the war.[47] It remained on grudging sale until at least the mid-1950s, pending completion of the civil National Grid 1:25,000 map:

> it cannot be classed amongst the series of maps upon which the reputation of Great Britain as the best-mapped country in the world is based... prospective users... are strongly urged to

---

[42] Yolande Hodson, 'MacLeod, MI4 and the Directorate of Military Survey 1919-1943', *Cartographic Journal*, 38 (2001), 155-172, pp.163-4.

[43] Hodson, 'MacLeod, MI4...', p.169, n.73, quoting 25A (11 November 1942) in PRO OS 1/161.

[44] See obituary of Willatts by Christopher Board in *Sheetlines* 57 (April 2000), 3-4.

[45] Willatts to Oliver, 15 August 1992 (in private possession). 'Steers' is J.A. Steers: the coastal survey formed the basis for his *The coastline of England and Wales*, Cambridge University Press, 1946.

[46] Memorandum by Cheetham on post-war programme, n.d. [October 1943], 1A in PRO OS 1/170.

[47] I am indebted to J.H. Andrews for an instance of this.

buy the new [civil Provisional Edition] sheets when published, or to await their publication.[48]

As a military map it was rendered obsolete in Britain by the adoption of the Transverse Mercator projection for military use in 1950, though in Ireland, at least in Northern Ireland, it seems to have continued in use for some time afterwards, and it seems to have continued in use in Britain as well by Officers Training Corps and similar bodies where there was no replacement 1:25,000 cover by GSGS 4627.[49] The Ordnance Survey in the Irish Free State had started work in the early 1930s on a 1:20,000 series, in 513 consecutively-numbered sheets covering all 32 counties, though only eight sheets appeared before the project was abandoned.[50] Like GSGS 2748 they were stylistically avant-garde, but unlike the British maps they were placed on sale. GSGS 3906 seems to have owed nothing to these Irish maps.

It should be noted that GSGS 3906 certainly did not have anything like the circulation of the military one-inch series, GSGS 3907 and GSGS 3908. By the turn of 1941-2 the standard first print-run for a Second War Revision printing of a GSGS 3907 sheet was 70,000 copies, and runs of 20,000 or 30,000 are routine on reprints; GSGS 3906 sheets with print codes (admittedly apparently a minority), often seem to have been printed in runs of only 2,000 or 3,000.

The civil 1:25,000: 1935-1954

The origins of the civil 1:25,000 map of Great Britain
In the early 1930s there was still strong Ordnance Survey opposition to putting GSGS 3906 on sale. Apart from the putative effect on six-inch sales there was the look of the map, which Winterbotham felt was 'ill adapted for civil use, competitive to the six-inch scale and, in effect, designed entirely for military purposes'.[51] In 1935 he was succeeded by M.N. MacLeod, who thought somewhat differently, and, better still, held office at exactly the right time to enable his ideas to be put into effect. The right time was the Davidson Committee, appointed early in 1935 to consider both tactical and strategic OS matters. The tactic, of getting more money to clear off arrears of revision and ensure that in future the large-scale maps be kept adequately up to date, was soon disposed of. The strategy, of devising future policy for the Ordnance Survey, and in particular how far it should relate to metric units, took somewhat longer, the Committee only made its Final Report in February 1938, and it was only published late that year. Amongst other things, it recommended adoption of a Transverse Mercator projection with a national grid based upon the international metre. The War Office Cassini projection and grid, used for both GSGS 2748, and GSGS 3906, had the advantage that it was easy to construct, but the disadvantage that it distorted angles: thus it was impossible for such purposes as invisible artillery 'map-shooting' to measure angles direct from the map. The Transverse Mercator projection, which had been adopted for small-scale civil mapping in the late 1920s, was orthomorphic: it had the property, particularly valuable for artillery purposes, that it did not distort angles. The Transverse Mercator metric grid was duly adopted, and was known successively as the Ordnance Survey Grid and (by 1944) the National Grid. Until 1951 the 100 km squares were designated by numbers derived from full co-ordinates, and thereafter by pairs of letters: the change was made because the military insisted on letters, and letters rather than numbers were used on pre-1951 military printings of National Grid mapping.[52]

---

[48] *A description of Ordnance Survey medium scale maps*, Chessington, Ordnance Survey, 1949, 17; cf *ibid*, 1955, 16.
[49] Information on O.T.C. use from Dr Christopher Board.
[50] See J.H. Andrews, *A paper landscape: the Ordnance Survey in nineteenth century Ireland*, Oxford University Press, 1975, 298-9.
[51] Winterbotham, 'Sidelights', 182.
[52] The development of the metric National Grid, and the various referencing systems intended or used in its early years is discussed in Richard Oliver, 'The evolution of the Ordnance Survey National Grid', *Sheetlines* 43 (August 1995), 25-46.

The Committee also recommended that experiments be made with 1:25,000 mapping, for which there had been some demand, with a view to its becoming a national series if successful. 'Whilst Davidson was certainly captain of the Committee boat, MacLeod was unquestionably the navigator', and its recommendations were largely based on his proposals.[53] MacLeod favoured replacing the six-inch altogether, by the 1:25,000 if possible; if not, he favoured recontouring the six-inch at 10 feet (3.28 metres) intervals, and to publish the contours on the 1:25,000 with metric values.[54] He had a particular interest in artillery survey,[55] and it is questionable whether anyone could have been in a better position to promote a map of possible civil use, but of definite interest to the War Office. However, the military representative on the Davidson Committee, Brigadier D.F. (later Lt-General Sir Desmond) Anderson, did not see matters in quite the same light, notwithstanding that the War Office would not (unlike with GSGS 3906) have to pay the preparation costs for the new series; this exasperated MacLeod, and may explain the lack of emphasis on the military applications in the Davidson Committee's Final Report.[56] It is, however, just to record that Anderson later became an enthusiastic supporter of the scale, and wrote warmly about it to Cheetham, MacLeod's successor, in 1949.[57] Not all senior officers were as unenthusiastic as Anderson. Rearmament was under way, and such a map was needed for Territorial Army training; 'unless they are accustomed to this scale in peace, they will not be able to get full value out of it in war', wrote Major-General A.F. Brooke.[58] Indeed, if possible the military would have liked to have complete cover of Great Britain at this scale within three or four years. However, the final report made no direct mention of the military requirement, 'no doubt as a matter of policy in times when the common sentiment was distinctly against the martial profession', beyond noting that the War Office had produced some sheets at this scale, and using one of them as an illustration.[59] (One wonders whether this was a tactic inspired by MacLeod, to circumvent the lack of enthusiasm for the scale in some quarters at the War Office, not least the head of GSGS, Col Percy Boulnois.)[60]

Instead, the report mentioned the possibilities of the new map for walkers and educational purposes.[61] Not mentioned was another civil interest group, the planners. All could make a case on the basis of existing mapping. The potential educational users were said to want a new intermediate-scale map, since the one-inch gave insufficient detail and the six-inch was too large a scale to give a general picture. Unlike the other potential users of a civil 1:25,000 they had no proper British prototype to point to: the nearest equivalents were the *Celtic Earthworks of Salisbury Plain* series, and the London three-inch of 1933, and overseas series such as the 1:20,000 of France and 1:25,000 of Switzerland. (Those with Territorial Army experience might perhaps have encountered GSGS 3906.)

The walkers had the advantage that, though their recreation was hardly novel, nonetheless it was unprecedentedly popular: it was epitomised by variously the 'mass trespasses', the Rights of Way Act of 1932, the formation of the Ramblers Association in 1935, and the 'hiker' cover for the OS one-inch Fifth Edition. There were a few examples of intermediate-scale mapping which walkers might have cited to support their case for the new scale. Starting in the early 1890s, a small number of footpath maps, usually at the two-inch or three-inch scales, had been printed for local footpath interest groups, occasionally by the Ordnance Survey (e.g. that of Oxford, 1933), but usually by private firms (e.g. Wirral and Warwickshire-Worcestershire groups). They mostly

---

[53] Yolande Hodson in Seymour, *A history*, 260.
[54] Seymour, *A history*, 264.
[55] Hodson, 'MacLeod, MI4...', pp 157, 158-9.
[56] Seymour, *A history*, 263. (Having been pushing for a national 1:100,000 for over twenty years, with a lack of support from some of those whom it might most benefit, the writer sympathises with MacLeod!)
[57] Anderson to Cheetham, 3 September 1949, in PRO OS 1/415.
[58] Quoted in Seymour, *A history*, 263.
[59] Yolande Hodson's comment, in Seymour, *A history*, 263.
[60] For more on Boulnois see Hodson, 'MacLeod, MI4...', pp 163-5.
[61] *Final Report of the Departmental Committee on the Ordnance Survey*, (London, H.M.S.O., 1938), p.13.

used photo-reduced OS six-inch material as a base, with the paths overprinted in red or green, but a few (e.g. of Watford, first published in 1902, and reissued as late as 1950) were newly drawn. The two-inch map of Cambridge of 1936 was unusual in that it was drawn privately (in a style derived from OS one-inch practice) but printed by the OS. These maps cannot have covered more than about one per cent of Great Britain, and they were perhaps mostly intended more as adjuncts to the OS one-inch than as substitutes for it, but they clearly demonstrated the potential for a properly-drawn national map series at a scale of around 1:25,000. There were also two-inch maps of Guernsey and Jersey (first published in 1901-2) and the Isles of Scilly (1933) and a three-inch of Guernsey (1933), but as the scale for each seems to have been determined by fitting a given land area onto a sheet of paper of given size, and none showed field boundaries, there are perhaps more closely related in spirit to the one-inch.

'The committee might further have mentioned that the new scale would be of use to Planning Authorities particularly in country districts', observed the OS in 1949.[62] Planning and administrative uses for the 1:25,000 were not mentioned in the Davidson Committee's Final Report. This seems curious at first, given that the demands of town and country planning were an important justification for catching up with the arrears of OS 1:2500 revision which represented the tactical side of the Davidson Committee's recommendations, but it also has a certain negative logic, in that there was no demand from this quarter recorded in the evidence presented to the Committee.[63] It was the dog which didn't bark in the night. There were at least two examples of mapping at somewhat similar scales which, even without the wisdom of hindsight, would have made at least as good a case for the 1:25,000 as did the footpath maps. These were the two-inch map of the London Passenger Transport Board (LPTB) area of 1935, which reproduced the two-inch drawings for the one-inch Fifth Edition at their original scale, and the three-inch mapping first produced by the OS in 1924 on behalf of Manchester and District Joint Town Planning Advisory Committee, by direct photo-reduction from the six-inch, in twenty sheets and covering 2,015 square miles (about 5,220 sq km).

A final justification for the 1:25,000 was MacLeod's suggestion that it might eventually replace the six-inch in mountainous areas, or indeed over the whole country, 'thereby effecting some economy!'[64]

The 'experiment': 'neck or nothing'

It was against this background that the Davidson Committee recommended that the 1:25,000 'should be tried out experimentally in certain selected areas, and, if successful, should be extended to cover the whole country in a National Series.'[65] However, according to MacLeod,

> The Committee's recommendation was couched in its tentative form only because one member wanted all work postponed until the arrears of 1/2500 revision had been overtaken, and was not to be satisfied with any assurances that the start of a 1/25,000 series would not affect field revision. I think the Committee were satisfied that there was need for a map intermediate between the 1 inch and the 6 inch. This does not mean that they thought it would be a 'commercial' success, but that it would be more suitable than either of these two scales for certain purposes for which these two scales have to be used because the 1/25,000 does not exist.
>
> The demand for the 1/25,000 was for education, for engineering, and for military training, to which may be added a general public need for a better map of our mountains.[66] The six-inch map in our mountainous areas consists almost entirely of 'rock' and 'rough pasture' symbols and is on too large a scale to be useful.

---

[62] *A description of medium scale Ordnance Survey maps*, Chessington, Ordnance Survey, 1949, 13, and *ibid*, 1955, 12.
[63] In Ordnance Survey library, Southampton, reference G.5186.
[64] MacLeod to Establishment and Finance Officer, 1 January 1940: minute 15 in PRO OS 1/383.
[65] *Final Report*, recommendations, 28.
[66] [The use for engineering was not mentioned in the *Final Report*.]

He observed:
> By the time we have discovered whether it is, or is not, wanted, we shall have incurred so much expense that it will be cheaper to go on than to draw back! It is, in fact, neck or nothing.[67]

The first extant mention of the proposed new scale in an OS civil context is found in a memorandum written by MacLeod for the Davidson Committee late in 1935:
> A new 1/25,000 map should be introduced and completed within 15 years, the style of this map to be based on the existing War Office map at this scale. ...
>
> The 1/25,000 map is a military necessity, and has been asked for by the War Office who, if the map is not produced by the Ordnance Survey, will have in any case to produce a number of sheets themselves. A map on this scale would undoubtedly have considerable civil value and should enjoy a good sale. It might indeed eventually replace the six-inch, with substantial saving in cost of revision.[68]

(The reference to 'style' should probably be interpreted as referring generally to colour-scheme and degree of generalisation, rather than to a proposal to perpetuate the specification of GSGS 3906 almost unchanged.) The trade was sounded out as to the possible demand for the scale: 'it was thought that the 1/25,000 would be a great success. There has been a good market waiting for it for some time.'[69]

Although the Davidson Committee would not report until early in 1938, MacLeod was evidently sanguine as to its outcome in this particular respect: preparations for the civil 1:25,000 were in hand by August 1936, when a specification for the new map was being worked out, based on a mixture of six-inch, one-inch and GSGS 3906 practice. A copy of GSGS 3906 sheet 47/16 SW (Andover) had hand-colour added to it (red and yellow road infills and green woods infill), giving a similar general result to the contemporary 'non-relief' one-inch; MacLeod liked the look of the map, but felt the colour scheme to be too elaborate.[70] He seems to have envisaged the map as a three-colour affair, similar to GSGS 3906, possibly with the addition of purple grid figures, but equally possibly with grid figures in blue, as on the Davidson report specimens and such contemporary series as GSGS 4041. Of printings of 500, 50 to 100 would be for civil use, and the rest for the military. It might be possible to sell it for a shilling if the education authorities would forgo their usual discount.[71] Such a map would serve both civil and military needs and, as there would be no road or wood infills, would presumably also avoid the need to produce a separate single-colour outline edition for specialised civil use. The approach would have been consistent with other maps associated with MacLeod, which combined the maximum of information with the minimum of printing, both of colour and of marginalia: the six-inch and 1:2500 specimens accompanying the Davidson Committee report, and the five-colour one-inch New Popular Edition in its original form. In 1950 he suggested saving paper by putting the legend and other information on the back of the 1:25,000 Regular Edition.[72] The ideal MacLeod map was a compact map.

Names and some ornament would have been produced by photonymograph (an early form of photo-typesetting), but there is no mention of the typeface which would have been employed.[73] Old Roman had been introduced a few years earlier for redrawn 1:2500 plans and was used on other contemporary mapping, including the 1:500,000 aviation series and marginalia for

---

[67] MacLeod to Establishment and Finance Officer, 1 January 1940: minute 15 in PRO OS 1/383. The important first sentence is not mentioned by W.A. Seymour in his account of the development of the 1:25,000, which cites this minute: see Seymour, *A history*, 292.
[68] Memorandum by MacLeod on acceleration of revision and scales and styles of maps, 30 December 1935, 1A in PRO OS 1/117.
[69] O.Maps to EO, 8 January 1936, in PRO OS 1/117.
[70] Minute 1, 12 August 1936, in PRO OS 1/355.
[71] Minutes 11-14, November-December 1939, in PRO OS 1/383.
[72] MacLeod to Burnett, 30 June 1950, 82A in PRO OS 1/416.
[73] Minute 1 in PRO OS 1/355.

specimens accompanying the Davidson report, and is used on the one surviving example of pre-1943 1:25,000 civil mapping.[74] This is a fragment of the West End of London (covering National Grid square TQ 2881), drawn on tracing paper some time in, or more probably some months before, August 1940, to demonstrate the effect which including street names would have on breaking the casing of blocks of buildings. Building infill is stippled, in place of the hatching used on GSGS 2748 and 3906 and on the first Provisional Edition sheets, prepared in the mid-1940s.[75]

The civil 1:25,000 was to be laid out in sheets covering an area 10 by 10 km (40 by 40 cm (15.74 by 15.74 inches) within the neat line), consonant with the Davidson Committee's recommendation that larger scale maps be square in shape. In December 1938 work was ordered to be put in hand on six sheets covering the country north-west of Salisbury, in an area which had long been covered by GSGS 2748 or 3906 mapping. They were sheets 4013, 4014, 4015, 4113, 4114 and 4115, corresponding to 41/03-05 and 41/13-15 in the numbering system used in 1944-51, and SU 03-05 and SU 13-15 in the post-1951 numbering system: three different numbering schemes were in use within a decade.[76] These maps would presumably have been 'provisional editions', produced by recasting GSGS 3906 material on Transverse Mercator metric grid sheet lines. Early in May 1939 a 1:25,000 drawing section was set up, and undertook some work on converting GSGS 3906 sheets to the new sheet lines; none of this was ever published.[77] In November 1939 one of MacLeod's subordinates argued that, from a sales point of view, it was 'absolutely essential' not to have small isolated groups of sheets:

> As I do not want to flog what may be, as a result of similar experience with the [one-inch] Fifth Edition, an already dead horse, I do not propose to give any other reasons unless required, beyond the utter impracticability of indicating to customers which sheets are or are not on sale.

This reason was evidently adequate: none other was asked for.[78] In April 1940 MacLeod decided that 'a reasonable block of sheets' should be completed as a stock job when drawing staff could be spared from war work.[79] After some three years of apparent inactivity, work was restarted in 1943: the 'reasonable block' would be much of Great Britain.

At this point it is necessary to consider the manner in which this new mapping would be made available. There were three possibilities. The first, which need not rule out either of the other two in the longer term, was the policy apparently pursued in 1938-40, of preparing immediately some 'provisional edition' sheets, using GSGS 3906 material rearranged on new Transverse Mercator National Grid sheet lines, which could be used to test the market. The second was to produce it on the basis of revised six-inch mapping, which in turn would depend on the output of revised 1:2500 mapping. Again, this would depend on how the 1:2500 was revised, whether by large blocks, e.g. counties, or small, e.g. towns and their immediate vicinities. (In May 1939 MacLeod wrote that the 1:2500 'overhaul' would start in Yorkshire, but whether this would have been of the whole county or of more limited areas is at present unclear.)[80] The third was a compromise between the two: not to wait for larger-scale revision, but to produce an interim edition which could be revised or replaced later once that larger-scale revision was available.

Geoffrey Cheetham: means, motive and opportunity

In the summer of 1943 the civil 1:25,000 project was revived. This immediately followed MacLeod's retirement in April 1943; he was succeeded by Major-General Geoffrey Cheetham but was on hand to give advice, and both men seem to have been of one mind as to the desirability of

---

[74] The designation 'Old Roman' appears in *Ordnance Survey Alphabets* [Southampton, Ordnance Survey, 1934].
[75] See items 22A-D in PRO OS 1/355.
[76] See minute 9 in PRO OS 1/355.
[77] See minutes 15 and 17, 16 May and 9 October 1939, in PRO OS 1/355, and minutes 3 and 16, May 1939 and 18 April 1940, in PRO OS 1/383.
[78] O.Maps to DG, 30 November 1939: minute 10 in PRO OS 1/383.
[79] Minute 17, 19 April 1940, in PRO OS 1/383.
[80] MacLeod to DMO & I, 3 May 1939, 6a in PRO OS 1/219.

the 1:25,000.[81] However, whereas in 1939 MacLeod had had nothing much more to work on than limited GSGS 3906 coverage, some evidence from potential users, and a qualified recommendation of experimental work from the Davidson Committee, by 1943 Cheetham had complete, if stylistically undistinguished, GSGS 3906 coverage, and a large drawing staff, who, in the intervals of war work, were available for civil production, precisely as envisaged by MacLeod in April 1940. And apart from the expected civil applications - educational and pedestrian - mentioned by the Davidson Committee, there was the dog which hadn't barked then, but did so now: planning. According to E.C. Willatts:

> ... I recall Cheetham discussing the post-war programme and I urged the need for a 1/25,000 map, pointing out that it had proved its worth already in the use we had made of the sheets I had obtained from the War Office. He readily accepted this & the scale was included in the programme.[82]

Although it is apparent that Willatts was knocking at an open door, it gave Cheetham a further justification for a proper civil map at this scale. After all, provided one overlooked GSGS 3906, postwar planning was a much more timely justification than either walking or education, both of which had apparently got along quite well with existing mapping. Further, at about the same time (the autumn of 1942) the Scott Committee on Land Utilisation in Rural Areas recommended that Government support should be given to the OS to make the 1:25,000 available for all planning purposes.[83] In October 1943 Cheetham wrote a long memorandum setting out a proposed post-war programme for the OS. It had long-term and short-term elements: the long-term programme was based on the Davidson Committee recommendations, and the short-term programme on immediate post-war needs. The most important component of the short-term programme was the production of a 1:25,000 map, which Cheetham justified at some length. Having noted the use made of GSGS 3906, he enumerated its drawbacks; it was illegible, it had the War Office Cassini grid, it could not be replaced sheet by sheet by mapping on the Transverse Mercator grid, and all work spent on improving it would be lost when the 'final' map was published.

> It is understood that the production of a good 1/25,000 map at a very early date is considered to be of great importance by planners. In addition it is now clear that this map will receive a hearty welcome from all those who are interested in education and by hikers. It therefore appears advisable to give a high order of priority to the production of a new 1/25,000 map. The proposal is that a provisional edition of the map on the final National Grid sheet lines should be produced. This proposal has the following advantages.
> ... there are a large number of draughtsmen now employed on war work who will before the end of the war reach periods when they are intermittently short of material to work on for war mapping. These draughtsmen will in about two years after demobilisation be required to draw the work which will be flowing in from the large scale surveyors in the Field. It is estimated that during this period those draughtsmen who are not suitable for the more highly skilled work required for the small scale maps will, with the addition of the post-war recruitment of about 280 men be able to complete this provisional map. We shall thus have a 1/25,000 completed within two years and should have a supply of draughtsmen available to take up the drawing of the large scale plans as fast as they are surveyed.
> By drawing this map on the final sheet lines, we shall have a map which will be able to incorporate sheet by sheet the results of the large scale survey which will be carried out in patches radiating out from the large towns.

He proposed that the final edition would have resurveyed contours: contours at both 25 feet and 10 metres interval were being surveyed experimentally in the Bournemouth area. He proposed 25 feet for the provisional map.

---

[81] Seymour, *A history*, 286, 292.
[82] Willatts to Oliver, 15 August 1992 (in private possession).
[83] *Report of the Committee on land utilisation in rural areas* (1942), paragraph 239. L. Dudley Stamp was vice-chairman of this Committee, which was set up jointly by the Ministries of Works and Agriculture & Fisheries. (I am indebted to Christopher Board for drawing this to my attention.)

> The drawing of one sheet has been put in hand to enable interested parties to criticise the style before work on the series is started in earnest. This sheet is in the Bournemouth area which is being covered by the experimental large scale survey and by the contouring experiment, so that a comparison may be made at an early date between the provisional and final maps.

If the resurveyed contours were at 25 feet intervals, then there would be discrepancies where they met old contours at sheet edges; thus it was for consideration whether the final map should be contoured at 10 metres interval. The adoption of the 1:25,000 also raised the question whether the Scottish highlands would be adequately served by this scale: he thought a properly contoured 1:25,000 of greater value than the present six-inch.

> To summarise:- It is proposed to produce a provisional map in three colours on the final sheet lines. It is hoped that the drawing of the outline, the water and the roads will be sufficiently good to be used in most cases for the final edition, and that the contours will have to be replaced as the new contour survey is completed. A decision is required as to whether the contour interval on the final map should be at 25 ft or at 10 metres vertical interval.[84]

In November 1943 the programme was approved in principle by the Ministry of Agriculture and Fisheries, which had departmental responsibility for the OS. The Permanent Secretary noted that it was

> of the utmost importance that the immediate needs of the planners... should be met. There would be a great and urgent demand for maps for many purposes connected with post-war reconstruction, e.g., housing, as soon as the war in Europe was over, and the Ordnance Survey must be ready to meet that demand when it arose. The Secretary said that he could not contemplate a situation in which urgent civilian work was held up for lack of up-to-date maps.

However, he thought that the resurveyed contours were the least important element, and no more was heard officially of metric contours until the 1960s.[85]

It is worth noting that, though Cheetham enthusiastically promoted the 1:25,000, he had private doubts about the scale, and preferred the 1:20,000. He enlarged on these in a letter to his successor, Major-General R.Ll. ('Bruno') Brown, in 1950. Noting that 1:25,000 was chosen for the civil map because it had been selected earlier for GSGS 3906, he continued:

> There are certain critical scales at which, in any given type of country, the power of showing particular types of detail changes. In Great Britain such a change occurs between 1/25,000 and 1/20,000... It is not now practical politics to consider changing to a scale of 1/20,000. It is however important to consider the points in which the 1/25,000 is at a serious disadvantage and to study how this disadvantage can be minimised by special care in design and compilation...

The larger scale was superior to the smaller in rural areas as roads could be drawn to scale rather than conventionally, as could the majority of field boundaries. In built-up areas many more names could be shown.[86] However, it is not known how far Cheetham was alive to these shortcomings when serious work began on the 1:25,000 in 1943, and how far they were the outcome of subsequent experience. Work was authorised on 14 June and was in hand by 17 August.[87]

---

[84] Memorandum by Cheetham, n.d. [October 1943], 1A in PRO OS 1/170.
[85] Record of meeting by D.B. T[oye], 6A in PRO OS 1/170.
[86] Cheetham to Brown, 28 November 1950, 158B in PRO OS 1/416.
[87] 'Chronology of 1:25 000 scale mapping', typescript [n.d., 1974] in OS Cartographic Library, citing Director-General's conference minutes.

Sheet 40/09 and the transition to Times Roman

The first sheet to be started was 40/09 (later SZ 09), which covered part of Bournemouth and Poole. As Cheetham explained in his memorandum on the post-war programme, this sheet was chosen because an experiment in 1:1250 resurvey was in hand in Bournemouth, and so it would be easy to compare provisional and final work.[88] In January 1944 a second pilot sheet, 33/89 (later SJ 89), of the centre of Manchester was started.[89] By 8 March 1944 a first proof of 40/09 was ready, further proofs followed, and in November specimens were distributed outside the Ordnance Survey.[90] 40/09 and 33/89 were the first two 1:25,000 sheets to go on sale, in November and December 1945 respectively; 33/89 was printed in three colours, and 40/09 in four. The surviving proofs, taken with written evidence in Public Record Office files, enable the development to be followed in some detail. All the 1944 proofs of 40/09 use the same three basic printings: outline in black, water in blue, and contours and main roads in reddish-orange-browny shades; one proof adds a fourth colour, green.[91]

Common to all these proofs, as it would be to the published maps, was the absence of any separate scale bars: instead, the outer border was graduated, in feet, yards, furlongs and miles, and graduation of the neat line supplied metric scales. This space-saving measure had been suggested by Lt-Col J.C.C. Craster in 1935, and was used on the six-inch and 1:2500 specimens in the Davidson Committee report.[92] (The principle was hardly a new one: a notable use was by John Bartholomew's, from about 1900 onwards on their half-inch and quarter-inch maps.)

The earliest proof seems to be that dated 3 March 1944. The names on the map face are a mixture of Old Roman, including those of the towns of Bournemouth, Poole and Wimborne Minster, and 'Roman Old Style no.5'. Roman Old Style no.5 had been used on 1:2500 mapping prepared between 1926 and 1932, and was also used on the 1:1250 National Grid pilot sheet, 40/0993 SW.[93] 'U.D.' is added beneath 'Wimborne Minster' to indicate that it is an Urban District. Woodland is shown by scattered black tree symbols, and it is not easy to decide what is close woodland, and what is scattered trees in relatively close proximity to each other. A and B roads are indicated in the legend and on the map, without numbers, which may be an oversight, given OS practice at other scales by this time. On railways, both *S.B.* (signal box) and *S.P.* (signal post) appear; neither had been shown on GSGS 2748 or GSGS 3906. Sandy foreshore is speckled, using the contour plate; mud is shown only by annotated 'white' between tide lines. Building infill is by hatching, and trigonometrical points on buildings are shown distinctively, as on GSGS 2748 and 3906. The military origins and uses of the map are thus still apparent. The contour plate is brown. Only county names are given in the border. The sheet number is given as '4009', top and bottom centre, and seems to follow the method used on the Davidson Committee specimens, of numbering based on shortening the full co-ordinate reference of the south-west corner of the map, in this case 400,000 metres east and 90,000 metres north of the false origin of the National Grid: a note appended to a diagram of 100 km squares states that the full kilometre reference of the south-west corner is 400090.

The next proof seems to be that dated 21 April 1944. The contour plate is a slightly darker brown. Some urban roads have been widened, names added inside the casing, and the infill

---

[88] Memorandum by Cheetham, n.d. [October 1943], 1A in PRO OS 1/170.

[89] 'Chronology', in OS Cartographic Library, citing Director-General's conference minutes.

[90] There is one in the 'specimen drawer' in the map room of the Royal Geographical Society.

[91] The main PRO files bearing on early development are OS 1/383 and OS 1/415. A set of eight proofs of sheet 40/09, dated March to November 1944, is at present held in the Charles Close Society-Ordnance Survey deposit in the map department of Cambridge University Library, and is the source for the following paragraphs. So far, no proof of sheet 33/89 has been found in a public collection, but one is known to exist in private hands. The placing of the proofs in the sequence given here was made after close study of changes to the black plate; it is not thought necessary to describe these in exhaustive detail here.

[92] Craster to MacLeod, in PRO OS 1/84B.

[93] 'Roman Old Style No.5' is referred to by implication in H.S.L. Winterbotham, *The national plans*, London, H.M.S.O., 1934, 105. There is a copy of 40/0993 SW in the Royal Geographical Society map room's 'OS specimen drawer'.

deleted: this has been effected by pushing the casing across any 'front gardens', without redrawing or moving the buildings behind. Parks are in a slightly heavier black stipple. Most of the parish names have been transferred from the map face to the border; the town name 'Bournemouth' is also deleted. The contour figures have been reset, in a slightly smaller point size. The grid figures in the margin have also been reset, in Roman Old Style no.5. Another proof is undated, and seems to be identical, except that the contour plate is printed in a reddish shade resembling 'Air Force Brown', which was used for contours and roads on printings of GSGS 3907, the military one-inch of England and Wales, at this time. Until shortly before, not only GSGS 3907, but other military series (e.g. GSGS 4040) had used orange for contours and roads, but the change to Air Force Brown had been made 'owing to the habit of both the Army and the Air Force of using amber torches'.[94]

The presumed next proof is also undated, and differs from the others in that it adds a fourth colour: green. This is used to print the tree symbols which had hitherto appeared in black and as a green stipple (in place of black stipple) for parkland which, until examined closely, looks like a solid tint. As a result (and this is thought to be unique in OS practice), parks appear far more prominently than do woods. Road infills have been restored to named roads in built-up areas.

The next proof is dated 15 August 1944. The sheet number is now '40/09'. The contour plate is a light, pinkish shade of brown. Numerous Old Roman names have been reset in Roman Old Style no.5. Park stipple is once more on the black plate, and is noticeably lighter than before. Various minor names and descriptions have been deleted. Only one urban road is now named. Contour figures have again been reset, in a smaller point size. Grid figures in the margin have also been reset, but in a larger point size, whereas all names in the margin have been reduced in size, and road destination distances have been added. Signal posts have been deleted from railways, but signal boxes are still indicated. Another proof, dated the following day, is identical, except that the contour plate is more 'brown'. A still darker brown was used on the next proof, dated 8 September 1944, which was otherwise unchanged.

The last of the proofs, dated 21 November 1944, has been completely relettered using Times Roman. Park stipple is lighter, and the contour plate uses a lighter (rather orangey) brown. It was this proof which was circulated in November 1944.[95] A modified version of 40/09 and a proof of 33/89 were sent out for further consultation in May 1945. One result of the consultation process was that the number of colours was increased from three to four, grey being used for field boundaries, building infill, and most vegetation symbols.

Publication progress: ten years rather than two

The original intention was to publish the Provisional Edition by large regional blocks, and by the summer of 1945 two were in hand, covering Lancashire, north Cheshire and part of the Lake District, and part of East Anglia. The original intention had been that the first block would be in Lancashire and Yorkshire, as 25 feet contours were already available for these counties, whereas they would have to be surveyed or interpolated in most other areas.[96] No reason is known why the East Anglian block (corresponding to most of National Grid square 62, later TM) was chosen, but it is notable that it was wholly covered by GSGS 3036 mapping, with its 10 and 20 feet contouring; in turn, this suggests that it was seen as nearly as convenient as Lancashire and Yorkshire.

> Publication by regional blocks soon broke down, as statutory demands prevented the progressive development over the country that was intended, and production was diverted to groups of sheets here and there covering many of the main towns and their immediate neighbourhoods.[97]

---

[94] Willis to Cheetham, 5 July 1945, 42 in PRO OS 1/415.
[95] The comments received are summarised in the notes on the conference on 1:25,000 design, 30 December 1944, 19/1 in PRO OS 1/383, but the comments themselves do not appear to have been preserved.
[96] Memorandum by Cheetham, n.d. [October 1943], 1A in PRO OS 1/170.
[97] *A description of Ordnance Survey medium scale maps*, Chessington, Ordnance Survey, 1949, 14.

The 'statutory demands' were presumably town and country planning and the Parliamentary Boundary Commission: the latter gave birth to the Administrative Areas Series, described later. Cheetham was anxious to complete the Provisional Edition as quickly as possible: in 1946 he wrote: 'Everything possible must be done to speed up the 1/25,000. We are aiming at 2 years, not 10.'[98] In March 1944 it was envisaged that 2,700 sheets (i.e. the whole of Great Britain) would be produced over three years.[99] The maps appeared much more slowly than had been intended, partly because expected post-war staffing levels were never reached, thus 'making nonsense of its intended function of meeting immediate needs'.[100] By January 1948 only 516 sheets had been completed, and though another 620 had been published by the time that Cheetham retired eighteen months later, and a total of 1,857 by October 1951, the Provisional Edition only reached its maximum extent of coverage in 1956.[101] By that time the first sheets intended to replace it were about to be published. The original intention not to publish in the Provisional Edition the sheets of certain mountain areas in Scotland and Wales was modified in January 1950, so that the final total of Provisional Edition sheets was 2,027.[102] Completion was delayed not, for once, because of shortages of money or staff, but because new 1:25,000 drawing was used for testing drawing trainees: the supply of new work therefore had to be eked out.[103] Even so, the Provisional Edition still excluded most of the Scottish highlands and islands, and the official estimate of the total number of sheets in the 'Regular Edition' which was intended to replace it was 2,632. This number would have been somewhat greater (2,859) had there not been a number of 10 by 12, 12 by 10 and 15 by 10 km sheets published in coastal areas.[104]

Popularity: a question of definition

The term 'Provisional' must not be read to imply any denigration of the quality of the map production. On the contrary, admiration for the map is widely expressed, and it is becoming increasingly popular.[105]

It would perhaps be unfair to describe this official version of events as 'slanted', but it is undeniable that, though a much cleaner-looking map than the contemporary one-inch New Popular Edition, the newcomer failed to sell in anything like the quantity of the smaller scale. Even by 1954, with the series nearly complete, the average sale for each sheet was only 81 copies per year.[106] No doubt the planners were pleased, but there were simply not enough planners to justify production of the series on these grounds alone, and it was apparent that the walking market was not being adequately tapped, either.

One difficulty may have been that the consultative process was overweighted towards 'professional' users, and thus gave lesser weight to 'leisure' users. In 1950 a consultation exercise was carried out in order to assist the design of the Regular Edition: those approached included a number of borough engineers, whose interests might have been thought to be served by much larger scales.

In 1948 the possibility of a publicity campaign using various journals and newspapers was investigated; whilst a review in a professional journal might draw the attention of a limited number of potential new customers, newspaper publicity was capable of reaching a much wider audience. The only difficulty was getting the newspapers to mention the subject in the first place. A press campaign was attempted to assist the launch of the eleven Regular Edition sheets of

---

[98] Minute by Cheetham, 30 May 1946, 19 in PRO OS 1/198.
[99] Walter to Director General, n.d., 52 in PRO OS 1/355.
[100] W.A. Seymour in Seymour, *A history*, 291.
[101] Seymour, *A history*, 291; OS Policy Statement 11, October 1951, 11A in PRO OS 11/29.
[102] Minute 83, PRO OS 1/383: the excluded areas are indicated on Plates VIII and IX of the 1949 *Description*.
[103] Minutes 143-145, 177, 178 in PRO OS 1/383.
[104] Minute 51 in PRO OS 1/355.
[105] *Description of ... medium scale maps*, 1949, 14, and *ibid*, 1955, 13.
[106] See minute by O.Maps, 24 January 1955, 225 in PRO OS 1/383. (Minute 6 in PRO OS 1/1123 quotes 17 [!] copies per sheet per year in 1953-4, which perhaps refers to folded issues.)

Plymouth and Dartmoor in 1956: review copies of maps were sent to ten papers, but the reaction was generally 'apathetic', with only one mention.[107] Publicity might perhaps have been a problem, but OS officers were sometimes prepared to show initiative: in December 1948 Brigadier H.E.M. Newman sent a specimen of the newly-published sheet 41/86 to his old school, Wellington College. He pointed out its advantages: the College was well situated in it (near the centre), and it would have advantages for field days, natural history societies, and Sunday walks. The school shop might stock them. The Bursar replied that he would go into it: with what effect, History, at any rate as represented by surviving documentation, does not record.[108] This was not the last time that the belief would be expressed that boys need not live by tuck alone, but could also feast on OS 1:25,000 maps.

There were also those who knew better than other users what the latter wanted. Mr Bosanquet of the Ministry of Agriculture and Fisheries wrote in June 1945:

> We all agree that any maps at this scale will be used *not* by motorists & hikers but by Govt. Depts., Local Authorities, and others concerned with planning [so] it is important that they shall be, as nearly as possible, photographic reductions of the 6" maps showing as much detail as possible, & having the minimum of colouring.[109]

Stanley F. Marriott, the Ramblers Association Honorary OS Adviser, took a somewhat different line. Whilst the new maps would not supplant the one-inch, they would supplement it: they would be 'popular with walkers who do most of their rambling in a small area', and such people as walk-leaders. Average walkers would not buy them 'wholesale', but might buy them in areas of close detail, such as around Leith Hill in Surrey. His Association thought that special sheets ought to be produced for holiday areas.[110] Col F.J. Salmon, a notable military surveyor, wrote that 'here we have the ideal scale for most map users who carry out their activities, whatever they may be, on foot'.[111]

'It remains something of a mystery why the series was ever made at all', observed Col Seymour. (By his time at the OS it had reached the 'awful map' stage.) He continued:

> Many of the planners working during the war on post-war reconstruction had certainly found the War Office 1:25,000 (GSGS 3906) very useful. This map... was very difficult to read, ... Cheetham referred to it as being 'nearly up to date'...; the wish for a 'good 1:25,000 map' might have been an acknowledgement of the value of the War Office map and an expression of the planners' hope for a more legible version of it.[112]

With the advantage of hindsight (and in particular in the light of the protests in later years when the 1:25,000 had started to make its mark with a wider public, but was still uneconomic) this verdict may be reasonable. However, there was some justification for Cheetham following the course which he did, even if one sets aside the fact that the 'planning requirement' was a convenient excuse for a course of action already decided on by MacLeod in 1939. In theory a civil 1:25,000, were there really a need for one, could have been produced in the 1940s much more easily in the way that GSGS 3036 had been produced thirty years earlier: by taking a direct photo-reduction of the six-inch, making some discreet retouchings, and rewriting the names. However, there was a subtle difference. In 1911-14 the OS had all its reproduction materials for the six-inch in pristine condition, and it was easy to get good results with a venture such as GSGS 3036. By 1943 most of the original material for the six-inch had been destroyed, and it survived only as prints, of variable quality, on enamelled zinc, or as printing plates. Had the method used for GSGS 3036 been repeated the results would have been nowhere as good, as the parent six-inch material would have been much inferior. Even in the 1930s the six-inch material tended to lack the crispness which had characterised it before 1914; this is exemplified by the two-inch Oxford

---

[107] Report on 1:25,000 Regular Edition, October 1956, 25A in PRO OS 1/1123.
[108] Newman to Bursar (Foley), and reply, 2 and 15 December 1948, 223A, 224A in PRO OS 1/415.
[109] Bosanquet note, 6 June 1945, 29C in PRO OS 1/415.
[110] Marriott to [OS], 15 June 1945, 30A in OS 1/415.
[111] Salmon to Cheetham, 19 March 1947, 97A in OS 1/415.
[112] Seymour, *A history*, 291-2.

footpath map of 1933, which was photo-reduced directly from the six-inch and is of similar quality to the GSGS 3906 sheets of 1940. There is no evidence that such a method was ever considered by MacLeod or Cheetham for a civil 1:25,000.

The short-term justification for the 1:25,000 in 1943 was that it kept drawing staff occupied in the intervals of war work, and until such time as National Grid 1:2500 and 1:1250 survey was sufficiently advanced to be taken up for drawing. The staff employed included not only regular OS staff but Royal Engineer sappers (until December 1946) and 'Junior Technical Civil Assistants', who were youths who had just left school and were filling in time before being called up for National Service.[113] By 1950, when the emphasis for the 1:25,000 moved from producing new sheets to revising existing ones, this work-creation for drawers was becoming a liability, and it was necessary to find a balance between the competing demands of drawing the new large-scale National Grid mapping and revising the Provisional Editions of the six-inch and 1:25,000.[114] 'With the advantage of hindsight', Seymour felt that the effort which was put into the 1:25,000 would have been better directed to the six-inch Provisional Edition, work on which had also begun in 1943: one cannot really quarrel with this, but by the time that it had become apparent that the 1:25,000 was both appearing much more slowly and proving much less successful than had been expected, so much work had been put into it that, for the time being, there could be no turning back.[115]

In 1940, MacLeod had written that the 1:25,000 was 'neck or nothing'. By the time that the venture had had a few years' trial, the OS neck was well and truly stuck out.

Compiling, drawing and revision

The Provisional Edition was initially produced by a fundamentally similar procedure to GSGS 3906: it was drawn at 1:15,000 on photo-reductions of the latest available six-inch material, but now printed in ferro-prussiate blue on enamelled zinc rather than on Whatman's paper. A complication was the division of what had been a single black 'detail' drawing for GSGS 3906 into separate drawings for the black and grey printings of the finished map. After the 'black' detail had been drawn, including names, it was photographed, and a combined ferro-prussiate blue print of the original six-inch compilation and the 'black' drawing made, on which the 'grey' detail was drawn. Once the four 1:15,000 drawings - outline and names (black), fields and buildings (grey), water (blue) and contours and roads (orange-brown) - had been examined, they were photographed, and negatives at 1:25,000 produced, from which printing plates could be made.

Revision could be carried out by several methods: (1) by starting again with a re-compilation at six-inch, and then following the same procedure as for a new sheet; (2) by correcting the original 1:15,000 drawings; (3) by correcting the 1:25,000 negatives; or (4) by a combination of (2) and (3). In 1952 it was reported that (4) was used for 90 per cent of sheets: two clear instances of (1) being used are the redrawn editions in the 'later' style of SD 42 and SJ 69. Having field boundaries in grey made revision more difficult.[116] Revision was often the occasion for replacing Roman Old Style no.5 or photo-written names with Times Roman: a few 'original style' sheets were wholly relettered.

The latest six-inch mapping was augmented by any other available revision, which might include one-inch material where it could be satisfactorily enlarged, bomb-damage surveys, some limited 1:2500 revision, scraps of surveys undertaken by military units during the war, and, particularly in south-east England, one-inch road revision of 1945-8.[117] As a result it was as up to date as it could be given the material available to the OS at the time of its compilation, but unfortunately there was a great disparity between the consequent paper landscape and the real

---

[113] Seymour, *A history*, 291.
[114] Minutes 62-68, 72-77 (March-October 1949) in PRO OS 1/383; revised Policy Statement 11, May 1959, 41A in PRO OS 11/29.
[115] Seymour, *A history*, 292.
[116] Minute, 12 March 1952, 110A-B in PRO OS 1/567.
[117] See minute 1A in PRO OS 36/7. 24 one-inch New Popular Edition sheets underwent road revision in 1945-8.

landscape perceived by its users, potential users and critics. For example, there was a complaint in 1949 that the Edinburgh sheet was twenty years out of date and whilst, like all the others, it would have incorporated the results of the rapid revision of new urban development carried out for Air Raid Precaution purposes in 1938-9, the accusation would certainly have been true of the less developed parts of the sheet and of minor detail generally.[118] The semantic distinction 'Provisional Edition' recognised this difficulty but did not remove it.

In 1950 work began on republishing the Provisional Edition with revision primarily collected for the one-inch Seventh Series. As the field-work was at six-inch there was no great problem involved in using it for both that scale and the 1:25,000, although it was only to one-inch standards of completeness, and thus excluded such details as field boundaries which were shown at all the larger scales. This was not a high priority: 1,013 of the 2,027 sheets had been republished by January 1957 and the publication of the revised sheets was only completed in 1965, though one further revised sheet, NS 27, appeared in 1970.[119] In 1958 a second one-inch revision cycle was started, but, with a few exceptions, thereafter only a limited amount of major road revision was added to the Provisional Edition.

The change in emphasis around 1949-50 from drawing new 1:25,000s to revising existing ones coincided with increasing pressure on OS drawing resources, and with a gradual reduction in establishment. Taken together, these implied that existing commitments such as the 1:25,000 Regular Edition would need to be kept under review.[120] No figures have been found for the time taken to compile and draw a new sheet, but in December 1950 the full revision of an average sheet took 65 man-days: 50 for drawing, 10 for examination, and 5 for proof examination. The 1:25,000 section had 77 staff, 50 on drawing and 20 on examination, but was 7 below establishment.[121]

There were various minor economies that could be made, such as printing in battery, whereby the need to 'wash down' a press between jobs in different colours could be reduced by printing, say, six sheets as a group: first, the black plate would be printed for all six sheets, then the press would be 'washed down', then the next plate would be printed and the press again washed down, and so on: only four washings-down, with the consequent expenditure of time and labour, would be needed, compared with twenty-four were all the sheets to be printed individually.[122] No such short cuts could be taken with revision, short of simply not drawing it. The pressure on drawing 1:25,000 revision was already noticeable in 1951, when it was decided not to add intermediate revision to a number of sheets expected to need reprinting in the near future. The OS Archaeology Officer, Charles W. Phillips, reacted to this by observing that

> 'The 1/25,000 series is a very popular one with archaeologists, and as the unrevised information is extremely out of date and frequently ludicrously wrong, it is suggested that the archaeological aspect should be an important factor in assessing whether or not a sheet... requires revision.' '... its initial warm reception has been much damaged by the unreliable and obsolete character of much of the information.'

The most that seems to have been achieved was that the archaeological state of a sheet was taken into account when assessing whether it was worth revising.[123] At this time, the aim was to revise about 300 1:25,000 sheets each year, which would have kept in step with the progress of one-inch revision; in the year to June 1953 only 103 were revised, though by September the rate of output had risen to 156, plus another 57 reprinted with corrections.[124] In 1955 it was believed that demand for the map was not greatly affected by its out-datedness (there being little difference in

---

[118] Seymour, *A history*, 291.
[119] Thompson to Hardy, 21 January 1957, 54A in PRO OS 1/697.
[120] For OS staffing at this period, see Seymour, *A history*, 290-1, 299.
[121] Memorandum, December 1950, 124A in PRO OS 1/383.
[122] Minute, 19 April 1955, 228 in PRO OS 1/383.
[123] Minutes by Phillips, 26 April and 17 May 1951, and by DMP, 11 June 1951, 141, 147 and 152 in PRO OS 1/383.
[124] Minutes 169, 183 and 191 in PRO OS 1/383.

sales of revised and unrevised sheets), but that it was desirable to continue revising the Provisional Edition from one-inch material.[125]

By early 1960 the demands of the National Grid mapping at six-inch and larger scales were such that it was decided in April that whilst completion of a first revision of the Provisional Edition by 1964 would be a priority, there could be no resources for a second comprehensive revision, i.e. using the material generated by the new revision for the one-inch Seventh Series which had begun in 1958.[126] Further revision of the Provisional Edition was largely confined to motorways, new main roads, the addition of buildings and runways at airfields, new reservoirs, and occasional conspicuous new building, such as the atomic power station at Dungeness on sheet TR 01. More thorough revision of the Provisional Edition was discussed again in 1965, but (except possibly such limited additions as the airfield structures) little seems to have come of it, probably because of the continuing pressure on drawing staff.[127]

'Restoring the 1:25,000 Provisional Series to a fully maintained status as an alternative to the Second Series' was considered in 1970, including issuing Provisional Edition material on Second Series sheet lines.[128] A combined print of the Provisional and Second Series sheets SK 49/59 was made, to test the accuracy of the older mapping; the fit was good, but not perfect.[129]

However the 1:25,000 still did better than the half-inch map, which was abandoned in 1961 after only a tenth of the sheets had been published, primarily so as to release drawing staff for the six-inch National Grid Regular Edition. As this in turn implied the 1:25,000 Regular Edition, one can almost say that the smaller scale and its users were sacrificed for the larger.[130]

The cartography and reliability of the 1:25,000 Provisional Edition

There is just enough evidence surviving - the experiment of 1940 referred to above and early proofs of sheet 40/09 with some names in Old Roman - to suggest that Cheetham started with MacLeod's ideas and modified them in the light of experience. Even the earliest proof, of March 1944, differs in that it represents a move away from MacLeod's three-colour map to include infill of main roads, using the contour plate. A development was the official 'original' style, which was not used on 40/09 in its published form, but was used for the 124 sheets comprising the Lancashire and East Anglian blocks. It was in three colours: black for outline, most names and grid, blue for water and water names and orange-brown for contours and road infills. Building infill was hatched black, field boundaries and vegetation ornament were also in black. Names were in Roman Old Style No.5, which had been replaced by Times Roman on 40/09 in November 1944. It was an alphabet which was effective at 1:2500 but which tended to look heavy and drab at 1:25,000: much more so, indeed, than Old Roman. Perhaps Cheetham felt the same, as sheet 40/09, in its final form, put on sale in November 1945, had lettering in Times Roman. Although most newly-drawn wartime series (e.g. GSGS 4250, France 1:50,000; GSGS 4416, Central Europe 1:100,000) were lettered in Egyptian, similar to the pre-war GSGS 3906, Times Roman started to be used for series in hand towards the end of the war (e.g. GSGS 4507, Germany 1:50,000).

However, that was not the only striking modification; as has been mentioned, a grey printing was added, for field boundaries, building infill and most vegetation, and details such as aqueduct alignments, hidden archaeological features, disused railways and canals. This can perhaps be traced to demands from the interested parties circulated in November 1944 for a more delicate style. Though green usually improves the look of a topographic map it did not really do much in its one appearance on 40/09 in its original form, perhaps because of its eccentric use; the

---

[125] Policy Statement 11, February 1955, 21A-B in PRO OS 11/29.
[126] Extract from Director-General's Conference 388, 33 in PRO OS 1/692.
[127] Minutes 81B, 82, 89, 91, 92, 3 August - 23 December 1965, in PRO OS 11/29.
[128] Supt SS3E, '1:25,000 2nd Series - Regular', December 1970: photocopy in possession of late J.B. Harley, 1985-6.
[129] Copy in OS Cartographic Library, December 1992.
[130] See papers in PRO OS 1/701.

extensive use of grey in the second version gave the map a much 'lighter' look and feel, although, as will be seen, using grey was to prove troublesome later and, at any rate with the printing technology of the 1940s and 1950s, was an unsatisfactory choice for a mass-production general-use map. However, the 'java grey' was only adopted after an experiment with 'broken black'.[131] Whatever the merits of the grey, the dominant orange-brown (the shade varied over the years) gave many sheets a decidedly 'dry' feel, which had been mitigated on wartime series - e.g. GSGS 3907 - by the use of green for woods. Also questionable was the indication of bench marks by small arrows, without the value being given; potential users were expected to refer to the official bench mark lists, but the engineers who might have been the main beneficiaries were sceptical from the beginning.[132] An outcome of the second consultation, in May 1945, was the change in position of the sheet number, from twice (centre top and bottom) to four times (in each corner).

Cheetham felt that critics of the first circulated proof of 40/09 had failed to realise the difference between the photographically-reduced GSGS 3906, 'a survey map', and the new 1:25,000, 'a conventional one'. One of those preferring a 'survey' approach was Dudley Stamp.[133] He wanted the actual widths of roads to be shown, and felt that all roads should be colour-infilled, rather than just those classified by the Ministry of Transport. This was important for questions of farm access.[134] A possible answer to this might have been to refer such critics to those sheets of GSGS 3906 which had been produced from GSGS 2748 drawings: narrower roads which had been legible on GSGS 2748 sometimes looked 'pinched' on GSGS 3906.

Colonel Sir Charles Close (DGOS 1911-22) called it 'a credit to the O.S.' and Brigadier E.M. Jack (DGOS 1922-30) thought it 'a beautifully clear map'.[135] Willatts wrote that he and his colleagues in the Maps Office of the Ministry of Town and Country Planning 'have looked at the map with increasing delight and admiration', though he thought that buildings should be slightly less generalised, so as to give a better idea of their character. (This was being put in hand, but Willatts had to wait until the advent of the Second Series nearly twenty years later for really detailed building depiction at this scale.)[136]

The use of orangey-brown was not popular with all users. 'Everyone who has studied the map dislikes the orange contours', wrote Col J.C.T. Willis (DGOS 1953-7) on behalf of the Directorate of Military Survey (DMS):

> The colour itself is unpleasant, and may well be quite overpowering in hilly districts. We are always, from the Service point of view, a little wary of orange as a colour owing to the habit of both the Army and the Air Force of using amber torches. In fact it was this reason which made us plump for 'Air Force Brown'. I presume that you are, for reasons of economy and speed, tied to four colours only. I think that the map would be enormously improved if you could manage to go to six, using a green for wood information and separating out contours and road fillings.[137]

Although the first and the second versions of Sheet 40/09 both looked radically different from anything previously seen at this scale, and indeed anything previously produced by the Ordnance Survey, there was some continuity of content with GSGS 2748 and GSGS 3906. A few road names were retained, usually on the periphery of built-up areas, and trigonometrical points on buildings continued to be shown, primarily for military use, but also so that surveyors, engineers and the like could see which were available in a given area of interest, and obtain the precise co-

---

[131] Conference on 1:25,000, 25 May 1945, 21/1A in PRO OS 1/383.
[132] Conference on 1:25,000 design, 30 December 1944, 19/1 in PRO OS 1/383.
[133] Cheetham to Houghton, 30 May 1945, 2A in PRO OS 1/415; comments by Stamp forwarded in Houghton to Cheetham, 17 Oct 1945, 29A in PRO OS 1/224.
[134] Stamp and Wilson memorandum, 6 June 1945, 29B in PRO OS 1/415.
[135] Close and Jack to Cheetham, 8 and 17 June 1945, 23A and 31A in PRO OS 1/415.
[136] Willatts to Cheetham, 4 July 1945, 40A in PRO OS 1/415; Willatts to Thackwell, 22 April 1963, 92A in PRO OS 1/1192.
[137] Willis to Cheetham, 5 July 1945, 42 in PRO OS 1/415.

ordinate values from published lists.[138] (Indeed, they continued to appear on new 1:25,000 sheets until the mid-1980s, by which time the original reason for them must have been long forgotten.) The showing of other trig. points was discussed in 1950 with DMS, who wanted all visible trig. marks to be shown, whether appropriate to the new triangulation begun in 1935 or to its predecessor, which stretched back into the nineteenth century. The OS thought this difficult to implement, but to be borne in mind as an aim, and adopted a policy of showing all visible points, old or new. As the Provisional Edition was derived from the existing six-inch, new trig. points did not usually appear on it.[139]

The railway symbols followed the example first set on the Western Front by GSGS 2742, and continued on GSGS 2748 and GSGS 3906 and the one-inch Fifth Edition and its derivatives, of using a chequer pattern, short-black-long-white for single track and long-black-short-white for multiple track. On the Fifth Edition there was greater contrast in the length of the chequers, making for much easier comprehension, but the only recorded criticism seems to be that received in 1950 from, appropriately enough, British Railways, who thought the chequers insufficiently distinct.[140] The one-inch Popular Edition's convention of chequers for single track and solid for multiple track had been adopted for new one-inch drawing by 1948, and was adopted by 1954 for the 1:25,000 Regular Edition.

Unlike its pre-1939 predecessors, the Provisional Edition included parish boundaries and names and National Trust areas. The parish and other administrative names were prominent; later, Cheetham criticised them for being too dominant.[141]

The end result of these modifications to the original concept of a three-colour map, and to the design of the pre-war GSGS 3906, was to push the balance of design away from a minimally-coloured derivative of large-scale mapping, towards a more colourful general-user map, akin to the relatively mass-market smaller scale mapping by which the OS had hitherto been known to most of the public, but with an element of 'specialist' content, such as the trig. points and the bench marks. (One might question what real need there was at this scale for the depiction of signposts by '*G.P.*', unless there was some residual military need, or habit.) Points of detail were sometimes a matter for consultation and advice: a notable instance was the definition of public buildings, on which advice was sought in 1947. Some of the suggestions were striking: Col Sanceau of DMS suggested various buildings as either militarily important, including schools, 'or (in the case of cathedrals) buildings the destruction or desecration of which should be avoided'.[142] A representative of the Ministry of Transport suggested that several public utility buildings similar to electricity works should be included for consistency: bus and tram depots were perhaps predictable, abattoirs less so.[143] Christie Willatts provided a comprehensive list, based on planning use zoning, which was privately dismissed later as 'quite impracticable'.[144]

Until 1948 infill was reserved for those roads which bore Ministry of Transport numbers; uncoloured roads might be anything from freshly tarred to the most viscous of rutted cart-tracks. Following public comment, colour infill was extended to unclassified roads, presumably in practice those which were infilled on the one-inch, and the original bold style of road-drawing was modified in about 1947-8 so as to reflect more accurately the irregularity of the boundaries of many roads.

Most of the sheets as first published followed established six-inch practice, and GSGS 2748 and 3036, in distinguishing footpaths ('*FP*') and bridle roads ('*BR*'). (Only footpaths seem to

---

[138] Cobb to Burnett, 29 June 1950, 74A in PRO OS 1/416.

[139] Minutes 109A-111, 114-117 in PRO OS 1/383. There is a good selection of old trig. points on TA 31/pt 41, B/ (1959): see e.g. 363193, 384196, 393184.

[140] Bennett to Burnett, 28 June 1950, 69A-B in PRO OS 1/416.

[141] Cheetham to Burnett, 20 June 1950, 29A in PRO OS 1/416.

[142] Sanceau to DGOS, 4 November 1947, 159A in OS 1/415.

[143] Phillips to Papworth, 31 October 1947, 157A in PRO OS 1/415.

[144] Willatts to Papworth, 6 December 1947, 162A in PRO OS 1/415; note on meetings on 1:25,000 content, 30 November & 5 December 1950, 97A in PRO OS 1/567.

have been distinguished on GSGS 3906.) Shorter pecks were used for footpaths and longer pecks for bridle roads. In 1950 Colonel Heap, then Deputy Director of Field Divisions, suggested that on the Regular Edition bridle roads should not be marked as such, and this seems to have been the occasion for removing them from published Provisional Edition mapping as well.[145] S.F. Marriott did not like footpaths running between double hedges to be shown only by the grey of the boundaries, without black pecks: this was a consequence of the map leaning towards a 'survey' rather than a 'conventional' style. Nothing seems to have been done about this on the Provisional Edition, and numerous examples of roads and paths indicated only by double ditches are to be found on the Second Series in its Explorer form.[146]

As has been noted already, bench marks had been shown from the first, without values, on the experimental versions of sheet 40/09, where they had had a mixed reception. Whilst bench marks with values had always appeared on the six-inch, they had not appeared on any of the Provisional Edition's military predecessors. They are another good example of the 1:25,000 being caught between the 'conventional' and 'survey' concepts. After 1945, bench mark values, like trig. values, were published in lists arranged by 10 km National Grid squares; these lists were usually much more up to date than the values published on the face of the County Series six-inch and 1:2500. One argument for showing bench marks without values on the 1:25,000 was similar to that for showing trigs: engineers, surveyors and others could see readily which ones were situated within their area of interest, though this time the argument came not from a surveyor of the Royal Engineers, but from that of the city of Lincoln.[147] One R.E. surveyor who thought they should be deleted was Major (later Col) Michael Cobb.[148] The Director of the Geological Survey, W.F.P. (later Sir William) McLintock, not only wanted the bench marks to be retained, but to be figured as well.[149] They were omitted from the Regular Edition, but there were some geologists who hoped, vainly, that they might be restored to the Second Series.[150]

If the comparative lack of recorded adverse comment is any guide, woodland was one of the more successful aspects of the map. However, it was not liked by MacLeod, who in 1935 had removed tree symbols (and thus the distinction between coniferous and non-coniferous woodland) from the one-inch Fifth Edition. In 1950 he described the tree symbols on the Provisional Edition as 'an ugly speckly effect', and suggested that on the Regular Edition woodland should be shown by the grey stipple used on the existing map for parks, and that parks, 'the shape of which has no significance', could be shown by a 'riband' (vignetting?). This would enable contours and other details in woodland to stand out better.[151] The suggestion does not appear to have been tried out.

In 1948 the bottom margin was redesigned. In its original form, this included a full legend, with a note on the projection and spheroid used for the National Grid; this necessitated using a cumbersome semi-Bender fold. From June 1948 the conventional signs and geodetic information were omitted, 'to save paper', and on folded copies the top margin was trimmed off and the bottom margin was tucked under. The Provisional Edition diligently continued to save paper in this fashion until its demise forty years later. (With the adoption of integral covers in 1970 it was possible to use an ordinary Bender fold.) Restoration of the conventional signs was considered in the consultation in 1950 as to Regular Edition design. Responses were mixed: some wanted them all to be shown, others thought that they could be omitted, as users would either know them anyway or else could refer to the separately-published conventional sign card, and there were those who thought that some should be shown. These last divided into those who wanted the more obvious signs, and those who wanted the less obvious ones. MacLeod wanted the lot, but printed on the back of the map so as to save paper, or else on the back of the cover, at present containing

---

[145] Heap to Burnett, 10 June 1950, 78A in PRO OS 1/416.
[146] Marriott to OS, 30 June 1950, 89A in PRO OS 1/416.
[147] Adlington to Burnett, 28 June 1950, and see also Watts to Burnett, n.d.: 70A and 98A in PRO OS 1/416.
[148] Cobb to Burnett, 29 June 1950, 74A in PRO OS 1/416.
[149] McLintock to OS, 30 June 1950, 90A in PRO OS 1/416.
[150] Wilson to Thackwell, 22 April 1963, 88A-B in PRO OS 1/1192.
[151] MacLeod to Burnett, 30 June 1950, 82A in PRO OS 1/416.

only 'a useless and meaningless device!' The signs did eventually appear on the back of the Provisional Edition, but only from 1970, a year after MacLeod's death.[152]

The simplification of the bottom margin also entailed removing the publication date after June 1948. So far, the Provisional Edition had followed the example of GSGS 2748 and 3906 in omitting any revision or compilation notes, which, given the mixed parentage of the map, might have been thought particularly necessary for this of all series. The official explanation was that there was no compilation note 'because of the very varied nature of the information used', and the publication date had been omitted 'as a result of queries from the public' (presumably as to why newly-published maps were so out of date).[153] One user asked for it to be made plain by a note on the map that it was based on old material and didn't include wartime and later changes. The official response was that the publication date was omitted because the material was not up to date.[154] The restoration of the publication date and the addition of a revision date seems to have been prompted by Lt-General Sir Desmond Anderson: their omission 'was a criticism made to me when I was extolling to someone the merits of the map'.[155] A proposed note, which was accepted, was drafted by the Director of Map Production and Publication, which gave the date ranges for the full revision of the constituent six-inch sheets and for 'partial systematic revision' (which included one-inch revision, 'air raid precautions' revision of 1938-9 and subsequent miscellaneous revision), and the date of any road revision.[156]

Assimilation of very out-of-date six-inch material to somewhat less-out-of-date one-inch material was not helped by the loss of revision documents during the war. In 1947 S.F. Marriott drew attention to a number of shortcomings on sheet 35/21, covering part of the Lake District, where the six-inch was based on a revision of 1900, with a later one-inch revision in 1920-1. Several important features, including some footpaths, appeared on the one-inch but not on the 1:25,000.[157] The Editor of the *Lakeland Rambler* (published by the local federation of the Ramblers Association) was disappointed that very few fell routes were shown. 'These are the routes mostly required by visitors who climb the fells', and the Federation would be quite happy to advise on them.[158] Yet another Lakeland complaint came from Mr Evans, the Chief Geologist of the Burmah Oil Company: on sheet 35/20 sometimes neither the contours nor the rock-drawing resembled the physical reality, and near 274019 some of the cliff-drawing was upside down - as indeed it was![159] Early in 1947 Major C.J.P. Thompson had visited the Swiss Federal Survey to investigate their methods of rock depiction, but there is no evidence that his visit had any effect on OS practice: the revelation that the Swiss 1:50,000 was still engraved on copper, which the OS had abandoned nearly 25 years earlier, can hardly have been encouraging.[160] Rock depiction was studied at intervals thereafter, with a view to drawing the Regular Edition rather than improving the Provisional Edition. In 1954 an experiment was tried with part of sheet SH 72. 'We all agree that the sheet concerned looks horrible; so, probably does the country itself, and in that respect at any rate the map may well be an adequate depiction of the true state of affairs!' wrote Willis. Half the rock and boulder ornament was deleted in the trial area, but the 'measles effect' remained.[161]

Otherwise, changes to design and content were few. A motorway symbol was added *circa* 1960. From 1957 onwards, as a security measure, industrial descriptions were bowdlerised to 'Works' and 'Mine', names and descriptions of reservoirs were deleted (with the result that some

---

[152] MacLeod to Burnett, 30 June 1950, 82A in PRO OS 1/415.
[153] Minute by Director of Map Production, 8 September 1949, 247 on PRO OS 1/415.
[154] Harvey to OS and Burnett to Harvey, 1 and 6 July 1949, 237A, 238A in PRO OS 1/415.
[155] Anderson to Cheetham, 3 September 1949, 243A in PRO OS 1/415.
[156] Minute by Director of Map Production and Publication, 13 October 1949, 253A on PRO OS 1/415.
[157] Marriott to OS, 14 January 1947, 84A in PRO OS 1/415.
[158] Taylor to OS, 6 December 1947, 163A in PRO OS 1/415.
[159] Evans to OS, 6 July 1948, 204A in PRO OS 1/415.
[160] See minutes 58-71 in PRO OS 1/355; the report is at 70A. For the abolition of engraving see Seymour, *A history*, 242-3, and Hodson, *Popular maps*, 35 ff.
[161] Minutes by Willis and DMP, 10 and 25 June 1954, 87 and 90 in PRO OS 1/355.

lakes adapted as reservoirs, such as Thirlmere, were unnamed for a time), and underground aqueducts, such as those from the Elan Valley and Thirlmere, were also deleted.[162] In 1950 Professor Daysh of the geography department of the University of Durham suggested that the proposed Regular Edition should enable users to assess 'something of the economic characteristics of the area represented', and should also carry information such as tidal ranges and tide times relative to London Bridge, and additional form lines for minor features. He did not wish the OS to go as far in these respects as did French and German mapping, but he did want to see 'the usage of industrial properties' and also 'the nature of shipping facilities in ports'.[163] No-one else then suggested either the physical or the economic information, and the latter would have fallen foul of the new security rules introduced in 1957.

However, Daysh's was not an isolated voice: the comparative lack of detail on OS mapping, particularly the 1:25,000, compared with European counterparts, was the subject of an article by D.R. MacGregor in the *Scottish Geographical Magazine* in 1960. It noted that 88 symbols were used on the British 1:25,000, compared with nearly 120 on the Swiss maps, and, 'the most astonishing figure', 300 on the French 1:20,000.

> The consequences of the trend towards greater content are considerable in their effect. Large-scale maps such as the French 1:20,000 series are becoming extraordinarily informative documents from which a concise and realistic conception of any landscape can be formed, although by contrast the British 1:25,000 map is a relatively crude instrument which fails to realise the potential value of such a scale. The contrast is probably due to the extensive use of aerial photographs by continental map-makers and the employment of geographers in map making. The use of such informative maps requires a high standard of map reading, higher in fact than that accepted of many geography students and graduates today, for the modern topographic map is employing an increasingly complicated code with which a map reader must become easily conversant, and to obtain satisfactory results the student must also master the technique of creating from the map-plan a three-dimensional view in the mind in both detail and colour. The growing content of continental topographic maps also suggests that British geographers might profitably start a critical examination of the empty spaces appearing on the OS 1:10,560 and 1:25,000 series. Information concerning the character of streams, the extent of floodable land, types of vegetation and the nature of industrial premises might all be improved, and would it not be practicable and valuable to show land classification on the *Six-Inch Map*?[164]

The OS attitude was, in effect, that in Britain the 1:25,000 was not a basic scale, and thus need not show so much detail. Maps at each scale had to be designed to suit 'the most general probable requirements': the six-inch was probably used by local authorities, not geographers, and the main 1:25,000 users were probably planners, walkers and schools. And anyway,

> It is very hard to find out exactly what the public wants. The most vocal elements are usually specialists with conflicting personal requirements. We have to aim to provide them with a base on which to add their special requirements without cluttering up the map for the general user.[165]

It is perhaps significant that these requests for a much more detailed 1:25,000 came from academic geographers who may have habitually used comparable overseas mapping in their departments, rather than from those such as walkers who regularly used it on the ground.

Although the introduction of grey had been welcomed in 1945 as helping to lighten to look of the map, it became apparent in time that it needed close control were an acceptable balance to be maintained between legibility of field boundaries on the one hand, and building infill not appearing too heavy on the other. A darker mix of grey had been used on earlier sheets; in 1951

---

[162] For the background to this see PRO OS 1/1479. In fact, very few underground aqueducts were shown.

[163] Daysh to Burnett, 30 June and 6 September 1950, 86A-B and 155A in PRO OS 1/416.

[164] D.R. MacGregor, 'Geographical reflections on modern mapping', *Scottish Geographical Magazine*, 76 (1960), 77-84; quotation from p.80.

[165] Minute 67, 21 October 1960, in PRO OS 1/697.

there was no enthusiasm for using different ink mixes on different sheets.[166] (It would presumably have complicated 'printing in battery'.) Complaints that the field boundaries were too faint appeared during the consultation exercise in 1950-1: S.F. Marriott suggested that green be tried.[167] Green was already regularly used in proof examination, and was substituted for grey on proofs of NY 83 (Weardale in County Durham) and SP 08 (Birmingham); the official view was that it was quite pleasing on NY 83, but 'overpowering' on SP 08. (The present writer felt it to be inadequate on NY 83, and about right on SP 08.) Marriott was told that it was impossible to find a shade of green suitable for both town and country.[168] In 1953 it was decided to use black for field boundaries on the Regular Edition, as it was more straightforward to draw.[169]

Recent building development was shown by grey stipple on a few sheets published in the early 1980s.[170] A similar procedure had been followed on an adaptation of the one-inch Popular Edition to cover the LPTB area, published in 1934; this was a relatively obscure series, and the later use may have been reinvention rather than revival.

These apart, the 1:25,000 Provisional Edition lasted to its end in 1989 with the least apparent design changes of any post-war OS map series. In 1968 it was officially redesignated the 'First Series': this appeared on revised reprints from 1974, but was trimmed off folded copies. It was the only one of the numerous twentieth century 'provisional editions' produced by the OS to be wholly newly compiled and newly drawn; otherwise, 'provisional edition' meant, at its most drastic, the recasting of existing mapping on new sheet lines, exemplified by part of the 1:63,360 New Popular Edition in 1944-7 and the 1:50,000 First Series in 1972-6.

Contours

The mapping was uniformly contoured at 25 feet (7.62 metres). As on GSGS 2748 and the pre-war GSGS 3906, most of these contours were interpolated rather than precisely surveyed; because of mid-nineteenth century doubts as to the value of the method, most of Britain was furnished with surveyed contours only at 50 and 100 feet, and thence at 100 feet intervals to 1000 feet (305 metres).[171] From 1913 the one-inch map had been furnished with contours at 50 feet intervals by the same interpolation method that would be used for GSGS 2748 and 3906 (and presumably post-1914 GSGS 3036 sheets, too).[172] Such interpolation was an art rather than a science, and sometimes those responsible came to somewhat different conclusions on different scales and at different dates. Additional surveyed contours had been recommended by the Davidson Committee in 1938, but they made no suggestions as to the vertical interval.[173] MacLeod had thought that new contours for the six-inch might be surveyed at 10 feet (3.28 metres) intervals, but published on the 1:25,000 with metric values. (His attempt to persuade the Committee to recommend full OS metrication was a failure, but he lived long enough to see it adopted as Government policy in 1965.)[174]

Contours were shown on the Provisional Edition with those every 100 feet emphasised as 'index contours'. They were shown by broken lines in built-up areas, being considered of little real value, and showing 'how the land lay before man disturbed it'.[175] The contours were popular with educationalists, and T.C. Warrington of the Geographical Association arranged for contour-

---

[166] Minutes by DMP, 14 September and 19 October 1951, 85 and 87 in PRO OS 1/567.

[167] Association of Municipal Corporations to Burnett, 4 July 1950, 96A in PRO OS 1/416; Houghton to Burnett, 16 January 1951, and Marriott to OS, 4 December 1951, 1A and 29A in PRO OS 1/697.

[168] Minute by DDSS, 18 April 1952 and accompanying maps, and OS to Marriott, 1 May 1952, 36 and 37A in PRO OS 1/697.

[169] Minute, 24 August 1953, 123 and 123A, in PRO OS 1/567.

[170] *e.g.* SE 80, edition B//* (1983); SJ 60, edition B///* (1982); SU 65, edition C//*/*/* (1982).

[171] Further details by county will be found in Richard Oliver, *Ordnance Survey maps: a concise guide for historians*, London, Charles Close Society, 1993, 123-171.

[172] Hodson, *Popular maps*, 99-107.

[173] *Final report*, 17.

[174] Seymour, *A history*, 261-4, 327.

[175] Newman to Peek, 16 February 1948, 170A in PRO OS 1/415.

and-water pulls to be supplied for the Association's members. He thought that 'there is great need for town children (and other folk for that matter) to get some picture of the relief of their own area and it is anything but easy for them on the ground. If it is at all possible, I should like to see the relief made more obvious.'[176] (Cheetham may have been glad to help, but one of those more closely concerned with producing these pulls regarded them as 'a perpetual source of worry'!)[177] They were perhaps less popular with some specialist users, as evidenced by criticisms in an article by Keith Clayton, a rising young geomorphologist, published in 1953.[178] Contours were only shown on the coloured edition, but there were periodical requests for them to be included on the outline edition as well, mostly from the geologists.[179]

The 1:25,000 and the mapping of towns

Cheetham's public espousal of the 1:25,000 and his private preference for 1:20,000 have already been described. The 1:25,000 was drawn at 1:15,000, and this may have given him the idea of producing a 1:15,000 series covering urban areas, which could include far more names of streets and public buildings than was possible at the smaller scale. The possibility was first mentioned in September 1946, in a letter covering a proof of sheet 51/28 (north-west London), which was sent out to several government departments for comments and suggestions as to what might be shown on the London sheets, including the treatment of underground railways.[180] Most of the respondents, including Stamp, Willatts and Brigadier Martin Hotine (head of the Directorate of Colonial Surveys, and widely regarded as 'the best Director-General that the OS never had'), were supportive: Willatts reported that within the MTCP there was not unanimity as to the usefulness of the 1:15,000, but that he and his staff thought that the idea 'should be accorded a hearty welcome'.[181]

Accordingly, preparations were made at the turn of 1946-7 for compiling lists of street names and public buildings: on 1 January nine men were working on the project, but little more was heard of it.[182] Over forty years later OS, in collaboration with George Philip, began publishing a series of county street atlases at 3½ inches to 1 mile (1:18,103), initially using photo-enlarged 1:50,000 material, which provided roads wide enough for general naming, so perhaps Cheetham was ahead of his time. The only practical outcome of the exercise seems to have been the adoption of a symbol for stations on underground railways.

Cheetham returned to the idea of improved urban mapping, this time at 1:25,000, after he retired. In response to the consultation exercise for the design of the 1:25,000 Regular Edition, he said that the 1:20,000 would have been a better choice of scale but that it was too late to change, and suggesting ways in which the 1:25,000 could be improved as a 'town map'. 'The object is to provide a good topographical map of an Urban area.' Names should be topographical and not repeated for administrative purposes, and prominent rather than public buildings should be emphasised, even if the OS was liable to attack for advertising the Plaza and not the Odeon, as the former was prominent and the latter was not. As the object would be to show sufficient detail to enable one to locate oneself quickly, street names should be shown wherever possible, as should well-known local names, including important bus stops (often named after public houses), and landmarks of all sorts, such as 'conspicuous public clocks'. Buildings might be treated in perhaps three categories: normal, prominent, and very prominent. It would be worth putting house outlines

---

[176] Warrington to Cheetham, 9 November 1946, 74A in PRO OS 1/415; for the contour-and-water pulls see correspondence from 24 October 1946 onwards, 73A in PRO OS 1/415.
[177] Minute by O.P., 9 August 1947, 141 in PRO OS 1/415.
[178] K.M.Clayton, 'A note on the twenty-five foot "contours" shown on the Ordnance Survey 1:25,000 map', *Geography*, 36 (2), 1953, 77-83.
[179] Welch to OS and reply, 27 April and 16 May 1948, 183, 185/1A in PRO OS 1/415; McLintock to OS, 30 June 1950, 90A in PRO OS 1/416; Advisory Committee on Survey and Mapping, 30 April 1964, 48 in PRO OS 1/1303.
[180] OS circular letter, 30 September 1946, 60A in PRO OS 1/415.
[181] Willatts to Papworth, 30 October 1946, 70A in PRO OS 1/415.
[182] Minutes 80, 81A, 83 in PRO OS 1/415.

on the grey plate, with prominent houses with black outlines, and very prominent ones infilled black. Landmarks such as churches would show up better against the grey, and names would stand out better. This depiction might be co-extensive with areas mapped at 1:1250. Cheetham admitted that the fieldwork involved would be much more demanding of surveyors than ordinary OS large-scale work.[183]

Some of these ideas were tried out on part of sheet 51/39: this involved removing most road and building casings, and adding street names. The result seemed harder to read, the distinction between terraces opening direct onto the street and houses with gardens was lost, it took about 30 per cent longer to prepare, and it joined the long list of OS experiments which have never reached the public.[184]

Cheetham returned to the theme later when writing the official *Textbook of Topographical Surveying*, published posthumously in 1965:

> On topographical maps it is desirable to show a well-distributed selection of landmarks. Such features are not always easy to find in towns. Churches are usually shown but these are not so conspicuous as they are in rural areas. Street names afford a ready means of checking position but it is not possible to show many at scales below 1/20,000. The best method is to show a selection of prominent buildings, particularly when these can be readily identified because they include such a feature as a large clock. The number of such landmarks which can be shown clearly is greatly increased if the filling for buildings and some minor detail is shown on a grey plate against which the landmarks, shown in black, will show up prominently.[185]

At least he tried.

Alternative formats: the outline edition

At first the Provisional Edition was offered in three forms: the standard coloured version, an outline version printed in grey, and an Administrative Areas Series. There was also a military edition of the standard coloured map, GSGS 4627. The Administrative and military editions are discussed below.

The outline edition was printed in grey. Black had hitherto been usual on outline editions of smaller-scale mapping and was tried on sheet 40/09, but may have been rejected because of a feeling that solid black buildings in urban areas would be far too 'heavy' in appearance.[186] Whilst grey was useful as a base for overprinting, for example the Administrative Areas edition, it was unsatisfactory when hand-tinting was applied, since the detail, already subdued, tended to disappear completely.[187] The choice of it may have been influenced by various 1:625,000 'planning maps' being produced in the mid-1940s, most of which used grey bases. Various alternatives were tried in 1948, without conclusive results.[188] 'Blue-grey' was suggested by the Director of the Geological Survey in 1950, but is not known to have been tried.[189]

By 1949 the OS was obliged to admit that only the lightest water-colour could be used with the grey; the shade was claimed to have been adopted on advice from the MTCP, but a few months later Willatts called it 'often too pale for easy reading'.[190] Despite this, the outline edition was 'quite popular'.[191] In 1949 average issues per sheet were 159.04 for the coloured version,

---

[183] Cheetham to Brown, 28 June 1950, 158A-B in PRO OS 1/416.

[184] Items 10, 10A, 11, 14 in PRO OS 1/697.

[185] [Geoffrey Cheetham], *Textbook of Topographical Surveying* [fourth edition], London, HMSO, 1965, 307-8.

[186] There is a 'black' specimen of sheet 40/09 in the collection of 1:25,000 specimens at present in the Charles Close Society-Ordnance Survey deposit in the map department of Cambridge University Library.

[187] Town Planning Institute to OS, 30 June 1950, 83A in PRO OS 1/416.

[188] See 176A, 178A, 180A, 182A in PRO OS 1/415, and 60A in PRO OS 1/416.

[189] McLintock to OS, 30 June 1950, 90A in PRO OS 1/416.

[190] OS to Stanford's, 2 November 1949, 79A in PRO OS 1/383; Willatts to Burnett, 4 July 1950, 109A in PRO OS 1/416.

[191] Note on meetings on 1:25,000 content, 30 November and 5 December 1950, 97A in PRO OS 1/567.

26.57 for the outline, and 13.99 for the Administrative Areas Series; in 1950 the figures were 131.9, 27.06 and 10.55 respectively. In 1954 22,800 outline sheets were issued, an average of 12 per published sheet.[192] In 1963-4 average annual issues per sheet were about 150 for the coloured edition, and about 11 for the outline.[193] At this time the main users of the outline edition were apparently local authorities and market researchers. The gradual decline in the ratio of outline to coloured issues by this time may account for the heavy 'cancellation' (scrapping) rate of 16-17 per cent for unrevised copies, which was officially attributed to printing excessive stock. This contributed to a small loss on outline publication, but at this time there was a 'presentation difficulty' in charging more for the outline than for the coloured edition.[194] This difficulty disappeared from the later 1970s, as the concept of 'special interest' mapping and associated special charging developed; by the time of the Provisional Edition's demise in 1989, the coloured version was £1:90 per sheet, and the outline £7:50. (The corresponding prices for the Second Series, usually offering twice the area, were £2:80 and £8:55; copies of the outline edition of both series were also available as film transparencies, as a special service and correspondingly higher price.)[195] Grey continued to be used until October 1973, when black was substituted. By that time probably few people cared either way.

The Administrative Areas Series[196]

The Administrative Areas Series originated in the need of the Parliamentary Boundary Commission for maps at this scale, and it was thought that there might also be a sale for them amongst local authorities, public utilities, 'and possibly the public generally'. It would be possible to show rural parish boundaries in much greater detail than on the one-inch, and publication of an administrative version of the 1:25,000 'may decide the very vexed question' of showing parish boundaries on the smaller scale, in favour of their omission. National Trust areas would be shown, including those not open to the public.[197]

Several interested parties were consulted, and Willatts thought that the series 'should prove a very valuable one', though 'the extension of the map to rural areas would seem to be an extravagance'. He was opposed to omitting parish boundaries from the one-inch.[198] Stanford's, the main OS agents, thought that the mapping would be in demand in rural as well as urban areas.[199] Other replies were less enthusiastic, though none were outright hostile, and the new map would receive 'a very warm welcome' from the Ministry of Health.[200] Work on the series went ahead; because of the Boundary Commission's requirements, various compromises were resorted to, including basing some sheets on GSGS 3906 material (e.g. sheet 36/37), and making available outline sheets without names, 'for *exhibition* purposes' (presumably displaying proposed boundary schemes to interested parties).[201] These makeshift maps were not published; those which were, were sometimes delayed by proposals to rearrange wards, as in Edinburgh and Glasgow.[202]

---

[192] Tabulated figures, 1949-50, and minute, 24 January 1955, 84A and 225 in PRO OS 1/383.

[193] These figures have been derived from those tabulated in 45A in PRO OS 1/776, and exclude an exceptional issue of outline sheets noted there.

[194] Minutes 45, 47B, 48 in PRO OS 1/776.

[195] OS Price List, November 1988.

[196] This was referred to as 'Administrative Areas Series' in the map headings, but as 'Administrative Areas Edition' elsewhere, notably in OS Central Registry files now in the PRO, and in the medium scale *Descriptions* of 1949 and 1955 (pp 16 and 15 respectively; plate XIV in both). Here, 'Series' is used, following the example of the maps.

[197] Item 1A, in PRO OS 1/224.

[198] Willatts to Cheetham, August 1945, 11A in PRO OS 1/224.

[199] Stanford to OS, 23 August 1945, 12A in PRO OS 1/224.

[200] White to Cheetham, 1 November 1945, 30A in PRO OS 1/224.

[201] Copy of sheet 36/37 at 40A in OS 1/224; for advance copies, see minutes 83 and 85, 9 and 18 January 1947, in PRO OS 1/224.

[202] Cheetham to Scottish Home Dept and reply, 5 and 24 April 1948, 50A, 51A in PRO OS 1/224.

In commenting on print runs and sales in October 1947, Cheetham wrote:
> In originating this series I was in some doubt as to whether it should be confined to urban areas or carried over the whole country. It was I think largely on the advice of the M.T.C.P. that we adopted the latter proposal.[203]

The written record suggests rather that the advice came from Stanford's, but from 1948 Administrative Areas publication was confined to sheets containing wards. By June 1950 there were doubts as to whether the series was worth maintaining, and those consulted as to Regular Edition design were specifically asked whether it was worth publishing.[204] Unsurprisingly, those directly concerned with administration rallied to its support: for example, the Town Planning Institute called it 'a most important map'.[205] Cheetham, now retired, thought that the series 'will eventually find a market', but if not, it should be dropped, 'and wait... for the squeals'. Demand might pick up if the Local Government Commission was reconstituted, and if the Conservatives were returned to power and amended the Representation of the People Act.[206] (They were, and they didn't.) Overall the replies did not indicate any increase in demand and, in meetings later in 1950 to consider comments elicited by the consultation, it was suggested that no administrative version be produced of the first 200 or 300 Regular sheets; any demand could be assessed from objections received.[207] Perhaps those most enthusiastic about this mapping overestimated the charm of administration for a wider public.

For the time being, the Administrative Areas Series continued to be printed, and a few revised sheets were issued. Problems were likely to arise when boundaries on the base-maps, which were standard outline edition stock, and might be several years out of date, were at odds with the revised overprint.[208] Sales were declining: 5,800 copies representing 980 published sheets had been issued in 1951 (an average of 5.9 copies of each sheet), 4,080 copies, representing 1,000 published sheets, in 1952 (average 4.1 copies per sheet), and 3,640 copies, of 1,004 published sheets, in January to October 1953 (average 3.6 copies per sheet); the cancellation rate was 88 per cent. Issues to OS agents declined from an average of 14 per sheet in 1949 to 2.7 in 1953.[209] 'The main users of this edition are Local Authorities and Political Organisations.'[210] It cost about £21 to draw a sheet and, because of problems with the drawing material used, revised sheets had to be completely redrawn.[211] The formal decision to stop production and reprinting of the whole series was taken on 11 May 1954: customers were to be told that it had been discontinued on account of the small demand.[212]

In 1958 the plates were cleaned off and the job files destroyed, but the remaining stock continued to be offered for sale.[213] 'Offered' seems to be the operative word, as in July 1963 112,096 copies were held: 450 had been issued in 1961 and 440 in 1962, and about half of those were in lieu of out-of-stock outline editions. Accordingly, the whole stock was withdrawn from sale and cancelled in September 1963.[214]

However, that was not quite the end of the story. In the spring of 1964 the Metropolitan Boroughs Joint Standing Committee noted that, because of the London Government Act of 1963, the boundaries shown on OS large-scale mapping of the capital would shortly be obsolete. As it would take a long time to revise these maps, the OS Director-General - now Major-General A.H.

---

[203] Minute by Cheetham, 13 October 1947, 33 in OS 1/383.
[204] Circular letters, 15A-M in PRO OS 1/416.
[205] Town Planning Institute to OS, 30 June 1950, 83A in PRO OS 1/416.
[206] Cheetham to Burnett, 20 June 1950, 29A in PRO OS 1/416.
[207] Note on meetings on 1:25,000 content, 30 November and 5 December 1950, 97A in PRO OS 1/567.
[208] Minutes, 5 and 11 November 1953, 193 and 195 in PRO OS 1/383.
[209] Minutes by ADP, 1 December 1953 and 30 April 1954, 199 and 208A in PRO OS 1/383.
[210] Minute by ADP, January 1954, 203 in OS 1/383.
[211] Minute by DDSS, 10 December 1953, 201 in PRO OS 1/383.
[212] Extract from minutes of Director-General's Conference 246: 210 in PRO OS 1/383.
[213] Minutes 238, 241 in PRO OS 1/383.
[214] Minutes 1, 4, 9 and 10 in PRO OS 1/1267.

Dowson - offered an overprinted 1:25,000 instead, which was accepted.[215] It was originally proposed that the mapping be published in 20 by 10 km sheets, so that Second Series mapping could be substituted for the base if necessary, but in the event 10 by 10 km sheet lines were used, with a few extended sheets to avoid publishing otherwise largely redundant sheets. No Treasury authority was sought, as none had been sought for abandoning the original Administrative Areas Series authorised in 1945.[216] The maps were duly published, and appear to have been withdrawn from sale in 1976.[217]

The military version: GSGS 4627
As recounted above, in the late 1930s MacLeod's original concept for the 1:25,000 had been of a map printed primarily for military use, but with a 'run-on' for civil purposes. The military continued to use the War Office Cassini gridded 1:25,000 and one-inch series, GSGS 3906-8, until 1950, but in 1947 work began on preparing military editions of the Transverse Mercator National Grid mapping at these scales. The military 1:25,000 was GSGS 4627, redesignated M.821 in 1957, following the introduction of NATO Standard Series Designations. Although GSGS 3906 had eventually covered the British Isles, the immediate justification for it was removed with the end of hostilities; GSGS 4627 was more limited in its cover, and output lagged behind that of the civil map. The military differed from the civil edition in that it added a purple overprint, mainly so as to enable grid figures to appear on the map face. At first this overprint was in purple; from October 1952 blue began to be substituted, but some sheets using purple continued to be printed until 1964. Preparation began at a time when the civil map still carried a full legend in the bottom margin, and this was retained for military sheets prepared by the OS in 1948-9 after it had been omitted from the civil map. This made extra work, and the War Office raised no objection in May 1949 when the OS proposed to omit the legend and thus bring GSGS 4627 in line with its civil counterpart.[218] Three sets of reproduction materials were prepared, one of which was controlled by the OS, one by DMS, and one of which was kept in the United States, in case of an 'emergency'.[219]

The early sheets with legends were all of military training areas in the south of England, including extensive cover of Aldershot and Salisbury Plain. The next priority seems to have been other training areas, ports, and other important potential enemy targets: it is difficult otherwise to understand why, for instance, the lower Humber estuary was comprehensively covered by 1951. Extension thereafter was rather patchy, and cover was only complete for East Anglia, central southern and south-east England. The justification for extending cover was that it was expected that the map would be used for civil defence work in the event of nuclear attack; as it was impossible to predict fall-out it was essential to have the whole country covered at this scale. It was felt to be adequate to meet military needs for 'some years' (!) and so there was no War Office pressure for replacement by the Regular Edition, though when Regular Edition mapping was available in Devon it was used.[220] National cover was evidently not so essential that there was any urgency, and there are striking gaps, including several industrial areas, parts of some coalfields, and much of central Scotland. The mapping was also used by the Royal Air Force for target practice, which may account for some coastal and other isolated sheets being published in GSGS 4627.[221] The apparent lack of urgency perhaps explains why only about a quarter of the sheets were issued in a second GSGS edition; 29 went into a third edition, and two into a fourth. Some sheets seem to have been republished as second editions purely because of the change in 1951 in the designation of 100 kilometre grid squares from numbers to letters, with a consequent

---

[215] Minutes 11, 11A, 12 in PRO OS 1/1267.
[216] Minutes of 21 and 23 September 1964, 21 and 22 in PRO OS 1/1267.
[217] They are mentioned in the *OS Map Catalogue 1976*, 47, but not in that for 1977.
[218] 13A and 15A in PRO OS 1/567.
[219] DMS to DGOS, 12 April 1950, item 36(1)A in PRO OS 1/567.
[220] DMS to DGOS, 4 November 1955, item 26A in PRO OS 1/1123.
[221] Air Ministry to OS, 3 July 1950, 95A in PRO OS 1/416.

effect on sheet numbers; this procedure was soon abandoned, presumably because it led to cancelling large stocks of otherwise perfectly good maps. 111 sheets were first published as second editions, without having an earlier edition, perhaps because of an intention that second edition sheets would be those with the letter designations of the grid squares. Some sheets were evidently not worth preparing in GSGS 4627 form, and the War Office drew civil stocks from OS, which could affect reprinting programmes.[222]

Whereas most printing of the military version of the one-inch (GSGS 4620 and GSGS 4639, later M.722 and M.725) was carried out by OS, much printing, and particularly reprinting, of GSGS 4627 was undertaken by the military, including some by the Royal Canadian Engineers, from the reproduction materials supplied by OS. Whilst this relieved OS of a burden, it had the disadvantage that it disguised the true extent to which the 1:25,000, taken as a whole, was being printed. It would appear that, whatever arrangements may have applied since, DMS did not at this time pay to OS any 'reproduction royalty' on its printings of GSGS 4627. In fact, the average runs for GSGS 4627 were substantially in excess of those for the civil map, and MacLeod's concept of a primarily military and only secondarily civil map was being realised, thought this went unperceived at the time. As there was a regular interchange of senior officers between OS and DMS - the usual pattern was for the Director of Military Survey to become the next Director-General of the Ordnance Survey, having previously served on the OS in a lesser capacity - it is strange that this apparently went unremarked on in OS circles, at any rate in writing. One reason may have been that the Director of Establishments and Finance was always a civilian. As will be related later, the DMS printing of GSGS 4627 was not the only 'hidden' 1:25,000 consumption statistic.

After 1,084 of the 2,027 sheets had been printed, the military lost interest in the erstwhile GSGS 4627, now M.821, as a national series, and it was retained only for a few training areas, on ad hoc sheet lines and with individual GSGS numbers. These usually had all the grey information on the parent mapping printed in black, and additional overprints in various colours. As a result of incomplete cover in Scotland, it was not always possible to use the Provisional Edition as a starting-point: thus *Garelochhead Military Training Area*, GSGS 4962, most of which lay outside the area mapped in the Provisional Edition, was produced by photo-enlarging one-inch Seventh Series material, and using the standard training-map colour-scheme, including orange roads and contours. The result looked like a modernist commentary on the Second War Revision of GSGS 3907.[223]

The start of the demise of GSGS 4627 can perhaps be dated to 1960. DMS reviewed its policy of mapping the United Kingdom at one-inch and 1:25,000 with a view to ensuring that up-to-date editions were available; the OS advised that it was not intended to incorporate the results of the new one-inch revision, begun in 1958, in the Provisional Edition.[224] Did DMS expect the nascent Second Series to serve instead? Perhaps: the first three Second Series sheets published of England were duly printed as M.822 in 1966, but they had no successors.[225]

Thematic maps

Provisional Edition material was also used for three thematic mapping series using OS film reproduction materials; two were printed but none were published by OS. These were the Second Land Utilisation Survey of Great Britain (LUS), the Geological Survey 'Classical areas of British Geology', and sheets of the Soil Survey. The LUS began issuing its 20 by 10 km sheets in 1961, and they appeared at irregular intervals until 1977: all but five of the 120 were published by 1971, and only one was a Scottish sheet. The sheets were printed in the three or four colours of the parent OS map, with the addition of several land-use colours. The distribution of the published

---

[222] Minute, November 1950, 119 in PRO OS 1/383; Willis to OS, 12 May 1950, 36(1) in PRO OS 1/567.
[223] Comments apply to the copy seen, which is edition 2, 1973.
[224] DMS to OS and reply, 1 June and 12 July 1960, 8A and 8A in PRO OS 1/776.
[225] These were SX 86/96, TQ 20/30 and TQ 21/31; copies of all three are in the Ministry of Defence Map Library, and there is a copy of SX 86/96 at 101A in PRO OS 1/1303.

sheets reflects varying degrees of local enthusiasm, with some county councils more willing than others to subsidise publication. The maps were printed by various commercial firms and were produced to a very high standard: some would say that there has been nothing like before or since.[226] Provisional Edition bases were used even when Second Series bases were at least theoretically available (e.g. TQ 20/30, Brighton, of 1968). The Geological Survey began publishing its 1:25,000 sheets in 1968; by 1988 some 38 had been published using Provisional Edition base maps. They were often formed either by adding part of another sheet to a published 10 by 10 km sheet, or else were independent of the standard sheet lines altogether. The Soil Survey sheets, of which, again, only specimen areas were published, used the standard sheet lines, but were sometimes prepared in two or three different versions, with monochrome base-maps in blue, green or brown. Both the geological and the soil series used a monochrome base derived from the outline edition, with the addition of contours.[227]

Sheet sizes, numbering and names

Although the more vocal specialist and professional users might complain of the out-datedness of the Provisional Edition, the map-using public at large, who had formed the main civil justification for the 1:25,000 at the time of the Davidson Committee, had two more quite different objections. They were related: the small sheet size, and the sheet arrangement and numbering. In the light of more recent developments there can be no doubt that these two causes were partly responsible for the long-continued low sales of the 1:25,000. Though the same geographical area could be covered much more economically at 1:25,000 than at six-inch, the small sheet size made the 1:25,000 an expensive buy compared with the one-inch. In the late 1940s both the one-inch New Popular and the 1:25,000 cost three shillings for the mounted-and-folded format, which represented 19.3 square inches of map area per penny for the one-inch against 6.9 square inches per penny for the 1:25,000. The wonder was less that so few 1:25,000s were sold, as that so many were. The relative proportion was not greatly different in the last months of the Provisional Edition in 1989: a standard 10 by 10 km sheet, costing £1:90, gave the purchaser 8.42 square centimetres of map or 0.52 square kilometre per (new) penny, compared with 20.64 and 5.16 respectively for the 1:50,000, selling at £3:10 per sheet. The 1:25,000's sheet size was of course the outcome of MacLeod's advice, eagerly accepted by the Davidson Committee, that in future OS maps at larger than one-inch scale should be laid out in square butt-jointed sheets on the National Grid. One could plead in mitigation of this that in 1939 it had been hoped to sell the 1:25,000 at one shilling per sheet, which would have made the price per square inch of map more comparable to the one-inch, and one might also plead that the National Grid may have been expected to catch on with the public in a way that it signally failed to do.[228] The argument that one could deduce the sheet number of a 1:25,000 or larger scale map simply from a grid reference was a sound one, but it rested on a very shaky foundation: the assumption that the grid would be generally understood and used. The National Grid was probably even less successful than the 1:25,000 map; the one-inch and smaller scale maps carried it but did not ultimately depend on it, and so for them the baleful influence of the Grid was effectively neutralised. But the 1:25,000 suffered. Complaints about the sheet numbering system and inconveniently small size made no impression with those responsible for Ordnance Survey policy; the confident prediction in 1950 that criticism of the sheet numbering system would 'evaporate' once the figure designations of the

---

[226] See Alice Coleman, 'The Second Land Use Survey: progress and Prospect', *Geographical Journal* 127 (1961), 170-86; Alice Coleman, 'Some cartographic aspects of the Second Series Land Use maps', *Geographical Journal* 130 (1964), 167-70; Alice Coleman, 'Some technical and economic limitations of cartographic colour representation on land use maps', *Cartographic Journal* 2 (1965), 90-4; Alice Coleman and W.G.V. Balchin, 'Land use maps', *Cartographic Journal* 16 (1979), 97-103.
[227] Only those thematic maps based on First Series material are noted in this volume.
[228] See minutes 10-15 in PRO OS 1/383.

100 km squares had been replaced by letters must rank among the more unfortunate in Ordnance Survey history.[229]

The small sheet size did not commend itself to all potential users. 'A walker on Dartmoor covering any distance would need quite a pocket-ful of these little sheets.'[230] The rigid sheet lines were a source of complaint from the very beginning: the Central London YHA Group called them a 'retrograde step'.[231]

Sheet names were for long resisted, but by 1964 there had been a change of heart, and they were allocated, usually on the basis of the place names appearing on the back-cover indexes then in use on folded copies of the maps, which had first been prepared in the later 1940s for the initial publications.[232] The process which led to their introduction to both the Provisional Edition and its successor was apparently started by Professor W.G.V. Balchin, who at a Map Users Conference in October 1962 asked if the 1:25,000 Second Series sheets would have the same names as the Land Utilisation Survey sheets then being published. The initial reaction was that this was a 'retrograde step', and it was noted that the LUS was having great difficulty in finding names for sheets, but also that they might help sales by indicating the locality.[233] The subject was raised again in April 1963 by Professor K.C. Edwards on behalf of the Geographical Association. The final decision to name may have been swayed by the Director-General, Major-General A.H. Dowson.[234]

Map covers[235]

Sales were probably not helped by the information given on, or omitted from, the covers. The first cover design had a diagram showing the sheet inside plus the twenty-four surrounding sheets, but the surrounding sheets were not numbered. This design was printed first on buff and then (from about 1949-50) on cream-coloured card. From about 1955 a modified design was used, in which the index was omitted (h.107). A later development was a series of regional indexes on the back cover, which could potentially help to locate sheets over a wide area of interest, although in practice the sheet inside often seemed to be close to the edge of the diagram. In 1965, as part of a wider restyling of OS map covers, a completely new design was introduced. It featured a 'magnifying glass' drawing attention to mapped detail (H.113). The area shown was the village of Berry Pomeroy in Devon (SX 828612), but all the text was fictionalised.[236] From 1970 hinged card covers were abandoned for the Provisional Edition, in favour of using the back of the map for an integral cover, with a standardised 'minimalist' design in blue (H.132.3) and the description 'First Series' (for the first time on a 1:25,000 cover), a national index (wonder of wonders!) and a legend. MacLeod would have approved.

The Provisional Edition was originally offered either paper flat, or mounted and folded; around 1950 it began to be offered paper folded. The mounted and folded format sold poorly, and was discontinued for new publications in 1954, although considerable stocks were still in store in 1964, by which time sales (of the declining number of unrevised sheets) were modest: in the previous year thirty had been sold, 200 had been returned by agents as superseded, 3,480 had been cancelled, and 14,500 remained in stock. Rather than cancel them, they were issued in the new 'magnifying glass' covers, and at paper-folded price.[237]

---

[229] Note on meetings on 1:25,000 content, 30 November and 5 December 1950, 97A in PRO OS 1/567.

[230] Harvey to OS, 1 July 1949, 237A in OS 1/415; see also Thompson to OS, 24 May 1948, 196A in OS 1/415.

[231] Marriott to OS, 15 June 1945; Central London YHA Group to OS, 28 February and 22 March 1946: 30A, 53A, 56A in PRO OS 1/415.

[232] Minutes 25, 25A in PRO OS 1/776.

[233] Minutes 68, 70, 71 in PRO OS 1/1192.

[234] Edwards to Thackwell, 20 April 1963, 89A-B in OS 1/1192; minute by Dowson, 42 in PRO OS 1/1303.

[235] The 'H' numbers cited here are references to Roger Hellyer's listing of Ordnance Survey cover designs in John Paddy Browne, *Map Cover Art*, Southampton, Ordnance Survey, 1991.

[236] See 'Puzzle corner' in *Sheetlines* 29 (January 1991), 29.

[237] Minute 10 in PRO OS 1/1123; minutes 55A-68 in PRO OS 1/776.

In the early days of the series, folding and covering maps followed publication 'haphazardly'.[238]

Print runs and sales

Unlike for the one-inch Seventh Series and the 1:50,000 First Series, no job files survive for the 1:25,000 Provisional Edition, and so it is impossible to say precisely how many were printed and how many were sold.[239] Such figures as do survive are to be found in OS files now in the Public Record Office which are primarily concerned with policy and design, and may not always be strictly comparable with each other. It was noted above that in 1939 an OS officer thought that publishing small isolated groups of sheets would hinder rather than help sales: whilst, because of the 'planning justification', this was the publication pattern in the early years of the series, there is no evidence whether this had an depressing effect on those sheets which had been published.

On the early sheets, printed in 1945-6, quantity-year print codes were employed, but these do not seem to have followed the same rule as for smaller-scale maps, so that whereas '15046' on a one-inch map can be taken to mean 15,000 copies printed in 1946, on a 1:25,000 it must surely signify 1500. In October 1947 print orders for all three styles - coloured, outline and administrative - were noted as 'hitherto ... based on guesswork tempered with discretion'. At this time it was thought that orders for a two-year stock should be 1,000 copies for areas of urban or rambling interest, 500 for small towns, and 125 for rural sheets. For the outline style, the figures should be 100 for urban and 50 for rural sheets, which was less than current orders, and for the administrative version 200 and 50 respectively, which were above current orders. Cheetham's reaction was that the sales figures on which these recommendations were based 'give cause for disappointment but not at present for despondency', and that surplus stock might be used by the War Office.[240]

Six months later sales seem to have been picking up: 125 was thought too low for rural sheets, as any significant sales usually necessitated a reprint, so runs for the three categories of sheet were increased to 1,200, 600 and 250 respectively.[241] In 1949 180,821 copies of the coloured edition were issued, an average of 159.04 copies of each published sheet; in 1950 this declined to 131.90. The average reprint run at this time was 850 copies; this represented anything between 250 and 2,000 copies for five years' stock.[242] Subsequently, reprinting was complicated both by orders by the War Office for copies of the civil version (of areas not covered by GSGS 4627), and by the juggling of revision and the call on drawing resources by other series, notably the six-inch, with a smooth flow of work through the presses.[243]

Sales could be unexpected. Sheet 53/44 (later TF 44) only contained about 25 square kilometres above high water mark (HWM), adjacent to the Wash; this extended sheet of 150 square kilometres mostly depicted foreshore and offshore sandbanks. As such, it was 'admittedly not of much use to the general public', but it sold 160 copies, nearly half the print-run, in the ten months following publication in June 1949.[244] No sales figures are available for that remarkable pair of sheets covering (quite literally) parts of Morecambe Bay: 34/35 (later SD 35), which contained about 1.5 square kilometres above HWM, in a sheet of 120 square kilometres (extended, as about 5 hectares of the marine component of the borough of Fleetwood lay outside the basic sheet), and 34/36 (later SD 36), which contained less than 10 hectares above HWM in the whole sheet. Both were republished in the 1950s in revised editions.

---

[238] Minute, 9 October 1947, 31 in PRO OS 1/383.

[239] The one-inch and 1:50,000 job files are in the Charles Close Society-Ordnance Survey deposit in the map department of Cambridge University Library.

[240] Minutes by OP and Cheetham, 9 and 13 October 1947, 31 and 33 in PRO OS 1/383.

[241] Minutes, 10-27 April 1948, 42-45 in PRO OS 1/383.

[242] Issue figures for 1949-50 and minute, 28 February 1950, 84A and 88 in PRO OS 1/383.

[243] Minutes 119, 124A, 167-169 in PRO OS 1/383.

[244] Minutes, 21 April and 12 May 1950, 39 and 40 in PRO OS 1/567.

In February 1951 a list of sheets awaiting reprint shows estimated annual consumption varying from 75 (sheet 22/40 (later SN 40), Kidwelly and Burry Port) to 480 (31/56 (later ST 56), southern fringe of Bristol).[245] In 1954 162,800 coloured sheets were issued; average issues were 81 per sheet, being 63 paper flat and 18 folded.[246] Costs at this time were such that, for an edition of 50 copies the cost per copy was 81.84d, for 200 copies 27.28d, for 500 copies 16.13d, for 1,000 copies 12.80d, and for 3,000 copies 10.34d. On average, 1,000 copies would give ten years' stock.[247] As in all printing, the shorter the run, the higher the proportion of 'make ready' costs each copy would have to bear. Costs were also affected by the number of superseded copies cancelled on publishing a new edition. In July 1951 five years' stock of sheet 42/56 was being ordered, which would entail 380 copies, or 71 per cent, of the existing stock being cancelled, but this was probably most exceptional, and the 1955 figures imply an expected cancellation rate of about 5 per cent.[248] In 1956 an average of 670 copies were printed of the coloured edition of the 11 experimental Regular Edition sheets of Plymouth and Dartmoor.[249]

In the late 1950s and early 1960s it was believed that revising the maps had had a beneficial effect on sales: the Plymouth group sold about 100 copies each in 1955-6 as Provisionals and about 200 each in 1958-9 as Regulars, the best seller in both years being SX 64, with 217 copies in 1955-6 and 472 in 1958-9. The average annual sale for revised sheets was thought in 1960 to be about 145.[250] In 1961 1080 copies were issued of the experimental large sheet 856, described below, compared with an average annual issue of 115.[251]

All these figures show variations, and may depend on how consumption is defined: whether confined to sales to agents, or including issues to government departments and for internal OS purposes. By 1963-4, when only a few sheets remained unrevised, total annual issues were 300,600, an average of 148.37 copies per sheet: of these, 196,000 (65 per cent) were for general sale, 65,000 (22 per cent) were sold at educational discount to 'schools' (which may include further and higher education), and 39,600 (13 per cent) went to government departments and for internal OS use. The 'government' issues included an unspecified proportion to the War Office for civil defence purposes.[252] 266,000 1:25,000 sheets were issued for sale in 1963-4 (an increase of 22,000, 11 per cent, over 1962-3), compared with 1,239,000 of the one-inch (an increase of 166,000, 15 per cent, over 1962-3). There were unspecified 'heavy' cancellations when old sheets were replaced by revised ones, whereas there were few cancellations when revised sheets were reprinted.[253] In 1970 375,000 1:25,000 sheets a year were issued, of which it was stated that certainly a third and possibly a half were for administration and education. At that time one-inch sales were over 1.5 million copies annually, and about 130,000 quarter-inch maps were issued, representing about 7 per cent of small-scale issues. The 1:25,000 figure includes Second Series sheets, including a number in Scotland which had no Provisional Edition predecessors, but after allowing for these there still seems to be a marked increase since 1964.[254]

The Ordnance Survey versus Coca Cola and baked beans: the tastes of youth
In 1964 it was believed that the main customers for paper folded 1:25,000 sheets were walkers, 'tourists', campers and 'motor rallies (occasional)', and for paper flat sheets local authorities,

---

[245] Consumption figures, 135A in PRO OS 1/383.
[246] Minute by O.Maps, 24 January 1955, 225 in PRO OS 1/383.
[247] Table and minute, 23 and 31 May 1955, 230A and 232 in PRO OS 1/383.
[248] Minutes, 30 July 1951 and 23 May 1955, 156 and 230A in PRO OS 1/383.
[249] Report on 1:25,000 Regular Edition, 25A in PRO OS 1/692.
[250] Review of 1:25,000 sheet size, 16 February 1960, 24A in PRO OS 1/1192.
[251] Minute by DMP, 15 January 1962, 42 in PRO OS 1/1192.
[252] Minute 45A in PRO OS 1/776; minute 25 in PRO OS 1/1303.
[253] Lundie to Burnett, 16 November 1964, 54 in PRO OS 1/776; minute 77, 6 November 1964, in PRO OS 1/1303.
[254] Background information on OS sales, 16B in PRO OS 1/1494.

'schools (for desk work)', the Territorial Army, Civil Defence, and 'examination purposes'.[255] It is interesting that scouts and guides are not mentioned. In the consultative exercise in 1950, the Boy Scout Association said that 10 by 10 km sheets were too small, and they seem to have favoured somewhat larger sheets centred on areas of tourist (and presumably scouting) interest. As it was, they usually used the one-inch, which covered a disproportionately larger area in much less detail.[256]

The possibilities of the scout market were raised again in 1964 by D.O'Donovan, a Treasury official, who wrote to his opposite number at the OS, G.D. Lundie:

> ... to my mind the 2½ inch maps have several advantages when it comes to teaching small boys how to find their way by map. There is at the Gilwell Park camping ground a shop where the boys can buy their Coca Cola and baked beans, and I suspect that a 2½ inch map, framed so as to embrace the Gilwell Park area and Epping Forest together, would command a ready sale there.

There were probably near 1,000 boys there every Whitsun, and smaller numbers at Easter and other weekends and throughout the summer. And there were Girl Guides, too.[257]

Lundie's reaction was sceptical: 'I am sure this letter has been written with the best of intentions even though you may feel there is an element of "teaching his grandmother to suck eggs" about it.'[258] The Deputy Director Small Scales produced some costs and observed:

> What worries me is the assumption that a large number of copies might be bought by Boy Scouts. It is my experience that boy scouts never spend a penny if they can avoid it, and at the best each troup or patrol leader might be prevailed upon to buy a copy of the new map. I note that Mr. O'Donovan stops short of guessing what the annual sales might be. In my opinion we would be lucky to sell 400 in the first year, and then drop to one or two hundred a year thereafter. It will not be an attractive map, unless we spend more money in preparation e.g., dropping in a green wood fill and providing some additional overprint information in red, e.g., site of main scout areas and other features of scouting interest. This sort of exercise might well put up our preparation costs by £200. Frankly I do not think the project is worth while.[259]

That settled it, and scouts at Gilwell Park and elsewhere were left free to spend their pocket-money as they always had done. Perhaps Mr O'Donovan had forgotten the penury and tastes of boys.

The 'Regular Edition', 1954-65

The concept of a 'regular edition'
As we have seen, in 1943 Cheetham envisaged that most of the drawing for the Provisional Edition, except for the contours, would be reused for the Regular Edition, which would be revised from the larger-scale resurvey which had begun shortly before. At some time before he retired in 1949 Cheetham appears to have changed his mind, and decided on a wholly redrawn Regular Edition; even if he did not, his successors certainly thought in these terms. Before the chequered history of the Regular Edition is recounted, it is useful to consider whether it was necessarily inevitable. Could the Provisional Edition have been maintained in its stead? There are two answers to this: the technical and sales justifications.

The essence of the technical justification was that the Provisional Edition was of substandard quality, and therefore unsuited as a 'permanent' map. One problem was that 'the sheets were not made to an accurate standard size'; another was that 'the method of reproduction

---

[255] Minute by SEP, 9 October 1964, 45 in PRO OS 1/776.
[256] Boy Scout Association to OS, 7 July 1950, 106A in PRO OS 1/416.
[257] O'Donovan to Lundie, 12 June 1964, 28A in PRO OS 1/776.
[258] Minute by Lundie to DMP, 18 June 1964, 29 in PRO OS 1/776.
[259] Minute by DDSS, 26 June 1964, 30 in PRO OS 1/776.

was rather hit and miss, the quality of the drawing leaves much to be desired...'[260] Whilst these may have been true, and the trained eye can sometimes detect, say, buildings that should be but are not perfectly rectangular, or slightly 'sausagey' lines, these do not affect either the general appearance or, more important, the functionality of the mapping. Redrawing would have permitted substantial redesign, but in practice redesign was a matter of tinkering rather than fundamental reconception. This was demonstrated, no doubt unintentionally, in 1972 when Provisional Edition material was used for the early Outdoor Leisure maps, described later. One cannot help feeling that the best was the enemy of the good, and that the OS was seeking an unnecessarily exacting technical standard. There is, however, a parallel: in 1943-6 those sheets of the one-inch New Popular Edition which were in Fifth Edition style were regarded as a 'final' edition, but by 1947 dissatisfaction with both their quality and their design had led to a decision to treat them as a temporary measure pending a wholly redrawn one-inch map, which was published as the Seventh Series. The sheer improvement of quality represented even by the pilot sheet for the series of 1949, which compared with the production mapping was rather conservative in appearance, must have inspired those who favoured redrawing, rather than revising an existing map.[261]

The sales justification was less straightforward. The nub of the matter was best expressed by Willis, commenting in February 1955 on a revision of 1:25,000 policy, and the intention to publish a few Regular sheets as an experiment:

> If an overwhelming success, then we have clear evidence for fulfilling our present policy of overall coverage of Great Britain. We may however, in the course of the experiment, gain further information as to whether the regular series is only required in towns and on their outskirts, i.e. within hiking distance of centres of population, and in 'tourist' areas, or perhaps whether the demand in the Plymouth area is such a flop that we shall be justified in assuming that we are filling an unfelt want.[262]

In other words, there might in fact be no sales justification for the Regular Edition. It is worth noting that the out-dated nature of much of the Provisional Edition does not seem to have been used as a strong argument in favour of replacing it by a Regular Edition.

The War Office and the 1:25,000 Regular Edition

Though in the early 1950s it was apparent that the 1:25,000 was not proving particularly successful in civil sales terms, there was one user who seemed to have few complaints: the army. Although they might not have supported MacLeod as wholeheartedly as he would have liked at the time of the Davidson Committee, when he was thinking of a primarily military map, by the mid 1950s, when it was ostensibly a civil map, they were probably its most enthusiastic users: in some ways, the military was the *eminence grise* behind the development of the 1:25,000 at this time. In the 1930s the War Office had been the most obvious customer for a national 1:25,000, and in the 1940s Cheetham had provided one, albeit in a 'Provisional Edition'.

The 1:25,000 was not the only instance where a tentative Davidson Committee recommendation had been elevated into a prominent OS activity. The Committee had recommended consideration of 1:1250 urban mapping once the arrears of 1:2500 revision had been cleared off, but by 1945 the OS had committed itself to the 1:1250, effectively as a substitute for further 1:2500 revision of larger urban areas, and from 1945 to 1960 most of the OS large-scale field survey effort went into the 1:1250. The output of National Grid rural 1:2500 and derived six-inch mapping was much more modest. In 1935-6 MacLeod had envisaged a fifteen-year programme, without the 1:1250 element. In 1943 Cheetham envisaged completing a geographically ambitious 1:1250 programme in ten years, and the 1:2500 in fifteen years, with

---

[260] Minute by DMP, 28 July 1954, 215 in PRO OS 1/383; minute by DMP, 18 March 1952, 111 in PRO OS 1/567.
[261] Richard Oliver, *A guide to the Ordnance Survey one-inch Seventh Series*, London, Charles Close Society, 1991, 5-7.
[262] Minute, Willis to Collins, 18 February 1955, 19 in PRO OS 11/29.

derived six-inch and 1:25,000 mapping following close behind.[263] In the event a somewhat less extensive 1:1250 programme was completed in the late 1960s and the 1:2500 in the early 1980s.

This larger-scale mapping would have provided the starting-point for the 1:25,000 Regular Edition, and the extended timescale for its appearance meant that the 1:25,000 Provisional Edition had a much longer life than its principal architect had intended. Though Cheetham was anxious to do something about starting the Regular Edition before he retired in 1949, using 36/27 (Edinburgh, later NT 27) as the pilot sheet, the replacement map was to develop much more slowly.[264] That it developed at all was probably due at least as much to the War Office as to civil users, actual or potential; that it developed along the rather erratic lines that it did was also due to the War Office.

The first stage was to circularise various government departments, recreational organisations such as the Ramblers and the Scouts, and a few individuals, to elicit their views; the letters, with specimens (a miscellany of stock copies), were sent in June 1950. The OS suggested amongst other things that it might be possible to offer the Regular Edition in 20 by 20 km sheets, by pasting together, mounting and folding four ordinary sheets; this idea had previously been floated by Cheetham in 1946.[265] This may possibly have reduced the number of complaints about the small sheet size, though there were still a few respondents who raised the point. Brigadier H.E.M. Newman, recently retired from the OS, even had the audacity to write that when at OS he had handed out the official line on sheet sizes, but 'I finished up convinced that the critics were justified'. Cheetham was also, unsurprisingly, in favour of the four-in-one scheme.[266] However, that was about the last that was heard of the combined sheet idea, and by 1954 a scheme was in hand for reducing the maximum size of Regular Edition sheets in coastal areas from 15 by 10 km to 12 by 10 km. The index at 1:625,000 showing the revised layout was duly published, but its proposals were overtaken almost immediately by a scheme for using integral covers on the Regular Edition, which (like those used for the Second Series from 1980 onwards) would have been printed as, in effect, an extension of the marginalia (and unlike those used for the Provisional Edition from 1970, which were printed on the back of the map).

In the 1950s the military were the most contented of the 1:25,000's users and potential users. They had no quarrels with the fundamental design of the map, they were apparently untroubled by the state of its revision, they were the one group of map-users who habitually employed the grid, and they loved the sheet size, which was extremely convenient for printing by mobile units in the field. It was unfortunate that the makers of the map were becoming particularly prey to doubt about the sheet size. By the early 1950s, when it was clearly apparent that the 1:25,000 was not the hoped-for success with the public, the Provisional Edition was nearly complete. Though it was policy that, once enough National Grid resurvey material became available, the Regular Edition would be started, it called for no great leap of imagination to toy with the idea of not proceeding with the Regular Edition, and allowing the Provisional Edition to wither away, at any rate as a civil map. If needs be, all the reproduction materials could be handed over to DMS, to do what they wished with them.[267] This would have the further advantage of reducing pressure on OS drawing resources.

As has been related above, in the early 1950s experiments using Provisional Edition material were made, which included field boundaries in black.[268] At this time there was no great desire for a radical redesign of the map. In 1952 'Bruno' Brown observed that Cheetham 'made a pretty good design in the Provisional Edition, and I see no evidence as yet which would

---

[263] Memoranda by MacLeod, 30 December 1935, 1A in PRO OS 1/117, and Cheetham, n.d. [October 1943], 1A in PRO OS 1/170.
[264] See early minutes on PRO OS 1/567.
[265] Circular letters, June 1950, 15A-M in PRO OS 1/416; Cheetham to Central London YHA Group, 5 March 1946, 54A in PRO OS 1/415.
[266] Newman to Burnett, [June 1950], and Cheetham to Burnett, 20 June 1950, 17A and 29A in PRO OS 1/416.
[267] See Willis to DMS, 4 October 1955, item 25A in PRO OS 1/1123.
[268] See minute 84, etc, in PRO OS 1/567.

conclusively force me very far from it'.[269] Because of the slow areal progress of the large-scale resurvey (mostly concentrated on 1:1250 urban work, with the only extensive rural work being a semi-experimental 'overhaul' in Devon), there was no great sense of urgency. The nearest approach to Regular drawing before Brown was succeeded by Major-General J.C.T. Willis in 1953 was an experiment with drawing scale in sheet NT 27: the north-west part was drawn at 1:17,500, and the south-west part at 1:25,000. It suggested that drawing at 1:25,000 might be practicable for open rural sheets, but not for close urban ones, and that anyway little progress could be made until the design of the six-inch Regular Edition was settled, as it was in 1953.[270]

Thus it was only in 1954-5 that the Regular Edition started to come to fruition; by now there was a preparedness to use green in order to make the map look more attractive.[271] This was no doubt egged on if not initiated by Willis, who had suggested a six-colour map when at DMS back in 1946. By this time the OS was under considerable financial pressure, and running-down the 1:25,000 seemed worth thinking about. An alternative was to cut costs, such as by omitting field boundaries, thereby saving 20 per cent of drawing costs.[272] One would have thought that this would have destroyed much of the rationale for the scale in the first place, but the suggestion was put to the Advisory Committee on Survey and Mapping at its meeting on 21 July 1955. 'There was a firm and reasonable requirement for the continuation of this series... The omission of field boundaries could, if necessary, be accepted though with reluctance.'[273]

In writing to DMS some ten weeks later, Willis went further: 'the field boundaries tend to clutter up the map, add little to its value, and involve considerable problems of maintenance'. Were there no other factors, he would consider a policy of redesigning the map for the civil market, in much larger sheet lines and without field boundaries and certain other minor detail, concentrating only on areas of likely or proven demand, and allow all other cover to 'waste away'. But there was of course, as he acknowledged, the military interest. Perhaps, if there was a 'negligible' civil requirement, the War Office would be prepared to take over the whole series?[274]

The eleven published Regular Edition sheets

Though the 1:25,000 might not be a complete success with the public, it was not a complete failure either, and any attempt to abandon it would not go unnoticed. It was decided to try a few Regular Edition sheets, to 'sound the market', and in July 1956 a block of eleven covering Plymouth and south-west Dartmoor was published (SX 45-47, 54-57, 64-67). The style of the published maps was broadly similar to the Provisional Edition: the main differences were the use of black for field boundaries, solid lines for multiple track railways, no park stipple, and a green infill for woods, which involved a fifth printing, but which was thought, rightly, to make the map more attractive. The economy of an integral cover was experimented with in 1955, but the published maps appeared in the customary hinged card covers.[275] As with Sheet 40/09 in 1944, some more adventurous colour-schemes were tried on proofs, including red road infill and green building infill.[276] A legend was once more provided, necessitating reversion to the semi-Bender fold.

A feature of the eleven sheets as first published, which was deleted from the revised editions published in 1959-60, was the provision of a convention (pecked line plus '*RW*') for public rights of way, which appeared in the legend, but not on the map. Though in 1949 local authorities had been given the duty of compiling Definitive Maps of public rights of way, the work went forward slowly, and if in 1955 the OS might have considered showing them on the 1:25,000, by 1956 it

---

[269] Minute by Brown, 9 April 1952, 112 in PRO OS 1/567.
[270] Minutes, report and specimens, 114, 114A, 120 in PRO OS 1/567.
[271] Minute, 11 February 1955, 20 in PRO OS 1/1123.
[272] Minute, 23 in PRO OS 1/1123; Collins to Willis, 19 May 1955, 36 in PRO OS 36/7.
[273] Minutes at 24 in PRO OS 1/1123. This committee was a precursor of the later consultative committee system.
[274] Willis to DMS, 4 October 1955, 25A in PRO OS 1/1123.
[275] Minutes 6, 11, 13, 16, 17, 21 in PRO OS 1/1123.
[276] Copies of these experiments are in PRO OS 36/7.

Ordnance Survey
Two-and-a-Half-Inch Map
Regular Edition

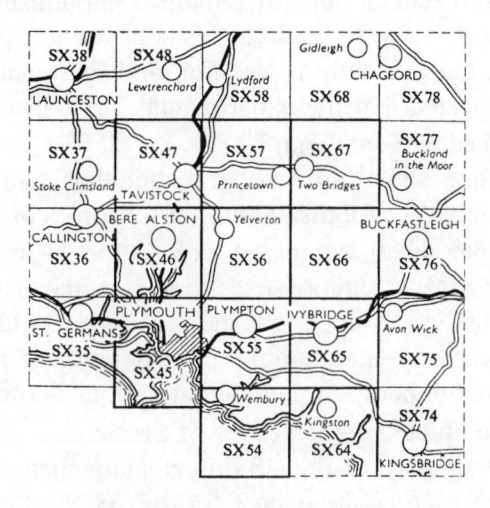

The Regular Edition of the new Ordnance Survey map at a scale of Two-and-a-Half Inches to One Mile has been published for the eleven sheets bounded by the thick lines (SX 45 · 46 · 47 · 54 · 55 · 56 · 57 · 64 · 65 · 66 & 67) in the index above.

Maps of areas outside the limits of the thickened lines have not been published in the Regular Edition, but are available in the Provisional Edition.

Figure 1. The leaflet pasted to the inside cover of Regular Edition sheets when first published in 1956. This is the only recorded published appearance of the expression Regular Edition *in connection with the 1:25,000 map. The wording on the spines of the same map covers reads* 2½ inch (Regular).

was thinking of showing them only on the one-inch, and made experiments with one-inch Seventh Series sheet 85 to this end. By this time the 1:25,000 was beginning to make some impression with walkers, and S.F. Marriott, the Rambler's Association mapping advisor, had an unsatisfactory correspondence with the OS, in which he urged that rights of way be shown on the 1:25,000, to which the OS response was that the 1:25,000 was an uneconomic map: '... demand... does not justify any money being spent on it', with sales 'apathetic'. Marriott's response was that the 1:25,000 would be much more popular were rights of way to be shown.[277] He might have added that in the past the Ramblers Association had been keen to supply the OS with information about paths, and had been issued with maps for this purpose.[278] In 1958 a rights of way experiment was made with the experimental large sheet 409 (Durham), with paths annotated with discreet sans-serif letters, but that remained unpublished.[279] At the same time it was intended to show rights of way on the revised sheets covering Epping Forest, provided that the information was available, but evidently it was not, and these sheets were published in the usual style. In 1960 it was expected that the experimental large sheet 856, would show public rights of way, but it was published without them.[280]

The eleven Regular Edition sheets sold rather better than had their Provisional Edition predecessors: in the period 23 July - 31 August, a total of 92 copies of the eleven sheets had been sold in 1955 (average of 8.36 per sheet), compared with 498 in 1956 (average 45.27 copies).[281] This helped to give the 1:25,000 a stay of execution. At the same time there were two contradictory developments. MacLeod's hope that the 1:25,000 would enable the six-inch to be abandoned in mountain areas was translated into post-war policy, and remained so until February 1956, when a meeting in Edinburgh between the OS and various Scottish map users showed that there was a strong desire for complete six-inch cover of Scotland.[282] It is possible that the Scots were pushing at a door which was ajar, but the OS immediately changed its policy, and the Scots were assured of their six-inch. A useful consequence for the OS was that one residual justification for the 1:25,000, that of being the largest scale at which the more remote areas would be published, was removed, though a disquieting feature of the Edinburgh meeting was that the planners were still keen on the scale. So were their counterparts in England.[283]

One big problem with the Regular Edition was that hitherto it had been assumed that it would have to be wholly redrawn, as the Provisional Edition material was felt not to be up to the best OS standards. At a time when the OS was short of drawing staff, this was a useful argument for not proceeding with new map series. What was now contemplated was in effect almost a reversion to the method which had been used forty years before for GSGS 3036 (but which might have been inspired by more recent recollections of GSGS 3906), whereby a single outline drawing would suffice for both the six-inch and the 1:25,000, though the provision of names would have to continue to be a separate operation for each scale. Several experiments were carried out in 1955-7, and eventually a satisfactory compromise between an over-bold six-inch and an unduly congested 1:25,000 was achieved.[284] Also tried, in 1958, was an early version of what twenty years later would be called four-colour or trichromatic printing (using the primary colours to give a much more elaborate colour-scheme); the results look good, but were not persisted with.[285] In 1960 it

---

[277] Collins to Marriott and reply, 27 Feb and 9 March 1956, 273A and 274A in PRO OS 1/700. For the development of rights of way depiction see PRO OS 1/700.
[278] 47A, 49A in PRO OS 1/415.
[279] Copy in PRO OS 1/1090.
[280] DDSS minute, 10 November 1958, 32A in PRO OS 1/692; note on Advisory Committee meeting, 28 April 1960, in OS 1/1192.
[281] Report on Regular Edition, October 1956, 25A in PRO OS 1/692.
[282] See PRO OS 1/798, particularly item 59A.
[283] Minute 28 in PRO OS 1/1123.
[284] The experiments are described in PRO OS 1/1241.
[285] Minute, 49 in PRO OS 1/1123; the only known surviving example is in the collection of 1:25,000 specimens at present in the Charles Close Society-Ordnance Survey deposit in the map department of Cambridge University Library.

was noted that 2,500 copies of a four-colour 20 by 15 km sheet would cost £1,298, compared with £1,167 for a conventional five-colour sheet; the four-colour sheet involved more elaborate preparation, including correction costs at least 20 per cent greater, and high cost on revision. 'This may be the reason why no other agency has yet adopted the system.'[286] The method was only adopted on a significant scale by the OS in 1977-8, and then only for the 1:50,000 and 1:250,000.

Rumours of discontinuance, and concealed popularity
At the same time as these experiments were taking place, rumours were circulating to the effect that the 1:25,000 was to be discontinued. It is in the nature of rumours that they do not lend themselves to the critical apparatus of academic writing, and so the origin and route of transmission of those of the demise of the 1:25,000 cannot be traced. They took written form on 9 January 1956 when, in a letter marked 'confidential', A.V. Hardy of the Cambridge Examinations Syndicate wrote:

> We understand that there is some risk of further work on the 2½" series... being discontinued. We very much hope that this is not the case since in our experience they are of very great use in the teaching of Geography and we consider them superior for this purpose to the 1" maps.[287]

The immediate reaction to this of Brigadier M.O. Collins, (Director of Map Publication, and former secretary to the Davidson Committee) was that

> His comments are interesting, but might not be so readily forthcoming if we put the conflict on a pure price basis, i.e. would he buy a much more expensive 1/25,000 map for the reason he gives...

The official reply was rather less forthright: the Provisional Edition was practically complete and was being revised, but no decision had been taken as to whether to produce a Regular Edition.[288]

Less than six months later, Alice Garnett wrote to express the Geographical Association's 'deep concern' that the 1:25,000 might be withdrawn. It was now built into examination syllabuses and geographical research and 'its withdrawal... would be a most serious and in some respects irreplaceable loss'. The following day a Mr Stevens of Brentwood, described as 'Agricultural Contractor', wrote that 'There seems to be an idea among map sellers that this series is being abandoned...' Both received similar replies to that sent to Hardy, who wrote again in January 1957, to express the Syndicate's appreciation of the mapping. The OS replied that, whilst it did not contemplate ceasing to publish, the Syndicate's views could be valuable if reconsideration was forced by 'economic factors outside our control'.[289] In October 1956 the Darlington Co-Operative Holidays Association and Holiday Fellowship Joint Group's representative wrote that they had 'heard rumours from different sources' that publication of these 'most useful' maps might cease, but at the same time had received a leaflet advertising a new edition with green woods. They were puzzled.[290]

In 1958-9 15 per cent of 1:25,000 sales were to schools, and in 1963-4 they were nearly 22 per cent, mostly in paper flat form.[291] However, these figures probably underestimate the real influence and reach of the 1:25,000. It has already been noted that many War Office printings of the military version, GSGS 4627, were made other than by the OS, and so no regard was paid to them in OS discussion of the 1:25,000's fortunes. There were two other categories of 1:25,000 which are also conspicuously absent from these discussions: they were extracts printed by the OS for the various examination boards, and in map-reading and interpretation textbooks, on which

---

[286] Report on production and sheet sizes, 2 March 1960, 26A in PRO OS 1/1192.
[287] Hardy to OS, 9 January 1956, 40A in PRO OS 1/697.
[288] Minute by Collins and OS to Hardy, 18 and 25 January 1956, 42 and 43A on PRO OS 1/697.
[289] Garnett and Stevens to OS and replies, 28 and 29 June and 10 July and 8 August 1956, Hardy to OS and reply, 11 and 21 January 1957, 44A, 45A, 47A, 50A, 54A and 55A in PRO OS 1/697.
[290] Barras to OS, 25 October 1956, 52A in PRO OS 1/697.
[291] Review of sheet size policy, 16 February 1960, 24A in PRO OS 1/1192; 1:25,000 issues, October 1964, 45A in PRO OS 1/776.

royalties would have been payable to the OS. No statistics are available for the output of either but, if the writer's experience in a rural grammar-school in the later 1960s is any guide, exam extracts were also a useful, and cheap, source for classroom teaching sets, particularly in schools funded by fiscally cautious local authorities, and they were carefully rounded up after the exams had been sat. The same applied to map-reading textbooks. Both were far easier to handle than full-size sheets. If this argument is correct, the short-term effect would be to depress sales of 1:25,000 mapping, but in the long term might encourage the map-using habit generally, and in particular to draw attention to the possibilities of the 1:25,000. Yolande Hodson has noted the encouragement of map-use in schools in the inter-war period and the consequent effect of increased map-consciousness on map sales.[292] Thus the humble extracts of the 1950s and 1960s were, all unbeknown to OS, cultivating a potential demand for the 1:25,000 in the 1970s and 1980s. The exiguous sales figures for the 1:25,000 suggest that they were already slowly increasing in the early 1960s.

However, the education market could work two ways, as the next episode demonstrates.

Sheet 856: a false dawn

Even as the eleven Regular Edition sheets were being drawn in 1954-6, the complaints from the public about the sheet size and numbering of the 1:25,000 were making an impression on the Ordnance Survey. Unfortunately for the public, whilst they only had to deal with the OS, the OS had to deal both with them and with the military, who strongly favoured the status quo. The fundamental difficulty was that DMS wanted a map which could be reproduced by mobile survey units, which implied a maximum sheet size of 15 by 10 km, which was inadequate to meet the complaints from the public.[293] The OS's first suggestion was for a 15 by 15 km sheet, with the possibility of so disposing the marginalia that the map could be printed in two halves. Willis remarked that, though the military version would only have proper margins round three sides,

> in the event of our having to use a tactical battle map in anger within this Island I think there will be so many more serious problems to be coped with that a little matter of a sheet lacking one internal margin would fade into insignificance...

and commercial firms might be used instead of mobile presses.[294] The eventual solution was a 20 by 15 km sheet and the suggestion came, ironically, from DMS: they were concerned that there were NATO and other agreements bearing on the matter, and if Britain produced a 1:25,000 larger than 'Princess' size then the Americans would probably produce a 1:25,000 of their own, which would not be as good, but which British forces would have to use. DMS claimed not to have the resources to produce their own 1:25,000 (though as they printed most copies of GSGS 4627 this seems a curious argument: perhaps they were thinking of revision), and would be practically bound to take whatever was provided by OS, though 'I don't think you should build on it'. If, however, OS was determined on a larger sheet size, then why not 20 by 15 km, which would be suitable for printing on a Quad Crown (45 by 33.5 inch, about 114 by 85 cm) machine, and double the size for a Princess machine (maximum size 30 by 22.5 inches, about 76 by 57 cm).[295] In 1955 the Advisory Committee had been prepared to accept a larger sheet size, provided that it was no larger than that of the one-inch Seventh Series; that it even more closely resembled that of the quarter-inch Fifth Series then in preparation seems to have been a coincidence.[296] Each 20 by 15 km sheet could be split in half, and reproduction materials for the consequent 10 by 15 km sheets supplied to the military. Two sheet line schemes were drawn up, the first of which envisaged 997

---

[292] Hodson, *Popular maps*, 182-5, 209.

[293] Minute by Willis, 10 April 1956, 29 in OS 1/1123.

[294] Willis to Carey, 5 March 1957, 32 in PRO OS 1/1123.

[295] Dowson to Carey, 2 July 1957, 42A in PRO OS 1/1123; paper sizes from Report, 2 March 1960, 26A in PRO OS 1/1192.

[296] Advisory Committee minutes, 21 July 1955, 24 in PRO OS 1/1123.

sheets, and the second of which reduced the total to 977.[297] The sheets were butt-jointed and consecutively numbered, and most large urban areas fell inconveniently across two or four sheets, but this was inevitable given the anxiety to have the minimum number of sheets in the series.

At this time (1958) it was briefly the intention to use the 20 by 15 km format not only for the Regular Edition, but also for the Provisional Edition, with 'the Second Provisional Series' starting production in 1961. It would have incorporated the fresh revision for the one-inch Seventh Series which was getting under way.[298] Four 20 by 15 km sheets based on Provisional material were definitely prepared, 409 (in County Durham), 856 (Ilfracombe and Lundy), and 950 and 951 (Isle of Wight), but only 856 was put on sale, apparently in response to an undertaking to George Philip & Son.[299] It had an integral cover with the name *Ilfracombe*, which was guillotined from paper-flat copies, a full legend in the bottom margin, and green infill for woods, but otherwise was similar to the Provisional Edition specification.

By the time that sheet 856 went on sale, the whole question of sheet size was once more open. In February 1959 Alice Garnett wrote again about a rumour, this time that the 1:25,000 sheets were to be replaced by a much larger size. The present size had 'great educational use', and she hoped that it could be retained after the larger sheet was adopted. The Director-General, now Major-General L.F. de Vic Carey, replied that it was anticipated that the new size would be more popular with the general public, and that only one series could be kept going.[300] An index to the large sheets was exhibited at a Geographical Association conference in January 1960, 'and it aroused considerable adverse criticism from the teaching profession, who particularly objected to the increased sheet size'. Association members visiting the OS stand at a publisher's exhibition in the same month expressed their 'grave misgivings', and said that if teachers lost interest then future sales might be seriously affected. (Presumably a flat 20 by 15 km sheet would be too large for most school desks, but it might have been folded in two or four. And what did teachers do about the one-inch?) A review of sheet sizes concluded that the effect of revision on increasing sales had not hitherto been taken into account and that, whereas 20 by 10 km might do, though 'ill-proportioned' and an unpleasing shape, 20 by 15 km was on balance unsuitable.[301]

A second 20 by 15 km sheet, of the Isles of Scilly, was published in 1964: it replaced three standard sheets, and the two-inch map which had first appeared in 1933.[302] As such, it had no connection with the large sheet series proposals of 1958-60. At about the same time sheet 856 was reprinted, now with its number replaced by the name *Ilfracombe and Lundy*.

The emergence of the 1:25,000 Second Series

Sheet sizes had implications for printing press capacity and replacement of existing machines, and the whole question was investigated in detail in February 1960. The hostility to the 20 by 15 km size from the Geographical Association perhaps confirmed growing doubts: the study suggested that the best option was either to retain 10 by 10 km, or adopt 20 by 10 km.[303] The case for 20 by

---

[297] The 997 sheet layout is in OS 1/1123, together with tracing paper overlays, which were manipulated to obtain the best fit. A 977 sheet layout index at 1:1,250,000 was printed (copy in Cartographic Library, OS), but it is not known if it was published.

[298] Extract from Director-General's Conference 336, 20 August 1958, 11 in PRO OS 1/1192.

[299] Extract from Director-General's Conference 388, 4 April 1960, 33 in OS 1/692, and Thackwell to Dowson, 15 January 1962, 42 in OS 1/1192; no other documentation for this seems to be extant. No copies of sheets 950 and 951 have so far been located, but there is a copy of sheet 409 in PRO OS 1/1090: it is printed in the standard four colours of the contemporary published Provisional Edition sheets, with the addition of discreet letters in black to show the status of public rights of way. 966 was also printed (as 'Sheet 971': copy at OS), but as a quasi-Regular sheet.

[300] Garnett to Carey and reply, 6 and 19 February 1959, 20 and 21A in PRO OS 1/1192.

[301] Review of sheet sizes, 16 February 1960, 24A in PRO OS 1/1192.

[302] See PRO OS 1/183.

[303] Report on production and sheet sizes, 2 March 1960, 26A in PRO OS 1/1192.

15 km sheets was further undermined by a shortage of drawing staff, and by the decision not to incorporate the post-1958 one-inch revision in the Provisional Edition.[304]

The question then seems to have been left until early in 1962, by which time sheet 856 had been on sale for over a year. Meanwhile, in 1961 the second Land Utilisation Survey had started publishing its work at 1:25,000 using Provisional Edition material in 20 by 10 km sheets, on OS advice.[305] On 26 February 1962 the 20 by 10 km size was officially decided on. As the 'halving' policy was still in force, the War Office was consulted, and raised no objection.[306] Work was put in hand immediately on a pilot sheet, TQ 21/31; it was first proved in March 1963. It was in four colours: improvements in screening meant that a 'grey' effect could be obtained for building infill on the black plate.[307] After various modifications, notably moving the legend and other notes from the bottom of the map to the left-hand side so as to facilitate folding, and replacing abbreviations for public rights of way by green symbols, the first fourteen Second Series sheets were published in December 1965. The Ramblers Association was delighted, particularly with the rights of way information, and their Secretary, Tom Stephenson, wrote to the OS that the Ramblers would do all in their power to promote the maps.[308]

Even then, however, progress was not smooth. From 1964 onwards the OS was under increasing pressure to recover more of its costs, and it was by no means clear whether the Second Series would ever be completed, notwithstanding that six-inch mapping drawn boldly - many thought over-boldly - with a view to 1:25,000 use was now being produced in some quantity. One problem was the long time-lag between 1:2500 and 1:1250 field survey and eventual 1:25,000 publication, 'with the embarrassing sequel that one of the early Second Series 1:25,000 sheets was found to be more out of date in some respects than the Provisional map it replaced'.[309] This was TQ 20/30, the Brighton sheet, which omitted several areas of building development which appeared on the Provisional Edition predecessors, using one-inch revision of 1957.[310] To make matters worse, sales of the map remained obstinately low (as many of the early Second Series sheets covered areas in the far north of Scotland, this is perhaps not surprising!) and the Ordnance Survey felt obliged to warn map users at the consultative meetings in 1969 and 1970 that it was doubtful if production could be continued, 'an announcement which at the time gave rise to no more than a murmur of dissent'.[311] As noted above, in 1970 the possibility was investigated of abandoning the Second Series and revising the Provisional Edition.

In 1968-9 changing to a 20 by 20 km sheet size was considered; estimates suggested an overall saving in preparation costs of about 5 per cent for the series, and the 80 by 80 cm basic sheet size would also be a good standard for metric 1:50,000 or 1:100,000 mapping, which was then being investigated, and indeed was adopted shortly after for the 1:50,000. A difficulty was that each sheet would have had to wait until the sixteen component six-inch Regular Edition sheets were available, and another was (unspecified) 'technical reasons' from the Drawing section, which may have decided the matter in favour of the status quo.[312] There is no evidence that the Provisional Edition would have been republished in this format.

The next development was metrication of the series: in 1969 further production of the six-inch Regular Edition was stopped in favour of the 1:10,000, with either relabelled imperial 25 foot interval contours or else proper metric 5 metre interval contours, and by 1971 the 1:25,000

---

[304] Extract from Director-General's conference 385, 4 April 1960, 31 in PRO OS 1/1192.
[305] Extract from Director-General's conference 425, 30 July 1962, 55 in PRO OS 1/1192.
[306] Minutes 42-45, 48, 49 in PRO OS 1/1192.
[307] 77A in PRO OS 1/1192.
[308] Stephenson to OS, 19 April 1966, 103A in PRO OS 1/1303.
[309] Seymour, *A history*, 328; for the effect of 'cost recovery' on the 1:25,000 see minute by Director of Establishments and Finance, 7 October 1964, 43A in PRO OS 1/776.
[310] Compare Provisional Edition sheets TQ 20, edition B, and TQ 30, edition C, with Second Series sheet TQ 20/30, edition A, around 228046, 250070, 273076, 297084, 300087 and 360062.
[311] Seymour, *A history*, 328.
[312] PRO OS 1/747, passim; Supt SS3E, '1:25,000 2nd Series - Regular', December 1970.

Second Series was being published in a similar style. In 1973 there was a firm proposal to abandon the Second Series: this led to strong complaints from walkers and others, and in October 1973 the retention of the series was announced.[313] This was satisfactory as far as it went, but the rate of output of the Second Series, at around 35 sheets a year, remained low; thanks to pressure from the Ramblers Association the OS increased output to about 80 sheets a year after 1975.[314] The Provisional Edition continued in print, but with minimal further revision. There was a further threat to the 1:25,000 in 1981-2 in the wake of the Ordnance Survey Review ('Serpell') Committee's findings, and its questioning the utility of the map. Once again there were vocal complaints from walkers and others, and the map was allowed to continue; by this time it was approaching the stage where it would be very difficult to abandon it, as so much had been published.

The Pathfinder and the Outdoor Leisure Maps

One practical reason for the pressure to complete the Second Series was the demand for public rights of way information. By the early 1970s Definitive Maps had been prepared for most of England and Wales and were being used to show rights of way on the one-inch Seventh Series and its 1:50,000 successor. The big drawback to these two smaller scales was that they did not show field boundaries, which were very desirable for those who wished to follow rights of way, but who did not have the map reading skills to follow those not clearly indicated on the ground, and not clearly referable to features shown on the one-inch or 1:50,000. Thus by the mid-1970s the argument was less for the 1:25,000 *per se* and more for a map which showed both field boundaries and public rights of way, conditions which the Second Series met very well. Admittedly, as Marriott had pointed out twenty years earlier, the 1:25,000 Provisional Edition with rights of way information would have served most practical purposes equally well, and, as an interim measure, in 1976-7 the OS overprinted fifteen Provisional Edition sheets with rights of way information, in response to a request from some county councils. Provisional Edition mapping was also used as a basis for a few independent series of rights-of-way maps, for example *Ramblers' maps of the Cotswolds*, published by the Gloucestershire area of the Ramblers Association in the mid 1970s. In 1979 the great interest of the walkers in the 1:25,000 Second Series was recognised by giving it the alternative name of 'Pathfinder'. The sheet numbering system came under review, and although the adoption of a consecutive numbering system as an alternative to National Grid numbering was announced in 1983, it was two or three years before it came into use.

Meanwhile, a series of Outdoor Leisure Maps had appeared, the first being *The Dark Peak* in 1972. The original concept was an economical reissue of whatever Provisional Edition or Second Series material happened to be available, on sheet lines designed to cover areas of tourist interest, with the addition of green wood infill where necessary, public rights of way in green and tourist information symbols in blue. The colour scheme was that of the Second Series, except that the 'black' plate on sheets using Provisional material (the majority at first) was the black and grey plates combined, and printed in the darkest of greys.[315] In the mid 1980s it was decided not to publish, or continue publishing, Second Series sheets wholly covered by Outdoor Leisure maps based on Second Series material. As a result, when the Pathfinder series was finally completed in December 1989, with the publication of Sheet 1362 (SX 54/63/64), which replaced two of the Regular Edition sheets of 1956, it was a hybrid depending on some coverage by Outdoor Leisure Maps to provide complete cover of Great Britain at 1:25,000 scale.

The Provisional Edition and the original Regular Edition thus came to a formal end in December 1989, and Ordnance Survey designated 1990 'the year of the Pathfinder'. There were still obstinate demands from sections of the public for a larger standard sheet size, and in 1992 two 20 by 10 km sheets were joined together to provide single-sheet cover of the Gower

---

[313] 'Chronology', in OS Cartographic Library, citing Director-General's conference minutes.
[314] Information from Michael Holroyd.
[315] OS Experiment 367, minute 33A: copy in OS Cartographic Library, December 1992.

peninsula; tourist information was added. The experiment was successful, and in March 1994 the Explorer series was introduced, for recreational areas not covered by Outdoor Leisure Maps. The first sheets were similar to the Gower sheet, but in 1995 the Outdoor Leisure colour scheme (using four-colour process printing and enabling with five road infill colours) was adopted, and early in 1997 it was announced that all the Pathfinders were to be replaced by a consecutively numbered national Explorer series: these were to be a mixture of 30 by 20 km single-sided and pairs of 20 by 20 km double-sided sheets. The conversion process would be greatly facilitated by digital scanning and editing, and the new series would be complete in the spring of 2003. The significance of this for the Provisional Edition is that those who argued, first, for a much larger sheet size and, second, for a more colourful map, have been proved 'right all along'.

By March 2002 most of the Pathfinder maps had been superseded, and the Outdoor Leisure maps were redesignated and renumbered as part of the Explorer series. Thus from the spring of 2003 Great Britain ought to be covered for the first time by a single, uniform 1:25,000 map. But even that may yet prove to be just another 'provisional edition', in function if not in name. In the spring of 2002 information was received from Ordnance Survey that work on a new specification for the Explorer series would begin in May 2003; presumably this has been prompted in part by the passage of the Countryside and Rights of Way Act 2000, with its 'access land' provisions and its implications for mapping. Taken together with the intention to begin producing the 1:50,000 series direct from the large-scale digital database by May 2004, this implies that a similar procedure can be expected in due course for the 1:25,000, which should reduce considerably problems of revision and maintenance at this scale, but will also mean another 'redrawing'.

'Beyond the best there is a better.'

Abbreviations used in the introductory essay

ADP	Assistant Director of Publication
BLML	British Library, Map Library
DDSS	Deputy Director, Small Scales
DGOS	Director General, Ordnance Survey
DMO & I	Director of Military Operations and Intelligence
DMP	Director of Map Publication
DMS	Directorate or Director of Military Survey
EO	[Officer in charge at OS office, 'Esher' (Hinchley Wood)]
GSGS	Geographical Section, General Staff
LPTB	London Passenger Transport Board
MoD	Ministry of Defence
MTCP	Ministry of Town and Country Planning
O.Maps	Officer-in-charge, Map Stores and Issues Branch
OP	Officer in charge of the Publications Department
PRO	Public Record Office, London
Trig.	Triangulation, triangulation point

Compiler's notes

Over 25,000 separate states of the 2,027 sheets are identified in this cartobibliography, but it must be admitted at the outset that this is an incomplete listing. As soon as it became evident that the job files, in which the Ordnance Survey maintained the record of work done on each sheet, had been destroyed, it was clearly impossible to determine whether all the printings of any sheet had been identified. In some cases those states that are still missing are unmistakable, usually from the sequence of print codes. A gap between "A" and "C" printings is obvious enough, so too that between "B/" and "B///", but until it is recorded, there is no way the researcher can be aware of the existence of a "B/" printing between "B" and "C". There may be many such missing here. The omission of such sheets is unfortunate because they would probably show an intermediate stage of revision. Less important are missing reprints that are identical within the neatline, where the only changes would be to details such as the price, the headings, and other marginalia - changes that provide evidence that additional printings took place, but otherwise nothing of importance. Here again many unrecorded reprints almost certainly exist - a known printing priced at 3/-, and thus made between 1958 and 1961, might be followed in the list by an unpriced sheet made after 1964. The possibility of a printing between the two is strong, but without the job files this cannot be confirmed unless recorded. The sequence of printings after 1964, once the price statement had been removed, is probably the least complete of all. After this date there were few changes to unrevised reprints, and the most significant of those occurred to the sheet headings, in particular the removal of the words "Provisional Edition". Readers will be aware, however, that most sheets of this period had an integral blue cover printed on the reverse, and are usually found folded. In this format the paper size was reduced - by the removal of the information above the map. Thus this telling evidence is usually destroyed. One should add that outline printings are almost always less well represented than coloured ones. There is good reason to believe that most coloured print runs would be accompanied by an outline one as well, but that the quantity printed would be only a tenth as many. Thus they turn up less frequently, the more so since most libraries that made collections of this map avoided the outline editions.

But enough of this negativity. Over 25,000 printings are listed - coloured and outline sheets, Regular and rights of way editions, military editions, and, in footnotes, Administrative Area, Geological Survey and Soil Survey issues, and more ephemeral types. What follows is a detailed description of the layout of this work, and the information readers may expect to find. Though the Provisional Edition of the 1:25,000 First Series began in the early days of the National Grid when its values, and therefore sheet numbers, were wholly numerical, the sheets are listed here according to the alpha-numeric system adopted in 1951, as is now standard practice.

The layout described

The entry for each sheet in the cartobibliography comprises a heading, and a list of printings.

The heading contains four elements:

1. Sheet number, shown in both its alpha-numeric and its earlier numeric styles. On sheets published too late to have had a numeric sheet number, only the alpha-numeric version is listed here.

2. Sheet title. This was never printed on the sheet itself. When taken from map covers titles are in italics; there may be minor textual variants between cover issues. Those with an asterisk added are taken from catalogues, those in square brackets have not been recorded and are supplied by the author. The earliest known published list appears in the Ordnance Survey Catalogue for 1974, by which time many sheets had been superseded by Second Series. Sheets already superseded may thus never have been allocated titles.

3. Sheet coverage (width by height), usually 10 by 10 km. There are many extended sheets around the coast.

4. Revision details. No revision dates were offered on early printings. Later it became the norm to offer two or three categories of revision information: that of the parent six-inch map, partial systematic revision, and often a road revision date, taken initially from the one-inch map. In many cases these dates were themselves revised during the lifetime of a sheet. In later years the dates of many additional categories may also be given. All revision dates displayed during the lifetime of every sheet are offered in the headings, and are given letter codes, as listed below, which correspond to those set against the relevant printings in column 7 of the lists. Occasionally no civilian printing is recorded with a specific revision date: these may be found on maps in the military list.

a, b: six-inch revision date
c, d, (and if necessary) e: partial systematic revision date
e, (or if already used) f: one-inch road revision date
f, g, h, i: other road revision dates
j, k, l, m: other types of revision, as specified

The list contains fourteen columns:

1. Sheet number, summarised as type 1, 2 or 3.
    1. numeric format - 42/25
    2. transitional format - SP 25 (42/25)
    3. alpha-numeric format - SP 25

2. Print code. The earliest codes (three sheets only) offered a print run, month and year, where printed (**E**sher, **C**rabwood or **C**hessington), in the form "1700/11/45 Cr", quickly superseded by a figure combining both print run and year of publication in the form "10046". From February 1947 these gave way to the unique letter system which itself had two different interpretations. Initially a letter was superseded by another for any reprint, in some instances even the initial outline printing being the "B" edition. Thus on some sheets the sequence quickly advanced to "D" or even further. Such reprints did not necessarily reach the copyright libraries. The system was later adjusted, so that an advance in letter signified a reprint with important revision. Reprints with less important change might have a "_" added under the parent letter, eg "B̲", listed here as "B/", and later, if these reprints were listed in the publication reports, it was customary to added an asterisk, eg "B•", listed here as "B/*". Confusion between the two systems led on occasions to two completely different editions with the same letter code, listed here "B [1]", "B [2]", or even a reversion to the earlier letter. Thus a sequence such as "A, B, A/" is known. In many cases reprints with minor change apparently failed to merit a change in print code, and the reverse could also be true, that the minor change apparent in an adjusted print code was not recorded in the marginal information on the sheet. Print code letters were usually in italic; from the late 1950s a change to roman character was made on revised editions, though not always on facsimile reprints (see also 3). Overprinted parts of print codes are given in curved brackets.

3. Coloured edition details, comprising three elements:
    First upper case letter: **C**oloured edition, some overprinted in **G**reen showing public rights of way[316].
    Second upper case letter: **P**rovisional Edition, **R**egular Edition, **0** - no heading, **F**irst Series.
    Lower case suffix letter - price codes:
    r: 2/-,3/-
    s: 2/6,3/6,4/6
    t: 2/6,4/6,6/-
    u: 2/6,4/6
    v: 2/6
    w: 3/-
    x: 4/-, in italic lettering
    y: 4/-, in Roman lettering
    z: no price

Only eleven sheets ever appeared in the Regular Edition, in square SX. This heading was not used on the maps themselves, though the spine writing of mid-1950s covers was altered from "2½ inch (Provisional)" to "2½ inch (Regular)" for such sheets. It also appeared on a descriptive index pasted inside some of these covers (see page 45). Only fifteen sheets were published overprinted in green with public rights of way information. "Provisional Edition" headings were removed from otherwise unaltered reprints from 1973, and on revised reprints the heading "First Series" replaced "Provisional Edition" from 1974. Sheets with different price statements appear in the same entry if no other specification changes occur. They are usually separated by commas. The full stop separator indicates the change from an italic to a roman print code letter. The "z" code may acquire suffix symbols: a "+" on extended sheets when the full sheet number also appears at the foot of the sheet, a "-" when "1:25,000" in the heading is altered to "1:25 000", a "o" when the notice about conventional signs centre bottom is removed. These marginalia alterations are for the most part not listed: they are noted when they are the only indication of additional printings once the price statement had been deleted.

---

[316] Public rights of way issues were also made for official use, overprinted, for example, in purple. This was probably not the work of the Ordnance Survey, and they are not covered in this volume.

4. Outline edition details, comprising three elements:
   First upper case letter: **O**utline edition, in grey, from 1973 in black.
   Second upper case letter: **P**rovisional Edition, **R**egular Edition, **0** - no heading, **F**irst Series, and in one instance, **A**dministrative Areas Edition.
   Lower case suffix letter - price codes:
   r: 2/-
   s: 2/9 paper flat
   t: 2/6
   u: 2/9,3/6
   v: 2/9
   w: 2/9,4/6
   x: 4/-, in italic lettering
   y: 4/-, in roman lettering
   z: no price

Administrative Areas editions usually appear in footnotes, being purely derivative of an outline edition printing. Further information may be found in the notes to the coloured edition [3].

5. Publication details, comprising two elements, the date and place of publication.
   1. Date of publication, or nd (no date). Crown copyright dates became a feature of later editions, and after their introduction publication dates were either never altered or were removed. In these cases the copyright date is used, in the form "1979ᶜ".
   2. po: Printed and Published by the Director General at the Ordnance Survey Office
      mo: Made and Published by the Director General at the Ordnance Survey Office
      pc: Printed and published by the Director General of the Ordnance Survey, Chessington, Surrey
      mc: Made and published by the Director General of the Ordnance Survey, Chessington, Surrey
      ms: Made and published by the Director General of the Ordnance Survey, Southampton
      m-s: Made and published by the Ordnance Survey, Southampton

6. Reprint information, comprising two elements, its date and the nature of the detail added.
   1. Date of reprint, or nd (no date).
   2. c: corrections
      mc: minor corrections
      ch: minor changes
      Ch: major changes

      a: airfields or aerodromes, some named
      b: building development or the extent of built up areas
      d: radio stations
      f: afforestation
      i: industrial development
      l: lakes or lochs, some named
      r: major roads
      t: tidal change
      u: urban development
      v: reservoirs, some named

Alternatively, sometimes additionally, republication dates appear in this column, in the form "1952rpb". These appeared on some sheets published in 1952 and 1953, signifying the alteration of the National Grid information to the new alpha-numeric format, with easting values printed upright rather than sideways.

7. Revision information, displayed by letters which refer to the revision dates listed in the heading (qv).

8. Magnetic variation date, using an abbreviated numerical format, or nd (no date).

9. Seven aspects of specification. This list is a selection from the many details of specification that could have been listed, though space did not permit. For instance the presence or absence centre bottom of footnotes concerning acreage and conventional signs might have been noted, but they were considered of marginal interest, and are rarely the sole indicators of change. They are in fact only noted when they are identified as such. An "o" is provided as a spacer where information is lacking or irrelevant. The several elements are:

1. The number of colours in the coloured edition – 3 (black, blue, brown), 4 (black, blue, brown, grey) or 5 (black, blue, brown, grey, green). The brown sometimes tends towards orange. This figure is retained even when referring only to an outline edition.
2. B: bench marks are present. The bench mark note on the map may be misleading.
3. E: easting values are printed sideways.
4. Relevant to those sheets with legends. See the illustrations on pages 68-71. The left hand legend on Provisional Edition sheets (1-4 below) contains two important variables: roads may be Ministry of Transport Class "A", "B" and "Other Roads (not classified by Ministry of Transport)", or "I", "II" and "Other Roads" subdivided as "Good, metalled" (now with pecked infill) and "Poor, or unmetalled", with the "not classified" expression usually omitted. In the Trigonometrical Station line churches may appear as "Church", or subdivided into "Church with Spire" and "without Spire". The minor variants in the right hand legend (Osier Bed or Beds, County of City for County Borough on Scottish sheets) are ignored here. Thus four permutations are possible:
    1. Roads MOT Class A, B, etc; Church
    2. Roads MOT Class A, B, etc; Churches subdivided
    3. Roads MOT Class I, II, etc; Church
    4. Roads MOT Class I, II, etc; Churches subdivided
        Also listed are the legends of the Regular Edition sheets:
    5. Rights of way information is present
    6. Rights of way information is absent
    7. Legends on the *Isles of Scilly* special sheet, 1964
5. X: an adjoining sheet index is present. Only those on the black plate are considered; those overprinted in purple or (from late 1952) blue on military printings are ignored.
6. The map frame. See the illustrations on page 67.
    1. The original frame on three-colour sheets with all figures in old Roman letterpress
    2. The three-colour map with National Grid figures phototypeset in Times Roman. This change usually occurred when eastings were redrawn upright
    3. The original frame on four-colour sheets, with small figures on the graduated outer frame
    4. The redesigned graduated outer frame with figures in Times Roman, on both three and four-colour sheets
    5. The frame used on the Regular Edition sheets
7. L: present only when the legend *Contours in Lochs are given in feet and are taken from the Bathymetrical Survey of Fresh Water Lochs of Scotland* appears.

10. Three further aspects of specification, left blank when not relevant.
    1. R: railway company names are present. They were usually removed after 1952.
    2. B: bridle roads are present. A few are represented by the pecked line symbol without the *B.R.* identifier, others have the *B.R.* identifier set against a double pecked line, or even something on the grey plate.
    3. C: canals are shown in solid blue. Normally canals appear in stippled blue: disused canals or flights of locks may also appear in solid blue.

11. The depiction of airfields, left blank when not relevant. See the illustrations on page 72. As many as three airfields have been identified on some sheets, though represented in some cases by a fragment of perimeter or blank area at the edge of a sheet. Their locations may be identified from their National Grid references given on pages 58-59, listed in reference order.
    0: full topography is shown with no sign of an airfield
    1: an airfield is represented by a blank space on the map
    2: perimeter detail is shown, perhaps with dispersals and buildings
    3: full runway layout is shown, perhaps with buildings

12. The date of appearance of the coloured edition in the Ordnance Survey Publication Report (OSPR).

13. The date of appearance of the outline edition in the Ordnance Survey Publication Report (OSPR).

14. Cross reference to notes.

Footnote references are usually applied to: Administrative Areas editions (abbreviated to AA), some with National Trust information (NT), the Special Administrative Area Edition Great London (SAAEGL), the military edition GSGS 4627, later M821 (M), the North Pennines Rural Development Board Areas (NPRDBA), Geological Survey editions (GS), Soil Survey editions (SS). It should be noted that the only Geological Survey and Soil Survey sheets to be listed here are those which use First Series base maps, and are on standard series sheet lines. Locations listed at the foot of the sheets, usually reservoirs, or airfields when named on the map, are also noted, also features of peculiar interest. Military edition notes provide cross references both to the list starting on page 302 and to the civilian printings used. It should be borne in mind that in most cases the plates had been altered since the most recent civilian printing. Typical alterations might be an adjustment in the magnetic variation date, the addition of the diagram of adjoining sheets, the removal of sheet prices. Such changes may be deduced from the information in the military list. Some military printings show clear evidence of civilian printings that have not been recorded, and appear in the principal lists, with an "M" in the notes section.

List of airfields

The following list of four-figure National Grid references supports the minimum detail provided in column 11 of the table. Airfields were usually not named, those which do appear are normally to be found in sheet footnotes. Often airfields are marked as "disused". The designation of airfields varies considerably and often differ from one edition to another; those noted, with the abbreviations used, include: flying boat base (b), aerodrome (d), air establishment (e), airfield (f), air park (k), airport (p), aquadrome (q), water aerodrome (w). It is not noted here when airfields are given which designation, or when none.

NC 96 9967f ▪ ND 26 2166f, ND 35 3652d, 3256f ▪ NH 66 6266f, 6567f, NH 75 7751d, NH 85 8551f, NH 87 8475f, NH 88 8282f ▪ NJ 06 0663d, NJ 15 1959, NJ 16 1960, 1968d, NJ 25 2059, NJ 26 2060, 2069d, 2665, NJ 36 3663d, NJ 66 6164f, NJ 81 8712d ▪ NK 04 0747f, NK 05 0657f, NK 06 0464f ▪ NO 00 0903f, NO 10 1003, NO 12 1528d, NO 22 2624f, NO 33 3936, NO 42 4620f, NO 43 4037f, NO 51 5611f, NO 53 5937, NO 60 6208f, NO 64 6243f, NO 65 6051f, NO 66 6269f, NO 77 7577 ▪ NS 20 2007f, NS 32 3627p, 3524f, NS 33 3535, NS 46 4766f,p, 4965, NS 56 5166p, NS 98 9380f ▪ NT 16 1265f, NT 17 1573f,p, NT 18 1584f, NT 47 4473, NT 48 4980, NT 57 5578f,d, NT 58 5081f, NT 74 7646f, NT 85 8950f, NT 93 9432f ▪ NU 12 1925, NU 20 2200f, NU 21 2513f, NU 22 2025f ▪ NX 06 Loch Ryan, NX 15 1054f, 1159, NX 16 1160, NX 45 4353f, NX 97 9978f ▪ NY 07 0078, NY 15 1253f,d, 1758f, NY 25 2254f, NY 26 2169f, NY 27 2170, NY 35 3053f, 3859, NY 45 4859, NY 46 4168f, 4860f,p ▪ NZ 06 0869f, NZ 07 0870, NZ 17 1971p,d, NZ 18 1681, NZ 19 1898f, NZ 20 2806f, 2400f, NZ 27 2071, NZ 29 2299, NZ 31 3713f, NZ 35 3458f, NZ 41 4515f, NZ 42 4928p, NZ 52 5028p, NZ 62 6222d ▪ SD 17 1772d, 1379f, SD 20 2907f, SD 30 3009f, SD 31 3010f, 3419, 3218, SD 33 3132d, 3335d, SD 37 3774f, SD 41 4211f, SD 42 4127f, SD 43 4336, SD 46 4666, SD 63 6231f,d ▪ SE 11 1113, SE 24 2241f,d, SE 28 2988, SE 29 2496f, SE 37 3871f, 3978f, SE 38 3781f, 3089f, SE 39 3090, SE 40 4905, 4303, SE 45 4552f, SE 46 4861f, 4867, SE 47 4079f 4175f, SE 50 5901, 5005, SE 52 5928, 5921, SE 53 5132f, 5238f, SE 54 5743f, 5349, SE 55 5350f, 5954f, SE 60 6806f, SE 62 6021f, 6028f, SE 63 6436f, SE 64 6748f, SE 66 6063f, SE 68 6782f, SE 70 7507f, SE 73 7134f, SE 74 7641f, 7848f, SE 75 7454f, SE 83 8235f, SE 90 9801f, SE 92 9425f, SE 95 9956f, SE 96 9964f ▪ SH 27 2975, SH 33 3333, SH 37 3075f, SH 45 4358, SH 47 4176f, SH 52 5626f ▪ SJ 32 3727f, SJ 35 3959, 3652f, SJ 36 3465f, 3269, 3960, SJ 37 3270, 3779, SJ 41 4217, SJ 42 4826f, SJ 45 4059, SJ 46 4060f, SJ 48 4183p, SJ 50 5004f, SJ 51 5710, SJ 52 5422f, SJ 53 5637f, SJ 55 5957d, SJ 58 5689, SJ 59 5690f, SJ 61 6018, SJ 62 6623f, 6626, SJ 63 6430f, SJ 65 6057, SJ 68 6583f, SJ 70 7904, SJ 72 7224, SJ 76 7369, SJ 77 7370, SJ 79 7497d, SJ 80 8902, 8600, SJ 81 8414f, SJ 82 8625d, SJ 88 8184p, 8982d, SJ 90 9002p,d, SJ 92 9926f. SJ 94 9341d, SJ 98 9082d ▪ SK 02 0026, SK 11 1412f, SK 12 1524, SK 13 1931, SK 14 1945f, 1742f, SK 23 2032f, 2830p, SK 24 2045, SK 31 3212, SK 42 4525f,p, SK 52 5221f,d, 5822f, SK 54 5246, SK 58 5380, SK 60 6501f,d, SK 61 6513, SK 63 6136d, SK 64 6841, SK 67 6976f, SK 68 6281f,d, SK 69 6698f, 6393f, SK 71 7415f, SK 73 7433f, SK 74 7247f, SK 76 7464, SK 77 7076f, SK 82 8626f, 8923, SK 84 8141f, 8149f, SK 85 8255f, 8950, 8150, SK 86 8862f, 8569f, SK 87 8570, SK 88 8888, SK 89 8695f, SK 90 9304f, 9406, SK 91 9015, 9513, SK 92 9422, SK 93 9334f, 9032, SK 94 9641f, 9949, SK 95 9050, 9950, 9959,

SK 96 9864f, 9369f, SK 97 9679f, 9978f, 9370f, SK 98 9680f, 9489f, 9683f, SK 99 9497f, 9490 ▪ SM 70 7906, SM 72 7925f, SM 80 8502f, 8006, SM 81 8311, SM 82 8425f, SM 90 9503b, SM 91 9519f ▪ SN 00 0502f, SN 01 0911, SN 11 1011, SN 30 3903, SN 40 4003 ▪ SO 35 3959, SO 36 3960f, SO 43 4237, SO 45 4059, SO 46 4060, SO 68 6188, SO 70 7909, SO 71 7910, SO 81 8816f, SO 82 8821f, SO 84 8944, SO 89 8290f, 8699f, SO 90 9100f, SO 92 9227, SO 94 9749f, 9044, SO 95 9750f ▪ SP 01 0412f, SP 10 1803, SP 11 1811, SP 14 1142f, 1648f, SP 15 1959, SP 16 1960, SP 17 1473, SP 18 1783p, SP 19 1491f, SP 20 2906f, 2406f, SP 21 2018d,f, SP 23 2133, SP 25 2151, 2655, SP 27 2373f, SP 30 3006, SP 31 3313, SP 32 3925f, 3225f, SP 34 3643d, SP 35 3554f, SP 37 3574p, SP 38 3982, SP 39 3796, SP 40 4700, 4105, SP 41 4615, SP 42 4026, 4926, SP 43 4434f, SP 44 4949, SP 45 4950f, SP 47 4473, SP 48 4087, 4082, SP 51 5319, SP 52 5026, 5320, 5924, SP 53 5436f, 5632f, SP 54 5049, SP 55 5050, SP 58 5988f, 5184f, SP 60 6309, SP 61 6310f, 6916, SP 62 6024, SP 63 6431f, 6138f, SP 64 6741, SP 68 6582f, 6089, SP 70 7309p, SP 71 7017f, SP 72 7729f, SP 73 7730f, SP 77 7777f, SP 78 7188f, SP 81 8711, SP 82 8523f, SP 86 8268f, 8061q, SP 88 8186f, SP 90 9904f, SP 91 9115f, SP 94 9442f, SP 95 9559, SP 96 9560f, 9968, SP 98 9181f, 9589, SP 99 9397f, 9590f ▪ SS 43 4934f, SS 53 5034, SS 59 5691, SS 60 6209f, SS 61 6210, SS 87 8379, SS 88 8380, SS 96 9968f, SS 97 9671 ▪ ST 06 0068, 0667p, ST 10 1307f, 1809f, ST 11 1810f, ST 21 2015, ST 27 2277p, ST 33 3634f, ST 35 3459, ST 36 3460d,p, ST 46 4965, ST 51 5315d, ST 52 5523f, ST 56 5968p, 5065d, ST 57 5979, ST 58 5980f, ST 68 6080, ST 71 7519, ST 72 7520, ST 76 7569, ST 77 7971f, 7570f, ST 87 8071f, 8576, ST 88 8981, ST 90 9405f, ST 95 9257, ST 97 9978, ST 98 9081f, ST 99 9099, 9596f, 9291 ▪ SU 07 0078f, 0570f, 0775, SU 09 0791f, 0598f, SU 10 1508f, SU 13 1433f, 1739f, SU 14 1649f, 1740, SU 15 1554, SU 17 1378f, SU 18 1888, SU 19 1498f, 1196, SU 21 2412, SU 23 2938, SU 24 2745d, SU 30 3400f, SU 33 3038d,f, 3938, SU 34 3245f, SU 37 3075, SU 38 3989f, SU 39 3293, 3990, SU 40 4707d, 4606w, SU 41 4516p, SU 43 4635f, SU 46 4964, SU 47 4174f, SU 48 4187, SU 49 4799f, 4197, SU 50 5602f, 5800f, SU 56 5064f, 5963p, SU 57 5477f, SU 58 5888, SU 59 5896f, 5395, SU 60 6603p, SU 64 6743f, SU 66 6063, SU 69 6291f, 6398f, SU 70 7602f, SU 74 7349d,f, SU 75 7958, SU 77 7773d, SU 80 8707, SU 85 8059p,f, 8553e, SU 87 8578, SU 89 8290f, SU 90 9060f, 9903f ▪ SV 91, *Isles of Scilly* 9110d,p ▪ SW 32 3728d, SW 61 6816, SW 62 6725f, SW 64 6646, SW 75 7352d, SW 86 8664f, 8768f, SW 87 8871f ▪ SX 08 0484, SX 18 1485f, SX 46 4960, SX 56 5167f, 5060p, SX 99 9993p ▪ SY 09 0093 ▪ SZ 19 1198p, 1892, SZ 29 2198f, SZ 49 4894d,f, SZ 58 5884p, SZ 68 6286, SZ 69 6090f ▪ TA 00 0802f, 0909f, TA 01 0413f, 0910f, TA 04 0243d,f, TA 05 0051f, 0056f, TA 11 1217f, 1010, TA 12 1121f, TA 14 1248f, TA 15 1358f, TA 16 1464f, TA 20 2702f, TA 30 3602f, 3108 ▪ TF 00 0302f, 0101, TF 02 0429, TF 03 0430f, TF 04 0149f, TF 05 0050f,d, 0356f, 0059, TF 06 0060f, 0961, TF 07 0472f, 0078, TF 08 0980, 0385, TF 10 1700, TF 16 1060f, TF 17 1471f, TF 18 1988f, 1080, TF 19 1996f, TF 24 2943d, TF 25 2256f, TF 26 2161f, TF 28 2088f, TF 29 2692f, 2095, TF 36 3461f, TF 38 3886f, TF 41 4819, TF 42 4820f, TF 46 4465f, TF 48 4480f, TF 56 5667, TF 60 6304f, TF 70 7208f, TF 71 7310, TF 72 7923, TF 73 7834f, TF 80 8406f, 8400, TF 82 8023f, 8424f, 8529, TF 83 8531f,d, 8938f, TF 90 9400f, 9807f, TF 91 9214f, 9918f, TF 93 9633f, 9038, TF 94 9942f ▪ TG 01 0018f, 0914, TG 02 0226f, TG 04 0042, TG 10 1500f, TG 11 1014f, TG 12, 1427f, 1320f, TG 13 1434f, TG 21 2113p, 2814f, TG 22 2622f, TG 31 3919, TG 32 3920, TG 41 4019f 5210 ▪ TL 00 0004f, 0900f, TL 05 0354f, 0459f, 0950, TL 06 0068f, 0460f, TL 07 0777f, 0970, TL 08 0986f, TL 09 0297f, TL 10 1503d, 1908, TL 12 1220p, TL 13 1636f, TL 14 1544d, TL 15 1852f, TL 16 1161f, 1069f, TL 17 1976, 1070f, TL 18 1886f, 1086, TL 19 1799, TL 20 2008p,f,d, TL 21 2612d, TL 24 2941, TL 25 2955f, TL 26 2364f, TL 27 2076f, 2874f, 2979f, TL 28 2684f, TL 34 3345f, 3041, TL 35 3359f, TL 36 3964, TL 37 3079, TL 40 4804f, TL 41 4617f, 4213, TL 43 4134f, TL 44 4645f, 4144f, TL 45 4858d, TL 46 4165f, 4867f, TL 47 4479f, TL 48 4480, TL 50 5806f, TL 51 5412f, TL 52 5322d,f,p, 5823f, TL 53 5634f, TL 54 5643f, TL 57 5178f, TL 62 6924f, 6022,TL 63 6135f, TL 64 6241f, 6449, TL 65 6350f, TL 66 6566, 6163, TL 67 6876f, TL 68 6988, TL 71 7312, TL 72 7024, TL 73 7731f, 7133, TL 74 7541, TL 75 7956f, 7251, TL 77 7571, TL 78 7481f, 7089f, TL 79 7393f, 7090, TL 82 8220f, 8427f, TL 84 8943f, TL 85 8952f, TL 86 8864f, TL 87 8975f, TL 88 8989, TL 89 8399, 8990, TL 91 9119f, TL 92 9120, 9229, TL 93 9230f, TL 94 9043, TL 95 9656f, 9052, TL 96 9966, TL 97 9679f, 9872f, 9075, TL 98 9089, 9988, TL 99 9499f, 9090 ▪ TM 00 0008f, TM 02 0129, TM 03 0130f, 0539f, TM 05 0251f, TM 06 0066f, TM 08 0785f, 0089f, TM 09 0298f, 0793f, 0090, TM 14 1941, TM 16 1263f, TM 17 1375f, 1972, TM 18 1488f, 1880f, 1983f, TM 19 1599, TM 23 2733, TM 24 2445f, TM 25 2353f, TM 27 2073f, TM 28 2083, 2589, TM 29 2490d,f, TM 34 3348f, TM 35 3553f, 3159f, TM 36 3261f, TM 37 3179f, 3979, TM 38 3286f, 3180, TM 39 3195f, TM 46 4264f, TM 47 4079f, TM 48 4588f, 4080 ▪ TQ 00 0003, TQ 03 0236f, TQ 05 0757f, TQ 06 0662f, 0062d, TQ 07 0775p, 0178f, TQ 08 0984p, 0388d, TQ 09 0999, TQ 10 1905, TQ 17 1177p, 1172k, 1076, TQ 18 1084f, TQ 19 1596d, TQ 20 2005p, TQ 23 2539, TQ 24 2840p, 2947, TQ 26 2963, TQ 28 2189, TQ 29 2190d,f, TQ 34 3047d, TQ 35 3357d, TQ 36 3063p, TQ 45 4259, TQ 46 4160, TQ 48 4589, TQ 49 4590f, 4997, TQ 58 5384, TQ 65 6755, TQ 67 6671f, TQ 70 7709, TQ 76 7464f, TQ 85 8159, TQ 88 8789p ▪ TR 02 0621p, TR 13 1135p, 1039, TR 36 3266f, 3767p.

The military list

The military list begins on page 302. As far as possible this follows the layout of the list of civilian printings, though the need to add extra details and omit others necessitates some rearrangement. The details in each column include:

1. The GSGS number, appearing in one of two forms:
   G: GSGS 4627
   M: M821 (beginning in 1957)

2. The sheet number style, as described in the civilian list column 1.

3. The sheet number, listed in alpha-numeric style. For its actual style see column 2. A "+" signifies an extended sheet, details of which may be found in the civilian list.

4. A combination of two elements:
   1. The civilian print code, deduced from other sources if in [ ], or nc (no code).
   2. The price code, using the same codes letters as the civilian list column 3.

5-11: 5 as civilian list column 5; 6 as 6; 7 as 7 (dates as in the civilian list); 8 as 8; 9 as 9; 10 as 10; 11 as 11.

12. Military edition details:
    WOE: War Office Edition
    1EG: First (etc) Edition-GSGS
    E1G: Edition 1-GSGS (etc)

13. A combination of two elements:
    1. Publication date of the military edition, or nd (no date).
    2. The colour of the military overprint: **purple** or **blue**

14. Print codes and printer codes of the military edition. These are sometimes situated separately on the map, but they are placed together here. The following is a list of printers' abbreviations:
    A: Adams Bros & Shardlow
    ASE RCE: Army Survey Establishment, Royal Canadian Engineers
    OS: Ordnance Survey
    SMS: School of Military Survey
    SPC RE [later 1 SPC RE]: [No 1] Survey Production Centre, Royal Engineers
    SPC (Air) [later 2 SPC (Air)]: [No 2] Survey Production Centre (Air)
    13 FSS RE: 13th Field Survey Squadron, Royal Engineers
    14 RE: 14th Company [later Squadron], Royal Engineers
    42 SER: 42nd Survey Engineers Regiment
    135 SER (TA): 135th Survey Engineers Regiment (Territorial Army)

Specification changes

Listed here is the sequence in which specification alterations were made to the 1:25,000 First Series. The potential value of such a chronology is enhanced when it is recalled, for the greater part of the life of the map, how few direct dating aids there are on the sheets themselves. These statistics are derived generally from the date of appearance of new editions in Ordnance Survey Publication Reports[317], so variant dates may be discovered among other reprints not so listed. In 1948 publication dates were removed, and when restored the date of initial publication was that usually displayed, itself by then misleading. Minor correction dates, and even new publication dates could quickly become out of date, and did not represent the actual date of printing when later sheet prices and other specification changes are considered. Such details were often not meant to be of interest to the buying public, though they may now be to the historian and map collector. Only a selection of the specification changes appear in the cartobibliography, since it was wholly impractical to contemplate a layout which would have run to more than a single line for each map state. Nonetheless enough is listed to identify almost all known reprints. Other alterations may be deduced from the following summary of specification change, with the caveat that most changes were in practice made when a revised reprint was undertaken. These were of course usually noted in the publication reports. Others, such as price changes, would have occurred at the date of printing of what in other respects may be a facsimile reprint. One specification change dating from about 1953 that is not listed, because the compiler frankly found himself incapable of confronting the prospect of examining the several states of over two thousand sheets for when it occurred, is when the full stop was removed from some map symbols, such as abbreviations like F.P. to FP, or road numbers, altered from A.127 to A127, the more so since these various categories of full stop were not necessarily removed from a sheet at the same time.

A.	Sheet number shown in National Grid numerical values	11/45 - 1/52
B.	Heading *1:25,000……Provisional Edition*	11/45
C.	Print code in print run/month/year/printer format	11/45 - 4/46
D.	Coloured map price "r" (2/-,3/-)	11/45
	Outline map price "r" (2/-)	11/45 - 8/50
E.	*Printed………Ordnance Survey Office,* date	11/45 - 8/48
F.	Copyright notice in words	11/45
G.	Legends present (on civilian printings)	11/45 - 6/48
	Legend "1". Roads MoT Classes A,B; triangulation station on Church	11/45
H.	Easting values printed sideways	11/45
J.	H.(L.)W.M.O.T. or H.(L.)W.M.O.S.T. (in Scotland)	11/45
K.	Notice *The full Kilometric reference...*	11/45
L.	Magnetic variation, abbreviated as *Mag Var^n*	11/45
	True North arrows are usually at the corners, sometimes 1 km in	
M.	Corner extensions on three-colour map	11/45
N.	Notice *To refer to a particular point...*	11/45
O.	Minor roads uncoloured	11/45
P.	Index to adjoining sheets	11/45
Q.	Frame with distance measurement in small figures	11/45
R.	Heights of Bench Marks notice	11/45
	Bench Marks present	11/45
S.	Railway company names	11/45
T.	Bridle Roads	12/45

The Rights of Way notice was standard throughout.
Mean Sea Level was measured at Newlyn, except on Isles of Scilly sheets, at Hugh Town, St Mary's.

C.	Print code in combined print run + year (eg 20046) format	4/46 - 1/47
K.	*The full Kilometric reference...* deleted	4/46
L.	*Mag Var^n* altered to *Mag Var*	4/46
M.	Corner extensions deleted from three-colour sheets	6/46
G.	Legend "2". Triangulation station on church with and without spire	8/46
N.	*To refer to a particular point...* deleted	8/46

---

[317] Dates derived from this source appear in the form "x/xx"; deduced dates are in the form "x.xx".

C.	Print code in letter format, located below sheet number	1/47 - 1/52
	Print code on outline editions sometimes advances to B	11/47
A.	The full sheet number of extended sheets given above the map	10/47
G.	Legend "4". Both changes (see 8/46, 12/47)	10/47
O.	Minor roads with coloured pecks introduced	10/47
Q.	Prototype frame with distance measurements in larger figures (type 4)	11/47
G.	Legend "3". Roads MoT Classes I, II	12/47
E.	*Printed............Ordnance Survey Office,* no date	5/48 - 9/50
G.	Legends removed from civilian printings	5/48
P.	Index to adjoining sheets removed	5/48
U.	Conventional signs notice added	5/48
V.	Reprint date with corrections added	1948
Q.	Frame with distance measurements in larger figures (type 4)	1/50?
G.	Legends removed from military printings in ST, SU, SX, TR	1.50
E.	*Printed and published...Ordnance Survey, Chessington, Surrey,* date	6/50 - 1/59
W.	Revision information (types "a", "c", "e") added	6/50
D.	Coloured map price "s" (2/6,3/6,4/6)	8/50 - 6/52
	Outline map price "s" (2/9 paper flat)	8/50 - 5/56
T.	Clearance of bridle roads begun	11/50
C.	Print code letter located above sheet number	8/51
A.	Alpha-numeric sheet numbers (numerical values in brackets)	11/51 - 12/53
H.	Easting values printed upright	11/51 (last sideways 1/52)
T.	Last new sheet including bridle roads issued	1/52
D.	Coloured map price "t" (2/6,4/6,6/-)	6/52 - 11/54
Q.	Three-colour map frame refigured	6/52
S.	Railway company names clearance begun	9/52 (last new 4/53)
P.	Index to adjoining sheets reinstated, with large letters	10/52
	Blue overprint colour on military series introduced	10.52
Y.	RAC and AA telephones introduced	11/52
X.	Republication date in use	1952 - 1953
P.	Index to adjoining sheets standardised	1/53
R.	Bench Marks and heights notice deleted	8/53
A.	Alpha-numeric sheet numbers (numeric system no longer displayed)	11/53
J.	H.(L.)W.M.M.T. added	4/54
D.	Outline map price "t" (2/6 paper flat)	6/54 - 10/55
	Coloured map price "u" (2/6,4/6)	10/54 - 6/56
D.	Outline map price "u" (2/9,3/6)	1/55
D.	Outline map price "v" (2/9)	5/56 - 2/61
	Coloured map price "v" (2/6)	5/56 - 8/58
Z.	Acreage notice added	10/56
V.	Reprinted with minor corrections	1956
	Military series number GSGS 4627 replaced by M821	1957
D.	Outline map price "w" (2/9,4/6)	9/57
	Coloured map price "w" (3/-)	8/58 - 1/61
E.	*Made and published..........Chessington, Surrey,* date	11/58 - 3/70
	NB no date later than 1965 until 1970.	
F.	Crown Copyright date as original publication	11/58 -

D.	Coloured map price "x" (4/-, italic lettering)	1/61 - 1/63
	Outline map price "x" (4/-, italic lettering)	1/61 - 1/63
C.	Print code in Roman lettering	2/61
W.	Road revision dates included	5/61
V.	Reprinted with minor changes	1961
U.	Conventional signs notice shortened	6/62
D.	Coloured map price "y" (4/-, Roman lettering)	3/63 - 2/64
	Outline map price "y" (4/-, Roman lettering)	2/63 - 2/64
D.	Coloured map price "z" (no price)	3/64
	Outline map price "z" (no price)	3/64
W.	More revision categories added	5/64
	Final military printings overprinted in purple	5.64
A.	Extended sheet number given also beneath the map	1/65
J.	M.H.W. or M.H.W.S.	3/65
	Last military printings (ST 70,71,80,81)	1.68
E.	*Made………Southampton,* date	3/68 - 11/70
U.	Conventional Signs notice deleted	9/70
E.	*Made………Southampton,* no date	12/70 - 1/82
B.	Heading *1:25 000………Provisional Edition*	1/72
B.	No heading, 1:25,000 or 1:25 000	2/73
	Outline editions in black	10/73
B.	Heading *1:25 000………First Series*	6/74
F.	Crown Copyright date updated	1/80 - 1/82
E.	*Made………* [no DG] *Southampton,* no date	1/82
F.	Crown Copyright date to left of sheet	1/82
V.	Reprint including SUSI derivatives	1/82
E.	*Made………* [no DG] *Southampton,* no date, sans serif lettering	6/86

Pages 64 to 66: 1:25,000 First Series index diagram. Figure 2: England (west) and Wales. Figure 3: England (east). Figure 4: Scotland and England (north). The original numerical and the later alphabetical methods of identifying the 100 km National Grid squares appears in the centre of each. For the full sheet numbers of the coastal combined sheets see the cartobibliography.

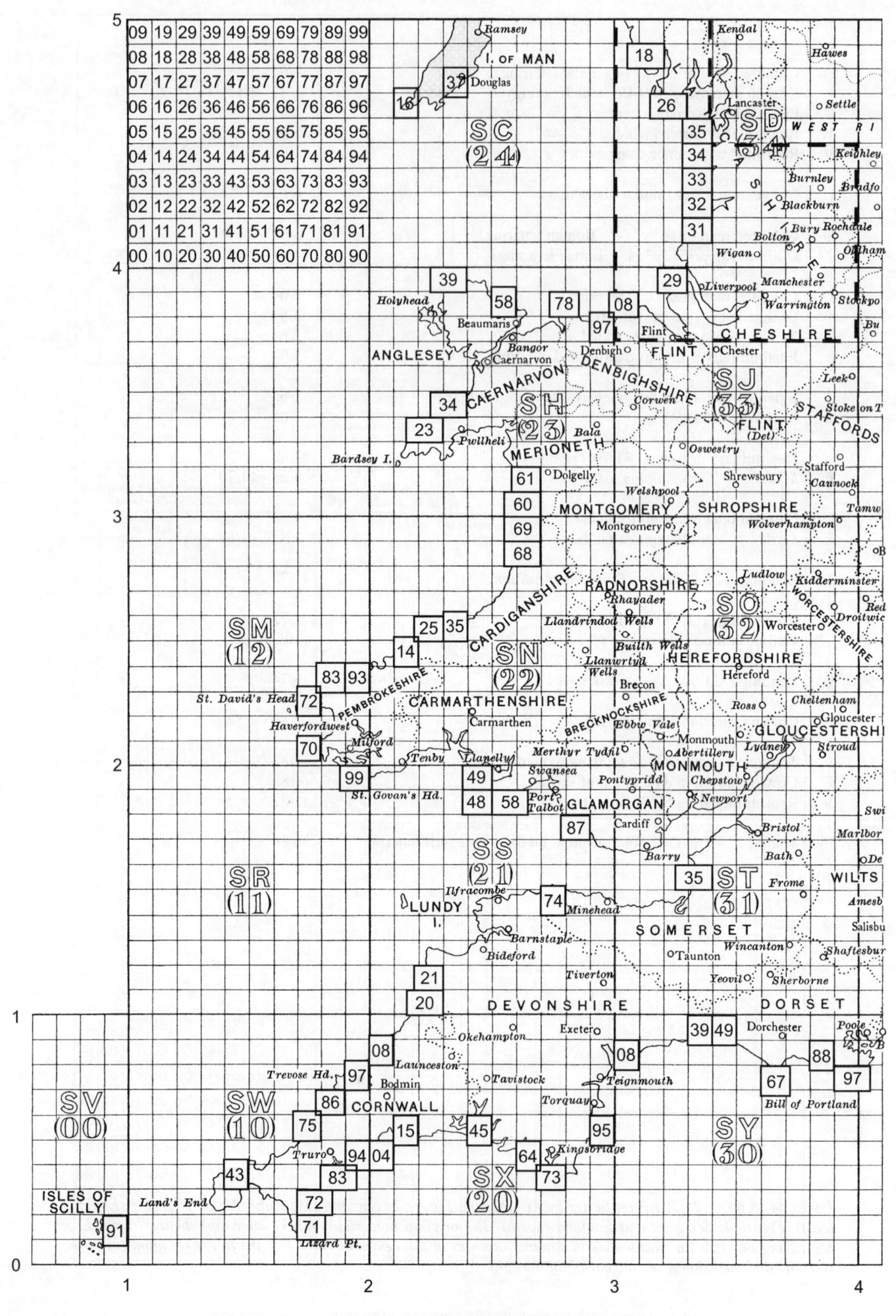

_ _ _   *Area within which sheets are in the original three-colour style*

INDEX

showing the sheet lines of the 1:25,000 First Series

- - - Area within which sheets are in the original three-colour style

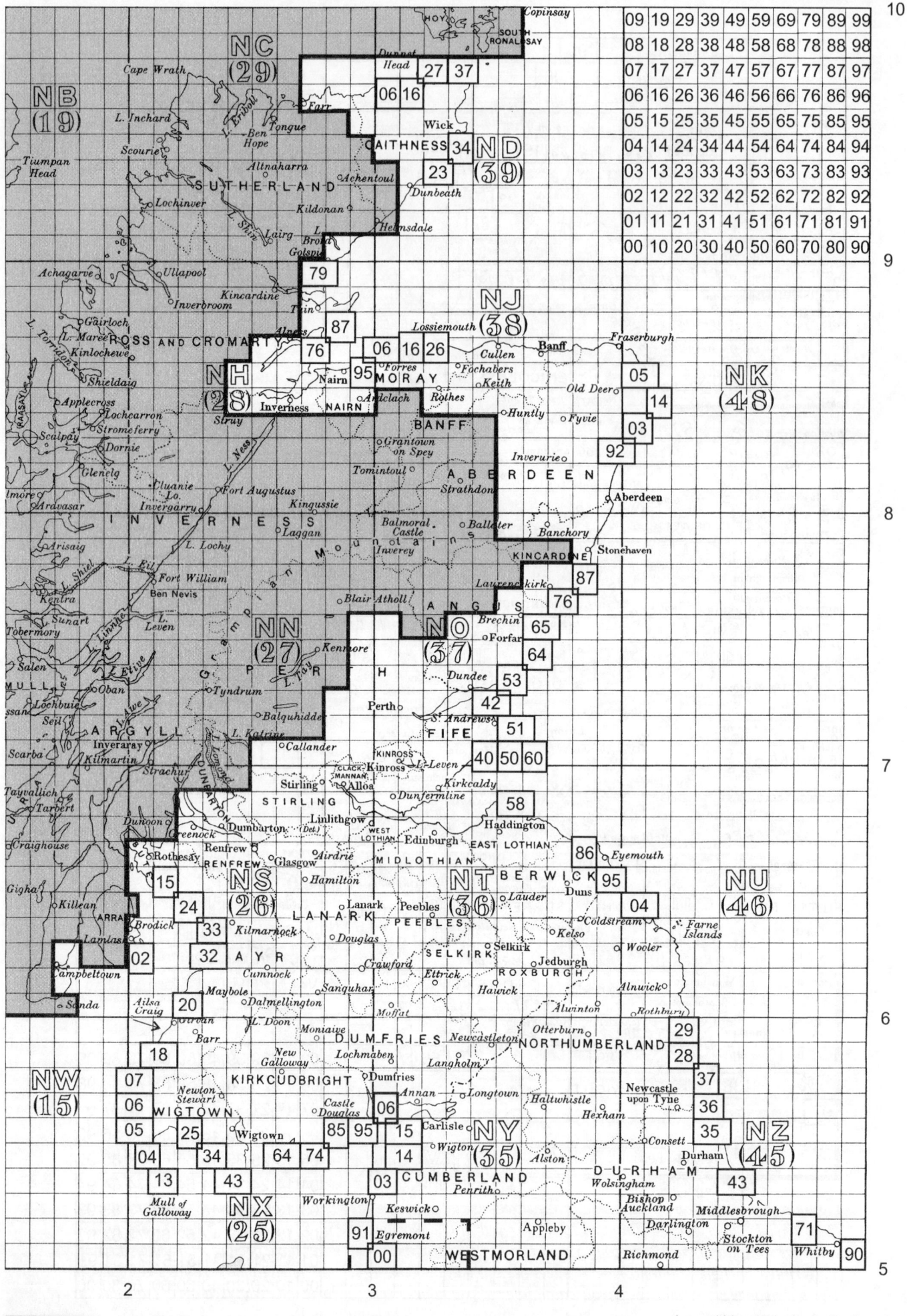

09	19	29	39	49	59	69	79	89	99
08	18	28	38	48	58	68	78	88	98
07	17	27	37	47	57	67	77	87	97
06	16	26	36	46	56	66	76	86	96
05	15	25	35	45	55	65	75	85	95
04	14	24	34	44	54	64	74	84	94
03	13	23	33	43	53	63	73	83	93
02	12	22	32	42	52	62	72	82	92
01	11	21	31	41	51	61	71	81	91
00	10	20	30	40	50	60	70	80	90

——— Limit of published mapping

– – – Area within which sheets are in the original three-colour style

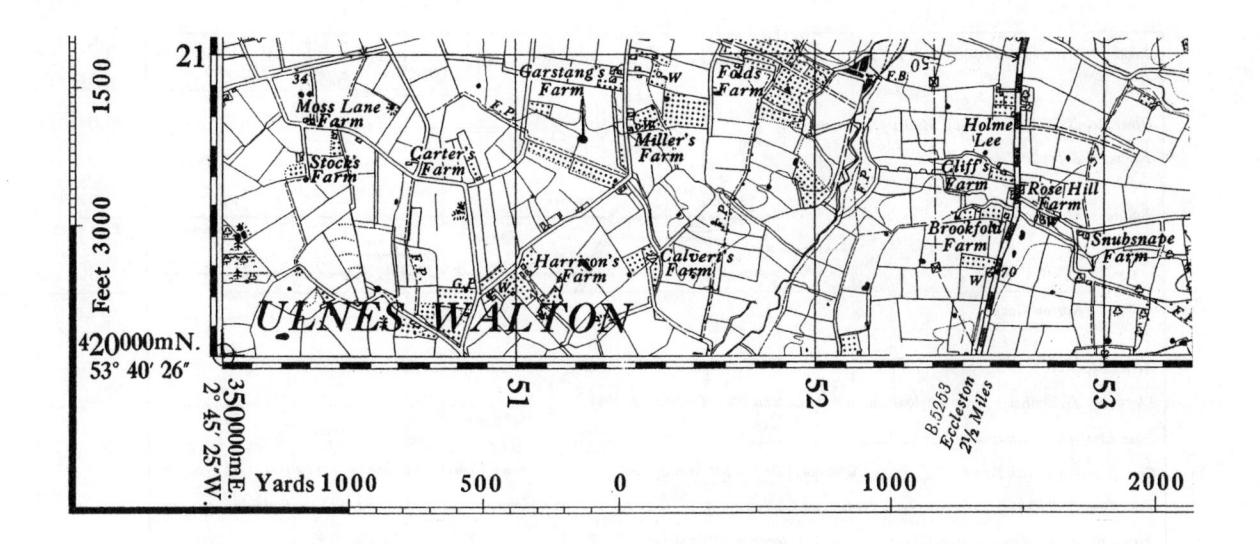

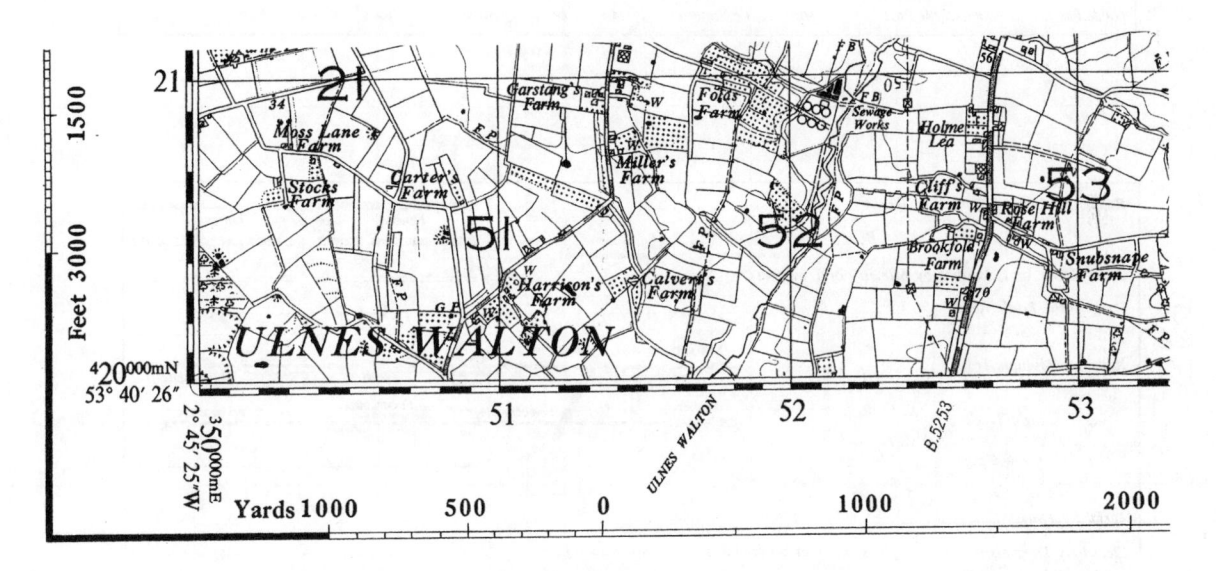

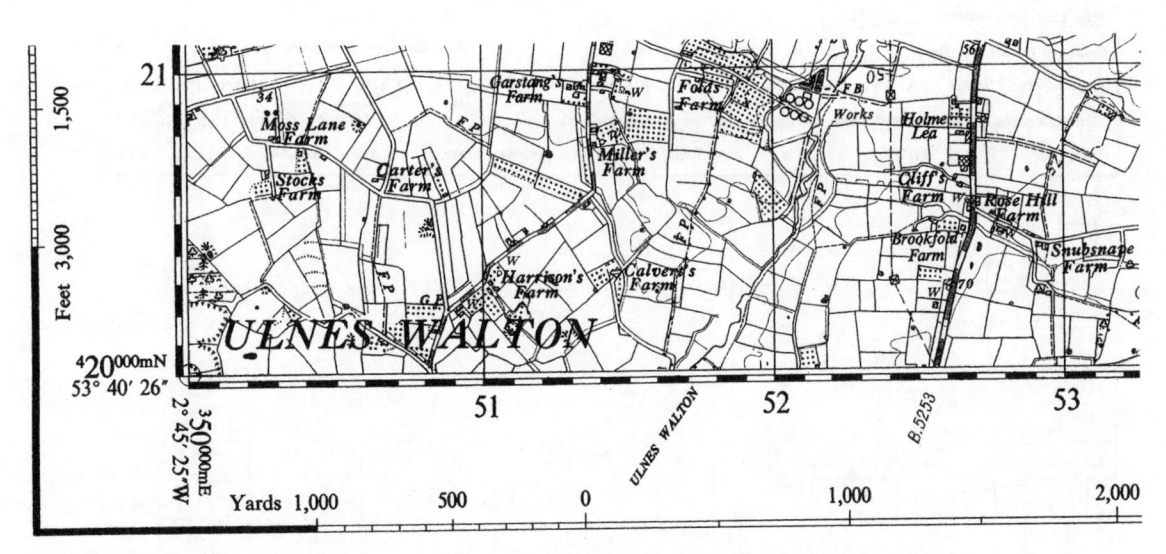

Figure 5. The evolution of the frame of the three-colour map. See the explanation of column 9.6 on page 57. (Top) type 1: sheet 34/52, edition A, 1947: lettered in Old Roman style no.5; easting grid figures are printed sideways. (Middle) type 2: sheet SD 52, edition C, 1955: easting grid figures have been reset in Times Roman, and printed upright. (Bottom) type 4: sheet SD 52, edition C//*, 1963: the outer frame text has been reset in Times Roman. This frame was also used on later four-colour sheets; the original style used on the four-colour map (type 3) is depicted on sheet 40/09 (see colour plate 3, right hand panel).

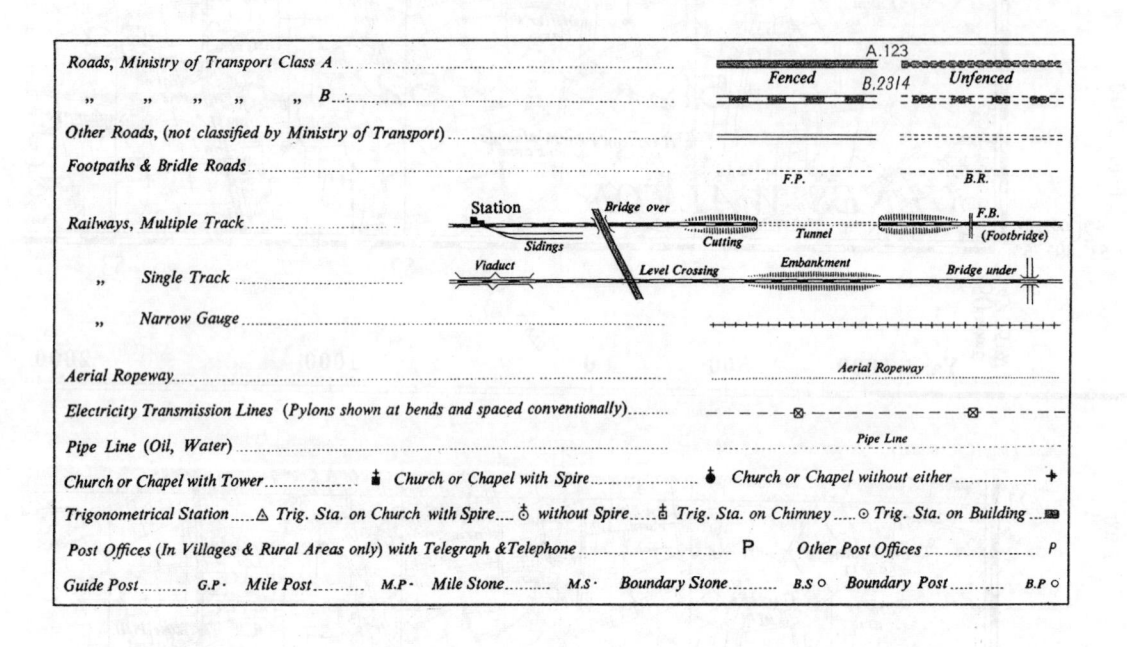

Figure 6. The left-hand legend on Provisional Edition sheets. See the explanation of column 9.4 on page 57 detailing the differences. (Top) type 1: sheet 41/67, 1947: roads MOT Class A, B, etc; Church. (Bottom) type 2: sheet 41/02, 1948: roads MOT Class A, B, etc; Churches subdivided.

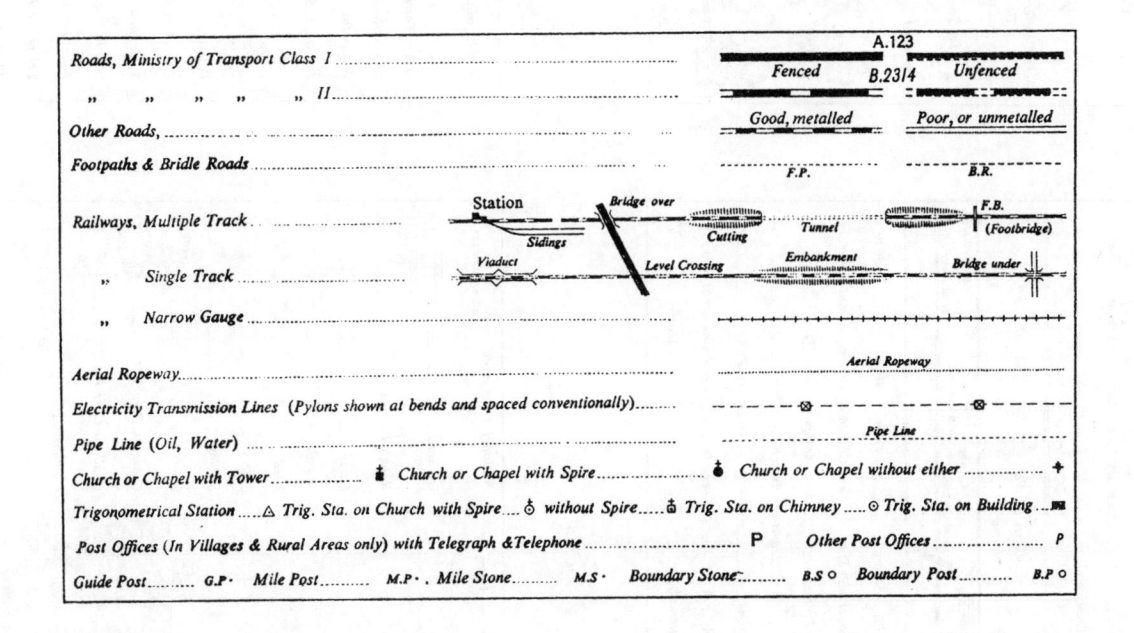

Figure 7. The left-hand legend on Provisional Edition sheets [cp figure 6]. (Top) type 3: sheet 34/95, 1948: roads MOT Class I, II, etc; Church. (Bottom) type 4: sheet 41/55, 1949: roads MOT Class I, II, etc; Churches subdivided.

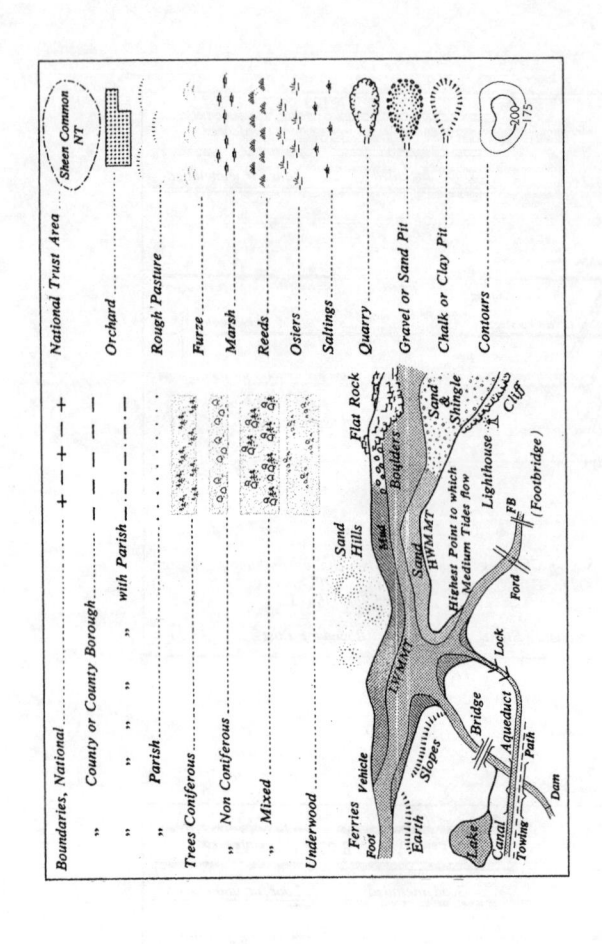

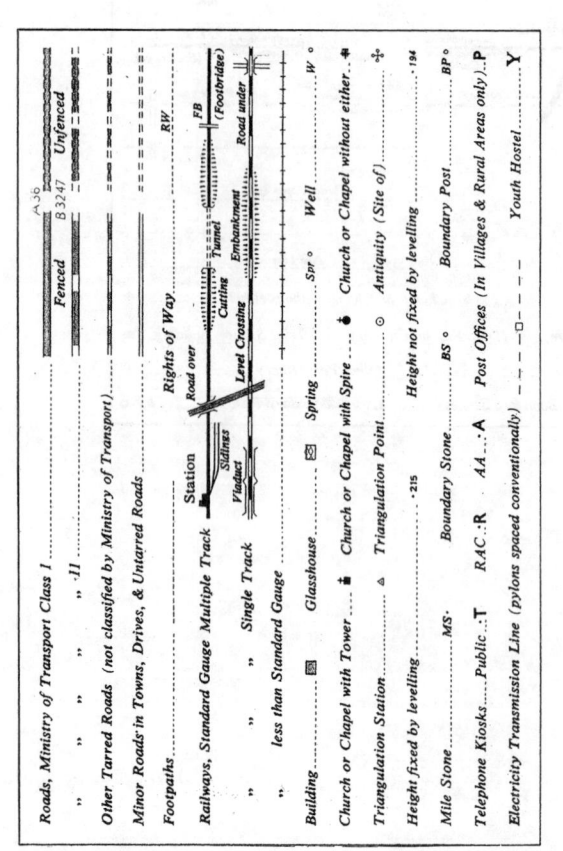

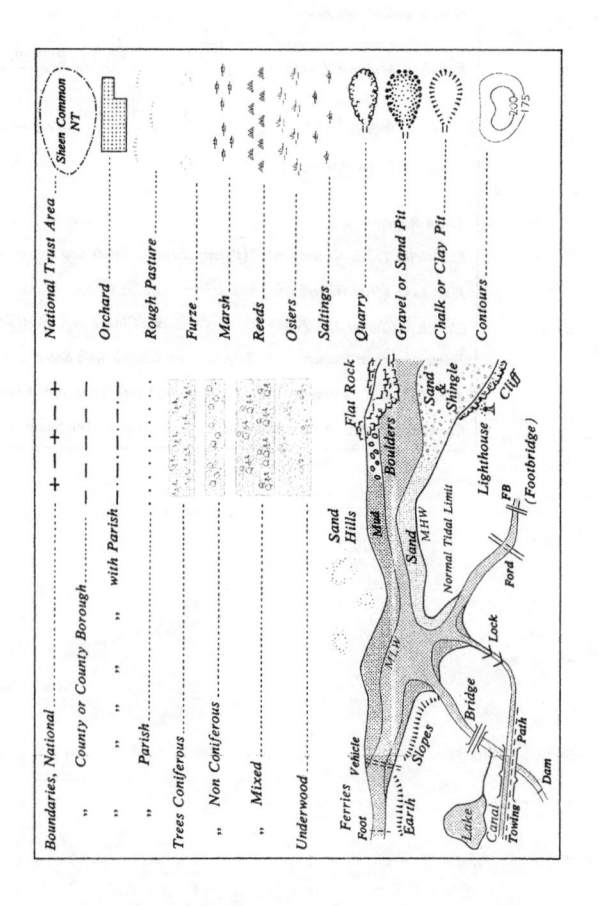

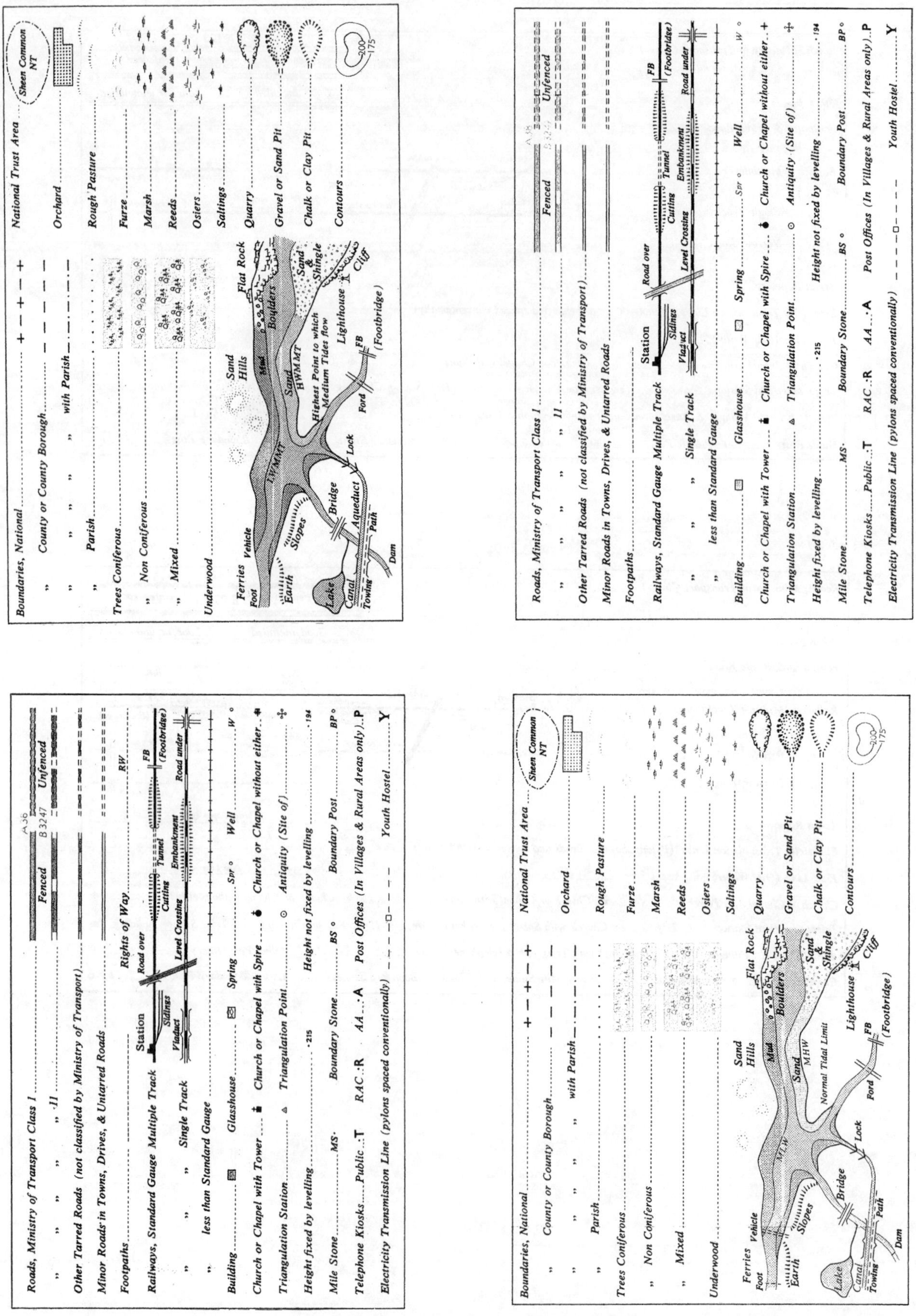

Figure 8. Legends on Regular Edition sheets. See the explanation of column 9.4 on page 57. (Top) type 5: sheet SX 45 & Part of SX 44, edition A, 1956: rights of way information is present. (Bottom) type 6: Sheet SX 45 & Part of SX 44, edition B/*/, 1967: rights of way information is absent.

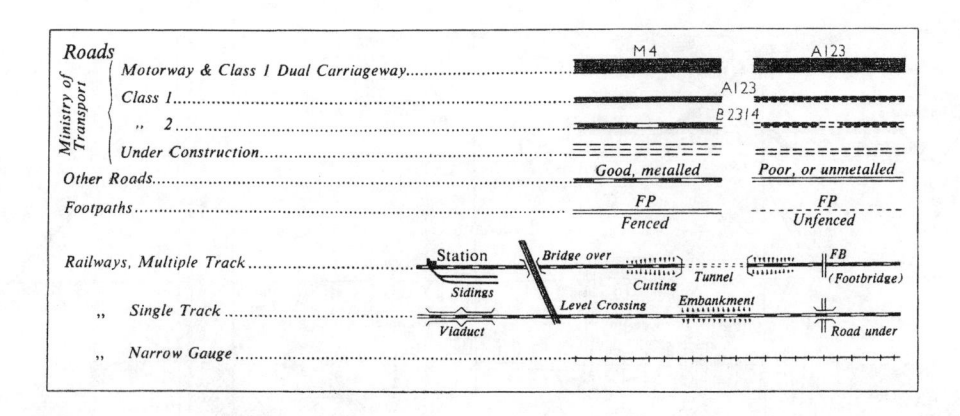

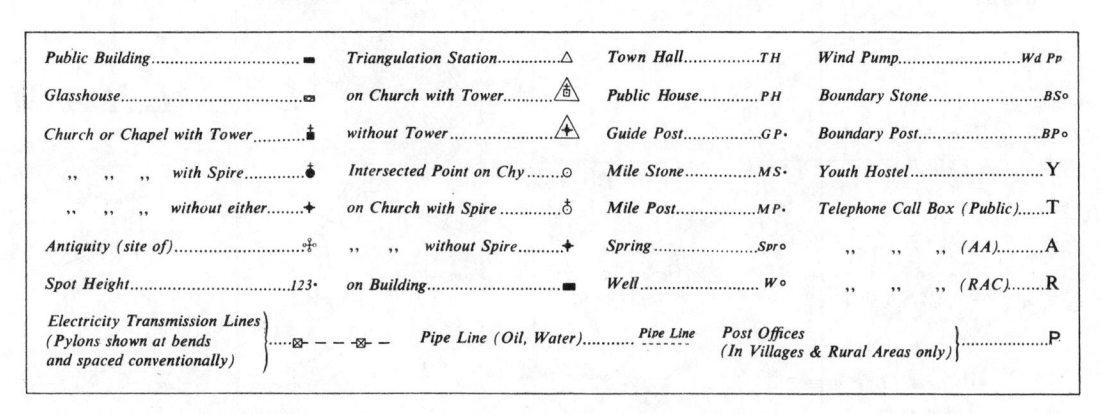

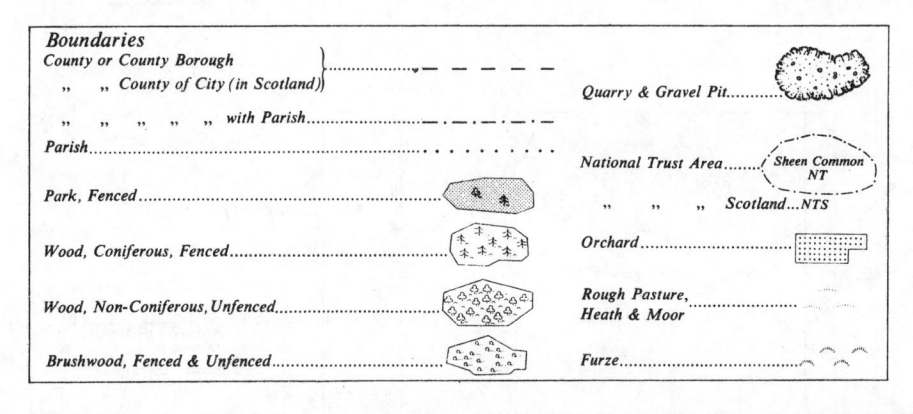

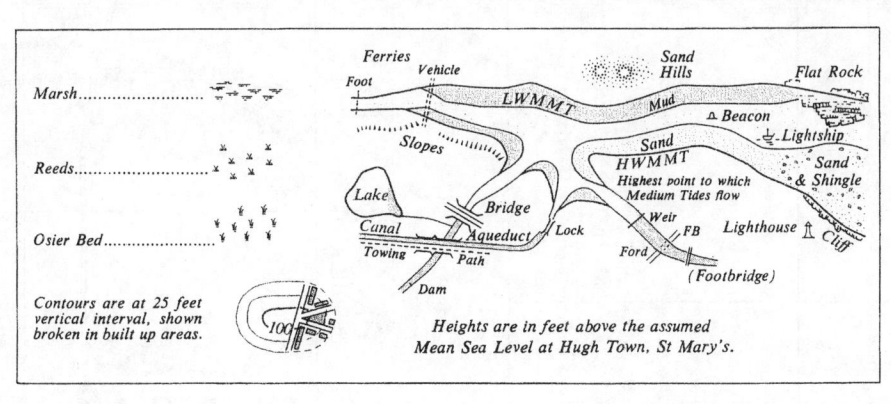

Figure 9. Legends for Isles of Scilly, *edition A, 1964 (type 7). This includes a motorway symbol.*

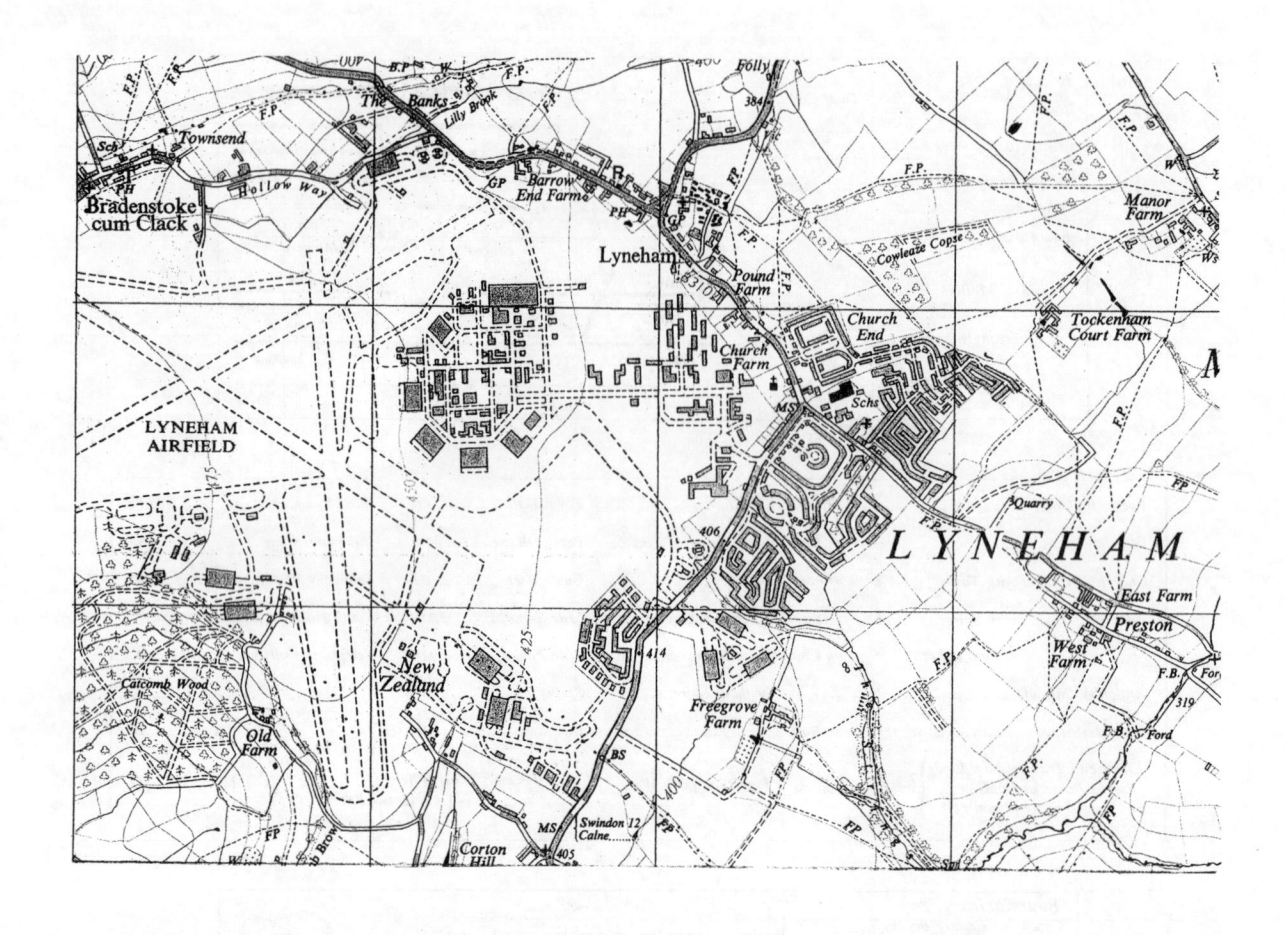

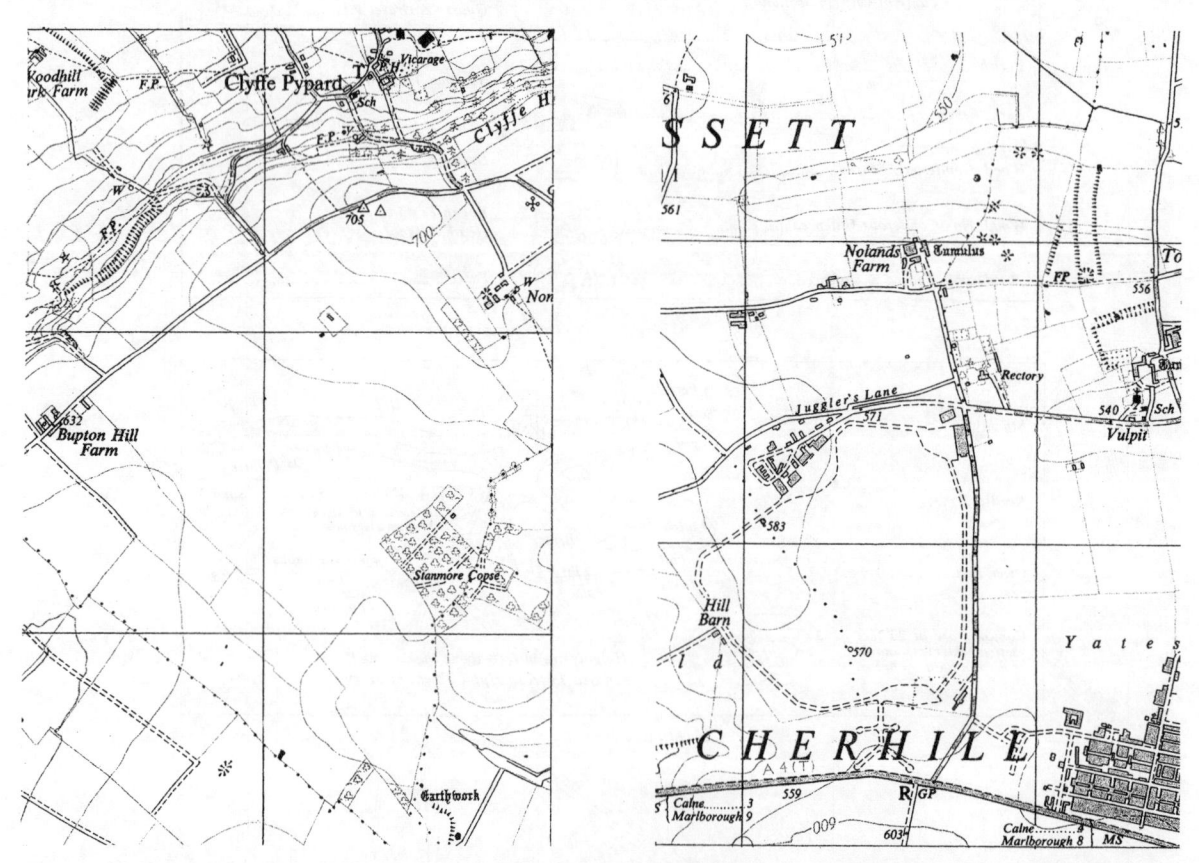

Figure 10. Depiction of airfields on sheet SU 07, edition B/, 1971. See the explanation of column 11 on page 57. Type 0 requires no illustration. (Bottom left) type 1: blank space. (Bottom right) type 2: perimeter track with dispersals and buildings only. (Top) type 3: full runway layout and buildings.*

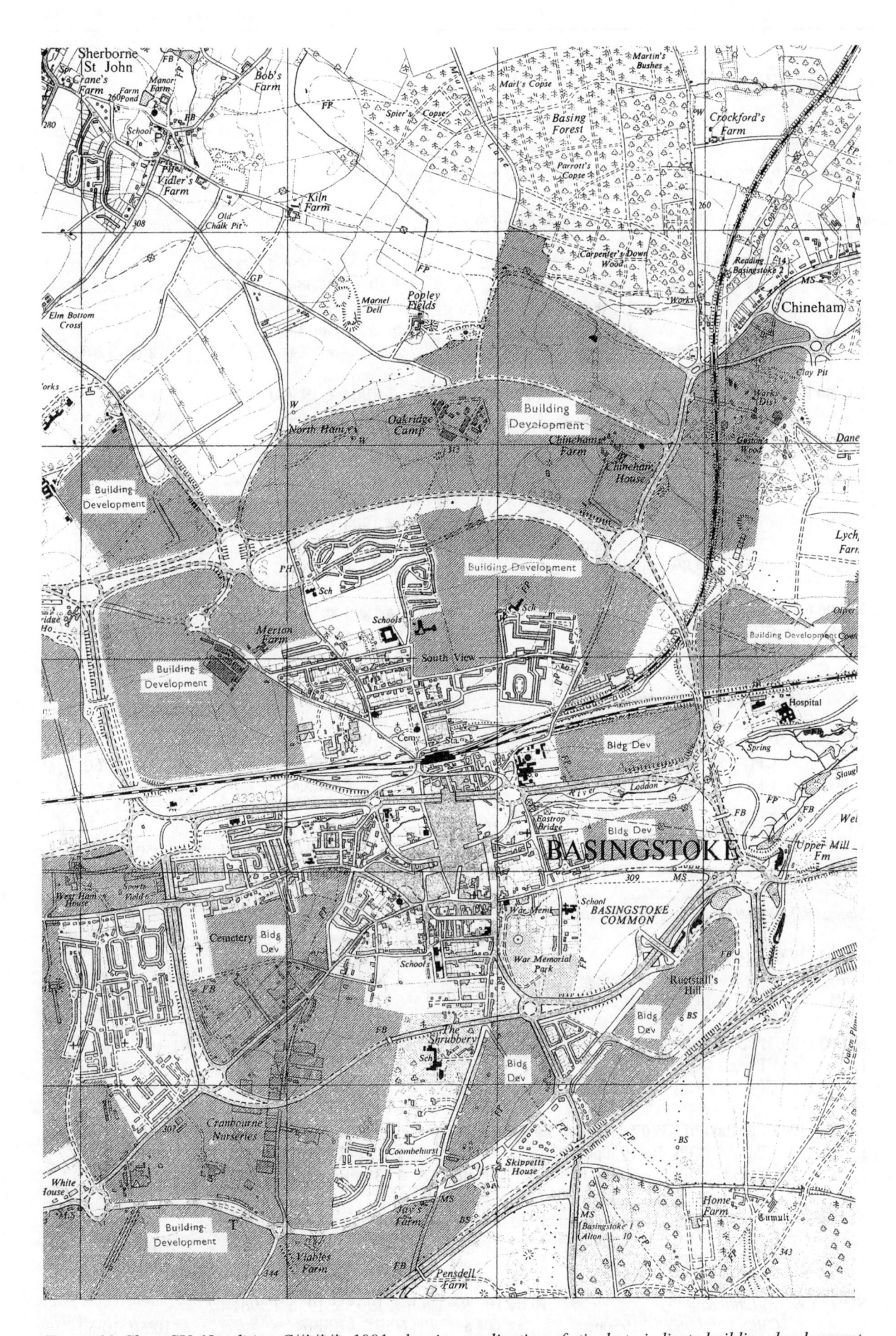

Figure 11. Sheet SU 65, edition C//*/*/*, 1981, showing application of stipple to indicate building development.

1   2   3   4   5   6   7   8   9   10   11   12   13   14

Cartobibliography

Sheets in HP, HT, HU, HW, HX, HY, HZ, NA and NB were not published.

Sheets NC 00 to 03, 10 to 14, 20 to 27, 30 to 37, 40 to 46, 50 to 56, 60 to 66, 70 to 75 were not published.

NC 76 (29/76) [Bettyhill]   10 by 10   Revision a 1904 c 1956 f 1961
1   A   CPs   OPs   1950pc   -   a   1.50   4BEoo4   8/50   8/50
3   A   CPt,v   1950pc   1953rpb a   6.53   4BooX4
3   A   CPw   1950mc   -   a   6.53   4BooX4
3   B   CPy   OPy   1963mc   -   acf   6.63   4oooX4   12/63   12/63

NC 80 (29/80) [Dunrobin]   10 by 10   Revision a 1904 c 1956 f 1961
1   A   CPs   OPs   1950pc   -   a   6.50   4BEoo4   R   9/50   10/50
3   B   CPy,z   OPy   1963mc   -   acf   6.63   4oooX4L   9/63   9/63

Sheets NC 81 to 85 were not published.

NC 86 (29/86) [Portskerra]   10 by 10   Revision a 1904 c 1956 f 1961
1   A   CPs   OPs   1950pc   -   a   6.50   4BEoo4   9/50   10/50
3   B   CPx   OPx   1963mc   -   acf   6.63   4oooX4   1/63   1/63

NC 90 (29/90) [Brora]   10 by 10   Revision a 1904 c 1956 f 1961
1   A   CPr   OPs   nd:po   -   -   6.50   4BEoo4   R   8/50   10/50
3   B   CPy,z   OPy   1963mc   -   acf   6.63   4oooX4   8/63   8/63

Sheets NC 91 to 94 were not published.

NC 95 (29/95) [Loch na Seilige]   10 by 10   Revision a 1904-05 c 1956 f 1961
1   A   CPr   OPs   nd:po   -   -   1.50   4BEoo4   7/50   8/50
3   B   CPy   OPy   1963mc   -   acf   6.63   4oooX4   8/63   8/63

NC 96 (29/96) [Reay]   10 by 10   Revision a 1904-05 c 1956 f 1961
1   A   CPr   OPs   1950pc   -   a   1.50   4BEoo4   0   7/50   8/50
3   B   CPy   OPy   1963mc   -   acf   1.63   4oooX4   3   8/63   8/63

Sheets ND 01 to 03 were not published.

ND 04 (39/04) *Altnabreac**   10 by 10   Revision a 1905 c 1946 d 1956
1   A   CPs   OPs   1950pc   -   ac   6.50   4BEoo4   RB   10/50   10/50
3   B   CPx   OPx   1963mc   -   ad   6.63   4oooX4   1/63   1/63

ND 05 (39/05) *Scotscalder**   10 by 10   Revision a 1905 c 1956 f 1961-62
1   A   CPr   OPs   nd:po   -   -   1.50   4BEoo4   RB   7/50   8/50
3   B   CPy   OPy   1963mc   -   acf   6.63   4oooX4   8/63   8/63

ND 06 (39/06) **& Part of ND 07** *Bridge of Forss**   10 by 12   Revision a 1904-05 c 1938 d 1938-56 f 1961
1   A   CPs   OPs   1950pc   -   ac   1.50   4BEoo3   8/50   9/50
3   B   CPy,z   OPy   1963mc   -   adf   6.63   4oooX3   11/63   11/63

ND 12 (39/12) *Berriedale**   10 by 10   Revision a 1905 c 1956 f 1961-62
1   A   CPr   OPs   1950pc   -   a   1.50   4BEoo4   B   8/50   10/50
3   B   CPx,z   OPx   1963mc   -   acf   6.63   4oooX4   1/63   1/63

ND 13 (39/13) *Latheron**   10 by 10   Revision a 1905 c 1956 f 1961-62
1   A   CPs   OPs   1950pc   -   a   6.50   4BEoo4   10/50   10/50
3   B   CPy,z   OPy   1963mc   -   acf   6.63   4oooX4   8/63   8/63

1	2	3	4	5	6	7	8	9	10	11	12	13	14

ND 14 (39/14) *Achavanich**   10 by 10   Revision a 1905 c 1956 f 1961
| 1 | A | CPr | OPs | 1950pc | - | a | 1.50 | 4BEoo4 | | | 8/50 | 10/50 | |
| 3 | B | CPx | OPx | 1962mc | - | acf | 6.62 | 4oooX4 | | | 12/62 | 12/62 | |

ND 15 (39/15) *Halkirk**   10 by 10   Revision a 1905 c 1956-57 f 1961-62
| 1 | A | CPr | OPs | 1950pc | - | a | 1.50 | 4BEoo4 | R | | 8/50 | 9/50 | |
| 3 | B | CPy,z | OPy | 1963mc | - | acf | 6.63 | 4oooX4L | | | 8/63 | 8/63 | |

ND 16 (39/16) **& Part of ND 17** *Thurso**   10 by 12   Revision a 1904-05 c 1938 d 1938-56 f 1961
| 1 | A | CPs | OPs | 1950pc | - | ac | 1.51 | 4BEoo3 | R | | 10/50 | 10/50 | |
| 3 | B | CPy,z | OPy | 1963mc | - | adf | 6.63 | 4oooX3 | | | 11/63 | 11/63 | |

ND 23 (39/23) **& Part of ND 33** *Lybster**   12 by 10   Revision a 1905 c 1956 f 1961
| 1 | A | CPs | OPs | 1950pc | - | a | 6.50 | 4BEoo3 | R | | 10/50 | 11/50 | |
| 3 | B | CPy,z | OPy | 1963mc | - | acf | 6.63 | 4oooX3 | | | 9/63 | 9/63 | |

ND 24 (39/24) *Camster**   10 by 10   Revision a 1905 c 1956 f 1961
| 1 | A | CPs | OPs | 1950pc | - | a | 6.50 | 4BEoo4 | | | 9/50 | 11/50 | |
| 3 | B | CPx | OPx | 1962mc | - | acf | 6.62 | 4oooX4 | | | 12/62 | 12/62 | |

ND 25 (39/25) *Watten**   10 by 10   Revision a 1905 b 1905-07 c 1956 f 1961
| 1 | A | CPs | OPs | 1950pc | - | a | 1.50 | 4BEoo4 R | | | 8/50 | 9/50 | |
| 3 | B | CPy,z | OPy | 1963mc | - | bcf | 6.63 | 4oooX4L | | | 11/63 | 11/63 | |

ND 26 (39/26) *Loch Heilen**   10 by 10   Revision a 1905 c 1956 f 1961
| 1 | A | CPs | OPs | 1950pc | - | a | 1.50 | 4BEoo4 | | 0 | 10/50 | 10/50 | |
| 3 | B | CPy,z | OPy | 1963mc | - | acf | 6.63 | 4oooX4L | | 3 | 11/63 | 11/63 | |

ND 27 (39/27) **& Part of ND 17** *Dunnett**   12 by 10   Revision a 1905 c 1956 f 1961
| 1 | A | CPs | OPs | 1950pc | - | a | 1.50 | 4BEoo3 | | | 11/50 | 12/50 | |
| 3 | B | CPy | OPy | 1963mc | - | acf | 6.63 | 4oooX3L | | | 9/63 | 9/63 | |

Sheet ND 29 was not published.

ND 34 (39/34) **& Part of ND 33** *Thrumster**   10 by 12   Revision a 1905 c 1938 d 1938-56 f 1961
| 1 | A | CPs | OPs | 1950pc | - | ac | 6.50 | 4BEoo3 | R | | 9/50 | 11/50 | |
| 3 | B | CPy,z | OPy | 1963mc | - | adf | 6.63 | 4oooX3L | | | 8/63 | 8/63 | |

AA 1.7.50 on A(A), 3/-, OSPR 11/50.

ND 35 (39/35) *Wick Caithness**   10 by 10   Revision a 1905 c 1938 d 1938-56 f 1961
| 1 | A | CPr | OPs | 1950pc | - | ac | 1.50 | 4BEoo4 | R | 00 | 8/50 | 9/50 | |
| 3 | B | CPy,z | OPy | 1963mc | - | adf | 6.63 | 4oooX4 | | 33 | 11/63 | 12/63 | 1 |

AA 1.4.50 on A(A), 3/-, OSPR 9/50. 1. Wick Aerodrome.

ND 36 (39/36) *Freswick**   10 by 10   Revision a 1905 c 1956 f 1961
| 1 | A | CPs | OPs | 1950pc | - | a | 1.50 | 4BEoo4 | | | 10/50 | 10/50 | |
| 3 | B | CPy | OPy | 1963mc | - | acf | 6.63 | 4oooX4 | | | 8/63 | 8/63 | |

ND 37 (39/37) **& Part of ND 47** *John o' Groats**   12 by 10   Revision a 1905 c 1956 f 1961
| 1 | A | CPs | OPs | 1950pc | - | a | 1.50 | 4BEoo3 | | | 11/50 | 12/50 | |
| 3 | B | CPy,z | OPy | 1963mc | - | acf | 6.63 | 4oooX3 | | | 12/63 | 12/63 | |

Sheets ND 38, 39, 48 and 49, sheets in NF and NG, sheets NH 00 to 43 were not published.

NH 44 (28/44) *Kilmorack*   10 by 10   Revision a 1901-04 c 1955 f 1960
| 1 | A | CPs | OPs | 1950pc | - | a | 1.50 | 4BEoo4 | B | | 10/50 | 10/50 | |
| 3 | B | CPx,z | OPx | 1961mc | - | acf | 6.61 | 4oooX4 | | | 8/61 | 7/61 | |

NH 45 (28/45) *Strathpeffer*   10 by 10   Revision a 1902-07 c 1955 f 1961
| 1 | A | CPr | OPs | nd:po | - | - | 1.50 | 4BEoo4 | RB | | 8/50 | 10/50 | |
| 3 | B | CPx,z | OPx | 1962mc | - | acf | 6.62 | 4oooX4L | | | 3/62 | 2/62 | |

1	2	3	4	5	6	7	8	9	10	11	12	13	14

Sheets NH 46 to 53 were not published.

NH 54 (28/54) *Beauly**   10 by 10   Revision a 1902-04 c 1955 f 1960
1 A   CPr   OPs   nd:po   -   -   1.50   4BEoo4   R   8/50   8/50
3 B   CPx,z   OPx   1961mc   -   acf   6.61   4oooX4   9/61   10/61

NH 55 (28/55) *Dingwall**   10 by 10   Revision a 1904 c 1938 d 1938-55 f 1961
1 A   CPr   OPr   1950pc   -   ac   1.50   4BEoo4   R   7/50   8/50
3 B   CPx,z,0z OPx   1962mc   -   adf   6.62   4oooX4L   5/62   5/62

NH 56 (28/56) *[Foulis]*   10 by 10   Revision a 1904 c 1938 d 1938-56 f 1962
1 A   CPs   OPs   1950pc   -   ac   6.50   4BEoo4   R   10/50 10/50
3 B   CPy,z   OPy   1963mc   -   adf   6.63   4oooX4   8/63   8/63

Sheets NH 57 to 63 were not published.

NH 64 (28/64) *Inverness** 10 by 10   Revision a 1902-29 c 1938-55 f 1961 g 1971
1 A   CPr   OPs   nd:po   -   -   1.50   4BEoo4   R   7/50   9/50
3 B   CPx,z   OPx   1962mc   -   acf   6.62   4oooX4   3/62   4/62
3 B/* CPz   OPz   1962^{c}ms nd:r acg 6.62 4oooX4   6/72   8/72
AA 1.3.50 on A(A), 3/-, OSPR 9/50.

NH 65 (28/65) *Munlochy**   10 by 10   Revision a 1904-07 c 1938-55 f 1960
1 A   CPr   OPs   nd:po   -   -   1.50   4BEoo4   R   7/50   8/50
3 B   CPx,z,0z OPx   1961mc   -   acf   6.61   4oooX4   10/61 10/61

NH 66 (28/66) *Alness**   10 by 10   Revision a 1904 c 1938-56 f 1961
1 A   CPs   OPs   1950pc   -   a   6.50   4BEoo4   R   00   9/50   10/50
3 B   CPy,z   OPy   1963mc   -   acf   6.63   4oooX4   32   8/63   8/63   1
1. Alness Airfield.

Sheets NH 67 to 73 were not published.

NH 74 (28/74) *Culloden Muir** 10 by 10   Revision a 1904-05 c 1955 f 1960
1 A   CPr   OPs   nd:po   -   -   1.50   4BEoo4   R   8/50   8/50
3 B   CPx,z   OPx   1961mc   -   acf   6.61   4oooX4   6/61   6/61
M on A.

NH 75 (28/75) *Fortrose**   10 by 10   Revision a 1902-04 c 1938-55 f 1960
1 A   CPr   OPs   nd:po   -   -   1.50   4BEoo4   R   0   8/50   10/50 1
3 B   CPx,z   OPx   1961mc   -   acf   6.61   4oooX4   3   1/62   12/61 2
M on A. 1. Fort George is not shown. 2. Inverness (Dalcross) Aerodrome.

NH 76 (28/76) **& Part of NH 86** *Cromarty**   12 by 10   Revision a 1904-38 c 1938 d 1955-56 f 1961-62
1 A   CPs   OPs   1950pc   -   ac   1.50   4BEoo3   R   8/50   9/50
3 B   CPy,z   OPy   1963mc   -   adf   6.63   4oooX3   8/63   8/63

NH 77 (28/77) *Ballchraggan**   10 by 10   Revision a 1904-05 c 1938-56 f 1961
1 A   CPr   OPs   nd:po   -   -   6.50   4BEoo4   R   8/50   10/50
3 B   CPy,z   OPy   1963mc   -   acf   6.63   4oooX4   8/63   8/63

NH 78 (28/78) *Tain Ross & Cromarty**   10 by 10   Revision a 1904-05 c 1938 d 1938-56 f 1961
1 A   CPs   OPs   1950pc   -   ac   6.50   4BEoo4   R   10/50 10/50
3 B   CPy,z   OPy   1963mc   -   adf   6.63   4oooX4   11/63 11/63

NH 79 (28/79) **& Part of NH 89** *Mound**   15 by 10   Revision a 1904-05 c 1938 d 1938-56 f 1961
1 A   CPs   OPs   1950pc   -   ac   6.50   4BEoo3   R   10/50 11/50
3 B   CPy,z   OPy   1963mc   -   adf   6.63   4oooX3   12/63 12/63

Sheets NH 80 to 83 were not published.

1	2	3	4	5	6	7	8	9	10	11	12	13	14

NH 84 (28/84) *Cawdor*   10 by 10 Revision a 1903 c 1955 f 1960
1 A   CPr   OPs   nd:po   -   -   1.50   4BEoo4   7/50 8/50
3 B   CPx,z OPx   1961mc -   acf   6.61   4oooX4   10/61 10/61
M on A.

NH 85 (28/85) *Nairn*   10 by 10 Revision a 1904-38 c 1955 f 1961
1 A   CPr   OPr   nd:po   -   -   1.50   4BEoo4   R   0   8/50 8/50
3 B   CPx,z OPx   1962mc -   acf   6.62   4oooX4   2   5/62 8/62
M on A.

NH 87 (28/87) **& Parts of NH 86 & NH 97** *Hill of Fearn* 12 by 12 Revision a 1904-05 c 1956 f 1961
1 A   CPs   OPs   1950pc -   a   6.50   4BEoo3   R   0   10/50 11/50
3 B   CPy,z OPy   1963mc -   acf   6.63   4oooX3L   3   12/63 12/63

NH 88 (28/88) *Inver Ross & Cromarty* 10 by 10 Revision a 1904-05 c 1956 f 1961
1 A   CPs   OPs   1950pc -   a   6.50   4BEoo4   R   0   10/50 11/50
3 B   CPy,z OPy   1963mc -   acf   6.63   4oooX4L   3   9/63 9/63

Sheets NH 90 to 93 were not published.

NH 94 (28/94) *Ferness*   10 by 10 Revision a 1903-04 c 1926-27 d 1955 f 1961
1 A   CPr   OPs   1950pc -   ac   1.50   4BEoo4   8/50 8/50
3 B   CPx   OPx   1962mc -   adf   6.62   4oooX4L   5/62 5/62
M on A.

NH 95 (28/95) **& Part of NH 96** *Auldearn* 10 by 12 Revision a 1904 c 1938-55 f 1960
1 A   CPr   OPs   nd:po   -   -   1.50   4BEoo3   R   7/50 9/50
3 B   CPx,z OPx   1962mc -   acf   6.62   4oooX3   5/62 6/62
M on A.

NH 98 (28/98) *Tarbat Ness*   10 by 10 Revision a 1903-04 c 1944-56 f 1961
1 A   CPr   OPs   nd:po   -   -   6.50   4BEoo4   8/50 10/50
3 B   CPy,z,0z OPy   1963mc -   acf   6.63   4oooX4   8/63 8/63

Sheets NJ 00 to 04 were not published.

NJ 05 (38/05) *Forres*   10 by 10 Revision a 1903-04 b 1904-06 c 1938 d 1938-55
1 A   CPs   OPs   1950pc -   ac   6.50   4BEoo4   R   10/50 10/50
3 B   CPv,z OPv   1957pc -   bd   6.57   4oooX4L   1/58   1
M on A. 1. OSPR (outline edition) not found.

NJ 06 (38/06) **& Part of NH 96** *Findhorn* 15 by 10 Revision a 1904 c 1938-55
1 A   CPr   OPs   nd:po   -   -   1.50   4BEoo3   R   0   8/50 9/50
3 B   CPv,z OPv   1957pc -   ac   6.57   4oooX3   1   3/58 7/58 1
M on A. 1. Kinloss Aerodrome.

Sheets NJ 10 to 14 were not published.

NJ 15 (38/15) *Kellas*   10 by 10 Revision a 1903-04 c 1938-55
1 A   CPs   OPs   1950pc -   a   6.50   4BEoo4   0   10/50 11/50
3 B   CPv.z OPv   1957pc -   ac   6.57   4oooX4   2   1/58   1
M on A. 1. OSPR (outline edition) not found.

NJ 16 (38/16) **& Part of NJ 17** *Burghead* 10 by 12 Revision a 1903-04 c 1938 d 1938-55 f 1965
1 A   CPs   OPs   1950pc -   ac   6.50   4BEoo3   R   00   12/50 1/51
3 B   CPv   OP?   1957pc -   ad   6.57   4oooX3   10   1/58   3/58
3 B/   CPz   OPz   1957mc 1966ch adf   6.57   4oooX3   13
M on A.

Sheets NJ 20 to 23 were not published.

1	2	3	4	5	6	7	8	9	10	11	12	13	14

NJ 24 (38/24) *Charlestown of Aberlour** 10 by 10 Revision a 1902-03 c 1938 d 1938-55
1   A   CPs   OPs   1950pc   -   ac   6.50   4BEoo4   R   10/50 11/50
3   B   CPv,z   OPv   1957pc   -   ad   6.57   4oooX4   1/58   10/57
M on A.

NJ 25 (38/25) *Glen of Rothes** 10 by 10 Revision a 1903-04 c 1938 d 1938-55
1   A   CPs   OPs   1950pc   -   ac   6.50   4BEoo4   R   0   10/50 10/50
3   B   CPv,z   OPv   1957pc   -   ad   6.57   4oooX4   2   12/57 10/57
M on A.

NJ 26 (38/26) **& Part of NJ 27** *Elgin** 10 by 12 Revision a 1904-06 b 1903-06 c 1938 d 1938-55
                                                       f 1974; j 1970 airfield
1   A   CPs   OPs   1950pc   -   ac   6.50   4BEoo3   R   000 12/50 12/50
3   B   CPv,x.z   OP?   1957pc   -   bd   6.57   4oooX3   200 2/58   2/58
3   B/*   CFz   OFz   1957^{c}ms -   bdfj   6.57   4oooX3   230 9/74   9/74   1
With an inset of Hallman Skerries (1 km square). M on A. 1. Lossiemouth Aerodrome.

Sheets NJ 30 to 33 were not published.

NJ 34 (38/34) *Dufftown** 10 by 10 Revision a 1900-03 b 1902-03 c 1938-55
1   A   CPs   OPs   1950pc   -   a   6.50   4BEoo4   R   10/50 10/50
3   B   CPv,z   OPv   1957pc   -   bc   6.57   4oooX4   1/58   -

NJ 35 (38/35) *Fochabers** 10 by 10 Revision a 1902-03 c 1938-55 f 1973
1   A   CPs   OPs   1951pc   -   a   1.51   4BEoo4   R   2/51   3/51
2   A   CPs   1951pc   1951   a   1.52   4Booo4   R
3   B   CPv,x   OP?   1957pc   -   ac   6.57   4oooX4   2/58   2/58
3   B/*   CPz   OPz   1957^{c}ms -   acf   6.57   4oooX4   9/73   12/73

NJ 36 (38/36) *Garmouth** 10 by 10 Revision a 1902-03 c 1938-55
1   A   CPs   OPs   1950pc   -   a   1.50   4BEoo4   R   0   10/50 10/50
2   A   CPs   1950pc   1952   a   1.52   4Booo4   R   0
3   B   CPv,z   OPv   1957pc   -   ac   6.57   4oooX4   3   9/57   9/57

Sheets NJ 40 to 43 were not published.

NJ 44 (38/44) *Cairnie** 10 by 10 Revision a 1900-24 c 1938-40 d 1938-55
1   A   CPs   OPs   1950pc   -   ac   1.51   4BEoo4   R   9/50   10/50
3   B   CPv,w,z   OPv   1957pc   -   ad   6.57   4oooX4   12/57 9/57

NJ 45 (38/45) *Keith** 10 by 10 Revision a 1902-03 c 1938-40 d 1938-55
1   A   CPs   OPs   1950pc   -   ac   6.50   4BEoo4   R   11/50 12/50
3   B   CPv,z   OPv   1957pc   -   ad   6.57   4oooX4   1/58   11/57

NJ 46 (38/46) *Buckie** 10 by 10 Revision a 1902-28 c 1938 d 1938-55
1   A   CPs   OPs   1951pc   -   ac   1.51   4BEoo4   R   1/51   1/51
3   B   CPv,z   OPv   1957pc   -   ad   6.57   4oooX4   9/57   9/57

NJ 50 (38/50) *Lumphanan** 10 by 10 Revision a 1899-1929 c 1942 d 1955 f 1960
1   A   CPs   OPs   1950pc   -   ac   6.50   4BEoo4   R   10/50 10/50
3   B   CPx,z   OPx   1961mc   -   adf   6.61   4oooX4   5/61   5/61

NJ 51 (38/51) *Alford Aberdeenshire** 10 by 10 Revision a 1899 c 1942 d 1942-55
1   A   CPs   OPs   1950pc   -   ac   6.50   4BEoo4   R   11/50 11/50
3   B   CPx,z   OPx   1961mc   -   ad   6.61   4oooX4   5/61   5/61

NJ 52 (38/52) *Kennethmont** 10 by 10 Revision a 1899 c 1927 d 1955 f 1960
1   A   CPs   OPs   1950pc   -   ac   6.50   4BEoo4   RB   10/50 10/50
3   B   CPx   OPx   1961mc   -   adf   6.61   4oooX4   10/61 10/61

1	2	3	4	5	6	7	8	9	10	11	12	13	14
NJ 53 (38/53) *Huntly (South)**					10 by 10	Revision a 1898-1925 c 1938 d 1938-55 f 1960							
1	A	CPs	OPs	1950pc	-	ac	6.50	4BEoo4	R		11/50	1/51	
3	B	CPx,z	OPx	1961mc	-	adf	6.61	4oooX4			7/61	6/61	
NJ 54 (38/54) *Huntly (North)**					10 by 10	Revision a 1900-03 b 1899-1925 c 1938 d 1938-55							
1	A	CPs	OPs	1951pc	-	ac	6.51	4BEoo4	R		1/51	1/51	
3	B	CPv,z	OPv	1957pc	-	bd	6.57	4oooX4			7/57	8/57	
NJ 55 (38/55) *Glen Barry**					10 by 10	Revision a 1900-03 c 1942 d 1955							
1	A	CPs	OPs	1951pc	-	ac	1.51	4BEoo4	R		1/51	1/51	
3	B	CPv,z	OPv	1957pc	-	ad	6.57	4oooX4			9/57	9/57	
NJ 56 (38/56) *Portsoy**					10 by 10	Revision a 1902-28 c 1938 d 1938-55							
1	A	CPs	OPs	1950pc	-	ac	6.50	4BEoo4	R		12/50	1/51	
3	B	CPv,z	OPv	1957pc	-	ad	6.57	4oooX4			12/57	9/57	
NJ 60 (38/60) *Torphins**					10 by 10	Revision a 1899 c 1942 d 1942-55 f 1960							
1	A	CPs	OPs	1950pc	-	ac	6.50	4BEoo4	R		9/50	11/50	
3	B	CPx,z	OPx	1961mc	-	adf	6.61	4oooX4			7/61	6/61	
NJ 61 (38/61) *Tillyfourie**					10 by 10	Revision a 1899 c 1942 d 1955 f 1960							
1	A	CPs	OPs	1950pc	-	ac	6.50	4BEoo4	R		11/50	12/50	
3	B	CPx,z	OPx	1961mc	-	adf	6.61	4oooX4			8/61	8/61	
NJ 62 (38/62) *Insch**					10 by 10	Revision a 1901-28 c 1954-55 f 1961							
1	A	CPs	OPs	1950pc	-	a	6.50	4BEoo4	R		10/50	10/50	
3	B	CPx,z	OPx	1962mc	-	acf	6.62	4oooX4			4/62	4/62	
NJ 63 (38/63) *Culsalmond**					10 by 10	Revision a 1899-1900 c 1955 f 1960							
1	A	CPs	OPs	1950pc	-	a	6.50	4BEoo4	R		11/50	1/51	
3	B	CPx,z	OPx	1961mc	-	acf	6.61	4oooX4			7/61	7/61	
NJ 64 (38/64) *Inverkeithny**					10 by 10	Revision a 1902-28 b 1900-23 c 1938 d 1938-55							
1	A	CPs	OPs	1950pc	-	ac	6.50	4BEoo4			10/50	11/50	
3	B	CPv,z	OPv	1957pc	-	bd	6.57	4oooX4			9/57	9/57	
NJ 65 (38/65) *Aberchirder**					10 by 10	Revision a 1904-05 b 1900-25 c 1927 d 1938-55							
1	A	CPs	OPs	1950pc	-	ac	6.50	4BEoo4			12/50	1/51	
3	B	CPv,z	OPv	1957pc	-	bd	6.57	4oooX4			9/57	9/57	
NJ 66 (38/66) *Banff**					10 by 10	Revision a 1901-29 c 1938 d 1938-55							
1	A	CPs	OPs	1951pc	-	ac	1.51	4BEoo4	R	0	1/51	1/51	
3	B	CPv,w,z	OPv	1957pc	-	ad	6.57	4oooX4		3	1/58	10/57	
NJ 70 (38/70) *Dunecht**					10 by 10	Revision a 1899-1901 c 1942 d 1955-56							
1	A	CPs	OPs	1950pc	-	ac	6.50	4BEoo4			10/50	11/50	
3	B	CPv,z	OPv	1957pc	-	ad	6.57	4oooX4			9/57	10/57	
NJ 71 (38/71) *Kintore**					10 by 10	Revision a 1899-1924 c 1938-42 d 1938-56							
1	A	CPs	OPs	1950pc	-	ac	6.50	4BEoo4	R		10/50	11/50	
3	B	CPv,z	OPv	1957pc	-	ad	6.57	4oooX4			9/57	10/57	
NJ 72 (38/72) *Inverurie**					10 by 10	Revision a 1899-1924 c 1938 d 1938-55							
1	A	CPs	OPs	1950pc	-	ac	6.50	4BEoo4	R		10/50	10/50	
3	A	CPu		1950pc	-	ac	6.50	4BEoo4	R				
3	B	CPv,w,z	OPv	1957pc	-	ad	6.57	4oooX4			9/57	10/57	
NJ 73 (38/73) *Fyvie**					10 by 10	Revision a 1899-1900 c 1955 f 1975							
1	A	CPs	OPs	1950pc	-	a	6.50	4BEoo4	R		10/50	11/50	
3	B	CPv.z	OPv	1957pc	-	ac	6.57	4oooX4			9/57	9/57	
3	B/*	CFz	OFz	1957^{c}ms	nd:r	acf	6.57	4oooX4			12/75	8/75	

1	2	3	4	5	6	7	8	9	10	11	12	13	14

NJ 74 (38/74) *Turriff (South)* *   10 by 10   Revision a 1899-1925 b 1900-25 c 1938 d 1938-55 f 1966

1	A	CPs	OPs	1950pc	-	ac	6.50	4BEoo4	R		10/50	11/50	
3	B	CPv	OPv	1957pc	-	bd	6.57	4oooX4			9/57	9/57	
3	B/	CPz	OPz	1957mc	1966ch	bdf	6.57	4oooX4					

NJ 75 (38/75) *Turriff (North)* *   10 by 10   Revision a 1900-25 c 1938 d 1938-55

| 1 | A | CPs | OPs | 1950pc | - | ac | 6.50 | 4BEoo4 | R | | 12/50 | 1/51 | |
| 3 | B | CPv,z | OPv | 1957pc | - | ad | 6.57 | 4oooX4 | | | 9/57 | 10/57 | |

NJ 76 (38/76) *Macduff* *   10 by 10   Revision a 1900-29 c 1938 d 1938-55

| 1 | A | CPs | OPs | 1950pc | - | ac | 6.50 | 4BEoo4 | R | | 11/50 | 11/50 | |
| 3 | B | CPv,z | OPv | 1957pc | - | ad | 6.57 | 4oooX4 | | | 9/57 | 9/57 | |

NJ 80 (38/80) [Peterculter]   10 by 10   Revision a 1899-1924 c 1938-42 d 1938-56

1	A	CPr	OPr	1947po	-	-	6.47	4BE1X3	R		4/47	8/47	
1	B	CPs		1950pc	-	ac	6.50	4BEoo3	R				
3	B		OPs	1950pc	-	ac	6.50	4BEoo3	R				
3	C	CPv,z	OPv	1957pc	-	ad	6.57	4oooX3			1/58	11/57	

AA 1.5.47 on A(A), 3/-, OSPR 8/47; AA 1.11.53 on B(B), 4/-, OSPR 4/55.

NJ 81 (38/81) [Dyce]   10 by 10   Revision a 1899-1924 c 1938-56

| 1 | A | CPr | OPr | nd:po | - | - | 1.49 | 4BEoo3 | R | 0 | 1/49 | 2/49 | |
| 3 | B | CPv.z | OPv | 1957pc | - | ac | 6.57 | 4oooX3 | | 3 | 9/57 | 8/57 | 1 |

M on A. 1. Aberdeen (Dyce) Aerodrome.

NJ 82 (38/82) *Oldmeldrum* *   10 by 10   Revision a 1899-1924 c 1938 d 1938-56

1	A	CPs	OPs	1950pc	-	ac	6.50	4BEoo4	R		11/50	11/50	
3	B	CPv	OPv	1957pc	-	ad	6.57	4oooX4			9/57	9/57	
3	B/	CP?x,z.z		1957mc	1961ch	ad	6.57	4oooX4					

NJ 83 (38/83) *Methlick* *   10 by 10   Revision a 1899 b 1899-1924 c 1955-56

| 1 | A | CPs | OPs | 1950pc | - | a | 6.50 | 4BEoo4 | | | 11/50 | 11/50 | |
| 3 | B | CPv.z | OPv | 1957pc | - | bc | 6.57 | 4oooX4 | | | 7/57 | 9/57 | |

NJ 84 (38/84) *New Deer* *   10 by 10   Revision a 1899-1900 c 1955

| 1 | A | CPs | OPs | 1950pc | - | a | 6.50 | 4BEoo4 | | | 11/50 | 11/50 | |
| 3 | B | CPv,z | OPv | 1957pc | - | ac | 6.57 | 4oooX4 | | | 5/57 | 6/57 | |

NJ 85 (38/85) *New Pitsligo* *   10 by 10   Revision a 1900-01 c 1944-55

| 1 | A | CPs | OPs | 1951pc | - | a | 1.51 | 4BEoo4 | | | 2/51 | 1/51 | |
| 3 | B | CPv,z | OPv | 1957pc | - | ac | 6.57 | 4oooX4 | | | 9/57 | 9/57 | |

NJ 86 (38/86) *New Aberdour* *   10 by 10   Revision a 1900-02 b 1900-25 c 1955

| 1 | A | CPs | OPs | 1950pc | - | a | 6.50 | 4BEoo4 | | | 12/50 | 1/51 | |
| 3 | B | CPv,z | OPv | 1957pc | - | bc | 6.57 | 4oooX4 | | | 1/58 | 12/57 | |

NJ 90 (38/90) [Aberdeen]   10 by 10   Revision a 1923-24 c 1938-50 d 1938-56 f 1968 j 1967 building

1	A	CPr	OPr	1947po	-	-	6.47	4BE1X3	R		6/47	8/47	1
1	B	CPs		1950pc	-	ac	6.50	4BEoo3	R		1/52	-	
2	B		OPu	1950pc	-	ac	6.50	4Booo3	R		-	1/55	2
3	C	CPv,y,z	OPv	1957pc	-	ad	6.57	4oooX3			1/58	10/57	
3	C/*	CPz	OPz	1957ms	1970b	adfj	6.57	4oooX3			7/70	8/70	

AA 1.5.47 on A(A), 3/-, OSPR 8/47; AA 1.11.53 on B(B), 4/-, OSPR 1/55. 1. Showing the Bridge of Don tramway (9409, 9509). 2. No other example of this outline price state has been recorded.

NJ 91 (38/91) [Bridge of Don (North)]   10 by 10   Revision a 1899-1924 c 1938-56 f 1965

| 1 | A | CPr | OPr | 1948po | - | - | 6.48 | 4BE2X3 | | | 4/48 | 4/48 | 1 |
| 1 | B | CPr | | nd:po | - | - | 6.48 | 4BEoo3 | | | | | |

1	2	3	4	5	6	7	8	9	10	11	12	13	14
3	C	CPv,y	OPv	1957pc	-	ac	6.57	4oooX3			9/57	9/57	
3	C/*	CPz	OPz	1957mc	1966r	acf	6.57	4oooX3			3/66	4/66	

1. Showing the base line measured in 1817 (9513 to 9819+), also the Bridge of Don to Balgownie Links tramway (9510, 9511, 9512).

NJ 92 (38/92) **& Part of NK 02** *Logierieve* 15 by 10 Revision a 1899-1924 c 1938 d 1938-56
1	A	CPs	OPs	1951pc	-	ac	1.51	4BEoo3	R		1/51	2/51	1
3	B	CPv,z.z+	OPv	1957pc	-	ad	6.57	4oooX3			9/57	8/57	

M on A, B. 1. Showing the base line measured in 1817 (9820, 9821).

NJ 93 (38/93) *Ellon** 10 by 10 Revision a 1899-1924 c 1938 d 1938-55
1	A	CPs	OPs	1950pc	-	ac	6.50	4BEoo4	R		12/50	1/51	
3	B	CPv.z	OPv	1957pc	-	ad	6.57	4oooX4			9/57	9/57	

NJ 94 (38/94) *Maud** 10 by 10 Revision a 1899-1900 b 1899-1901 c 1955
1	A	CPs	OPs	1950pc	-	a	6.50	4BEoo4	R		12/50	12/50	
3	B	CPv,z	OPv	1957pc	-	bc	6.57	4oooX4			9/57	9/57	

NJ 95 (38/95) *Strichen** 10 by 10 Revision a 1899-1901 c 1955
1	A	CPs	OPs	1950pc	-	a	6.50	4BEoo4	R		12/50	1/51	
3	B	CPv,z	OPv	1957pc	-	ac	6.57	4oooX4			9/57	8/57	

NJ 96 (38/96) *Fraserburgh** 10 by 10 Revision a 1900-25 c 1938 d 1938-55
1	A	CPs	OPs	1950pc	-	ac	6.50	4BEoo4			12/50	1/51	
3	B	CPv,z	OPv	1957pc	-	ad	6.57	4oooX4			1/58	11/57	

M on A, B.

NK 03 (48/03) **& Parts of NK 02 & NK 13** *Cruden Bay** 12 by 12 Revision a 1899-1900 b 1899 c 1938-56
1	A	CPs	OPs	1950pc	-	a	6.50	4BEoo3	R		12/50	1/51	1
3	B	CPv	OPv	1957pc	-	bc	6.57	4oooX3			1/58	11/57	
3	B/	CPw.z+		1957mc	1961mc	bc	6.57	4oooX3					

1. Showing the Cruden Bay Hotel Tramway, from the south side of the station at 083366 into the hotel.

NK 04 (48/04) *Longside** 10 by 10 Revision a 1899-1924 c 1938-55
1	A	CPs	OPs	1951pc	-	a	6.51	4BEoo4	R	0	1/51	1/51	1
3	B	CPv.z	OPv	1957pc	-	ac	6.57	4oooX4		3	1/58	-	

1. There is no sign of the Lenabo airship base (0242), nor the railway to it.

NK 05 (48/05) **& Part of NK 15** *Rattray Head** 15 by 10 Revision a 1899-1901 c 1955
1	A	CPs	OPs	1950pc	-	a	1.50	4BEoo3	R	0	12/50	1/51	
3	B	CPv.z,0z	OPv	1957pc	-	ac	6.57	4oooX3		3	3/58	12/57	

NK 06 (48/06) *Inverallochy** 10 by 10 Revision a 1900-25 c 1938 d 1938-55
1	A	CPs	OPs	1950pc	-	ac	6.50	4BEoo4	R	0	10/50	11/50	
3	B	CPv,z.z	OPv	1957pc	-	ad	6.57	4oooX4		3	1/58	-	

NK 14 (48/14) **& Part of NK 13** *Peterhead** 10 by 12 Revision a 1899-1924 c 1938 d 1938-55
1	A	CPs	OPs	1950pc	-	ac	6.50	4BEoo3	R		12/50	1/51	1
3	B	CPv.z,0z	OPv	1957pc	-	ad	6.57	4oooX3			1/58	12/57	

1. Showing the Stirling Hill-Burnhaven tramway, with its extension along the Peterhead South Breakwater. There is another down the north Breakwater. The Peterhead harbour branch is closed.

Sheets in NL and NM, sheets NN 00 to 49 were not published.

NN 50 (27/50) *Aberfoyle** 10 by 10 Revision a 1898-99 c 1954 f 1972; j 1970 reservoir
1	A	CPr	OPr	nd:po	-	-	6.49	4BEoo3	R		1/50	3/50	
3	B	CPu,x,z	OPv	1955pc	-	ac	6.55	4oooX3			1/56	10/57	
3	B/*	CPz	OPz	1955^{c}ms	nd:rv	acfj	6.55	4oooX3			9/73	9/73	1

1. Glen Finglas Reservoir.

Sheets NN 51 to 59 were not published.

1	2	3	4	5	6	7	8	9	10	11	12	13	14

NN 60 (27/60) *Callander**   10 by 10   Revision a 1898-99 c 1938-54
1	A	CPr	OPr	nd:po	-	-	6.49	4BEoo3	R		1/50	3/50	
3	B	CPu	OPv	1956pc	-	ac	6.56	4oooX3			1/56	10/57	
3	B/	CPw.z		1956mc	1961mc	ac	6.56	4oooX3					

Sheets NN 61 to 69 were not published.

NN 70 (27/70) *Dunblane**   10 by 10   Revision a 1899-1930 c 1947-54 f 1975
1	A	CPr	OPr	nd:po	-	-	1.50	4BEoo3	RB		1/50	3/50	
3	B	CPu	OPv	1956pc	-	ac	6.56	4oooX4			3/56	10/57	
3	B/	CP?w,y.z		1956mc	1961mc	ac	6.56	4oooX4					
3	B//*	CFz	OFz	1956ᶜms	nd:r	acf	6.56	4oooX4			11/75	9/75	

Sheets NN 71 to 79 were not published.

NN 80 (27/80) *Strathallan**   10 by 10   Revision a 1896-1930 c 1947-54 f 1972
1	A	CPr	OPr	nd:po	-	-	1.49	4BEoo3	R		1/50	3/50	1
3	B	CPu	OPv	1956pc	-	ac	6.56	4oooX3			3/56	5/57	
3	B/	CPw.z		1956mc	1959mc	ac	6.56	4oooX3					
3	B//*	CPz	OPz	1956ᶜms	nd:r	acf	6.56	4oooX3			7/73	12/73	

1. Showing the Royal Caledonian Curling Pond (8609) and the Royal Curling Club Station at 869090.

NN 81 (27/81) *Muthill**   10 by 10   Revision a 1898-99 c 1954
1	A	CPr	OPr	nd:po	-	-	1.50	4BEoo3	R		3/50	3/50	
3	B	CPu		1956pc	-	ac	6.56	4oooX3L			4/56	-	
3	B/	CPw.z		1956mc	1959mc	ac	6.56	4oooX3L					

NN 82 (27/82) *Crieff**   10 by 10   Revision a 1898-1930 c 1938-54 f 1975
                                                   j 1971 [reservoir]
1	A	CPr	OPr	nd:po	-	-	1.50	4BEoo3	R		3/50	3/50	
3	B	CPu,x.z		1956pc	-	ac	6.56	4oooX3L			4/56	-	
3	B/*	CFz	OFz	1956ᶜms	nd:rv	acfj	6.56	4oooX3L			7/76	7/76	1

1. Loch Turret Reservoir.

Sheets NN 83 to 89 were not published.

NN 90 (27/90) *Glen Devon*   10 by 10   Revision a 1894-1920 c 1948-54 f 1969
                                                   j 1967 afforestation k 1967 reservoir
1	A	CPr	OPr	nd:po	-	-	6.49	4BEoo3	R		8/49	9/49	
	B												
3	C	CPu	OPv	1956pc	-	ac	6.56	4oooX3			2/56	10/57	
3	C/	CPw,x		1956mc	1960mc	ac	6.56	4oooX3					
3	C//*	CPz	OPz	1956ᶜms	nd:fv	acfjk	6.56	4oooX3			12/70	2/71	1

1. Upper Glendevon Reservoir.

NN 91 (27/91) *Auchterarder**   10 by 10   Revision a 1899-1930 c 1938-54
| 1 | A | CPr | OPr | nd:po | - | - | 6.49 | 4BEoo3 | R | | 9/49 | 9/49 | |
| 3 | B | CPu,x.z | OPv | 1956pc | - | ac | 6.56 | 4oooX3 | | | 3/56 | 6/57 | 1 |

1. The Gleneagles Hotel is shown, together with the railway approach 930116 to 916115.

NN 92 (27/92) *Keillour Forest**   10 by 10   Revision a 1899-1930 c 1938-54
| 1 | A | CPr | OPr | nd:po | - | - | 6.49 | 4BEoo3 | R | | 8/49 | 9/49 | |
| 3 | B | CPu,x,z | OPv | 1956pc | - | ac | 6.56 | 4oooX3 | | | 3/56 | 6/57 | |

NN 93 (27/93) *Upper Strathbraan** 10 by 10   Revision a 1899 c 1954
| 1 | A | CPr | OPr | nd:po | - | - | 6.48 | 4BEoo3 | | | 8/48 | 8/48 | |
| 3 | B | CPw,x.z | OPv | 1958mc | - | ac | 6.58 | 4oooX4 | | | 2/59 | - | |

NN 94 (27/94) *Craigvinean Forest** 10 by 10 Revision a 1899 c 1954
| 1 | A | CPr | OPr | 1948po | - | - | 1.48 | 4BEoo3 | R | | 8/48 | 8/48 | |
| 3 | B | CPw,z | OPv | 1959mc | - | ac | 6.59 | 4oooX4L | | | 4/59 | 5/59 | |

1	2	3	4	5	6	7	8	9	10	11	12	13	14

NN 95 (27/95) *Pitlochry*   10 by 10 Revision a 1898-99 c 1938-54
1 A   CPr   OPr   nd:po   -   -   1.49 4BEoo3 R   11/48 11/48
    B
3 C   CPw,z   OPv   1959mc   -   ac   6.59 4oooX3   6/59   9/59

Sheets NN 96 to 99 were not published.

NO 00 (37/00) *Crook of Devon*   10 by 10 Revision a 1894-1913 c 1938-54 f 1970
1 A   CPr   OPr   nd:po   -   -   6.48 4BEoo3 R   0   9/48   10/48
3 B   CPu,x,z OPv   1956pc   -   ac   6.56 4oooX4   1   1/56   6/57
3 B/   CPz   OPz   1956^{c}ms nd:ch acf 6.56 4oooX4   3   1
1. WT station and masts are shown.

NO 01 (37/01) *Dunning*   10 by 10 Revision a 1894-1930 c 1954
1 A   CPr   OPr   nd:po   -   -   1.48 4BEoo3 RB   12/48 1/49
3 B   CPu,w,z OPv   1956pc   -   ac   6.56 4oooX3   3/56   5/57

NO 02 (37/02) *Perth (West)*   10 by 10 Revision a 1899-1930 c 1954-55
1 A   CPr   OPr   nd:po   -   -   6.48 4BEoo3 R   8/48   10/48
3 B   CPu   OPv   1956pc   -   ac   6.56 4oooX4   4/56   6/57
3 B/   CP?w,y,z   1956mc   1961mc ac   6.56 4oooX4
AA 1.6.48 on A(A), 3/-, OSPR 10/48.

NO 03 (37/03) *Bankfoot*   10 by 10 Revision a 1898-1901 c 1938-55
1 A   CPr   OPr   nd:po   -   -   1.49 4BEoo3 R   9/49   9/49
3 B   CPw,y,z OPv   1958mc   -   ac   6.58 4oooX4   2/59   3/59

NO 04 (37/04) *Dunkeld*   10 by 10 Revision a 1898-99 c 1938-55; j 1982 selected
1 A   CPr   OPr   nd:po   -   -   6.49 4BEoo3 R   11/49 11/49
3 B   CPv,x,z OPv   1957pc   -   ac   6.57 4oooX3L   6/57   8/57 1
3 B/*   CFz   OFz   1982^{c}m-s nd:r   acj 6.57 4oooX3L   7/82   7/82
1. A 1956 proof is recorded, with the grey plate coloured green.

NO 05 (37/05) *Creag nam Mial*   10 by 10 Revision a 1898-99 c 1954
1 A   CPr   OPr   1948po   -   -   6.48 4BEoo3   8/48   8/48
3 B   CPw,z   OPv   1959mc   -   ac   6.59 4oooX4L   4/59   7/59

Sheets NO 06 to 09 were not published.

NO 10 (37/10) *Kinross*   10 by 10 Revision a 1894-1919 c 1938-54 f 1975
1 A   CPr   OPr   nd:po   -   -   1.48 4BEoo3 R   0   10/48 11/48
3 B   CPu   OPv   1955pc   -   ac   6.55 4oooX3   1   1/56   11/58
3 B/   CPw,z   1955mc   1960mc ac   6.55 4oooX3   1
3 B//* CFz   OFz   1955^{c}ms nd:r   acf 6.55 4oooX3   3   4/76   11/75

NO 11 (37/11) *Bridge of Earn*   10 by 10 Revision a 1894-1931 c 1938-54
1 A   CPr   OPr   nd:po   -   -   1.48 4BEoo3 R   9/48   10/48
    B
3 C   CPu,z.z OPv   1956pc   -   ac   6.56 4oooX3   1/56   -

NO 12 (37/12) *Perth (East)*   10 by 10 Revision a 1898-1931 c 1938-54 f 1981
1 A   CPr   OPr   nd:po   -   -   6.48 4BEoo3 R   0   9/48   10/48
    B
3 C   CPu.z   OPv   1956pc   -   ac   6.56 4oooX3   2   2/56   5/57 1
3 C/*   CFz   OFz   1981^{c}ms nd:r   acf 6.56 4oooX3   2   1/82   1/82
AA 1.5.48 on A(A), 3/-, OSPR 10/48. 1. Perth Aerodrome.

NO 13 (37/13) *Stanley Perthshire*   10 by 10 Revision a 1898-1900 c 1938-55
1 A   CPr   OPr   nd:po   -   -   1.49 4BEoo3 R   10/49 10/49
3 B   CPw,z   OPv   1959mc   -   ac   6.59 4oooX3   4/59   5/59

1	2	3	4	5	6	7	8	9	10	11	12	13	14

NO 14 (37/14) *Blairgowrie**   10 by 10   Revision a 1898-99 c 1938-55
| 1 | A | CPr | OPr | nd:po | - | - | 1.49 | 4BEoo3 | R | | 10/49 | 11/49 |
| 3 | B | CPw,z | OPv | 1959mc | - | ac | 6.59 | 4oooX3L | | | 4/59 | 5/59 |

Sheets NO 15 to 19 were not published.

NO 20 (37/20) *Glenrothes**   10 by 10   Revision a 1912-19 c 1938-54 f 1967
1	A	CPr	OPr	1948po	-	-	6.48	4BEoo3	R		5/48	6/48
	B											
3	C	CPu	OPv	1956pc	-	ac	6.56	4oooX3			4/56	4/57
3	C/	CPw		1956mc	1959mc	ac	6.56	4oooX3				
3	C//*	CPz	OPz	1956mc	1968r	acf	6.56	4oooX3			1/68	1/68

NO 21 (37/21) *Newburgh Fife**   10 by 10   Revision a 1898-1913 c 1938-54
| 1 | A | CPr | OPr | 1948po | - | - | 6.48 | 4BEoo3 | RB | | 5/48 | 6/48 |
| 3 | B | CPu,y,z | OPv | 1956pc | - | ac | 6.56 | 4oooX3L | | | 4/56 | 5/57 |

NO 22 (37/22) *Errol**   10 by 10   Revision a 1898-1912 c 1954 f 1965; j 1965 airfield
1	A	CPr	OPr	nd:po	-	-	1.49	4BEoo3	R	0	10/49	10/49
3	B	CPu	OPv	1956pc	-	ac	6.56	4oooX3		1	2/56	-
3	B/	CPw		1956mc	1960mc	ac	6.56	4oooX3		1		
3	B//*	CPz,0z	OPz	1956mc	1966r	acfj	6.56	4oooX3		3	7/66	9/66

NO 23 (37/23) *Hallyburton Forest**   10 by 10   Revision a 1898-1921 c 1938-55
| 1 | A | CPr | OPr | nd:po | - | - | 1.49 | 4BEoo3 | R | | 10/49 | 11/49 |
| 3 | B | CPw,z | OPv | 1959mc | - | ac | 6.59 | 4oooX3L | | | 4/59 | 7/59 |

NO 24 (37/24) *Alyth**   10 by 10   Revision a 1898-1922 c 1938-55
| 1 | A | CPr | OPr | nd:po | - | - | 6.49 | 4BEoo3 | R | | 10/49 | 11/49 |
| 3 | B | CPw,z | OPv | 1959mc | - | ac | 6.59 | 4oooX3L | | | 4/59 | 7/59 |

Sheets NO 25 to 29 were not published.

NO 30 (37/30) *Leven Fife**   10 by 10   Revision a 1912-13 c 1938-46 d 1938-54
1	A	CPr	OPr	1948po	-	-	1.48	4BEoo3	R		5/48	7/48
1	B	CPs		1951pc	-	ac	1.51	4BEoo3	R		1/52	-
3	C	CPu	OPv	1956pc	-	ad	6.56	4oooX3			1/56	-
3	C/	CP?w,z		1956mc	1959mc	ad	6.56	4oooX3				
AA 1.3.48 on A(A), 3/-, OSPR 7/48.

NO 31 (37/31) *Cupar**   10 by 10   Revision a 1912-13 c 1938-54
1	A	CPr	OPr	nd:po	-	-	1.48	4BEoo3	R		8/48	8/48
	B											
3	C	CPu,y.z	OPv	1955pc	-	ac	6.55	4oooX3			1/56	8/57

NO 32 (37/32) *Tay Bridge**   10 by 10   Revision a 1898-1921 c 1938 d 1938-54
1	A	CPr	OPr	1947po	-	-	6.47	4BE2X3	RB		6/47	10/47
1	B	CPs		1951pc	-	ac	1.51	4BEoo3	RB		1/52	-
3	C	CPu	OPv	1956pc	-	ad	6.56	4oooX3			2/56	8/57
3	C/	CPw,z.z		1956mc	1959mc	ad	6.56	4oooX3				
AA 1.5.47/NT 1.7.47 on A(A), 3/-, OSPR 10/47.

NO 33 (37/33) *Dundee (West)*   10 by 10   Revision a 1898-1921 c 1938-55 f 1967
1	A	CPr	OPr	1947po	-	-	1.47	4BE2X3	R	0	3/47	8/47
3	B	CPw,z	OPv	1959mc	-	ac	6.59	4oooX3L		3	6/59	10/59
3	B/*	CPz	OPz	1959mc	1967r	acf	6.59	4oooX3L		3	10/67	10/67
AA 1.5.47 on A(A), 3/-, OSPR 8/47.

NO 34 (37/34) *Glamis**   10 by 10   Revision a 1898-1922 c 1955 f 1966
1	A	CPr	OPr	nd:po	-	-	1.49	4BEoo3	R		11/49	11/49
3	B	CPw	OPv	1959mc	-	ac	6.59	4oooX3			4/59	7/59
3	B/	CPz		1959mc	1966ch	acf	6.59	4oooX3				

84

1	2	3	4	5	6	7	8	9	10	11	12	13	14

NO 35 (37/35) *Kirriemuir**   10 by 10   Revision a 1921-22 c 1938-55
| 1 | A | CPr | OPr | nd:po | - | - | 1.50 | 4BEoo3 | R | | 5/50 | 6/50 | |
| 3 | B | CPw,z | OPv | 1959mc | - | ac | 6.59 | 4oooX3 | | | 4/59 | 4/59 | |

Sheets NO 36 to 39 were not published.

NO 40 (37/40) **& Part of NT 49** *Largo*   10 by 12   Revision a 1911-13 c 1938-54
1	A	CPr	OPr	nd:po	-	-	1.48	4BEoo3	R		10/48	10/48	
3	B	CPu	OPv	1955pc	-	ac	6.55	4oooX3L			1/56	6/57	
3	B/	CPw.z		1955mc	1960mc	ac	6.55	4oooX3L					

NO 41 (37/41) *Guard Bridge**   10 by 10   Revision a 1912 c 1938-54
1	A	CPr	OPr	nd:po	-	-	6.48	4BEoo3	R		8/48	9/48	1
?	B	CP?		nd:po	-	-	?	4BEoo3	R				M
3	C	CPu		1956pc	-	ac	6.56	4oooX3			4/56	-	
3	C/	CPw,z,0z		1956mc	1960mc	ac	6.56	4oooX3					

AA 1.5.48 on A(A), 3/-, OSPR 9/48. M on B, C/. 1. Cameron Reservoir is in course of construction.

NO 42 (37/42) **& Part of NO 52** *Newport-on-Tay*   15 by 10   Revision a 1912-21 c 1938-54 f 1970
                                         j 1967 airfield
1	A	CPr	OPr	1947po	-	-	6.47	4BE2X3	RB	0	7/47	10/47	
3	B	CPu		1956pc	-	ac	6.56	4oooX3		1	4/56	-	1
3	B/	CPw	OPv	1956mc	1960mc	ac	6.56	4oooX3		1	-	2/58	2
3	B//*	CPz	OPz	1956ᶜms	nd:ar	acfj	6.56	4oooX3		3	2/71		3

AA 1.5.47 on A(A), 3/-, OSPR 10/47. M on A, B/. 1. Showing observation towers around the coast, not present on the A printing. 2. Leuchars Airfield. 3. Showing the road bridge. OSPR (outline edition) not found.

NO 43 (37/43) *Dundee (East)**   10 by 10   Revision a 1901-21 b 1920-21 c 1924-38 d 1937-55
                                         f 1967
1	A	CPr	OPr	1947po	-	-	6.47	4BE2X3	R	0	5/47	8/47	
2	B	CPt		1953pc	-	ac	6.53	4BooX3		0	10/53	-	
3	C	CPv.y,z	OPv	1957pc	-	bd	6.57	4oooX3		3	9/57	9/57	
3	C/*	CPz	OPz	1957mc	1968r	bdf	6.57	4oooX3		3	1/68	1/68	

AA 1.5.47 on A(A), 3/-, OSPR 8/47.

NO 44 (37/44) *Inverarity**   10 by 10   Revision a 1922-27 c 1938-55
1	A	CPr	OPr	nd:po	-	-	1.50	4BEoo3	R		6/50	8/50	
3	B	CPv	OPv	1957pc	-	ac	6.57	4oooX3			3/57	5/57	
3	B/	CPw,z		1957mc	1961mc	ac	6.57	4oooX3					

NO 45 (37/45) *Forfar**   10 by 10   Revision a 1922 c 1938-55 f 1969
1	A	CPr	OPr	nd:po	-	-	6.49	4BEoo3	R		11/49	11/49	
3	B	CPv,z	OPv	1957pc	-	ac	6.57	4oooX3L			5/57	8/57	
3	B/	CPz	OPz	1957ᶜms	nd:ch	acf	6.57	4oooX3L					

Sheets NO 46 to 49 were not published.

NO 50 (37/50) **& Part of NT 59** *Anstruther**   10 by 12   Revision a 1912-38 c 1954
| 1 | A | CPr | OPr | nd:po | - | - | 6.48 | 4BEoo3 | R | | 9/48 | 9/48 | |
| 3 | B | CPu,z | OP? | 1956pc | - | ac | 6.56 | 4oooX3 | | | 2/56 | 2/58 | |

NO 51 (37/51) **& Part of NO 61** *St Andrews*   15 by 10   Revision a 1912 c 1938-54
1	A	CPr	OPr	nd:po	-	-	1.48	4BEoo3	R	0	9/48	10/48	
3	B	CPu	OPv	1956pc	-	ac	6.56	4oooX3		2	1/56	11/57	
3	B/	CPw,z.z		1956mc	1960mc	ac	6.56	4oooX3		2			

AA 1.6.48 on A(A), 3/-, OSPR 10/48.

NO 53 (37/53) **& Parts of NO 52 & NO 63** *Carnoustie**   12 by 12   Revision a 1920-21 c 1938-55
| 1 | A | CPr | OPr | nd:po | - | - | 1.49 | 4BEoo3 | R | 0 | 2/49 | 3/49 | |
| 3 | B | CPv,z.z | OPv | 1957pc | - | ac | 6.57 | 4oooX3 | | 3 | 12/57 | 12/57 | |

M on A, B.

1	2	3	4	5	6	7	8	9	10	11	12	13	14

NO 54 (37/54) *Letham Angus**   10 by 10   Revision a 1921-22 c 1955
1   A   CPr   OPr   nd:po   -   -   1.50   4BEoo3   R   6/50   8/50
3   B   CPv,w,z OPv   1957pc   -   ac   6.57   4oooX3   3/57   5/57

NO 55 (37/55) *Aberlemno**   10 by 10   Revision a 1921-23 c 1938-55
1   A   CPr   OPr   nd:po   -   -   1.50   4BEoo3   R   6/50   8/50
3   B   CPv,z   OPv   1957pc   -   ac   6.57   4oooX3L   3/57   6/57

NO 56 (37/56) *Brechin (North West)**   10 by 10   Revision a 1900-22 c 1938-55
1   A   CPr   OPr   nd:po   -   -   1.50   4BEoo3   R   3/50   4/50
3   B   CPv,z   OPv   1957pc   -   ac   6.57   4oooX3   3/57   6/57

Sheets NO 57 and 58 were not published.

NO 59 (37/59) *Aboyne**   10 by 10   Revision a 1899-1929 c 1955 f 1960
1   A   CPs   OPs   nd:po   -   -   1.50   4BEoo4   R   9/50   10/50
3   B   CPx,z   OPx   1961mc   -   acf   6.61   4oooX4L   7/61   7/61

NO 60 (37/60) **& Part of NT 69** *Crail**   10 by 12   Revision a 1912 c 1938-54
1   A   CPr   OPr   nd:po   -   -   1.48   4BEoo3   R   0   8/48   9/48
1   B   CPr   nd:po   -   -   1.48   4BEoo3   R   0
3   C   CPu,z.z OP?   1955pc   -   ac   6.55   4oooX3   1   1/56   2/58

NO 64 (37/64) **& Parts of NO 63 & NO 74** *Arbroath**   12 by 12   Revision a 1921-22 c 1938-55 f 1969
1   A   CPr   OPs   nd:po   -   -   1.50   4BEoo3   R   0   7/50   8/50
3   B   CPv,y,z OPv   1957pc   -   ac   6.57   4oooX3   0   3/58   7/58
3   B/ CPz   OPz   1957ms   1970ch acf 6.57   4oooX3   3
AA 1.3.50 on A(A), 3/-, OSPR 8/50.

NO 65 (37/65) **& Part of NO 75** *Montrose*   15 by 10   Revision a 1922-23 c 1938-55
1   A   CPr   OPr   nd:po   -   -   1.50   4BEoo3   R   0   3/50   3/50 1
3   B   CPv,z.z OPv   1957pc   -   ac   6.57   4oooX3   3   3/58   11/57
1. Showing the Ordnance Depot and camping ground at Broomfield.

NO 66 (37/66) *Marykirk**   10 by 10   Revision a 1922-23 c 1938-55
1   A   CPr   OPr   nd:po   -   -   1.50   4BEoo3   R   0   5/50   5/50
3   B   CPv,w   OPv   1957pc   -   ac   6.57   4oooX3   1   6/57   8/57
3   B/ CPw,z   1957mc   1961mc ac 6.57   4oooX3   1

NO 67 (37/67) *Fettercairn**   10 by 10   Revision a 1901-22 b 1901-23 c 1938 d 1938-55
1   A   CPs   OPs   1950pc   -   ac   1.50   4BEoo4   B   10/50 10/50
3   B   CPv,w   OPv   1957pc   -   bd   6.57   4oooX4   9/57   8/57

Sheet NO 68 was not published.

NO 69 (37/69) *Banchory (W) Kincardine**   10 by 10   Revision a 1899-1923 c 1938-42 d 1938-55 f 1960
1   A   CPs   OPs   1950pc   -   ac   6.50   4BEoo4   R   10/50 10/50
3   B   CPx,z   OPx   1961mc   -   adf   6.61   4oooX4   10/61 10/61

NO 76 (37/76) **& Part of NO 86** *St Cyrus**   12 by 10   Revision a 1901-23 c 1938-55
1   A   CPr   OPr   nd:po   -   -   1.50   4BEoo3   R   7/50   8/50
3   B   CPv,z   OPv   1957pc   -   ac   6.57   4oooX3   1/58   11/57 1
1. Showing the Ordnance Depot and camping ground at Broomfield.

NO 77 (37/77) *Laurencekirk**   10 by 10   Revision a 1901-23 c 1938 d 1938-55
1   A   CPs   OPs   1950pc   -   ac   6.50   4BEoo4   R   0   10/50 11/50
3   B   CPv,z   OPv   1957pc   -   ad   6.57   4oooX4   3   7/57   8/57

Sheet NO 78 was not published.

1	2	3	4	5	6	7	8	9	10	11	12	13	14

NO 79 (37/79) *Banchory (E) Kincardine**   10 by 10   Revision a 1899-1923   c 1942   d 1938-56
| 1 | A | CPs | OPs | 1950pc | - | ac | 1.50 | 4BEoo4 | R | | 10/50 | 11/50 | |
| 3 | B | CPv,z | OPv | 1957pc | - | ad | 6.57 | 4oooX4 | | | 6/57 | 8/57 | |

NO 87 (37/87) **& Part of NO 86** *Inverbervie**   10 by 12   Revision a 1901-23   c 1938-55
| 1 | A | CPr | OPs | nd:po | - | - | 1.50 | 4BEoo3 | R | | 8/50 | 10/50 | |
| 3 | B | CPv,z | OPv | 1957pc | - | ac | 6.57 | 4oooX3 | | | 6/57 | 6/57 | |

NO 88 (37/88) *Stonehaven**   10 by 10   Revision a 1901-23   c 1938-42   d 1938-55
| 1 | A | CPs | OPs | 1950pc | - | ac | 6.50 | 4BEoo4 | R | | 9/50 | 9/50 | |
| 3 | B | CPv,z | OPv | 1957pc | - | ad | 6.57 | 4oooX4 | | | 9/57 | 9/57 | |

NO 89 (37/89) *Maryculter**   10 by 10   Revision a 1901-28   c 1956
| 1 | A | CPr | OPs | nd:po | - | - | 1.50 | 4BEoo4 | R | | 7/50 | 8/50 | |
| 3 | B | CPv,x.z | OPv | 1957pc | - | ac | 6.57 | 4oooX4 | | | 7/57 | 9/57 | |

NO 99 (37/99) *Portlethen**   10 by 10   Revision a 1902-23   c 1956
| 1 | A | CPr | OPs | nd:po | - | - | 1.50 | 4BEoo4 | R | | 8/50 | 10/50 | |
| 3 | B | CPv,z.z | OPv | 1957pc | - | ac | 6.57 | 4oooX4 | | | 9/57 | 9/57 | |

Sheets NR 15, 24 to 27, 34 to 39, 44 to 49, 56 to 70 were not published.

NR 71 (16/71) *Campbeltown (South)*   10 by 10   Revision a 1897   c 1914-15   d 1914-54   f 1960
| 1 | A | CPs | OPs | 1950pc | - | ac | 6.50 | 4BEoo4 | | | 12/50 | 1/51 | |
| 3 | B | CPx,z | OPx | 1961mc | - | adf | 6.61 | 4oooX4 | | | 10/61 | 9/61 | |

NR 72 (16/72) *Campbeltown (North)**   10 by 10   Revision a 1914-15   c 1938   d 1938-54   f 1960
| 1 | A | CPs | OPs | 1951pc | - | ac | 1.51 | 4BEoo4 | | | 3/51 | 5/51 | |
| 3 | B | CPz | OPy | 1964mc | - | adf | 6.64 | 4oooX4 | | | 3/64 | 2/64 | |

Sheets NR 73 to 79, 83 to 89, 92 to 99 were not published.

NS 02 (26/02) **& Part of NS 01** *Whiting Bay**   10 by 12   Revision a 1914-15   c 1953   f 1963
1	A	CPr	OPr	nd:po	-	-	6.49	4BEoo3			3/50	3/50	
1	A	CPs		1950pc	1951	-	1.51	4BEoo3					
3	B	CPz,0z	OPz	1964mc	-	acf	6.64	4oooX3			4/64	4/64	

NS 03 (26/03) *Brodick**   10 by 10   Revision a 1914-15   c 1953   f 1960
| 1 | A | CPr | OPr | nd:po | - | - | 1.50 | 4BEoo3 | | | 1/50 | 3/50 | |
| 3 | B | CPz | OPz | 1964mc | - | acf | 6.64 | 4oooX3 | | | 3/64 | 3/64 | |

Sheet NS 04 was not published.

NS 05 (26/05) *Sound of Bute**   10 by 10   Revision a 1915   c 1954
| 1 | A | CPr | OPr | nd:po | - | - | 6.49 | 4BEoo3 | | | 1/50 | 2/50 | |
| 3 | B | CPu,w | OPv | 1955pc | - | ac | 6.55 | 4oooX3 | | | 9/55 | 11/57 | |

NS 06 (26/06) *Rothesay**   10 by 10   Revision a 1897-1915   c 1938-54
1	A	CPr	OPr	nd:po	-	-	1.50	4BEoo3			1/50	3/50	1
3	B	CPu	OPv	1955pc	-	ac	6.55	4oooX3L			9/55	3/58	
3	B/	CPw,z		1955mc	1959mc	ac	6.55	4oooX3L					

AA 1.9.49 on A(A), 3/-, OSPR 3/50. 1. Showing a narrow gauge tramway across the peninsula (0366 to 0666).

Sheets NS 07 to 09 were not published.

NS 15 (26/15) **& Part of NS 14** *The Cumbraes**   10 by 12   Revision a 1895-1935   c 1938-54
1	A	CPr	OPr	nd:po	-	-	6.48	4BEoo3			4/49	5/49	
3	B	CPu	OPv	1955pc	-	ac	6.55	4oooX3			9/55	6/57	
3	B/	CP?w,z.z		1955mc	1960mc	ac	6.55	4oooX3					

1	2	3	4	5	6	7	8	9	10	11	12	13	14

NS 16 (26/16) *Skelmorlie** 10 by 10 Revision a 1909-15 c 1938-54
1   A   CPr   OPr   nd:po   -   -   6.48   4BEoo3   R   2/49   4/49
3   B   CPu   1955pc   -   ac   6.55   4oooX3   9/55   -
3   B/   CPw,z   OP?   1955pc   1958mc   ac   6.55   4oooX3   -   2/58
AA 1.12.48 on A(A), 3/-, OSPR 4/49.

Sheets NS 17 to 19 were not published.

NS 20 (26/20) **& Part of NS 10** *Dailly** 12 by 10 Revision a 1907-08 c 1938-53 f 1965
1   A   CPr   OPr   nd:po   -   -   6.49   4BEoo3   R   0   10/49 12/49
3   B   CPv,y   OPv   1956pc   -   ac   6.56   4oooX3   1   5/56   11/57
3   B/   CPz   OPz   1956mc   1966ch   acf   6.56   4oooX3   3   1
1. This is the airfield on Turnberry golf links.

NS 21 (26/21) *Dunure** 10 by 10 Revision a 1907-08 c 1938-53
1   A   CPr   OPr   nd:po   -   -   6.49   4BEoo3   R   10/49 11/49
3   B   CPu,z   OPv   1956pc   -   ac   6.56   4oooX3   2/56   11/57

NS 24 (26/24) **& Parts of NS 14 & NS 23** *Saltcoats** 12 by 12 Revision a 1908-09 c 1938-54 f 1967 g 1975
1   A   CPr   OPr   nd:po   -   -   6.49   4BEoo3   R   5/49   5/49   1
3   B   CPu   OPv   1955pc   -   ac   6.55   4oooX3   12/55 11/57
3   B/   CPw   1955mc   1959mc   ac   6.55   4oooX3
3   B//   CPw,z   1955mc   1960mc   ac   6.55   4oooX3
3   B///*   CPz   OPz   1955ms   1968r   acf   6.55   4oooX3   3/68   5/68
3   B///*/* CFz   OFz   1955ᶜms   nd:r   acg   6.55   4oooX3   6/76   6/76
1. The Stevenston explosives works area is blank.

NS 25 (26/25) *Largs** 10 by 10 Revision a 1908-09 c 1938-54
1   A   CPr   OPr   nd:po   -   -   1.49   4BEoo3   R   4/49   5/49
3   B   CPu   OPv   1955pc   -   ac   6.55   4oooX3   11/55 11/57
3   B/   CP?w,x,z.z 1955mc   1959mc   ac   6.55   4oooX3
AA 1.1.49 on A(A), 3/-, OSPR 5/49.

NS 26 (26/26) *Largs (North)** 10 by 10 Revision a 1908-12 c 1938-54
1   A   CPr   OPr   nd:po   -   -   1.48   4BEoo3   2/49   4/49
3   B   CPu   OPv   1955pc   -   ac   6.55   4oooX3   10/55 11/57
3   B/   CPw,z   1955mc   1959mc   ac   6.55   4oooX3
AA 1.12.48 on A(A), 3/-, OSPR 4/49. M on B.

NS 27 (26/27) *Greenock** 10 by 10 Revision a 1909-14 c 1954
1   A   CPr   OPr   nd:po   -   -   6.48   4BEoo3   R   2/49   5/49
3   B   CPu   OPv   1955pc   -   ac   6.55   4oooX3   11/55 11/57
3   B/   CPw   1955mc   1960mc   ac   6.55   4oooX3   1
AA 1.12.48 on A(A), 3/-, OSPR 5/49. M on A. 1. The reservoirs are all named.

NS 27 & Part of NS 17 *Greenock** 10 by 10 Revision a 1909-14 d 1954-66 f 1970
3   C   CPz   OPz   1970ms   -   adf   6.70   4oooX3   5/70   5/70
The NS 17 area is shown as an extrusion. This is the first time it appeared in this series. This is the last fully revised new edition in the 1:25,000 First Series, and the first since TQ 23 in July 1965.

NS 28 (26/28) *Helensburgh (West)** 10 by 10 Revision a 1898-1914 c 1938-54
1   A   CPr   OPr   nd:po   -   -   6.49   4BEoo3   R   10/49 12/49
3   B   CPu   OPv   1955pc   -   ac   6.55   4oooX3   11/55 11/57 1
3   B/   CP?w,z   1955mc   1960mc   ac   6.55   4oooX3
AA 16.10.49 on A(A), 3/-, OSPR 12/49. M on A. 1. Showing Faslane Bay dock and railway system.

Sheet NS 29 was not published.

NS 30 (26/30) *Kirkmichael Ayrshire** 10 by 10 Revision a 1907-08 c 1938-53
1   A   CPr   OPr   nd:po   -   -   1.50   4BEoo3   R   10/49 11/49
3   B   CPv,z   OPv   1957pc   -   ac   6.57   4oooX3   9/57   8/57

1	2	3	4	5	6	7	8	9	10	11	12	13	14

NS 31 (26/31) *Dalrymple**   10 by 10 Revision a 1908 c 1938-54 f 1972
1 A   CPr   OPr   nd:po   -   -   6.48   4BEoo3   R   3/49   5/49
3 B   CPv,x,z   OPv   1957pc   -   ac   6.57   4oooX3L   3/57   6/57
3 B/*   CPz   OPz   1957^{c}ms nd:r   acf   6.57   4oooX3L   5/72   8/72
AA 1.12.48 on A(A), 3/-, OSPR 5/49.

NS 32 (26/32) **& Part of NS 22** *Ayr**   15 by 10 Revision a 1908 c 1936-48 d 1936-54
                                                  f 1962 g 1965 h 1972; j 1962 airfield
1 A   CPr   OPr   nd:po   -   -   1.48   4BEoo3   R   30   6/49   8/49
2 B   CPt   1952pc   -   ac   6.52   4Booo3   R   31   9/52   -   1
3 C   CPv   OPv   1957pc   -   ad   6.57   4oooX3   31   9/57   6/57
3 C/
3 C//*   CPz   OPz   1957mc   1964r   adfj   6.57   4oooX3   31   10/64 2/65
3 C//*/*   CPz   OPz   1957mc   1966r   adgj   6.57   4oooX3   33   9/66   10/66
3 C//*/*/* CPz   OPz   1957^{c}ms nd:r   adhj   6.57   4oooX3   33   7/72   12/72
AA 1.12.48 on A(A), 3/-, OSPR 8/49. 1. Prestwick Airport.

NS 33 (26/33) **& Part of NS 23** *Troon Ayrshire** 12 by 10 Revision a 1908 c 1937-46 d 1937-54
1 A   CPr   OPr   nd:po   -   -   1.49   4BEoo3   R   0   5/49   6/49   1
2 B   CPt   1952pc   -   ac   6.52   4Booo3   R   3   1/53   -
3 C   CPv,y.z   OPv   1957pc   -   ad   6.57   4oooX3   0   1/58   11/57
AA 1.1.49 on A(A), 3/-, OSPR 6/49. M on C. 1. The sheet number is incorrect on the coloured edition. The Stevenston explosives works area is blank.

NS 34 (26/34) *Kilwinning**   10 by 10 Revision a 1907-08 c 1938-55 f 1968
1 A   CPr   OPr   nd:po   -   -   1.49   4BEoo3   R   5/49   6/49
3 B   CPw   OPv   1958pc   -   ac   6.58   4oooX3   11/58 12/58
3 B/   CPw   OPz   1958mc   1960mc   ac   6.58   4oooX3
3 B//*   CPz   OPz   1958ms   1969r   acf   6.58   4oooX3   7/69   8/69
AA 1.1.49 on A(A), 3/-, OSPR 6/49. M on B, B/.

NS 35 (26/35) *Kilbirnie*   10 by 10 Revision a 1908-09 b 1908-12 c 1954-55
1 A   CPr   OPr   nd:po   -   -   1.49   4BEoo3   RB   5/49   5/49
3 B   CPv   OP?u,z 1958pc   -   bc   6.58   4oooX3L   3/58   3/58
3 B[/]   CPw   1958mc   1960mc   bc   6.58   4oooX3L
3 B/   CPz   1958mc   1960mc   bc   6.58   4oooX3L
M on A.

NS 36 (26/36) *Bridge of Weir**   10 by 10 Revision a 1910-40 b 1909-40 c 1949-55
1 A   CPr   OPr   nd:po   -   -   6.48   4BEoo3   RB   4/49   4/49
3 B   CPv,w,z   OPv   1957pc   -   bc   6.57   4oooX3   1/58   3/58
M on A.

NS 37 (26/37) *Port Glasgow**   10 by 10 Revision a 1910-38 c 1955
1 A   CPr   OPr   nd:po   -   -   6.48   4BEoo3   R   5/49   7/49
3 B   CPv   OP?   1958pc   -   ac   6.57   4oooX3   3/58   3/58
3 B/   CPw,z   1958mc   1960mc   ac   6.57   4oooX3
AA 1.1.49 on A(A), 3/-, OSPR 7/49. M on A.

NS 38 (26/38) *Helensburgh (East)**   10 by 10 Revision a 1914 c 1937-38 d 1954-55
1 A   CPr   OPr   nd:po   -   -   6.49   4BEoo3   R   11/49 12/49
3 B   CPw.x,z   OPv   1959mc   -   ad   6.59   4oooX3L   11/59 10/59
AA 16.10.49 on A(A), 3/-, OSPR 12/49. M on A.

Sheet NS 39 was not published.

NS 40 *Dalmellington**   10 by 10 Revision a 1907-08 c 1938 d 1938-53
2 A   CPs   OPs   1952pc   -   ac   1.52   4Booo4   R   3/52   4/52
3 B   CPv,z   OPv   1957pc   -   ad   6.57   4oooX4   5/57   8/57

1	2	3	4	5	6	7	8	9	10	11	12	13	14

NS 41 (26/41) *Drongan**   10 by 10   Revision a 1908 c 1938-54
1 A   CPr   OPr   nd:po   -   -   6.49   4BEoo3   R   10/49 11/49
3 B   CPv.x,z   OPv   1957pc   -   ac   6.57   4oooX3L   6/57   8/57

NS 42 (26/42) *Tarbolton**   10 by 10   Revision a 1908 c 1938-48 d 1938-54
1 A   CPr   OPr   nd:po   -   -   6.49   4BEoo3   R   10/49 11/49
2 B   CPt   OPs   1952pc   -   ac   6.52   4Booo3   R   10/52 10/52
3 C   CPv,y,z   OPv   1957pc   -   ad   6.57   4oooX3   3/57   6/57

NS 43 (26/43) *Kilmarnock**   10 by 10   Revision a 1908 c 1937-46 d 1937-54 f 1973
1 A   CPr   OPr   nd:po   -   -   6.48   4BEoo3   R   5/49   7/49
2 B   CPt   1952pc   -   ac   6.52   4Booo3   R   12/52
3 C   CPv   OPv   1957pc   -   ad   6.57   4oooX3   9/57   9/57
3 C/   CPw,z   1957mc   1960mc   ad   6.57   4oooX3
3 C//*   CPz   OPz   1957^{c}ms nd:r   adf   6.57   4oooX3   9/73   10/73
AA 1.1.49 on A(A), 3/-, OSPR 7/49. M on C.

NS 44 (26/44) *St Warton**   10 by 10   Revision a 1908-09 c 1938-55
1 A   CPr   OPr   nd:po   -   -   6.48   4BEoo3   R   2/49   5/49
3 B   CPv   OP?   1957pc   -   ac   6.57   4oooX3   2/58   2/58
3 B/   CPw.z   1957mc   1960mc   ac   6.57   4oooX3
AA 1.12.48 on A(A), 3/-, OSPR 5/49. M on B.

NS 45 (26/45) *Neilston**   10 by 10   Revision a 1908-12 b 1908-09 c 1938 d 1938-55
1 A   CPr   OPr   1947po   -   -   6.47   4BE2X3   R   11/47 8/48
1 B   CPr   1947pc   1950c   ac   6.50   4BEoo3   R
3 C   CPv,x.z   OPv   1958pc   -   bd   6.58   4oooX3   3/58   7/58
AA 1.5.48 on A(A), 3/-, OSPR 8/48. M on B.

NS 46 (26/46) *Paisley**   10 by 10   Revision a 1909-14 c 1937-50 d 1937-55
                                          f 1963 g 1968 h 1972
1 A   CPr   OPr   1948po   -   -   6.48   4BE4X3   R   00   4/48   5/48
1 B   CPs   nd:po   -   -   6.48   4BEoo3   R   00   M
2 C   CPt,u   OPs   1952pc   -   ac   6.52   4Booo3   R   13   11/52 11/52
3 D   CPw,x   OPv   1959mc   -   ad   6.59   4oooX3   13   11/59 11/59 1
3 D/   CPz   OPz   1959mc   1965ch   adf   6.59   4oooX3   33
3 D//*   CPz   OPz   1959ms   1968r   adg   6.59   4oooX3   30   9/68   9/68
3 D//*/* CPz   OPz   1959^{c}ms nd:r   adh   6.59   4oooX3   30   3/73   4/73   2
AA 1.1.48 on A(A), 3/-, OSPR 5/48. M on B. 1. Abbotsinch Airfield. 2. Abbotsinch renamed Glasgow Airport.

NS 47 (26/47) *Kilpatrick**   10 by 10   Revision a 1909-33 b 1896-14 c 1938-39 d 1933-50
                                          e 1933-55 (psr); f 1972
1 A   CPr   1947po   -   -   6.47   4BE2X3   R   12/47 -
1 B   OPr   nd:po   -   -   6.47   4BE2X3   R   -   10/48
1 C   CPs   1951pc   -   ac   1.51   4BEoo3   R   M
2 D   CPt   1952pc   -   bd   6.52   4Booo3   R   9/52   -
3 E   CPv   OPv   1958pc   -   be   6.58   4oooX3   6/58   7/58
3 E/   CPx,z   1958mc   1961ch   be   6.58   4oooX3
3 E//*   CPz   OPz   1958^{c}ms nd:r   bef   6.58   4oooX3   9/73   10/73
AA 1.7.48 on B(A), 3/-, OSPR 10/48. M on C.

NS 48 (26/48) *Drymen**   10 by 10   Revision a 1914 c 1954-55 f 1965
1 A   CPr   OPr   nd:po   -   -   1.50   4BEoo3   R   10/49 11/49
3 A/   CPw   1949pc   1958mc   a   1.50   4BooX3
3 B   CPw   OPv   1960mc   -   ac   6.60   4oooX3L   12/60 12/60
3 B/   CPz   OPz   1960mc   1966ch   acf   6.60   4oooX3L
M on A.

Sheet NS 49 was not published.

1	2	3	4	5	6	7	8	9	10	11	12	13	14

NS 50 *High Pennyvenie**   10 by 10 Revision a 1907-08 c 1953
3   A   CPv,z   OPv   1956pc   -   ac   6.56   4oooX4L   7/56   8/56

NS 51 *Cumnock (South)**   10 by 10 Revision a 1907-08 c 1938 d 1938-53
2   A   CPs   OPs   1952pc   -   ac   1.52   4Booo4   R   4/52   6/52
3   B   CPv,x,z OPv   1957pc   -   ad   6.57   4oooX4   6/57   8/57

NS 52 *Catrine**   10 by 10 Revision a 1908 c 1938 d 1938-54 f 1962
2   A   CPt   OPs   1953pc   -   ac   1.53   4BooX4   R   3/53   3/53
3   B   CPv   OPv   1957pc   -   ad   6.57   4oooX4   7/57   9/57
3   B/
3   B//   CPz,0z   O0z   1957mc   1964ch   adf   6.57   4oooX4

NS 53 (26/53) *Newmilns and Greenholm**   10 by 10 Revision a 1856-1908 c 1938-54
1   A   CPr   OPr   nd:po   -   -   1.50   4BEoo4   R   7/50   8/50
3   B   CPv,w   OPv   1957pc   -   ac   6.57   4oooX4   5/57   8/57
M on B.

NS 54 (26/54) *Dunwan Dam**   10 by 10 Revision a 1908-12 c 1938 d 1938-55
1   A   CPr   OPr   nd:po   -   -   1.50   4BEoo3   5/50   6/50
3   A   CPu   1950pc   -   -   1.50   4BEoo3
3   B   CPv,z   OP?   1958pc   -   ad   6.58   4oooX3   3/58   3/58
M on A.

NS 55 (26/55) *Newton Mearns*   10 by 10 Revision a 1909-40 c 1938-55
1   10047   CPr   1947po   -   -   1.47   4BE2X3   R   1/47   -
1   A   OPr   nd:po   -   -   1.47   4BE2X3   R   -   10/48
1   B   CPr   OPs   nd:po   -   -   1.47   4BEoo3   R
3   C   CPw,y,z,0z OPv 1959mc   -   ac   6.59   4oooX3   9/59   7/59
AA 1.5.48 on A(A), 3/-, OSPR 10/48. M on B.

NS 56 (26/56) *Glasgow (West)**   10 by 10 Revision a 1909-35 c 1938-50 d 1938-55
                                       f 1969 g 1974
1   A   CPr   OPr   1947po   -   -   1.47   4BE2X3   R   0   3/47   10/48
1   B   CPr   OPs   nd:po   -   -   1.47   4BEoo3   R   0
2   C   CPt   1952pc   -   ac   6.52   4Booo3   R   3   12/52 -   1
3   C/   OPv   1952pc   1956c   ac   6.52   4BooX3   3
3   D   CPw,z   OPv   1958pc   -   ad   6.58   4oooX3   3   9/58   11/58
3   D/*   CPz   OPz   1958ms   1970r   adf   6.58   4oooX3   0   11/70 12/70 2
3   D/*/*   CFz   OFz   1958^{c}ms nd:r   adg   6.58   4oooX3   0   9/75   5/75
AA 1.8.48 on A(A), 3/-, OSPR 10/48; AA 1.5.51 on B(B), 3/-, OSPR 8/51. M on B. 1. Renfrew Airport. 2. Renfrew Airport is deleted.

NS 57 (26/57) *Milngavie**   10 by 10 Revision a 1910-35 c 1937-50 d 1937-55
1   A   CPr   1947po   -   -   1.47   4BE2X3   R   1/47   -
1   A   OPr,s nd:po   -   -   1.47   4BE2X3   R   -   10/48
2   B   CPt   1952pc   -   ac   6.52   4Booo3   R   9/52   -
3   B/   OPv   1952pc   1956mc   ac   6.52   4BooX3   -   2/57
3   C   CPw,z,0z OPv   1958pc   -   ad   6.58   4oooX3   9/58   10/58
AA 1.9.48 on A(A), 3/-, OSPR 10/48. M on A.

NS 58 (26/58) *Killearn**   10 by 10 Revision a 1914 c 1955
1   A   CPr   OPr   nd:po   -   -   6.49   4BEoo3   R   10/49 11/49
1   B   CPs,t   1949pc   1950c   a   6.50   4BEoo3   R
3   C   CPw,z.z OPv   1960mc   -   ac   6.60   4oooX3   2/60   2/60
M on B.

NS 59 (26/59) *Buchlyvie**   10 by 10 Revision a 1898-1914 c 1954
1   A   CPr   OPr   nd:po   -   -   6.49   4BEoo3   R   11/49 11/49
3   B   CPu   OPv   1956pc   -   ac   6.56   4oooX3   2/56   11/57
3   B/   CPx,z   1956mc   1961ch   ac   6.56   4oooX3

91

1	2	3	4	5	6	7	8	9	10	11	12	13	14

NS 60 *Afton Reservoir**   10 by 10   Revision a 1898-1908 c 1953
3   A   CPv,z   OPv   1956pc   -   ac   6.56   4oooX4   10/56   10/56

NS 61 *New Cumnock**   10 by 10   Revision a 1908 c 1938 d 1938-53
2   A   CPs   OPs   1952pc   -   ac   1.52   4Booo4   R   3/52   4/52
3   B   CPv,x   OPv   1957pc   -   ad   6.57   4oooX4   5/57   8/57

NS 62 *Muirkirk**   10 by 10   Revision a 1908 c 1923 d 1923-53
2   A   CPs   OPs   1952pc   -   ac   1.52   4Booo4   R   3/52   4/52
3   B   CPv,x.z   OPv   1957pc   -   ad   6.57   4oooX4   5/57   6/57

NS 63 (26/63) *Glengavel Reservoir**   10 by 10   Revision a 1908-13 c 1938-54
1   A   CPr   OPr   nd:po   -   -   1.50   4BEoo3   R   1/50   2/50
3   B   CPv   OPv   1957pc   -   ac   6.57   4oooX3   5/57   8/57
3   B/   CPx.z   1957mc   1961ch   ac   6.57   4oooX3

NS 64 (26/64) *Strathaven (West)**   10 by 10   Revision a 1909-10 c 1946-55
1   A   CPr   OPr   nd:po   -   -   6.49   4BEoo3   R   3/50   3/50
1   B   CPs   1950pc   1951c   a   1.51   4BEoo3   R
3   C   CPv,x,z   OPv   1957pc   -   ac   6.57   4oooX3   1/58   10/57
M on B.

NS 65 (26/65) *East Kilbride**   10 by 10   Revision a 1910-40 c 1938-55 f 1967
1   A   CPr   1947po   -   -   1.47   4BE2X3   R   1/47   -
1   A   OPr   nd:po   -   -   1.47   4BE2X3   R   -   9/48
3   B   CPv   OP?   1958pc   -   ac   6.58   4oooX3   3/58   3/58
3   B/   CP?w,z   1958mc   1960mc   ac   6.58   4oooX3
3   B//   CPz   OPz   1958ms   1968ch   acf   6.58   4oooX3
AA 1.5.48 on A(A), 3/-, OSPR 9/48. M on A.

NS 66 (26/66) *Glasgow (East)*   10 by 10   Revision a 1910-36 c 1938 d 1938-55 f 1972
1   A   CPr   1947po   -   -   1.47   4BE2X3   R   4/47   -
1   A   OPr   nd:po   -   -   1.47   4BE2X3   R   -   10/48
1   B   CPr   1947pc   1950c   ac   6.50   4BEoo3   R
3   B/   CPw   1947mc   1959c   ac   6.50   4BooX3
3   C   CPw,x   OPv   1961mc   -   ad   6.61   4oooX4   11/60   1/61
3   C/*   CPz,0z   OPz   1961ᶜms   nd:r   adf   6.61   4oooX4   2/73   3/73   1
AA 1.5.48 on A(A), 3/-, OSPR 10/48. M on B. 1. Showing the M73 under construction.

NS 67 (26/67) *Kirkintilloch**   10 by 10   Revision a 1910-35 c 1938-55
1   A   CPr   1947po   -   -   1.47   4BE2X3   R   2/47   -
1   A   OPr   nd:po   -   -   1.47   4BE2X3   R   -   10/48
1   B   CPs   nd:po   -   -   1.51   4BEoo3   R   M
3   C   CPw   OPv   1958pc   -   ac   6.58   4oooX3   11/58   1/59
3   C/   CPw,z,0z   1958mc   1960mc   ac   6.58   4oooX3
AA 1.5.48 on A(A), 3/-, OSPR 10/48. M on B.

NS 68 (26/68) *Fintry**   10 by 10   Revision a 1913-14 c 1938-46 d 1938-55
1   A   CPr   OPr   nd:po   -   -   6.49   4BEoo3   10/49   11/49
3   A/   CPw   1949mc   1958mc   ac   6.49   4BooX3
3   B   CPw,z   OPv   1960mc   -   ad   6.60   4oooX3   9/60   6/60
M on A.

NS 69 (26/69) *Kippen**   10 by 10   Revision a 1898-1914 c 1946-54
1   A   CPr   OPr   nd:po   -   -   6.49   4BEoo3   R   10/49   11/49
1   A   CPr   1949pc   1950   -   6.50   4BEoo3   R
3   B   CPu   OPv   1956pc   -   ac   6.56   4oooX3   3/56   11/57
3   B/   CPw,z   1956mc   1959mc   ac   6.56   4oooX3

NS 70 *Sanquhar**   10 by 10   Revision a 1898-99 c 1938-54
3   A   CPu,z   OPs   1955pc   -   ac   6.55   4oooX4   9/55   11/55

1	2	3	4	5	6	7	8	9	10	11	12	13	14

NS 71 *Kirkconnel Dumfriesshire**   10 by 10   Revision a 1898-1908 c 1938-54
3 A   CPu,z   OPs   1955pc   -   ac   6.55   4oooX4   11/55 1/56

NS 72 *Glenbuck**   10 by 10   Revision a 1898-1909 c 1938-54
3 A   CPu,y,z OPs   1955pc   -   ac   6.55   4oooX4   9/55   10/55

NS 73 (26/73) *Logan Water**   10 by 10   Revision a 1908-09 c 1941-54
1 A   CPr   OPr   nd:po   -   -   1.50   4BEoo4   R   5/50   7/50
3 B   CPu,x,z OPv   1955pc   -   ac   6.55   4oooX4   9/55   6/57

NS 74 (26/74) *Strathaven (East)**   10 by 10   Revision a 1909-10 c 1939-54 f 1967
1 A   CPr   OPr   nd:po   -   -   1.49   4BEoo3   R   4/49   4/49
3 B   CPu,x   OPv   1956pc   -   ac   6.56   4oooX3   3/56   5/57
3 B/*   CPz   OPz   1956ms   1968r   acf   6.56   4oooX3   3/68   2/68
M on B.

NS 75 (26/75) *Motherwell**   10 by 10   Revision a 1909-36 c 1935-54 f 1963 g 1968 h 1973
                                                       j 1966 railway
1 A   CPr   OPr   1948po   -   -   6.48   4BE2X3   R   3/48   5/48
1 B   CPr   nd:po   -   -   6.48   4BEoo3   R
3 C   CPv   OPv   1956pc   -   ac   6.56   4oooX3   10/56 11/56
3 C/   CP?w,x   1956mc   1958mc   ac   6.56   4oooX3
3 C//   CPz   OPz   1956mc   1965ch   acf   6.56   4oooX3
3 C///*   CPz   OPz   1956ms   1969r   acgj   6.56   4oooX3   3/69   5/69
3 C///*/* CPz   OPz   1956ᶜms   nd:r   achj   6.56   4oooX3   9/73   10/73
AA 1.1.48 on A(A), 3/-, OSPR 5/48. M on B.

NS 76 (26/76) *Airdrie Lanarkshire**   10 by 10   Revision a 1910-36 c 1938-54 f 1965 g 1971
                                                       j 1963 industrial k 1965 railway l 1971 railway
1 A   CPr   1947po   -   -   6.47   4BE2X3   R   3/47   -
1 A   OPr   nd:po   -   -   6.47   4BE2X3   R   -   10/48
1 B   CPr   nd:po   -   -   6.47   4BEoo3   R
1 C   CPs   nd:po   -   -   1.51   4BEoo3   R   M
3 D   CPv   OP?   1956pc   -   ac   6.56   4oooX3   7/56   3/58
3 D/   CP?w,y   OPz   1956mc   1960mc   ac   6.56   4oooX3
3 D//*   CPz   OPz   1956mc   1966ir   acfjk 6.56   4oooX3   10/66 10/66
3 D//*/*   CPz,0z   OPz   1956ᶜms   nd:r   acgjl 6.56   4oooX3   7/72   7/72
AA 1.8.48 on A(A), 3/-, OSPR 10/48. M on C.

NS 77 (26/77) *Kilsyth**   10 by 10   Revision a 1910-35 c 1940-55 f 1964 g 1968 h 1973
                                                       j 1963-67 building k 1966 railway
1 A   CPr   OPr   1948po   -   -   1.48   4BE2X3   R   4/48   5/48
1 B   CPr   nd:po   -   -   1.48   4BEoo3   R
3 C   CPv   OPv   1956pc   -   ac   6.56   4oooX3   6/56   6/57
3 C/   CPw   1956mc   1960mc   ac   6.56   4oooX3
3 C//*   CPz   OPz   1956mc   1965r   acf   6.56   4oooX3   4/65   3/65
3 C//*/*   CPz   OPz   1956ms   1969br   acgjk 6.56   4oooX3   3/69   1/69
3 C//*/*/* CPz   OPz   1956ᶜms   nd:r   achjk 6.56   4oooX3   4/74   6/74
AA 1.1.48 on A(A), 3/-, OSPR 5/48. M on B.

NS 78 (26/78) *Carron Bridge**   10 by 10   Revision a 1896-1943 c 1946-55
1 A   CPr   OPr   nd:po   -   -   6.49   4BEoo3   1/50   2/50
3 B   CPu   OPv   1956pc   -   ac   6.56   4oooX3   6/56   11/57
3 B/   CPw,z   1956mc   1958mc   ac   6.56   4oooX3
M on B.

NS 79 (26/79) *Stirling**   10 by 10   Revision a 1899-43 c 1914-54 d 1946-54
                                                       f 1967 g 1971 h 1976
1 A   CPr   OPr   nd:po   -   -   1.50   4BEoo3   R   3/50   4/50
3 B   CPu   OPv   1956pc   -   ac   6.56   4oooX3   2/56   11/57
3 B/   CPw,z   1956mc   1960mc   ad   6.56   4oooX3

1	2	3	4	5	6	7	8	9	10	11	12	13	14
3	B//*	CPz	OPz	1956mc	1968r	adf	6.56	4oooX3			1/68	1/68	
3	B//*/*	CPz	OPz	1956^{c}ms	nd:r	adg	6.56	4oooX3			6/72	8/72	
3	B//*/*/*	CFz	OFz	1956^{c}ms	nd:r	adh	6.56	4oooX3					1

AA 1.11.49 on A(A), 3/-, OSPR 4/50. 1. OSPR not found.

NS 80 *Upper Nithsdale** 10 by 10 Revision a 1898-1909 c 1938-54
3 A CPu.z OPs 1955pc - ac 6.55 4oooX4   9/55 10/55

NS 81 *Leadhills** 10 by 10 Revision a 1898-1909 c 1938-54
3 A CPu OPs 1955pc - ac 6.55 4oooX4   10/55 11/55
3 A/ CPw,z   1955mc 1959mc ac 6.55 4oooX4

NS 82 *Crawfordjohn** 10 by 10 Revision a 1898-1909 c 1954 f 1969
3 A CPu,x OPs 1955pc - ac 6.55 4oooX4   10/55 11/55
3 A/ CPz   1955^{c}ms nd:ch acf 6.55 4oooX4

NS 83 (26/83) *Douglas Water** 10 by 10 Revision a 1909 c 1941-54 f 1964
1 A CPr OPr nd:po - - 1.50 4BEoo4 R 6/50 6/50
3 B CPu OPv 1955pc - ac 6.55 4oooX4   11/55 10/57
3 B/ CPw   1955mc 1960mc ac 6.55 4oooX4
3 B//* CPz OPz 1955mc 1965r acf 6.55 4oooX4   10/65 11/65

NS 84 (26/84) *Lanark** 10 by 10 Revision a 1909-41 c 1938-54 f 1968
1 A CPr OPr nd:po - - 1.49 4BEoo3 R 4/49 4/49
3 B CPv OPv 1956pc - ac 6.56 4oooX4   6/56 10/57
3 B/ CPw,z   1956mc 1959mc ac 6.56 4oooX4
3 B// CPz OPz 1956ms 1968ch acf 6.56 4oooX4
M on B.

NS 85 (26/85) *Carluke** 10 by 10 Revision a 1910-14 c 1938-54 f 1965 g 1974
1 A CPr OPr nd:po - - 1.48 4BEoo3 R 10/48 11/48
2 A CPs,t OPs 1948pc - - 1.52 4Booo3 R
3 B CPu OPv 1956pc - ac 6.56 4oooX3   3/56 11/57
3 B/ CP?w,x   1956mc 1959mc ac 6.56 4oooX3
3 B// CPz OPz 1956mc 1966ch acf 6.56 4oooX3
3 B///* CFz OFz 1956^{c}ms nd:r acg 6.56 4oooX3   1/75 1/75
AA 1.6.51 on A(A), 3/-, OSPR 4/52.

NS 86 (26/86) [*Shotts*] 10 by 10 Revision a 1910-36 c 1938-55 f 1968; j 1966 railway
1 A CPr OPr nd:po - - 1.50 4BEoo3 R 10/49 11/49
3 A CPu   1949pc - - 1.50 4BEoo3 R
3 B CPu OPv 1956pc - ac 6.56 4oooX3   4/56 10/57
3 B/ CP?x,z   1956mc 1961mc ac 6.56 4oooX3
3 B//* CPz OPz 1956ms 1969r acfj 6.56 4oooX3   3/69 3/69
M on A.

NS 87 (26/87) *Falkirk (South)* 10 by 10 Revision a 1910-44 c 1946-55
1 A CPr OPr nd:po - - 1.49 4BEoo3 R 4/49 5/49
3 B CPv,x,z OPv 1956pc - ac 6.56 4oooX3   7/56 6/57
AA 1.1.49 on A(A), 3/-, OSPR 5/49. M on A, B.

NS 88 (26/88) [*Falkirk (North)*] 10 by 10 Revision a 1913-44 c 1946-55 f 1964 g 1968
                                                    j 1966 railway
1 A CPr OPr nd:po - - 6.49 4BEoo3 R 2/50 4/50
3 B CPv OPv 1956pc - ac 6.56 4oooX3   10/56 10/56
3 B/ CPw   1956mc 1960mc ac 6.56 4oooX3
3 B//* CPz OPz 1956mc 1965r acf 6.56 4oooX3   6/65 6/65
3 B//*/* CPz OPz 1956ms 1969r acgj 6.56 4oooX3   4/69 3/69
AA 1.9.49 on A(A), 3/-, OSPR 4/50. M on B.

1	2	3	4	5	6	7	8	9	10	11	12	13	14

NS 89 (26/89) *Alloa* 10 by 10 Revision a 1899-1943 c 1938-54
1	A	CPr	OPr	nd:po	-	-	6.49	4BEoo3	R		1/50	2/50	
3	B	CPv	OPv	1956pc	-	ac	6.56	4oooX3			7/56	6/57	
3	B/	CP?x,y,z		1956mc	1961mc	ac	6.56	4oooX3					

AA 1.9.49 on A(A), 3/-, OSPR 2/50. M on B.

NS 90 *Daer Reservoir** 10 by 10 Revision a 1898-1909 c 1954 f 1972; j 1968 reservoir
3	A	CPu	OPs	1955pc	-	ac	6.55	4oooX4			9/55	9/55	
3	A/*	CPz	OPz	1955^{c}ms	nd:v	acfj	6.55	4oooX4			4/73	6/73	1

1. Daer Reservoir.

NS 91 *Elvanfoot** 10 by 10 Revision a 1898-1909 c 1954
3	A	CPu,y.z	OPs	1955pc	-	ac	6.55	4oooX4			11/55	11/55	

NS 92 *Abington Lanarkshire** 10 by 10 Revision a 1909 c 1954 f 1960 g 1965
3	A	CPu	OPs	1955pc	-	ac	6.55	4oooX4			11/55	11/55	
3	A/*	CPx	OPx	1955mc	1961r	acf	6.55	4oooX4			9/61	9/61	
3	A/*/*	CPz	OPz	1955mc	1966r	acg	6.55	4oooX4			10/66	10/66	

NS 93 (26/93) *Tinto Hills** 10 by 10 Revision a 1909 c 1941-54
1	A	CPr	OPr	nd:po	-	-	1.50	4BEoo3	R		3/50	3/50	
1	B	CPs		1951pc	-	a	1.51	4BEoo3	R				
3	C	CPu	OPv	1955pc	-	ac	6.55	4oooX3			11/55	10/57	
3	C/	CPw,z		1955mc	1960mc	ac	6.55	4oooX3					

NS 94 (26/94) *Carstairs** 10 by 10 Revision a 1909-10 c 1938-54
1	A	CPr	OPr	nd:po	-	-	6.48	4BEoo3	R		3/49	4/49	
3	B	CPu	OPv	1956pc	-	ac	6.56	4oooX3			5/56	5/57	
3	B/	CPw,z		1956mc	1961mc	ac	6.56	4oooX3					

NS 95 (26/95) *Forth** 10 by 10 Revision a 1905-14 c 1941-54
1	A	CPr	OPr	nd:po	-	-	6.48	4BEoo3	R		3/49	4/49	
3	B	CPu	OPv	1956pc	-	ac	6.56	4oooX3			1/56	2/57	
3	B/	CPx,z		1956mc	1961mc	ac	6.56	4oooX3					

NS 96 (26/96) [Bathgate] 10 by 10 Revision a 1905-14 c 1938-55 f 1966; j 1966 industrial
1	A	CPr	OPr	nd:po	-	-	1.50	4BEoo3	R		10/49	11/49	
3	B	CPu	OPv	1956pc	-	ac	6.56	4oooX3			4/56	10/57	
3	B/	CPw		1956mc	1961mc	ac	6.56	4oooX3					
3	B//*	CPz	OPz	1956mc	1967ir	acfj	6.56	4oooX3			1/67	1/67	

M on A.

NS 97 (26/97) *Linlithgow (West)** 10 by 10 Revision a 1913-44 c 1938-55 f 1966 g 1972
1	A	CPr	OPr	nd:po	-	-	6.48	4BEoo3	R		5/49	6/49	
2	A	CPu		1949pc	-	-	6.48	4BooX3	R				
3	B	CPv	OPv	1956pc	-	ac	6.56	4oooX3L			6/56	10/57	
3	B/	CP?w,y		1956mc	1961mc	ac	6.56	4oooX3L					
3	B//	CPz	OPz	1956mc	1967ch	acf	6.56	4oooX3L					
3	B///*	CPz	OPz	1956^{c}ms	nd:r	acg	6.56	4oooX3L			4/73	9/73	

AA 1.1.49 on A(A), 3/-, OSPR 6/49. M on A, B/.

NS 98 (26/98) [Grangemouth] 10 by 10 Revision a 1913-44 c 1938-55 f 1963
1	A	CPr	OPr	nd:po	-	-	1.50	4BEoo3	R	0	11/49	12/49	
1	B [1]	CPs		nd:po	-	-	1.51	4BEoo3	R	0			M
3	B [2]	CPv	OPv	1956pc	-	ac	6.56	4oooX3L		2	8/56	12/57	
3	B/	CPw		1956mc	1960mc	ac	6.56	4oooX3L		2			
3	B//	CPz	OPz	1956mc	1965ch	acf	6.56	4oooX3L		0			

AA 1.10.49/NT 12.2.49 on A(A), 3/-, OSPR 12/49. M on B [1].

1	2	3	4	5	6	7	8	9	10	11	12	13	14

NS 99 (26/99) *Dollar**   10 by 10   Revision a 1899-1920 c 1946-54
1	A	CPr	OPr	nd:po	-	-	1.50	4BEoo3	R		1/50	1/50	
3	B	CPu	OPv	1956pc	-	ac	6.56	4oooX3			3/56	10/57	
3	B/	CP?w,z		1956mc	1960mc	ac	6.56	4oooX3					

NT 00 *Moffat*   10 by 10   Revision a 1898-1909 c 1938-54 f 1967
| 3 | A | CPu | OPs | 1955pc | - | ac | 6.55 | 4oooX4 | | | 10/55 | - | |
| 3 | A/ | CPz,0z | | 1955mc | 1968ch | acf | 6.55 | 4oooX4 | | | | | |

NT 01 [Tweed's Well]   10 by 10   Revision a 1898-1909 c 1954
| 3 | A | CPu | OPs | 1955pc | - | ac | 6.55 | 4oooX4 | | | 11/55 | 11/55 | |

NT 02 [Camps Reservoir]   10 by 10   Revision a 1906-09 c 1954
| 3 | A | CPu | OPs | 1955pc | - | ac | 6.55 | 4oooX4 | | | 10/55 | 11/55 | |

NT 03 (36/03) *Biggar Lanarkshire**   10 by 10   Revision a 1906-41 c 1954
1	A	CPr	OPr	nd:po	-	-	6.50	4BEoo4	R		7/50	8/50	
3	B	CPu	OPv	1955pc	-	ac	6.55	4oooX4			11/55	5/57	
3	B/	CPw,y,z		1955mc	1960mc	ac	6.55	4oooX4					

NT 04 (36/04) *Elsrickle*   10 by 10   Revision a 1906-10 c 1954
| 1 | A | CPr | OPr | nd:po | - | - | 1.50 | 4BEoo3 | R | | 3/50 | 3/50 | |
| 3 | B | CPu,x,z | OPv | 1956pc | - | ac | 6.56 | 4oooX3 | | | 4/56 | 9/57 | |

NT 05 (36/05) *Cobbinshaw*   10 by 10   Revision a 1905-10 c 1954
| 1 | A | CPr | OPr | nd:po | - | - | 6.49 | 4BEoo3 | R | | 2/50 | 3/50 | |
| 3 | B | CPu,x,z | OPv | 1956pc | - | ac | 6.56 | 4oooX3 | | | 2/56 | 9/57 | |

NT 06 (36/06) *The Calders*   10 by 10   Revision a 1905-14 c 1938-55 f 1972
1	A	CPr	OPr	nd:po	-	-	1.48	4BEoo3	R		8/48	8/48	
1	B	CPr		nd:po	-	-	1.49	4BEoo3	R				
3	C	CPu	OPv	1956pc	-	ac	6.56	4oooX3			3/56	2/57	
3	C/	CPw.z		1956mc	1960mc	ac	6.56	4oooX3					
3	C//*	CPz	OPz	1956^{c}ms	nd:r	acf	6.56	4oooX3			2/73	3/73	

M on B.

NT 07 (36/07) *Linlithgow (East)*   10 by 10   Revision a 1913-14 c 1938-55 f 1971
1	A	CPr	OPr	nd:po	-	-	1.48	4BEoo3	R		9/48	9/48	
1	B	CPr		nd:po	-	-	6.48	4BEoo3	R				
3	C	CPv	OPv	1956pc	-	ac	6.56	4oooX3L			6/56	2/57	
3	C/	CPx	OPx	1956mc	1961mc	ac	6.56	4oooX3L					
3	C//*	CPz	OPz	1956^{c}ms	nd:r	acf	6.56	4oooX3L			7/72	8/72	

M on B.

NT 08 (36/08) *Dunfermline (West)*   10 by 10   Revision a 1913-25 c 1938-55 f 1963
1	A	CPr	OPr	nd:po	-	-	1.49	4BEoo3	R		12/48	2/49	1
3	B	CPv		1956pc	-	ac	6.56	4oooX3			8/56	-	
3	B/	CPv		1956pc	1957mc	ac	6.56	4oooX3					
3	B//	CPz		1956mc	1964ch	acf	6.56	4oooX3					2

AA 1.9.48 on A(A), 3/-, OSPR 2/49. M on A. 1. Rosyth Dockyard is blank. 2. Rosyth Docks and Barracks are shown.

NT 09 (36/09) *Saline*   10 by 10   Revision a 1899-1943 c 1938-54
| 1 | A | CPr | OPr | 1948po | - | - | 6.48 | 4BEoo3 | RB | | 6/48 | 7/48 | |
| 3 | B | CPu.x,z | OPv | 1956pc | - | ac | 6.56 | 4oooX3 | | | 1/56 | 6/57 | |

AA 1.4.48 on A(A), 3/-, OSPR 7/48. M on B.

NT 10 *Loch Fell*   10 by 10   Revision a 1898 c 1938-53
| 3 | A | CPu,z | OPs | 1954pc | - | ac | 6.54 | 4oooX4 | | | 12/54 | 1/55 | |

1	2	3	4	5	6	7	8	9	10	11	12	13	14

NT 11 [Loch Skeen]   10 by 10   Revision a 1897-1906 c 1953

| 3 | A | CPu,x | OPs | 1954pc | - | ac | 6.54 | 4oooX4L | | | 11/54 | 11/54 | |

NT 12 [Talla Reservoir]   10 by 10   Revision a 1897-1906 c 1953

| 3 | A | CPu | OPs | 1954pc | - | ac | 6.54 | 4oooX4 | | | 11/54 | 11/54 | |

NT 13 *Calzeat*   10 by 10   Revision a 1906 c 1953

| 3 | A | CPu | OPs | 1954pc | - | ac | 6.54 | 4oooX4 | | | 12/54 | 1/55 | |
| 3 | A/ | CPw,z | | 1954mc | 1961mc | ac | 6.54 | 4oooX4 | | | | | |

NT 14 (36/14) *Drochil*   10 by 10   Revision a 1908-12 c 1953

| 1 | A | CPr | OPr | nd:po | - | - | 6.49 | 4BEoo3 | R | | 12/49 | 1/50 | |
| 3 | B | CPu,x,z | OPv | 1955pc | - | ac | 6.55 | 4oooX3 | | | 5/55 | 2/57 | |

M on the 1958 index, but not recorded.

NT 15 (36/15) *West Linton*   10 by 10   Revision a 1893-1906 c 1953-54 f 1967
j 1966 reservoir

1	A	CPr	OPr	nd:po	-	-	6.49	4BEoo3			12/49	1/50	
3	B	CPu	OPv	1955pc	-	ac	6.55	4oooX3			7/55	2/57	
3	B/	CP?w.z		1955mc	1960mc	ac	6.55	4oooX3					
3	B//*	CPz	OPz	1955ms	1968v	acfj	6.55	4oooX3			3/68	3/68	1

M on B. 1. West Water Reservoir.

NT 16 (36/16) *Balerno*   10 by 10   Revision a 1905-32 c 1938-54 f 1971

1	A	CPr		1947po	-	-	6.47	4BE2X3	R	0	8/47	-	
1	A		OPr	nd:po	-	-	6.47	4BE2X3	R	0	-	10/48	
3	B	CPu	OPv	1955pc	-	ac	6.55	4oooX3		1	8/55	2/57	
3	B/	CPw,y,z		1955mc	1958mc	ac	6.55	4oooX3		1			
3	B//*	CPz		1955^{c}ms	nd:r	acf	6.55	4oooX3		1	9/72		

AA 1.6.48 on A(A), 3/-, OSPR 10/48. M on A.

NT 17 (36/17) *Queensferry West Lothian** 10 by 10   Revision a 1895-1932 c 1938-54 f 1963-64 g 1970
j 1963-64 airfield k 1964 airfield

1	A	CPr		1947po	-	-	6.47	4BE2X3	R	0	5/47	-	
1	A		OPr	nd:po	-	-	6.47	4BE2X3	R	0	-	10/48	
1	B	CPr		nd:po	-	-	6.47	4BEoo3	R	0			
3	C	CPu	OPv	1955pc	-	ac	6.55	4oooX3		1	11/55	2/57	
3	C/													
3	C//*	CPz	OPz	1955mc	1964ar	acfj	6.55	4oooX3		3	12/64	12/64	1
3	C//*/*	CPz,0z	OPz	1955^{c}ms	nd:r	acgk	6.55	4oooX3		3	10/71	1/72	

AA 1.6.48 on A(A), 3/-, OSPR 10/48. M on B. 1. Edinburgh (Turnhouse) Airport.

NT 18 (36/18) *Dunfermline (East)*   10 by 10   Revision a 1913-25 c 1938-48 d 1938-54
f 1967 g 1972; j 1983 selected k SUSI

1	A	CPr	OPr	nd:po	-	-	1.49	4BEoo3	R	0	11/48	1/49	1
1	B	CPs		1951pc	-	ac	1.51	4BEoo3	R	0	6/51	-	
2	B		OPs	1951pc	-	ac	1.52	4Booo3	R	0	-	4/52	
3	C	CPu	OPv	1956pc	-	ad	6.56	4oooX3		1	6/56	9/57	
3	C/	CP?w,x		1956mc	1959c	ad	6.56	4oooX3		1			2
3	C//*	CPz	OPz	1956mc	1968r	adf	6.56	4oooX3		1	3/68	1/68	
3	C//*/*	CPz	OPz	1956^{c}ms	nd:r	adg	6.56	4oooX3		1	9/73	10/73	
3	C//*/*/*	CFz	OFz	1984^{c}m-s	nd:br	adgjk	6.85	4oooX3		1	9/84	9/84	3

AA 1.9.48 on A(A), 3/-, OSPR 1/49. M on B. 1. Rosyth Dockyard is blank. Showing the Fordell Railway. 2. Donibristle Airfield. 3. A building development area lies across the Rosyth Dockyard.

NT 19 (36/19) *Cowdenbeath*   10 by 10   Revision a 1913-43 c 1938-54 f 1973

1	A	CPr	OPr	nd:po	-	-	1.48	4BEoo3	R		11/48	1/49	
3	B	CPu	OPv	1956pc	-	ac	6.56	4oooX3L			5/56	9/57	
3	B/	CPw,z		1956mc	1960mc	ac	6.56	4oooX3L					
3	B//*	CPz	OPz	1956^{c}ms	nd:r	acf	6.56	4oooX3L			9/73	10/73	

AA 1.5.48 on A(A), 3/-, OSPR 1/49. M on B.

1	2	3	4	5	6	7	8	9	10	11	12	13	14

NT 20 [Davington]   10 by 10   Revision a 1897-1918 c 1953
3   A   CPu   OPs   1955pc   -   ac   6.55   4oooX4   1/55   1/55

NT 21 [Ettrickhall]   10 by 10   Revision a 1897-98 c 1953-54
3   A   CPu   OPs   1954pc   -   ac   6.54   4oooX4   10/54   11/54

NT 22 [St Mary's Loch]   10 by 10   Revision a 1897-1906 c 1953
3   A   CPu   OPs   1954pc   -   ac   6.54   4oooX4   10/54   11/54

NT 23 [Kings Muir]   10 by 10   Revision a 1897-1906 c 1938-53 f 1966
3   A   CPu,y   OPs   1954pc   -   ac   6.54   4oooX4   12/54   1/55
3   A/   CPz   1954mc   1967ch   acf   6.54   4oooX4

NT 24 (36/24) [Peebles]   10 by 10   Revision a 1905-06 c 1938-54
1   A   CPr   OPr   nd:po   -   -   6.50   4BEoo3   RB   12/49   1/50
3   B   CPu   OPv   1955pc   -   ac   6.55   4oooX3L   5/55   9/57
3   B/   CP?w,y   1955mc   1960mc   ac   6.55   4oooX3L

NT 25 (36/25) *Leadburn*   10 by 10   Revision a 1905-32 c 1953-54
1   A   CPr   OPr   nd:po   -   -   1.50   4BEoo3   R   12/49   1/50
3   B   CPu,x,z   OPv   1955pc   -   ac   6.55   4oooX3L   7/55   2/57
M on B.

NT 26 (36/26) *Loanhead*   10 by 10   Revision a 1905-32 c 1938-54 f 1970
1   A   CPr   1947po   -   -   1.47   4BE2X3   R   5/47   -
1   A   CPr   1947po   -   -   1.47   4BE2X3   R   10/47   -   1
1   A   OPr   nd:po   -   -   1.47   4BE2X3   R   -   10/48
1   B   CPr   nd:po   -   -   1.47   4BEoo3   R
3   C   CPu   OPv   1955pc   -   ac   6.55   4oooX3   11/55   2/57
3   C/   CPx   1955pc   -   ac   6.55   4oooX3
3   C//*   CPz,0z   OPz   1955^{c}ms   nd:br   acf   6.55   4oooX3   5/72   8/72
AA 1.6.48 on A(A), 3/-, OSPR 10/48. M on C. 1. The second state, recorded in OSPR 10/47, may be identified with the words "Edinburgh" and "Midlothian" crossed through in the top margin.

NT 27 (36/27) [Edinburgh]   10 by 10   Revision a 1895-1932 c 1938-54
1   A   CPr   1947po   -   -   1.47   4BE2X3   R   8/47   -
1   A   OPr   nd:po   -   -   1.47   4BE2X3   R   -   10/48
2   A   CPt   OPt   1947pc   1952   -   1.52   4Booo3   R
3   B   CPw,x,z   OPv,z   1959mc   -   ac   6.59   4oooX3L   9/59   10/59
AA 1.6.48 on A(A), 3/-, OSPR 10/48; AA 1.11.53 on A(B), 4/-, OSPR 6/54. M on A.

NT 28 (36/28) *Burntisland**   10 by 10   Revision a 1913 c 1938-54
1   A   CPr   nd:po   -   -   6.48   4BEoo3   R   9/48   -
1   B [1]   OPr   nd:po   -   -   6.48   4BEoo3   R   -   11/48
3   B [2]   CPu,x   OPv   1956pc   -   ac   6.56   4oooX3L   4/56   2/57
AA 1.7.48 on B(A), 3/-, OSPR 11/48. M on B [1], B [2].

NT 29 (36/29) [Kirkcaldy]   10 by 10   Revision a 1912-13 c 1938-54 f 1963 g 1965
                                                j 1965 railway
1   A   CPr   OPr   nd:po   -   -   1.49   4BEoo3   R   11/48   1/49
3   B   CPu   OPv   1956pc   -   ac   6.56   4oooX3L   5/56   ?
3   B/
3   B//   CPz   1956mc   1964ch   acf   6.56   4oooX3L
3   B///*   CPz   OPz   1956mc   1966r   acgj   6.56   4oooX3L   6/66   6/66
AA 1.9.48 on A(A), 3/-, OSPR 1/49.

NT 30 [Craik Forest]   10 by 10   Revision a 1897-1918 c 1938-53
3   A   CPu   OPs   1954pc   -   ac   6.54   4oooX4   11/54   11/54

NT 31 [Buccleuch]   10 by 10   Revision a 1897-1918 c 1938-53
3   A   CPu   OPs   1954pc   -   ac   6.54   4oooX4   11/54   11/54

1	2	3	4	5	6	7	8	9	10	11	12	13	14

NT 32 [Ettrickbridge End]   10 by 10   Revision a 1897-1906 c 1953
3   A   CPu   OPs   1955pc   -   ac   6.55   4oooX4   1/55   1/55

NT 33 (36/33) [Innerleithen]   10 by 10   Revision a 1897-1906 c 1938-53
1   A   CPr   OPr   nd:po   -   -   1.50   4BEoo3   R   3/50   3/50
3   B   CPu   OPv   1955pc   -   ac   6.55   4oooX3   3/55   12/57
3   B/   CPw   1955mc   1959mc   ac   6.55   4oooX3

NT 34 (36/34) [Glentress]   10 by 10   Revision a 1897-1906 c 1938-54
1   A   CPr   OPr   nd:po   -   -   1.50   4BEoo3   B   3/50   3/50
3   B   CPu   OPv   1955pc   -   ac   6.55   4oooX3   4/55   2/57

NT 35 (36/35) *Middleton Midlothian*   10 by 10   Revision a 1905-06 c 1953-54
1   A   CPr   OPr   nd:po   -   -   6.49   4BEoo3   RB   12/49   2/50
3   B   CPu,z   OPv   1955pc   -   ac   6.55   4oooX3   6/55   2/57
M on B.

NT 36 (36/36) *Dalkeith*   10 by 10   Revision a 1905-32 c 1938-54
1   A   CPr   1947po   -   -   1.48   4BE4X3   R   11/47   -
1   A   OPr   nd:po   -   -   1.48   4BE4X3   R   -   10/48
3   B   CPu,x,z   OPv   1955pc   -   ac   6.55   4oooX3   7/55   5/57
AA 1.6.48 on A(A), 3/-, OSPR 10/48. M on A, B.

NT 37 (36/37) [Musselburgh]   10 by 10   Revision a 1912-32 c 1947-54 f 1966; j 1966 industrial
1   A   CPr   1947po   -   -   6.47   4BE2X3   R   9/47   -
1   A   OPr   nd:po   -   -   6.47   4BE2X3   R   -   10/48
    B
3   C   CPu   OPv   1955pc   -   ac   6.55   4oooX3   9/55   9/57
3   C/   CP?w,z   1955mc   1960mc   ac   6.55   4oooX3
3   C//*   CPz   OPz   1955mc   1967ir   acfj   6.55   4oooX3   2/67   2/67   1
AA 1.6.48 on A(A), 3/-, OSPR 10/48. M on A, C/. 1. Showing the Monktonhall railway freight terminal.

NT 39 (36/39) [Methil]   10 by 10   Revision a 1912-13 c 1938-54
1   A   CPr   OPr   1948po   -   -   6.48   4BEoo3   R   7/48   8/48
3   B   CPu   OPv   1956pc   -   ac   6.56   4oooX3   1/56   6/57
3   B/   CP?w.y   1956mc   1961mc   ac   6.56   4oooX3
AA 1.5.48 on A(A), 3/-, OSPR 8/48.

NT 40 *Teviothead*   10 by 10   Revision a 1916-18 c 1938-53
3   A   CPu,v,z   OPs   1954pc   -   ac   6.54   4oooX4   12/54   1/55   1
M on A. 1. Stobs Camp.

NT 41 (36/41) *Hawick (West) Roxburghshire*   10 by 10   Revision a 1897-1918 c 1938-53
1   A   CPr   OPr   nd:po   -   -   1.50   4BEoo4   B   5/50   6/50
3   B   CPu   OPv   1955pc   -   ac   6.55   4oooX4   3/55   6/57
3   B/   CPw,z.z   1955mc   1961mc   ac   6.55   4oooX4
AA 1.1.50 on A(A), 3/-, OSPR 6/50. M on A.

NT 42 (36/42) *Selkirk*   10 by 10   Revision a 1897-1930 c 1938-53 f 1975
1   A   CPr   OPr   nd:po   -   -   6.50   4BEoo4   RB   6/50   6/50
3   B   CPu   OPv   1955pc   -   ac   6.55   4oooX4   1/55   9/57
3   B/   CPw,z   1955mc   1961mc   ac   6.55   4oooX4
3   B//*   CFz   OFz   1955^{c}ms   nd:r   acf   6.55   4oooX4   7/76   7/76

NT 43 (36/43) *Galashiels (West)*   10 by 10   Revision a 1897-1930 c 1938-53 f 1970
1   A   CPr   OPr   nd:po   -   -   1.50   4BEoo3   RB   1/50   2/50
3   B   CPu   OPv   1955pc   -   ac   6.55   4oooX3   2/55   9/57
3   B/   CPw,z   1955mc   1959mc   ac   6.55   4oooX3
3   B//   CPz   1955^{c}ms   nd:ch   acf   6.55   4oooX3
AA 1.10.49 on A(A), 3/-, OSPR 2/50.

1	2	3	4	5	6	7	8	9	10	11	12	13	14

NT 44 (36/44) *Stow Midlothian* 10 by 10 Revision a 1897-1919 c 1953
1 A   CPr   OPr   nd:po   -   -   1.50   4BEoo3   R   12/49 2/50
3 B   CPu,z   OPv   1955pc   -   ac   6.55   4oooX3   4/55   2/57

NT 45 (36/45) *Oxton Berwickshire** 10 by 10 Revision a 1908-09 c 1953-54
1 A   CPr   OPr   nd:po   -   -   1.50   4BEoo3   RB   12/49 1/50
3 B   CPu   OPv   1955pc   -   ac   6.55   4oooX3   3/55   2/57
3 B/   CPw,z   1955mc   1960mc   ac   6.55   4oooX3

NT 46 (36/46) *Ormiston* 10 by 10 Revision a 1906-32 c 1938-54 f 1966
1 A   CPr   OPr   nd:po   -   -   1.49   4BEoo3   R   12/48 12/48
3 B   CPu   OPv   1955pc   -   ac   6.55   4oooX3   5/55   9/57
3 B/   CPw   1955mc   1960mc   ac   6.55   4oooX3
3 B//   CPz   1955mc   1966ch   acf   6.55   4oooX3
M on A, B/.

NT 47 (36/47) *Tranent** 10 by 10 Revision a 1906-32 c 1938-54
1 A   CPr   OPr   nd:po   -   -   1.49   4BEoo3   R   0   11/48 12/48
3 B   CPu,z   OPv   1955pc   -   ac   6.55   4oooX3   2   7/55   9/57
M on A, B.

NT 48 (36/48) *Gullane** 10 by 10 Revision a 1906 c 1952-54
1 A   CPr   OPr   nd:po   -   -   1.48   4BEoo3   R   0   9/48   9/48
3 B   CPu,z   OPv   1955pc   -   ac   6.55   4oooX3   1   11/55 10/57
M on A, B.

NT 50 *Shankend* 10 by 10 Revision a 1916-17 c 1938-53
3 A   CPt   OPs   1954pc   -   ac   6.54   4oooX4   8/54   8/54
3 A/   CPw,z   1954mc   1959mc   ac   6.54   4oooX4
M on A.

NT 51 (36/51) *Hawick (East) Roxburghshire* 10 by 10 Revision a 1916-17 c 1938-53
1 A   CPr   OPr   nd:po   -   -   1.50   4BEoo3   R   2/50   4/50
3 B   CPt   OPv   1954pc   -   ac   6.54   4oooX3   10/54 9/57
3 B/   CPw,z.z   1954mc   1959mc   ac   6.54   4oooX3
AA 1.9.49 on A(A), 3/-, OSPR 4/50. M on A.

NT 52 (36/52) *Liliesleaf** 10 by 10 Revision a 1897-1930 c 1938-53
1 A   CPr   OPr   nd:po   -   -   1.50   4BEoo4   RB   4/50   5/50
3 B   CPt   OPv   1954pc   -   ac   6.54   4oooX4   10/54 9/57
3 B/   CPw,z   1954mc   1959mc   ac   6.54   4oooX4
M on B/.

NT 53 (36/53) *Melrose Roxburgh* 10 by 10 Revision a 1906-19 c 1930-53 f 1976
1 A   CPr   OPr   nd:po   -   -   1.50   4BEoo3   RB   12/49 2/50
   B
3 C   CPt   1954pc   -   ac   6.54   4oooX3   9/54   -
3 C/   CPv   1954pc   1957mc   ac   6.54   4oooX3
3 C//   CPw,z   1954mc   1961mc   ac   6.54   4oooX3
3 C///* CFz   OFz   1954^{c}ms nd:r   acf   6.54   4oooX3   2/77   2/77
AA 1.10.49 on A(A), 3/-, OSPR 2/50.

NT 54 (36/54) *Lauder* 10 by 10 Revision a 1906-19 c 1938-52 f 1975
1 A   CPr   OPr   nd:po   -   -   1.50   4BEoo3   RB   12/49 1/50
3 B   CPu   OPv   1954pc   -   ac   6.54   4oooX3   10/54 9/57
3 B/   CPw,z   1954mc   1960mc   ac   6.54   4oooX3
3 B//*   CFz   OFz   1954^{c}ms nd:r   acf   6.54   4oooX3   11/75 11/75

NT 55 (36/55) *Hogs Law* 10 by 10 Revision a 1906 c 1952
1 A   CPr   OPr   nd:po   -   -   1.50   4BEoo3   RB   3/50   3/50
3 B   CPt,w,0z OPv   1954pc   -   ac   6.54   4oooX3   8/54   9/57
M added by hand on an official copy of the 1958 index, but not recorded.

1	2	3	4	5	6	7	8	9	10	11	12	13	14

NT 56 (36/56) *Gifford* 10 by 10 Revision a 1906 c 1952
1 A   CPr   OPr   nd:po   -   -   6.48 4BEoo3 B   11/48 12/48
3 B   CPu   OPv   1954pc -   ac   6.54 4oooX3   11/54 5/57
3 B/ CPw,z   1954mc 1960mc ac   6.54 4oooX3
M on A, B/.

NT 57 (36/57) *Haddington East Lothian** 10 by 10 Revision a 1906 c 1938-52 f 1965
1 A   CPr   OPr   nd:po   -   -   1.49 4BEoo3 R   0   1/49   1/49
3 B   CPu   OPv   1954pc -   ac   6.54 4oooX3   1   10/54 9/57
3 B/ CPw   1954mc 1959mc ac   6.54 4oooX3   1
3 B// CPz   OPz   1954mc 1966ch acf   6.54 4oooX3   3
M on A, B/.

NT 58 (36/58) **& Part of NT 68** *North Berwick** 15 by 10 Revision a 1906 c 1938-52
1 A   CPr   OPr   nd:po   -   -   1.48 4BEoo3 R   0   9/48   10/48
3 B   CPu   OP?   1954pc -   ac   6.54 4oooX3   1   11/54 2/58
3 B/ CPw,z   1954mc 1960mc ac   6.54 4oooX3   1
M on A, B/.

NT 60 *Southdean* 10 by 10 Revision a 1896-1920 c 1948-52
3 A   CPt   OPs   1954pc -   ac   6.54 4oooX4   9/54   10/54
3 A/ CPw,z   1954mc 1958mc ac   6.54 4oooX4
M on A. NPRDBA 63 on A/.

NT 61 (36/61) *Oxnam* 10 by 10 Revision a 1896-1918 c 1938-53
1 A   CPr   OPr   nd:po   -   -   1.50 4BEoo4 B   6/50   6/50
3 B   CPt   OPv   1954pc -   ac   6.54 4oooX4   8/54   9/57
3 B/ CPw,z   1954mc 1959mc ac   6.54 4oooX4
M on B.

NT 62 (36/62) *Jedburgh** 10 by 10 Revision a 1917-19 c 1938-53
1 A   CPr   OPr   nd:po   -   -   1.50 4BEoo4 R   5/50   6/50
3 B   CPt   OPv   1954pc -   ac   6.54 4oooX4   10/54 12/57
3 B/ CPw,z.z   1954mc 1960mc ac   6.54 4oooX4
M on B/.

NT 63 (36/63) *Smailholm** 10 by 10 Revision a 1906-19 c 1938-53
1 A   CPr   OPr   nd:po   -   -   1.50 4BEoo4 R   2/50   3/50
   B
3 C   CPt   OPv   1954pc -   ac   6.54 4oooX4   9/54   9/57
3 C/ CPx,z   1954mc 1961mc ac   6.54 4oooX4

NT 64 (36/64) *Gordon* 10 by 10 Revision a 1906 c 1952
1 A   CPr   OPr   nd:po   -   -   1.50 4BEoo3 RB   2/50   3/50
3 B   CPt   OPv   1954pc -   ac   6.54 4oooX3   8/54   9/57
3 B/ CPw,z   1954pc -   ac   6.54 4oooX3

NT 65 (36/65) *Longformacus* 10 by 10 Revision a 1906 c 1952 f 1971
1 A   CPr   OPr   nd:po   -   -   1.49 4BEoo3 B   3/50   3/50
3 B   CPt   OPv   1954pc -   ac   6.54 4oooX3   8/54   9/57
3 B/ CPw,z   1954mc 1959mc ac   6.54 4oooX3
3 B// CPz   1954^c ms nd:ch acf   6.54 4oooX3
M added by hand on an official copy of the 1958 index, but not recorded.

NT 66 (36/66) *Cranshaws* 10 by 10 Revision a 1906 c 1952 f 1972; j 1972 reservoir
1 A   CPr   OPr   nd:po   -   -   1.50 4BEoo3 B   3/50   4/50
3 B   CPt   OPv   1954pc -   ac   6.54 4oooX3   8/54   9/57
3 B/ CPw,z   1954mc 1961mc ac   6.54 4oooX3
3 B//* CPz   OPz   1954^c ms nd:rv acfj   6.54 4oooX3   9/73   10/73 1
M on A, B/. 1. Whiteadder Reservoir.

1	2	3	4	5	6	7	8	9	10	11	12	13	14

NT 67 (36/67) *Dunbar* — 10 by 10 Revision a 1906 c 1932-52; j 1983 selected
1	A	CPr	OPr	nd:po	-	-	6.48	4BEoo3	RB		1/49	2/49	
3	B	CPt	OPv	1954pc	-	ac	6.54	4oooX3			8/54	-	
3	B/	CPw.z		1954mc	1959mc	ac	6.54	4oooX3					
3	B//*	CFz	OFz	1983^{c}m-s	nd:r	acj	6.84	4oooX3			1/84	1/84	

M on A, B.

NT 70 *Catcleugh* — 10 by 10 Revision a 1896-1917 c 1952-53 f 1969
3	A	CPt	OPs	1954pc	-	ac	6.54	4oooX4			9/54	10/54	
3	A/	CPw		1954mc	1959mc	ac	6.54	4oooX4					
3	A//	CPz	OPz	1954ms	1970ch	acf	6.54	4oooX4					

M on A. NPRDBA 64 on A/.

NT 71 *Hownam* — 10 by 10 Revision a 1896-1918 c 1952-53
| 3 | A | CPu.z | OPs | 1954pc | - | ac | 6.54 | 4oooX4 | | | 11/54 | 11/54 | |

M on A. NPRDBA 65 on A.

NT 72 (36/72) *Morebattle** — 10 by 10 Revision a 1892-1918 c 1938-53
| 1 | A | CPr | OPr | nd:po | - | - | 1.50 | 4BEoo4 | B | | 6/50 | 7/50 | |
| 3 | B | CPt,w,z | OPv | 1954pc | - | ac | 6.54 | 4oooX4 | | | 10/54 | 9/57 | |

NT 73 (36/73) *Kelso** — 10 by 10 Revision a 1906-19 c 1938-53
1	A	CPr	OPr	nd:po	-	-	1.50	4BEoo3	R		2/50	4/50	
	B												
3	C	CPt	OPv	1954pc	-	ac	6.54	4oooX3			9/54	9/57	
3	C/	CPw.z		1954mc	1960mc	ac	6.54	4oooX3					

AA 1.10.49 on A(AA), 3/-, OSPR 4/50.

NT 74 (36/74) *Greenlaw Berwickshire** — 10 by 10 Revision a 1897-1906 c 1952
1	A	CPr	OPr	nd:po	-	-	1.50	4BEoo3	RB	0	3/50	3/50	
3	B	CPu	OPv	1954pc	-	ac	6.54	4oooX3		1	10/54	10/57	
3	B/	CPw,z		1954mc	1961mc	ac	6.54	4oooX3		1			

NT 75 (36/75) *Duns* — 10 by 10 Revision a 1906 c 1938-52
1	A	CPr	OPr	nd:po	-	-	1.50	4BEoo3	RB		2/50	3/50	
1	A	CPs		1950pc	1951	-	1.51	4BEoo3	RB				
3	B	CPu,x	OPv	1954pc	-	ac	6.54	4oooX4			11/54	?	
3	B/	CPz		1954pc	-	ac	6.54	4oooX4					

NT 76 (36/76) *Abbey St Bathans* — 10 by 10 Revision a 1906 c 1940 d 1952
1	A	CPr	OPr	nd:po	-	-	1.50	4BEoo3	RB		2/50	3/50	
1	?	CP		1950pc	1951c	ac		4BEoo3	RB				M
	B												
3	C	CPt,w.z	OPv	1954pc	-	ad	6.54	4oooX3			8/54	-	

M on ?A, C.

NT 77 (36/77) *Innerwick** — 10 by 10 Revision a 1906 c 1938-52
1	A	CPr	OPr	nd:po	-	-	6.49	4BEoo3	R		1/50	2/50	
3	B	CPt	OPv	1954pc	-	ac	6.54	4BooX3			4/54	9/57	
3	B/	CPw,z		1954mc	1959mc	ac	6.54	4BooX3					

M on A, B/.

NT 80 *Ridlees* — 10 by 10 Revision a 1896-1921 c 1953
| 3 | A | CPt | OPs | 1954pc | - | ac | 6.54 | 4oooX4 | | | 8/54 | 8/54 | |
| 3 | A/ | CPw,z | | 1954mc | 1959mc | ac | 6.54 | 4oooX4 | | | | | |

M on A.

NT 81 *Windy Gyle* — 10 by 10 Revision a 1896-1921 c 1952-53
| 3 | A | CPt | OPs | 1954pc | - | ac | 6.54 | 4oooX4 | | | 10/54 | 10/54 | |
| 3 | A/ | CPv,z.0z | OPz | 1954pc | 1958mc | ac | 6.54 | 4oooX4 | | | | | |

M on A. NPRDBA 66 on A/.

```
 1   2          3         4       5        6        7      8      9          10    11   12      13      14
```

NT 82 *Yetholm*   10 by 10 Revision a 1896-1922 c 1952-53
3   A   CPt   OPs   1954pc   -   ac   6.54   4oooX4   9/54   10/54
3   A/   CP?w,x,z,0z   1954mc   1960mc   ac   6.54   4oooX4
NPRDBA 67 on A/.

NT 83 (36/83) *Coldstream**   10 by 10 Revision a 1906-22 c 1938-53
1   A   CPr   OPr   nd:po   -   -   1.50   4BEoo3   R   3/50   4/50
3   B   CPt,w.x,z OPv   1954pc   -   ac   6.54   4oooX3   8/54   9/57
AA 1.9.49 on A(AA), 3/-, OSPR 4/50. NPRDBA 1 on B.

NT 84 (36/84) *Norham (West)*   10 by 10 Revision a 1906-22 c 1938-52
1   A   CPr   nd:po   -   -   1.48   4BEoo3   R   12/48   -
1   B   OPr   nd:po   -   -   1.48   4BEoo3   R   -   2/49
3   C   CPt,u,z   OPv   1954pc   -   ac   6.54   4oooX3   8/54   5/57
AA 1.8.48 on B(AA), 3/-, OSPR 2/49.

NT 85 (36/85) *Chirnside*   10 by 10 Revision a 1906 c 1938-52
1   A   CPr   OPr   nd:po   -   -   1.48   4BEoo3   R   0   12/48   1/49
3   B   CPt   OPv   1954pc   -   ac   6.54   4oooX3   1   8/54   9/57
3   B/   CPw.z   1954mc   1959mc   ac   6.54   4oooX3   1

NT 86 (36/86) **& Part of NT 87** *Reston** 10 by 12 Revision a 1906 c 1952 f 1974
1   A   CPr   OPr   nd:po   -   -   1.50   4BEoo3   R   3/50   3/50
1   B   CPs   1950pc   1951c   a   1.51   4BEoo3   R
3   C   CPt   OP?   1954pc   -   ac   6.54   4oooX3   7/54   2/58
3   C/   CPw.z,z+   1954mc   1960mc   ac   6.54   4oooX3
3   C//*   CFz   OFz   1954^{c}ms nd:r   acf   6.54   4oooX3   5/76   1/75
M on A, C/.

NT 90 *Harbottle*   10 by 10 Revision a 1896-1921 c 1952
3   A   CPt   OPs   1954pc   -   ac   6.54   4oooX4   10/54   10/54
3   A/   CPw.z   1954mc   1961mc   ac   6.54   4oooX4
M on A. NPRDBA 7 on A/.

NT 91 *Alnham*   10 by 10 Revision a 1896-1921 c 1952
3   A   CPt   OPs   1954pc   -   ac   6.54   4oooX4   8/54   8/54
3   A/ CPv,x,z,0z.Pz   1954pc   1956mc   ac   6.54   4oooX4
NPRDBA 6 on A/.

NT 92 *Wooler*   10 by 10 Revision a 1896-1922 c 1952
3   A   CPt   OPs   1954pc   -   ac   6.54   4oooX4   9/54   10/54
3   A/   CP?w,x,z,0z   1954mc   1960mc   ac   6.54   4oooX4
NPRDBA 3 on A/.

NT 93 (36/93) *Milfield**   10 by 10 Revision a 1892-1922 c 1952
1   A   CPr   OPr   nd:po   -   -   6.49   4BEoo3   RB   0   2/50   2/50
3   B   CPt   1954pc   -   ac   6.54   4oooX3   1   8/54   -
3   B/   CPv,z.z   1954pc   1957mc   ac   6.54   4oooX3   1
NPRDBA 2 on B/.

NT 94 (36/94) *Norham (East)*   10 by 10 Revision a 1892-1922 c 1938-52
1   A   CPr   OPr   nd:po   -   -   1.48   4BEoo3   R   11/48   2/49
3   B   CPt   OPv   1954pc   -   ac   6.54   4oooX3   6/54   11/57
3   B/   CPw,z   1954mc   1959c   ac   6.54   4oooX3
AA 1.9.48 on A(AA), 3/-, OSPR 2/49.

NT 95 (36/95) **& Part of NU 05** *Berwick-upon-Tweed* 12 by 10 Revision a 1906-23 c 1938-53 f 1974
1   A   CPr   OPr   nd:po   -   -   1.48   4BEoo3   R   10/48   12/48
3   B   CPu   OP?   1954pc   -   ac   6.54   4oooX3   11/54   2/58
3   B/   CPw,z.z   1954mc   1959mc   ac   6.54   4oooX3
3   B//*   CFz   OFz   1954^{c}ms nd:r   acf   6.54   4oooX3   8/74   8/74
AA 1.8.48 on A(AA), 3/-, OSPR 12/48.

	1	2	3	4	5	6	7	8	9	10	11	12	13	14

NT 96 (36/96) *Eyemouth**   10 by 10   Revision a 1906 c 1938-53
1	A	CPr	OPr	nd:po	-	-	6.49	4BEoo3	R		2/50	3/50	
3	B	CPt		1954pc	-	ac	6.54	4oooX3			8/54	-	
3	B/	CPw,z		1954pc	1958mc	ac	6.54	4oooX3					

M on A, B/.

NU 00 (46/00) *Rothbury*   10 by 10   Revision a 1921 c 1952
| 1 | A | CPr | OPr | nd:po | - | - | 6.49 | 4BEoo3 | RB | | 2/50 | 3/50 | |
| 3 | B | CPt,v,x,z,0z| OPv | 1954pc | - | ac | 6.54 | 4oooX3 | | | 9/54 | 1/57 | |

NPRDBA 9 on B.

NU 01 (46/01) *Whittingham*   10 by 10   Revision a 1921-22 c 1952
1	A	CPr	OPr	nd:po	-	-	1.50	4BEoo3	RB		2/50	3/50	
3	B	CPt	OPv	1954pc	-	ac	6.54	4oooX3			9/54	9/57	
3	B/	CPw,x,z		1954mc	1958mc	ac	6.54	4oooX3					

NPRDBA 5 on B/.

NU 02 (46/02) *Chatton**   10 by 10   Revision a 1921-22 c 1946-52
1	A	CPr	OPr	nd:po	-	-	6.49	4BEoo3	RB		2/50	3/50	
3	B	CPt	OPv	1954pc	-	ac	6.54	4oooX3			8/54	9/57	
3	B/	CP?w,x,z		1954mc	1959mc	ac	6.54	4oooX3					

NPRDBA 4 on B/.

NU 03 (46/03) *Lowick Northumberland**   10 by 10   Revision a 1892-1922 c 1952
1	A	CPr	OPr	nd:po	-	-	6.49	4BEoo3	RB		2/50	2/50	
3	B	CPt	OPv	1954pc	-	ac	6.54	4oooX3			8/54	9/57	
3	B/	CPw,z		1954mc	1960mc	ac	6.54	4oooX3					

NU 04 (46/04) **& Part of NU 14**   15 by 10   Revision a 1892-1922 c 1938-53 f 1974
Holy Island Northumberland
1	A	CPr	OPr	nd:po	-	-	1.50	4BEoo3	R		2/50	3/50	
3	B	CPt	OPv	1954pc	-	ac	6.54	4oooX3			9/54	8/57	
3	B/	CPw,x,z.z		1954mc	1958mc	ac	6.54	4oooX3					
3	B//*	CFz	OFz	1954^{c}ms	nd:r	acf	6.54	4oooX3			11/74	11/74	

NU 10 (46/10) *Felton Northumberland*   10 by 10   Revision a 1921 c 1942-52
| 1 | A | CPr | OPr | nd:po | - | - | 1.50 | 4BEoo3 | B | | 2/50 | 3/50 | |
| 3 | B | CPt,v.x,z,0z | OPv | 1954pc | - | ac | 6.54 | 4oooX3 | | | 8/54 | 5/57 | |

NU 11 (46/11) *Alnwick*   10 by 10   Revision a 1921-22 c 1938-52 f 1976
1	A	CPr	OPr	nd:po	-	-	1.50	4BEoo3	RB		3/50	3/50	
3	B	CPt,v	OPv	1954pc	-	ac	6.54	4oooX3			8/54	8/57	
3	B/	CP?w,z		1954mc	1960mc	ac	6.54	4oooX3					
3	B//*	CFz	OFz	1954^{c}ms	nd:r	acf	6.54	4oooX3			3/77	3/77	

NU 12 (46/12) *Ellingham Northumberland*   10 by 10   Revision a 1921-22 c 1952
1	A	CPr	OPr	nd:po	-	-	1.50	4BEoo3	RB	0	2/50	3/50	
3	B	CPt	OPv	1954pc	-	ac	6.54	4oooX3		1	8/54	9/57	
3	B/	CPw,z		1954mc	1959c	ac	6.54	4oooX3		1			

NU 13 (46/13) *Belford*   10 by 10   Revision a 1896-1922 c 1922-53
1	A	CPr	OPr	nd:po	-	-	1.50	4BEoo3	RB		3/50	3/50	
3	B	CPt	OPv	1954pc	-	ac	6.54	4oooX3			8/54	9/57	
3	B/	CPw,z,0z		1954mc	1958mc	ac	6.54	4oooX3					

NU 20 (46/20) *Amble**   10 by 10   Revision a 1921-22 c 1938-53 f 1965
1	A	CPr	OPr	nd:po	-	-	1.50	4BEoo3	RB	0	3/50	5/50	
3	B	CPt	OPv	1954pc	-	ac	6.54	4oooX3		1	9/54	9/57	
3	B/	CPw	OPw	1954mc	1959c	ac	6.54	4oooX3		1			1
3	B//	CPz	OPz	1954mc	1966ch	acf	6.54	4oooX3		3			

M on A, B. 1. Acklington Airfield.

104

1	2	3	4	5	6	7	8	9	10	11	12	13	14

NU 21 (46/21) *Alnmouth* 10 by 10 Revision a 1921-22 c 1938-53 f 1966 g 1972

1	A	CPr	OPr	nd:po	-	-	1.50	4BEoo3	R	0	2/50	3/50	
3	B	CPt	OPv	1954pc	-	ac	6.54	4oooX3		1	9/54	1/57	
3	B/	CPw,y		1954mc	1959c	ac	6.54	4oooX3		1			
3	B//	CPz	OPz	1954mc	1966ch	acf	6.54	4oooX3		3			
3	B///*	CPz,0z	OPz	1954^{c}ms	nd:r	acg	6.54	4oooX3		3	9/73	12/73	

NU 22 (46/22) *Embleton Northumberland* 10 by 10 Revision a 1922 c 1952

1	A	CPr	OPr	nd:po	-	-	6.49	4BEoo3	R	0	2/50	3/50	
3	B	CPt	OPv	1954pc	-	ac	6.54	4oooX3		1	8/54	8/57	
3	B/	CPw,z.z,0z		1954mc	1958mc	ac	6.54	4oooX3		1			

NU 23 (46/23) *Farne Islands* 10 by 10 Revision a 1896-1922 c 1952-53

1	A	CPr	OPr	nd:po	-	-	6.49	4BEoo3	R		2/50	3/50	
3	B	CPt	OPv	1954pc	-	ac	6.54	4oooX3			9/54	8/57	
3	B/	CP?w,z.z		1954mc	1960mc	ac	6.54	4oooX3					

NX 04 (25/04) **& Part of NX 14** *Mull of Logan** 10 by 10 Revision a 1906 c 1946-51

1	A	CPr	OPr	nd:po	-	-	1.50	4BEoo4			5/50	6/50	
3	B	CPw.z	OPv	1960mc	-	ac	6.60	4oooX4			12/60	1/61	

Sheet lines are offset, at 203 km E to 213 km E, 540 km N to 550 km N.

NX 05 (25/05) **& Part of NW 95** *Lochans** 15 by 10 Revision a 1906-07 c 1938-51

1	A	CPr	OPr	1950pc	-	a	6.50	4BEoo3	R		7/50	8/50	
2	B	CPt	OPv	1953pc	-	ac	6.53	4BooX3			6/53	9/57	

NX 06 (25/06) **& Part of NW 96** *Stranraer** 15 by 10 Revision a 1906-07 c 1938-53

1	A	CPr	OPr	nd:po	-	-	6.50	4BEoo3	R	0	7/50	8/50	
2	B	CPt	OPv	1953pc	-	ac	6.53	4BooX3		0	12/53	11/57	1
3	B/	CPw.z		1953mc	1959mc	ac	6.53	4BooX3		0			

1. Showing the Cairnryan railway system.

NX 07 (25/07) **& Part of NW 97** *Mouth of Loch Ryan** 15 by 10 Revision a 1906-07 c 1951-53

1	A	CPr	OPr	nd:po	-	-	1.50	4BEoo3			7/50	8/50	
3	B	CPw.z	OPv	1961mc	-	ac	6.61	4oooX3			12/60	12/60	

NX 13 (25/13) **& Parts of NX 03 & NX 23** *Mull of Galloway** 12 by 10 Revision a 1906 c 1951

1	A	CPr	OPr	1950pc	-	a	6.50	4BEoo3			7/50	7/50	
3	B	CPw,x	OPv	1960mc	-	ac	6.60	4oooX3			2/60	2/60	

With an inset of the Scares (225 km E to 227 km E, 533 km N to 535 km N).

NX 15 (25/15) *Luce Sands** 10 by 10 Revision a 1906-07 c 1951

1	A	CPr	OPr	nd:po	-	-	1.50	4BEoo4	R	00	5/50	6/50	
3	B	CPx,z	OPx	1961mc	-	ac	6.61	4oooX4		11	5/61	5/61	1

1. West Freugh Airfield.

NX 16 (25/16) *New Luce** 10 by 10 Revision a 1906-07 c 1951-53 f 1972

1	A	CPr	OPr	nd:po	-	-	1.50	4BEoo4	R	0	5/50	6/50	
3	B	CPw,z	OPx	1961mc	-	ac	6.61	4oooX4		0	1/61	2/61	
3	B/	CPz		1961^{c}ms	nd:ch	acf	6.61	4oooX4		0			

NX 17 *Glenwhilly** 10 by 10 Revision a 1907-08 c 1951

2	A	CPt	OPs	1953pc	-	ac	1.53	4BooX4			4/53	4/53	
3	A	CPw		1953mc	-	ac	1.53	4BooX4					

NX 18 (25/18) **& Part of NX 08** *Ballantrae** 15 by 10 Revision a 1907-08 c 1946-51

1	A	CPr	OPr	nd:po	-	-	1.50	4BEoo3	R		3/50	4/50	
3	B	CPt		1953pc	-	ac	6.53	4BooX3			1/54	-	
3	B/	CPw		1953mc	1959mc	ac	6.53	4BooX3					

1	2	3	4	5	6	7	8	9	10	11	12	13	14

NX 19 (25/19) & Parts of NX 09 & NS 00 *Girvan** 10 by 10 Revision a 1907-08 c 1938-53
1 A   CPr   OPr   nd:po   -   -   1.50 4BEoo3   R   5/50 6/50
3 B   CPu.z OP?   1956pc -   ac   6.56 4oooX3   4/56 2/58
With an inset of Ailsa Craig (201 km E to 203 km E, 599 km N to 601 km N).

NX 25 (25/25) & Part of NX 24 *Castle Loch Wigtownshire** 10 by 12 Revision a 1906-07 c 1951 f 1959
1 A   CPr   OPr   nd:po   -   -   1.50 4BEoo3   R   5/50 7/50
3 B   CPx,z OPx   1961mc -   acf 6.61 4oooX3   7/61 7/61

NX 26 (25/26) *Loch Ronald**   10 by 10 Revision a 1906-07 c 1951
1 A   CPr   OPr   nd:po   -   -   6.50 4BEoo4   RB   5/50 6/50
3 B   CPx.z OPx   1961mc -   ac   6.61 4oooX4   6/61 3/61

NX 27 *Drumlamford**   10 by 10 Revision a 1907 c 1951
2 A   CPt   OPs   1953pc -   ac   6.53 4BooX4   5/53 5/53
3 A   CPw.0z   1953mc -   ac   6.53 4BooX4

NX 28 *Barrhill**   10 by 10 Revision a 1907-08 c 1951
2 A   CPt   OPs   1953pc -   ac   1.53 4BooX4   4/53 4/53
3 A   CPz   1953pc -   ac   1.53 4BooX4

NX 29 *Barr**   10 by 10 Revision a 1907-08 c 1938-53
3 A   CPt   OPs   1954pc -   ac   6.54 4oooX4   8/54 9/54
3 A/ CPw,z   1954mc 1960mc ac   6.54 4oooX4

NX 34 (25/34) & Part of NX 24 *Port William** 12 by 10 Revision a 1906-07 c 1951
1 A   CPr   OPs   nd:po   -   -   1.50 4BEoo3   8/50 9/50
3 B   CPw   OPv   1959mc -   ac   6.59 4oooX3   7/59 9/59

NX 35 (25/35) *Clugston Loch**   10 by 10 Revision a 1906-07 c 1951
1 A   CPr   OPr   nd:po   -   -   1.50 4BEoo4   R   6/50 8/50
3 B   CPx.z OPx   1961mc -   ac   6.61 4oooX4   6/61 5/61

NX 36 (25/36) *Kirkcowan**   10 by 10 Revision a 1906-07 c 1951 f 1959
1 A   CPr   OPr   nd:po   -   -   1.50 4BEoo4   R   5/50 6/50
3 B   CPx,z OPx   1961mc -   acf 6.61 4oooX4   8/61 8/61

NX 37 *Bargrennan**   10 by 10 Revision a 1907 c 1946-51
2 A   CPt   OPs   1953pc -   ac   6.53 4BooX4   9/53 10/53
3 A   CPz   1953pc -   ac   6.53 4BooX4

NX 38 *Glentrool Forest**   10 by 10 Revision a 1907 c 1951 f 1969
3 A   CPt   OPs   1953pc -   ac   6.53 4BooX4   12/53 12/53
3 A/ CPz   OPz   1953ᶜms nd:ch acf 6.53 4BooX4

NX 39 *Balloch Ayrshire**   10 by 10 Revision a 1907-08 c 1951-53
3 A   CPv,z OPv   1956pc -   ac   6.56 4oooX4   11/56 11/56

NX 43 (25/43) & Part of NX 33 *Burrow Head** 15 by 10 Revision a 1906 c 1951
1 A   CPr   OPs   1950pc -   a   1.50 4BEoo3   8/50 9/50
2 B   CPt   OPv   1953pc -   ac   6.53 4BooX3   10/53 11/57
3 B/ CPw.z   1953mc 1960mc ac   6.53 4BooX3

NX 44 (25/44) *Garlieston**   10 by 10 Revision a 1906-07 c 1951 f 1959
1 A   CPr   OPr   1950pc -   a   1.50 4BEoo4   R   6/50 8/50
3 B   CPx,z OPx   1961mc -   acf 6.61 4oooX4   8/61 7/61

NX 45 (25/45) *Wigtown**   10 by 10 Revision a 1906-07 c 1938-53
1 A   CPs   OPs   1950pc -   a   6.50 4BEoo4   R   0 8/50 8/50
3 B   CPt   OPv   1953pc -   ac   6.53 4BooX4   1 12/53 11/57
3 B/ CPw,z   1953mc 1959mc ac   6.53 4BooX4   1

1	2	3	4	5	6	7	8	9	10	11	12	13	14

NX 46 (25/46) *Newton Stewart**   10 by 10   Revision a 1907 c 1938 d 1938-51 f 1965
1 A   CPs   OPs   1950pc   -   ac   6.50   4BEoo4   R   10/50 10/50
3 B   CPt   1953pc   -   ad   6.53   4BooX4   11/53 -
3 B/   CPw   1953mc   1959mc   ad   6.53   4BooX4
3 B// CPz   OPz   1953mc   1966ch   adf 6.53   4BooX4

NX 47 *Loch Dee*   10 by 10   Revision a 1907 c 1946-51
2 A   CPt   OPs   1953pc   -   ac   6.53   4BooX4   8/53   8/53
3 A   CPz   1953pc   -   ac   6.53   4BooX4

NX 48 *Merrick**   10 by 10   Revision a 1907 c 1946-51
2 A   CPt   OPs   1953pc   -   ac   6.53   4BooX4   5/53   5/53
3 A   CPz   1953mc   -   ac   6.53   4BooX4

NX 49 *Loch Doon**   10 by 10   Revision a 1907-08 c 1951-53
3 A   CPv   OPv   1956pc   -   ac   6.56   4oooX4L   11/56 11/56

NX 55 (25/55) *Water of Fleet**   10 by 10   Revision a 1907 c 1942 d 1942-51
1 A   CPr   OPs   1950pc   -   ac   6.50   4BEoo4   8/50   8/50
3 B   CPt   OPv   1953pc   -   ad   6.53   4BooX4   11/53 11/57
3 B/   CPw,z   1953mc   1959mc   ad   6.53   4BooX4

NX 56 (25/56) *Big Water of Fleet**   10 by 10   Revision a 1907 c 1942 d 1942-51 f 1970
1 A   CPr   OPs   1950pc   -   ac   1.50   4BEoo4   R   7/50   8/50
3 B   CPx,z OPx   1961mc   -   ad   6.61   4oooX4L   3/61   5/61
3 B/   CPz   1961^{c}ms nd:ch adf 6.61 4oooX4L

NX 57 *Clatteringshaws*   10 by 10   Revision a 1907 c 1951
3 A   CPt,z OPs   1953pc   -   ac   6.53   4BooX4   11/53 11/53

NX 58 *Polharrow Burn**   10 by 10   Revision a 1907 c 1951
3 A   CPt   OPs   1953pc   -   ac   6.53   4BooX4   11/53 11/53
3 A/   CP?w,z   1953mc   1960mc   ac   6.53   4BooX4

NX 59 *Carsphairn**   10 by 10   Revision a 1907 c 1951-53
3 A   CPv   OPv   1956pc   -   ac   6.56   4oooX4   10/56 11/56

NX 64 (25/64) **& Part of NX 54** *Kirkcudbright Bay**   15 by 10 Revision a 1907 c 1938-51
1 A   CPr   OPs   nd:po   -   -   1.50   4BEoo3   8/50   9/50
3 B   CPt   OPv   1953pc   -   ac   6.53   4BooX3   12/53 9/57
3 B/   CPw,z.z   1953mc   1960mc   ac   6.53   4BooX3
M on A.

NX 65 (25/65) *Kirkcudbright**   10 by 10   Revision a 1907 c 1938-42 d 1938-51
1 A   CPs   OPs   1950pc   -   ac   1.50   4BEoo4   R   8/50   8/50
3 B   CPt   OPv   1953pc   -   ad   6.53   4BooX4   11/53 11/57
3 B/   CPw,z   1953mc   1960mc   ad   6.53   4BooX4

NX 66 (25/66) *Laurieston Kirkcudbright**   10 by 10 Revision a 1907 c 1951 f 1972; j 1972 loch
1 A   CPr   OPr   nd:po   -   -   1.50   4BEoo4   R   7/50   8/50
3 B   CPx,z OPx   1961mc   -   ac   6.61   4oooX4L   1/61   3/61
3 B/* CPz   OPz   1961^{c}ms nd:lr acfj 6.61 4oooX4L   9/73   1
1. River Dee is enlarged into Loch Ken. OSPR (outline edition) not found.

NX 67 (25/67) *New Galloway**   10 by 10   Revision a 1907 c 1946-51
1 A   CPr   OPr   nd:po   -   -   1.50   4BEoo4   R   8/50   8/50
3 B   CPt,v,z OP? 1954pc   -   ac   6.54   4BooX4   1/54   2/58

NX 68 *St John's Town of Dalry** 10 by 10   Revision a 1907 c 1946-51
2 A   CPt   OPs   1953pc   -   ac   6.53   4BooX4   7/53   7/53
3 A/   CPw   1953mc   -   ac   6.53   4BooX4

1   2   3   4   5   6   7   8   9   10   11   12   13   14

NX 69 *Water of Ken**   10 by 10 Revision a 1893-1907 c 1951-53
3   A   CPv   OPv   1956pc   -   ac   6.56   4oooX4   10/56   10/56

NX 74 (25/74) **& Part of NX 84** *Dunbrennan** 12 by 10 Revision a 1907 c 1938-53
1   A   CPr   OPs   nd:po   -   -   1.50   4BEoo3   8/50   9/50
3   B   CPt.z   OPv   1954pc   -   ac   6.54   4BooX3   6/54   9/57
M on A, B.

NX 75 (25/75) *Gelston Kirkcudbright**   10 by 10 Revision a 1907 c 1938-42 d 1946-53
1   A   CPs   OPs   1950pc   -   ac   6.50   4BEoo4   R   9/50   9/50
3   B   CPt.z   OP?   1954pc   -   ad   6.54   4oooX4   7/54   2/58

NX 76 (25/76) *Castle Douglas**   10 by 10 Revision a 1907 c 1938-42 d 1938-53 f 1972
                                                         j 1972 loch
1   A   CPs   OPs   1950pc   -   ac   6.50   4BEoo4   R   9/50   11/50
3   B   CPt   OPv   1954pc   -   ad   6.54   4oooX4   8/54   11/57
3   B/   CPw,z   1954mc   1959mc   ad   6.54   4oooX4
3   B//*   CPz   OPz   1954^{c}ms nd:lr   adfj   6.54   4oooX4   3/73   6/73   1
1. River Dee is enlarged into Loch Ken.

NX 77 (25/77) *Corsock**   10 by 10 Revision a 1907 c 1942 d 1942-53
1   A   CPs   OPs   1950pc   -   ac   6.50   4BEoo4   8/50   9/50
3   B   CPt,w   OPv   1954pc   -   ad   6.54   4oooX4   8/54   11/57

NX 78 *Castlefairn**   10 by 10 Revision a 1893-1907 c 1951
3   A   CPt,z   OPs   1954pc   -   ac   6.54   4BooX4   3/54   4/54

NX 79 *Moniaive**   10 by 10 Revision a 1898-1907 c 1951-54
3   A   CPu,z   OPs   1955pc   -   ac   6.55   4oooX4   7/55   9/55

NX 85 (25/85) **& Part of NX 84** *Dalbeattie Forest* 10 by 12 Revision a 1907 c 1942 d 1942-53
1   A   CPs   OPs   1950pc   ac   6.50   4BEoo3   8/50   10/50
3   B   CPt   1954pc   ad   6.54   4oooX3   8/54   -
3   B/   CPw,z.z   OPv   1954pc   ad   6.54   4oooX3   -   10/58

NX 86 (25/86) *Dalbeattie**   10 by 10 Revision a 1907 c 1938-42 d 1938-53 f 1972
1   A   CPs   OPs   1950pc   -   ac   6.50   4BEoo4   R   9/50   10/50
3   B   CPt,z   OPv   1954pc   -   ad   6.54   4oooX4   6/54   11/57
3   B/*   CPz   OPz   1954^{c}ms nd:r   adf   6.54   4oooX4   5/73   9/73

NX 87 (25/87) *Crocketford**   10 by 10 Revision a 1907 c 1942 d 1942-53
1   A   CPs   OPs   1950pc   -   ac   1.50   4BEoo4   R   8/50   9/50
3   B   CPt   OPv   1954pc   -   ad   6.54   4oooX4   8/54   11/57
3   B/   CPw,x   1954mc   1960mc   ad   6.54   4oooX4

NX 88 (25/88) *Dunscore**   10 by 10 Revision a 1898-1907 c 1942 d 1942-51
1   A   CPs   OPs   1950pc   -   ac   6.50   4BEoo4   R   11/50   12/50
3   B   CPt,y   OPv   1954pc   -   ad   6.54   4oooX4   8/54   11/57

NX 89 (25/89) *Thornhill Dumfriesshire**   10 by 10 Revision a 1898-99 c 1951-54
1   A   CPs   OPs   1950pc   -   a   6.50   4BEoo4   R   11/50   11/50
3   B   CPu   OPv   1955pc   -   ac   6.55   4oooX4   9/55   12/57

NX 91 (25/91) **& Part of NX 90** *Whitehaven** 10 by 12 Revision a 1922-23 c 1938-51
1   15046   CPr   1946po   -   -   1.47   3BE1X1   R   9/46   -
1   A   OPr   1946po   -   -   1.47   3BE1X1   R   -   1/47
2   B   CPt   1952pc   -   ac   6.52   3Booo2   R   11/52   -
3   B/   CPw,x.z,0z   1952mc   1958mc   ac   6.52   3BooX2
AA 1.6.46 on A(A), 3/-, OSPR 1/47.

108

1	2	3	4	5	6	7	8	9	10	11	12	13	14

NX 92 (25/92) *Workington** 10 by 10 Revision a 1923 c 1938-51
1 A CPr OPr nd:po - - 1.50 4BEoo4 R   5/50 6/50
2 B CPt OPs 1952pc - ac 6.52 4Booo4 R   9/52 7/54
3 B/ CP?w,x,z,0z   1952mc 1959mc ac 6.52 4BooX4
AA 1.4.50 on A(A), 3/-, OSPR 6/50.

NX 95 (25/95) **& Part of NY 05** *Kirkbean** 12 by 10 Revision a 1907 c 1942 d 1942-53
1 A CPs OPs 1950pc - ac 1.50 4BEoo3   9/50 9/50
3 B CPt,u,z OPv 1954pc - ad 6.54 4oooX3   6/54 9/57

NX 96 (25/96) *New Abbey** 10 by 10 Revision a 1907-08 b 1898-1908 c 1942 d 1942-53
1 A CPs OPs 1950pc - ac 6.50 4BEoo4   9/50 9/50
3 B CPt,w OPv 1954pc - bd 6.54 4BooX4   4/54 11/57
3 B CPw,z   1954mc - bd 6.54 4BooX4

NX 97 (25/97) *Dumfries** 10 by 10 Revision a 1899-1929 c 1938-53 f 1975
1 A CPr OPr nd:po - - 1.50 4BEoo4 R 0 7/50 7/50
3 B CPt OPv 1954pc - ac 6.54 4oooX4   1 6/54 11/57
3 B/ CP?w,y,z   1954mc 1960mc ac 6.54 4oooX4   1
3 B//* CFz OFz 1954^{c}ms nd:r acf 6.54 4oooX4   1 1/76 12/75
AA 1.2.50 on A(A), 3/-, OSPR 7/50.

NX 98 (25/98) *Lower Nithsdale** 10 by 10 Revision a 1898-1907 c 1946-53
1 A CPr OPr nd:po - - 1.50 4BEoo4 R   6/50 6/50
3 B CPt OPv 1954pc - ac 6.54 4oooX4   7/54 11/57
3 B/ CPw,z   1954mc 1959mc ac 6.54 4oooX4

NX 99 *Loch Ettrick** 10 by 10 Revision a 1898-99 c 1947-54 f 1970
3 A CPu OPs 1955pc - ac 6.55 4oooX4   8/55 9/55
3 A/ CPz OPz 1955^{c}ms nd:ch acf 6.55 4oooX4

NY 00 (35/00) **& Part of NX 90** 12 by 10 Revision a 1897-1924 c 1948-51 f 1967 g 1974
*Gosforth Cumberland**   j 1962 industrial
1 A CPr OPr 1947po - - 6.47 3BE1X1 RB   7/47 11/47
2 B CPt OPv 1952pc - ac 6.52 3Booo2 R   11/52 8/57
3 B/ CPw,x   1952mc 1960mc ac 6.52 3BooX2
3 B//* CPz OPz 1952ms 1968i acfj 6.52 3BooX2   5/68 4/68
3 B//*/ CFz   1952^{c}ms nd:ch acgj 6.52 3BooX2
AA 1.4.47 on A(A), 3/-, OSPR 11/47.

NY 01 (35/01) *Cleator Moor* 10 by 10 Revision a 1923-24 c 1938-51 f 1974
1 10046 CPr   1946po - - 6.46 3BE1X1 RB   9/46 -
1 2546   OPr 1946po - - 6.46 3BE1X1 RB   - 11/46
2 B CPt,u OPv 1952pc - ac 6.52 3Booo2 R   12/52 10/57
3 B CPx   1952pc - ac 6.52 3Booo2 R
3 B/ CPz OPz 1952pc - ac 6.52 3Booo2
3 B// CFz   1952^{c}ms nd:ch acf 6.52 3Booo2
AA 1.5.46 on 2546(2546), 3/-, OSPR 11/46.

NY 02 (35/02) *Distington* 10 by 10 Revision a 1923-24 c 1938-48 d 1938-51
    f 1970 g 1974
1 A CPr OPs 1950pc - ac 6.50 4BEoo4 R   7/50 8/50
3 B CPw,x,z OPv 1958pc - ad 6.58 4oooX4   8/58 10/58
3 B/ CPz OPz 1958^{c}ms nd:ch adf 6.58 4oooX4
3 B// CFz   1958^{c}ms nd:ch adf 6.58 4oooX4
AA 1.7.50 on A(A), 3/-, OSPR 8/50.

NY 03 (35/03) **& Part of NX 93** *Maryport** 12 by 10 Revision a 1923 c 1938-47 d 1938-51
1 A CPr OPs 1950pc - ac 6.50 4BEoo3 R   7/50 9/50
2 B CPt OPs 1953pc - ad 1.53 4Booo3 R   1/53 4/55

1	2	3	4	5	6	7	8	9	10	11	12	13	14
3	B/	CPv,x,z		1953pc	1957mc	ad	1.53	4BooX3					
3	B/	CPz		1953pc	1956mc	ad	1.53	4BooX3					

AA 1.5.50 on A(A), 3/-, OSPR 9/50.

NY 06 (35/06) **& Part of NY 05** *Blackshaw** 10 by 12 Revision a 1894-1907 c 1945-52
| 1 | A | CPs | OPs | nd:po | - | - | 1.50 | 4BEoo3 | R | | 9/50 | 9/50 | |
| 3 | B | CPt,x,z | OPv | 1954pc | - | ac | 6.54 | 4oooX3 | | | 6/54 | 10/57 | |

NY 07 (35/07) *Collin** 10 by 10 Revision a 1898-1929 c 1946-52
1	A	CPr	OPr	nd:po	-	-	1.50	4BEoo3	RB	0	8/50	7/50	
3	B	CPt		1954pc	-	ac	6.54	4BooX3		1	4/54	-	
3	B/	CPw,z		1954mc	1959mc	ac	6.54	4BooX3		1			

NY 08 (35/08) *Lochmaben* 10 by 10 Revision a 1898-99 c 1938-53
1	A	CPr	OPr	nd:po	-	-	1.50	4BEoo3	R		6/50	8/50	
3	B	CPt	OPv	1954pc	-	ac	6.54	4oooX3			6/54	10/57	
3	B/	CPw,z,0z		1954mc	1960mc	ac	6.54	4oooX3					

NY 09 *St Ann's* 10 by 10 Revision a 1898 c 1947-54 f 1967
| 3 | A | CPu,x | OPs | 1955pc | - | ac | 6.55 | 4oooX4 | | | 9/55 | 10/55 | |
| 3 | A/ | CPz,0z | O0z | 1955ms | 1968ch | acf | 6.55 | 4oooX4 | | | | | |

NY 10 (35/10) *Wast Water* 10 by 10 Revision a 1897-1924 c 1948-51 f 1974
1	10047	CPr		1947po	-	-	1.47	3BE1X1	RB		1/47	-	
1	A		OPr	1947po	-	-	1.47	3BE1X1	RB		-	4/47	
1	B	CPr		nd:po	-	-	1.47	3BEoo1	RB				
3	C	CPx,z.z	OPx	1961mc	-	ac	6.61	3oooX2			5/61	5/61	
3	C/	CFz		1961ᶜms	nd:ch	acf	6.61	3oooX2					

AA 1.6.46 on A(A), 3/-, OSPR 4/47.

NY 11 (35/11) *Buttermere Cumberland* 10 by 10 Revision a 1898-1924 c 1946-51 f 1974
1	15046	CPr		1946po	-	-	6.46	3BE1X1	B		8/46	-	
1	2546		OPr	1946po	-	-	6.46	3BE1X1	B		-	11/46	
2	B	CPt,u	OPv	1952pc	-	ac	6.52	3Booo2			12/52	2/57	
3	B/	CPx,z		1952mc	1961ch	ac	6.52	3BooX2					
3	B//	CFz	OFz	1952ᶜms	nd:ch	acf	6.52	3BooX2					

AA 1.5.46 on 2546(2546), 3/-, OSPR 11/46.

NY 12 (35/12) *Lorton Vale* 10 by 10 Revision a 1898-1924 c 1947-51 f 1974
1	A	CPr	OPr	nd:po	-	-	1.50	4BEoo4	RB		4/50	5/50	
2	B	CPt		1953pc	-	ac	1.53	4Booo4	R		1/53	-	
3	B/	CPv	OP?	1953pc	1957mc	ac	1.53	4BooX4			-	9/57	
3	B//	CPx,z.z		1961ᶜmc	1961mc	ac	1.53	4BooX4					
3	B///	CFz		1953ᶜms	nd:ch	acf	1.53	4BooX4					

NY 13 (35/13) *Cockermouth* 10 by 10 Revision a 1923 c 1938-48 d 1938-51 f 1974
1	A	CPr	OPr	1950pc	-	ac	6.50	4BEoo4	RB		7/50	7/50	
3	B	CPv	OPv	1957pc	-	ad	6.57	4oooX4			1/58	9/57	
3	B/	CPw,x,z.z		1957mc	1960mc	ad	6.57	4oooX4					
3	B//	CFz		1957ᶜms	nd:ch	adf	6.57	4oooX4					

NY 14 (35/14) **& Part of NY 04** *Aspatria* 15 by 10 Revision a 1923-24 c 1946-51
1	A	CPs	OPs	1950pc	-	a	6.50	4BEoo3	RB		9/50	9/50	
2	B	CPt	OP?	1953pc	-	ac	6.53	4BooX3			5/53	2/58	
3	B/	CPw,y,z		1953mc	1959mc	ac	6.53	4BooX3					

NY 15 (35/15) **& Part of NY 05** *Silloth* 15 by 10 Revision a 1899-1924 b 1898-1924 c 1938 d 1938-52
1	A	CPs	OPs	1950pc	-	ac	6.50	4BEoo3	R	00	8/50	9/50	
3	B	CPt	OPv	1954pc	-	bd	6.54	4oooX3		11	7/54	11/57	
3	B/	CP?w,y,z		1954pc	1958mc	bd	6.54	4oooX3		11			1

1. Silloth Aerodrome, a name possibly not present on the unrecorded CPw state.

1	2	3	4	5	6	7	8	9	10	11	12	13	14

NY 16 (35/16) *Annan**   10 by 10   Revision a 1898-1924 b 1898-1929 c 1938 d 1938-52
1	A	CPr	OPs	1950pc	-	ac	6.50	4BEoo4	R		8/50	10/50
3	B	CPt		1954pc	-	bd	6.54	4BooX4			4/54	-
3	B/	CPw,y,z		1954mc	1958mc	bd	6.54	4BooX4				

NY 17 (35/17) *Ecclefechan**   10 by 10   Revision a 1898-99 c 1930-52 f 1965 g 1967
1	A	CPs	OPs	1950pc	-	a	6.50	4BEoo3	R		11/50	12/50
3	B	CPt	OPv	1954pc	-	ac	6.54	4oooX3			6/54	10/57
3	B/	CPw		1954mc	1959mc	ac	6.54	4oooX3				
3	B//*	CPz	OPz	1954mc	1966r	acf	6.54	4oooX3			1/66	1/66
3	B//*/*	CPz	OPz	1954ms	1968r	acg	6.54	4oooX3			5/68	4/68

NY 18 (35/18) *Lockerbie*   10 by 10   Revision a 1898-1930 c 1938 d 1938-53 f 1964
1	A	CPs	OPs	1950pc	-	ac	6.50	4BEoo3	RB		9/50	10/50
3	B	CPt		1954pc	-	ad	6.54	4BooX3			5/54	-
3	B/	CPv		1954pc	1957mc	ad	6.54	4BooX3				
3	B//*	CPz,0z	OPz	1954mc	1965r	adf	6.54	4BooX3			6/65	2/65

NY 19 *Boreland*   10 by 10   Revision a 1898 c 1948-53 f 1965
| 3 | A | CPu | OPs | 1954pc | - | ac | 6.54 | 4oooX4 | | | 10/54 | 10/54 |
| 3 | A/* | CPz | OPz | 1954mc | 1965r | acf | 6.54 | 4oooX4 | | | 11/65 | 11/65 |

NY 20 (35/20) *Sca Fell*   10 by 10   Revision a 1897-1913 b 1897-98 c 1951 f 1974
1	10046	CPr		1946po	-	-	1.47	3BE1X1	B		9/46	-
1	A		OPr	1946po	-	-	1.47	3BE1X1	B		-	1/47
1	B	CPr		nd:po	-	-	1.47	3BEoo1	B			
2	B	CPt		1948pc	1952	-	1.47	3Booo2	B			
3	B	CPu		1948pc	1952	-	1.47	3Booo2	B			
3	B/	CPw		1948mc	1960mc	a	1.47	3BooX2	B			
3	C	CPw.x,z	OPv	1960mc	-	bc	6.60	3oooX4			8/60	11/60
3	C/	CFz	OFz	1960^{c}ms	nd:ch	bcf	6.60	3oooX4				

AA 1.6.46 on A(A), 3/-, OSPR 1/47.

NY 21 (35/21) *Borrowdale Cumberland*   10 by 10   Revision a 1897-1924 c 1951 f 1962 g 1974
1	2046	CPr		1946po	-	-	1.47	3BE1X1	B		10/46	-
1	2547		OPr	1947po	-	-	1.47	3BE1X1	B		-	1/47
1	B	CPr		nd:po	-	-	1.47	3BEoo1	B			
3	C	CPv	OPv	1956pc	-	ac	6.56	3oooX2			11/56	2/57
3	C/	CPw		1956mc	1961ch	ac	6.56	3oooX2				
3	C//	CPy,z		1956mc	1963ch	acf	6.56	3oooX2				
3	C///	CFz	OFz	1956^{c}ms	nd:ch	acg	6.56	3oooX2				

AA 1.6.46 on 2547(2547), 3/-, OSPR 1/47.

NY 22 (35/22) *Keswick Cumberland*   10 by 10   Revision a 1897-23 c 1938-51 f 1962 g 1965 h 1974
1	A	CPr	OPs	nd:po	-	-	1.50	4BEoo3	RB		7/50	8/50
2	B	CPt	OPv	1952pc	-	ac	6.52	4Booo3	R		11/52	8/57
3	B/	CPv		1952pc	1957mc	ac	6.52	4BooX3				
3	B//	CPw		1952mc	1960mc	ac	6.52	4BooX3				
3	B///	CPy		1952mc	1963ch	acf	6.52	4BooX3				
3	B////	CPz	OPz	1952mc	1966ch	acg	6.52	4BooX3				
3	B/////	CFz		1952^{c}ms	nd:ch	ach	6.52	4BooX3				

AA 1.3.50/NT 17.4.46 on A(A), 3/-, OSPR 8/50.

NY 23 (35/23) *Bassenthwaite**   10 by 10   Revision a 1898-1923 c 1951
| 1 | A | CPr | OPr | nd:po | - | - | 1.50 | 4BEoo3 | RB | | 4/50 | 5/50 |
| 3 | B | CPw,x,z | OPv | 1959mc | - | ac | 6.59 | 4oooX3 | | | 3/59 | 4/59 |

NY 24 (35/24) *Wigton**   10 by 10   Revision a 1923-24 c 1938-51
1	A	CPr	OPr	nd:po	-	-	1.50	4BEoo3	R		5/50	5/50
			OP?								-	9/57
3	B	CPw,x,z	OPv	1960mc	-	ac	6.60	4oooX3			6/60	6/60

1	2	3	4	5	6	7	8	9	10	11	12	13	14

NY 25 (35/25) *Kirkbride** 10 by 10 Revision a 1924 c 1938-52
1	A	CPr	OPr	nd:po	-	-	1.50	4BEoo3	RB	0	5/50	5/50	
3	B	CPt	OPv	1954pc	-	ac	6.54	4BooX3		1	5/54	9/57	
3	B/	CPw,z		1954mc	1958mc	ac	6.54	4BooX3		1			

NY 26 (35/26) *Eastriggs* 10 by 10 Revision a 1898-1929 c 1938-52 f 1972
1	A	CPr	OPs	nd:po	-	-	1.50	4BEoo3	R	0	7/50	8/50	1
3	B	CPt		1954pc	-	ac	6.54	4oooX3		1	6/54	-	
3	B/	CPw	OPv	1954pc	1958mc	ac	6.54	4oooX3		1			
3	B//	CPw,z		1954mc	1960mc	ac	6.54	4oooX3		1			
3	B///*	CPz,0z	OPz	1954^{c}ms	nd:r	acf	6.54	4oooX3		1	5/73	9/73	

1. The Gretna armaments complex is not shown, nor its railway system.

NY 27 (35/27) *Kirtlebridge* 10 by 10 Revision a 1898-1930 b 1898 c 1938 d 1946-52 f 1968 g 1972
1	A	CPs	OPs	1950pc	-	ac	6.50	4BEoo3	R	0	9/50	9/50	
3	B	CPt,x	OPv	1954pc	-	bd	6.54	4BooX3		1	5/54	2/57	
3	B/	CPz	OPz	1954ms	1968ch	bdf	6.54	4BooX3		1			
3	B//*	CPz	OPz	1954^{c}ms	nd:r	bdg	6.54	4BooX3		1	9/73	10/73	

NY 28 (35/28) *Paddockhole** 10 by 10 Revision a 1898 c 1938-43 d 1943-52
| 1 | A | CPs | OPs | 1950pc | - | ac | 1.50 | 4BEoo3 | | | 8/50 | 9/50 | |
| 3 | B | CPt,w,z | OPv | 1954pc | - | ad | 6.54 | 4BooX3 | | | 3/54 | 10/57 | |

NY 29 *Eskdalemuir** 10 by 10 Revision a 1898 c 1947-53
| 3 | A | CPu,w,z | OPs | 1954pc | - | ac | 6.54 | 4oooX4 | | | 10/54 | 10/54 | |

NY 30 (35/30) *Ambleside* 10 by 10 Revision a 1897-1923 c 1951 f 1974
1	A	CPr	OPr	1947po	-	-	6.47	3BE1X1	B		9/47	12/47	
1	B	CPr		nd:po	-	-	6.47	3BEoo1	B				
2	B	CPt		1947pc	1949c	-	6.47	3Booo2					
3	C	CPw,x,z.z	OPv	1959mc	-	ac	6.59	3oooX2			10/59	11/59	
3	C/	CFz		1959^{c}ms	nd:ch	acf	6.59	3oooX2					

AA 1.5.47/NT 26.11.47 on A(A), 3/-, OSPR 12/47.

NY 31 (35/31) *Thirlmere* 10 by 10 Revision a 1897-1923 c 1951 f 1974
1	10047	CPr		1947po	-	-	1.47	3BE1X1	B		1/47	-	
1	A		OPr	1947po	-	-	1.47	3BE1X1	B		-	3/47	
1	B	CPr		nd:po	-	-	1.47	3BEoo1	B				
1	C	CPs,t		1946pc	1950c	a	6.50	3BEoo1	B				
2	D	CPt		1952pc	-	ac	6.52	3Booo2			12/52	-	
3	D/	CPw		1952pc	1958mc	ac	6.52	3BooX2					
3	D//	CPw,x,z.z		1952mc	1960mc	ac	6.52	3BooX2					
3	D///	CFz		1952^{c}ms	nd:ch	acf	6.52	3BooX2					

AA 1.9.46 on A(A), 3/-, OSPR 3/47.

NY 32 (35/32) *Threlkeld* 10 by 10 Revision a 1897-1923 c 1938-51 f 1968 g 1974
1	A	CPr	OPr	nd:po	-	-	1.50	4BEoo3	R		4/50	5/50	
3	A	CPt	OP?	1950pc	-	-	1.50	4BEoo3	R		-	9/57	
3	B	CPw.x,z	OPv	1958mc	-	ac	6.58	4oooX3			11/58	2/59	
3	B/*	CPz	OPz	1958ms	1968r	acf	6.58	4oooX3			11/68	8/68	
3	B/*/	CFz		1958^{c}ms	nd:ch	acg	6.58	4oooX3			8/75		

NY 33 (35/33) *Caldbeck* 10 by 10 Revision a 1898-99 c 1946-51
1	A	CPr	OPr	nd:po	-	-	1.50	4BEoo3	B		5/50	5/50	
2	B	CPt		1952pc	-	ac	6.52	4Booo3			11/52	-	
3	B/	CPv,x,z		1952pc	1957mc	ac	6.52	4BooX3					

NY 34 (35/34) *Sebergham* 10 by 10 Revision a 1898-1924 c 1951
| 1 | A | CPr | OPr | nd:po | - | - | 1.50 | 4BEoo3 | RB | | 4/50 | 5/50 | |
| 3 | B | CPx,z | OPx | 1961mc | - | ac | 6.61 | 4oooX3 | | | 6/61 | 5/61 | |

M on A.

1	2	3	4	5	6	7	8	9	10	11	12	13	14

NY 35 (35/35) *Carlisle (West)*   10 by 10   Revision a 1899-1924 c 1937-38 d 1937-52 f 1971
1	A	CPr	OPr	nd:po	-	-	1.49	4BEoo3	R	00	10/48	11/48	
1	B	CPs		1951pc	-	ac	1.51	4BEoo3	R	00	7/51	-	
3	C	CPt	OPv	1954pc	-	ad	6.54	4oooX3		11	6/54	2/57	
3	C/	CPw,y,z		1954pc	1958mc	ad	6.54	4oooX3		11			
3	C//*	CPz,0z	OPz	1954^{c}ms	nd:r	adf	6.54	4oooX3		11	10/72	12/72	

AA 1.7.48 on A(A), 3/-, OSPR 11/48. M on C/.

NY 36 (35/36) *Gretna*   10 by 10   Revision a 1898-1924 c 1938-52 f 1972
1	A	CPr	OPr	1948po	-	-	6.48	4BE4X3	R		4/48	4/48	1
1	B	CPr		nd:po	-	-	6.48	4BEoo3	R				
3	C	CPt	OPv	1954pc	-	ac	6.54	4oooX3			6/54	5/57	
3	C/	CPw	OPv	1954mc	1959mc	ac	6.54	4oooX3					
3	C//*	CPz	OPz	1954^{c}ms	nd:r	acf	6.54	4oooX3			4/73	5/73	

M on C, C/. 1. The military railway system is not shown.

NY 37 (35/37) *Canonbie*   10 by 10   Revision a 1898-1900 c 1945-52
1	A	CPs	OPs	1950pc	-	a	6.50	4BEoo3	R		9/50	9/50
3	B	CPt	OPv	1954pc	-	ac	6.54	4BooX3			5/54	9/57
3	B/	CPw,z		1954mc	1960mc	ac	6.54	4BooX3				

NPRDBA 58 on B/.

NY 38 (35/38) *Langholm*   10 by 10   Revision a 1898 c 1938 d 1938-52 f 1969
1	A	CPs	OPs	1950pc	-	ac	6.50	4BEoo3	R		11/50	12/50
3	B	CPt	OPv	1954pc	-	ad	6.54	4BooX3			4/54	9/57
3	B/	CPz,0z	OPz	1954ms	1969ch	adf	6.54	4BooX3				

NY 39 *Westerkirk*   10 by 10   Revision a 1898-1917 c 1947-53 f 1969
3	A	CPt	OPs	1954pc	-	ac	6.54	4oooX4			9/54	10/54
3	A/	CPw		1954mc	1960mc	ac	6.54	4oooX4				
3	A//	CPz		1954ms	1969ch	acf	6.54	4oooX4				

M on A.

NY 40 (35/40) *Troutbeck*   10 by 10   Revision a 1897-1913 c 1951 f 1974
1	A	CPr	OPr	1947po	-	-	6.47	4BE1X3	B		9/47	11/47
1	B	CPr		nd:po	-	-	6.47	4BEoo3	B			
3	C	CPw,y,z	OPx	1961mc	-	ac	6.61	4oooX4			1/61	2/61
3	C/	CFz	OFz	1961^{c}ms	nd:ch	acf	6.61	4oooX4				

AA 1.7.47/NT 6.7.47 on A(A), 3/-, OSPR 11/47.

NY 41 (35/41) *Martindale*   10 by 10   Revision a 1897-1913 c 1938-51 f 1974
1	A	CPr	OPr	1947po	-	-	1.47	4BE1X3	B		9/47	11/47
1	B	CPr		nd:po	-	-	1.47	4BEoo3	B			
3	C	CPv,x.z	OPv	1957pc	-	ac	6.57	4oooX3			1/58	11/57
3	C/	CFz		1957^{c}ms	nd:ch	acf	6.57	4oooX3				

AA 1.12.46/NT 23.9.47 on A(A), 3/-, OSPR 11/47.

NY 42 (35/42) *Ullswater (North)*   10 by 10   Revision a 1897-1923 c 1951 f 1973
1	A	CPr	OPr	nd:po	-	-	1.50	4BEoo3	RB		5/50	5/50	
3	B	CPw,z	OPv	1960mc	-	ac	6.60	4oooX3			11/60	11/60	
3	B/*	CFz	OFz	1960^{c}ms	nd:r	acf	6.62	4oooX3			11/74	6/74	1

1. An alteration of magnetic variation date at this stage is unusual.

NY 43 (35/43) *Skelton Cumberland*   10 by 10   Revision a 1898-1923 c 1938-51 f 1972
1	A	CPr	OPr	nd:po	-	-	1.50	4BEoo3	R		4/50	5/50
2	B	CPt		1952pc	-	ac	6.52	4Booo3	R		11/52	-
3	B/	CP?v,y,z		1952pc	1957mc	ac	6.52	4BooX3				
3	B//*	CPz	OPz	1952^{c}ms	nd:r	acf	6.52	4BooX3			9/73	10/73

AA 1.3.50 on A(A), 3/-, OSPR 5/50.

1	2	3	4	5	6	7	8	9	10	11	12	13	14

NY 44 (35/44) *Southwaite*   10 by 10   Revision a 1898-99 c 1951 f 1972
1	A	CPr	OPr	nd:po	-	-	1.50	4BEoo3	RB		3/50	4/50	
2	B	CPt	OPv	1953pc	-	ac	1.53	4BooX3	R		1/53	9/57	
3	B/	CPw		1953mc	1958mc	ac	1.53	4BooX3					
3	B/	CPz		1960^{c}mc	1958mc	ac	1.53	4BooX3					
3	B//*	CPz,0z	OPz	1953^{c}ms	nd:r	acf	1.53	4BooX3			3/73	7/73	

M on B/.

NY 45 (35/45) *Carlisle (East)*   10 by 10   Revision a 1899-1924 c 1937-52 f 1972
1	A	CPr	OPr	1948po	-	-	6.48	4BEoo3	R	0	6/48	8/48	
3	B	CPt	OPv	1954pc	-	ac	6.54	4BooX3		1	4/54	5/57	
3	B/	CPv,y,z		1954pc	1958mc	ac	6.54	4BooX3		1			
3	B//*	CPz	OPz	1954^{c}ms	nd:r	acf	6.54	4BooX3		3	9/73	10/73	
3	B//*/	CFz	OFz	1979^{c}ms	nd:ch	acf	6.54	4BooX3		3			

AA 1.4.48 on A(A), 3/-, OSPR 8/48. M on B/.

NY 46 (35/46) *Scaleby*   10 by 10   Revision a 1899-1924 c 1938 d 1937-52 f 1970
1	A	CPr	OPr	1948po	-	-	6.48	4BEoo3	B	00	7/48	8/48	
1	B	CPs		1949pc	1950c	ac	6.50	4BEoo3	B	00			
3	C	CPt	OPv	1954pc	-	ad	6.54	4BooX3		11	5/54	2/57	
3	C/	CP?w,z		1954mc	1960mc	ad	6.54	4BooX3		11			
3	C//	CPz,0z	OPz	1954^{c}ms	nd:ch	f	6.54	4BooX3		33			1

M on C, C/. NPRDBA 56 on C/. 1. Carlisle Airport.

NY 47 (35/47) *Rowanburn*   10 by 10   Revision a 1896-1924 c 1916-52
1	A	CPr	OPr	nd:po	-	-	1.50	4BEoo3	RB		6/50	6/50	
3	B	CPt	OPv	1954pc	-	ac	6.54	4BooX3			2/54	9/57	
3	B/	CP?w.z		1954mc	1959mc	ac	6.54	4BooX3					

NPRDBA 57 on B/.

NY 48 (35/48) *Newcastleton*   10 by 10   Revision a 1899-1916 b 1898-1917 c 1938-52
1	A	CPs	OPs	1950pc	-	a	6.50	4BEoo3	R		9/50	10/50	
3	B	CPt	OPv	1954pc	-	bc	6.54	4BooX3			1/54	9/57	
3	B/	CPw,z		1954mc	1959mc	bc	6.54	4BooX3					

NPRDBA 59 on B/.

NY 49 *Gorrenberry**   10 by 10   Revision a 1898-1917 c 1948-53
| 3 | A | CPt,z | OPs | 1954pc | - | ac | 6.54 | 4oooX4 | | | 9/54 | 10/54 | |

M on A.

NY 50 (35/50) *Shap Fells*   10 by 10   Revision a 1911-13 c 1951 f 1972 g 1974
1	A	CPr	OPr	1947po	-	-	6.47	4BE1X3	RB		9/47	11/47	
2	B	CPt	OPv	1952pc	-	ac	6.52	4Booo3	R		11/52	9/57	
3	B/	CP?w,z		1952mc	1959mc	ac	6.52	4BooX3					
3	B//*	CPz	OPz	1952^{c}ms	nd:r	acf	6.52	4BooX3			3/73	7/73	
3	B//*/	CFz		1952^{c}ms	nd:ch	acg	6.52	4BooX3					

AA 1.7.47 on A(A), 3/-, OSPR 11/47. NPRDBA 48 on B/.

NY 51 (35/51) *Shap*   10 by 10   Revision a 1911-13 c 1942 d 1951
                                        f 1968 g 1970 h 1974; j 1966 reservoir
1	A	CPr	OPr	1947po	-	-	6.47	4BE1X3	RB		2/47	5 /47	
1	B	CPs	OPv	1951pc	-	ac	1.51	4BEoo3	RB		6/51	9/57	
3	C	CPw,y	OP?	1960mc	-	ad	6.60	4oooX4			2/60	4/60	
3	C/*	CPz	OPz	1960ms	1969v	adfj	6.60	4oooX4			9/69	8/69	1
3	C/*/*	CPz	OPz	1960^{c}ms	nd:r	adgj	6.60	4oooX4			6/71	7/71	
3	C/*/*/	CFz		1960^{c}ms	nd:ch	adhj	6.60	4oooX4					

AA 1.10.46 on A(A), 3/-, OSPR 5/47. NPRDBA 49 on C. 1. Wet Sleddale Reservoir.

```
 1  2        3        4     5         6        7     8     9        10   11  12      13    14
```

NY 52 (35/52) *Lowther*　　　　　　　10 by 10　Revision a 1913-23 c 1938-51 f 1970 g 1974
1　A　　　CPr　　OPr　nd:po　　　-　　　-　　1.50　4BEoo3　RB　　　5/50　5/50
2　B　　　CPt　　　　　1953pc　　-　　　ac　1.53　4Booo3　R　　　1/53　-
3　B/　　 CPw,x,z　　　 1953pc　　1958mc　ac　1.53　4oooX3
3　B//*　　CPz　　OPz　1953ᶜms　nd:r　acf　1.53　4oooX3　　　　　8/71　9/71
3　B//*/　 CFz　　　　　1953ᶜms　nd:ch　acg　1.53　4oooX3
AA 1.4.50 on A(A), 3/-, OSPR 5/50. NPRDBA 50 on B/.

NY 53 (35/53) *Penrith*　　　　　　　10 by 10　Revision a 1898-1923 c 1938-51 f 1970 g 1974
1　A　　　CPr　　OPr　nd:po　　　-　　　-　　1.50　4BEoo3　RB　　　5/50　5/50
2　B　　　CPt　　　　　1952pc　　-　　　ac　6.52　4Booo3　R　　　1/53　-
3　B/　　 CPv,x　OPv　1952pc　　1956mc　ac　6.52　4BooX3　　　　　-　　5/57
3　B//*　　CPz　　OPz　1952ᶜms　nd:r　acf　6.52　4BooX3　　　　　6/71　7/71
3　B//*/　 C0z　　　　　1952ᶜms　nd:ch　acg　6.52　4BooX3　　　　　12/80
AA 1.4.50 on A(A), 3/-, OSPR 5/50. SS: Soil, Land Use Capability, coloured, 1977 on B//*/. NPRDBA 52 on B/.

NY 54 (35/54) *Kirkoswald Cumberland**　10 by 10　Revision a 1898-1924 c 1951
1　A　　　CPr　　OPr　nd:po　　　-　　　-　　1.50　4BEoo3　RB　　　4/50　5/50
3　B　　　CPw.x,z　OPv　1960mc　　-　　　ac　6.60　4oooX3　　　　　4/60　4/60
NPRDBA 53 on B.

NY 55 (35/55) *Castle Carrock**　　　10 by 10　Revision a 1898-1924 c 1946-52
1　A　　　CPr　　OPr　nd:po　　　-　　　-　　1.50　4BEoo3　RB　　　5/50　5/50
3　B　　　CPt.x,z　OPs　1954pc　　-　　　ac　6.54　4BooX3　　　　　2/54　2/57
NPRDBA 54 on B.

NY 56 (35/56) *Brampton Cumberland**　10 by 10　Revision a 1899-1924 c 1938-52
1　A　　　CPr　　OPr　nd:po　　　-　　　-　　1.50　4BEoo3　R　　　4/50　5/50
3　B　　　CPt　　　　　1954pc　　-　　　ac　6.54　4BooX3　　　　　2/54　-
3　B/　　 CPv,x,z　　　 1954pc　　1957mc　ac　6.54　4BooX3
NPRDBA 55 on B/. SS: Soil, coloured, 1983 on B/.

NY 57 *Bewcastle*　　　　　　　　　10 by 10　Revision a 1899 c 1946-52
3　A　　　CPt　　OPs　1954pc　　-　　　ac　6.54　4BooX4　　　　　4/54　4/54
3　A/　　 CPw.z　　　 1954mc　　1959mc　ac　6.54　4BooX4
GS: Solid, 1969 on A/; Drift, 1969 on A/.

NY 58 *Newcastleton Forest*　　　　　10 by 10　Revision a 1895-1917 c 1920-52
3　A　　　CPt　　OPs　1954pc　　-　　　ac　6.54　4BooX4　　　　　4/54　4/54
3　A/　　 CPw,x,z　　 1954mc　　1961mc　ac　6.54　4BooX4
3　A//　　CPz　　O0z　1954ᶜms　nd:ch　ac　6.54　4BooX4
NPRDBA 60 on A/.

NY 59 *Saughtree**　　　　　　　　　10 by 10　Revision a 1895-1920 c 1945-53
3　A　　　CPt,v,z　OPs　1954pc　　-　　　ac　6.54　4oooX4　　　　　8/54　11/54
M on A. NPRDBA 61 on A.

NY 60 (35/60) *Tebay*　　　　　　　　10 by 10　Revision a 1911-13 c 1948-51 f 1959 g 1971 h 1975
1　A　　　CPr　　OPr　1947po　　-　　　-　　1.48　4BE2X3　RB　　　11/47　1/48
1　B　　　CPs　　　　　1950pc　　-　　　a　　6.50　4BEoo3　RB　　　11/50　-
3　C　　　CPx　　OPx　1961mc　　-　　　acf　6.61　4oooX3　　　　　6/61　6/61
3　C/*　　 CPz　　OPz　1961ᶜms　nd:r　acg　6.61　4oooX3　　　　　12/72　2/73
3　C/*/*　CFz　　OFz　1961ᶜms　nd:r　ach　6.61　4oooX3　　　　　7/76　7/76
AA 1.9.47 on A(A), 3/-, OSPR 1/48.

NY 61 (35/61) *Great Asby*　　　　　10 by 10　Revision a 1911-14 c 1951
1　A　　　CPr　　OPr　1947po　　-　　　-　　1.48　4BE1X3　RB　　　10/47　12/47
1　B　　　CPr　　　　　nd:po　　　-　　　-　　1.48　4BEoo3　RB
3　C　　　CPx.z,0z　OPx　1961mc　-　　　ac　6.61　4oooX4　　　　　1/61　2/61
AA 1.8.47 on A(A), 3/-, OSPR 12/47.

1	2	3	4	5	6	7	8	9	10	11	12	13	14

NY 62 (35/62) *Appleby Westmorland* 10 by 10 Revision a 1897-1923 c 1938-51
1 A   CPr   OPr   nd:po   -   -   1.50 4BEoo4 RB   5/50   5/50
2 B   CPt   1953pc   -   ac   1.53 4BooX4   5/53   -
3 B   CPv,x,z OPv   1953pc   -   ac   1.53 4BooX4   -   8/57
AA 1.4.50 on A(A), 3/-, OSPR 5/50. NPRDBA 51 on B.

NY 63 (35/63) *Skirwith* 10 by 10 Revision a 1897-1923 c 1951
1 A   CPr   OPr   nd:po   -   -   1.50 4BEoo3 B   5/50   5/50
3 B   CPv,z,0z OPv   1957pc   -   ac   6.57 4oooX3   10/57 9/57

NY 64 (35/64) *Gilderdale Forest* 10 by 10 Revision a 1895-98 b 1895-1919 c 1951
1 A   CPs   OPs   1950pc   -   a   6.50 4BEoo3 RB   8/50   9/50
3 B   CPw,x,z,0z OPv 1959mc -   bc   6.59 4oooX3   5/59   4/59

NY 65 (35/65) *Slaggyford* 10 by 10 Revision a 1898-1920 b 1895-1924 c 1946-52
1 A   CPs   OPs   1950pc   -   a   6.50 4BEoo3 RB   11/50 11/50
3 B   CPt   1954pc   -   bc   6.54 4BooX3   1/54   -
3 B/   CPv,x.z.0z   1954pc   1957mc bc   6.54 4BooX3

NY 66 (35/66) *Greenhead* 10 by 10 Revision a 1895-1924 c 1946-52 f 1969
1 A   CPr   OPr   nd:po   -   -   1.50 4BEoo3 RB   4/50   5/50 1
3 B   CPt   1954pc   -   ac   6.54 4BooX3   3/54   -
3 B/   CPv,x   1954pc   1957mc ac   6.54 4BooX3
3 B//   CPz,0z   OPz   1954ms   1970ch acf 6.54 4BooX3
1. Several Roman alignments - Maiden Way, Stanegate, Hadrian's Wall and the vallum - are on the grey plate, and other (?exposed) sections on the black. This affects other sheets as well, eg NY 56, NY 89.

NY 67 *Spadeadam Forest* 10 by 10 Revision a 1895-1924 c 1946-52 f 1969
                                                        j 1962 industrial k 1962 afforestation
3 A   CPt   OPs   1954pc   -   ac   6.54 4BooX4   4/54   4/54
3 A/   CPw,z   1954mc   1958mc ac   6.54 4BooX4
3 A//*   CPz,0z   OPz   1954ms   1970fi acfjk 6.54 4BooX4   2/70   3/70

NY 68 *Wellhaugh Moor** 10 by 10 Revision a 1895-1920 c 1945-52
3 A   CPt,x.z   OPs   1954pc   -   ac   6.54 4BooX4   3/54   3/54

NY 69 (35/69) *Kielder* 10 by 10 Revision a 1896-1920 c 1945-53 f 1969
1 A   CPr   OPr   nd:po   -   -   1.49 4BEoo3 RB   8/48   9/48
3 B   CPt   OPv   1954pc   -   ac   6.54 4oooX3   10/54 9/57
3 B/   CPw,y   1954mc   1960mc ac   6.54 4oooX3
3 B//   CPz,0z   OPz   1954ms   1969ch acf 6.54 4oooX3
NPRDBA 62 on B//.

NY 70 (35/70) *Kirkby Stephen* 10 by 10 Revision a 1911-13 c 1950-51
1 10046 CPr   1946po   -   -   6.46 4BE1X3 RB   9/46   -
1 2547   OPr   1947po   -   -   6.46 4BE1X3 RB   -   1/47
2 B   CPt,u   OPv   1952pc   -   ac   6.52 4Booo3 R   12/52 9/57
3 B/   CPx,z.0z   1952mc   1961mc ac   6.52 4BooX3
AA 1.5.46 on 2547(2547), 3/-, OSPR 1/47.

NY 71 (35/71) *Warcop* 10 by 10 Revision a 1911-13 c 1951
1 A   CPr   1947po   -   -   1.48 4BE1X3 RB   11/47 -
1 B   OPr   1947po   -   -   1.48 4BE1X3 RB   -   2/48
2 C   CPt   1953pc   -   ac   1.53 4Booo3 R   1/53   -
3 C/   CPv,z.z   1953pc   1957mc ac   1.53 4BooX3
AA 1.8.47 on B(A), 3/-, OSPR 2/48. M on B, C/.

NY 72 (35/72) *Murton Fell Westmorland* 10 by 10 Revision a 1896-1913 c 1951
1 A   CPr   OPr   nd:po   -   -   1.50 4BEoo3 B   7/50   7/50
2 A   CPt   1950pc   1952   -   1.50 4Booo3 B
3 B   CPw,x.z.z,0z OPv 1958pc -   ac   6.58 4oooX3   7/58   8/58
M on A, B.

1	2	3	4	5	6	7	8	9	10	11	12	13	14

NY 73 (35/73) *Tyne Head*   10 by 10   Revision a 1895-1911   c 1951   f 1972; j 1972 reservoir

1	A	CPr	OPr	nd:po	-	-	1.50	4BEoo3	B		5/50	7/50	
2	A	CPt		1950pc	1952rpb	-	1.50	4BooX3	B				
3	B	CPw,x,z	OPv	1960mc	-	ac	6.60	4oooX3			11/60	11/60	
3	B/*	CPz,0z	OPz	1960^{c}ms	nd:v	acfj	6.60	4oooX3			5/73	6/73	1

1. Cow Green Reservoir.

NY 74 (35/74) *Alston Cumberland*   10 by 10   Revision a 1898   b 1895-1920   c 1948-51

1	A	CPr	OPs	1950pc	-	a	6.50	4BEoo3	RB		7/50	8/50	
2	B	CPt	OPv	1952pc	-	bc	6.52	4Booo3	R		11/52	2/57	
3	B/	CPw,y,z,0z		1952mc	1958mc	bc	6.52	4BooX3					

NY 75 (35/75) *Whitfield Northumberland*   10 by 10   Revision a 1898-1920   c 1946-52

| 1 | A | CPs | OPs | 1950pc | - | a | 6.50 | 4BEoo3 | B | | 8/50 | 9/50 | |
| 3 | B | CPw,z,0z | OPx | 1961mc | - | ac | 6.61 | 4oooX3 | | | 1/61 | 2/61 | |

NY 76 (35/76) *Haltwhistle*   10 by 10   Revision a 1920   b 1895-1920   c 1946-52   f 1966

1	A	CPs	OPs	1950pc	-	a	6.50	4BEoo3	RB		11/50	12/50	
3	B	CPt		1954pc	-	bc	6.54	4BooX3			2/54	-	
3	B/	CPw,x		1954mc	1958mc	bc	6.54	4BooX3					
3	B//*	CPz,0z		1954mc	1967r	bcf	6.54	4BooX3			12/66	12/66	

NY 77 *Black Fell*   10 by 10   Revision a 1895-1920   c 1946-52   f 1971

3	A	CPt	OPs	1954pc	-	ac	6.54	4BooX4			2/54	3/54	
3	A/	CPw,x,z		1954mc	1958mc	ac	6.54	4BooX4					
3	A//	CPz	OPz	1954^{c}ms	nd:ch	acf	6.54	4BooX4					

NY 78 *Falstone*   10 by 10   Revision a 1895-1921   c 1946-52   f 1970

3	A	CPt	OPs	1954pc	-	ac	6.54	4BooX4			4/54	6/54	
3	A/	CPw,x		1954mc	1958mc	ac	6.54	4BooX4					
3	A//	CPz	OPz	1954^{c}ms	nd:ch	acf	6.54	4BooX4					

NY 79 *Emblehope Moor*   10 by 10   Revision a 1896-1921   c 1948-52

| 3 | A | CPt | OPs | 1954pc | - | ac | 6.54 | 4oooX4 | | | 7/54 | 8/54 | |
| 3 | A/ | CPv,x,z | | 1954pc | 1957mc | ac | 6.54 | 4oooX4 | | | | | |

NY 80 (35/80) *Keld Yorkshire*   10 by 10   Revision a 1910-13   c 1920   d 1950-51

1	A	CPr	OPr	1947po	-	-	1.48	4BE1X3	RB		10/47	12/47	
1	B	CPs		1951pc	-	ac	1.51	4BEoo3	RB		1/52	-	
2	B	CPt	OPs	1951pc	-	ac	1.51	4BooX3	RB		-	7/53	
3	C	CPw,y,z,0z	OPv	1960mc	-	ad	6.60	4oooX4			7/60	7/60	

AA 1.8.47 on A(A), 3/-, OSPR 12/47.

NY 81 (35/81) *Stainmore*   10 by 10   Revision a 1911-13   c 1951   f 1965

1	10046	CPr		1946po	-	-	6.46	4BE1X3	RB		8/46	-	
1	2546		OPr	1946po	-	-	6.46	4BE1X3	RB		-	11/46	
2	B	CPt	OPv	1952pc	-	ac	6.52	4Booo3	R		9/52	9/57	
3	B/	CPw		1952mc	1959mc	ac	6.52	4BooX3					
3	B//	CPz,0z		1952mc	1966ch	acf	6.52	4BooX3					

AA 1.5.46 on 2546(2546), 3/-, OSPR 11/46. M on B.

NY 82 (35/82) *Lune Forest*   10 by 10   Revision a 1895-1912   c 1951   f 1962   g 1972
     j 1962 reservoir   k 1972 reservoir

1	A	CPr	OPr	nd:po	-	-	1.50	4BEoo3	B		5/50	7/50	
2	A	CPt		1950pc	1952	-	1.50	4Booo3	B				
3	B	CPw	OPv	1960mc	-	ac	6.60	4oooX3			2/60	2/60	
3	B/*	CPz	OPz	1960mc	1964ch v	acfj	6.60	4oooX3			8/64	9/64	1
3	B/*/*	CPz	OPz	1960^{c}ms	nd:v	acgk	6.60	4oooX3			9/73	10/73	2

M on A, B. 1. Selset Reservoir. 2. Cow Green Reservoir.

1	2	3	4	5	6	7	8	9	10	11	12	13	14

NY 83 (35/83) *St John's Chapel*   10 by 10   Revision   a 1895-1922   b 1895-1919   c 1951   f 1975
j 1973 reservoir

1	A	CPr	OPr	1950pc	-	a	1.50	4BEoo3	RB		7/50	7/50	
2	B	CPt		1952pc	-	bc	6.52	4Booo3	R		8/52	-	
3	B/	CPv,x,z		1952pc	1956mc	bc	6.52	4BooX3					
3	B//*	CFz	OFz	1952^{c}ms	nd:rv	bcfj	6.52	4BooX3			7/76	7/76	1

1. Cow Green Reservoir.

NY 84 (35/84) *Allenheads*   10 by 10   Revision   a 1895-1922   c 1951   f 1970

1	A	CPs	OPs	1950pc	-	a	6.50	4BEoo3	B		8/50	9/50	
3	A	CPu		1950pc	-	a	6.50	4BEoo3	B				
3	B	CPw,y	OPv	1958pc	-	ac	6.58	4oooX3			6/58	8/58	
3	B/	CPz,0z	OPz,0z	1958^{c}ms	nd:ch	acf	6.58	4oooX3					

NY 85 (35/85) *Allendale Town**   10 by 10   Revision   a 1919-20   b 1895-1920   c 1919-52

| 1 | A | CPs | OPs | 1950pc | - | a | 6.50 | 4BEoo3 | RB | | 11/50 | 12/50 | |
| 3 | B | CPt,v.x,z | OPv | 1954pc | - | bc | 6.54 | 4BooX3 | | | 1/54 | 2/57 | |

NY 86 (35/86) *Haydon Bridge**   10 by 10   Revision   a 1920   c 1946-52

1	A	CPs	OPs	1950pc	-	a	6.50	4BEoo3	RB		11/50	1/51	
3	B	CPt,v	OPv	1953pc	-	ac	6.53	4BooX3			1/54	9/57	
3	B/	CPw,z.z,0z		1953mc	1961mc	ac	6.53	4BooX3					

AA 1.11.50/NT 30.10.50 on A(A), 3/-, OSPR 1/51. NPRDBA 16 on B/.

NY 87 (35/87) *Wark Northumberland**   10 by 10   Revision   a 1920-21   c 1952   f 1959

| 1 | A | CPs | OPs | 1951pc | - | a | 1.51 | 4BEoo4 | RB | | 4/51 | 5/51 | |
| 3 | B | CPw,x,z | OPv | 1959mc | - | ac | 6.59 | 4oooX4 | | | 9/59 | 10/59 | |

NPRDBA 14 on B.

NY 88 (35/88) *Bellingham*   10 by 10   Revision   a 1895-1921   c 1945-52

1	A	CPr	OPr	nd:po	-	-	6.50	4BEoo3	RB		5/50	5/50	
3	B	CPt	OPv	1954pc	-	ac	6.54	4BooX3			2/54	9/57	
3	B/	CPw,z,0z		1954mc	1959c	ac	6.54	4BooX3					

M on A.

NY 89 (35/89) *Otterburn*   10 by 10   Revision   a 1896-1921   c 1945-52

1	A	CPr	OPr	nd:po	-	-	1.50	4BEoo4	B		5/50	5/50	
3	B	CPt	OPv	1954pc	-	ac	6.54	4oooX4			8/54	10/57	
3	B/	CPw,z,0z		1954mc	1959c	ac	6.54	4oooX4					

M on A.

NY 90 (35/90) *Whaw*   10 by 10   Revision   a 1910-12   c 1950-51

1	10046	CPr		1946po	-	-	6.46	4BE1X3	B		8/46	-	
1	2546		OPr	1946po	-	-	6.46	4BE1X3	B		-	10/46	
2	B	CPt		1952pc	-	ac	1.52	4Booo3			6/52	-	
3	B/	CPv,x.z,0z		1952mc	1957c	ac	1.52	4BooX3					

AA 1.5.46 on 2546(2546), 3/-, OSPR 10/46.

NY 91 (35/91) *Bowes*   10 by 10   Revision   a 1912-13   c 1951   f 1967; j 1965 reservoir

1	A	CPr	OPr	1947po	-	-	1.48	4BE1X3	RB		11/47	2/48	
2	B	CPt	OPv	1952pc	-	ac	6.52	4Booo3	R		7/52	2/57	
3	B/	CPw,x		1952pc	1958mc	ac	6.52	4BooX3					
3	B//*	CPz,0z	OPz	1952ms	1968v	acfj	6.52	4BooX3			7/68	7/68	1

AA 1.9.47 on A(A), 3/-, OSPR 2/48. M on A. 1. Balderhead Reservoir.

NY 92 (35/92) *Middleton in Teesdale*   10 by 10   Revision   a 1896-1919   c 1950-51   f 1962
j 1962 reservoir

1	A	CPr	OPr	nd:po	-	-	1.50	4BEoo3	RB		3/50	4/50	
3	B	CPw	OPv	1961mc	-	ac	6.61	4oooX3			1/61	1/61	
3	B/*	CPz	OPz	1961mc	1964ch,v	acfj	6.61	4oooX3			8/64	9/64	1

M on A. 1. Selset Reservoir.

1	2	3	4	5	6	7	8	9	10	11	12	13	14

NY 93 (35/93) *Stanhope*   10 by 10 Revision a 1895-1920 b 1895-96 c 1951
1 A   CPr   OPr   nd:po   -   -   1.50 4BEoo3   R   5/50   5/50
3 B   CPw,y,z,0z OPv   1960mc   -   bc   6.60 4oooX3   11/60 11/60
M on A.

NY 94 (35/94) *Rookhope*   10 by 10 Revision a 1895-1922 c 1951
1 A   CPr   OPr   nd:po   -   -   1.50 4BEoo3   RB   5/50   5/50   1
3 A/   CPw   1950pc   1958mc   a   1.50 4BooX3   B
3 B   CPw,z,0z   OPv   1959mc   -   ac   6.59 4oooX3   9/59   9/59
1. The Middlehope line is erroneously shown as LNER. Showing Crawley and Weatherhill stationary engines.

NY 95 (35/95) *Blanchland* 10 by 10 Revision a 1895-1920 c 1952 f 1967; j 1966 [reservoir]
1 A   CPr   OPr   nd:po   -   -   1.50 4BEoo3   B   3/50   3/50
3 A/   CPw   1950mc   1958mc   a   1.50 4BooX3   B
3 B   CPw.y   OPv   1960mc   -   ac   6.60 4oooX3   12/59 1/60
3 B/*   CPz   OPz   1960mc   1968v   acfj 6.60 4oooX3   1/68   1/68
NPRDBA 17 on B/*.

NY 96 (35/96) *Hexham*   10 by 10 Revision a 1920 c 1938-52
1 A   CPr   OPr   nd:po   -   -   1.50 4BEoo3   RB   2/50   3/50
3 B   CPt   OPv   1954pc   -   ac   6.54 4BooX3   2/54   2/57
3 B/ CP?w,x,z.z,0z   1954mc   1958mc   ac   6.54 4BooX3
AA 1.12.49 on A(A), 3/-, OSPR 3/50. M added by hand on an official copy of the 1958 index, but not recorded.
NPRDBA 15 on B/.

NY 97 (35/97) *Humshaugh*   10 by 10 Revision a 1920-21 c 1941-52
1 A   CPr   OPr   nd:po   -   -   1.50 4BEoo3   RB   5/50   5/50
3 B   CPt   OPv   1954pc   -   ac   6.54 4BooX3   1/54   2/57
3 B/   CPw,z   1954mc   1960mc   ac   6.54 4BooX3
M on A. NPRDBA 13 on B/.

NY 98 (35/98) *Kirkwhelpington* 10 by 10 Revision a 1920-21 c 1952 f 1959
1 A   CPr   OPr   nd:po   -   -   6.49 4BEoo3   RB   5/50   7/50
3 B   CPx,z,0z   OPx   1961mc   -   acf   6.61 4oooX3   6/61   6/61
M on A. NPRDBA 12 on B.

NY 99 (35/99) *Elsdon*   10 by 10 Revision a 1896-1921 c 1945-52
1 A   CPr   OPr   nd:po   -   -   1.50 4BEoo3   B   4/50   5/50
3 B   CPt   OPv   1954pc   -   ac   6.54 4oooX3   5/54   6/57
3 B/   CPw,z,0z   OPz   1954mc   1960mc   ac   6.54 4oooX3
M on A. NPRDBA 8 on B/.

NZ 00 (45/00) *Langthwaite* 10 by 10 Revision a 1910-27 c 1950-51
1 A   CPr   OPr   1948po   -   -   1.48 4BE3X3   B   3/48   4/48
2 B   CPs   OPv   1952pc   -   ac   1.52 4Booo3   5/52   8/57
3 B/   CPw,y,0z   1952mc   1959c   ac   1.52 4BooX3
AA 1.1.48 on A(A), 3/-, OSPR 4/48. M on A, B, B/.

NZ 01 (45/01) *Barnard Castle* 10 by 10 Revision a 1896-1919 c 1938-51
1 A   CPr   OPr   1948po   -   -   6.48 4BE1X3   RB   4/48   5/48
1 B   CPr   nd:po   -   -   6.50 4BEoo3   RB   M
2 C   CPt   1952pc   -   ac   6.52 4Booo3   R   11/52 -
3 C/ CPv,w,z.z,0z   OPv   1952pc   1956mc   ac   6.52 4BooX3   -   2/57
AA 1.2.48 on A(A), 3/-, OSPR 5/48. M on B. NPRDBA 25 on C/.

NZ 02 (45/02) *Woodland Durham* 10 by 10 Revision a 1896-1919 c 1951
1 A   CPr   OPr   nd:po   -   -   1.49 4BEoo3   RB   10/49 10/49
2 B   CPt   OPs   1952pc   -   ac   6.52 4Booo3   R   9/52   10/52
3 B/   CPv,w,y,z,0z   1952pc   1957mc   ac   6.52 4BooX3
M on A. NPRDBA 24 on B/.

1	2	3	4	5	6	7	8	9	10	11	12	13	14

NZ 03 (45/03) *Wolsingham*   10 by 10   Revision a 1896-1920 c 1939-51

1	A	CPr	OPr	nd:po	-	-	1.49	4BEoo3	RB		10/49	10/49	
2	B	CPt	OPs	1952pc	-	ac	6.52	4Booo3	R		9/52	10/52	
3	B/	CPv,z		1952pc	1956mc	ac	6.52	4BooX3					

M on B/. NPRDBA 22 on B/.

NZ 04 (45/04) *Castleside Durham*   10 by 10   Revision a 1916 c 1951

1	A	CPr	OPr	nd:po	-	-	1.49	4BEoo3	RB		10/49	11/49	
3	A/	CPv		1949pc	1957mc	a	1.49	4BooX3	B				
3	B	CPw,x,z,0z	OPv	1958pc	-	ac	6.58	4oooX3			1/59	2/59	

AA 1.9.49 on A(A), 3/-, OSPR 11/49. NPRDBA 19 on B.

NZ 05 (45/05) *Consett (West)*   10 by 10   Revision a 1914-20 c 1946-52 f 1966; j 1966 reservoir

1	A	CPr	OPr	nd:po	-	-	1.48	4BEoo3	RB		8/48	10/48	
3	B	CPt	OPv	1954pc	-	ac	6.54	4BooX3			4/54	9/57	
3	B/	CPw,z		1954mc	1959mc	ac	6.54	4BooX3					
3	B//*	CPz,0z	OPz	1954mc	1968rv	acfj	6.54	4BooX3			1/68	3/68	

AA 1.5.48/NT 30.7.48 on A(A), 3/-, OSPR 10/48. M on B, B/. NPRDBA 18 on B//*.

NZ 06 (45/06) [*Prudhoe*]   10 by 10   Revision a 1913-20 c 1938-52

1	A	CPr	OPr	1948po	-	-	6.48	4BEoo3	RB	0	7/48	9/48	
3	B	CPt	OPv	1954pc	-	ac	6.54	4BooX3		1	3/54	8/57	
3	B/	CPw,z		1954mc	1959c	ac	6.54	4BooX3		1			1

AA 1.6.48 on A(A), 3/-, OSPR 9/48. M on A, B. 1. Ouston Airfield.

NZ 07 (45/07) [*Stamfordham*]   10 by 10   Revision a 1913-20 c 1938-47 d 1938-52

1	A	CPr	OPr	1948po	-	-	1.48	4BEoo3	B	0	7/48	7/48	
1	B	CPs		1950pc	-	ac	6.50	4BEoo3	B	0	11/50	-	
3	C	CPx	OPx	1961mc	-	ad	6.61	4oooX3		1	4/61	3/61	

M on B.

NZ 08 (45/08) *Hartburn**   10 by 10   Revision a 1920-21 c 1938-52

1	A	CPr	OPr	nd:po	-	-	6.49	4BEoo3	RB		2/50	2/50	
3	B	CPt	OPv	1954pc	-	ac	6.54	4BooX3			4/54	9/57	
3	B/	CPw,z		1954mc	1960mc	ac	6.54	4BooX3					

NPRDBA 11 on B/.

NZ 09 (45/09) *Nunnykirk*   10 by 10   Revision a 1921 c 1947-52

1	A	CPr	OPr	nd:po	-	-	1.50	4BEoo3	RB		2/50	3/50	
3	B	CPt	OPv	1954pc	-	ac	6.54	4BooX3			5/54	9/57	
3	B/	CPw,y,z		1954mc	1958mc	ac	6.54	4BooX3					

NPRDBA 10 on B/.

NZ 10 (45/10) *Richmond Yorkshire*   10 by 10   Revision a 1911-27 c 1938-51

1	A	CPr	OPr	1948po	-	-	6.48	4BEoo3	RB		5/48	7/48	
2	B	CPt	OPs	1952pc	-	ac	6.52	4Booo3	R		11/52	12/53	
3	B/	CPv.x,z,0z	OPz	1952pc	1957mc	ac	6.52	4BooX3					

AA 1.3.48 on A(A), 3/-, OSPR 7/48. M on A. NPRDBA 27 on B/.

NZ 11 (45/11) *Gainford*   10 by 10   Revision a 1912-19 c 1938-51

1	10046	CPr		1946po	-	-	1.47	4BE1X3	RB		11/46	-	
1	A		OPr	1946po	-	-	1.47	4BE1X3	RB		-	2/47	
2	B	CPt		1953pc	-	ac	1.53	4Booo3	R		1/53	-	
3	B/	CPv,y,z		1953pc	1957mc	ac	1.53	4BooX3					

AA 1.9.46 on A(A), 3/-, OSPR 2/47. M on A. NPRDBA 26 on B/.

NZ 12 (45/12) *West Auckland*   10 by 10   Revision a 1914-19 c 1938-51

1	A	CPr	OPr	nd:po	-	-	1.49	4BEoo3	RB		10/49	11/49	1
2	B	CPt	OPv	1953pc	-	ac	1.53	4BooX3			4/53	-	
3	B/	CPw,x,z,0z		1953mc	1959mc	ac	1.53	4BooX3					

AA 1.9.49 on A(A), 3/-, OSPR 11/49. M on A. NPRDBA 23 on B/. 1. The private Woodland Colliery line, dismantled in about 1923, is erroneously shown as LNER property.

1	2	3	4	5	6	7	8	9	10	11	12	13	14

NZ 13 (45/13) *Crook Durham* 10 by 10 Revision a 1915-19 c 1938-51 f 1972
1	A	CPr		nd:po	-	-	6.49	4BEoo3	RB		9/49	-	
1	B		OPr	nd:po	-	-	6.49	4BEoo3	RB		-	11/49	
2	C	CPt		1953pc	-	ac	1.53	4BooX3			4/53	-	
3	C/	CPv.x,z		1953pc	1957mc	ac	1.53	4BooX3					
3	C//*	CPz,0z	OPz	1953ᶜms	nd:r	acf	1.53	4BooX3			3/73	3/73	

AA 1.5.49 on B(A), 3/-, OSPR 11/49. M on C/. NPRDBA 21 on C/.

NZ 14 (45/14) *Lanchester* 10 by 10 Revision a 1915-40 b 1915-19 c 1947-48 d 1947-51 f 1959
1	A	CPr	OPr	nd:po	-	-	1.49	4BEoo3	RB		10/49	11/49	
2	A	CPu		1949pc	-	-	1.49	4Booo3	RB				
3	B	CPx,z,0z	OPx	1961mc	-	bdf	6.61	4oooX4			9/61	9/61	

AA 1.4.49 on A(A), 3/-, OSPR 11/49. M on A. NPRDBA 20 on B.

NZ 15 (45/15) *Consett (East)* 10 by 10 Revision a 1914-19 c 1938-50 f 1975
1	A	CPr	OPr	1948po	-	-	6.48	4BEoo3	RB		7/48	9/48	1
2	A	CPs		1948pc	1952	-	1.52	4Booo3	RB				
3	B	CPt,v,x,z.z	OPv	1954pc	-	ac	6.54	4oooX3			6/54	9/57	2
3	B/*	CFz	OFz	1954ᶜms	nd:r	acf	6.54	4oooX3			12/75	11/75	

AA 1.4.48 on A(A), 3/-, OSPR 9/48. M on B. 1. Five place names are in the large Times Roman font reserved for village names, a number only exceeded on SS 99. 2. Only Annfield Plain remains.

NZ 16 (45/16) *Blaydon* 10 by 10 Revision a 1913-15 c 1938-48
1	A	CPr	OPr	1947po	-	-	6.47	4BE2X3	RB		8/47	11/47	
1	B	CPr		1947po	-	-	6.47	4BE2X3	RB				1
1	C	CPs		1951pc	-	ac	1.51	4BEoo3	RB		12/51	-	
3	C/	CPv	OPv	1951pc	1956mc	ac	1.51	4BooX3	B				
3	C//	CPw,y,z		1951mc	1960mc	ac	6.51	4BooX3	B				

AA 1.4.47 on A(A), 3/-, OSPR 11/47. M on C, C//. 1. One of only three coloured edition B sheets with legends.

NZ 17 (45/17) [*Ponteland*] 10 by 10 Revision a 1913-21 c 1938-48
1	A	CPr	OPr	1948po	-	-	6.48	4BEoo3	RB	0	5/48	6/48	
1	B	CPr		nd:po	-	-	6.48	4BEoo3	R	3			
3	B/	CPw,x.z		1948mc	1959c	ac	6.48	4BooX3		3			1

M on B, B/. 1. Newcastle (Woolsington) Aerodrome.

NZ 18 (45/18) *Morpeth (West)* 10 by 10 Revision a 1920-21 f 1971
1	A	CPr	OPr	nd:po	-	-	1.50	4BEoo3	RB	0	2/50	3/50	
3	A/	CPw		1950mc	1959mc	a	1.50	4BooX3	B	0			
3	A//	CPw,y,z		1950mc	1960mc	a	1.50	4BooX3	B	0			
3	A///*	CPz	OPz	1950ᶜms	nd:r	af	1.50	4BooX3	B	0	4/72	8/72	

NZ 19 (45/19) *Longhorsley** 10 by 10 Revision a 1921 c 1947-52 f 1965
1	A	CPr	OPr	nd:po	-	-	1.50	4BEoo3	B	0	2/50	3/50	
3	B	CPt	OPv	1954pc	-	ac	6.54	4oooX3		1	9/54	9/57	
3	B/	CPw		1954mc	1958mc	ac	6.54	4oooX3		1			
3	B//	CPz		1954mc	1966ch	acf	6.54	4oooX3		3			

NZ 20 (45/20) *Scotch Corner** 10 by 10 Revision a 1912-27 c 1938-51 f 1965 j 1958-60 airfield
1	A	CPr		1947po	-	-	1.48	4BE1X3	RB	00	11/47	-	
1	B		OPr	1947po	-	-	1.48	4BE1X3	RB	00	-	2/48	
1	C	CPr		nd:po	-	-	1.48	4BEoo3	RB	00			
2	D	CPt	OPs	1952pc	-	ac	6.52	4Booo3	R	11	11/52	11/52	
3	D/	CPw,x,z		1952pc	1958mc	ac	6.52	4BooX3		11			
3	D//*	CPz,0z	OPz	1952mc	1966ar	acfj	6.52	4BooX3		33	5/66	5/66	

AA 1.9.47 on B(A), 3/-, OSPR 2/48. M on C.

121

1	2	3	4	5	6	7	8	9	10	11	12	13	14

NZ 21 (45/21) *Darlington (West)**   10 by 10   Revision a 1912-14 c 1938-50 f 1965
1	A	CPr	OPr	1948po	-	-	1.48	4BE3X3	RB		1/48	4/48	
1	B	CPr,s		nd:po	-	-	1.48	4BEoo3	R				
3	C	CPx,z	OPx	1961mc	-	ac	6.61	4oooX3			6/61	5/61	
3	C/*	CPz	OPz	1961mc	1966r	acf	6.61	4oooX3			6/66	7/66	

AA 1.11.47 on A(A), 3/-, OSPR 4/48. M on B.

NZ 22 (45/22) *Bishop Auckland*   10 by 10   Revision a 1913-15 c 1938-50 f 1966 g 1968
1	A	CPr	OPr	1948po	-	-	1.48	4BE2X3	RB		3/48	6/48	
2	A	CPt		1948pc	1952rpb	-	1.52	4BooX3	RB				
2	B	CPt	OPs	1953pc	-	ac	1.53	4BooX4			5/53	8/54	1
3	B/												
3	B//	CPx		1953mc	1961ch	ac	1.53	4BooX4					
3	B///*	CPz	OPz	1953mc	1967r	acf	1.53	4BooX4			1/67	1/67	
3	B///*/*	CPz	OPz	1953ms	1969r	acg	1.53	4BooX4			8/69	9/69	

AA 1.11.47 on A(A), 3/-, OSPR 6/48. M on A. 1. Aycliffe Industrial Estate, built as a munitions factory during the Second World War, is shown, complete with railway connections (not present on edition A).

NZ 23 (45/23) [Spennymoor]   10 by 10   Revision a 1914-15 c 1938-50
| 1 | A | CPr | OPr | nd:po | - | - | 1.50 | 4BEoo3 | RB | | 10/49 | 11/49 | |
| 3 | B | CPw,x,z | OPv | 1958pc | - | ac | 6.58 | 4oooX3 | | | 10/58 | 10/58 | |

AA 1.10.49 on A(A), 3/-, OSPR 11/49.

NZ 24 (45/24) [Durham]   10 by 10   Revision a 1915 c 1938-50 f 1963
1	A	CPr	OPr	nd:po	-	-	6.49	4BEoo3	RB		6/49	8/49	
2	B	CPt,u	OPv	1953pc	-	ac	6.53	4BooX3			10/53		
3	B/	CPw		1953mc	1961ch	ac	6.53	4BooX3					
3	B//	CPy,z		1953mc	1963ch	acf	6.53	4BooX3					

AA 1.7.49 on A(A), 3/-, OSPR 8/49. M on B, B/.

NZ 25 (45/25) [Chester-le-Street]   10 by 10   Revision a 1914-15 c 1938-52
1	A	CPr	OPr	1947po	-	-	6.47	4BE2X3	RB		7/47	10/47	
3	B	CPt	OPv	1954pc	-	ac	6.54	4BooX3			3/54	-	
3	B/												
3	B//	CPw.z		1954mc	1960mc	ac	6.54	4BooX3					

AA 1.5.47 on A(A), 3/-, OSPR 10/47. M on A, ?B//.

NZ 26 (45/26) *Newcastle upon Tyne*   10 by 10   Revision a 1912-14 c 1938-49 f 1962
1	A	CPr	OPr	1947po	-	-	6.47	4BE2X3	R		5/47	7/47	
1	B	CPs		1951pc	-	ac	1.51	4BEoo3	R		12/51	-	
2	B	CPu	OPs	1951pc	-	ac	1.51	4Booo3	R		-	7/52	
3	B/	CPw		1951mc	1960mc	ac	1.51	4BooX3					
3	B//	CPy,z		1951mc	1963ch	acf	1.51	4BooX3					1

AA 1.4.47 on A(A), 3/-, OSPR 7/47. M on A. 1. A copy in the Geography Department, University of Newcastle-upon-Tyne, has SP 70 (D, CPz), on the reverse.

NZ 27 (45/27) [Seghill]   10 by 10   Revision a 1913-21 c 1938-51 f 1966
1	A	CPr	OPr	1948po	-	-	1.48	4BE2X3	RB	0	3/48	5/48	
1	A	CPs		1948pc	1951	-	6.51	4BEoo3	RB	0			
3	B	CPt	OPv	1954pc	-	ac	6.54	4BooX3		0	1/54	9/57	
3	B/	CPw,y		1954mc	1960mc	ac	6.54	4BooX3		0			
3	B//	CPz	OPz	1954mc	1967ch	acf	6.54	4BooX3		3			

AA 1.4.48 on A(A), 3/-, OSPR 5/48. M on B.

NZ 28 (45/28) **& Part of NZ 38** *Blyth**   12 by 10   Revision a 1920-21 c 1938-47 f 1965
1	A	CPr		1948po	-	-	6.48	4BE4X3	R		6/48	-	
1	B		OPr	1948po	-	-	6.48	4BE4X3	R		-	6/48	
1	C	CPs	OPv	1951pc	-	ac	6.51	4BEoo3	R		11/51	8/57	
3	C/	CPw.x		1951mc	1958c	ac	6.51	4BooX3					
3	C//	CPz	OPz	1951mc	1966ch	acf	6.51	4BooX3					

AA 1.6.46 on B(A), 3/-, OSPR 6/48.

1	2	3	4	5	6	7	8	9	10	11	12	13	14

NZ 29 (45/29) **& Part of NZ 39** *Lynemouth** 12 by 10 Revision a 1921 c 1938-53 f 1964
1	A	CPr	OPr	nd:po	-	-	6.49	4BEoo3	RB	0	2/50	3/50	
3	B	CPt		1954pc	-	ac	6.54	4oooX3		0	11/54	-	
3	B/	CPw	OP?	1954mc	1958mc	ac	6.54	4oooX3		0	-	3/58	
3	B//	CPz		1954mc	1964ch	acf	6.54	4oooX3		1			

NZ 30 (45/30) *Great Smeaton** 10 by 10 Revision a 1911-13 c 1950
1	A	CPr	OPr	1948po	-	-	6.48	4BEoo3	RB		6/48	8/48	
2	A	CPs		1948pc	1952	-	1.52	4Booo3	B				
3	A/	CPv		1948pc	1957mc	a	1.52	4BooX3	B				
3	B	CPw,z	OPv	1958mc	-	ac	6.58	4oooX3			1/59	12/58	

AA 1.5.48 on A(A), 3/-, OSPR 8/48.

NZ 31 (45/31) *Darlington (East)** 10 by 10 Revision a 1912-14 c 1938-50 f 1965 g 1969
                                                    j 1969 airfield
1	A	CPr	OPr	1948po	-	-	6.48	4BEoo3	RB	0	5/48	7/48	
2	B	CPt		1953pc	-	ac	1.53	4BooX3	R	1	3/53	-	
3	B/	CPv		1953pc	1957mc	ac	1.53	4BooX3		1			
3	B//	CPw		1953mc	1961ch	ac	1.53	4BooX3		1			
3	B///	CPz	OPz	1953mc	1966ch	acf	1.53	4BooX3		3			
3	B////*	CPz	OPz	1953ms	1969r	acgj	1.53	4BooX3		3	2/70	2/70	1

AA 1.3.48 on A(A), 3/-, OSPR 7/48. 1. Tees-side Airport.

NZ 32 (45/32) *Sedgefield* 10 by 10 Revision a 1913-15 c 1948-50 d 1938-50 f 1971
1	A	CPr	OPr	1948po	-	-	1.48	4BE2X3	RB		3/48	5/48	
1	B	CPs		1951pc	-	ac	6.51	4BEoo3	RB		7/51	-	
2	B		OPs	1951pc	-	ac	6.51	4Booo3	RB		-	7/52	
2	C	CPt	OPv	1952pc	-	ad	6.52	4Booo3	R		12/52	-	
3	C/	CPw,x		1952mc	1960mc	ad	6.52	4BooX3					
3	C//*	CPz	OPz	1952ᶜms	-	adf	6.52	4BooX3			6/72	8/72	

M on C, C/.

NZ 33 (45/33) [Coxhoe] 10 by 10 Revision a 1914-15 c 1938-50
1	A	CPr	OPr	nd:po	-	-	1.48	4BEoo3	RB		8/48	8/48	
1	A	CPt		1948pc	1952	-	1.52	4Booo3	RB				
2	B	CPt,u	OPv	1953pc	-	ac	1.53	4BooX3	R		4/53	1/57	
3	B	CPy,z		1953pc	-	ac	1.53	4BooX3	R				

M on B.

NZ 34 (45/34) [Hetton le Hole] 10 by 10 Revision a 1914-19 b 1914-15 c 1938-48 d 1938-50
1	A	CPr	OPr	1948po	-	-	1.48	4BE2X3	RB		3/48	6/48	
1	B	CPs		1951pc	-	ac	6.51	4BEoo3	RB		7/51	-	
2	B		OPs	1951pc	-	ac	6.51	4BooX3	RB		-	2/53	
2	C	CPt	OPv	1953pc	-	bd	1.53	4BooX3			5/53	1/57	
3	C/												
3	C//	CPw,x,z		1953mc	1959mc	bd	1.53	4BooX3					

AA 1.4.48 on A(A), 3/-, OSPR 6/48. M on C, C//.

NZ 35 (45/35) **& Part of NZ 45** [Sunderland] 15 by 10 Revision a 1913-15 c 1938-47 d 1938-48 f 1963
1	A	CPr	OPr	1947po	-	-	1.47	4BE2X3	RB	0	5/47	9/47	
1	B	CPs	OPs	1951pc	-	ac	6.51	4BEoo3	RB	0	9/51	9/51	
2	B	CPt		1951pc	1953rpb	ac	1.51	4BooX3	RB	0			
3	C	CPt		1954pc	-	ad	6.54	4BooX3		1	4/54	-	
3	C/												
3	C//	CPz	OPz	1954mc	1964ch	adf	6.54	4BooX3		1			

AA 1.10.46/NT 1.7.47 on A(A), 3/-, OSPR 9/47. M on B.

NZ 36 (45/36) **& Part of NZ 46** [South Shields] 12 by 10 Revision a 1912-20 c 1938-48 f 1962 g 1967
1	A	CPr	OPr	1947po	-	-	1.47	4BE2X3	RB		6/47	11/47	
1	B	CPr		nd:po	-	-	1.47	4BEoo3	RB				
2	C		OPs	1952pc	-	ac	1.52	4Booo3	RB		-	9/52	

1	2	3	4	5	6	7	8	9	10	11	12	13	14
2	D	CPt	OPs	1953pc	-	ac	1.53	4BooX3	RB		3/53	1/56	
3	D/	CPw		1953mc	1960mc	ac	1.53	4BooX3	B				
3	D//	CPy,z		1953mc	1963ch	acf	1.53	4BooX3	B				
3	D///*	CPz	OPz	1953ms	1968r	acg	1.53	4BooX3	B		7/68	7/68	

AA 1.10.47 on A(A), 3/-, OSPR 11/47; AA 1.3.52 on C(B), 4/-, OSPR 9/52. M on B.

NZ 37 (45/37) **& Part of NZ 38** *Whitley Bay* 10 by 12 Revision a 1913-21 c 1938-48 f 1966

1	A	CPr	OPr	1947po	-	-	1.48	4BE2X3	RB		12/47	3/48	
3	B	CPt		1954pc	-	ac	6.54	4BooX3			1/54	-	
3	B/	CPv,x		1954pc	1957c	ac	6.54	4BooX3					
3	B//*	CPz	OPz	1954mc	1967r	acf	6.54	4BooX3			2/67	2/67	

AA 1.9.47 on A(A), 3/-, OSPR 3/48. M on B, B/.

NZ 40 (45/40) *Crathorne* 10 by 10 Revision a 1910-13 c 1950 f 1976

1	A	CPr	OPr	1948po	-	-	1.48	4BE3X3	RB		12/47	2/48	1
3	B	CPv	OP?	1958pc	-	ac	6.58	4oooX3			5/58	5/58	
3	B/	CPw,z		1958mc	1960mc	ac	6.58	4oooX3					
3	B//*	CFz	OFz	1958^{c}ms	nd:r	acf	6.58	4oooX3					2

AA 1.10.47/NT 5.1.48 on A(A), 3/-, OSPR 2/48. 1. The mounted and folded price is missing, thus the 2/- is present, not the 3/- price. 2. OSPR not found.

NZ 41 (45/41) *Stockton-on-Tees* 10 by 10 Revision a 1912-31 c 1938-51 f 1973

1	A	CPr	OPr	1947po	-	-	1.48	4BE1X3	RB	0	11/47	12/47	
2	B	CPt		1953pc	-	ac	1.53	4BooX3	R	1	3/53	-	
3	B/	CPv.x,z		1953pc	1957mc	ac	1.53	4BooX3		1			
3	B//*	CPz	OPz	1953^{c}ms	nd:r	acf	1.53	4BooX3		1	12/73	5/74	

AA 1.6.47 on A(A), 3/-, OSPR 12/47. M on A.

NZ 42 (45/42) *Billingham* 10 by 10 Revision a 1913-14 b 1913-28 c 1938-50 d 1938-51 f 1972

1	A	CPr	OPr	1947po	-	-	6.47	4BE2X3	RB	0	6/47	8/47	
1	B	CPs		1951pc	-	ac	1.51	4BEoo3	RB	0	6/51	-	
2	B		OPs	1951pc	-	ac	1.51	4Booo3	RB	0	-	9/52	
2	C	CPt		1953pc	-	bd	1.53	4BooX3	R	2	3/53	-	1
3	C/	CPv		1953pc	1956mc	bd	1.53	4BooX3		2			
3	C//	CPx,z		1953mc	1961ch	bd	1.53	4BooX3		2			
3	C///*	CPz,0z	OPz	1953^{c}ms	nd:r	bdf	1.53	4BooX3		2	9/73	9/73	

AA 1.3.47 on A(A), 3/-, OSPR 8/47. M on A. 1. West Hartlepool Civic Airport.

NZ 43 (45/43) **& Part of NZ 53** *West Hartlepool* 15 by 10 Revision a 1914-15 c 1938-50 f 1974

1	A	CPr	OPr	1948po	-	-	6.48	4BEoo3	RB		6/48	6/48	
2	B	CPt		1953pc	-	ac	1.53	4BooX3	R		2/53	-	
3	B/	CPv,y,z		1953pc	1957c	ac	1.53	4BooX3					
3	B//*	CFz	OFz	1953^{c}ms	nd:r	acf	1.53	4BooX3			6/75	5/75	

AA 1.2.48 on A(A), 3/-, OSPR 6/48. M on B, B/.

NZ 44 (45/44) *Seaham** 10 by 10 Revision a 1914-15 c 1938-50

1	A	CPr	OPr	1947po	-	-	6.47	4BE2X3	RB		6/47	10/47	
2	B	CPt		1953pc	-	ac	1.53	4BooX3	R		2/53	-	
3	B/	CPv,x,z	OPv	1953pc	1956mc	ac	1.53	4BooX3			-	2/57	

AA 1.4.47 on A(A), 3/-, OSPR 10/47. M on A, B.

NZ 50 (45/50) *Stokesley** 10 by 10 Revision a 1910-27 c 1938-50 f 1966 g 1972

1	A	CPr	OPr	1947po	-	-	6.47	4BE1X3	RB		11/47	2/48	
1	B	CPr		1947pc	1950c	a	6.50	4BEoo3	RB				
	B		OP?								-	11/52	
3	C	CPt	OPv	1954pc	-	ac	6.54	4BooX3			5/54	1/57	
3	C/	CPw,x,y		1954mc	1960mc	ac	6.54	4BooX3					
3	C//*	CPz	OPz	1954mc	1967r	acf	6.54	4BooX3			2/67	2/67	
3	C//*/*	CPz,0z	OPz,0z	1954^{c}ms	nd:r	acg	6.54	4BooX3			4/73	7/73	

AA 1.8.47 on A(A), 3/-, OSPR 2/48.

1	2	3	4	5	6	7	8	9	10	11	12	13	14

NZ 51 (45/51) *Eston*   10 by 10 Revision a 1913-28 c 1938-52 f 1965 g 1968
                                       j 1958-63 building
1	A	CPr		1947po	-	-	1.48	4BE1X3	RB		8/47	-	
1	B		OPr	1947po	-	-	1.48	4BE1X3	RB		-	11/47	
3	C	CPt	OPv	1954pc	-	ac	6.54	4BooX3			5/54	9/57	
3	C/	CPw,x		1954pc	1958mc	ac	6.54	4BooX3					
3	C//*	CPz	OPz	1954mc	1966r	acf	6.54	4BooX3			5/66	6/66	
3	C//*/*	CPz	OPz	1954ms	1970br	acgj	6.54	4BooX3			4/70	5/70	

AA 1.4.47 on B(A), 3/-, OSPR 11/47. M on B.

NZ 52 (45/52) *Tees Mouth*   10 by 10 Revision a 1853-1927 b 1913-39 c 1940-53
                                            f 1963 g 1974
1	A	CPr	OPr	1947po	-	-	6.47	4BE2X3	R	0	5/47	8/47	
1	B	CPs		1947pc	1950c	a	6.50	4BEoo3	R	0			
3	C	CPt	OPv	1954pc	-	bc	6.54	4oooX4		1	6/54	9/57	1
3	C/	CPw.x		1954mc	1959c	bc	6.54	4oooX4		1			
3	C//	CPz	OPz	1954mc	1965ch	bcf	6.54	4oooX4		0			
3	C///*	CFz	OFz	1954^{c}ms	nd:r	bcg	6.54	4oooX4		0	11/75	7/75	

AA 1.3.47 on A(A), 3/-, OSPR 8/47. M on A. 1. West Hartlepool Civic Airport.

NZ 60 (45/60) *Castleton Yorkshire*   10 by 10 Revision a 1910-27 c 1938-50
1	A	CPr	OPr	1947po	-	-	6.47	4BE1X3	RB		11/47	2/48	
1	B	CPr		nd:po	-	-	6.47	4BEoo3	RB				
3	C	CPt	OPv	1954pc	-	ac	6.54	4BooX3			4/54	1/57	
3	C/	CPw,y.z,0z		1954mc	1960mc	ac	6.54	4BooX3					

AA 1.9.47 on A(A), 3/-, OSPR 2/48.

NZ 61 (45/61) *Guisborough**   10 by 10 Revision a 1926-27 c 1938-50
1	A	CPr	OPr	1947po	-	-	1.47	4BE1X3	RB		9/47	12/47	
3	B	CPt	OPv	1954pc	-	ac	6.54	4BooX4			3/54	1/57	
3	B/	CPv,x,z	O0z	1954pc	1957c	ac	6.54	4BooX4					

AA 1.12.46 on A(A), 3/-, OSPR 12/47. M on A, B.

NZ 62 (45/62) *Redcar*   10 by 10 Revision a 1927 c 1938-50
1	50046	CPr		1946po	-	-	6.46	4BE1X3	RB	0	6/46	-	
1	5046		OPr	1946po	-	-	6.46	4BE1X3	RB	0	-	11/46	
3	B	CPt,v,y,z	OPv	1954pc	-	ac	6.54	4BooX3		0	4/54	5/57	1

AA 1.4.46 on 5046(2546), 3/-, OSPR 11/46. M on A, B. 1. The Aerodrome (Depot for Engineering Accessories).

NZ 70 (45/70) *Glaisdale*   10 by 10 Revision a 1910-27 c 1920 d 1950
1	A	CPr	OPr	1947po	-	-	1.47	4BE1X3	RB		4/47	7/47	
1	B	CPs		1951pc	-	ac	6.51	4BEoo3	RB		8/51	-	
3	C	CPt	OPv	1954pc	-	ad	6.54	4BooX3			4/54	9/57	
3	C/	CPw,y,z.z,0z	O0z	1954mc	1960mc	ad	6.54	4BooX3					

AA 1.12.46 on A(A), 3/-, OSPR 7/47.

NZ 71 (45/71) **& Part of NZ 72** *Loftus**   10 by 12 Revision a 1910-27 c 1938-50 f 1962; j 1962 reservoir
1	10047	CPr		1947po	-	-	1.47	4BE1X3	RB		12/46	-	1
1	A		OPr	1947po	-	-	1.47	4BE1X3	RB		-	5/47	
3	B	CPt	OPv	1954pc	-	ac	6.54	4BooX3			4/54	12/57	
3	B/	CPw		1954mc	1958mc	ac	6.54	4BooX3					
3	B//*	CPy,z	OPy,0z	1954mc	1963r	acfj	6.54	4BooX3			9/63	10/63	2

AA 1.9.46 on A(A), 3/-, OSPR 5/47. M on A, B. 1. Showing the Port Mulgrave tramway system, including its tunnel, also the Skinningrove railway zig-zag. 2. Scaling Reservoir.

NZ 80 (45/80) *Sleights*   10 by 10 Revision a 1910-27 c 1938-50
1	A	CPr	OPr	1948po	-	-	6.48	4BE1X3	RB		2/48	6/48	
3	B	CPt	OPs	1954pc	-	ac	6.54	4BooX3			3/54	11/55	
3	B/	CPw,x,z,0z	OPz	1954mc	1958mc	ac	6.54	4BooX3					

AA 1.12.47 on A(A), 3/-, OSPR 6/48. M on A, B.

1	2	3	4	5	6	7	8	9	10	11	12	13	14

NZ 81 (45/81) *Whitby Yorkshire*   10 by 10   Revision a 1910-27 c 1938-50 f 1966
1	A	CPr	OPr	1947po	-	-	1.48	4BE1X3	RB		11/47	1/48	
3	B	CPt	OPv	1954pc	-	ac	6.54	4BooX3			3/54	1/57	
3	B/	CPw,x,z		1954mc	1958mc	ac	6.54	4BooX3					
3	B//	CPz,0z	OPz	1954mc	1967ch	acf	6.54	4BooX3					

AA 1.8.47 on A(A), 3/-, OSPR 1/48. M on A, B.

NZ 90 (45/90) **& Part of NZ 91** *Robin Hood's Bay* 10 by 12   Revision a 1910-27 c 1938-50 f 1973
1	A	CPr	OPr	1947po	-	-	6.47	4BE1X3	RB		10/47	3/48	
3	B	CPt		1954pc	-	ac	6.54	4BooX3			2/54	-	
3	B/	CPv.x,z		1954pc	1957mc	ac	6.54	4BooX3					
3	B//*	CPz,0z	OPz,0z	1954ᶜms	nd:r	acf	6.54	4BooX3				7/74	6/74

AA 1.8.47/NT 12.1.48 on A(A), 3/-, OSPR 3/48. M on A, B.

SD 09 (34/09) *Ravenglass**   10 by 10   Revision a 1897-98 c 1948-51 f 1974
1	12546	CPr		1946po	-	-	6.46	3BE1X1	R		6/46	-	
1	2546		OPr	1946po	-	-	6.46	3BE1X1	R		-	5/47	
2	B	CPt	OPv	1952pc	-	ac	6.52	3Booo2	R		6/52	10/57	
3	B/	CPw,x,z		1952mc	1958mc	ac	6.52	3BooX2					
3	B//	CFz	OFz	1952ᶜms	nd:ch	acf	6.52	3BooX2					

AA 1.4.46 on 2546(2546), 3/-, OSPR 5/47.

SD 17 (34/17) *Barrow-in-Furness (North West)** 10 by 10 Revision a 1910-31 c 1938-51 f 1965
1	10046	CPr		1946po	-	-	6.46	3BE1X1	R	00	4/46	-	
1	2546		OPr	1946po	-	-	6.46	3BE1X1	R	00	-	4/46	
2	B	CPt		1952pc	-	ac	6.52	3Booo2	R	31	6/52	-	1
3	B/	CPw	OPv	1952mc	1960c	ac	6.52	3BooX2		31	-	2/60	
3	B//	CPz	OPz	1952mc	1966ch	acf	6.52	3BooX2		33	-	-	2

AA 1.3.46 on 2546(2546), 3/-, OSPR 7/46. M on A. 1. Barrow (Walney Island) Aerodrome. The other airfield is noted over the topography. 2. Walney Aerodrome (Private).

SD 18 (34/18) **& Part of SD 08** *Millom (North)** 15 by 10 Revision a 1897-1924 c 1950-51
1	15046	CPr		1946po	-	-	6.46	3BE1X1	R		7/46	-	
1	2546		OPr	1946po	-	-	6.46	3BE1X1	R		-	5/47	
2	B	CPt		1952pc	-	ac	6.52	3Booo2	R		8/52	-	
3	B/	CPv,z.z		1952pc	1956mc	ac	6.52	3BooX2					

AA 1.4.46 on 2546(2546), 3/-, OSPR 5/47. M on B/.

SD 19 (34/19) *Muncaster**   10 by 10   Revision a 1897-1912 c 1948-51
1	15046	CPr		1946po	-	-	6.46	3BE1X1	RB		6/46	-	
1	2546		OPr	1946po	-	-	6.46	3BE1X1	RB		-	9/46	
2	B	CPt,u.x,z	OPv	1952pc	-	ac	6.52	3Booo2	R		9/52	10/57	

AA 1.4.46 on 2546(2546), 3/-, OSPR 9/46.

SD 20 (34/20) **& Part of SD 21** *Formby (West)** 10 by 12 Revision a 1925-38 c 1925-53 f 1969
1	10046	CPr		1946po	-	-	6.46	3BE1X1	R	0	5/46	-	
1	2546		OPr	1946po	-	-	6.46	3BE1X1	R	0	-	5/47	
2	B	CPt		1953pc	-	ac	6.53	3BooX2		1	10/53	-	
3	B	CPv	OPv	1953pc	-	ac	6.53	3BooX2		1	-	8/57	
3	B/	CPz	OPz	1953pc	-	ac	6.53	3BooX2		1			
3	B//	CPz		1953ms	1970ch	acf	6.53	3BooX2		3	-	-	1

AA 1.3.46 on 2546(2546), 3/-, OSPR 5/47. M on A, B. 1. Woodvale Airfield.

SD 26 (34/26) **& Part of SD 16** *Barrow-in-Furness (South East)* 15 by 10 Revision a 1910-32 c 1938-51
1	15046	CPr		1946po	-	-	6.46	3BE1X1	R		5/46	-	
1	2546		OPr	1946po	-	-	6.46	3BE1X1	R		-	7/46	
2	B	CPt		1952pc	-	ac	6.52	3Booo2	R		8/52	-	
3	B/	CPv.z		1952pc	1956mc	ac	6.52	3BooX2					

AA 1.3.46 on 2546(2546), 3/-, OSPR 7/46. M on A.

1	2	3	4	5	6	7	8	9	10	11	12	13	14

SD 27 (34/27) *Dalton-in-Furness**   10 by 10   Revision a 1910-32 c 1938-51
1 15046   CPr   1946po   -   -   6.46 3BE1X1   R   10/46 -
1 A   OPr   1946po   -   -   6.46 3BE1X1   R   -   1/47
2 B   CPt   OPv   1952pc   -   ac   6.52 3Booo2   R   10/52 10/57
3 B/   CP?w,x,z   1952mc   1960mc ac   6.52 3BooX2   R
AA 1.8.46 on A(A), 3/-, OSPR 1/47. M on A.

SD 28 (34/28) *Broughton in Furness*   10 by 10   Revision a 1911-23 c 1951 f 1966
1 10046   CPr   1946po   -   -   6.46 3BE1X1   R   9/46   -
1 2546   OPr   1946po   -   -   6.46 3BE1X1   R   -   11/46
2 B   CPt,u OPv   1953pc   -   ac   1.53 3Booo2   R   1/53   5/57
3 B   CPx   1953pc   -   ac   1.53 3Booo2   R
3 B/   CPz,0z OPz   1953mc   1967ch acf 1.53 3BooX2
AA 1.6.46 on 2546(2546), 3/-, OSPR 11/46. M on B.

SD 29 (34/29) *Torver**   10 by 10   Revision a 1897-1912 c 1950 d 1951
1 10046   CPr   1946po   -   -   1.47 3BE1X1   RB   11/46 -
1 A   OPr   1946po   -   -   1.47 3BE1X1   RB   -   1/47
1 B   CPs   OPv   1950pc   -   ac   1.50 3BEoo1   RB   -   5/57
3 C   CPx,y,z OPx   1961mc   -   ad   6.61 3oooX4   5/61   5/61
AA 1.8.46 on A(A), 3/-, OSPR 1/47.

SD 30 (34/30) *Formby (East)**   10 by 10   Revision a 1925-26 c 1937-53 f 1966
1 15046   CPr   1946po   -   -   1.47 3BE1X1   R   0   9/46   -
1 2546   OPr   1946po   -   -   1.47 3BE1X1   R   0   -   11/46
2 B   CPt   OPs   1953pc   -   ac   6.53 3BooX2   1   7/53   7/53
3 B/   CPv,x   1953pc   1958mc ac   6.53 3BooX2   1
3 B//   CPz   OPz   1953mc   1966ch acf 6.53 3BooX2   3   1
AA 1.7.46 on 2546(2546), 3/-, OSPR 11/46. M on A. 1. Woodvale Airfield.

SD 31 (34/31) **& Part of SD 21** *Southport**   12 by 10   Revision a 1925-27 c 1938-53 f 1966
1 30046   CPr   1946po   -   -   6.46 3BE1X1   R   000 7/46 -
1 5046   OPr   1946po   -   -   6.46 3BE1X1   R   000 -   11/46
3 B   CPt   1953pc   -   ac   6.53 3BooX4   100 12/53 -
3 B/   CPw,x OP?   1953pc   1958mc ac   6.53 3BooX4   100 -   2/58
3 B//   CPz   OPz   1953mc   1967ch acf 6.53 3BooX4   300   1
AA 1.6.46 on 5046(2546), 3/-, OSPR 11/46. M on B, B/. 1. Woodvale Airfield.

SD 32 (34/32) **& Part of SD 22** *Lytham St Anne's* 12 by 10 Revision a 1908-30 c 1937-51
1 20046   CPr   1946po   -   -   6.46 3BE1X1   R   6/46   -
1 2546   OPr   1946po   -   -   6.46 3BE1X1   R   -   9/46
3 B   CPu   OPs   1954pc   -   ac   6.54 3oooX2   11/54 1/55
3 B/   CPw.z   1954mc   1960mc ac   6.54 3oooX2
AA 1.4.46 on 2546(2546), 3/-, OSPR 11/46. M on B.

SD 33 (34/33) **& Part of SD 23** *Blackpool** 12 by 10 Revision a 1909-31 c 1937-51
1 50046   CPr   1946po   -   -   6.46 3BE1X1   R   11   6/46 -   1
1 5046   OPr   1946po   -   -   6.46 3BE1X1   R   11   -   11/46
3 B   CPu   1955pc   -   ac   6.55 3oooX2   31   1/55 -   2
3 B/   CPw,x,z.0z   1955pc   1958mc ac   6.55 3oooX2   31
AA 1.4.46 on 5046(5046), 3/-, OSPR 11/46. Showing the Blackpool & Fleetwood tramway. 1. Blackpool Aerodrome. 2. Blackpool (Squire's Gate) Aerodrome.

SD 34 (34/34) **& Part of SD 24** *Fleetwood* 12 by 10 Revision a 1930-31 c 1937-51
1 20046   CPr   1946po   -   -   6.46 3BE1X1   R   9/46   -
1 2546   OPr   1946po   -   -   6.46 3BE1X1   R   -   12/46
3 B   CPu   1954pc   -   ac   6.54 3oooX2   11/54 -
3 B/   CPv.y,z,z+   1954pc   1957mc ac   6.54 3oooX2
AA 1.6.46 on 2546(2546), 3/-, OSPR 12/46. Showing the Blackpool & Fleetwood tramway.

1	2	3	4	5	6	7	8	9	10	11	12	13	14

SD 35 (34/35) **& Part of SD 25** *Point of Lune** 12 by 10 Revision a 1910-31 c 1938-51
1	250/3/46 E	CPr		1946po			6.46	3oE1X1			4/46	-	
1	1046		OPr	1946po			6.46	3oE1X1			-	4/46	1
3	A	CPt		1946pc			6.46	3oEoo1					
3	B	CPt	OPv	1954pc		ac	6.54	3oooX2			11/54	3/57	
3	B/	CP?w,z		1954mc	1960mc	ac	6.54	3oooX2					

1. The outline edition appears also in OSPR 7/46.

SD 36 (34/36) *Yeoman Wharf**   10 by 10 Revision a 1889 b 1889-1931 c 1938-51
1	2546	CPr		1946po			6.46	3BE1X1			4/46	-	
1	1046		OPr	1946po			6.46	3BE1X1			-	4/46	1
2	A	CPs		1946pc	1952	-	1.52	3Booo2					
3	A	CPv		1946pc	1952	a	1.52	3BooX2					
3	B	CPx,z	OPx	1961mc	-	bc	6.61	3oooX2			3/61	2/61	

1. The outline edition also appears in OSPR 7/46.

SD 37 (34/37) *Cartmel*   10 by 10 Revision a 1910-32 c 1938-51 f 1970
1	15046	CPr		1946po			6.46	3BE1X1	RB	0	4/46	-	
1	2546		OPr	1946po			6.46	3BE1X1	RB	0	-	4/46	
2	B	CPt,u	OPv	1952pc	-	ac	6.52	3Booo2	R	1	9/52	10/57	
3	B	CPz		1952pc	-	ac	6.52	3Booo2	R	1			
3	B/	CPz	OPz	1952pc	-	ac	6.52	3Booo2		1			
3	B//	CPz,0z	OPz	1952^{c}ms	nd:ch	acf	6.52	3Booo2		3			1

AA 1.3.46 on 2546(2546), 3/-, OSPR 7/46. 1. Cark Airfield.

SD 38 (34/38) *Newby Bridge*   10 by 10 Revision a 1910-12 c 1938-51 f 1965; j 1963 tide lines
1	A	CPr	OPr	1947po			1.47	3BE1X1	R	4/47	7/47
2	B	CPt,u	OPv	1952pc	-	ac	6.52	3Booo2	R	9/52	5/57
3	B	CPx		1952pc	-	ac	6.52	3Booo2	R		
3	B/*	CPz	OPz	1952mc	1966rt	acfj	6.52	3Booo2		12/66	12/66

AA 1.9.46 on A(A), 3/-, OSPR 7/47. The base map sheet price is lacking.

SD 39 (34/39) *Hawkshead**   10 by 10 Revision a 1911-12 c 1938-51
1	A	CPr	OPr	1947po			1.48	3BE1X1		11/47	2/48
1	B	CPr		nd:po			1.48	3BEoo1			
2	C	CPt		1952pc	-	ac	1.52	3Booo2		6/52	-
3	C	CPu	OPv	1952pc	-	ac	1.52	3Booo2		-	5/57
3	C/	CPw,y,z		1952mc	1960mc	ac	1.52	3BooX2			

AA 1.8.47/NT 23.12.47 on A(A), 3/-, OSPR 2/48.

SD 40 (34/40) *Ormskirk**   10 by 10 Revision a 1925-26 c 1938-50 f 1969; j 1968 railway
1	15046	CPr		1946po			1.47	3BE1X1	RB	9/46	-	1
1	2546		OPr	1946po			1.47	3BE1X1	RB	-	12/46	
2	B	CPt,u	OPv	1953pc	-	ac	1.53	3BooX2	R	2/53	6/57	
3	B	CPw,z		1953pc	-	ac	1.53	3BooX2	R			
3	B/*	CPz	OPz	1953^{c}ms	nd:r	acfj	1.53	3BooX2		11/70	12/70	

AA 1.7.46 on 2546(2546), 3/-, OSPR 12/46. M on B. 1. Showing the Inland Sorting Depot (3240-3340).

SD 41 (34/41) *Bursclough**   10 by 10 Revision a 1926-27 c 1938-48 f 1965
1	10046	CPr		1946po			1.47	3BE1X1	R	0	11/46	-	
1	A		OPr	1946po			1.47	3BE1X1	R	0	-	1/47	
1	no code		OPs	1947pc			1.47	3BEoo1	R	0			1
2	B	CPt,u	OPs	1953pc	-	ac	6.53	3BooX2	R	1	2/53	1/55	
3	B	CPx		1953pc	-	ac	6.53	3BooX2	R	1			
3	B/	CPz	OPz	1953mc	1966ch	acf	6.53	3BooX2		3			

AA 1.8.46 on A(A), 3/-, OSPR 1/47; AA 1.6.51/NT 26.4.51 on no code(B), 3/-, OSPR 8/51. M on B. 1. Not recorded as an outline edition, but used as the base to the Administrative Areas edition B.

1	2	3	4	5	6	7	8	9	10	11	12	13	14

SD 42 (34/42) *Hesketh Bank*   10 by 10   Revision a 1908-30 c 1938-51 f 1962 g 1967
1	15046	CPr		1946po			6.46	3BE1X1	R	0	6/46	-	
1	2546		OPr	1946po			6.46	3BE1X1	R	0	-	9/46	
3	B	CPu	OPs	1955pc		ac	6.55	4oooX4		1	1/55	1/55	1
3	B/												
3	B//	CPy		1955mc	1963ch	acf	6.55	4oooX4		1			
3	B///	CPz	OPz	1955ms	1968ch	acg	6.55	4oooX4		1			

AA 1.4.46 on 2546(2546), 3/-, OSPR 9/46. M on B. 1. One of two sheets remade to four-colour specification.

SD 43 (34/43) *Kirkham Lancashire**   10 by 10   Revision a 1909-30 c 1937-51
1	10046	CPr		1946po			1.47	3BE1X1	R	0	11/46	-
1	A		OPr	1946po			1.47	3BE1X1	R	0	-	1/47
3	B	CPu	OPv	1954pc		ac	6.54	3oooX2		3	12/54	5/57
3	B/	CPw,x,z		1954mc	1960mc	ac	6.54	3oooX2		3		

AA 1.8.46 on A(A), 3/-, OSPR 1/47.

SD 44 (34/44) *Garstang*   10 by 10   Revision a 1910-30 c 1937-51
1	A			1947po			1.48	3BE1X1	R		9/47	-
1	B		OPr	1947po			1.48	3BE1X1	R		-	11/47
1	C	CPr		nd:po			6.48	3BEoo1	R			
3	D	CPu	OPv	1954pc		ac	6.54	3oooX2			12/54	10/57
3	D/	CP?w,x,y,z		1954mc	1959mc	ac	6.54	3oooX2				

AA 1.12.46 on B(A), 3/-, OSPR 11/47.

SD 45 (34/45) *Galgate**   10 by 10   Revision a 1910-31 c 1937-51 f 1962 g 1965
                                         j 1958 industrial
1	A	CPr	OPr	1947po			1.47	4BE1X3	R		11/47	1/48
1	B	CPr		nd:po			1.47	4BEoo3	R			
3	C	CPu	OPv	1954pc		ac	6.54	4oooX3			12/54	10/57
3	C/											
3	C//*	CPx	OPx	1954mc	1962r	acf	6.54	4oooX3			8/62	8/62
3	C//*/*	CPz	OPz	1954mc	1966ir	acgj	6.54	4oooX3			5/66	7/66

AA 1.9.47 on A(A), 3/-, OSPR 1/48.

SD 46 (34/46) *Lancaster**   10 by 10   Revision a 1910-31 c 1938-50 f 1962
1	A	CPr	OPr	1947po			1.48	4BE1X3	R	0	11/47	2/48
	B											
2	C	CPt,u		1952pc		ac	6.52	4Booo3	R	0	11/52	-
3	C/*	CPx,z	OPx	1952mc	1963r	acf	6.52	4BooX3		0	1/63	1/63

AA 1.9.47 on A(A), 3/-, OSPR 2/48. M on C.

SD 47 (34/47) *Grange-over-Sands*   10 by 10   Revision a 1910-32 c 1938-51
1	A	CPr	OPr	1947po			1.47	4BE1X3	RB		1/47	4/47
2	B	CPt,u	OPv	1952pc		ac	6.52	4Booo3	R		9/52	5/57
3	B/	CPx,z		1952mc	1961mc	ac	6.52	4BooX3				

AA 1.9.46 on A(A), 3/-, OSPR 4/47. M on B/.

SD 48 (34/48) *Milnthorpe*   10 by 10   Revision a 1896-1912 c 1951-55
1	A	CPr	OPr	1948po		6.48	4BEoo3	RB		5/48	7/48
3	A	CPt		1948pc		6.48	4BEoo3	RB			
3	B	CPw,x,y.z	OPv	1960mc	ac	6.60	4oooX3			4/60	4/60

AA 1.4.48 on A(A), 3/-, OSPR 7/48.

SD 49 (34/49) *Windermere*   10 by 10   Revision a 1911-12 c 1938-51 f 1971 g 1974
1	A	CPr	OPr	1948po			1.48	4BEoo3	RB		5/48	7/48
2	B	CPt		1952pc		ac	6.52	4Booo3	R		12/52	-
3	B/	CPv,x		1952pc	1958mc	ac	6.52	4BooX3				
3	B//*	CPz	OPz	1952^{c}ms	nd:r	acf	6.52	4BooX3			12/72	3/73
3	B//*/	CFz	OFz	1952^{c}ms	nd:ch	acg	6.52	4BooX3				

AA 1.4.48/NT 13.5.48 on A(A), 3/-, OSPR 7/48.

1	2	3	4	5	6	7	8	9	10	11	12	13	14

SD 50 (34/50) *Wigan**   10 by 10   Revision a 1925-39 c 1937-48 f 1963
1	A	CPr	OPr	1947po	-	-	1.47	3BE1X1	R		2/47	5/47	
2	B	CPt	OPs	1953pc	-	ac	6.53	3BooX2			8/53	12/53	
3	B	CPu.z		1953pc	-	ac	6.53	3BooX2					
3	B/*	CPz	OPz	1953mc	1964r	acf	6.53	3BooX2			8/64	10/64	

AA 1.11.46 on A(A), 3/-, OSPR 5/47. M on B.

SD 51 (34/51) *Chorley Lancashire*   10 by 10   Revision a 1927-28 c 1938-49 f 1964 g 1972
1	A	CPr	OPr	1947po	-	-	6.47	3BE1X1	R		2/47	5/47	
1	B	CPs		1950pc	-	ac	6.50	3BEoo1	R		12/50	-	
3	B	CPu	OPv	1950pc	-	ac	6.50	3BEoo1	R		-	2/57	
3	C	CPx	OPx	1961mc	-	ac	6.61	3oooX4			5/61	5/61	
3	C/*	CPz	OPz	1961mc	1965r	acf	6.61	3oooX4			5/65	4/65	
3	C/*/*	CPz	OPz	1961^{c}ms	nd:r	acg	6.61	3oooX4			12/72	6/73	

AA 1.12.46 on A(A), 3/-, OSPR 5/47. M on B.

SD 52 (34/52) *Preston (South)**   10 by 10   Revision a 1927-39 c 1938-51 f 1963 g 1967 h 1972
1	A	CPr	OPr	1947po	-	-	1.47	3BE1X1	R		2/47	5/47	
1	B	CPr		nd:po	-	-	6.50	3BEoo1	R				
3	C	CPu	OPv	1955pc	-	ac	6.55	3oooX2			2/55	2/57	
3	C/	CPw		1955mc	1959c	ac	6.55	3oooX2					
3	C//*	CPw	OPv	1955mc	1960r	ac	6.55	3oooX4				10/60	1
3	C//*/	CPy		1955mc	1963ch	acf	6.55	3oooX4					
3	C//*//	CPz	OPz	1955ms	1968ch	acg	6.55	3oooX4					
3	C//*///*	CPz	OPz	1955^{c}ms	nd:r	ach	6.55	3oooX4			9/73	10/73	

AA 1.11.46 on A(A), 3/-, OSPR 5/47. M on C. 1. OSPR (coloured edition) not found.

SD 53 (34/53) *Preston (North)**   10 by 10   Revision a 1910-30 c 1937-51 f 1962 g 1965
                                                                   j 1958 railway
1	A	CPr	OPr	1947po	-	-	1.47	3BE1X1	RB		2/47	5/47	
3	B	CPu	OPv	1955pc	-	ac	6.55	3oooX2			1/55	11/57	
3	B/	CPw		1955mc	1958mc	ac	6.55	3oooX2					
3	B//*	CPx	OPx	1955mc	1962r	acf	6.55	3oooX2			12/62	12/62	
3	B//*/*	CPz	OPz	1955mc	1966r	acgj	6.55	3oooX2			10/66	10/66	

AA 1.9.46 on A(A), 3/-, OSPR 5/47. M on B.

SD 54 (34/54) *Bleasdale*   10 by 10   Revision a 1907-30 c 1951 f 1968
1	15046	CPr		1946po	-	-	6.46	3BE1X1	R		6/46	-	
1	2546		OPr	1946po	-	-	6.46	3BE1X1	R		-	9/46	
3	B	CPu		1955pc	-	ac	6.55	3oooX2			1/55	-	
3	B/	CPv,x	OPv	1955pc	1956mc	ac	6.55	3oooX2			-	2/57	
3	B//*	CPz,0z	OPz	1955ms	1968r	acf	6.55	3oooX2			8/68	8/68	

AA 1.4.46 on 2546(2546), 3/-, OSPR 9/46.

SD 55 (34/55) *Dolphinholme*   10 by 10   Revision a 1889-1931 c 1938-51 f 1963 g 1966
1	10046	CPr		1946po	-	-	6.46	4BE1X3			8/46	-	
1	2546		OPr	1946po	-	-	6.46	4BE1X3			-	11/46	
3	B	CPu		1955pc	-	ac	6.55	4oooX3			1/55	-	
3	B/	CPv		1955pc	1957mc	ac	6.55	4oooX3					
3	B//	CPx		1955mc	1961mc	ac	6.55	4oooX3					
3	B///	CPz		1955mc	1964ch	acf	6.55	4oooX3					1
3	B////*	CPz,0z	OPz	1955mc	1967r	acg	6.55	4oooX3			5/67	4/67	

AA 1.5.46 on 2546(2546), 3/-, OSPR 11/46. 1. Showing the M6 under construction.

SD 56 (34/56) *Caton*   10 by 10   Revision a 1910-31 c 1938-51 f 1962 g 1963
1	A	CPr	OPr	1948po	-	-	6.48	4BE1X3	RB		3/48	5/48	
1	B	CPr		nd:po	-	-	6.48	4BEoo3	RB				
2	C	CPt		1952pc	-	ac	6.52	4Booo3	R		9/52	-	
3	C/	CPv	OPv	1952pc	1956mc	ac	6.52	4BooX3			-	2/57	
3	C//*	CPx	OPx	1952mc	1962r	acf	6.52	4BooX3			6/62	12/62	
3	C//*/*	CPz	OPz	1952mc	1965r	acg	6.52	4BooX3			10/65	10/65	

AA 1.1.48 on A(A), 3/-, OSPR 5/48.

1	2	3	4	5	6	7	8	9	10	11	12	13	14

SD 57 (34/57) *Burton Westmorland** 10 by 10 Revision a 1910-11 c 1938-51 f 1960 g 1970
1	A	CPr	OPr	1947po	-	-	6.47	4BE1X3	RB		9/47	11/47	
2	B	CPs,t,u	OPv	1952pc	-	ac	6.52	4Booo3	R		9/52	10/57	
3	B/*	CPx,y	OPx	1952mc	1961r	acf	6.52	4BooX3			9/61	9/61	
3	B/*/*	CPz,0z	OPz	1952^{c}ms	nd:r	acg	6.52	4BooX3			12/71	5/72	

AA 1.6.47 on A(A), 3/-, OSPR 11/47. NPRDBA 45 on B/*.

SD 58 (34/58) *Sedgwick** 10 by 10 Revision a 1910-12 c 1938-51 f 1971 g 1975
1	A	CPr	OPr	1947po	-	-	1.48	4BE1X3	RB		10/47	1/48	
1	B	CPr		nd:po	-	-	1.48	4BEoo3	RB				
3	C	CPw.z	OPv	1960mc	-	ac	6.60	4oooX4			11/60	1/61	
3	C/*	CPz	OPz	1960^{c}ms	nd:r	acf	6.60	4oooX4			9/72	12/72	
3	C/*/*	CFz		1960^{c}ms	nd:r	acg	6.60	4oooX4			9/76		

AA 1.8.47 on A(A), 3/-, OSPR 1/48. SS: Soil, outline, 1972 on C. NPRDBA 46 on C.

SD 59 (34/59) *Kendal** 10 by 10 Revision a 1911-12 c 1938-51 f 1971 g 1974
1	A	CPr	OPr	1947po	-	-	6.47	4BE2X3	RB		8/47	11/47	
1	B	CPr		nd:po	-	-	6.47	4BEoo3	RB				
2	C	CPt	OPv	1952pc	-	ac	6.52	4Booo3	R		11/52	10/57	
3	C/	CP?w.y		1952mc	1960mc	ac	6.52	4oooX3					
3	C//*	CPz	OPz	1952^{c}ms	nd:r	acf	6.52	4oooX3			5/72	8/72	
3	C//*/	CFz		1952^{c}ms	nd:ch	acg	6.52	4oooX3					

AA 1.2.47 on A(A), 3/-, OSPR 11/47. NPRDBA 47 on C/.

SD 60 (34/60) *Leigh (North) Lancashire* 10 by 10 Revision a 1926-28 c 1936-48
1	20046	CPr		1946po	-	-	6.46	3BE1X1	R		8/46	-	
1	2547		OPr	1947po	-	-	6.46	3BE1X1	R		-	1/47	
2	B	CPt,u	OPs	1953pc	-	ac	1.53	3BooX4	R		2/53	2/53	
3	B	CPy		1953pc	-	ac	1.53	3BooX4	R				
3	B/	CPz	OPz	1953pc	-	ac	1.53	3BooX4					

AA 1.6.46 on 2547(2547), 3/-, OSPR 1/47. M on B.

SD 61 (34/61) *Horwich* 10 by 10 Revision a 1927-28 c 1938-48
1	A	CPr	OPr	1947po	-	-	1.47	3BE1X1	R		3/47	6/47	
1	B	CPr		nd:po	-	-	1.47	3BEoo1	R				
3	C	CPu	OPs	1954pc	-	ac	6.54	3oooX2			10/54	1/55	
3	C/	CPw,x.z		1954mc	1959mc	ac	6.54	3oooX2					

AA 1.12.46 on A(A), 3/-, OSPR 6/47. M on C.

SD 62 (34/62) *Blackburn** 10 by 10 Revision a 1927-29 c 1937-51
1	20046	CPr		1946po	-	-	1.47	3BE1X1	R		9/46	-	
1	5047		OPr	1947po	-	-	1.47	3BE1X1	R		-	1/47	
3	B	CPu	OPv	1955pc	-	ac	6.55	3oooX2			3/55	11/57	
3	B/	CPw,y,z		1955mc	1959mc	ac	6.55	3oooX2					

AA 1.6.46 on 5047(2547) 3/-, OSPR 1/47. M on B.

SD 63 (34/63) *Longridge Lancashire** 10 by 10 Revision a 1928-30 c 1937-51 f 1970
1	10047	CPr		1947po	-	-	6.46	3BE1X1	R	0	12/46	-	
1	A		OPr	1947po	-	-	6.46	3BE1X1	R	0	-	5/47	
3	B	CPu	OPv	1954pc	-	ac	6.54	3oooX2		1	10/54	10/57	
3	B/	CPw.x,y,z	OPy	1954mc	1958mc	ac	6.54	3oooX2		1			
3	B//	CPz	OPz	1954^{c}ms	nd:ch	acf	6.54	3oooX2		3			1

AA 1.3.47 on A(A), 3/-, OSPR 5/47. M on B. 1. Samlesbury Aerodrome.

SD 64 (34/64) *Chipping* 10 by 10 Revision a 1906-30 c 1951
1	15046	CPr		1946po	-	-	6.46	3BE1X1	B		6/46	-	
1	2546		OPr	1946po	-	-	6.46	3BE1X1	B		-	10/46	
3	B	CPu	OPv	1954pc	-	ac	6.54	3oooX2			12/54	-	
3	B/	CPw		1954mc	1958mc	ac	6.54	3oooX2					
3	B//	CPx,z,0z	OPx	1954mc	1958mc	ac	6.54	3oooX2					

AA 1.4.46 on 2546(2546), 3/-, OSPR 10/46.

1	2	3	4	5	6	7	8	9	10	11	12	13	14

SD 65 (34/65) *Whitendale Fell*   10 by 10   Revision a 1907-10 c 1950-51
1	A	CPr	OPr	1948po	-	-	1.48	4BE3X3	B		12/47	2/48	
1	B	CPr		nd:po	-	-	1.48	4BEoo3	B				
2	B		OPs	1947pc	1949c	-	1.52	4Booo3	B				
3	C	CPu		1955pc	-	ac	6.55	4oooX3			1/55	-	
3	C/	CPv,x.x,0z	OPx.z	1955pc	1957mc	ac	6.55	4oooX3					

AA 1.11.47 on A(A), 3/-, OSPR 2/48.

SD 66 (34/66) *Bentham Yorkshire**   10 by 10   Revision a 1907-10 c 1950-51
1	A	CPr		1948po	-	-	1.48	4BE3X3	RB		1/48	-	
1	B		OPr	1948po	-	-	1.48	4BE3X3	RB		-	3/48	
1	B	CPr	OPr	nd:po	-	-	1.48	4BEoo3	RB				
3	C	CPw,y,z	OPx	1961mc	-	ac	6.61	4oooX4			12/60	2/61	

AA 1.11.47 on B(A), 3/-, OSPR 3/48.

SD 67 (34/67) *Kirkby Lonsdale*   10 by 10   Revision a 1889-1920 c 1938-51
1	A	CPr	OPr	1948po	-	-	1.48	4BE3X3	RB		2/48	4/48	
1	B	CPr		nd:po	-	-	1.48	4BEoo3	RB				
2	C	CPt,u,x.z	OPv	1952pc	-	ac	6.52	4Booo3	R		11/52	2/57	
3	C/	CPz	OPz	1952pc	-	ac	6.52	4Booo3					

AA 1.12.47 on A(A), 3/-, OSPR 4/48. NPRDBA 44 on C/.

SD 68 (34/68) *Middleton Fell*   10 by 10   Revision a 1907-13 c 1950-51
1	A	CPr	OPr	1947po	-	-	1.48	4BE1X3	RB		11/47	2/48	
3	B	CPr		nd:po	-	-	1.48	4BEoo3	RB				
3	C	CPv,x,z.z	OPv	1958pc	-	ac	6.58	4oooX3			5/58	7/58	

AA 1.8.47 on A(A), 3/-, OSPR 2/48.

SD 69 (34/69) *Sedbergh*   10 by 10   Revision a 1896-1912 c 1951 f 1972
1	A	CPr	OPr	1948po	-	-	6.48	4BE4X3	RB		3/48	4/48	
1	B	CPs		1948pc	1951c	a	1.51	4BEoo3	RB				
3	C	CPx,z	OPx	1961mc	-	ac	6.61	4oooX3			7/61	5/61	
3	C/*	CPz	OPz	1961^{c}ms	nd:r	acf	6.61	4oooX3			3/73	4/73	
3	C/*/	CFz		1961^{c}ms	nd:r	acf	6.61	4oooX3					

AA 1.1.48 on A(A), 3/-, OSPR 4/48.

SD 70 (34/70) *Bolton (South)**   10 by 10   Revision a 1926-32 c 1936-50 f 1959
1	A	CPr	OPr	1947po	-	-	1.47	3BE1X1	R		4/47	7/47	
1	B	CPr		nd:po	-	-	1.47	3BEoo1	R				
2	C	CPt	OPs	1952pc	-	ac	6.52	3Booo2	R		9/52	3/55	
3	C/	CPv		1952pc	1956mc	ac	6.52	3BooX2					
3	C//	CPx.x,z	OPx	1952mc	1961ch	acf	6.52	3BooX2					

AA 1.5.46 on A(A), 3/-, OSPR 7/47. M on B.

SD 71 (34/71) *Bolton (North)**   10 by 10   Revision a 1927-39 c 1938 d 1938-50 e 1947
1	A	CPr	OPr	1947po	-	-	1.47	3BE1X1	R		6/47	9/47	
1	B	CPr		nd:po	-	-	1.47	3BEoo1	R				
3	B	CPu,w,x	OPx	1950pc	-	-	1.47	3BEoo1	R				
3	B/	CPz	OPz	1950mc	1964ch	ad	1.47	3BooX2					

AA 1.10.46/NT 1.7.47 on A(A), 3/-, OSPR 9/47. M on B.

SD 72 (34/72) *Accrington**   10 by 10   Revision a 1927-29 c 1937-50 f 1971
1	15046	CPr		1946po	-	-	6.46	3BE1X1	RB		11/46	-	
1	A		OPr	1946po	-	-	1.47	3BE1X1	R		-	4/47	
3	B	CPu	OPs	1954pc	-	ac	6.54	3oooX2			12/54	7/55	
3	B/	CP?v.x,z		1954pc	1957c	ac	6.54	3oooX2					
3	B//*	CPz	OPz	1954^{c}ms	nd:r	acf	6.54	3oooX2			2/72	7/72	

AA 1.3.47 on A(A), 3/-, OSPR 4/47. M on B.

1	2	3	4	5	6	7	8	9	10	11	12	13	14

SD 73 (34/73) *Great Harwood**   10 by 10   Revision a 1907-30 c 1937-50
1	15047	CPr		1947po	-	-	6.46	3BE1X1	R		12/46	-	1
1	A		OPr	1947po	-	-	6.46	3BE1X1	R		-	5/47	
3	B	CPu	OPv	1955pc	-	ac	6.55	3oooX2			1/55	6/57	
3	B/	CPw,x,z	OPz	1955pc	1958c	ac	6.55	3oooX2					

AA 1.3.47 on A(A), 3/-, OSPR 5/47. 1. Showing Calderstones Institution, with its railway connection.

SD 74 (34/74) *Clitheroe*   10 by 10   Revision a 1883-1930 c 1929-50 f 1965 g 1970
1	15046	CPr		1946po	-	-	6.46	3BE1X1	RB		9/46	-	
1	2546		OPr	1946po	-	-	6.46	3BE1X1	RB		-	12/46	
3	B	CPt		1954pc	-	ac	6.54	3oooX2			9/54	-	
3	B/	CPv,x		1954pc	1957mc	ac	6.54	3oooX2					
3	B//	CPz	OPz	1954mc	1966ch	acf	6.54	3oooX2					
3	B///*	CPz,0z	OPz	1954^{c}ms	nd:r	acg	6.54	3oooX2			5/72	8/72	

AA 1.7.46 on 2546(2546), 3/-, OSPR 12/46.

SD 75 (34/75) *Slaidburn*   10 by 10   Revision a 1907 c 1938-50 f 1969
1	A	CPr	OPr	1948po	-	-	6.48	4BEoo3	B		6/48	7/48	
3	B	CPu		1954pc	-	ac	6.54	4oooX3			11/54	-	
3	B/	CP?w,y		1954pc	1958mc	ac	6.54	4oooX3					
3	B//	CPz,0z	OPz	1954ms	1970ch	acf	6.54	4oooX3					

AA 1.4.48 on A(A), 3/-, OSPR 7/48.

SD 76 (34/76) *Clapham Yorkshire*   10 by 10   Revision a 1907 c 1950
1	15046	CPr		1946po	-	-	6.46	4BE1X3	RB		4/46	-	
1	2546		OPr	1946po	-	-	6.46	4BE1X3	RB		-	4/46	1
2	B	CPt,u	OPv	1952pc	-	ac	1.52	4Booo3	R		6/52	10/57	
3	B	CPx,z		1952pc	-	ac	1.52	4Booo3	R				
3	B/	CPz,0z	OPz	1952pc	-	ac	1.52	4Booo3					

AA 1.3.46 on 2546(2546), 3/-, OSPR 7/46. NPRDBA 42 on B. 1. Also listed in OSPR 7/46.

SD 77 (34/77) *Ingleborough Hill*   10 by 10   Revision a 1907-28 c 1938-51 d 1950
1	10046	CPr		1946po	-	-	6.46	4BE1X3	RB		10/46	-	
1	A		OPr	1946po	-	-	6.46	4BE1X3	RB		-	1/47	
1	B	CPr		nd:po	-	-	6.46	4BEoo3	RB				
3	B	CPu		1950po	-	-	6.50	4BEoo3	RB				
3	B/	CPw		1950mc	1958mc	ac	6.46	4BooX3	B				
3	C	CPw,x.x,y,z	OPv	1959mc	-	ad	6.59	4oooX4			8/59	8/59	

AA 1.7.46 on A(A), 3/-, OSPR 1/47. NPRDBA 43 on C.

SD 78 (34/78) *Dent*   10 by 10   Revision a 1889-1928 c 1950
1	A	CPr	OPr	1947po	-	-	6.47	4BE1X3	RB		10/47	1/48	
1	B	CPr		nd:po	-	-	6.47	4BEoo3	RB				
3	C	CPw,x,z.0z	OPv	1960mc	-	ac	6.60	4oooX4			11/60	11/60	

AA 1.7.47 on A(A), 3/-, OSPR 1/48.

SD 79 (34/79) *Baugh Fell*   10 by 10   Revision a 1907-13 c 1950
1	A	CPr	OPr	1947po	-	-	1.48	4BE1X3	RB		11/47	2/48	
1	B	CPr		nd:po	-	-	1.48	4BEoo3	RB				
3	C	CPv,x,z.x,0z	OP?	1958pc	-	ac	6.58	4oooX3			3/58	3/58	

AA 1.9.47 on A(A), 3/-, OSPR 2/48.

SD 80 (34/80) *Middleton Lancashire**   10 by 10   Revision a 1927-33 c 1936-49 f 1973
1	15047	CPr		1947po	-	-	1.47	3BE1X1	RB		1/47	-	
1	A		OPr	1947po	-	-	1.47	3BE1X1	RB		-	5/47	
1	B	CPr		1950pc	-	ac	6.50	3BEoo1	RB				
3	B		OPs	1950pc	-	ac	6.50	3BEoo1	RB		-	5/56	
3	B/	CPx,z	OPv	1950mc	1960mc	ac	6.50	3BooX2	B				
3	B//*	CPz	OPz	1950^{c}ms	nd:r	acf	6.50	3BooX2	B		1/74	3/74	

AA 1.10.46 on A(A), 3/-, OSPR 5/47. M on B, B/.

133

1	2	3	4	5	6	7	8	9	10	11	12	13	14

SD 81 (34/81) *Bury*   10 by 10   Revision a 1905-33 c 1938 d 1938-49 f 1963 g 1974
                                             j 1981 selected k SUSI

1	A	CPr	OPr	1947po	-	-	1.47	3BE1X1	R		3/47	6/47
1	B	CPr		nd:po	-	-	1.47	3BEoo1	R			
	C											
3	D	CPr		nd:po	-	-	1.47	3BEoo1	R			
3	D/	CPw,x		1950mc	1959mc	ac	1.52	3BooX2				
3	D//	CPz	OPz	1950mc	1964ch	adf	1.52	3BooX2				
3	D///*	CFz	OFz	1950^{c}ms	nd:r	adg	1.52	3BooX2			4/76	4/76
3	D///*/*	CFz	OFz	1982^{c}m-s	nd:br	adgjk	1.52	3BooX2			6/82	6/82

AA 1.2.46 on A(A), 3/-, OSPR 6/47. M on D.

SD 82 (34/82) *Rawtenstall**   10 by 10   Revision a 1905-28 c 1938-50 f 1970

1	20046			1946po	-	-	6.46	3BE1X1	RB		9/46	-
1	2546		OPr	1946po	-	-	6.46	3BE1X1	RB		-	11/46
3	B	CPu	OPv	1954pc	-	ac	6.54	3oooX2			10/54	3/55
3	B/		OP?								-	2/57
3	B//	CP?w,x		1954mc	1960c	ac	6.54	3oooX2				
3	B///	CPz	OPz,Oz	1954^{c}ms	nd:ch	acf	6.54	3oooX2				

AA 1.6.46 on 2546(2546), 3/-, OSPR 11/46. M on B.

SD 83 (34/83) *Burnley**   10 by 10   Revision a 1905-30 c 1938-50 f 1964

1	50046	CPr		1946po	-	-	6.46	3BE1X1	R		7/46	-
1	5046		OPr	1946po	-	-	6.46	3BE1X1	R		-	11/46
3	B	CPu	OPv	1954pc	-	ac	6.54	3oooX2			12/54	11/57
3	B/	CP?w,x	OPz	1954mc	1959mc	ac	6.54	3oooX2				
3	B//	CPz	OPz	1954mc	1965ch	acf	6.54	3oooX2				

AA 1.4.46 on 5046(2546), 3/-, OSPR 11/46.

SD 84 (34/84) *Barnoldswick*   10 by 10   Revision a 1884-1930 c 1938-50

1	10046	CPr		1946po	-	-	1.47	3BE1X1	RB		12/46	-
1	A		OPr	1946po	-	-	1.47	3BE1X1	RB		-	4/47
3	B	CPu		1954pc	-	ac	6.54	3oooX2			10/54	-
3	B/	CP?w,x,z OPz		1954pc	1958mc	ac	6.54	3oooX2				

AA 1.10.46 on A(A), 3/-, OSPR 4/47.

SD 85 (34/85) *Hellifield*   10 by 10   Revision a 1907 c 1938-50 f 1974

1	A	CPr	OPr	1947po	-	-	1.48	4BE1X3	RB		11/47	2/48
1	B	CPr		nd:po	-	-	1.48	4BEoo3	RB			
3	C	CPu	OPv	1954pc	-	ac	6.54	4oooX3			10/54	6/57
3	C/	CPv,y,z.z		1954pc	1958mc	ac	6.54	4oooX3				
3	C//	CFz		1954^{c}ms	nd:ch	acf	6.54	4oooX3				

AA 1.9.47 on A(A), 3/-, OSPR 2/48. NPRDBA 40 on C/.

SD 86 (34/86) *Settle*   10 by 10   Revision a 1907 c 1950 f 1974

1	A	CPr	OPr	1947po	-	-	1.48	4BE1X3	RB		11/47	2/48
	B											
	C											
3	D	CPt,v	OPv	1954pc	-	ac	6.54	4BooX3			4/54	10/56
3	D/	CPw.y,z		1954mc	1960mc	ac	6.54	4BooX3				
3	D//	CFz		1954^{c}ms	nd:ch	acf	6.54	4oooX3				

AA 1.9.47/NT 4.12.47 on A(A), 3/-, OSPR 2/48. NPRDBA 41 on D/.

SD 87 (34/87) *Horton in Ribblesdale*   10 by 10   Revision a 1907 c 1950

1	A	CPr	OPr	1948po	-	-	1.48	4BE3X3	RB		2/48	3/48
1	B	CPs		1948pc	1951c	a	1.51	4BEoo3	RB			
3	B/	CPv		1948pc	1957c	a	1.51	4BooX3	B			
3	C	CPv,x,z,Oz	OPv	1957pc	-	ac	6.57	4oooX3			9/57	9/57

AA 1.11.47 on A(A), 3/-, OSPR 3/48.

1	2	3	4	5	6	7	8	9	10	11	12	13	14

SD 88 (34/88) *Hawes*   10 by 10 Revision a 1907-10 c 1950 f 1974
1	A	CPr	OPr	1947po	-	-	1.48	4BE1X3	RB		9/47	11/47	1
2	B	CPs,t,x	OPv	1952pc	-	ac	1.52	4Booo3	R		5/52	10/57	2
3	B	CPz		1952pc	-	ac	1.52	4Booo3	R				
3	B/	CPz	OPz	1952pc	-	ac	1.52	4Booo3					
3	B//	CFz		1952^{c}ms	nd:ch	acf	1.52	4Booo3					

AA 1.10.47 on A(A), 3/-, OSPR 11/47. 1. The railway east of Hawes is incorrectly marked LMS. 2. The railway is corrected to LNER.

SD 89 (34/89) *Thwaite Yorkshire*   10 by 10 Revision a 1907-13 c 1950
1	A	CPr	OPr	1947po	-	-	1.48	4BE1X3	RB		11/47	2/48
1	B	CPr		nd:po	-	-	1.48	4BEoo3	RB			
3	C	CPw,x,z.0z	OPv	1958pc	-	ac	6.58	4oooX4			9/58	10/58

AA 1.9.47 on A(A), 3/-, OSPR 2/48.

SD 90 (34/90) *Oldham*   10 by 10 Revision a 1904-33 c 1932-49 f 1971
1	A	CPr	OPr	1947po	-	-	6.47	3BE1X1	R		8/47	11/47
1	B											
1	C	CPs		1951pc	-	ac	1.51	3BEoo1	R		6/51	-
2	C	CPs		1951pc	-	ac	1.51	3BEoo1	R			
2	C		OPs	1951pc	-	ac	1.51	3BooX1	R		-	11/53
3	C/	CP?w,x	OPx	1951mc	1960mc	ac	1.51	3BooX2				
3	C//	CPx,z	OPz	1951mc	1960mc	ac	1.51	3BooX2				
3	C///*	CPz	OPz,0z	1951^{c}ms	nd:r	acf	1.51	3BooX2			6/72	8/72

AA 1.5.46/NT 1.9.47 on A(A), 3/-, OSPR 11/47. M on C.

SD 91 (34/91) *Littleborough Lancashire*   10 by 10 Revision a 1904-29 c 1936-53 f 1972
1	A	CPr	OPr	1947po	-	-	1.47	3BE1X1	R		3/47	5/47
1	B	CPr		nd:po	-	-	1.47	3BEoo1	R			
3	C	CPu	OPv	1954pc	-	ac	6.54	3oooX2			12/54	11/57
3	C/	CPw		1954mc	1959c	ac	6.54	3oooX2				
3	C//	CPw,z		1954mc	1961c	ac	6.54	3oooX2				
3	C///*	CPz,0z	OPz	1954^{c}ms	nd:r	acf	6.54	3oooX2			10/73	11/73

AA 1.12.46 on A(A), 3/-, OSPR 5/47. M on C.

SD 92 (34/92) *Todmorden*   10 by 10 Revision a 1904-34 c 1938-50
1	15046	CPr		1946po	-	-	6.46	3BE1X1	R		7/46	-	
1	2546		OPr	1946po	-	-	6.46	3BE1X1	R		-	10/46	
3	B	CPt	OPs	1954pc	-	ac	6.54	3oooX2			8/54	10/54	
3	B/	CPw,x,z.z	OPv	1954mc	1960mc	ac	6.54	3oooX2					
3	B/	G0z		1954mc	1960mc	ac	6.54	3oooX2			5/77	-	1

AA 1.5.46 on 2546(2546), 3/-, OSPR 10/46. M on B, B/. 1. Overprinted in green with public rights of way information to February 1976.

SD 93 (34/93) *Trawden*   10 by 10 Revision a 1890-1934 c 1938-50
1	10046	CPr		1946po	-	-	1.47	3BE1X1	B		12/46	-
1	A		OPr	1946po	-	-	1.47	3BE1X1	B		-	4/47
3	B	CPu	OPv	1954pc	-	ac	6.54	3oooX2			12/54	2/57
3	B/	CPw,x,z		1954mc	1960mc	ac	6.54	3oooX2				

AA 1.9.46 on A(A), 3/-, OSPR 4/47.

SD 94 (34/94) *Earby*   10 by 10 Revision a 1891-1935 c 1910-50
1	20046	CPr		1946po	-	-	6.46	3BE1X1	RB		9/46	-
1	A		OPr	1946po	-	-	6.46	3BE1X1	RB		-	1/47
3	B	CPu		1954pc	-	ac	6.54	3oooX2			10/54	-
3	B/	CPv,x,z	OPv	1954pc	1958mc	ac	6.54	3oooX2				

AA 1.7.46 on A(A), 3/-, OSPR 1/47.

1	2	3	4	5	6	7	8	9	10	11	12	13	14

SD 95 (34/95) *Skipton*   10 by 10   Revision a 1907 c 1938-50 f 1974; j 1981 selected

1	A	CPr	OPr	1948po	-	-	6.48	4BE3X3	RB		4/48	5/48	
1	B	CPr		nd:po	-	-	6.48	4BEoo3	RB				
3	C	CPu	OPv	1954pc	-	ac	6.54	4oooX3			12/54	1/57	
3	C/	CPv,x,z		1954pc	1958mc	ac	6.54	4oooX3					
3	C//	CFz		1954ᶜms	nd:ch	acf	6.54	4oooX3					
3	C///*	CFz	OFz	1981ᶜms	nd:r	acfj	6.54	4oooX3			1/82	1/82	1

AA 1.3.48 on A(A), 3/-, OSPR 5/48. NPRDBA 39 on C/. 1. Showing the Skipton bypass under construction.

SD 96 (34/96) *Conistone*   10 by 10   Revision a 1907 c 1950

1	A	CPr		1948po	-	-	1.48	4BEoo3	RB		6/48	-	
1	B		OPr	1948po	-	-	1.48	4BEoo3	RB		-	7/48	
1	B	CPr		nd:po	-	-	1.48	4BEoo3	RB				
1	C	CPs,t		1948pc	1950c	a	6.50	4BEoo3	RB				
3	D	CPv,x,z	OPv	1957pc	-	ac	6.57	4oooX3			1/58	11/57	

AA 1.4.48 on B(A), 3/-, OSPR 7/48.

SD 97 (34/97) *Buckden Yorkshire*   10 by 10   Revision a 1907-10 c 1950 f 1974

1	15046	CPr		1946po	-	-	1.47	4BE1X3	B		7/46	-	
1	2546		OPr	1946po	-	-	1.47	4BE1X3	B		-	9/46	
2	B	CPs		1952pc	-	ac	1.52	4Booo3			5/52	-	
3	B	CPv	OPv	1952pc	-	ac	1.52	4BooX3			-	1/57	
3	B/	CPw,x,z		1952mc	1960mc	ac	1.52	4BooX3					
3	B//	CFz		1952ᶜms	nd:ch	acf	1.52	4oooX3					

AA 1.5.46 on 2546(2546), 3/-, OSPR 9/46.

SD 98 (34/98) *Stalling Busk*   10 by 10   Revision a 1907-29 c 1950

| 1 | A | CPr | OPr | 1948po | - | - | 1.48 | 4BE3X3 | RB | | 2/48 | 5/48 | |
| 3 | B | CPw.x,z,0z | OPv | 1960mc | - | ac | 6.60 | 4oooX3 | | | 5/60 | 4/60 | |

AA 1.12.47 on A(A), 3/-, OSPR 5/48.

SD 99 (34/99) *Askrigg*   10 by 10   Revision a 1910-11 c 1950

1	A	CPr	OPr	1948po	-	-	6.48	4BEoo3	RB		7/48	8/48	
1	B	CPs		1948pc	1950c	a	6.50	4BEoo3	RB				
3	C	CPw,z,0z	OPv	1960mc	-	ac	6.60	4oooX3			10/60	10/60	

AA 1.5.48 on A(A), 3/-, OSPR 8/48.

SE 00 (44/00) *Saddleworth Moor*   10 by 10   Revision a 1889-1930 c 1938-52 f 1967 g 1971
j 1966 reservoir

1	A	CPr	OPr	1947po	-	-	1.47	4BE1X3	R		1/47	5/47	
1	B	CPr		nd:po	-	-	1.47	4BEoo3	R				
3	C	CPu,x.z	OPv.y,z	1955pc	-	ac	6.55	4oooX3			5/55	11/56	
3	C/*	CPz	OPz	1955mc	1968v	acfj	6.55	4oooX3			1/68	1/68	
3	C/*/	CPz,0z	OPz,0z	1955ᶜms	nd:ch	acgj	6.55	4oooX3			6/72		1

AA 1.9.46 on A(A), 3/-, OSPR 5/47. M on C. 1. Dover Stone Reservoir. OSPR (outline edition) not found.

SE 01 (44/01) *Marsden*   10 by 10   Revision a 1904-30 c 1938-51 f 1970 g 1978
j 1958 reservoir

1	A	CPr	OPr	1947po	-	-	6.47	4BE1X3	R		10/47	11/47	
2	B	CPt		1947pc	1953rpb	-	6.53	4BooX3					
3	C	CPu	OPv	1955pc	-	ac	6.55	4oooX3			4/55	9/57	
3	C/	CPw,x,z	OPv	1955pc	1958mc	ac	6.55	4oooX3					
3	C//*	CPz	OPz	1955ᶜms	nd:rv	acfj	6.55	4oooX3			6/72	8/72	1
3	C//*/*	CFz	OFz	1979ᶜms	nd:r	acgj	6.55	4oooX3			2/80	2/80	

AA 1.4.47 on A(A), 3/-, OSPR 11/47. M on C, C/. 1. Baitings Reservoir.

SE 02 (44/02) *Halifax*   10 by 10   Revision a 1904-34 c 1938-48 d 1938-51

1	A	CPr	OPr	1947po	-	-	6.47	4BE1X3	RB		10/47	1/48	
1	B	CPs		1951pc	-	ac	1.51	4BEoo3	RB		11/51	-	
3	C	CPu	OPv	1955pc	-	ad	6.55	4oooX4			5/55	10/57	
3	C/	CPw,x,z		1955pc	1958mc	ad	6.55	4oooX4					

AA 1.6.47 on A(A), 3/-, OSPR 1/48. M on B.

1	2	3	4	5	6	7	8	9	10	11	12	13	14

SE 03 (44/03) *Denholme*   10 by 10   Revision a 1932-34 c 1938 d 1938-50
1 A   CPr   OPr   1947po   -   -   1.48   4BE1X3   RB   9/47   11/47
1 B   CPs   1950pc   -   ac   6.50   4BEoo3   RB   11/50   -
2 C   CPt   1952pc   -   ad   6.52   4Booo3   R   11/52   -
3 C/   CPv,x,z,0z   1952pc   1957mc ad 6.52 4BooX3
AA 1.6.47 on A(A), 3/-, OSPR 11/47. M on B.

SE 04 (44/04) *Keighley*   10 by 10   Revision a 1907-35 c 1938-50
1 A   CPr   OPr   1948po   -   -   6.48   4BE1X3   RB   5/48   6/48
1 B   CPr   nd:po   -   -   6.48   4BEoo3   RB
2 C   CPt   OPv   1952pc   -   ac   6.52   4Booo4   R   11/52   9/57
3 C/   CPv,y,z   1952pc   1958mc ac 6.52 4BooX4
AA 1.3.48/NT 22.4.48 on A(A), 3/-, OSPR 6/48. M on C. NPRDBA 37 on C/.

SE 05 (44/05) *Bolton Abbey*   10 by 10   Revision a 1906-07 c 1938 d 1938-50 f 1975
1 A   CPr   1947po   -   -   6.47   4BE1X3   RB   9/47   -
1 B   OPr   1947po   -   -   6.47   4BE1X3   RB   -   11/47
1 C   CPr   nd:po   -   -   6.47   4BEoo3   RB
1 D   CPs   OPs   1947pc   1951c   ac 1.51 4BEoo3   RB
3 E   CPv   OPv   1957pc   -   ad   6.57   4oooX3   4/57   6/57
3 E/   CPw,z   1957pc   1958mc ad 6.57 4oooX3
3 E//* CFz   OFz   1957^{c}ms nd:ch adf 6.57 4oooX3   12/75   11/75
AA 1.1.47 on B(A), 3/-, OSPR 11/47. M on E, E/. NPRDBA 38 on E/.

SE 06 (44/06) *Grassington*   10 by 10   Revision a 1907 c 1950 f 1974
1 A   CPr   1948po   -   -   6.48   4BE1X3   B   3/48   -
1 B   OPr   1948po   -   -   6.48   4BE1X3   B   -   4/48
1 C   CPr   nd:po   -   -   6.48   4BEoo3   B
2 D   CPt   1952pc   -   ac   6.52   4Booo3   6/52   -
2 D   OPs   1952pc   -   ac   6.52   4BooX3   -   11/53
3 D/   CPv   1952pc   1956mc ac 6.52 4BooX3
3 D// CPw,y,z   1952mc   1960mc ac 6.52 4BooX3
3 D/// CFz   1952^{c}ms nd:ch acf 6.52 4BooX3
AA 1.1.48 on B(A), 3/-, OSPR 4/48.

SE 07 (44/07) *Middlesmoor*   10 by 10   Revision a 1907-27 c 1950 f 1974
1 A   CPr   OPr   1948po   -   -   6.48   4BEoo3   B   6/48   7/48
3 B   CPv,x,z OPv   1957pc   -   ac   6.57   4oooX3   2/58   2/58
3 B/   CFz   1957^{c}ms nd:ch acf 6.57 4oooX3
AA 1.4.48 on A(A), 3/-, OSPR 7/48.

SE 08 (44/08) *West Witton*   10 by 10   Revision a 1907-30 c 1950 f 1961
1 A   CPr   OPr   1948po   -   -   1.48   4BE1X3   RB   2/48   6/48
1 B   CPr   nd:po   -   -   1.48   4BEoo3   RB
2 B   CPu   1949pc   -   -   1.48   4Booo3   RB
3 C   CPx,z,0z OPx 1961mc   -   acf 6.61   4oooX3   8/61   7/61
AA 1.4.48 on A(A), 3/-, OSPR 6/48.

SE 09 (44/09) *Reeth*   10 by 10   Revision a 1910-28 c 1950
1 A   CPr   OPr   1948po   -   -   6.48   4BEoo3   RB   6/48   7/48
3 B   CPx,y,z,0z OPx 1961mc   -   ac   6.61   4oooX4   5/61   6/61
AA 1.4.48 on A(A), 3/-, OSPR 7/48. M on A.

SE 10 (44/10) *Holmfirth*   10 by 10   Revision a 1903-29 c 1938-51 f 1967; j 1959 reservoir
1 A   CPr   OPr   1947po   -   -   6.47   4BE1X3   RB   7/47   11/47
1 B   CPr   nd:po   -   -   6.47   4BEoo3   RB
3 C   CPu   OPv   1955pc   -   ac   6.55   4oooX3   6/55   10/57
3 C/   CP?w,x,z OPv 1955mc   1960mc ac 6.55 4oooX3
3 C//* CPz   OPz   1955ms   1968v   acfj 6.55 4oooX3   9/68   9/68
3 C//*/ CPz,0z OPz   1955^{c}ms nd:ch acfj 6.55 4oooX3   7/72   1
AA 1.12.46 on A(A), 3/-, OSPR 11/47. M on C, C/. 1. OSPR (outline edition) not found.

1	2	3	4	5	6	7	8	9	10	11	12	13	14

SE 11 (44/11) *Huddersfield* 10 by 10 Revision a 1929-31 c 1938-51 f 1961 g 1973
1	A	CPr	OPr	1947po	-	-	6.47	4BE1X3	R	0	6/47	8/47	
1	B	CPr		nd:po	-	-	6.47	4BEoo3	R	0			
3	C	CPu	OPv	1955pc	-	ac	6.55	4oooX3		0	5/55	7/56	
3	C/	CPw		1955mc	1958mc	ac	6.55	4oooX3		0			
3	C//	CPx,z		1955mc	1963ch	acf	6.55	4oooX3		0			
3	C///*	CFz	OFz	1955^{c}ms	nd:r	acg	6.55	4oooX3		0	3/74	6/74	

AA 1.1.47 on A(A), 3/-, OSPR 8/47. M on B.

SE 12 (44/12) *Brighouse* 10 by 10 Revision a 1930-38 c 1938-51 f 1974
1	20046	CPr		1946po	-	-	1.47	4BE1X3	R		11/46	-	
1	A		OPr	1946po	-	-	1.47	4BE1X3	R		-	2/47	
3	B	CPu,v	OPv	1955pc	-	ac	6.55	4oooX3			4/55	9/57	
3	B	CPx,z		1955mc	-	ac	6.55	4oooX3					
3	B/*	CFz	OFz	1955^{c}ms	nd:r	acf	6.55	4oooX3			3/74	7/74	

AA 1.7.46 on A(A), 3/-, OSPR 2/47. M on A.

SE 13 (44/13) *Bradford** 10 by 10 Revision a 1932-34 c 1938 d 1938-50 f 1971
1	A	CPr	OPr	1947po	-	-	6.47	4BE2X3	R		8/47	11/47	
1	B	CPr		nd:po	-	-	6.47	4BEoo3	R				
1	C	CPs		1947pc	1950c	ac	6.47	4BEoo3	R				M
2	D	CPt	OPv	1952pc	-	ad	6.52	4Booo3			12/52	11/56	
3	D/	CPw,x,z		1952pc	1958mc	ad	6.52	4BooX3					
3	D//*	CPz	OPz	1952^{c}ms	nd:r	adf	6.52	4BooX3			1/72	4/72	

AA 1.3.47 on A(A), 3/-, OSPR 11/47. M on C.

SE 14 (44/14) *Ilkley* 10 by 10 Revision a 1906-34 c 1938-50
1	A	CPr	OPr	1947po	-	-	6.47	4BE1X3	RB		6/47	10/47	
1	B	CPr		nd:po	-	-	6.47	4BEoo3	RB				
2	C	CPt,u	OPs	1952pc	-	ac	6.52	4Booo3	R		12/52	11/55	
3	C/	CPw,x,z.z,0z		1952mc	1959mc	ac	6.52	4BooX3					

AA 1.2.47 on A(A), 3/-, OSPR 10/47. M on C, C/. NPRDBA 36 on C/.

SE 15 (44/15) *Blubberhouses* 10 by 10 Revision a 1906-07 c 1938-50 f 1966; j 1966 reservoir
1	A	CPr	OPr	1947po	-	-	1.48	4BE1X4	B		11/47	2/48	1
2	B	CPt,u	OPs	1952pc	-	ac	6.52	4Booo4			9/52	10/52	
3	B/	CPw,y		1952mc	1960mc	ac	6.52	4BooX4					
3	B//*	CPz,0z	OPz	1952mc	1967v	acfj	6.52	4BooX4			12/66	12/66	2

AA 1.9.47 on A(A), 3/-, OSPR 2/48. M on B, B/. NPRDBA 34 on B//*. 1. A prototype sheet with Times Roman figures in the frame. Most yard values lack commas. It was not used again until 1950. 2. Thruscross Reservoir.

SE 16 (44/16) *Pateley Bridge* 10 by 10 Revision a 1907-08 c 1951
1	A	CPr	OPr	1948po	-	-	6.48	4BEoo3	RB		5/48	7/48	
1	B	CPr		nd:po	-	-	6.48	4BEoo3	RB				
2	C	CPt,u	OPv	1952pc	-	ac	6.52	4Booo3	R		9/52	9/57	
3	C/	CPw.z,0z		1952mc	1960mc	ac	6.52	4BooX3					

AA 1.4.48 on A(A), 3/-, OSPR 7/48. NPRDBA 33 on C/.

SE 17 (44/17) *Ramsgill** 10 by 10 Revision a 1907-28 c 1950-51
1	A	CPr	OPr	1948po	-	-	1.48	4BE1X3	B		2/48	3/48	
1	B	CPr		nd:po	-	-	1.48	4BEoo3	B				
1	B	CPs		1948pc	1951	-	1.51	4BEoo3	B				
3	C	CPw,y,z,0z	OPv	1959mc	-	ac	6.59	4oooX3			11/59	11/59	

AA 1.11.47 on A(A), 3/-, OSPR 3/48.

SE 18 (44/18) *Middleham* 10 by 10 Revision a 1910-30 c 1938-51
1	A	CPr	OPr	1948po	-	-	1.48	4BE1X3	RB		1/48	3/48	
1	B	CPr		nd:po	-	-	1.48	4BEoo3	RB				
3	B/	CPw		1948mc	1958mc	a	1.48	4BooX3	B				
3	C	CPw,z,0z	OPv	1960mc	-	ac	6.60	4oooX3			9/60	5/60	

AA 1.10.47/NT 2.2.48 on A(A), 3/-, OSPR 3/48. NPRDBA 29 on C.

1	2	3	4	5	6	7	8	9	10	11	12	13	14

SE 19 (44/19) *Catterick Camp*   10 by 10   Revision a 1910-32 c 1938-51
1 A   CPr   OPr   1948po   -   -   6.48   4BEoo3   RB   7/48   8/48   1
2 B   CPt   OPv   1952pc   -   ac   6.52   4Booo3   R   9/52   9/57
3 B/   CPw,z,0z   1952mc   1958mc   ac   6.52   4BooX3
AA 1.5.48 on A(A), 3/-, OSPR 8/48. M on A, B, B/. NPRDBA 28 on B/. 1. Camp features shown include the Signal Training Centre, barracks, ranges and hospitals. The camp railway system is shown, as LNER property.

SE 20 (44/20) [Penistone]   10 by 10   Revision a 1929-30 c 1938-51
1 A   CPr   1948po   -   -   1.48   4BE1X3   RB   2/48   -
1 B   OPr   1948po   -   -   1.48   4BE1X3   RB   -   6/48
1 C   CPr,t   nd:po   -   -   1.48   4BEoo3   RB   1
3 D   CPu   OPs   1955pc   -   ac   6.55   4oooX3   4/55   1/56
3 D/   CPw,x,z   OPz   1955mc   1958mc   ac   6.55   4oooX3
AA 1.12.47 on B(A), 3/-, OSPR 6/48. M on D, D/. 1. Some copies have the erroneous "t" price state "2/6 net paper flat, 6/- net Mounted & folded".

SE 21 (44/21) [Horbury]   10 by 10   Revision a 1929-31 c 1937-51
1 A   CPr   OPr   1948po   -   -   1.47   4BE1X3   RB   12/47 3/48   1
1 B   CPr   nd:po   -   -   1.47   4BEoo3   RB
3 C   CPu   OPs   1955pc   -   ac   6.55   4oooX3   3/55   3/56
3 C/   CPv,x.z   1955pc   1958c   ac   6.55   4oooX3
AA 1.1.47 on A(A), 3/-, OSPR 3/48. M on B. 1. Showing the Flockton Colliery line, with its zig-zag.

SE 22 (44/22) *Dewsbury*   10 by 10   Revision a 1930-39 c 1938-51 f 1971 g 1974
1 20046   CPr   1946po   -   -   1.47   4BE1X3   R   11/46 -
1 A   OPr   1946po   -   -   1.47   4BE1X3   R   -   2/47
3 B   OPs   1946pc   1951c   -   1.47   4BEoo3   R
3 C   CPu   OPv   1955pc   -   ac   6.55   4oooX3   5/55   11/56
3 C/   CPw.x,z   1955mc   1959c   ac   6.55   4oooX3
3 C//*   CPz   OPz   1955cms   nd:r   acf   6.55   4oooX3   9/72   12/72
3 C//*/* CFz   OFz   1955cms   nd:r   acg   6.55   4oooX3   7/75   1
AA 1.8.46 on A(A), 3/-, OSPR 2/47. M on A. 1. OSPR (coloured edition) not found.

SE 23 (44/23) *Leeds (West)**   10 by 10   Revision a 1932-34 c 1938-50 f 1959 g 1964
1 A   CPr   OPr   1947po   -   -   1.47   4BE1X3   RB   2/47   6/47
1 B   CPr   nd:po   -   -   1.47   4BEoo3   RB   M
2 C   CPt   OPs   1953pc   -   ac   1.53   4BooX3   R   2/53   11/54
3 C/
3 C//*   CPx   OPx   1953mc   1961r   acf   1.53   4BooX3   12/61 12/61
3 C//*/   CPz,0z   OPz   1953mc   1965ch   acg   1.53   4BooX3
AA 1.9.46 on A(A), 3/-, OSPR 6/47; AA 1.12.53 on C(B), 4/-, OSPR 11/54. M on B.

SE 24 (44/24) *Otley Yorkshire**   10 by 10   Revision a 1906-34 c 1938-50
1 10047   CPr   1947po   -   -   1.47   4BE1X3   RB   0   1/47   -
1 A   OPr   1947po   -   -   1.47   4BE1X3   RB   1   -   4/47   1
1 B   CPr   nd:po   -   -   1.47   4BEoo3   RB   1
2 C   CPt   OPs   1952pc   -   ac   6.52   4Booo3   R   3   11/52 3/56   2
3 C/   CP?v,x.z.z   1952pc   1957mc   ac   6.52   4BooX3   3
AA 1.9.46 on A(A), 3/-, OSPR 4/47; AA 1.12.53 on C(B), 4/-, OSPR 4/54. M on C. 1. Leeds-Bradford Municipal Airfield. 2. Leeds, Bradford (Yeadon) Aerodrome.

SE 25 (44/25) *Harrogate (West)*   10 by 10   Revision a 1906-32 c 1938-50
1 15046   CPr   1946po   -   -   1.47   4BE1X3   RB   6/46   -
1 2546   OPr   1946po   -   -   1.47   4BE1X3   RB   -   9/46
2 B   CPt   1952pc   -   ac   6.52   4Booo3   R   11/52 -
3 B   CPu,x   OPv   1952pc   -   ac   6.52   4Booo3   R   -   11/56
3 B/   CPz,0z   OPz   1952pc   -   ac   6.52   4BooX3
AA 1.4.46 on 2546(2546), 3/-, OSPR 9/46. M on B. NPRDBA 35 on B.

1	2	3	4	5	6	7	8	9	10	11	12	13	14

SE 26 (44/26) *Ripley Yorkshire*   10 by 10   Revision a 1907-32 c 1938-51

1	15046	CPr		1946po	-	-	6.46	4BE1X3	RB		7/46	-	
1	2546		OPr	1946po	-	-	6.46	4BE1X3	RB		-	9/46	
2	B	CPt	OPv	1952pc	-	ac	6.52	4Booo3	R		9/52	9/57	
3	B/	CPw,z,0z		1952pc	1958mc	ac	6.52	4BooX3					

AA 1.4.46 on 2546(2546), 3/-, OSPR 9/46. NPRDBA 32 on B/.

SE 27 (44/27) *West Tanfield*   10 by 10   Revision a 1907-28 c 1938-51

1	A	CPr	OPr	1948po	-	-	6.48	4BEoo3	RB		7/48	8/48	
2	B	CPt		1952pc	-	ac	6.52	4Booo3	R		9/52	-	
3	B/	CPw.x,z,0z		1952pc	1958mc	ac	6.52	4BooX3					

AA 1.5.48 on A(A), 3/-, OSPR 8/48. M on A. NPRDBA 31 on B/.

SE 28 (44/28) *Bedale*   10 by 10   Revision a 1911-28 c 1950-51 f 1967

1	A	CPr		1948po	-	-	6.48	4BEoo3	RB	0	7/48	-	
1	B		OPr	1948po	-	-	6.48	4BEoo3	RB	0	-	8/48	
	C												
2	D	CPt	OPv	1953pc	-	ac	1.53	4BooX4	R	1	2/53	9/57	
3	D/	CPw		1953mc	1958mc	ac	1.53	4BooX4		1			
3	D//*	CPz,0z	OPz	1953mc	1968r	acf	1.53	4BooX4		1	2/68	2/68	

AA 1.5.48 on B(A), 3/-, OSPR 8/48. NPRDBA 30 on D//*.

SE 29 (44/29) *Catterick* *   10 by 10   Revision a 1911-32 c 1932-51 f 1963

1	A	CPr	OPr	1948po	-	-	6.48	4BEoo3	RB	1	6/48	7/48	
2	B	CPt		1952pc	-	ac	6.52	4Booo3	R	1	9/52	-	
3	B/	CPv		1952pc	1957mc	ac	6.52	4BooX3		1			
3	B//*	CPz	OPz	1952mc	1965r	acf	6.52	4BooX3		1	1/65	2/65	

AA 1.4.48 on A(A), 3/-, OSPR 7/48. M on A, B.

SE 30 (44/30) [*Barnsley*]   10 by 10   Revision a 1929-30 b 1929-31 c 1938-48 d 1938-51 e 1948 f 1968

1	A	CPr	OPr	1948po	-	-	6.48	4BEoo3	RB		6/48	8/48	
1	B	CPr		nd:po	-	-	6.48	4BEoo3	RB				
1	B	CPs		1948pc	1951c	ace	1.51	4BEoo3	RB				
3	C	CPu	OPs	1955pc	-	bd	6.55	4oooX3			8/55	9/55	
3	C/	CPw,x,z	OPz	1955mc	1960mc	bd	6.55	4oooX3					
3	C//*	CPz	OPz	1955ms	1969r	bdf	6.55	4oooX3			3/70	2/70	

AA 1.1.48 on A(A), 3/-, OSPR 8/48. M on C, C/.

SE 31 (44/31) [*Wakefield (South)*]   10 by 10   Revision a 1929-31 c 1937-53

1	A	CPr	OPr	1947po	-	-	6.47	4BE1X3	RB		11/47	3/48	
2	nc [?B]		OPt	1947pc	1953c	-	6.53	4BooX3	RB				
3	C	CPu	OPs	1955pc	-	ac	6.55	4oooX3			8/55	4/56	
3	C/	CPw,y,z		1955mc	1960mc	ac	6.55	4oooX3					

AA 1.7.47 on A(A), 3/-, OSPR 3/48; AA 1.4.53 on no code(B), 4/-, OSPR 2/54. M on A.

SE 32 (44/32) *Wakefield (North)*   10 by 10   Revision a 1905-32 c 1938-53 f 1967 g 1971 h 1975

1	A	CPr	OPr	1947po	-	-	6.47	4BE1X3	RB		6/47	11/47	
1	B	CPr		nd:po	-	-	6.47	4BEoo3	RB				
3	C		OPt	1949pc	1951c	-	6.47	4BEoo3	RB				
3	D	CPu	OPv	1955pc	-	ac	6.55	4oooX4			8/55	9/57	
3	D/	CPw		1955mc	1959c	ac	6.55	4oooX4					
3	D//	CPw		1955mc	1961ch	ac	6.55	4oooX4					
3	D///*	CPz	OPz	1955ms	1968r	acf	6.55	4oooX4			6/68	3/68	
3	D///*/*	CPz,z-	OPz	1955^{c}ms	nd:r	acg	6.55	4oooX4			9/72	12/72	
3	D///*/*/*	CFz	OFz	1955^{c}ms	nd:r	ach	6.55	4oooX4					1

AA 1.2.47 on A(A), 3/-, OSPR 11/47; AA 1.12.53 on C(B), 4/-, OSPR 8/54. M on B. 1. OSPR not found.

1	2	3	4	5	6	7	8	9	10	11	12	13	14

SE 33 (44/33) *Leeds (East)**   10 by 10   Revision a 1905-34 c 1938-50 f 1968; j 1964 building
1 A   CPr   OPr   1947po   -   -   1.47 4BE1X3 RB   3/47   6/47
1 B   CPr   nd:po   -   -   1.47 4BEoo3 RB
2 C   CPt   OPs   1953pc   -   ac   1.53 4BooX4   5/53   2/55
3 C/   CPw,x,z   1953pc   1958mc   ac   1.53 4BooX4
3 C//* CPz,0z   OPz   1953ms   1970br   acfj 1.53 4BooX4   3/70   5/70
AA 1.12.46 on A(A), 3/-, OSPR 6/47; AA 1.12.53 on C(B), 4/-, OSPR 2/55. M on B.

SE 34 (44/34) *Harewood**   10 by 10   Revision a 1906-33 c 1937-38 d 1937-50
1 10046 CPr   1946po   -   -   1.47 4BE1X3 RB   11/46 -
1 A   OPr   1946po   -   -   1.47 4BE1X3 RB   -   3/47
1 B   CPs   1951pc   -   ac   1.51 4BEoo3 RB   4/51   -
3 B   OPs   1951pc   -   ac   1.51 4BEoo3 RB   -   11/56
3 C   CPx,z   OPx   1961mc   -   ad   6.61 4oooX3   7/61   5/61
AA 1.9.46 on A(A), 3/-, OSPR 3/47. M on B.

SE 35 (44/35) *Harrogate (East)**   10 by 10   Revision a 1907-32 c 1938-50 f 1965
1 A   CPr   OPr   1948po   -   -   6.48 4BEoo3 RB   6/48   8/48
2 B   CPt   1952pc   -   ac   6.52 4Booo4   12/52 -
3 B/   CPv,x   1952pc   1957mc   ac   6.52 4BooX4
3 B//   CPz   OPz   1952mc   1966ch   acf   6.52 4BooX4
AA 1.4.48 on A(A), 3/-, OSPR 8/48. M on B, B/.

SE 36 (44/36) *Boroughbridge*   10 by 10   Revision a 1907-32 c 1938-51 f 1962 g 1966
                                                     j 1966 railway
1 A   CPr   OPr   1947po   -   -   1.48 4BE1X3 RB   11/47 2/48
1 B   CPr   nd:po   -   -   1.48 4BEoo3 RB
2 C   CPt   OPs   1952pc   -   ac   6.52 4Booo3 R   11/52 8/54
3 C/   CPw   1952mc   1958mc   ac   6.52 4BooX3
3 C//   CPx   1952mc   1963ch   acf   6.52 4BooX3   1
3 C///* CPz,0z   OPz   1952mc   1967r   acgj 6.52 4BooX3   2/67   1/67
AA 1.8.47 on A(A), 3/-, OSPR 2/48. M on B. SS: Soil, outline, 1975 on C///*. 1. The A1 is under construction.

SE 37 (44/37) *Ripon*   10 by 10   Revision a 1907-28 c 1938-51 f 1969
1 A   CPr   1948po   -   -   6.48 4BEoo3 RB   00   6/48   -
1 B   OPr   1948po   -   -   6.48 4BEoo3 RB   00   -   7/48
2 C   CPt   OPv   1953pc   -   ac   1.53 4BooX3 R   11   3/53   11/56
3 C/   CPw,y   1953mc   1959mc   ac   1.53 4BooX3   11   1
3 C//   CPz,0z   OPz   1953ms   1970ch   acf   1.53 4BooX3   33
AA 1.4.48/NT 31.5.48 on B(A), 3/-, OSPR 7/48. M on A, C. 1. Dishforth and Topcliffe Airfields are named.

SE 38 (44/38) *Pickhill*   10 by 10   Revision a 1910-28 c 1951 f 1963
1 A   CPr   OPr   1948po   -   -   1.48 4BE1X3 RB   00   2/48   4/48
1 B   CPr   nd:po   -   -   1.48 4BEoo3 RB   00
2 C   CPt   OPs   1952pc   -   ac   6.52 4Booo3 R   11   12/52 10/54
3 C/   CPw   1952pc   1958mc   ac   6.52 4BooX3   11
3 C//* CPz,0z   OPz   1952mc   1965r   acf   6.52 4BooX3   31   3/65   4/65
AA 1.11.47 on A(A), 3/-, OSPR 4/48.

SE 39 (44/39) *Northallerton*   10 by 10   Revision a 1911-27 c 1938-51
1 A   CPr   OPr   1948po   -   -   1.48 4BE1X3 RB   0   12/47 2/48
1 B   CPr   nd:po   -   -   1.48 4BEoo3 RB   0
2 C   CPt,x   1952pc   -   ac   6.52 4Booo3 R   1   10/52 -
3 C   CPz   OPv   1952pc   -   ac   6.52 4Booo3 R   1   -   11/56
3 C/   CPz   OPz   1952pc   -   ac   6.52 4Booo3   1
AA 1.11.47 on A(A), 3/-, OSPR 2/48. M on A, C. SS: Soil, Land Use Capability, coloured, 1981 on C/.

SE 40 (44/40) [*Wath upon Dearne*]   10 by 10   Revision a 1928-30 c 1938-51
1 A   CPr   OPr   1947po   -   -   1.47 4BE1X3 RB   00   1/47   5/47
2 B   CPt   OPs   1952pc   -   ac   6.52 4Booo3 R   11   11/52 8/54
3 B/   CPv,z   OPz   1952pc   1956mc   ac   6.52 4BooX3   11
AA 1.9.46 on A(A), 3/-, OSPR 5/47. M on B/.

1	2	3	4	5	6	7	8	9	10	11	12	13	14

SE 41 (44/41) *Hemsworth** 10 by 10 Revision a 1904-31 c 1938-51 f 1965
1	15046	CPr		1946po	-	-	6.46	4BE1X3	RB		10/46	-	
1	A		OPr	1946po	-	-	6.46	4BE1X3	RB		-	2/47	
2	B	CPt	OPs	1953pc	-	ac	1.53	4BooX3	R		2/53	12/55	
3	B/	CPw,x		1953pc	1958mc	ac	1.53	4oooX3					
3	B//*	CPz,0z	OPz	1953mc	1966r	acf	1.53	4oooX3			1/66	12/65	

AA 1.7.46 on A(A), 3/-, OSPR 2/47. M on B, B/.

SE 42 (44/42) *Castleford* 10 by 10 Revision a 1905-32 c 1938-52 f 1963 g 1969
1	15046	CPr		1946po	-	-	6.46	4BE1X3	R		8/46	-	
1	2546		OPr	1946po	-	-	6.46	4BE1X3	R		-	12/46	
3	B	CPt	OPv	1954pc	-	ac	6.54	4BooX3			5/54	1/57	
3	B/	CPw		1954mc	1958c	ac	6.54	4BooX3					
3	B//	CPy,z		1954mc	1964ch	acf	6.54	4BooX3					
3	B///	CPz	OPz	1954^{c}ms	nd:ch	acg	6.54	4BooX3					

AA 1.5.46 on 2546(2546), 3/-, OSPR 12/46. M on B, B/.

SE 43 (44/43) [Garforth] 10 by 10 Revision a 1905-06 c 1938-50 f 1964 g 1968
1	A	CPz	OPr	1947po	-	-	6.47	4BE1X3	RB		10/47	1/48	1
1	B	CPr		nd:po	-	-	6.47	4BEoo3	RB				
3	C	CPt	OPv	1954pc	-	ac	6.54	4BooX3			3/54	1/57	
3	C/	CP?w		1954mc	1959mc	ac	6.54	4BooX3					M
3	C//*	CPz	OPz	1954mc	1965r	acf	6.54	4BooX3			1/65	2/65	
3	C//*/*	CPz,z-	OPz	1954ms	1969r	acg	6.54	4BooX3			5/69	1/69	

AA 1.8.47 on A(A), 3/-, OSPR 1/48. M on C, C/. 1. The absence of a price statement on the coloured edition is no doubt an error.

SE 44 (44/44) *Tadcaster** 10 by 10 Revision a 1906-07 c 1937-50 f 1961 g 1966
                                            j 1966 industrial
1	A	CPr	OPr	1948po	-	-	1.48	4BE1X3	RB		12/47	3/48	
1	B	CPr		nd:po	-	-	1.48	4BEoo3	RB				
3	C	CPt,u	OPv	1954pc	-	ac	6.54	4BooX3			1/54	9/57	
3	C/*	CPy	OPy	1954mc	1963r	acf	6.54	4BooX3			5/63	5/63	
3	C/*/*	CPz	OPz	1954mc	1967ir	acgj	6.54	4BooX3			1/67	1/67	

AA 1.10.47 on A(A), 3/-, OSPR 3/48. M on C. The wartime industrial complex at Thorpe Arch is not shown.

SE 45 (44/45) *Hammerton* 10 by 10 Revision a 1907 c 1950 f 1968
1	A	CPr	OPr	1948po	-	-	1.48	4BE1X3	RB	0	1/48	3/48	
3	B	CPt		1954pc	-	ac	6.54	4BooX3		1	1/54	-	
3	B/	CPw		1954pc	1958mc	ac	6.54	4BooX3		1			
3	B//*	CPx	OPx	1954pc	1961r	ac	6.54	4BooX3		1	12/61	12/61	
3	B//*/	CPz	OPz	1954ms	1968ch	acf	6.54	4BooX3		1			

AA 1.11.47 on A(A), 3/-, OSPR 3/48. M on B.

SE 46 (44/46) *Tholthorpe* 10 by 10 Revision a 1907-10 c 1950-51 f 1970; j 1970 airfield
1	A	CPr	OPr	1948po	-	-	1.48	4BE1X3	RB	00	2/48	6/48	
1	B	CPr		nd:po	-	-	1.48	4BEoo3	RB	00			
1	B	CPs	OPs	1948pc	1951	-	1.51	4BEoo3	RB	00			
3	C	CPx	OPx	1961mc	-	ac	6.61	4oooX4		13	10/61	6/61	1
3	C/*	CPz	OPz	1961^{c}ms	nd:a	acfj	6.61	4oooX4		33	6/71	7/71	

AA 1.12.47 on A(A), 3/-, OSPR 6/48; AA 1.6.51 on B(A), 3/-, OSPR 8/51. M on B. 1. Linton-on-Ouse Airfield.

SE 47 (44/47) *Dalton Yorkshire* 10 by 10 Revision a 1907-10 c 1951 f 1978
1	A	CPr	OPr	1948po	-	-	1.48	4BE1X3	RB	00	2/48	6/48	
2	B	CPt		1952pc	-	ac	6.52	4Booo3	R	11	11/52	-	
3	B/	CPv,z		1952pc	1957mc	ac	6.52	4BooX3		11			
3	B/*	CFz		1952^{c}ms	nd:r	acf	6.52	4BooX3		11			1

AA 1.11.47 on A(A), 3/-, OSPR 6/48. M on B/. 1. The print code is illogical. OSPR not found.

142

1	2	3	4	5	6	7	8	9	10	11	12	13	14

SE 48 (44/48) *Thirsk*   10 by 10   Revision a 1909-26 b 1909-27 c 1951

1	A	CPr	OPr	1948po	-	-	1.48	4BE3X3	RB		12/47	2/48	
1	B	CPs		1948pc	1950c	a	6.50	4BEoo3	RB				
2	C	CPt		1952pc	-	bc	6.52	4Booo3	R		11/52	-	
3	C	CPx	OPv	1952pc	-	bc	6.52	4Booo3	R		-	1/57	
3	C/	CPz,0z	O0z	1952pc	-	bc	6.52	4Booo3					

AA 1.10.47 on A(A), 3/-, OSPR 2/48.

SE 49 (44/49) *Osmotherley Yorkshire*   10 by 10   Revision a 1910-27 c 1950-51 f 1960 g 1972
j 1960 reservoir

1	A	CPr	OPr	1948po	-	-	6.48	4BE3X3	B		3/48	4/48	1
3	B	CPx,z	OPx	1961mc	-	acf	6.61	4oooX3			6/61	5/61	
3	B/*	CPz,0z	OPz,0z	1961^{c}ms	nd:rv	acgj	6.61	4oooX3			5/73	6/73	

AA 1.1.48/NT 3.3.48 on A(A), 3/-, OSPR 4/48. 1. The alignment of the Kepwick Railway is on the grey plate.

SE 50 (44/50) [*Doncaster*]   10 by 10   Revision a 1928-30 c 1938-48 d 1938-51 f 1962

1	A	CPr	OPr	1948po	-	-	6.48	4BEoo3	RB	00	5/48	7/48	
1	B												
1	C	CPs		1951pc	-	ac	1.51	4BEoo3	RB	10	9/51	-	
2	D	CPt	OPs	1952pc	-	ad	6.52	4Booo3	R	11	12/52	-	
3	D/	CPw		1952mc	1960mc	ad	6.52	4BooX3		11			
3	D//*	CPy,z	OPy,z	1952mc	1963r	adf	6.52	4BooX3		11	3/63	2/63	

AA 1.3.48 on A(A), 3/-, OSPR 7/48. M on D, D/.

SE 51 (44/51) *Askern**   10 by 10   Revision a 1904-30 c 1938-51

1	15047	CPr		1947po	-	-	1.47	4BE1X3	RB		1/47	-	
1	2547		OPr	1947po	-	-	1.47	4BE1X3	RB		-	1/47	
2	B	CPt	OPv	1952pc	-	ac	6.52	4Booo3	R		11/52	9/57	
3	B/	CP?w,x,z		1952mc	1958mc	ac	6.52	4BooX3					

AA 1.7.46 on 2547(2547), 3/-, OSPR 1/47. M on B, B/.

SE 52 (44/52) *Knottingley**   10 by 10   Revision a 1904-31 c 1938-50 f 1965

1	A	CPr		1948po	-	-	6.48	4BE1X3	RB	00	3/48	-	
1	B		OPr	1948po	-	-	6.48	4BE1X3	RB	00	-	4/48	
1	C	CPr		nd:po	-	-	6.48	4BEoo3	RB	00			
3	D	CPt	OPv	1954pc	-	ac	6.54	4BooX3		11	4/54	8/57	
3	D/	CPw		1954pc	1958mc	ac	6.54	4BooX3		11			
3	D//	CPw		1954mc	1961ch	ac	6.54	4BooX3		11			
3	D///	CPz,0z	OPz,0z	1954mc	1966ch	acf	6.54	4BooX3		33			

AA 1.1.48 on B(A), 3/-, OSPR 4/48. M on D/, D//.

SE 53 (44/53) [*Church Fenton*]   10 by 10   Revision a 1905-06 c 1938-50

1	A	CPr	OPr	1947po	-	-	6.47	4BE1X3	RB	00	9/47	1/48	
1	A	CPt		1947po	-	-	6.47	4BEoo3	RB	00			
3	B	CPt		1954pc	-	ac	6.54	4BooX4		11	1/54	-	
3	B/	CPv,x,z		1954pc	1957mc	ac	6.54	4BooX4		11			

AA 1.7.47 on A(A), 3/-, OSPR 1/48. M on A, B.

SE 54 (44/54) *Askham**   10 by 10   Revision a 1906-29 c 1937-50

1	10046	CPr		1946po	-	-	6.46	4BE1X3	RB	00	7/46	-	
1	2546		OPr	1946po	-	-	6.46	4BE1X3	RB	00	-	11/46	
1	A	CPt		1946po	-	-	6.46	4BEoo3	RB	00			
3	B	CPt,v,x,z	OPv	1954pc	-	ac	6.54	4BooX3		11	1/54	2/57	

AA 1.4.46 on 2546(2546), 3/-, OSPR 11/46. M on B.

SE 55 (44/55) *York (West)*   10 by 10   Revision a 1907-38 c 1950 f 1970

1	10046	CPr		1946po	-	-	1.47	4BE1X3	RB	00	12/46	-	
1	A		OPr	1946po	-	-	1.47	4BE1X3	RB	00	-	4/47	
3	B	CPt		1954pc	-	ac	6.54	4BooX4		13	1/54	-	
3	B/	CPv,y,z		1954pc	1957c	ac	6.54	4BooX4		11			
3	B//	CPz,z-	OPz	1954^{c}ms	nd:ch	acf	6.54	4BooX4		33			

AA 1.9.46 on A(A), 3/-, OSPR 4/47. M on B, B/.

1	2	3	4	5	6	7	8	9	10	11	12	13	14

SE 56 (44/56) *Tollerton Yorkshire*   10 by 10   Revision a 1907-28 c 1938-51
1 A   CPr   OPr   1948po   -   -   1.48 4BE1X3   RB   0   1/48   3/48
1 B   CPr   nd:po   -   -   1.48 4BEoo3   RB   0
3 C   CPt,x,z   OPv   1954pc   -   ac   6.54 4BooX3   1   3/54   1/57
AA 1.9.47 on A(A), 3/-, OSPR 3/48. M on C.

SE 57 (44/57) *Coxwold*   10 by 10   Revision a 1909-10 c 1938-50
1 A   CPr   OPr   1948po   -   -   6.48 4BE3X3   RB   2/48   6/48
1 A   CPs   1948pc   1951   -   1.51 4BEoo3   RB
3 B   CPt   OPv   1954pc   -   ac   6.54 4BooX4   2/54   9/57
3 B/   CPw,y.z,0z   1954mc   1959mc ac   6.54 4BooX4
AA 1.12.47 on A(A), 3/-, OSPR 6/48. M on B/.

SE 58 (44/58) *Rievaulx*   10 by 10   Revision a 1909-10 c 1950-51
1 A   CPr   OPr   1948po   -   -   6.47 4BE1X3   B   12/47 3/48
1 B   CPr   nd:po   -   -   6.47 4BEoo3   B
3 C   CPt   OPv   1954pc   -   ac   6.54 4BooX3   5/54   9/57
3 C/   CPw,z.z,0z O0z   1954mc   1958mc ac   6.54 4BooX3
AA 1.9.47 on A(A), 3/-, OSPR 3/48. SS: Soil, Land Use Capability, outline, 1979 on C/.

SE 59 (44/59) *Bilsdale**   10 by 10   Revision a 1910 c 1950
1 A   CPr   OPr   1948po   -   -   1.48 4BE1X3   B   2/48   4/48
3 B   CPt   OPv   1954pc   -   ac   6.54 4BooX3   3/54   1/57
3 B/   CPw,z,0z OPv,0z 1954mc   1960mc ac   6.54 4BooX3
AA 1.6.47 on A(A), 3/-, OSPR 4/48.

SE 60 (44/60) *Armthorpe*   10 by 10   Revision a 1901-30 c 1937-51 f 1965 g 1979
1 A   CPr   OPr   1947po   -   -   6.47 4BE1X3   RB   0   10/47 12/47
2 B   CPt   OPv   1952pc   -   ac   6.52 4Booo3   R   1   12/52 9/57
3 B/   CPx   OPx   1952mc   1961ch ac   6.52 4BooX3   1
3 B//   CPz   OPz   1952mc   1966ch acf   6.52 4BooX3   3   1
3 B///*   CFz   OFz   1979^{c}ms nd:r   acg   6.52 4BooX3   3   9/80   9/80
AA 1.7.47 on A(A), 3/-, OSPR 12/47. M on A, B. SS: Soil, Land Use Capability, outline, 1971 on B//. 1. Lindholme Airfield.

SE 61 (44/61) *Thorne Yorkshire**   10 by 10   Revision a 1904-30 c 1948-51
1 A   CPr   OPr   1947po   -   -   6.47 4BE2X3   R   11/47 2/48
1 B   CPr   OPs   nd:po   -   -   6.47 4BEoo3   R
2 C   CPt   OPv   1952pc   -   ac   6.52 4Booo3   11/52 9/57
3 C/   CPw,z.z   1952mc   1960mc ac   6.52 4BooX3
AA 1.9.47 on A(A), 3/-, OSPR 2/48; AA 1.5.51 on B(B), 3/-, OSPR 8/51. M on C/.

SE 62 (44/62) *Snaith**   10 by 10   Revision a 1904-05 c 1938-50 f 1970
1 A   CPr   OPr   1948po   -   -   1.48 4BE3X3   RB   00   2/48   3/48
3 B   CPt,x   OPv   1954pc   -   ac   6.54 4BooX4   11   3/54   1/57
3 B/   CPz   OPz   1954^{c}ms nd:ch acf   6.54 4BooX4   33   1
AA 1.10.47 on A(A), 3/-, OSPR 3/48. M on B. 1. Drax Power Station and it railways are shown.

SE 63 (44/63) *Selby*   10 by 10   Revision a 1905-08 c 1938-50
1 A   CPr   1947po   -   -   1.47 4BE1X3   RB   0   10/47 -
1 B   OPr   1947po   -   -   1.47 4BE1X3   RB   0   -   12/47
3 C   CPt   OPv   1954pc   -   ac   6.54 4BooX3   1   5/54   9/57
3 C/   CPw,z,0z   1954mc   1959mc ac   6.54 4BooX3   1
AA 1.7.47 on B(A), 3/-, OSPR 12/47. M on C, C/.

SE 64 (44/64) *Esrick*   10 by 10   Revision a 1906-38 c 1950
1 A   CPr   OPr   1947po   -   -   1.47 4BE1X3   RB   0   1/47   4/47
3 B   CPt   OPv   1954pc   -   ac   6.54 4BooX4   1   3/54   9/57
3 B/   CPw,z.z,0z   1954mc   1960mc ac   6.54 4BooX4   1   1
AA 1.9.46 on A(A), 3/-, OSPR 4/47. M on B, B/. SS: Soil, outline, 1971 on B/. 1. Elvington Airfield.

1	2	3	4	5	6	7	8	9	10	11	12	13	14

SE 65 (44/65) *York (East)*   10 by 10   Revision a 1909-29 c 1936-50
1   A   CPr   OPr   1947po   -   -   1.47   4BE1X3   RB   1/47   4/47
3   B   CPt   OPv   1954pc   -   ac   6.54   4BooX3   4/54   9/57
3   B/   CPw,y,z   1954mc   1960mc   ac   6.54   4BooX3
AA 1.10.46/NT 9.1.47 on A(A), 3/-, OSPR 4/47. M on B, B/. SS: Soil, outline, 1971 on B/.

SE 66 (44/66) *Strensall*   10 by 10   Revision a 1909-28 c 1936-50
1   A   CPr   OPr   1947po   -   -   1.48   4BE1X3   RB   0   11/47   2/48
1   B   CPs   1950pc   -   a   6.50   4BEoo3   RB   0   11/50   -
3   C   CPt,v,z.z   OPv   1954pc   -   ac   6.54   4BooX4   1   3/54   2/57
AA 1.9.47 on A(A), 3/-, OSPR 2/48. M on C.

SE 67 (44/67) *Hovingham*   10 by 10   Revision a 1909-10 c 1950
1   10047   CPr   1947po   -   -   1.47   4BE1X3   RB   1/47   -
1   A   OPr   1947po   -   -   1.47   4BE1X3   RB   -   4/47
1   B   CPr   OPr   nd:po   -   -   6.48   4BEoo3   RB
3   C   CPt,u,z.z   OPv   1954pc   -   ac   6.54   4BooX3   2/54   1/57
AA 1.9.46 on A(A), 3/-, OSPR 4/47. M on C.

SE 68 (44/68) *Helmsley*   10 by 10   Revision a 1910 c 1950
1   A   CPr   1947po   -   -   6.47   4BE1X3   RB   0   10/47   -
1   B   OPr   1947po   -   -   6.47   4BE1X3   RB   0   -   12/47
3   C   CPt   OPv   1954pc   -   ac   6.54   4BooX4   1   4/54   1/57
3   C/   CPw,z,0z   1954mc   1960mc   ac   6.54   4BooX4   1
AA 1.6.47 on B(A), 3/-, OSPR 12/47.

SE 69 (44/69) *Rudland Rigg*   10 by 10   Revision a 1910 c 1950
1   A   CPr   OPr   1947po   -   -   6.47   4BE1X3   RB   9/47   12/47
3   B   CPt   OPv   1954pc   -   ac   6.54   4BooX3   3/54   8/57
3   B/   CPw,z,0z   O0z   1954mc   1960mc   ac   6.54   4BooX3
AA 1.7.47 on A(A), 3/-, OSPR 12/47.

SE 70 (44/70) *Epworth*   10 by 10   Revision a 1901-30 c 1946-53
1   10046   CPr   1946po   -   -   6.46   4BE1X3   R   0   9/46   -
1   2546   OPr   1946po   -   -   6.46   4BE1X3   R   0   -   11/46
2   B   CPt   1953pc   -   ac   6.53   4BooX3   1   9/53   -
3   B   CPv.y,z   OPv   1953pc   -   ac   6.53   4BooX3   1   -   7/57
AA 1.6.46 on 2546(2546), 3/-, OSPR 11/46. M on B.

SE 71 (44/71) *Crowle Lincolnshire**   10 by 10   Revision a 1904-30 c 1938-51
1   A   CPr   OPr   1947po   -   -   1.47   4BE1X3   RB   10/47   12/47
2   B   CPt   1953pc   -   ac   1.53   4Booo3   R   1/53   -
3   B   CPu   OPv   1953pc   -   ac   1.53   4Booo3   R   -   11/56
3   B/   CPz,0z   OPz,0z   1953pc   -   ac   1.53   4BooX3
AA 1.6.47 on A(A), 3/-, OSPR 12/47. M on B.

SE 72 (44/72) *Goole**   10 by 10   Revision a 1904-08 c 1938-51 f 1966
1   A   CPr   OPr   1947po   -   -   1.48   4BE1X3   RB   10/47   12/47
3   B   CPt   OPv   1954pc   -   ac   6.54   4BooX4   2/54   8/57
3   B/   CPw,x   1954mc   1959mc   ac   6.54   4BooX4
3   B//   CPz,0z   OPz   1954mc   1967ch   acf   6.54   4BooX4
AA 1.7.47 on A(A), 3/-, OSPR 12/47. M on A.

SE 73 (44/73) *Bubwith**   10 by 10   Revision a 1905-08 c 1938-50 f 1969
1   A   CPr   OPr   1947po   -   -   1.47   4BE1X3   RB   0   10/47   2/48
3   B   CPt   OPs   1954pc   -   ac   6.54   4BooX4   1   2/54   9/55
3   B/   CPv,z   OPz   1954pc   1956mc   ac   6.54   4BooX4   1
3   B//   CPz   OPz   1954^{c}ms   nd:ch   acf   6.54   4BooX4   3
AA 1.8.47 on A(A), 3/-, OSPR 2/48. M on B/.

1	2	3	4	5	6	7	8	9	10	11	12	13	14

SE 74 (44/74) *Barmby Moor*   10 by 10   Revision a 1908-26 c 1950 f 1965
1 A   CPr   OPr   1948po   -   -   1.48   4BE3X3   RB   00   12/47 2/48
3 B   CPt,u   OPv   1953pc   -   ac   6.53   4BooX4   11   12/53 11/56
3 B/   CPz,0z   OPz   1953mc   1965ch   acf 6.53   4BooX4   33
AA 1.10.47 on A(A), 3/-, OSPR 2/48. M on B. SS: Soil, outline, 1971 on B/.

SE 75 (44/75) *Stamford Bridge**   10 by 10   Revision a 1909-26 c 1950
1 A   CPr   OPr   1947po   -   -   1.47   4BE1X3   RB   0   11/47 1/48
3 B   CPt,u,z   OPv   1954pc   -   ac   6.54   4BooX3   1   3/54   1/57
AA 1.7.47 on A(A), 3/-, OSPR 1/48. M on B.

SE 76 (44/76) *Westow*   10 by 10   Revision a 1909-26 c 1938-50
1 15046   CPr   1946po   -   -   6.46   4BE1X3   RB   6/46   -
1 2546   OPr   1946po   -   -   6.46   4BE1X3   RB   -   5/47
3 B   CPt,u   OPv   1954pc   -   ac   6.54   4BooX3   3/54   9/57
3 B/   CP?w,z.z,0z   1954mc   1960mc   ac   6.54   4BooX3
AA 1.5.46 on 2546(2546), 3/-, OSPR 5/47. M on B, B/. SS: Soil, outline, 1973 on B/.

SE 77 (44/77) *Malton*   10 by 10   Revision a 1909-27 c 1938-50
1 15046   CPr   1946po   -   -   6.46   4BE1X3   RB   7/46   -
1 2546   OPr   1946po   -   -   6.46   4BE1X3   RB   -   5/47
3 B   CPt   OPv   1954pc   -   ac   6.54   4BooX3   2/54   9/57
3 B/   CPw,z,0z   1954mc   1958mc   ac   6.54   4BooX3
AA 1.5.46 on 2546(2546), 3/-, OSPR 5/47. M on B/.

SE 78 (44/78) *Pickering (West)*   10 by 10   Revision a 1910-27 c 1938-50
1 A   CPr   1947po   -   -   6.47   4BE1X3   RB   10/47 -
1 B   OPr   1947po   -   -   6.47   4BE1X3   RB   -   12/47
3 C   CPt   OPv   1954pc   -   ac   6.54   4BooX3   3/54   1/57
3 C/   CPx.z   1954mc   1961ch   ac   6.54   4BooX3
AA 1.6.47 on B(A), 3/-, OSPR 12/47.

SE 79 (44/79) *Rosedale Abbey*   10 by 10   Revision a 1910-27 c 1950 f 1969
1 10046   CPr   1946po   -   -   1.47   4BE1X3   RB   9/46   -
1 2547   OPr   1947po   -   -   1.47   4BE1X3   RB   -   1/47
3 B   CPt   OPv   1954pc   -   ac   6.54   4oooX4   6/54   9/57
3 B/   CPw,z   OPz   1954mc   1958mc   ac   6.54   4oooX4
3 B//   CPz,0z   OPz   1954^{c}ms nd:ch acf 6.54   4oooX4
AA 1.8.46 on 2547(2547), 3/-, OSPR 1/47.

SE 80 (44/80) *Scunthorpe (South West)*   10 by 10   Revision a 1905-06 c 1938-51; j 1983 selected k SUSI
1 10046   CPr   1946po   -   -   1.47   4BE1X3   B   7/46   -
1 2546   OPr   1946po   -   -   1.47   4BE1X3   B   -   11/46
2 B   CPt   1953pc   -   ac   1.53   4BooX3   1/53   -
3 B/   CPv,x,z.z   1953pc   1956mc   ac   1.53   4BooX3
3 B//*   CFz   OFz   1983^{c}m-s nd:rb acjk 6.84   4BooX3   11/83 11/83 1
AA 1.5.46 on 2546(2546), 3/-, OSPR 11/46. 1. The areas of building development are shaded grey.

SE 81 (44/81) *Scunthorpe (North West)**   10 by 10   Revision a 1904-06 c 1938-51
1 10046   CPr   1946po   -   -   1.47   4BE1X3   RB   10/46 -
1 A   OPr   1946po   -   -   1.47   4BE1X3   RB   -   2/47
2 B   CPt   OPv   1953pc   -   ac   1.53   4BooX3   R   3/53   9/57
3 B/   CPw,z.z   1953mc   1959c   ac   1.53   4BooX3
AA 1.8.46 on A(A), 3/-, OSPR 2/47.

SE 82 (44/82) *Broomfleet*   10 by 10   Revision a 1904-08 c 1938-50
1 10046   CPr   1946po   -   -   1.47   4BE1X3   RB   8/46   -
1 2546   OPr   1946po   -   -   1.47   4BE1X3   RB   -   10/46
3 B   CPt   1953pc   -   ac   6.53   4BooX3   11/53 -
3 B/   CP?w.z   1953mc   1960mc   ac   6.53   4BooX3
AA 1.5.46 on 2546(2546), 3/-, OSPR 10/46. M on A.

1	2	3	4	5	6	7	8	9	10	11	12	13	14

SE 83 (44/83) *North Cave** 10 by 10 Revision a 1908 c 1948-50 f 1966
1 10046 CPr   1946po   -   -   1.47 4BE1X3 RB 0 9/46 -
1 2546   OPr 1946po   -   -   1.47 4BE1X3 RB 0 -   12/46
3 B   CPt,u   1953pc   -   ac   6.53 4BooX3   1 11/53 -
3 B/   CPz   OPz 1953mc 1966ch acf 6.53 4BooX3   3
AA 1.6.46 on 2546(2546), 3/-, OSPR 12/46. M on B.

SE 84 (44/84) *Market Weighton* 10 by 10 Revision a 1908-26 c 1950 f 1967
1 A   CPr   OPr 1947po   -   -   1.47 4BE1X3 RB   4/47 6/47
3 B   CPt   1953pc   -   ac   6.53 4BooX3   11/53 -
3 B/   CP?w,z   1953pc 1958mc ac   6.53 4oooX3
3 B//   CPz   OPz,0z 1953mc 1967ch acf 6.53 4oooX3
AA 1.9.46 on A(A), 3/-, OSPR 6/47. M on B/.

SE 85 (44/85) *Fridaythorpe* 10 by 10 Revision a 1909-26 c 1946-50
1 A   CPr   OPr 1947po   -   -   6.47 4BE1X3 B   10/47 1/48
3 B   CPt   1953pc   -   ac   6.53 4BooX3   11/53 -
3 B/   CPw,z,0z   1953mc 1959mc ac   6.53 4BooX3
AA 1.2.47 on A(A), 3/-, OSPR 1/48. M on A. SS: Soil, outline, 1985 on B/.

SE 86 (44/86) *North Grimston* 10 by 10 Revision a 1909-26 c 1938-50
1 10046 CPr   1946po   -   -   1.47 4BE1X3 RB   7/46 -
1 2546   OPr 1946po   -   -   1.47 4BE1X3 RB   -   9/46
3 B   CPt   1954pc   -   ac   6.54 4BooX4   1/54 -
3 B/   CPw,z,0z   1954mc 1958mc ac   6.54 4BooX4
AA 1.5.46 on 2546(2546), 3/-, OSPR 9/46. M on 10046.

SE 87 (44/87) *Rillington** 10 by 10 Revision a 1909-27 c 1950
1 10046 CPr   1946po   -   -   1.47 4BE1X3 RB   9/46 -
1 2547   OPr 1947po   -   -   1.47 4BE1X3 RB   -   1/47
3 B   CPt,u,y,z OPz 1954pc -   ac   6.54 4BooX4   2/54
AA 1.6.46 on 2547(2547), 3/-, OSPR 1/47. M on 10046.

SE 88 (44/88) *Thornton Dale* 10 by 10 Revision a 1909-27 c 1938-50
1 10046 CPr   1946po   -   -   1.47 4BE1X3 RB   9/46 -
1 2546   OPr 1946po   -   -   1.47 4BE1X3 RB   -   12/46
3 B   CPt,x,z   1954pc   -   ac   6.54 4BooX3   4/54 -
AA 1.6.46 on 2546(2546), 3/-, OSPR 12/46. M on 10046, B.

SE 89 (44/89) *Levisham* 10 by 10 Revision a 1910-27 c 1950
1 10046 CPr   1946po   -   -   1.47 4BE1X3 RB   9/46 -
1 2546   OPr 1946po   -   -   1.47 4BE1X3 RB   -   12/46
1 [B]   CPr   1946po   1.47 4BEoo3 RB   M
3 C   CPt,w,x,z OPv 1954pc -   ac   6.54 4BooX3   4/54 10/58
3 C   CPz,0z   O0z 1954mc -   ac   6.54 4BooX3
AA 1.5.46 on 2546(2546), 3/-, OSPR 12/46. M on ?B, C.

SE 90 (44/90) *Scunthorpe (South East)* 10 by 10 Revision a 1905-06 c 1938-53 f 1966; j 1980 selected
1 A   CPr   OPr 1948po   -   -   1.48 4BE3X3 RB 0 1/48 3/48
1 B   CPr   nd:po   -   -   1.48 4BEoo3 RB 0
2 C   CPt   1953pc   -   ac   6.53 4BooX3   1 9/53 -
3 C/   CP?w,z OPv 1953mc 1959mc ac   6.53 4BooX3   1
3 C//   CPz   OPz 1953mc 1966ch acf 6.53 4BooX3   3
3 C///* CFz   OFz 1980^{c}ms nd:r acfj 6.53 4BooX3   3 12/80 1/81
AA 1.10.47 on A(A), 3/-, OSPR 3/48.

SE 91 (44/91) *Scunthorpe (North East)** 10 by 10 Revision a 1905-06 c 1938-51
1 A   CPr   OPr 1947po   -   -   1.47 4BE1X3 RB   11/47 1/48
   B
2 C   CPt   1953pc   -   ac   1.53 4BooX3 R   2/53 -
3 C/   CPw.z,0z   1953mc 1960mc ac   1.53 4BooX3
AA 1.9.47 on A(A), 3/-, OSPR 1/48. M on C.

1	2	3	4	5	6	7	8	9	10	11	12	13	14

SE 92 (44/92) *Brough Yorkshire**   10 by 10   Revision a 1904-26   c 1938-50   f 1963   g 1973
1	A	CPr	OPr	1947po	-	-	6.47	4BE2X3	R	0	6/47	9/47	
3	B	CPt		1954pc	-	ac	6.54	4BooX3		1	12/53	-	
3	B/												
3	B//	CPz	OPz	1954mc	1964ch	acf	6.54	4BooX3		1			
3	B///*	CPz	OPz	1954ᶜms	nd:r	acg	6.54	4BooX3		1	9/73	10/73	

AA 1.4.47 on A(A), 3/-, OSPR 9/47. M on A.

SE 93 (44/93) *South Cave**   10 by 10   Revision a 1908-26   c 1938-50   f 1972
1	10046	CPr		1946po	-	-	1.47	4BE1X3	RB		11/46	-	
1	A		OPr	1946po	-	-	1.47	4BE1X3	RB		-	5/47	
3	B	CPt		1953pc	-	ac	6.53	4BooX3			11/53	-	
3	B/	CPw,x		1953mc	1958mc	ac	6.53	4BooX3					
3	B//*	CPz,z-	OPz	1953ᶜms	nd:r	acf	6.53	4BooX3			3/73	6/73	

AA 1.9.46 on A(A), 3/-, OSPR 5/47. M on B/.

SE 94 (44/94) *South Dalton**   10 by 10   Revision a 1908-26   c 1938-50
1	A	CPr	OPr	1947po	-	-	6.47	4BE1X3	RB		11/47	1/48	
3	B	CPt	OPs	1953pc	-	ac	6.53	4BooX3			12/53	3/55	
3	B	CPw,z		1953mc	-	ac	6.53	4BooX3					

AA 1.8.47 on A(A), 3/-, OSPR 1/48. M on A, B.

SE 95 (44/95) *Wetwang*   10 by 10   Revision a 1908-26   c 1946-53
1	A	CPr		1947po	-	-	1.47	4BE1X3	RB	0	10/47	-	
1	B		OPr	1947po	-	-	1.47	4BE1X3	RB	0	-	12/47	
3	C	CPt		1953pc	-	ac	6.53	4BooX3		1	11/53	-	
3	C/	CPw,z,0z		1953mc	1960mc	ac	6.53	4BooX3		1			1

AA 1.1.47 on B(A), 3/-, OSPR 12/47. M on B. 1. Driffield Airfield.

SE 96 (44/96) *Sledmere*   10 by 10   Revision a 1909-26   c 1946-50
1	1046	CPr		1946po	-	-	6.46	4BE1X3	RB	0	9/46	-	
1	A		OPr	1946po	-	-	6.46	4BE1X3	RB	0	-	1/47	
3	B	CPt		1954pc	-	ac	6.54	4BooX3		1	2/54	-	
3	B/	CPw,z,0z		1954mc	1960mc	ac	6.54	4BooX3		1			

AA 1.6.46 on A(A), 3/-, OSPR 1/47. M on A.

SE 97 (44/97) *Sherburn Yorkshire**   10 by 10   Revision a 1909   c 1950
| 1 | A | CPr | OPr | 1947po | - | - | 6.47 | 4BE1X3 | RB | | 10/47 | 12/47 | |
| 3 | B | CPt,u,z | OPz | 1954pc | - | ac | 6.54 | 4BooX3 | | | 2/54 | - | |

AA 1.7.47 on A(A), 3/-, OSPR 12/47. M on A, B.

SE 98 (44/98) *Wykeham Yorkshire*   10 by 10   Revision a 1909-27   c 1938-50
1	A	CPr	OPr	1947po	-	-	1.47	4BE1X3	RB		10/47	1/48	
2	A	CPs		1947pc	1951	-	6.51	4Booo3	RB				
3	B	CPt		1954pc	-	ac	6.54	4BooX3			2/54	-	
3	B/	CPv,x,z,0z		1954pc	1957mc	ac	6.54	4BooX3					

AA 1.12.46 on A(A), 3/-, OSPR 1/48. M on A, B.

SE 99 (44/99) **& Part of TA 09** *Scalby*   15 by 10   Revision a 1910-27   c 1929-38   d 1938-50
1	A	CPr	OPr	1948po	-	-	1.48	4BE2X3	RB		3/48	6/48	
1	B	CPs		1951pc	-	ac	1.50	4BEoo3	RB		1/51	-	
3	C	CPt	OPv	1954pc	-	ad	6.54	4BooX3			3/54	9/57	
3	C/	CPw,y,z.z,0z	O0z	1954mc	1960mc	ad	6.54	4BooX3					

AA 1.12.47 on A(A), 3/-, OSPR 6/48. M on A, B.

SH 12 (23/12) *Aberdaron*   10 by 10   Revision a 1914   c 1949
| 1 | A | CPr | OPs | nd:po | - | - | 1.50 | 4BEoo3 | B | | 7/50 | 8/50 | |
| 3 | A/ | CPw,y,z | | 1950pc | 1958mc | ac | 1.50 | 4BooX3 | B | | | | |

Some front cover sheet indexes show incorrect sheet lines.

1	2	3	4	5	6	7	8	9	10	11	12	13	14

SH 22 (23/22) *Porth Neigwl**   10 by 10   Revision a 1914 c 1949-50
1 A   CPr   OPr   nd:po   -   -   6.49 4BEoo3   6/50 7/50
3 B   CPu.x,z   OPv   1955pc   -   ac   6.55 4oooX3   1/56 6/57

SH 23 (23/23) **& Part of SH 13** *Botwnnog** 15 by 10 Revision a 1914 c 1948-49
1 A   CPs   OPs   1950pc   -   ac   6.50 4BEoo3   11/50 12/50
3 A/   CPv,x,z.z   1950pc   1957mc ac   6.50 4BooX3

SH 27 (23/27) *Trearddur Bay**   10 by 10   Revision a 1923 c 1949-50
1 A   CPr   OPr   nd:po   -   -   1.50 4BEoo3   R   0 5/50 5/50
3 B   CPu   OPv   1956pc   -   ac   6.56 4oooX3   0 4/56 12/57
3 B/   CP?w,y,z   1956mc   1960mc ac   6.56 4oooX3   1

SH 28 (23/28) *Holyhead*   10 by 10   Revision a 1922-23 c 1949 f 1972
1 A   CPr   OPr   nd:po   -   -   1.50 4BEoo3   R   3/50 5/50
3 B   CPu   OPv   1955pc   -   ac   6.55 4oooX3   1/56 6/57
3 B/   CPw,z   1955mc   1960mc ac   6.55 4oooX3
3 B//*   CPz   OPz   1955ᶜms nd:r   acf   6.55 4oooX3   7/73 11/73
AA 1.11.49 on A(A), 3/-, OSPR 5/50.

SH 32 (23/32) *Abersoch*   10 by 10   Revision a 1914 c 1949-50
1 A   CPr   OPr   nd:po   -   -   1.50 4BEoo3   6/50 7/50
3 B   CPu   OPv   1955pc   -   ac   6.55 4oooX3   1/56 6/57
3 B   CPw,z   1955mc   -   ac   6.55 4oooX3

SH 33 (23/33) *Pwllheli**   10 by 10   Revision a 1913-14 c 1948-50
1 A   CPr   OPr   nd:po   -   -   1.50 4BEoo3   R   2 6/50 7/50
3 B   CPu   OPv   1956pc   -   ac   6.56 4oooX3   2 2/56 6/57
3 B/   CP?w,z   1956mc   1960mc ac   6.56 4oooX3   2
AA 1.5.50 on A(A), 3/-, OSPR 8/50. M on B, B/.

SH 34 (23/34) **& Part of SH 24** *Nefyn* 15 by 10 Revision a 1913-14 c 1948-50
1 A   CPr   OPr   nd:po   -   -   1.50 4BEoo3   7/50 8/50
3 B   CPu,x.z   OPv   1956pc   -   ac   6.56 4oooX3   4/56 9/57

SH 36 (23/36) *Aberffraw**   10 by 10   Revision a 1913-15 c 1949-50 f 1967
1 A   CPr   OPr   nd:po   -   -   1.50 4BEoo3   R   3/50 5/50
3 B   CPu   OPv   1956pc   -   ac   6.56 4oooX3   4/56 12/57
3 B/   CPz   OPz   1956mc   1968ch acf   6.56 4oooX3

SH 37 (23/37) *Rhosneigr*   10 by 10   Revision a 1915-23 c 1949-50
1 A   CPr   OPr   nd:po   -   -   1.50 4BEoo3   R   0 3/50 4/50
3 B   CPu,y,z   OPy   1956pc   -   ac   6.56 4oooX3   1 2/56 -

SH 38 (23/38) *Llanfaethlu*   10 by 10   Revision a 1915-23 c 1949-50 f 1966; j 1965 reservoir
1 A   CPr   OPr   nd:po   -   -   1.50 4BEoo3   3/50 5/50
3 B   CPu,y   OPy   1955pc   -   ac   6.55 4oooX4   12/55 -
3 B/*   CPz,0z   OPz   1955mc   1967v   acfj 6.55 4oooX4   5/67 5/67 1
1. Alaw Reservoir.

SH 39 (23/39) **& Part of SH 29** *Cemaes Bay* 15 by 10 Revision a 1899-1922 c 1948-49 f 1968
1 A   CPr   OPr   nd:po   -   -   1.50 4BEoo3   6/50 7/50
3 B   CPu   OPv   1955pc   -   ac   6.55 4oooX3   9/55 9/57
3 B/   CPx   1955pc   -   ac   6.55 4oooX3
3 B//   CPz,0z   OPz   1955mc   1969ch acf   6.55 4oooX3

SH 43 (23/43) *Chwilog**   10 by 10   Revision a 1913-14 c 1948-50
1 A   CPr   OPr   nd:po   -   -   1.50 4BEoo3   R   6/50 7/50
3 B   CPv   1956pc   -   ac   6.56 4oooX3   6/56 -
3 B/   CPw,z   1956mc   1960mc ac   6.56 4oooX3
M on B, B/.

1	2	3	4	5	6	7	8	9	10	11	12	13	14

SH 44 (23/44) *Clynnog**   10 by 10   Revision   a 1913-14   c 1938-49
1   A   CPs   OPs   1951pc   -   ac   6.51   4BEoo4   R   11/51   11/51
3   A/   CPv,z   1951pc   1958mc   ac   6.51   4BooX4

SH 45 (23/45) *Pen-y-groes Caernarvonshire*   10 by 10   Revision   a 1913-14   c 1948-50
1   A   CPr   OPr   nd:po   -   -   1.50   4BEoo3   R   0   5/50   6/50
3   B   CPv   OPv   1956pc   -   ac   6.56   4oooX3   1   10/56   3/59
3   B/   CP?w,z   1956mc   1959mc   ac   6.56   4oooX3   1

SH 46 (23/46) *Caernarvon*   10 by 10   Revision   a 1913-15   c 1938-50
1   A   CPr   OPr   nd:po   -   -   1.50   4BEoo3   R   3/50   5/50
3   B   CPv   OPv   1956pc   -   ac   6.56   4oooX3   11/56   1/57
3   B/   CPw.x,z   1956mc   1960mc   ac   6.56   4oooX3
AA 1.11.49/NT 28.11.49 on A(A), 3/-, OSPR 5/50.

SH 47 (23/47) *Llangefni*   10 by 10   Revision   a 1913-15   c 1938-49   f 1967; j 1959 reservoir
1   A   CPr   OPr   nd:po   -   -   1.50   4BEoo3   R   3/50   5/50
3   B   CPu   1956pc   -   ac   6.56   4oooX3   1   2/56   -
3   B/   CPx   1956mc   1961ch   ac   6.56   4oooX3   1
3   B//*   CPz   OPz   1956mc   1968v   acfj   6.56   4oooX3   1   2/68   3/68

SH 48 (23/48) *Llanerchymedd**   10 by 10   Revision   a 1915-22   c 1949-51   f 1968; j 1966 [reservoir]
1   A   CPr   nd:po   -   -   6.50   4BEoo3   R   5/50   -
1   B [1]   OPr   nd:po   -   -   6.50   4BEoo3   R   -   7/50
3   B [2]   CPu.x   1956pc   -   ac   6.56   4oooX3   4/56   -
3   B/*   CPz   OPz   1956ms   1968rv   acfj   6.56   4oooX3   8/68   6/68   1
AA 1.5.50 on B(A), 3/-, OSPR 7/50. 1. Alaw Reservoir.

SH 49 (23/49) *Amlwch*   10 by 10   Revision   a 1922   c 1938-49
1   A   CPr   OPr   nd:po   -   -   1.50   4BEoo3   R   5/50   7/50
3   B   CPu   1956pc   -   ac   6.56   4oooX3   3/56   -
3   B/   CPw.z,0z   1956mc   1961mc   ac   6.56   4oooX3
AA 1.5.50 on A(A), 3/-, OSPR 7/50.

SH 52 *Llanbedr Merionethshire*   10 by 10   Revision   a 1899-1900   c 1949-50   f 1965
2   A   CPt   OPs   1953pc   -   ac   6.53   4BooX4   1   11/53   11/53
3   A   CPu,x   1953pc   -   ac   6.53   4BooX4   1
3   A/   CPz   OPz   1953mc   1966ch   acf   6.53   4BooX4   0
3   A/   G0z   1953mc   1966ch   acf   6.53   4BooX4   0   5/77   1
M on A. 1. Overprinted in green with public rights of way information to May 1965.

SH 53 (23/53) *Portmadoc*   10 by 10   Revision   a 1899-1914   c 1938-49
1   A   CPr   OPs   nd:po   -   -   1.50   4BEoo3   R   7/50   9/50
2   A   CPt   1950pc   1952   -   1.50   4Booo3   R
3   B   CPv   1956pc   -   ac   6.56   4oooX3   6/56   -
3   B/   CPw,x,z   1956mc   1960mc   ac   6.56   4oooX3
3   B/   G0z   1956mc   1960mc   ac   6.56   4oooX3   5/77   -   1
AA 1.5.50/NT 11.5.50 on A(A), 3/-, OSPR 9/50. M on A. 1. Overprinted in green with public rights of way information to May 1965.

SH 54 *Beddgelert*   10 by 10   Revision   a 1912-14   c 1949
2   A   CPt   OPs   1953pc   -   ac   6.53   4BooX4   8/53   10/53
3   A   CPu   1953pc   -   ac   6.53   4BooX4
3   A/   CP?w,x.z   1953mc   1960mc   ac   6.53   4oooX4
AA 1.6.53/NT 28.8.53 on A(A), 4/-, OSPR 10/53. M on A.

SH 55 *Llyn Cwellyn*   10 by 10   Revision   a 1912-13   c 1946-49
2   A   CPt   OPs   1953pc   -   ac   6.53   4BooX4   R   4/53   5/53
3   A/   CPv,x.z   1953pc   1957c   ac   6.53   4BooX4   R
M on A.

1	2	3	4	5	6	7	8	9	10	11	12	13	14

SH 56 (23/56) *Port Dinorwic** 10 by 10 Revision a 1912-13 c 1938-47
1 A CPr OPs 1950pc - ac 6.50 4BEoo3 R 7/50 8/50
  A/
3 A// CPx.y,z   1950mc 1961ch ac 6.50 4BooX3
AA 1.5.50 on A(A), 3/-, OSPR 8/50. M on A.

SH 57 (23/57) *Bangor Caernarvonshire* 10 by 10 Revision a 1913-15 c 1948-50 f 1961
1 A CPr OPr nd:po - - 1.50 4BEoo3 R 3/50 5/50
3 B CPu OP? 1956pc - ac 6.56 4oooX3   2/56 2/58
3 B/ CPw   1956mc 1960mc ac 6.56 4oooX3
3 B// CPy,z OPz 1956mc 1963ch acf 6.56 4oooX3
AA 1.12.49/NT 16.2.50 on A(A), 3/-, OSPR 5/50.

SH 58 (23/58) **& Part of SH 59** *Red Wharf Bay** 10 by 12 Revision a 1914-16 c 1949
1 A CPr OPr nd:po - - 1.50 4BEoo3 R 4/50 5/50
3 B CPu.z OPv 1956pc - ac 6.56 4oooX3   2/56 6/57

SH 60 & Part of SH 50 *Bryn-crug* 15 by 10 Revision a 1900 c 1948-52
3 A CPt,u OPs 1953pc - ac 6.53 4BooX3   11/53 2/54
3 A/ CPw,y,z.0z   1953mc 1960mc ac 6.53 4BooX3   1
AA 1.9.53 on A(A), 4/-, OSPR 2/54. M on A/. 1. Showing the alignment of the Towyn-Bryneglwys tramway.

SH 61 & Part of SH 51 *Barmouth* 12 by 10 Revision a 1899-1900 c 1938-52
3 A CPt,u OPs 1953pc - ac 6.53 4BooX3   12/53 1/54
3 A/ CPw,x,z.z   1953mc 1960mc ac 6.53 4BooX3
3 A[/] G0z   1953mc 1960mc ac 6.53 4BooX3   . 1
M on A, A/. 1. Overprinted in green with public rights of way information to May 1965.

SH 62 *Y Llethr* 10 by 10 Revision a 1899-1900 c 1949
2 A CPt OPs 1953pc - ac 6.53 4BooX4   8/53 9/53
3 A/ CPv   1953pc 1956mc ac 6.53 4BooX4
3 A// CP?w.y,z   1953mc 1960mc ac 6.53 4BooX4
3 A//[/*] GPz   1953mc 1960mc ac 6.53 4BooX4   5/77 - 1
M on A. 1. Overprinted in green with public rights of way information to May 1965.

SH 63 *Penrhyndeudraeth* 10 by 10 Revision a 1899-1913 c 1949
2 A CPt OPs 1953pc - ac 6.53 4BooX4   9/53 10/53
3 A CPv   1953pc - ac 6.53 4BooX4
3 A/ CPw.y,z   1953mc 1960mc ac 6.53 4BooX4
3 A/ G0z   1953mc 1960mc ac 6.53 4BooX4   1
AA 1.6.53 on A(A), 4/-, OSPR 10/53. M on A. 1. Overprinted in green with public rights of way information to May 1965.

SH 64 *Blaenau Ffestiniog (West)* 10 by 10 Revision a 1899-1914 c 1949 f 1964
2 A CPt,u OPs 1953pc - ac 6.53 4BooX4   10/53 1/54
3 A/
3 A// CPz OPz 1953mc 1965ch acf 6.53 4BooX4
3 A// G0z   1953mc 1965ch acf 6.53 4BooX4   5/77 - 1
AA 1.6.53/NT 6.11.53 on A(A), 4/-, OSPR 1/54. M on A. 1. Overprinted in green with public rights of way information to May 1965.

SH 65 *Pass of Llanberis* 10 by 10 Revision a 1911-14 c 1948-49
2 A CPt OPs 1953pc - ac 6.53 4BooX4   7/53 7/53
3 A/ CPv   1953pc 1957mc ac 6.53 4BooX4
3 A// CPw,y.z   1953mc 1961ch ac 6.53 4BooX4
M on A.

SH 66 *Bethesda* 10 by 10 Revision a 1911-13 c 1938-48
2 A CPt,u OPs 1953pc - ac 6.53 4BooX4   10/53 11/53
3 A/ CPw,x,z.z,0z   1953mc 1958mc ac 6.53 4BooX4
AA 1.6.53/NT 19.9.53 on A(A), 4/-, OSPR 11/53. M on A.

```
  1  2       3       4     5         6        7     8     9         10   11  12     13    14
```

SH 67 (23/67) *Llanfairfechan**   10 by 10 Revision a 1911-14 c 1938-51
1 A   CPr   OPs   nd:po   -   -   1.50 4BEoo3   R   8/50   9/50
3 B   CPu   OPv   1956pc   -   ac   6.56 4BooX3   3/56   6/57
3 B/   CPw,z   1956mc   1960mc   ac   6.56 4BooX3
AA 1.10.46 on A(A), 3/-, OSPR 9/50. M on A.

SH 68 (23/68) *Penmon*   10 by 10 Revision a 1913-14 c 1948-49
1 A   CPr   OPr   nd:po   -   -   1.50 4BEoo3   3/50   4/50
3 B   CPu,z,0z OPv 1956pc   -   ac   6.56 4oooX3   3/56   8/57

SH 70 *Machynlleth*   10 by 10 Revision a 1900 c 1938-49
3 A   CPt   OPs   1954pc   -   ac   6.54 4BooX4   2/54   4/54 1
3 A/   CPw,z.0z   1954mc   1960mc   ac   6.54 4oooX4
AA 1.12.53 on A(A), 4/-, OSPR 4/54. 1. Showing the Corris Tramway alignment.

SH 71 *Dolgellau*   10 by 10 Revision a 1900 c 1938-49
3 A   CPt,u   OPs   1954pc   -   ac   6.54 4BooX4   1/54   1/54
3 A/   CPx.z   1954mc   1961ch   ac   6.54 4BooX4
3 A/(/*) GPz   1954mc   1961ch   ac   6.54 4BooX4   5/77   -   1
1. Overprinted in green with public rights of way information to May 1965.

SH 72 *Ganllwyd*   10 by 10 Revision a 1899-1900 c 1938-49 f 1970
2 A   CPt   OPs   1953pc   -   ac   6.53 4BooX4   9/53   10/53
3 A   CPw   1953mc   -   ac   6.53 4BooX4
3 A/   CP?w,x,z   1953mc   1958mc   ac   6.53 4BooX4
3 A//   CPz   OPz   1953^{c}ms nd:ch   acf   6.53 4BooX4
3 A//   G0z   1953^{c}ms nd:ch   acf   6.53 4BooX4   6/77   -   1
M on A. 1. Overprinted in green with public rights of way information to May 1965.

SH 73 *Trawsfynydd*   10 by 10 Revision a 1899-1900 c 1949 f 1965 g 1972
2 A   CPt   OPs   1953pc   -   ac   6.53 4BooX4   8/53   8/53
3 A   CPv   1953pc   -   ac   6.53 4BooX4
3 A/   CPw   1953mc   1960mc   ac   6.53 4BooX4
3 A//*   CPz   OPz   1953mc   1966r   acf   6.53 4BooX4   3/66   3/66
3 A//*/   CPz,0z OPz   1953^{c}ms nd:ch   acg   6.53 4BooX4
M on A.

SH 74 *Blaenau Ffestiniog (East)*   10 by 10 Revision a 1899-1914 c 1949
2 A   CPt   1953pc   -   ac   6.53 4BooX4   10/53 -
3 A   OPs   1953pc   -   ac   6.53 4BooX4   -   1/54
3 A/   CPv,x,z   1953pc   1957mc   ac   6.53 4BooX4
3 A/   G0z   1953pc   1957mc   ac   6.53 4BooX4   5/77   -   1
AA 1.6.53/NT 6.11.53 on A(A), 4/-, OSPR 1/54. M on A. 1. Overprinted in green with public rights of way information to May 1965.

SH 75 *Betws-y-Coed*   10 by 10 Revision a 1910-14 c 1946-49
2 A   CPt,u   OPs   1953pc   -   ac   6.53 4BooX4   8/53   8/53
3 A/   CPw,y,z.z   1953mc   1960mc   ac   6.53 4BooX4
M on A. GS: *Capel Curig & Betws-y-Coed,* Solid and Drift, 1976 on A/.

SH 76 *Dolgarrog**   10 by 10 Revision a 1910-12 c 1938-49
2 A   CPt,u   OPs   1953pc   -   ac   6.53 4BooX4   8/53   8/53
3 A/   CPx.y,z   1953mc   1961ch   ac   6.53 4BooX4
M on A. GS: Solid and Drift, 1981 on A/.

SH 77 (23/77) *Conway*   10 by 10 Revision a 1911-39 c 1937-49
1 A   CPr   nd:po   -   -   1.50 4BEoo3   R   7/50   -
1 B   OPs   nd:po   -   -   1.50 4BEoo3   R   -   9/50
2 B   CPt   1950pc   1953rpb -   1.50 4BooX3   R
3 B/   CPv,x,z OPv   1950pc   1956c   ac   1.50 4BooX3   -   2/57
M on B. AA 1.6.50 on B(A), 3/-, OSPR 9/50.

1	2	3	4	5	6	7	8	9	10	11	12	13	14

SH 78 (23/78) **& Part of SH 88** *Llandudno* 12 by 10 Revision a 1911-39 c 1948-51
1	A	CPr		nd:po	-	-	1.50	4BEoo3	R	3/50	-	1
1	B		OPr	nd:po	-	-	1.50	4BEoo3	R	-	5/50	
3	B	CPu.z		1956pc	-	ac	6.56	4ooX3		4/56	-	

Sheet lines are offset, at 273 km E to 285 km E. Some front cover sheet indexes show incorrect sheet lines. AA 1.3.50 on B(A), 3/-, OSPR 5/50. 1. Showing the Great Ormes Head tramway.

SH 80 *Cemmaes* 10 by 10 Revision a 1900 c 1948-49 f 1965
2	A	CPt	OPs	1953pc	-	ac	6.53	4BooX4	10/53	11/53
3	A/	CPw		1953mc	1960mc	ac	6.53	4BooX4		
3	A//	CPz	OPz	1953mc	1966mc	acf	6.53	4BooX4		

SH 81 *Mallwyd* 10 by 10 Revision a 1900 c 1946-49 f 1965 g 1970
3	A	CPt	OPs	1953pc	-	ac	6.53	4BooX4	11/53	12/53
3	A/	CPw,x		1953mc	1958mc	ac	6.53	4BooX4		
3	A//	CPz		1953mc	1966ch	acf	6.53	4BooX4		
3	A///	CPz,0z	OPz	1953^{c}ms	nd:ch	acg	6.53	4BooX4		

SH 82 *Aran Fawddwy* 10 by 10 Revision a 1899-1900 c 1949
| 2 | A | CPt | OPs | 1953pc | - | ac | 6.53 | 4BooX4 | 7/53 | 8/53 |
| 3 | A/ | CPv,x.z,0z | | 1953pc | 1957mc | ac | 6.53 | 4BooX4 | | |

M on A.

SH 83 *Arennig Fawr* 10 by 10 Revision a 1899-1900 c 1949 f 1965
j 1965 major water k 1965 railway
2	A	CPt	OPs	1953pc	-	ac	6.53	4BooX4	7/53	7/53	
3	A/	CPv		1953pc	1957mc	ac	6.53	4BooX4			
3	A//*	CPz	OPz	1953mc	1966v	acfjk	6.53	4BooX4	7/66	8/66	1
3	A//*	G0z		1953mc	1966v	acfjk	6.53	4BooX4	5/77	-	2

M on A. 1. Llyn Celyn. 2. Overprinted in green with public rights of way information to May 1965.

SH 84 *Ysbyty Ifan* 10 by 10 Revision a 1899-1914 c 1949 f 1965 g 1969
j 1965 reservoir k 1965 railway
2	A	CPt	OPs	1953pc	-	ac	6.53	4BooX4	8/53	8/53	
3	A/	CPu		1953pc	1956mc	ac	6.53	4BooX4			
3	A//*	CPz	OPz	1953mc	1966v	acfjk	6.53	4BooX4	10/66	10/66	1
3	A//*/	CPz,0z	OPz,0z	1953ms	1970ch	acgjk	6.53	4BooX4		9/73	2

1. Llyn Celyn. 2. OSPR (coloured edition) not found.

SH 85 *Pentrefoelas* 10 by 10 Revision a 1910-11 c 1949
| 3 | A | CPt | OPs | 1953pc | - | ac | 6.53 | 4BooX4 | 11/53 | 12/53 |
| 3 | A/ | CPw,y.z | | 1953pc | 1958mc | ac | 6.53 | 4BooX4 | | |

SH 86 *Llangernyw* 10 by 10 Revision a 1910-12 c 1938-49
2	A	CPt	OPs	1952pc	-	ac	6.52	4Booo4	R	8/52	9/52
3	A/	CPv,x		1952pc	1958c	ac	6.52	4BooX4			
3	A//	CPz		1952pc	1958c	ac	6.52	4BooX4			

SH 87 (23/87) *Colwyn Bay* 10 by 10 Revision a 1910-39 c 1949
| 1 | A | CPr | OPr | nd:po | - | - | 1.50 | 4BEoo3 | R | 5/50 | 7/50 | |
| 3 | A/ | CPw,x,z | | 1950mc | 1959mc | ac | 1.50 | 4BooX3 | | | | 1 |

AA 1.1.50 on A(A), 3/-, OSPR 7/50. 1. Ordnance Survey hold 1967 experimental copies of this printing with "Contours are at 5 metre V.I. in N.W. Quadrant" and "Contours are at 10 metre V.I. in N.W. Quadrant".

SH 90 *Talerddig* 10 by 10 Revision a 1900 c 1949
| 2 | A | CPt | OPs | 1953pc | - | ac | 1.53 | 4BooX4 | 4/53 | 4/53 |
| 3 | A/ | CPw,z,0z | | 1953mc | 1959mc | ac | 1.53 | 4BooX4 | | |

SH 91 *Foel* 10 by 10 Revision a 1900 c 1949 f 1969
2	A	CPt	OPs	1953pc	-	ac	1.53	4BooX4	1/53	2/53
3	A	CPx,z		1953pc	-	ac	1.53	4BooX4		
3	A/	CPz,0z		1953^{c}ms	nd:ch	acf	1.53	4BooX4		

1	2	3	4	5	6	7	8	9	10	11	12	13	14

SH 92 *Lake Vyrnwy (West)*   10 by 10   Revision a 1899-1900 c 1949
2   A   CPt   OPs   1953pc   -   ac   6.53   4BooX4   1/53   2/53
3   A   CPz   1953pc   -   ac   6.53   4BooX4

SH 93 *Bala*   10 by 10   Revision a 1899-1900 c 1949
2   A   CPt   OPs   1953pc   -   ac   6.53   4BooX4   7/53   7/53
3   A/   CPw,z.z   1953mc   1958mc   ac   6.53   4BooX4
3   A/   G0z   1953mc   1958mc   ac   6.53   4BooX4   6/77   -   1
1. Overprinted in green with public rights of way information to May 1965.

SH 94 *Cerrigydrudion*   10 by 10   Revision a 1899-1910 c 1946-49
2   A   CPt   OPs   1953pc   -   ac   6.53   4BooX4   4/53   4/53
3   A/   CPv,z,0z   1953pc   1957mc   ac   6.53   4BooX4

SH 95 *Alwen Reservoir*   10 by 10   Revision a 1910 c 1949 f 1970
2   A   CPt   OPs   1952pc   -   ac   6.52   4Booo4   7/52   7/52
3   A/   CPw   1952mc   1960mc   ac   6.52   4BooX4
3   A//   CPz,0z   O0z   1952^{c}ms   nd:ch   acf   6.52   4BooX4

SH 96 *Llansannan*   10 by 10   Revision a 1910 c 1948-49
2   A   CPs,u   OPs   1951pc   -   ac   1.51   4Booo4   2/52   3/52
3   A   CPz   1951pc   -   ac   1.51   4Booo4
M on A.

SH 97 (23/97) **& Part of SH 98** *Abergele*   10 by 12   Revision a 1910-37 c 1949-50
1   A   CPr   OPr   nd:po   -   -   1.50   4BEoo3   R   6/50   7/50
3   A   CPt   1950pc   -   -   1.50   4BEoo3   R
3   A/   CPv,z.z   1950pc   1958mc   ac   1.50   4BooX3
AA 1.1.50 on A(A), 3/-, OSPR 8/50. M on A/.

SJ 00 *Llanllugan*   10 by 10   Revision a 1900-01 c 1946-49
2   A   CPt,w   OPs   1952pc   -   ac   6.52   4Booo4   9/52   10/52
3   A   CPz   1952pc   -   ac   6.52   4Booo4

SJ 01 *Llanwddyn*   10 by 10   Revision a 1900 c 1938-49 f 1969
2   A   CPt   OPs   1952pc   -   ac   6.52   4Booo4   9/52   10/52
3   A/   CPw,z   1952mc   1960mc   ac   6.52   4BooX4
3   A//   CPz,0z   1952ms   1970ch   acf   6.52   4BooX4

SJ 02 *Llangynog Montgomeryshire*   10 by 10   Revision a 1900-10 c 1938-49
2   A   CPt   OPs   1952pc   -   ac   6.52   4Booo4   R   11/52   1/53
3   A/   CPw,y,z.0z   1952mc   1959mc   ac   6.52   4BooX4

SJ 03 *Llandrillo*   10 by 10   Revision a 1899-1910 c 1949 f 1969
2   A   CPt   OPs   1952pc   -   ac   6.52   4Booo4   R   10/52   10/52
3   A/   CPw,z   1952pc   1958mc   ac   6.52   4BooX4
3   A//   CPz,0z   1952ms   1970ch   acf   6.52   4BooX4

SJ 04 (33/04) *Corwen**   10 by 10   Revision a 1899-1910 c 1949 f 1966
1   A   CPr   OPr   nd:po   -   -   6.49   4BEoo3   RB   1/50   1/50
3   B   CPv   OPv   1956pc   -   ac   6.56   4oooX3   6/56   10/57
3   B/   CP?w,z   1956mc   1960mc   ac   6.56   4oooX3
3   B//   CPz,0z   OPz   1956mc   1966ch   acf   6.56   4oooX3

SJ 05 *Gyffylliog*   10 by 10   Revision a 1899-1910 c 1949 f 1969
2   A   CPt   OPs   1952pc   -   ac   6.52   4Booo4   R   11/52   11/52
3   A/   CPw,x   1952mc   1959mc   ac   6.52   4BooX4
3   A//   CPz   OPz   1952ms   1969ch   acf   6.52   4BooX4

SJ 06 (33/06) *Denbigh**   10 by 10   Revision a 1910-11 c 1938-49 f 1966
1   A   CPr   OPr   nd:po   -   -   1.50   4BEoo3   R   3/50   3/50

1	2	3	4	5	6	7	8	9	10	11	12	13	14
3	B	CPu,x	OPv	1956pc	-	ac	6.56	4oooX3			1/56	5/57	
3	B/	CPz	OPz	1956mc	1967ch	acf	6.56	4oooX3					

M on B.

SJ 07 (33/07) *St Asaph** 10 by 10 Revision a 1910-11 c 1938-49

1	A	CPr	OPr	1947po	-	-	1.48	3BE1X1	R		11/47	2/48	
1	B	CPs		1950pc	-	ac	6.50	3BEoo1	R				
3	B/	CPv.z	OPv	1950pc	1957mc	ac	6.50	3BooX2			-	5/57	

AA 1.9.47 on A(A), 3/-, OSPR 2/48. M on B/.

SJ 08 (33/08) & Part of SH 98 *Rhyl** 12 by 10 Revision a 1910-11 c 1938-49

1	A	CPr		1947po	-	-	6.47	3BE1X1	R		8/47	-	
1	B		OPr	1947po	-	-	6.47	3BE1X1	R		-	11/47	
		CP?									1/55	-	
3	C	CPu	OPv	1956pc	-	ac	6.56	3oooX1			3/56	6/57	
3	C/	CPw,x		1956mc	1959mc	ac	6.56	3oooX1					
3	C//	CPz,z+		1956mc	1959mc	ac	6.56	3oooX1					

AA 1.2.47 on B(A), 3/-, OSPR 11/47.

SJ 10 *Llanfair Caereinion* 10 by 10 Revision a 1900-01 c 1949

| 2 | A | CPt | OPs | 1952pc | - | ac | 6.52 | 4Booo4 | R | | 10/52 | 1/53 | |
| 3 | A/ | CPw.z,0z | | 1952mc | 1960mc | ac | 6.52 | 4BooX4 | | | | | |

AA 1.8.52 on A(A), 4/-, OSPR 1/53.

SJ 11 *Llanfyllin* 10 by 10 Revision a 1900-01 c 1938-49

| 2 | A | CPt | OPs | 1952pc | - | ac | 6.52 | 4Booo4 | R | | 8/52 | 9/52 | |
| 3 | A/ | CPw,z,0z | | 1952mc | 1959mc | ac | 6.52 | 4BooX4 | | | | | |

AA 1.4.52 on A(A), 4/-, OSPR 9/52.

SJ 12 *Llanrhaeadr-ym-Mochnant* 10 by 10 Revision a 1900-10 c 1938-49 f 1966

2	A	CPt	OPs	1952pc	-	ac	6.52	4Booo4	R		11/52	1/53	
3	A/	CP?w,y		1952pc	1958mc	ac	6.52	4BooX4					
3	A//	CPz,0z		1952mc	1966ch	acf	6.52	4BooX4					

SJ 13 *Llanarmon Dyffryn Ceiriog* 10 by 10 Revision a 1899-1910 c 1949 f 1969

2	A	CPt	OPs	1952pc	-	ac	6.52	4Booo4			8/52	9/52	
3	A/	CPw		1952mc	1960mc	ac	6.52	4BooX4					
3	A//	CPz,0z	OPz	1952^{c}ms	nd:ch	acf	6.52	4BooX4					

SJ 14 (33/14) *Horseshoe Pass* 10 by 10 Revision a 1899-1910 c 1949

1	A	CPr	OPr	nd:po	-	-	1.49	4BEoo3	R		1/50	1/50	
3	B	CPu,x	OPv	1956pc	-	ac	6.56	4oooX3			4/56	1/57	
3	B/	CPz,0z	OPz	1956pc	-	ac	6.56	4oooX3					

SJ 15 (33/15) *Ruthin* 10 by 10 Revision a 1910 c 1938-49

1	A	CPr	OPr	nd:po	-	-	6.49	4BEoo3	RB		12/49	1/50	
3	B	CPu	OPv	1956pc	-	ac	6.56	4oooX3			5/56	5/57	
3	B/	CPw,z		1956mc	1961mc	ac	6.56	4oooX3					

SJ 16 (33/16) *Cilcain** 10 by 10 Revision a 1910 c 1938-49

1	A	CPr	OPr	nd:po	-	-	6.48	4BEoo3	R		10/49	10/49	
2	A	CPt		1949pc	1952	-	6.48	4Booo3	R				
3	B	CPu	OPv	1956pc	-	ac	6.56	4oooX3			3/56	12/57	
3	B/	CPw,y,z.z		1956mc	1960mc	ac	6.56	4oooX3					

SJ 17 (33/17) *Holywell Flintshire** 10 by 10 Revision a 1910 c 1949

1	A	CPr	OPr	1947po	-	-	6.47	3BE1X1	R		8/47	11/47	
3	B	CPt		1947pc	1954c	-	6.54	3BooX2					
3	C	CPv		1956pc	-	ac	6.56	3oooX2			8/56	-	
3	C/	CPv,z		1956pc	1958mc	ac	6.56	3oooX2					

AA 1.4.47 on A(A), 3/-, OSPR 11/47. SS: Soil, Land Use Capability, coloured, 1978 on C/.

1	2	3	4	5	6	7	8	9	10	11	12	13	14

SJ 18 (33/18) *Mostyn**   10 by 10   Revision a 1910-25 c 1949-51

1	A	CPr	OPr	1947po	-	-	1.48	3BE1X1	R		10/47	12/47	
3	B	CPu	OPv	1955pc	-	ac	6.55	3oooX2			1/55	5/57	
3	B/	CPx,z	OPx	1955mc	1961mc	ac	6.55	3oooX2					

AA 1.10.47 on A(A), 3/-, OSPR 12/47.

SJ 20 (33/20) *Welshpool*   10 by 10   Revision a 1900-01 c 1938-49

| 1 | A | CPr | OPr | nd:po | - | - | 6.49 | 4BEoo3 | R | | 2/50 | 4/50 | |
| 3 | B | CPv,x,z | OPv,z | 1957pc | - | ac | 6.57 | 4oooX3 | | | 7/57 | 8/57 | |

AA 1.9.49 on A(A), 3/-, OSPR 4/50.

SJ 21 (33/21) *Arddleen*   10 by 10   Revision a 1900 c 1949

1	A	CPr		nd:po	-	-	1.50	4BEoo3	R		3/50	-	
1	B [1]		OPr	nd:po	-	-	1.50	4BEoo3	R		-	5/50	
3	B [2]	CPv	OPv	1957pc	-	ac	6.57	4oooX3			9/57	9/57	
3	B/	CPw,x,z,0z	OPv	1957mc	1959mc	ac	6.57	4oooX3					

AA 1.1.50 on B(A), 3/-, OSPR 5/50. SS: Soil, coloured, 1982 on B/.

SJ 22 (33/22) *Oswestry (South)*   10 by 10   Revision a 1900-25 c 1938-49 f 1966

1	A	CPr	OPr	nd:po	-	-	6.49	4BEoo3	RB		1/50	2/50	
3	B	CPv	OPv,z	1957pc	-	ac	6.57	4oooX4			1/58	3/58	
3	B	CPw.z		1957mc	-	ac	6.57	4oooX4					
3	B/	CPz,0z	OPz	1957mc	1966ch	acf	6.57	4oooX4					

AA 1.11.49 on A(A), 3/-, OSPR 2/50.

SJ 23 (33/23) *Chirk*   10 by 10   Revision a 1899-1924 c 1938-49

1	A	CPr	OPr	nd:po	-	-	6.49	4BEoo3	R		1/50	2/50	
3	B	CPv,x,z	OPv,y	1957pc	-	ac	6.57	4oooX3			5/57	8/57	
3	B/	CPz,0z	OPz	1957pc	-	ac	6.57	4oooX3					

AA 1.11.49 on A(A), 3/-, OSPR 2/50.

SJ 24 (33/24) *Llangollen*   10 by 10   Revision a 1909-10 c 1938-49 f 1965

1	A	CPr	OPr	nd:po	-	-	6.48	4BEoo3	RC		12/48	1/49	
1	B	CPs	OPs	1951pc	-	ac	1.51	4BEoo3	RC		8/51	9/51	
3	B	CPu		1951pc	-	ac	1.51	4BEoo3	RC				
3	B/	CP?w,z	OPv,x	1951mc	1959mc	ac	1.51	4BooX3	C				
3	B//	CPz,0z	OPz	1951mc	1966ch	acf	1.51	4BooX3	C				

SJ 25 (33/25) *Treuddyn**   10 by 10   Revision a 1909-10 c 1938-49 f 1966

1	A	CPr	OPr	nd:po	-	-	1.49	4BEoo3	R		12/48	1/49	
1	B	CPs		1951pc	-	ac	1.51	4BEoo3	R		7/51	-	
3	B	CPu	OPv	1951pc	-	ac	1.51	4BEoo3	R		-	5/57	
3	B/	CPw,y	OPv	1951mc	1959mc	ac	1.51	4BooX3					
3	B//	CPz		1951mc	1966ch	acf	1.51	4BooX3					

SJ 26 (33/26) *Mold*   10 by 10   Revision a 1909-10 c 1938-49 f 1966

1	A	CPr	OPr	nd:po	-	-	1.50	4BEoo3	R		2/50	5/50	
2	B	CPt		1953pc	-	ac	6.53	4BooX3			9/53	-	
3	B/	CPv	OPv	1953pc	1956mc	ac	6.53	4BooX3			-	5/57	
3	B//	CPw	OPx	1953mc	1961mc	ac	6.53	4BooX3					
3	B///	CPz	OPz	1953mc	1967ch	acf	6.53	4BooX3					

AA 1.3.50 on A(A), 3/-, OSPR 5/50.

SJ 27 (33/27) *Flint**   10 by 10   Revision a 1909-10 c 1936-53

1	20046	CPr		1946po	-	-	6.46	3BE1X1	R		8/46	-	
1	2546		OPr	1946po	-	-	6.46	3BE1X1	R		-	11/46	
2	B	CPt		1953pc	-	ac	6.53	3BooX2			9/53	-	
3	B/	CPv		1953pc	1956mc	ac	6.53	3BooX2					
3	B//	CP?w,y,z	OPx,z	1953mc	1959mc	ac	6.53	3BooX2					

AA 1.6.46 on 2546(2546), 3/-, OSPR 11/46. M on B.

1	2	3	4	5	6	7	8	9	10	11	12	13	14

SJ 28 (33/28) [Hoylake]   10 by 10   Revision a 1909-36 c 1935-51 f 1965
1 15046   CPr   1946po   -   -   1.47   3BE1X1   R   11/46 -
1 A   OPr   1946po   -   -   1.47   3BE1X1   R   -   2/47
2 B   CPt   1953pc   -   ac   6.53   3BooX2   9/53   -
3 B/   CPv,x OPv   1953pc   1957mc ac   6.53   3BooX2
3 B//   CPz   OPz   1953mc   -   acf   6.53   3BooX2
AA 1.9.46 on A(A), 3/-, OSPR 2/47. M on A.

SJ 29 (33/29) **& Part of SJ 19** *Wallasey**   12 by 10   Revision a 1925-36 c 1938-51
1 10046   CPr   1946po   -   -   6.46   3BE1X1   R   5/46   -
1 2546   OPr   1946po   -   -   6.46   3BE1X1   R   -   7/46
2 B   CPt   OPv   1953pc   -   ac   6.53   3BooX2   R   4/53   9/57
3 B/   CPv,y,z.z OPv,z 1953pc 1957mc ac   6.53   3BooX2
AA 1.3.46 on 2546(2546), 3/-, OSPR 7/46. M on 10046.

SJ 30 (33/30) *Minsterley*   10 by 10   Revision a 1901 c 1949
1 A   CPr   OPr   nd:po   -   -   6.49   4BEoo3   RB   1/50   1/50   1
3 B   CPv   OPv   1957pc   -   ac   6.57   4oooX3   3/57   5/57
3 B/   CPx,z OPx   1957mc   1961mc ac   6.57   4oooX3
1. The Snailbeach District Railway is shown and named.

SJ 31 (33/31) *Nesscliff**   10 by 10   Revision a 1900-01 c 1949 f 1965
1 A   CPr   OPr   nd:po   -   -   1.50   4BEoo3   RB   2/50   3/50
3 B   CPv   OPv   1956pc   -   ac   6.56   4oooX3   8/56   10/57 1
3 B/   CPw,x   1956mc   1959mc ac   6.56   4oooX3
3 B//   CPz,0z OPz   1956mc   1966ch acf   6.56   4oooX3
1. The munitions sites at 3716 and 3519 are left blank.

SJ 32 (33/32) *Ruyton-XI-Towns**   10 by 10   Revision a 1900-25 c 1949
1 A   CPr   OPr   nd:po   -   -   6.49   4BEoo3   RB   0   1/50   2/50
3 B   CPv   OP?   1957pc   -   ac   6.57   4oooX3   3   2/58   2/58
3 B   CPw,z,0z   1957mc   -   ac   6.57   4oooX3   3
AA 1.8.49 on A(A), 3/-, OSPR 2/50.

SJ 33 (33/33) *St Martin's Shropshire*   10 by 10   Revision a 1899-1924 c 1949
1 A   CPr   OPr   nd:po   -   -   6.49   4BEoo3   R   1/50   2/50
3 A   CPv   1949pc   -   -   1.52   4Booo3   R
3 B   CPv,y,z.z,0z OPv 1957pc -   ac   6.57   4oooX3   7/57   8/57
AA 1.11.49 on A(A), 3/-, OSPR 2/50.

SJ 34 (33/34) *Wrexham (South)**   10 by 10   Revision a 1909-24 c 1949 f 1974
1 A   CPr   OPr   nd:po   -   -   1.48   4BEoo3   RB   10/48 11/48
2 A   CPt   OPr   1948pc   1952rpb -   1.48   4BooX3   RB
3 B   CPu,x,z OPv   1956pc   -   ac   6.56   4oooX3   5/56   5/57
3 B/*   CFz   OFz   1956^{c}ms nd:r acf   6.56   4oooX3   5/75   5/75
AA 1.8.48 on A(A), 3/-, OSPR 11/48.

SJ 35 (33/35) *Wrexham (North)**   10 by 10   Revision a 1909-10 c 1938-53 f 1965 g 1974
1 A   CPr   nd:po   -   -   6.48   4BEoo3   RB   00   10/48 -
1 A   CP   nd:po   -   -   6.48   4BEoo3   RB   00   1
1 B   OPr   nd:po   -   -   6.48   4BEoo3   RB   00   -   11/48
2 C   CPt   OPv   1953pc   -   ac   6.53   4BooX3   11   9/53   7/56
3 C/   CPv,x   1953pc   1957c ac   6.53   4BooX3   11
3 C//   CPz   OPz   1953mc   1966ch acf   6.53   4BooX3   33
3 C///*   CFz   OFz   1953^{c}ms nd:r acg   6.53   4BooX3   33   5/75   5/75
AA 1.6.48 on B(A), 3/-, OSPR 11/48. SS: Soil, outline, 1977 on C///*. 1. Overprinted for military use in GSGS Misc.1605/3 *Exercise Ballista III,* priced 2/6, 6/-.

157

	1	2	3	4	5	6	7	8	9	10	11	12	13	14

SJ 36 (33/36) *Chester (West)**   10 by 10   Revision a 1908-09   c 1938-49   f 1962   g 1965
1	A		CPr		1948po			1.48	4BE2X3	R	000	2/48	-
1	B			OPr	1948po			1.48	4BE2X3	R	000	-	6/48
2	C		CPs	OPs	1951pc		ac	6.51	4Booo3	R	010	1/52	4/54
3	C/		CPw	OPv	1951pc	1958mc	ac	6.51	4BooX3		010		
3	C//		CPy		1951mc	1963ch	acf	6.51	4BooX3		010		
3	C///		CPz	OPz	1951mc	1966ch	acg	6.51	4BooX3		310		

AA 1.4.48 on B(A), 3/-, OSPR 6/48.

SJ 37 (33/37) *Ellesmere Port (West)**   10 by 10   Revision a 1908-37   c 1938-49   f 1972; j 1971 industrial
1	10046	CPr		1946po			6.46	3BE1X1	R	10	11/46	-
1	A		OPr	1946po			6.46	3BE1X1	R	10	-	4/47
1	B	CPs		1951pc		ac	1.52	3BEoo1	R	10	12/51	-
3	B	CPu	OPv	1951pc		ac	1.52	3BooX2	R	10	-	11/57
3	B/	CP?w,x,z OPz	1951mc	1959mc	ac	1.52	3BooX2		10		
3	B//*	CPz	OPz	1951ᶜms	nd:ir	acfj	1.52	3BooX2		10	11/73	2/74

AA 1.9.46/NT 14.11.46 on A(A), 3/-, OSPR 4/47; AA 1.9.46/NT 1.7.47 on A(B), 3/-, OSPR 10/47. M on B. SS: Soil, outline, 1973 on B/.

SJ 38 (33/38) *Bebington*   10 by 10   Revision a 1908-36   c 1938-49   f 1964
1	20046	CPr		1946po			6.46	3BE1X1	RB	12/46	-
1	A		OPr	1946po			6.46	3BE1X1	RB		2/47
1	B	CPs		1951pc		ac	1.51	3BEoo1	RB	12/51	-
2	B		OPs	1951pc		ac	1.52	3Booo2	RB	-	7/52
3	B/	CPw,x	OPv	1951pc	1958c	ac	1.52	3BooX2	B		
3	B//	CPz	OPz	1951mc	1965ch	acf	1.52	3BooX2	B		

AA 1.5.46 on A(A), 3/-, OSPR 2/47; AA 1.11.53 on B(B), 4/-, OSPR 11/54. M on B, B/.

SJ 39 (33/39) *Bootle**   10 by 10   Revision a 1924-36   c 1938-50
1	30046	CPr		1946po			6.46	3BE1X1	R	7/46	-
1	5046		OPr	1946po			6.46	3BE1X1	R	-	12/46
2	B	CPt	OPs	1953pc		ac	6.53	3BooX2		9/53	8/54
3	B/	CPw,x,z	OPv	1953pc	1958mc	ac	6.53	3BooX2			

AA 1.5.46 on 5046(2546), 3/-, OSPR 12/46; AA 1.11.53 on B(B), 4/-, OSPR 8/54. M on 30046.

SJ 40 (33/40) *Dorrington Shropshire*   10 by 10   Revision a 1900-25   c 1949
| 1 | A | CPr | OPr | 1948po | | 6.48 | 4BEoo3 | R | 7/48 | 8/48 |
| 3 | B | CPv.x,z | OPv,z | 1957pc | ac | 6.57 | 4oooX3 | | 9/57 | 8/57 |

AA 1.2.48 on A(A), 3/-, OSPR 8/48.

SJ 41 (33/41) *Shrewsbury (West)*   10 by 10   Revision a 1900-25   c 1938-49
1	A	CPr	OPr	1948po			6.48	4BEoo3	RB	0	6/48	8/48
1	B	CPs		1951pc		ac	6.51	4BEoo3	RB	0	9/51	-
3	B/	CPv.x,z		1951pc	1957mc	ac	1.52	4BooX3	B	0		

AA 1.6.48 on A(A), 3/-, OSPR 8/48.

SJ 42 (33/42) *Myddle*   10 by 10   Revision a 1900-24   c 1949   f 1965
1	A	CPr	OPr	nd:po			6.49	4BEoo3	RB	0	1/50	1/50
3	A	CPu		1950pc			6.49	4BEoo3	RB	0		
3	B	CPv.y,z	OPv	1957pc		ac	6.57	4oooX3		1	5/57	6/57
3	B/	CPz	OPz	1957mc	1966ch	acf	6.57	4oooX3		3		

SJ 43 (33/43) *Welshampton*   10 by 10   Revision a 1909-29   c 1949
| 1 | A | CPr | OPr | nd:po | | 6.49 | 4BEoo3 | R | 2/50 | 3/50 |
| 3 | B | CPv,x,z.z | OPv | 1957pc | ac | 6.57 | 4oooX3 | | 7/57 | 8/57 |

SJ 44 (33/44) *Malpas*   10 by 10   Revision a 1909   c 1949   f 1961
1	A	CPr	OPr	1948po		6.48	4BEoo3	R	5/48	6/48
1	B	CPr		nd:po		6.48	4BEoo3	R		
3	C	CPx,z,0z	OPx	1961mc	acf	6.61	4oooX3		12/61	11/61

158

1	2	3	4	5	6	7	8	9	10	11	12	13	14

SJ 45 (33/45) *Farndon Cheshire**   10 by 10   Revision a 1908-09 c 1949
1	A	CPr	OPr	1948po	-	-	6.48	4BE2X3	R	0	3/48	4/48	
1	B	CPr		nd:po	-	-	6.48	4BEoo3	R	0			
1	B	CP		nd:po	-	-	6.48	4BEoo3	R	0			1
2	B	CPu		1949pc	-	-	6.48	4BooX3	R	0			
3	C	CPx,z,z-	OPx	1961mc	-	ac	6.61	4oooX3		1	5/61	5/61	

M on C. 1. Overprinted for military use in GSGS Misc.1605/3 *Exercise Ballista III,* priced 2/6, 6/-.

SJ 46 (33/46) *Chester (East)**   10 by 10   Revision a 1908-37 c 1938-53 f 1966
1	A	CPr	OPr	1948po	-	-	1.48	4BE2X3	RB	0	1/48	3/48	
1	B	CPr		nd:po	-	-	1.48	4BEoo3	RB	0			
2	C	CPt		1953pc	-	ac	6.53	4BooX3		1	9/53	-	
3	C/	CPw,x,z	OPv	1953pc	1958c	ac	6.53	4BooX3		1			
3	C//	CPz	OPz	1953mc	1966ch	acf	6.53	4BooX3		3			

AA 1.11.47 on A(A), 3/-, OSPR 3/48.

SJ 47 (33/47) *Ellesmere Port (East)*   10 by 10   Revision a 1908-37 c 1938-49 f 1965 g 1974
j 1963-64 tide lines k 1964 industrial
1	10046	CPr		1946po	-	-	1.47	3BE1X1	R		11/46	-	
1	A		OPr	1946po	-	-	1.47	3BE1X1	R		-	4/47	
1	B	CPs		1952pc	-	ac	1.52	3BEoo1	R		1/52	-	
2	B/	CPv		1952pc	1957c	ac	1.52	3BooX2					
3	B//	CPx	OPz	1952mc	1961c	ac	1.52	3BooX2					
3	B///*	CPz	OPz	1952mc	1966it	acfjk	1.52	3oooX4			10/66	8/66	
3	B///*/*	CFz	OFz	1952^{c}ms	nd:r	acgjk	1.52	3oooX4			6/74	7/74	

AA 1.10.46 on A(A), 3/-, OSPR 4/47. M on B.

SJ 48 (33/48) *Liverpool (South East)**   10 by 10   Revision a 1924-36 c 1938-49 f 1966
1	15046	CPr		1946po	-	-	6.46	3BE1X1	R	1	12/46	-	1
1	A		OPr	1946po	-	-	6.46	3BE1X1	R	1	-	4/47	
1	B	CPs		1951pc	-	ac	1.51	3BEoo1	R	3	12/51	-	
3	B		OPs	1951pc	-	ac	1.51	3BEoo1	R	3	-	2/54	
3	B		OPv,w	1951pc	-	ac	1.51	3BEoo1	R	3	-	9/57	
3	B/	CP?w.x,y		1951mc	1960mc	ac	1.51	3BooX1		3			
3	B//*	CPz	OPz	1951mc	1967r	acf	1.51	3BooX2		3	4/67	4/67	

AA 1.10.46/NT 21.10.46 on A(A), 3/-, OSPR 4/47; AA 1.11.53/NT 21.10.46 on B(B), 4/-, OSPR 1/55. M on B, B/. 1. Liverpool Airport.

SJ 49 (33/49) [*Prescot*]   10 by 10   Revision a 1925 c 1937-52
1	15046	CPr		1946po	-	-	6.46	3BE1X1	R		8/46	-	
1	2546		OPr	1946po	-	-	6.46	3BE1X1	R		-	12/46	
2	B	CPt	OPs	1953pc	-	ac	6.53	3BooX2			9/53	11/54	
3	B/	CPw,x,z	OPv	1953pc	1958mc	ac	6.53	3BooX2					

AA 1.5.46 on 2546(2546), 3/-, OSPR 12/46; AA 1.11.53 on B(B), 4/-, OSPR 11/54. M on 15046, B/.

SJ 50 (33/50) *Cressage*   10 by 10   Revision a 1900-25 c 1938-49 f 1970
1	A	CPr	OPr	nd:po	-	-	1.48	4BEoo3	R	0	8/48	9/48	
2	A	CPu		1948pc	-	-	1.52	4Booo3	R	0			
3	B	CPv	OPv	1956pc	-	ac	6.56	4oooX3		1	1/57	1/57	
3	B/	CPw,z		1956mc	1959mc	ac	6.56	4oooX3		1			
3	B//	CPz	OPz	1956^{c}ms	nd:ch	acf	6.56	4oooX3		3			

AA 1.6.48 on A(A), 3/-, OSPR 9/48.

SJ 51 (33/51) *Shrewsbury (East)**   10 by 10   Revision a 1900-25 c 1938-49 f 1963
1	A	CPr	OPr	1948po	-	-	6.48	4BEoo3	RB	0	6/48	8/48	
1	B	CPs	OPs	1951pc	-	ac	1.51	4BEoo3	RB	0	10/51	11/51	
3	B/		OPv	1951pc	1957mc	ac	1.51	4BooX3	B	0			
3	B//	CPy,z	OPz	1951mc	1963ch	acf	1.51	4BooX3	B	0			

AA 1.6.48 on A(A), 3/-, OSPR 8/48.

1	2	3	4	5	6	7	8	9	10	11	12	13	14

SJ 52 (33/52) *Wem*   10 by 10   Revision a 1900-24 c 1949 f 1963

1	A	CPr	OPr	nd:po	-	-	6.49	4BEoo3	RB	0	1/50	3/50	
3	B	CPv	OPv	1956pc	-	ac	6.56	4oooX3		1	2/57	2/57	
3	B/	CPy,z,0z		1956mc	1963ch	acf	6.56	4oooX3		1			

SJ 53 (33/53) *Prees*   10 by 10   Revision a 1900-24 c 1949

| 1 | A | CPr | OPr | nd:po | - | - | 6.49 | 4BEoo3 | R | 0 | 1/50 | 1/50 | |
| 3 | B | CPv,x,z,0z | OP? | 1957pc | - | ac | 6.57 | 4oooX4 | | 3 | 2/58 | 3/58 | |

SJ 54 (33/54) *Whitchurch Shropshire**   10 by 10   Revision a 1908-24 c 1949

| 1 | A | CPr | OPr | nd:po | - | - | 6.48 | 4BEoo3 | RB | | 7/48 | 10/48 | |
| 3 | B | CPv,x,z.z,0z | OPv | 1957pc | - | ac | 6.57 | 4oooX3 | | | 1/57 | 2/57 | |

AA 1.7.48 on A(A), 3/-, OSPR 10/48.

SJ 55 (33/55) *Bulkeley*   10 by 10   Revision a 1908-09 c 1949

1	A	CPr	OPr	1948po	-	-	6.48	4BEoo3	RB	0	6/48	7/48	
1	B	CPr		nd:po	-	-	6.48	4BEoo3	RB	0			
3	B/	CPw		1948pc	1958mc	a	6.48	4BooX3	B	0			
3	C	CPw,x,z	OPv	1959mc	-	ac	6.59	4oooX3		3	10/59	10/59	

M on C.

SJ 56 (33/56) *Kelsall*   10 by 10   Revision a 1908 c 1938-49 f 1969

1	A	CPr	OPr	1948po	-	-	1.48	4BE2X3	RB		1/48	4/48	
1	B	CPr		1950pc	-	ac	6.50	4BEoo3	RB				
3	B/	CPw,x	OPv,z	1950pc	1958mc	ac	6.50	4BooX3					
3	B//	CPz	OPz	1950^{c}ms	nd:ch	acf	6.50	4BooX3					

AA 1.3.48 on A(A), 3/-, OSPR 4/48.

SJ 57 (33/57) *Frodsham**   10 by 10   Revision a 1908-09 c 1938-49 f 1970 g 1974

1	A	CPr	OPr	1947po	-	-	1.47	3BE1X1	RB		1/47	4/47	
2	B	CPt		1953pc	-	ac	6.53	3BooX2			8/53	-	
3	B/	CPv,x,z	OPx,z	1953pc	1958mc	ac	6.53	3BooX2					
3	B//*	CPz	OPz	1953^{c}ms	nd:r	acf	6.53	3BooX2			1/72	5/72	
3	B//*/*	CFz	OFz	1953^{c}ms	nd:r	acg	6.53	3BooX2			12/75	1/75	

AA 1.10.46 on A(A), 3/-, OSPR 4/47. M on A.

SJ 58 (33/58) *Widnes*   10 by 10   Revision a 1905-26 c 1937-49 d 1937-52 f 1961

1	A	CPr	OPr	1947po	-	-	6.47	3BE1X1	R	0	6/47	2/48	
1	B	CPs		1950pc	-	ac	6.50	3BEoo1	R	0	11/50	-	
2	C	CPt		1953pc	-	ad	6.53	3BooX2		1	7/53	-	
3	C/	CPw		1953mc	1958mc	ad	6.53	3BooX2		1			
3	C//*	CPx,z	OPx	1953mc	1962r	adf	6.53	3BooX4		1	7/62	9/62	

AA A(A) not found; AA 1.4.47 on A(B), 3/-, OSPR 2/48. M on B.

SJ 59 (33/59) [*St Helens*]   10 by 10   Revision a 1925-26 c 1938-49 f 1964; j 1964 airfield

1	15046	CPr		1946po	-	-	1.47	3BE1X1	R	0	10/46	-	
1	A		OPr	1946po	-	-	1.47	3BE1X1	R	0	-	2/47	
2	B	CPt		1953pc	-	ac	6.53	3BooX2		1	4/53	-	
3	B/	CPv,x	OPv	1953pc	1958c	ac	6.53	3BooX2		1			
3	B//*	CPz	OPz	1953mc	1965ar	acfj	6.53	3BooX2		3		4/65	1

AA 1.7.46 on A(A), 3/-, OSPR 2/47. M on B. 1. Burtonwood Airfield. OSPR (coloured edition) not found.

SJ 60 (33/60) *The Wrekin*   10 by 10   Revision a 1900-25 c 1938-49 f 1966
                                                  j 1982 selected k SUSI

1	A	CPr	OPr	nd:po	-	-	6.49	4BEoo3	RB		1/50	2/50	
3	B	CPv	OPv	1957pc	-	ac	6.57	4oooX3			3/58	7/58	
3	B/	CPw,x,z		1957mc	1958mc	ac	6.57	4oooX3					
3	B//	CPz	OPz	1957mc	1966ch	acf	6.57	4oooX3					
3	B///*	CFz	OFz	1982^{c}m-s	nd:brvy	acfjk	6.57	4oooX3			3/83	3/83	1

AA 1.8.49 on A(A), 3/-, OSPR 2/50. 1. The areas of building development are shaded grey.

1	2	3	4	5	6	7	8	9	10	11	12	13	14

SJ 61 (33/61) *Wellington Shropshire*   10 by 10   Revision a 1900-25 c 1938-49; j 1981 selected k SUSI
1	A	CPr	OPr	nd:po	-	-	1.50	4BEoo3	RB	0	10/49	11/49	
3	B	CPv,y,z,0z	OPv	1957pc	-	ac	6.57	4oooX3		0	9/57	9/57	
3	B/*	CFz	OFz	1982^{c}m-s	nd:br	acjk	6.57	4oooX3		0	4/82	4/82	

AA 1.9.49 on A(A), 3/-, OSPR 11/49.

SJ 62 (33/62) *Hodnet*   10 by 10   Revision a 1900-25 c 1949
1	A	CPr	OPr	nd:po	-	-	1.49	4BEoo3	RB	00	1/50	3/50	
3	B	CPv	OPv	1957pc	-	ac	6.57	4oooX4		12	7/57	8/57	
3	B/	CPw,y,z		1957mc	1960mc	ac	6.57	4oooX4		12			

SJ 63 (33/63) *Market Drayton**   10 by 10   Revision a 1900-25 c 1930-49 f 1966
1	A	CPr	OPr	nd:po	-	-	1.49	4BEoo3	RB	0	2/49	3/49	
3	B	CPv		1956pc	-	ac	6.56	4oooX3		1	1/57	-	
3	B/	CPw,y		1956pc	1958mc	ac	6.56	4oooX3		1			
3	B//	CPz	OPz	1956mc	1966ch	acf	6.56	4oooX3		3			

SJ 64 (33/64) *[Audlem]*   10 by 10   Revision a 1908-24 c 1949
1	A	CPr	OPr	1948po	-	-	1.48	4BE2X3	RB		4/48	5/48	
1	B	CPr		nd:po	-	-	1.48	4BEoo3	RB				
3	C	CPv,x.y	OPv	1957pc	-	ac	6.57	4oooX4			1/58	-	

AA 1.3.48 on A(A), 3/-, OSPR 5/48.

SJ 65 (33/65) *Crewe (West)*   10 by 10   Revision a 1907-09 c 1938-49
1	A	CPr		1948po	-	-	1.48	4BE2X3	R	0	3/48	-	
1	B		OPr	1948po	-	-	1.48	4BE2X3	R	0	-	6/48	
1	C	CPs	OPs	1951pc	-	ac	1.51	4BEoo3	R	0	9/51	-	
3	C		OPs	1951pc	-	ac	1.51	4BEoo3	R	0	-	10/54	
3	C/	CPv.x,z,z-	OPz	1951pc	1956mc	ac	1.51	4BooX3					

AA 1.3.48 on B(A), 3/-, OSPR 6/48. SS: Soil, outline, 1971 on C.

SJ 66 (33/66) *Winsford Cheshire*   10 by 10   Revision a 1907-09 c 1938-49
1	A	CPr	OPr	1948po	-	-	1.48	4BE2X3	R		3/48	4/48	
	B												
2	C	CPs,t	OPs	1952pc	-	ac	1.52	4Booo3	R		5/52	6/52	
3	C/	CPw,x,z.z,0z	OPv	1952mc	1958mc	ac	1.52	4BooX3					

AA 1.11.47 on A(A), 3/-, OSPR 4/48.

SJ 67 (33/67) *Northwich*   10 by 10   Revision a 1907-09 c 1938-49 f 1964
1	15046	CPr		1946po	-	-	1.47	3BE1X1	RB		9/46	-	
1	A		OPr	1946po	-	-	1.47	3BE1X1	RB		-	1/47	
2	B	CPt		1952pc	-	ac	6.52	3Booo2	R		9/52	-	
3	B/	CPv		1952pc	1956mc	ac	6.52	3BooX2					
3	B//	CPx	OPx	1952mc	1961ch	ac	6.52	3BooX2					
3	B///*	CPz	OPz	1952mc	1965r	acf	6.52	3BooX2			7/65	9/65	

AA 1.8.46 on A(A), 3/-, OSPR 1/47. M on B.

SJ 68 (33/68) *Warrington**   10 by 10   Revision a 1905-26 c 1937-49 f 1962 g 1966
1	A	CPr	OPr	1947po	-	-	6.47	3BE1X1	RB	0	11/47	2/48	
	B												
2	C	CPt		1952pc	-	ac	6.52	3Booo2	R	1	8/52	-	
3	C/	CPv	OPv	1952pc	1956mc	ac	6.52	3BooX2		1			
3	C//	CPw		1952mc	1961mc	ac	6.52	3BooX2		1			
3	C///*	CPy	OPy	1952mc	1963r	acf	6.52	3BooX2		1	7/63	7/63	1
3	C///*/*	CPz	OPz	1952mc	1967r	acg	6.52	3BooX2		1	6/67	6/67	

AA 1.9.47 on A(A), 3/-, OSPR 2/48. M on C. 1. With the M6 under construction.

1	2	3	4	5	6	7	8	9	10	11	12	13	14

SJ 69 (33/69) *Leigh (South) Lancashire**   10 by 10   Revision a 1904-39 c 1937-49 f 1964
1	15046	CPr		1946po		-	6.46	3BE1X1	R		4/46	-	
1	2546		OPr,s	1946po		-	6.46	3BE1X1	R		-	4/46	
2	B	CPt		1953pc		ac	1.53	4BooX4	R		1/53	-	1
3	B/	CPv,w		1953pc	1956mc	ac	1.53	4BooX4					
3	B//*	CPz	OPz	1953mc	1964r	acf	1.53	4BooX4			11/64	12/64	

AA 1.3.46 on 2546(2546), 3/-, OSPR 9/46. M on B, B/. 1. One of the two sheets remade to four-colour specification. Showing in great detail the layout of the Ordnance Factory at Risley, complete with its railway system.

SJ 70 (33/70) *Shifnal*   10 by 10   Revision a 1901-25 c 1938-49 f 1974
1	A	CPr	OPr	nd:po	-	-	6.49	4BEoo3	R	0	5/49	6/49
1	B	CPs,w	OPv	1951pc	-	ac	6.51	4BEoo3	R	0	1/52	5/57
3	B	CPy	OPz	1951pc	-	ac	6.51	4BEoo3	R	0		
3	B/	CPz	OPz	1951pc	-	ac	6.51	4BooX3		0		
3	B//*	CFz	OFz	1951^{c}ms	nd:r	acf	6.51	4BooX3		0	1/76	7/75

M on B.

SJ 71 (33/71) *Newport Shropshire*   10 by 10   Revision a 1900-24 c 1938-49
1	A	CPr		nd:po	-	-	6.49	4BEoo3	RB		5/49	-
1	B [1]		OPr	nd:po	-	-	6.49	4BEoo3	RB		-	6/49
1	B [2]	CPs		1951pc	-	ac	6.51	4BEoo3	R		12/51	-
3	B	CPu	OPv	1951pc	-	ac	6.51	4BEoo3	R		-	8/57
3	B	CPx,z		1951pc	-	ac	6.51	4BooX3	R			
3	B/	CPz,0z	OPz	1951pc	-	ac	6.51	4BooX3				

AA 1.4.49 on B(A), 3/-, OSPR 6/49. M on B [2].

SJ 72 (33/72) *Norbury Staffordshire**   10 by 10   Revision a 1900-25 c 1949
1	A	CPr	OPr	nd:po	-	-	6.49	4BEoo3	B	0	5/49	6/49
3	A	CPt		1949pc	-	-	6.49	4BEoo3	B	0		
3	B	CPw,z	OPv	1960mc	-	ac	6.60	4oooX3		1	11/60	11/60

SJ 73 (33/73) *Ashley*   10 by 10   Revision a 1899-1925 c 1949
1	A	CPr	OPr	1948po	-	-	6.48	4BE2X3	RB		4/48	5/48
1	B [1]	CPr		nd:po	-	-	6.48	4BEoo3	RB			
2	B [2]		OPs	1949pc	-	-	1.52	4Booo3	RB			
3	B/	CPw		1949mc	1959mc	a	1.52	4BooX3	B			
3	C	CPw,z	OPv	1960mc	-	ac	6.60	4oooX3			8/60	11/60

SJ 74 (33/74) *[Madeley]*   10 by 10   Revision a 1899-1924 c 1938-49 f 1966
1	A	CPr	OPr	1948po	-	-	6.48	4BE2X3	RB		4/48	4/48
1	B	CPr		nd:po	-	-	6.48	4BEoo3	RB			
3	C	CPw,x	OPv	1958pc	-	ac	6.58	4oooX3			9/58	10/58
3	C/*	CPz	OPz	1958mc	1967r	acf	6.58	4oooX3			2/67	2/67

M on B.

SJ 75 (33/75) *Crewe (East)*   10 by 10   Revision a 1897-1922 c 1938-49 f 1966
1	A	CPr	OPr	1948po		-	1.48	4BE2X3	R		4/48	5/48
	B											
2	C	CPt,u	OPv	1952pc	-	ac	6.52	4Booo3	R		8/52	11/57
3	C/	CPw,x		1952mc	1960mc	ac	6.52	4BooX3				
3	C//*	CPz	OPz	1952mc	1967r	acf	6.52	4BooX3			3/67	4/67

AA 1.6.46 on A(A), 3/-, OSPR 5/48. M on C, C/.

SJ 76 (33/76) *Sandbach**   10 by 10   Revision a 1907-08 c 1938-49 f 1963
1	A	CPr	OPr	1948po	-	-	1.48	4BE2X3	R	0	1/48	4/48
1	B	CPs		1950pc	-	ac	6.50	4BEoo3	R	2		
3	B/	CPw		1950mc	1960mc	ac	6.50	4BooX3		2		
3	B//*	CPz	OPz	1950mc	1964r	acf	6.50	4BooX3		2	4/64	3/64

AA 1.6.46 on A(A), 3/-, OSPR 4/48.

1	2	3	4	5	6	7	8	9	10	11	12	13	14

SJ 77 (33/77) *Knutsford* — 10 by 10 Revision a 1907-08 c 1938-49 f 1964

1	10046	CPr		1946po	-	-	1.47	3BE1X1	R	0	11/46	-	
1	A		OPr	1946po	-	-	1.47	3BE1X1	R	0	-	2/47	
2	B	CPt,u	OPv	1952pc	-	ac	1.52	3Booo2	R	1	6/52	5/57	
3	B/	CP?w,x		1952mc	1959	ac	1.52	3BooX2		1			
3	B//*	CPz,0z	OPz	1952mc	1965r	acf	1.52	3BooX2		1	2/65	3/65	

AA 1.9.46 on A(A), 3/-, OSPR 2/47. M on B.

SJ 78 (33/78) *Altrincham* — 10 by 10 Revision a 1904-37 c 1934-49 f 1963 g 1972

1	10046	CPr		1946po	-	-	1.47	3BE1X1	R		11/46	-	
1	A		OPr	1946po	-	-	1.47	3BE1X1	R		-	4/47	
1	B	CPs	OPv	1951pc	-	ac	1.51	3BEoo1	R		1/52	5/57	
3	B/	CPw		1951mc	1959mc	ac	1.51	3BooX2					
3	B//		OPv	1951mc	1960mc	ac	1.51	3BooX2					
3	B///*	CPz	OPz	1951mc	1964r	acf	1.51	3BooX2			10/64	11/64	
3	B///*/*	CPz	OPz	1951^{c}ms	nd:r	acg	1.51	3BooX2			5/73	6/73	

AA 1.5.46 on A(A), 3/-, OSPR 4/47. M on B, B//.

SJ 79 (33/79) *Eccles* — 10 by 10 Revision a 1904-35 c 1936-49 f 1961 g 1972

1	A	CPr	OPr	1947po	-	-	1.47	3BE1X1	R	0	3/47	6/47	
1	B	CPr		nd:po	-	-	1.47	3BEoo1	R	0			
2	C	CPt	OPs	1953pc	-	ac	6.52	3Booo2	R	1	1/53	11/53	1
3	C/	CPw	OPv	1953pc	1958c	ac	6.52	3BooX2		1	-	9/57	
3	C//*	CPx,z	OPx,z	1953mc	1961r	acf	6.52	3BooX2		1			2
3	C//*/*	CPz	OPz	1953^{c}ms	nd:r	acg	6.52	3BooX2		1	5/73	9/73	

AA 1.5.46 on A(A), 3/-, OSPR 6/47. M on B. 1. Barton Aerodrome. 2. OSPR not found.

SJ 80 (33/80) [Wolverhampton (North West)] — 10 by 10 Revision a 1900-21 c 1937-49 f 1966

1	A	CPr	OPr	1947po	-	-	1.47	4BE2X3	RB	00	1/47	5/47	
2	B	CPs,u	OPv	1952pc	-	ac	1.52	4Booo3	R	20	4/52	10/56	
3	B/	CP?w,x		1952mc	1959c	ac	1.52	4BooX3		20			
3	B//*	CPz	OPz	1952mc	1967r	acf	1.52	4BooX3		20	5/67	5/67	

AA 1.10.46 on A(A), 3/-, OSPR 5/47. M on B.

SJ 81 (33/81) [Wheaton Aston] — 10 by 10 Revision a 1900-21 c 1938 d 1949 f 1960

| 1 | A | CPr | OPr | nd:po | - | - | 1.49 | 4BEoo3 | RB | 0 | 5/49 | 5/49 | |
| 3 | B | CPx,z | OPx,z | 1961mc | - | adf | 6.61 | 4oooX3 | | 3 | 6/61 | 6/61 | |

M on A.

SJ 82 (33/82) *Eccleshall* — 10 by 10 Revision a 1921-22 c 1938 d 1938-49 f 1963

1	A	CPr	OPr	nd:po	-	-	1.49	4BEoo3	RB	0	1/49	1/49	
3	A/	CPw		1949mc	1959mc	ac	1.49	4BooX3	B	0			
3	B	CPw	OPv	1960mc	-	ad	6.60	4oooX3		3	8/60	8/60	
3	B/*	CPz	OPz	1960mc	1965r	adf	6.60	4oooX3		3	7/65	7/65	

SS: Soil, outline, Land Use Capability, outline, 1975 on B.

SJ 83 (33/83) *Stone (West)* — 10 by 10 Revision a 1921-22 c 1937-49 f 1963

1	A	CPr	OPr	1947po	-	-	1.47	4BE2X3	RB		4/47	7/47	1
1	B	CPr		nd:po	-	-	1.47	4BEoo3	RB				
2	C	CPs	OPs	1952pc	-	ac	1.52	4Booo3	R		4/52	6/52	2
3	C/	CPw		1952mc	1959mc	ac	1.52	4BooX3					
3	C//*	CPz	OPz	1952mc	1964r	acf	1.52	4BooX3			4/64	4/64	3
3	C//*/	CPz,0z	O0z	1952mc	1964r	acf	1.52	4BooX3					

AA 1.12.46 on A(A), 3/-, OSPR 7/47. M on C. 1. The ROF industrial complex at Cold Meece is not shown, nor its railway system. 2. Halls of residence are shown. 3. Every phone box on the M6 has an "R" symbol.

SJ 84 (33/84) [Stoke on Trent] — 10 by 10 Revision a 1922-23 c 1938-49 f 1965

1	A	CPr	OPr	1947po	-	-	6.47	4BE2X3	R		7/47	11/47	
1	B	CPs		1951pc	-	ac	1.51	4BEoo3	R		3/51	-	
3	B/	CPv,x	OPv,z	1951pc	1956c	ac	1.51	4BooX3			-	2/57	
3	B//*	CPz	OPz	1951mc	1966r	acf	1.51	4BooX3			5/66	7/66	

AA 1.4.47 on A(A), 3/-, OSPR 11/47. M on B.

1	2	3	4	5	6	7	8	9	10	11	12	13	14

SJ 85 (33/85) *Kidsgrove*   10 by 10   Revision a 1908-23 c 1938-50 f 1966 g 1974
                                            j 1965 railway

1	A	CPr	OPr	1947po	-	-	6.47	4BE2X3	R		7/47	11/47	
1	B [1]		OPs	1947po	-	-	6.47	4BE2X3	R				
2	B [2]	CPt	OPv	1952pc	-	ac	6.52	4Booo3	R		11/52	5/56	
3	B/	CPx	OPz	1952pc	1956mc	ac	6.52	4BooX3					
3	B//*	CPz,0z	OPz	1952mc	1967r	acfj	6.52	4BooX3			2/67	2/67	
3	B//*/*	CFz	OFz	nd:ms	nd:r	acgj	6.52	4BooX3			5/75	1/75	

AA 1.4.47/NT 7.11.47 on A(A), 3/-, OSPR 11/47. M on B [2].

SJ 86 (33/86) *Congleton*   10 by 10   Revision a 1907-22 c 1938-49 f 1962

1	A	CPr		1948po	-	-	6.48	4BEoo3	R		5/48	-	
1	B		OPr	1948po	-	-	6.48	4BEoo3	R		-	7/48	
1	B	CPr		nd:po	-	-	6.48	4BEoo3	R				
2	C	CPt,u	OPs	1952pc	-	ac	6.52	4Booo3	R		8/52	9/52	
3	C/	CPw	OPv	1952mc	1959mc	ac	6.52	4BooX3					
3	C//	CPy,z	OPz	1952mc	1963ch	acf	6.52	4BooX3					

AA 1.6.48 on B(A), 3/-, OSPR 7/48.

SJ 87 (33/87) *Alderley Edge**   10 by 10   Revision a 1907-45 c 1938-49

1	A	CPr	OPr	1947po	-	-	1.48	3BE1X1	R		10/47	12/47	
1	B	CPr		nd:po	-	-	1.48	3BEoo1	R				
2	B	CPt		1949pc	1952	-	1.52	3Booo2	R				
3	C	CPv	OPv	1957pc	-	ac	6.57	3oooX2			9/57	9/57	
3	C/	CPw,y,z	OPv,z	1957mc	1960mc	ac	6.57	3oooX2					

AA 1.8.47 on A(A), 3/-, OSPR 12/47. M on C.

SJ 88 (33/88) *Wilmslow*   10 by 10   Revision a 1904-36 c 1938-49 f 1974; j 1973 airfield

1	15046	CPr		1946po	-	-	1.47	3BE1X1	R	11	11/46	-	1
1	A		OPr	1946po	-	-	1.47	3BE1X1	R	11	-	4/47	
1	B	CPr		1947po	-	-	1.47	3BE1X1	R	11	4/47	-	2
1	C	CPs		1951pc	-	ac	1.51	3BEoo1	R	00	10/51	-	
3	C		OPv	1951pc	-	ac	1.51	3BooX2	R	00	-	10/56	
3	C/	CPw		1951mc	1959mc	ac	1.51	3BooX2		00			
3	C//	CPw,z.z	OPz	1951mc	1960ch	ac	1.51	3BooX2		00			
3	C///*	CFz	OFz	nd:ms	nd:ar	acfj	1.51	3BooX2		30	3/75	1/75	3

AA 1.5.46 on A(A), 3/-, OSPR 4/47. M on C, C//. 1,3. Manchester Airport. 2. One of only three coloured edition B sheets with legends.

SJ 89 (33/89) *Manchester*   10 by 10   Revision a 1908-35 c 1939-55 f 1972

1	1700/11/45 Cr	CPr		1945po	-	-	1.46	3BE1X1	RBC		12/45	-	1
1	5046		OPr	1945po	-	-	1.46	3BE1X1	RBC		-	5/46	
1	B	CPr,t		nd:po	-	-	1.46	3BEoo1	RBC				
3	B	CPt		1948pc	-	-	1.46	3BEoo1	RBC				
3	C	CPw,x,z	OPv,x	1959mc	-	ac	6.59	3oooX4	C		2/59	3/59	
3	C/*	CPz	OPz	1959^{c}ms	nd:r	acf	6.59	3oooX4	C		6/73	11/73	

AA 1.2.46 on 5046(2546), 3/-, OSPR 5/46. M on B. For the only recorded pre-publication printing see the introductory essay. 1. The second sheet to be published. As on the first, 40/09 (SZ 09), sheet prices are situated bottom left, but here the print code is below the frame above the sheet number.

SJ 90 (33/90) [Wolverhampton (North East)]   10 by 10   Revision a 1913-38 c 1946-49 f 1967

1	A	CPr	OPr	1947po	-	-	6.47	4BE2X3	R	1	5/47	9/47	
1	B	CPs		1950pc	-	ac	6.50	4BEoo3	R	2	1/51	-	
3	B		OPs	1950pc	-	ac	6.52	4BooX3	R	2	-	6/54	
3	B/												
3	B//	CPx	OPz	1950pc	1961ch	ac	6.52	4BooX3		2			1
3	B///*	CPz	OPz	1950mc	1968r	acf	6.52	4BooX3		2	3/68	3/68	2

AA 1.2.47 on A(A), 3/-, OSPR 9/47. M on B. 1. Wolverhampton Aerodrome. 2. Wolverhampton Airport.

1	2	3	4	5	6	7	8	9	10	11	12	13	14

SJ 91 (33/91) [Cannock]   10 by 10   Revision a 1915-21 c 1938-49 f 1964
1	A	CPr	OPr	nd:po	-	-	1.49	4BEoo3	R		2/49	4/49	
2	B	CPt	OPs	1952pc	-	ac	1.52	4Booo3	R		6/52	11/55	
3	B/	CPv,w		1952pc	1956mc	ac	1.52	4BooX3					
3	B//*	CPz	OPz	1952mc	1965r	acf	1.52	4BooX3			11/65	11/65	

AA 1.2.49 on A(A), 3/-, OSPR 4/49. M on B.

SJ 92 (33/92) *Stafford**   10 by 10   Revision a 1921-22 c 1938-49 f 1963 g 1965
1	A	CPr	OPr	nd:po	-	-	1.49	4BEoo3	RB	0	2/49	4/49	
2	B	CPt,u		1952pc	-	ac	6.52	4Booo3	R	1	8/52	-	
3	B		OPv	1952pc	-	ac	6.52	4BooX3	R	1	-	1/57	
3	B/	CPy	OPz	1952mc	1964ch	acf	6.52	4BooX3	R	1			
3	B//	CPz	OPz	1952mc	1966ch	acg	6.52	4BooX3	R	3			

AA 1.2.49 on A(A), 3/-, OSPR 4/49. M on B.

SJ 93 (33/93) *Stone (East) Staffordshire*   10 by 10   Revision a 1921-22 c 1938-49
1	A	CPr	OPr	1947po	-	-	1.47	4BE2X3	RB		2/47	5/47	
2	B	CPt	OPs	1952pc	-	ac	1.52	4Booo3	R		6/52	3/54	
3	B/												
3	B//	CPx,z.0z	OPz	1952mc	1961mc	ac	1.52	4BooX3					

AA 1.12.46 on A(A), 3/-, OSPR 5/47. M on B.

SJ 94 (33/94) [Longton]   10 by 10   Revision a 1922-37 c 1937-50
1	A	CPr	OPr	1947po	-	-	6.47	4BE2X3	RB	0	5/47	7/47	1
2	B	CPt	OPs	1952pc	-	ac	6.52	4Booo3	R	3	11/52	12/55	2
3	B/	CPv,x,z	OPz	1952pc	1957mc	ac	6.52	4BooX3		3	-	-	3

AA 1.2.47 on A(A), 3/-, OSPR 7/47. M on B. 1. Longton is labelled as a village on the coloured edition, and a town on the outline. 2. Stoke on Trent Municipal Aerodrome. 3. Meir Aerodrome (Stoke-on-Trent Corporation).

SJ 95 (33/95) *Leek Staffordshire*   10 by 10   Revision a 1922-23 c 1938-49
1	A	CPr	OPr	1947po	-	-	6.47	4BE2X3	RB		4/47	6/47	
1	B	CPr,s		1950pc	-	ac	6.50	4BEoo3	RB				
3	B		OPv	1950pc	-	ac	6.50	4BEoo3	RB			6/57	
3	B/	CPw	OPx	1950mc	1961mc	ac	6.50	4BooX3					
3	B//	CPz.z,0z	OPz	1950mc	1961mc	ac	6.50	4BooX3					

AA 1.12.46 on A(A), 3/-, OSPR 6/47. M on B.

SJ 96 (33/96) *Bosley*   10 by 10   Revision a 1897-1923 c 1938-49
1	A	CPr	OPr	1948po	-	-	1.48	4BEoo3	RB		5/48	7/48	
2	A	CPs		1948pc	1951	-	6.51	4Booo3	RB				
3	B	CPw.x,z	OPv,x,z	1958mc	-	ac	6.58	4oooX3			3/59	4/59	

AA 1.5.48 on A(A), 3/-, OSPR 7/48.

SJ 97 (33/97) *Macclesfield**   10 by 10   Revision a 1907-10 c 1949 e no date f 1966
1	A	CPr	OPr	1947po	-	-	6.47	3BE1X1	R		10/47	2/48	
1	B	CPr	OPs	1950pc	-	ace	6.50	3BEoo2	R				
3	B	CPu	OP?	1950pc	-	ace	6.50	3BEoo2	R		-	11/55	
3	B/	CP?w,y	OPy	1950mc	1960mc	ace	6.50	3BooX2					
3	B//	CPz	OPz	1950mc	1966ch	acef	6.50	3BooX2					

AA 1.8.47/NT 5.12.47 on A(A), 3/-, OSPR 2/48.

SJ 98 (33/98) *Marple*   10 by 10   Revision a 1896-1919 c 1934-49 f 1970
j 1966 airfield k 1966 building
1	15047	CPr		1947po	-	-	1.47	3BE1X1	RB	0	12/46	-	
1	A		OPr	1947po	-	-	1.47	3BE1X1	RB	0	-	6/47	
2	B	CPt		1952pc	-	ac	6.52	3Booo2	R	1	8/52	-	
3	B/	CPx,z	OPv,x,z	1952pc	1956mc	ac	6.52	3BooX2		1	-	5/57	
3	B//*	CPz,0z	OPz	nd:ms	nd:ab	acfjk	6.52	3BooX2		3	9/70	12/70	

AA 1.4.46 on A(A), 3/-, OSPR 6/47. M on B.

1	2	3	4	5	6	7	8	9	10	11	12	13	14

SJ 99 (33/99) *Ashton-under-Lyne* 10 by 10 Revision a 1906-34 c 1938-49 f 1978
1	A		CPr	OPr	1947po		-	1.47	3BE1X1	RC		8/47	11/47	
1	B		CPs		1951pc		ac	1.51	3BEoo1	RC		7/51	-	
2	B		CPu	OPs	1951pc		ac	1.52	3BEoo1	RC		-	11/54	
3	B/		CPw,x,z	OPv,x,z	1951mc	1958c	ac	1.52	3BooX2	C				
3	B//*		CFz	OFz	1951^{c}ms	nd:r	acf	-	3BooX2	C		4/80	4/80	

AA 1.4.46 on A(A), 3/-, OSPR 11/47. M on B, B/.

SK 00 (43/00) *Brownhills* 10 by 10 Revision a 1912-21 c 1938-49 f 1968; j 1959 building
1	A		CPr	OPr	1947po		-	6.47	4BE2X3	R		6/47	10/47	
1	B		CPs		1951pc		ac	1.51	4BEoo3	R		8/51	-	
3	B			OPs	1951pc		ac	1.51	4BEoo3	R		-	11/55	
3	B/		CPw		1951pc	1958mc	ac	1.51	4BooX3					
3	B//		CPw,z.z	OPz	1951mc	1960mc	ac	1.51	4BooX3					
3	B///*		CPz	OPz	1951ms	1969br	acfj	1.51	4BooX3			4/69	5/69	

AA 1.4.47/NT 1.8.47 on A(A), 3/-, OSPR 10/47. M on B.

SK 01 (43/01) *Rugeley* 10 by 10 Revision a 1915-22 c 1938-49 f 1975
1	A		CPr	OPr	nd:po		-	6.48	4BEoo3	RB		1/49	2/49	
2	B		CPt	OPv	1952pc		ac	6.52	4Booo3	R		9/52	9/55	
3	B/		CPw,x.y,z	OPz	1952mc	1960mc	ac	6.52	4BooX3					
3	B//*		CFz	OFz	1952^{c}ms	nd:r	acf	6.52	4BooX3			7/76	7/76	

AA 1.8.48 on A(A), 3/-, OSPR 2/49. M on B.

SK 02 (43/02) *Abbot's Bromley** 10 by 10 Revision a 1920-22 c 1949 f 1965 g 1977
  j 1959 reservoir k 1959 railway
1	A		CPr	OPr	nd:po		-	1.49	4BEoo3	RB	0	1/49	2/49	
3	B		CPw	OPx	1961mc		ac	6.61	4oooX3		3	1/61	2/61	
3	B/*		CPz	OPz	1961mc	1967v	acfjk	6.61	4oooX3		3	1/67	2/67	1
3	B/*/*		CFz	OFz	1961^{c}ms	nd:r	acgjk	6.61	4oooX3		3			2

M on A. 1. Blithfield Reservoir (OSPR). 2. OSPR not found.

SK 03 (43/03) *Uttoxeter* 10 by 10 Revision a 1920-22 c 1937-49
1	A		CPr	OPr	nd:po		-	1.49	4BEoo3	RB		4/49	5/49	
2	B		CPt	OPs	1952pc		ac	6.52	4Booo3	R		7/52	8/54	
3	B/			OPv	1952pc	1956mc	ac	6.52	4BooX3					
3	B//		CPx.z,0z		1952mc	1961ch	ac	6.52	4BooX3					

SK 04 (43/04) *Cheadle* 10 by 10 Revision a 1920-22 c 1949
1	A		CPr	OPr	nd:po		-	1.49	4BEoo3	RB		6/49	6/49	
1	B		CPs		1951pc		ac	1.51	4BEoo3	RB		8/51	-	
3	B			OPv	1951pc		ac	1.51	4BEoo3	RB		-	2/57	
3	B/		CPw.y,z,0z	OPx,z	1951mc	1961mc	ac	1.51	4BooX3	B				

SK 05 (43/05) *Onecote* 10 by 10 Revision a 1919-23 c 1937-49
1	A		CPr	OPr	nd:po		-	6.48	4BEoo3	RB		1/49	3/49	
2	B		CPs,u	OPv	1951pc		ac	6.51	4Booo3	R		2/52	2/57	
3	B/		CPw,x,z,0z	OPz	1951mc	1959mc	ac	6.51	4BooX3					

AA 1.1.49/NT 24.11.48 on A(A), 3/-, OSPR 3/49. M on B/. SS: Soil, outline, 1975 on B/.

SK 06 (43/06) *Longnor Staffordshire* 10 by 10 Revision a 1897-1923 c 1938-49
1	A		CPr	OPr	nd:po		-	1.49	4BEoo3	RB		2/49	4/49	
1	B		CPs	OPs	1950pc		ac	6.50	4BEoo3	R		12/50	5/56	
3	B/		CP?w,x,z.0z		1950mc	1958mc	ac	6.50	4BooX3					

AA 1.2.49 on A(A), 3/-, OSPR 4/49. M on B/. GS: *The Roaches & Upper Dove Valley,* Solid and Drift, 1975 on B/.

1	2	3	4	5	6	7	8	9	10	11	12	13	14

SK 07 (43/07) *Buxton Derbyshire*   10 by 10   Revision a 1907-19 c 1938-49 f 1972; j 1968 reservoir
1	A	CPr	OPr	1948po	-	-	1.48	4BE1X3	RB		1/48	3/48
1	B	CPs	OPs	1951pc	-	ac	1.51	4BEoo3	RB		7/51	8/51
	B/											
3	B//	CPw,x,z		1951mc	1960mc	ac	1.51	4BooX3	B			
3	B///*	CPz,0z	OPz	1951ᶜms	nd:rv	acfj	1.51	4BooX3	B		4/73	6/73

AA 1.11.47 on A(A), 3/-, OSPR 3/48. GS: Solid and Drift, 1975 on B///*.

SK 08 (43/08) *Whaley Bridge*   10 by 10   Revision a 1897-19 b 1896-38 c 1938-49 d 1938-51 f 1974
1	A	CPr	OPr	1948po	-	-	6.48	4BE3X3	RB		4/48	6/48
1	B	CPs		1951pc	-	ac	1.51	4BEoo3	RB		2/51	-
2	C	CPt		1951pc	1953c	ac	6.53	4BooX3	B			
3	D	CPu	OPv	1955pc	-	bd	6.55	4oooX3			5/55	2/57
3	D/	CPw,x,z.z		1955mc	1959mc	bd	6.55	4oooX3				
3	D//	CFz	OFz	1955ᶜms	nd:ch	bdf	6.55	4oooX3				

AA 1.2.48/NT 5.4.48 on A(A), 3/-, OSPR 6/48. M on D.

SK 09 (43/09) *Glossop*   10 by 10   Revision a 1896-1936 c 1938-51 f 1973
1	A	CPr	OPr	1947po	-	-	6.47	4BE2X3	RB		6/47	8/47
1	B	CPr		nd:po	-	-	6.47	4BEoo3	RB			
2	B	CPt		1949pc	1952	-	6.47	4Booo3	RB			
3	C	CPu	OPv	1955pc	-	ac	6.55	4oooX3			6/55	2/57
3	C/	CPw,x,z.z		1955mc	1960mc	ac	6.55	4oooX3				
3	C//	CPz	OPz	1955ᶜms	nd:ch	acf	6.55	4oooX3				

AA 1.3.47 on A(A), 3/-, OSPR 8/47. M on C, C/.

SK 10 (43/10) *Lichfield*   10 by 10   Revision a 1912-21 c 1938-49 f 1961 g 1971
1	A	CPr	OPr	1947po	-	-	6.47	4BE2X3	RB		6/47	9/47
1	B	CPs		1951pc	-	ac	1.51	4BEoo3	RB		9/51	-
3	B		OPv	1951pc	-	ac	1.51	4BEoX3	RB		-	6/56
3	B/	CPv		1951pc	1956mc	ac	1.51	4BooX3	B			
3	B//	CPx,z		1951mc	1961ch	acf	1.51	4BooX3	B			
3	B///*	CPz,0z	OPz	1961ᶜms	nd:r	acg	1.51	4BooX3	B		5/72	9/72

AA 1.5.46 on A(A), 3/-, OSPR 9/47. M on B.

SK 11 (43/11) *Alrewas**   10 by 10   Revision a 1920-22 c 1949 f 1959 g 1974
1	A	CPr	OPr	nd:po	-	-	6.48	4BEoo3	RB	0	9/48	10/48
3	B	CPx,y,z	OPx	1961mc	-	acf	6.61	4oooX3		3	7/61	3/61
3	B/*	CFz	OFz	1961ᶜms	nd:r	acg	6.61	4oooX3		3	10/74	6/74

AA 1.9.48 on A(A), 3/-, OSPR 10/48. M on A.

SK 12 (43/12) *Draycott in the Clay*   10 by 10   Revision a 1920-22 c 1949 f 1959
1	A	CPr	OPr	nd:po	-	-	6.48	4BEoo3	B	0	3/49	4/49
3	A	CPu		1949pc	-	a	6.48	4BooX3	B	0		
3	B	CPx,z,0z	OPx,z	1961mc	-	acf	6.61	4oooX4		3	7/61	5/61

M on A.

SK 13 (43/13) *Sudbury Derbyshire*   10 by 10   Revision a 1920 c 1949 f 1969 g 1973
1	A	CPr	OPr	1948po	-	-	6.49	4BEoo3	RB	0	5/48	6/48
3	B	CPw,x,z	OPv,z	1960mc	-	ac	6.60	4oooX3		3	10/60	11/60
3	B/	CPz	OPz	1960ms	1970ch	acf	6.60	4oooX3		3		
3	B//*	CPz,0z	OPz	1960ᶜms	nd:r	acg	6.60	4oooX3		3	12/73	12/73

SK 14 (43/14) *Ashbourne*   10 by 10   Revision a 1920 c 1938-49 f 1966
1	A	CPr	OPr	nd:po	-	-	1.49	4BEoo3	RB	00	2/49	3/49
1	B	CPs		1951pc	-	ac	1.51	4BEoo3	R	00	5/51	-
3	B		OPs	1951pc	-	ac	1.51	4BEoo3	R	00	-	10/55
3	B/	CPw,y,z	OPv	1951mc	1958mc	ac	1.51	4BooX3		0?		
3	B//	CPz,0z	OPz	1951mc	1966ch	acf	1.51	4BooX3		33		

167

1	2	3	4	5	6	7	8	9	10	11	12	13	14

SK 15 (43/15) *Alstonfield** 10 by 10 Revision a 1919-20 c 1949
1	A	CPr	OPr	nd:po	-	-	1.49	4BEoo3	RB		3/49	4/49	
1	B	CPs		1950pc	-	ac	6.50	4BEoo3	RB		12/50	-	
3	B		OPs	1950pc	-	ac	6.50	4BEoo3	RB		-	1/55	
3	B/	CPw		1950mc	1958mc	ac	6.50	4BooX3	B				
3	B//	CPx,y,z,zo	OPz	1950mc	1961ch	ac	6.50	4BooX3	B				

GS: *Dovedale,* Solid and Drift, 1980.

SK 16 (43/16) *Monyash* 10 by 10 Revision a 1919-20 c 1938-49
1	A	CPr	OPr	nd:po	-	-	1.49	4BEoo3	RB		1/49	1/49	
1	B	CPs	OPs	1951pc	-	ac	1.51	4BEoo3	RB		9/51	10/51	
3	B	CPu		1951pc	-	ac	1.51	4BEoo3	RB				
3	B/	CPx,z.z,z		1951mc	1961mc	ac	1.51	4BooX3	B				1

GS: Solid and Drift, 1977 on B/. 1. The bar under the B is lower on the additional unpriced state.

SK 17 (43/17) *Tideswell* 10 by 10 Revision a 1919-20 c 1949
1	A	CPr	OPr	1948po	-	-	6.48	4BE1X3	RB		4/48	5/48	
1	B	CPr		nd:po	-	-	6.48	4BEoo3	RB				
1	B	CPs		1949pc	1951	-	6.51	4BEoo3	RB				
3	C	CPv	OPv	1958pc	-	ac	6.58	4oooX3			3/58	3/58	
3	C/	CP?x,z,0z	OPy	1958mc	1961ch	ac	6.58	4oooX3					

AA 1.2.48/NT 24.3.48 on A(A), 3/-, OSPR 5/48. GS: *Miller's Dale,* Solid and Drift, 1976 on C/. SS: Soil, Land Use Capability, outline, 1971 on C/.

SK 18 (43/18) *Castleton Derbyshire* 10 by 10 Revision a 1896-1920 c 1938-49 f 1971
1	A	CPr	OPr	1948po	-	-	1.48	4BE1X3	RB		1/48	3/48	
1	B	CPr		nd:po	-	-	1.48	4BEoo3	RB				
1	C	CPs		1951pc	-	ac	1.51	4BEoo3	RB		4/51	-	
2	C	CPt		1951pc	-	ac	1.52	4Booo3	RB				
2	C	CPt		1951pc	1952	ac	1.52	4Booo3	RB				
3	C/	CPv		1951pc	1957c	ac	1.51	4BooX3	B				
3	C//	CPw,y,z		1951mc	1960c	ac	1.51	4BooX3	B				
3	C///	CPz,0z	OPz	1951^{c}ms	nd:ch	acf	1.51	4BooX3	B				

AA 1.9.47/NT 10.1.48 on A(A), 3/-, OSPR 3/48. GS: Solid and Drift, 1975 on C///.

SK 19 (43/19) *Derwent Dale (North)* 10 by 10 Revision a 1896-1929 c 1948-51 f 1974
1	10046	CPr		1946po	-	-	6.46	4BE1X3	B		10/46	-	
1	A		OPr	1946po	-	-	6.46	4BE1X3	B		-	1/47	
3	B	CPu	OPv	1955pc	-	ac	6.55	4oooX3			3/55	2/57	
3	B/	CPw,y,z		1955mc	1958mc	ac	6.55	4oooX3					
3	B//	CFz		1955^{c}ms	nd:ch	acf	6.55	4oooX3					

AA 1.7.46 on A(A), 3/-, OSPR 1/47. M on B, B/.

SK 20 (43/20) [Tamworth] 10 by 10 Revision a 1920-22 c 1937-50
1	A	CPr	OPr	nd:po	-	-	1.49	4BEoo3	RB		12/48	2/49	
2	B	CPs	OPs	1952pc	-	ac	1.52	4Booo3	R		4/52	3/54	
3	B		OP?	1952pc	-	ac	1.52	4Booo3	R			9/57	
3	B/	CPw,x,z		1952mc	1960mc	ac	1.52	4BooX3					

AA 1.10.48 on A(A), 3/-, OSPR 2/49. M on B, B/.

SK 21 (43/21) [Church Gresley] 10 by 10 Revision a 1920-21 c 1938-49
1	A	CPr	OPr	1947po	-	-	6.47	4BE2X3	RB		8/47	11/47	
2	B	CPt	OPv	1952pc	-	ac	6.52	4Booo3	R		7/52	1/57	
3	B/	CPv,z		1952pc	1957mc	ac	6.52	4BooX3					

AA 1.4.47 on A(A), 3/-, OSPR 11/47. M on B, B/.

SK 22 (43/22) *Burton upon Trent* 10 by 10 Revision a 1920-21 c 1937-50 f 1968
1	A	CPr	OPr	1947po	-	-	6.47	4BE2X3	RB		7/47	10/47	
2	B	CPt	OPv	1952pc	-	ac	6.52	4Booo3	R		9/52	2/57	
3	B/	CPw,x,z	OPz	1952mc	1960mc	ac	6.52	4BooX3					
3	B//*	CPz	OPz	1952ms	1969r	acf	6.52	4BooX3			4/69	5/69	

AA 1.5.47 on A(A), 3/-, OSPR 10/47.

1	2	3	4	5	6	7	8	9	10	11	12	13	14

SK 23 (43/23) *Etwall*   10 by 10 Revision a 1913-20 c 1948-50
1   A   CPr   OPr   1948po   -   -   6.48   4BEoo3   RB   00   5/48   5/48
3   B   CPw.z,0z OPv   1960mc   -   ac   6.60   4oooX3   32   10/60 11/60 1
M on A. 1. Derby Airport.

SK 24 (43/24) *Brailsford*   10 by 10 Revision a 1913-20 c 1949-50
1   A   CPr   OPr   1948po   -   -   1.48   4BE4X3   RB   0   3/48   4/48
3   A   CPt   1948pc   -   -   1.48   4BEoo3   RB   0
3   B   CPw,y,z.z OPv   1960mc   -   ac   6.60   4oooX3   3   10/60 11/60
M on A.

SK 25 (43/25) *Wirksworth**   10 by 10 Revision a 1920 c 1938-49
1   A   CPr   OPr   nd:po   -   -   1.49   4BEoo3   RB   2/49   2/49
1   B   CPs   1951pc   -   ac   6.51   4BEoo3   RB   9/51   -
3   B/   CPv,z.z,0z   1951pc   1957mc   ac   6.51   4BooX3   B

SK 26 (43/26) *Bakewell*   10 by 10 Revision a 1919-20 c 1949
1   A   CPr   OPr   nd:po   -   -   1.49   4BEoo3   RB   2/49   3/49
3   A   CPv   1949pc   -   -   1.49   4BEoo3   RB
3   B   CPv.x,z,0z OP?v,z 1957pc   -   ac   6.57   4oooX4   1/58   1/58
M on A. GS: Solid and Drift, 1982 on B.

SK 27 (43/27) *Calver*   10 by 10 Revision a 1914-20 c 1945-49
1   A   CPr   OPr   1947po   -   -   1.47   4BE1X3   RB   2/47   5/47
1   B   CPr   nd:po   -   -   1.47   4BEoo3   RB
1   C   CPs   1951pc   -   ac   1.51   4BEoo3   RB   12/51   -
3   C   OPv   1951pc   -   ac   1.51   4BooX3   RB   -   5/56
3   C/   CPv,x,z.z,0z   1951pc   1956mc   ac   1.51   4BooX3   B
AA 1.8.46/NT 1.1.47 on A(A), 3/-, OSPR 5/47. M on C.

SK 28 (43/28) *Hathersage*   10 by 10 Revision a 1901-35 c 1934-49
1   A   CPr   OPr   1947po   -   -   6.47   4BE1X3   RB   4/47   1
1   B   CPr   nd:po   -   -   6.47   4BEoo3   RB
1   C   CPs   1951pc   -   ac   1.51   4BEoo3   RB   3/51   -
2   C   OPs   1951pc   -   ac   1.51   4BooX3   RB   -   2/53
3   C   CPu   1951pc   -   ac   1.51   4BooX3   RB
3   C/   CPw,y,0z.0z OPv,z.0z 1951mc 1960mc ac   1.51   4BooX3   B
AA A(A) not found; AA 1.11.53/NT 1.8.47 on C(B), 4/-, OSPR 2/55. M on C, C/. 1. OSPR (outline edition) not found.

SK 29 (43/29) *Stocksbridge*   10 by 10 Revision a 1901-35 c 1938-51
1   A   CPr   OPr   1948po   -   -   1.48   4BE1X3   RB   2/48   6/48
1   B   CPr   nd:po   -   -   1.48   4BEoo3   RB
3   C   CPu   OPv   1955pc   -   ac   6.55   4oooX3   3/55   9/57
3   C/   CPw,x,z,0z   1955mc   1958mc   ac   6.55   4oooX3
AA 1.12.47 on A(A), 3/-, OSPR 6/48. M on C, C/.

SK 30 (43/30) *[Shackerstone]* 10 by 10 Revision a 1901-27 c 1950
1   A   CPr   OPr   nd:po   -   -   1.49   4BEoo3   RB   2/49   3/49
1   B   CPs   1951pc   -   ac   1.51   4BEoo3   RB   12/51   -
3   B   OPs   1951pc   -   ac   1.51   4BEoo3   RB   -   12/55
3   B/   CPv   1951pc   1957mc   ac   1.51   4BooX3   B
3   B//   CPw,y,z   OPv   1951mc   1960mc   ac   1.51   4BooX3
M on B, B//.

SK 31 (43/31) *[Swadlincote]*   10 by 10 Revision a 1920-27 c 1938-50
1   A   CPr   OPr   1948po   -   -   1.48   4BE2X3   RB   0   2/48   5/48
1   B   CPs   OP?   1951pc   -   ac   1.51   4BEoo3   R   1   6/51   8/51
3   B/   CPw,x,z   OPv   1951mc   1960mc   ac   1.51   4BooX3   1
AA 1.4.46 on A(A), 3/-, OSPR 5/48.

169

1	2	3	4	5	6	7	8	9	10	11	12	13	14

SK 32 (43/32) *Melbourne Derbyshire* 10 by 10 Revision a 1920-21 c 1938-50 f 1965 g 1972
j 1959 industrial k 1959 reservoir

1	A	CPr	OPr	1948po	-	-	1.48	4BE1X3	RB		2/48	7/48	
1	B	CPs		1951pc	-	ac	1.51	4BEoo3	RB		4/51	-	
3	B/	CPw,x	OPv	1951pc	1958mc	ac	1.51	4BooX3	B				
3	B//*	CPz	OPz	1951mc	1966iv	acfjk	1.51	4BooX3	B		2/66	2/66	
3	B//*/	CPz	OPz	1951^{c}ms	nd:ch	acgjk	1.51	4BooX3	B			10/73	1

AA 1.5.48 on A(A), 3/-, OSPR 7/48. M on B, B/. 1. OSPR (coloured edition) not found.

SK 33 (43/33) *Derby* 10 by 10 Revision a 1913-20 c 1937-50 f 1974

1	A	CPr	OPr	1947po	-	-	6.47	4BE2X3	RB		7/47		1
1	B [1]	CPs		1947po	-	-	1.51	4BEoo3	RB				M
2	B [2]	CPt	OPv	1952pc	-	ac	6.52	4Booo3	R		1/53	2/57	
3	B/	CPw,y,z	OPv,z	1952mc	1960mc	ac	6.52	4BooX3					
3	B//*	CFz	OFz	1952^{c}ms	nd:r	acf	6.52	4BooX3			5/75	12/74	

M on B [1], B/. 1. OSPR (outline edition) not found.

SK 34 (43/34) *Belper** 10 by 10 Revision a 1913-20 c 1938-50

1	A	CPr	OPr	nd:po	-	-	1.49	4BEoo3	RB		12/48	2/49	
2	B	CPs		1951pc	-	ac	6.51	4Booo3	R		3/52	-	
3	B/		OPv	1951pc	1957mc	ac	6.51	4BooX3			-	2/57	
3	B//	CPx,z,0z		1951mc	1961ch	ac	6.51	4BooX3					

AA 1.1.49/NT 10.12.48 on A(A), 3/-, OSPR 2/49. M on B, B//.

SK 35 (43/35) *Crich* 10 by 10 Revision a 1913-20 c 1937-49 f 1965; j 1960 reservoir

1	A	CPr	OPr	nd:po	-	-	1.49	4BEoo3	RBC		2/49	4/49	
1	B	CPs		1951pc	-	ac	1.51	4BEoo3	RBC		4/51	-	
3	B	CPu		1951pc	-	ac	1.51	4BEoo3	RBC				
3	B/	CPw,x		1951mc	1959c	ac	1.51	4BooX3	BC				
3	B//*	CPz,0z	OPz	1951mc	1966rv	acfj	1.51	4BooX3	BC		5/66	5/66	1

AA 1.3.49/NT 1.3.49 on A(A), 3/-, OSPR 4/49. 1. Ogston Reservoir.

SK 36 (43/36) *Clay Cross** 10 by 10 Revision a 1913-20 c 1938-49 f 1963

1	A	CPr	OPr	1948po	-	-	6.48	4BEoo3	RB		7/48	8/48	
2	B	CPs,t	OPv	1951pc	-	ac	6.51	4Booo3	R		2/52	2/57	
3	B/	CPw	OPv	1951mc	1960mc	ac	6.51	4BooX3					
3	B//	CPy,z,0z	OPz	1951mc	1964ch	acf	6.51	4BooX3					

AA 1.7.48 on A(A), 3/-, OSPR 8/48. M on B, B/.

SK 37 (43/37) *Chesterfield Derbyshire* 10 by 10 Revision a 1914-39 c 1938-49

1	A	CPr	OPr	1947po	-	-	6.47	4BE1X3	RB		7/47	10/47	
1	B	CPr		nd:po	-	-	6.47	4BEoo3	RB				
2	C	CPt		1952pc	-	ac	6.52	4Booo4	R		9/52	-	
3	C/	CPv	OPv	1952pc	1957mc	ac	6.52	4BooX4					
3	C//	CPw.x,z,0z		1952mc	1960mc	ac	6.52	4BooX4					

AA 1.10.46 on A(A), 3/-, OSPR 10/47. M on C.

SK 38 (43/38) *Sheffield* 10 by 10 Revision a 1914-36 c 1938-50 f 1971

1	A	CPr	OPr	1947po	-	-	6.47	4BE1X3	RB		9/47	11/47	
1	B	CPs		1951pc	-	ac	6.51	4BEoo3	RB		8/51	-	
3	B		OPs	1951pc	-	ac	1.52	4Booo3	RB		-	1/55	
3	B/	CPw,x,z		1951mc	1958c	ac	1.52	4BooX3	B				
3	B//	CPz,0z	OPz	1951^{c}ms	nd:ch	acf	1.52	4BooX3	B				

AA 1.10.46 on A(A), 3/-, OSPR 11/47; AA 1.11.53 on B(B), 4/-, OSPR 1/55. M on B.

SK 39 (43/39) *Sheffield (North)* 10 by 10 Revision a 1920-35 c 1938-53 f 1970

1	A	CPr	OPr	1947po	-	-	6.47	4BE1X3	RB		6/47	8/47	
1	B	CPr		nd:po	-	-	6.47	4BEoo3	RB				
3	B		OPs	1949pc	-	-	6.47	4BEoo3	RB				
3	C	CPu	OPv	1955pc	-	ac	6.55	4oooX3			9/55	5/57	
3	C/	CPw,x,z		1955pc	1958mc	ac	6.55	4oooX3					

170

1	2	3	4	5	6	7	8	9	10	11	12	13	14
3	C//*	CPz,0z	OPz	1955^{c}ms	nd:r	acf	6.55	4oooX3			4/71	6/71	

AA 1.8.46 on A(A), 3/-, OSPR 8/47; AA 1.11.53 on B(B), 4/-, OSPR 1/55. M on B.

SK 40 (43/40) *Bagworth* 10 by 10 Revision a 1901-28 c 1938-50 f 1965
1	A	CPr	OPr	nd:po	-	-	1.49	4BEoo3	RB		2/49	2/49	
2	B	CPs	OPv	1951pc	-	ac	1.52	4Booo3	R		2/52	4/57	
3	B/	CPw,z		1951mc	1959mc	ac	1.52	4BooX3					
3	B//*	CPz	OPz	1951mc	1965r	acf	1.52	4BooX3			9/65	9/65	

SK 41 (43/41) *Coalville** 10 by 10 Revision a 1919-28 c 1938-50 f 1965 g 1967 h 1971
j 1966 railway
1	A	CPr	OPr	nd:po	-	-	1.48	4BEoo3	RB		1/49	2/49	
2	B	CPs,t	OPv	1951pc	-	ac	6.51	4Booo3	R		12/51	4/57	
3	B/	CPw,z		1951mc	1959mc	ac	6.51	4BooX3					
3	B//*	CPz	OPz	1951mc	1965r	acf	6.51	4BooX3			10/65	10/65	1
3	B//*/*	CPz	OPz	1951mc	1968r	acgj	6.51	4BooX3			3/68	3/68	
3	B//*/*/*	CPz	OPz	1951^{c}ms	nd:r	achj	6.51	4BooX3			4/73	5/73	

AA 1.1.49 on A(A), 3/-, OSPR 2/49. 1. Showing the M1 under construction.

SK 42 (43/42) *Kegworth* 10 by 10 Revision a 1919-20 c 1950 f 1959 g 1965 h 1968
i 1972; j 1965 airfield k 1959 industrial
1	A	CPr	OPr	nd:po	-	-	6.48	4BEoo3	RB	1	8/48	10/48	
2	B	CPt		1952pc	-	ac	6.52	4Booo3	R	1	7/52	7/52	
3	B/	CPv		1952pc	1956mc	ac	6.52	4BooX3		1			
3	B//	CPx		1952mc	1961ch	acf	6.52	4BooX3		1			
3	B///*	CPz	OPz	1952mc	1966ar	acgj	6.52	4BooX3		3	4/66	4/66	1
3	B///*/*	CPz	OPz	1952ms	1969ir	achk	6.52	4BooX3		3	4/69	4/69	
3	B///*/*/*	CPz	OPz	1952^{c}ms	nd:r	acik	6.52	4BooX3		3	9/72	12/72	

AA 1.9.48 on A(A), 3/-, OSPR 10/48. M on B, B//. 1. East Midlands Airport.

SK 43 (43/43) *Long Eaton** 10 by 10 Revision a 1912-20 c 1937-48 d 1937-50
f 1960 g 1966 h 1971
1	A	CPr	OPr	1947po	-	-	6.47	4BE2X3	RB		4/47	7/47	
1	B	CPs		1950pc	-	ac	6.50	4BEoo3	RB				
2	C	CPs,u	OPv	1952pc	-	ad	1.52	4Booo3	R		5/52	5/57	
3	C/	CPx		1952mc	1961ch	adf	1.52	4BooX3					
3	C//*	CPz	OPz	1952mc	1967r	adg	1.52	4BooX3			8/67	8/67	
3	C//*/	CPz	OPz	1952^{c}ms	nd:ch	adh	1.52	4BooX3					

AA 1.8.46 on A(A), 3/-, OSPR 7/47. M on C.

SK 44 (43/44) [Ilkeston] 10 by 10 Revision a 1913-14 c 1938-50 f 1967
1	A	CPr	OPr	1948po	-	-	6.48	4BEoo3	RB		7/48	9/48	
1	B	CPs		1951pc	-	ac	6.51	4BEoo3	RB		11/51	-	
3	B		OPv	1951pc	-	ac	6.51	4BooX3	RB		-	9/57	
3	B/	CPw,y		1951mc	1958mc	ac	6.51	4BooX3	B				1
3	B//*	CPz	OPz	1951mc	1968r	acf	6.51	4BooX3	B		3/68	3/68	

AA 1.7.48 on A(A), 3/-, OSPR 9/48. M on B, B/. 1. Some bridle roads have been deleted.

SK 45 (43/45) [Sutton in Ashfield] 10 by 10 Revision a 1913-14 c 1937-50
1	A	CPr	OPr	nd:po	-	-	6.48	4BEoo3	RB		10/48	11/48	
1	B	CPr		nd:po	-	-	6.48	4BEoo3	RB				
2	C	CPs		1951pc	-	ac	6.51	4Booo3	R		1/52	-	
3	C/	CPv,x,z	OPv	1951pc	1956mc	ac	6.51	4BooX3			-	4/57	

AA 1.10.48 on A(A), 3/-, OSPR 11/48. M on C.

SK 46 (43/46) [Tibshelf] 10 by 10 Revision a 1913-14 c 1938-50
1	A	CPr	OPr	1948po	-	-	1.48	4BE4X3	RB		3/48	4/48	
1	B	CPs	OPs	1951pc	-	ac	1.51	4BEoo3	RB		8/51	11/51	
3	B/	CPw,x		1951mc	1960mc	ac	1.51	4BooX3	B				

AA 1.3.48 on A(A), 3/-, OSPR 4/48. M on B, B/.

1	2	3	4	5	6	7	8	9	10	11	12	13	14

SK 47 (43/47) [Staveley] 10 by 10 Revision a 1874-75 b 1914-22 c 1938-50 d 1938-51 f 1968; j 1967 railway

1	A	CPr	OPr	1947po	-	-	1.47	4BE2X3	RB		10/47	2/48	
1	B	CPs		1951pc	-	ac	1.51	4BEoo3	R		5/51	-	
2	C	CPt	OPv	1953pc	-	bd	1.53	4Booo3	R		1/53	5/57	
3	C/	CPw,y		1953mc	1958mc	bd	1.53	4BooX3					
3	C//*	CPz	OPz	1953ms	1969r	bdfj	1.53	4BooX3			1/69	1/69	

AA 1.12.46 on A(A), 3/-, OSPR 2/48. M on C, C/.

SK 48 (43/48) [Woodhouse] 10 by 10 Revision a 1914-35 c 1938-51 f 1968

1	15046	CPr		1946po	-	-	6.46	4BE1X3	RB		9/46	-	
1	2547		OPr	1947po	-	-	6.46	4BE1X3	RB		-	1/47	
2	B	CPt	OPs	1953pc	-	ac	1.53	4BooX4			5/53	-	
3	B/	CPv		1953pc	1956mc	ac	1.53	4BooX4					
3	B//	CPw.x	OPz	1953mc	1960mc	ac	1.53	4BooX4					
3	B///*	CPz	OPz	1953ms	1968r	acf	1.53	4BooX4			12/68	12/68	

AA 1.7.46 on 2547(2547), 3/-, OSPR 1/47; AA 1.11.53 on B(B), 4/-, OSPR ? . M on A. The canal is coloured dark blue through Norwood Locks.

SK 49 (43/49) [Rotherham] 10 by 10 Revision a 1928-35 c 1938-51 f 1967

1	2046	CPr		1946po	-	-	6.46	4BE1X3	RB		10/46	-	
1	A		OPr	1946po	-	-	6.46	4BE1X3	RB		-	1/47	
2	B	CPt		1953pc	-	ac	1.53	4BooX3			4/53	-	
3	B/	CPv,x	OPv	1953pc	1956mc	ac	1.53	4BooX3			-	5/57	
3	B//*	CPz	OPz	1953ms	1968r	acf	1.53	4BooX3			4/68	5/68	

AA 1.7.46 on A(A), 3/-, OSPR 1/47. M on A.

SK 50 (43/50) *Leicester (West)* 10 by 10 Revision a 1927-28 c 1938-50 f 1961 g 1965

1	A	CPr	OPr	1947po	-	-	1.48	4BE2X3	RBC		9/47	11/47	
2	B	CPt	OPs	1952pc	-	ac	6.52	4Booo3	RC		10/52	12/53	
3	B/	CPv		1952pc	1957mc	ac	6.52	4BooX3	C				
3	B//	CPx		1952mc	1962ch	acf	6.52	4BooX3	C				
3	B///*	CPz	OPz	1952mc	1966r	acg	6.52	4BooX3	C		12/65	12/65	
3	B///*/	CPz	OPz	1952mc	1965r	acg	6.52	4BooX3	C				

AA 1.10.47 on A(A), 3/-, OSPR 11/47. M on B, B//.

SK 51 (43/51) *Loughborough (South)* 10 by 10 Revision a 1919-28 c 1938-50 f 1962 g 1965

1	A B	CPr	OPr	nd:po	-	-	1.49	4BEoo3	RB		12/48	2/49	
2	C	CPs	OPv	1951pc	-	ac	6.51	4Booo3	R		2/52	5/57	
3	C/	CPw		1951pc	1958mc	ac	6.51	4BooX3					
3	C//	CPy		1951mc	1963ch	acf	6.51	4BooX3					
3	C///	CPz,0z	OPz	1951mc	1966ch	acg	6.51	4BooX3					

AA 1.10.48 on A(A), 3/-, OSPR 2/49. M on C, C/.

SK 52 (43/52) *Loughborough (North)** 10 by 10 Revision a 1916-19 c 1938-49 d 1938-50 f 1965 g 1972

1	A	CPr	OPr	1948po	-	-	1.48	4BE2X3	RB	00	2/48	6/48	
1	B	CPs		1950pc	-	ac	6.50	4BEoo3	RB	00			
2	C	CPs		1951pc	-	ad	1.52	4Booo3	R	30	1/52	-	
3	C/	CPu		1951pc	1956mc	ad	1.52	4BooX3		30			1
3	C//	CPw,x	OPv	1951mc	1961mc	ad	1.52	4BooX3		30			
3	C///	CPz	OPz	1951mc	1966ch	adf	1.52	4BooX3		33			
3	C////*	CPz	OPz	1951^{c}ms	nd:r	adg	1.52	4BooX3		33	12/72	2/73	

AA 1.5.48 on A(A), 3/-, OSPR 6/48. 1. Loughborough Aerodrome.

SK 53 (43/53) *Beeston Nottinghamshire* 10 by 10 Revision a 1912-19 c 1938-50 f 1958 g 1965 h 1972 j 1962 industrial k 1966 building

| 1 | A | CPr | OPr | 1947po | - | - | 6.47 | 4BE2X3 | RB | | 3/47 | 6/47 | |
| 1 | B | CPr | | nd:po | - | - | 6.47 | 4BEoo3 | RB | | | | |

172

1	2	3	4	5	6	7	8	9	10	11	12	13	14
2	C	CPt	OPt	1952pc	-	ac	6.52	4Booo3	R		7/52	6/54	1
3	C/												
3	C//	CPx		1952mc	1961ch	acf	6.52	4oooX3					
3	C///*	CPz	OPz	1952mc	1966ir	acgj	6.52	4oooX3			7/66	7/66	
3	C///*/*	CPz	OPz	1952^{c}ms	nd:br	achjk	6.52	4oooX3			1/73	12/72	

AA 1.8.46 on A(A), 3/-, OSPR 6/47; AA 1.11.53 on C(B), 4/-, OSPR 6/54. M on C. 1. The outline copy in the Ordnance Survey Record Map Library has the 2/6 price blocked out and "2/9 net Paper flat" printed beneath.

SK 54 (43/54) [Nottingham]   10 by 10   Revision a 1913-19 c 1938-50 f 1967

1	A	CPr	OPr	1947po	-	-	6.47	4BE2X3	RB	1	4/47	6/47	
1	B	CPr		nd:po	-	-	6.47	4BEoo3	RB	1			
2	C	CPs	OPt	1951pc	-	ac	6.51	4Booo3	R	1	2/52	6/54	
3	C	CPu		1951pc	-	ac	6.51	4Booo3	R	1			
3	C/	CPw,x,z	OPz	1951mc	1959mc	ac	6.51	4BooX3		1			
3	C//*	CPz	OPz	1951mc	1968r	acf	6.51	4BooX3		1	1/68	1/68	

AA 1.8.46 on A(A), 3/-, OSPR 6/47; AA 1.11.53 on C(B), 4/-, OSPR 6/54. M on C.

SK 55 (43/55) [Mansfield (South)]   10 by 10   Revision a 1913-14 c 1938-50

1	A	CPr	OPr	1948po	-	-	1.48	4BE4X3	RB		4/48	6/48	
1	B	CPr		nd:po	-	-	1.48	4BEoo3	RB				
1	C	CPs		1951pc	-	ac	6.51	4BEoo3	R		1/52	-	
3	C		OPs	1951pc	-	ac	6.51	4BooX3	R		-	5/56	
3	C/	CPv		1951pc	1956mc	ac	6.51	4BooX3					
3	C//	CPw,y,z		1951mc	1960mc	ac	6.51	4BooX3					

AA 1.4.48 on A(A), 3/-, OSPR 6/48. M on C.

SK 56 (43/56) [Mansfield (North)]   10 by 10   Revision a 1913-14 c 1938-50

1	A	CPr	OPr	1948po	-	-	6.48	4BEoo3	RB		5/48	7/48	
	B												
2	C	CPs	OPv	1952pc	-	ac	6.52	4Booo3	R		6/52	9/57	
3	C/	CPw,z		1952pc	1958c	ac	6.52	4BooX3					

AA 1.5.48 on A(A), 3/-, OSPR 7/48. M on C, C/.

SK 57 (43/57) [Worksop]   10 by 10   Revision a 1914-28 c 1938-51

1	A	CPr	OPr	1947po	-	-	1.47	4BE1X3	RB		2/47	5/47	
2	B	CPt	OPs	1952pc	-	ac	6.52	4Booo3	R		1/53	11/55	
3	B/	CPv,x,z		1952pc	1956mc	ac	6.52	4BooX3					

AA 1.9.46 on A(A), 3/-, OSPR 5/47. M on B, B/.

SK 58 (43/58) [Dinnington]   10 by 10   Revision a 1914-28 c 1946-51

1	A	CPr	OPr	1947po	-	-	1.48	4BE1X3	RB	0	11/47	2/48	
1	B	CPr		nd:po	-	-	1.48	4BEoo3	RB	0			
2	C	CPt		1952pc	-	ac	6.52	4Booo3	R	0	11/52	-	
3	C/	CPv,y,z	OPv	1952pc	1956mc	ac	6.52	4BooX3		0	-	5/57	

AA 1.9.47 on A(A), 3/-, OSPR 2/48. M on C.

SK 59 (43/59) [Maltby]   10 by 10   Revision a 1918-29 c 1937-51 f 1963

1	20046	CPr		1946po	-	-	6.46	4BE1X3	RB		9/46	-	
1	A		OPr	1946po	-	-	6.46	4BE1X3	RB		-	1/47	
2	B	CPt	OPv	1952pc	-	ac	6.52	4BooX3	R		1/53	5/57	
3	B/	CPw		1952mc	1958mc	ac	6.52	4BooX3					
3	B//*	CPy,z	OPy	1952mc	1963r	acf	6.52	4BooX3			11/63	11/63	

AA 1.6.46 on A(A), 3/-, OSPR 1/47. M on B.

SK 60 (43/60) [Leicester (East)]   10 by 10   Revision a 1902-39 c 1938-50 f 1965; j 1969 airfield

1	A	CPr	OPr	1947po	-	-	6.47	4BE2X3	RB	0	-	8/47	
2	B	CPs,u	OPv	1951pc	-	ac	6.51	4Booo4	RB	0	1/52	11/56	
3	B/	CP?w,x		1951mc	1959c	ac	6.51	4BooX4	B	1			1
3	B//*	CPz	OPz	1951mc	1966ar	acfj	6.51	4BooX4	B	3	11/66	12/66	2

AA 1.6.47 on A(A), 3/-, OSPR 8/47. M on B. 1. Leicester East Airfield. 2. Leicester East Aerodrome.

1	2	3	4	5	6	7	8	9	10	11	12	13	14

SK 61 (43/61) *Rearsby*   10 by 10   Revision   a 1901-28   c 1938-50   f 1970

1	A	CPr	OPr	1948po	-	-	6.48	4BEoo3	RB	0	6/48	7/48	
1	B	CPs		1951pc	-	ac	1.51	4BEoo3	RB	0	12/51	-	
3	B/	CPv,z		1951pc	1957mc	ac	1.51	4BooX3	B	0			
3	B//*	CPz,0z	OPz	1951^{c}ms	nd:r	acf	1.51	4BooX3	B	0	6/71	7/71	

M on B/.

SK 62 (43/62) *Broughton*   10 by 10   Revision   a 1915-19   b 1902-19   c 1950

1	A	CPr	OPr	1948po	-	-	1.48	4BE2X3	RB		2/48	5/48	
1	B	CPr		nd:po	-	-	1.48	4BEoo3	RB				
2	B	CPt		1949pc	1952rpb	-	1.48	4BooX3	RB				
3	B/	CPv		1949pc	1957mc	a	1.48	4BooX3	B				
3	C	CPv,z	OPv	1957pc	-	bc	6.57	4oooX3			1/58	3/58	

M on B.

SK 63 (43/63) *Radcliffe on Trent**   10 by 10   Revision   a 1912-19   c 1938-50   f 1971

1	A	CPr	OPr	1947po	-	-	1.47	4BE2X3	RB	0	1/47	5/47	
2	B	CPt		1952pc	-	ac	6.52	4Booo3	R	0	7/52	-	
3	B/	CPu,x,z		1952pc	1956mc	ac	6.52	4BooX3		3			1
3	B//*	CPz	OPz,0z	1952^{c}ms	nd:r	acf	6.52	4BooX3		3	5/72	8/72	

AA 1.8.46/NT 1.1.47 on A(A), 3/-, OSPR 5/47. M on B/. 1. Nottingham (Tollerton) Aerodrome.

SK 64 (43/64) *Carlton*   10 by 10   Revision   a 1912-19   c 1938-50

1	A	CPr	OPr	1947po	-	-	6.47	4BE2X3	RB	0	3/47	6/47	
	B												
2	C	CPs,t	OPv	1951pc	-	ac	6.51	4Booo3	R	0	4/52	8/57	
3	C/	CPw,x,z,0z	OPx,z	1951mc	1958mc	ac	6.51	4BooX3		0			

AA 1.8.46 on A(A), 3/-, OSPR 6/47. M on C, C/.

SK 65 (43/65) *Oxton (Nottinghamshire)**   10 by 10   Revision   a 1912-19   c 1939-50   f 1958

1	A	CPr	OPr	1948po	-	-	6.48	4BE4X3	RB		3/48	6/48	
3	A	CPu		1948pc	-	-	6.48	4BEoo3	RB				
3	B	CPx,z.z	OPx	1961mc	-	acf	6.61	4oooX3			7/61	5/61	

AA 1.1.48 on A(A), 3/-, OSPR 6/48. M on A.

SK 66 (43/66) *Ollerton (Nottinghamshire)**   10 by 10   Revision   a 1913-15   c 1938-50

1	A	CPr	OPr	1948po	-	-	6.48	4BE2X3	RB		4/48	4/48	
1	B	CPs		1951pc	-	ac	1.51	4BEoo3	RB		10/51	-	1
2	B		OPs	1951pc	-	ac	1.51	4BooX3	RB		-	2/53	
3	B/	CPu,y		1951pc	1956mc	ac	1.51	4BooX3	B				
3	B/	CPz		1951ms	1956mc	ac	1.51	4BooX3	B				

M on B, B/. SS: Soil, outline, 1971 on B/. 1. The Mid-Nottinghamshire Joint line (north to Ollerton) and colliery lines at Ollerton are added.

SK 67 (43/67) *Clumber Park*   10 by 10   Revision   a 1914-16   c 1938-53   f 1965

1	A	CPr		nd:po	-	-	1.49	4BEoo3	RB	0	10/48	-	
1	B		OPr	nd:po	-	-	1.49	4BEoo3	RB	0	-	11/48	
2	C	CPt,v	OPv	1953pc	-	ac	6.53	4BooX4		1	9/53	2/57	
3	C	CPy		1953pc	-	ac	6.53	4BooX4		1			
3	C/	CPz,0z	OPz	1953mc	1966ch	acf	6.53	4BooX4		3			

AA 1.7.48 on B(A), 3/-, OSPR 11/48. M on C.

SK 68 (43/68) *Ranskill*   10 by 10   Revision   a 1914-18   c 1938-51   f 1962   g 1965   h 1972   j 1959 airfield

1	A	CPr		1947po	-	-	1.47	4BE1X3	RB	0	9/47	-	
1	B		OPr	1947po	-	-	1.47	4BE1X3	RB	0	-	12/47	1
1	C	CPr		nd:po	-	-	1.47	4BEoo3	RB	0			
2	D	CPt		1952pc	-	ac	6.52	4Booo4	R	1	10/52	-	
3	D		OPv	1852pc	-	ac	6.52	4Booo4	R	1	-	11/56	

174

1	2	3	4	5	6	7	8	9	10	11	12	13	14
3	D/	CPw		1952pc	1958mc	ac	6.52	4BooX4		1			2
3	D//	CPy		1952mc	1963ch	acf	6.52	4BooX4		1			
3	D///*	CPz	OPz	1952mc	1966ar	acgj	6.52	4BooX4		3	10/66	10/66	
3	D///*/*	CPz	OPz	1952^{c}ms	nd:r	achj	6.52	4BooX4		3	3/73	4/73	

AA 1.2.47 on A(A), 3/-, OSPR 12/47. M on D/. 1. The print code letter A is deleted. 2. Worksop Aerodrome.

SK 69 (43/69) *Bawtry* 10 by 10 Revision a 1919-29 c 1938-53 f 1961 g 1965 j 1965 airfield

1	A	CPr	OPr	1947po	-	-	1.48	4BE1X3	RB	00	11/47	1/48	
2	B	CPt	OPv	1953pc	-	ac	6.53	4BooX4		11	8/53	7/56	
3	B	CPv		1953pc	-	ac	6.53	4BooX4		11			
3	B/*	CPx	OPx	1953mc	1961r	acf	6.53	4BooX4		11	12/61	12/61	1
3	B/*/*	CPz,0z	OPz	1953mc	1966a	acgj	6.53	4BooX4		12	7/66	7/66	

AA 1.8.47 on A(A), 3/-, OSPR 1/48. M on B. 1. Finningley Airfield.

SK 70 (43/70) [*Billesdon*] 10 by 10 Revision a 1902-28 c 1949 f 1965

1	A	CPr	OPr	nd:po	-	-	6.49	4BEoo3	RB		6/49	6/49	
1	B	CPs	OPv	1951pc	-	ac	1.51	4BEoo3	RB		6/51	8/57	
3	B	CPy		1951pc	-	ac	1.51	4BEoo3	RB				
3	B/	CPz	OPz	1951mc	1966ch	acf	1.51	4BooX3					

M on B.

SK 71 (43/71) *Melton Mowbray* 10 by 10 Revision a 1902-28 c 1938-50 f 1975 j 1981 selected k SUSI

1	A B	CPr	OPr	nd:po	-	-	1.49	4BEoo3	RB	0	5/49	6/49	
3	C	CPu		1955pc	-	ac	6.55	4oooX3		1	1/55	-	
3	C/	CPv,z		1955pc	1957c	ac	6.55	4oooX3		3			
3	C//*	CFz	OFz	1955^{c}ms	nd:r	acf	6.55	4oooX3		3	7/76	7/76	
3	C//*/*	CFz	OFz	1981^{c}m-s	nd:b	acfjk	6.55	4oooX3		3	1/82	1/82	

M on C/.

SK 72 (43/72) *Scalford* 10 by 10 Revision a 1915-29 c 1938-50

1	A	CPr	OPr	1948po	-	-	1.48	4BE2X3	RB		2/48	5/48	
1	B	CPr		nd:po	-	-	1.48	4BEoo3	RB				
3	C	CPu	OPv	1954pc	-	ac	6.54	4oooX3			11/54	9/57	
3	C/	CPw,z,0z		1954mc	1959mc	ac	6.54	4oooX3					

M on C, C/.

SK 73 (43/73) *Whatton** 10 by 10 Revision a 1912-29 c 1950

1	A	CPr	OPr	1948po	-	-	1.48	4BE2X3	RB	0	4/48	5/48	
1	B	CPr		nd:po	-	-	1.48	4BEoo3	RB	0			
3	C	CPu	OPv	1954pc	-	ac	6.54	4oooX3		1	12/54	9/57	
3	C/	CPw,z.0z	OPv.0z	1954mc	1960mc	ac	6.54	4oooX3		1			1

M on C, C/. 1. Langar Airfield.

SK 74 (43/74) *Flintham* 10 by 10 Revision a 1912-15 c 1950 f 1965

1	A	CPr	OPr	1948po	-	-	6.48	4BE2X3	RB	0	4/48	5/48	
1	B	CPr		nd:po	-	-	1.48	4BEoo3	RB	0			
3	C	CPu	OPv	1955pc	-	ac	6.55	4oooX3		1	8/55	11/56	
3	C/	CPv	OPv	1955pc	1958mc	ac	6.55	4oooX3		1			
3	C//	CPz,0z	OPz	1955mc	1966ch	acf	6.55	4oooX3		3			

M on C/.

SK 75 (43/75) *Newark-on-Trent (West)** 10 by 10 Revision a 1915 c 1938-50 f 1967

1	A	CPr	OPr	1948po	-	-	6.48	4BE4X3	RB		4/48	6/48	
2	B	CPs		1951pc	-	ac	6.51	4Booo3	RB		1/52	-	
3	B/	CPv,y		1951pc	1956mc	ac	6.51	4BooX3	B				
3	B//*	CPz	OPz	1951ms	1968r	acf	6.51	4BooX3	B		5/68	3/68	

AA 1.4.48 on A(A), 3/-, OSPR 6/48. M on B/.

1	2	3	4	5	6	7	8	9	10	11	12	13	14

SK 76 (43/76) *Ossington**　　　　10 by 10　Revision a 1903-16 c 1950 f 1969
1	A	CPr	OPr	1948po	-	-	6.48	4BEoo3	RB	0	5/48	5/48	
3	B	CPw,x	OPv	1959mc	-	ac	6.59	4oooX3		3	3/59	5/59	
3	B/*	CPz	OPz	1959ms	1970r	acf	6.59	4oooX3		2	3/70	5/70	

M on B.

SK 77 (43/77) *East Markham*　　　10 by 10　Revision a 1915-16 c 1938-50 d 1938-51
　　　　　　　　　　　　　　　　　　　　　　　　f 1963 g 1970; j 1969 airfield
1	A	CPr	OPr	nd:po	-	-	1.49	4BEoo3	RB	0	10/48	11/48	
1	B	CPs		1951pc	-	ac	1.51	4BEoo3	RB	0	7/51	-	
2	C	CPt		1953pc	-	ad	1.53	4BooX3	R	1	1/53	-	
3	C/	CPv		1953pc	1957mc	ad	1.53	4BooX3		1			
3	C/	CPz		1953mc	1964ch	adf	1.53	4BooX3		1			1
3	C//*	CPz	OPz	1953mc	1964r	adf	1.53	4BooX3		1	6/64	8/64	2
3	C//*/*	CPz	OPz	1953^{c}ms	nd:ar	adgj	1.53	4BooX3		3	6/71	6/71	

AA 1.7.48 on A(A), 3/-, OSPR 11/48. M on C/. 1. Not found. The existence of a state with this maverick edition code is improbable, but see the next. 2. This state has "Reprinted with minor changes 1964" blocked out, and replaced with the roads note. The print code C/ is also blocked out, replaced by C//* below the sheet number.

SK 78 (43/78) *East Retford*　　　　10 by 10　Revision a 1915-18 c 1938-51 f 1965; j 1980 selected
1	A	CPr		1947po	-	-	1.48	4BE1X3	RB		10/47	-	
1	B		OPr	1947po	-	-	1.48	4BE1X3	RB		-	2/48	
2	C	CPt		1953pc	-	ac	6.53	4BooX3			9/53	-	
3	C	CPu,x	OPv	1953pc	-	ac	6.53	4BooX3			-	8/57	
3	C/	CPz	OPz	1953mc	1966ch	acf	6.53	4BooX3					1
3	C/	CPz,0z	OPz	1953mc	1966ch	acf	6.53	4BooX3					
3	C//*	CFz	OFz	1980^{c}ms	nd:r	acfj	6.53	4BooX3			2/81	3/81	

AA 1.8.47 on B(A), 3/-, OSPR 2/48. M on C. 1. West Burton (7986) parish name is omitted; it is on the next.

SK 79 (43/79) *Misterton Nottinghamshire*　10 by 10　Revision a 1918-40 c 1948-51
1	A	CPr	OPr	1947po	-	-	6.47	4BE1X3	RB		10/47	12/47	
2	B	CPt,u	OPv	1953pc	-	ac	1.53	4BooX3			5/53	8/57	
3	B	CPx,z	OPz	1953pc	-	ac	1.53	4BooX3					

AA 1.12.46 on A(A), 3/-, OSPR 12/47. M on B.

SK 80 (43/80) *Oakham*　　　　　　10 by 10　Revision a 1902-28 c 1949-50 f 1977; j 1977 reservoir
1	A	CPr	OPr	nd:po	-	-	6.49	4BEoo3	RB		6/49	6/49	
3	B	CPu,v,z	OPv	1954pc	-	ac	6.54	4oooX3			11/54	2/57	
3	B/*	CFz	OFz	1954^{c}ms	nd:rv	acfj	6.54	4oooX3			11/78	11/78	1

1. Rutland Water.

SK 81 (43/81) *Wymondham Leicestershire*　10 by 10　Revision a 1901-28 c 1949-50
1	A	CPr	OPr	nd:po	-	-	6.49	4BEoo3	RB		6/49	7/49	
1	B	CPs		1950pc	-	ac	6.50	4BEoo3	RB		11/50	-	
2	B		OPs	1950pc	-	ac	6.50	4BooX3	RB		-	2/53	
3	B/	CP?w,z.0z		1950mc	1960mc	ac	6.50	4BooX3	B				

SK 82 (43/82) *Croxton Kerrial*　　　10 by 10　Revision a 1902-29 c 1950 f 1966
1	A	CPr	OPr	nd:po	-	-	6.49	4BEoo3	B	00	6/49	6/49	1
3	B	CPu	OPv	1954pc	-	ac	6.54	4oooX3		11	11/54	10/56	2
3	B/	CP?w,z		1954mc	1959mc	ac	6.54	4oooX3		11			
3	B//	CPz,0z	OPz	1954mc	1967ch	acf	6.54	4oooX3		31			

1. The railway to Sproxton stops at the field boundary at 877252. 2. The continuation of the railway is restored.

SK 83 (43/83) *Sedgebrook*　　　　10 by 10　Revision a 1902-29 c 1938-50 f 1963
1	A	CPr	OPr	nd:po	-	-	1.49	4BEoo3	RB		6/49	7/49	1
3	B	CPu		1955pc	-	ac	6.55	4oooX3			3/55		
3	B/	CPv		1955pc	1958mc	ac	6.55	4oooX3					
3	B//*	CPz,0z	OPz	1955mc	1966r	acf	6.55	4oooX3			1/66	4/66	

AA 1.2.49 on A(A), 3/-, OSPR 7/49. M on B/. 1. Also overprinted red, black and blue for use in a Western Command pre-Staff College course (GSGS Misc.498/3, 100/11/50 SPC RE).

1	2	3	4	5	6	7	8	9	10	11	12	13	14

SK 84 (43/84) *Long Bennington*   10 by 10   Revision a 1902-29 c 1946-51 f 1968 g 1978

1	A	CPr	OPr	nd:po	-	-	1.49	4BEoo3	RB	00	2/49	2/49	
3	B	CPu	OPv	1955pc	-	ac	6.55	4oooX3		11	7/55	11/56	
3	B/	CPw		1955mc	1959mc	ac	6.55	4oooX3		11			
3	B//*	CPz	OPz	1955ms	1969r	acf	6.55	4oooX3		11	3/69	4/69	
3	B//*/*	CFz	OFz	1955^{c}ms	nd:r	acg	6.55	4oooX3		11	11/78	11/78	

M on B/.

SK 85 (43/85) *Newark-on-Trent (East)*   10 by 10   Revision a 1903-15 c 1938-53 f 1967 g 1976

1	A	CPr	OPr	nd:po	-	-	1.49	4BEoo3	RB	000	2/49	4/49	
2	B	CPt		1953pc	-	ac	6.53	4BooX3		111	9/53	-	
3	B	CPu,x	OPv	1953pc	-	ac	6.53	4BooX3		111	-	1/57	
3	B/*	CPz	OPz	1953ms	1968r	acf	6.53	4BooX3		111	5/68	3/68	?1
3	B/*/*	CFz	OFz	1953^{c}ms	nd:r	acg	6.53	4BooX3		111	5/77	5/77	1

AA 1.12.48 on A(A), 3/-, OSPR 4/49. M on B. SS: Soil, Land Use Capability, outline, 1974 on B/*. 1. The railway curve is added between the Lincoln and London lines.

SK 86 (43/86) *Collingham*   10 by 10   Revision a 1904-15 c 1947-53 f 1966
j 1961 airfield k 1961 tidal change

1	A	CPr	OPr	1948po	-	-	1.48	4BEoo3	RB	00	5/48	6/48	
1	B	CPr		nd:po	-	-	1.48	4BEoo3	RB	00			
2	C	CPt	OPs	1953pc	-	ac	6.53	4BooX3		11	9/53	6/55	
3	C	CPx		1953pc	-	ac	6.53	4BooX3		11			
3	C/*	CPz,0z	OPz	1953mc	1967rt	acfjk	6.53	4BooX3		33	1/67	2/67	

SK 87 (43/87) *Newton on Trent*   10 by 10   Revision a 1915-16 c 1938-48 d 1938-53 f 1965

1	A	CPr	OPr	nd:po	-	-	1.49	4BEoo3	RB	0	9/48	10/48	
1	B	CPs		1951pc	-	ac	1.51	4BEoo3	RB	0	9/51	-	
3	C	CPw,x,z	OPv,z	1959mc	-	ad	6.59	4oooX3		1	9/59	9/59	
3	C/	CPz,z-	OPz	1959mc	1966ch	adf	6.59	4oooX3		3			

SK 88 (43/88) *Gainsborough (South)*   10 by 10   Revision a 1904-18 c 1947-51

1	A	CPr	OPr	1947po	-	-	6.47	4BE1X3	RB	0	11/47	2/48	
2	B	CPt,u	OPv	1953pc	-	ac	6.53	4BooX3		0	9/53	9/57	
3	B/	CPw,z		1953mc	1960mc	ac	6.53	4BooX3		0			

AA 1.8.47 on A(A), 3/-, OSPR 2/48.

SK 89 (43/89) *Gainsborough (North)*   10 by 10   Revision a 1905-40 c 1938-53

1	10046	CPr		1946po	-	-	1.47	4BE1X3	RB	0	9/46	-	
1	2546		OPr	1946po	-	-	1.47	4BE1X3	RB	0	-	11/46	
2	B	CPt	OPv	1953pc	-	ac	6.53	4BooX4		1	9/53	1/57	
3	B	CPv		1953pc	-	ac	6.53	4BooX4		1			
3	B/	CPw,z.z	OPv	1953mc	1961mc	ac	6.53	4BooX4		1			

AA 1.6.46 on 2546(2546), 3/-, OSPR 11/46.

SK 90 (43/90) *Ketton*   10 by 10   Revision a 1899-1929 c 1938-50 f 1960 g 1977
j 1977 reservoir

1	A	CPr	OPr	nd:po	-	-	6.49	4BEoo3	RB	00	6/49	7/49	
1	B	CPs		1951pc	-	ac	1.51	4BEoo3	RB	01	10/51	-	
3	B		OPv	1951pc	-	ac	1.51	4BooX3	RB	01	-	11/56	
3	B/	CPx,z		1951mc	1961ch	acf	1.51	4BooX3	B	01			
3	B//*	CFz	OFz	1951^{c}ms	nd:rv	acgj	1.51	4BooX3	B	31	8/78	8/78	

M on B.

SK 91 (43/91) *Stretton*   10 by 10   Revision a 1902-28 c 1949-50 f 1972

1	A	CPr	OPr	nd:po	-	-	1.49	4BEoo3	RB	00	7/49	7/49	
1	B	CPs	OPv	1951pc	-	ac	1.51	4BEoo3	RB	00	1/52	11/56	
3	B/	CP?w,z		1951mc	1960mc	ac	1.51	4BooX3	B	00			
3	B//*	CPz,0z	OPz	1951^{c}ms	nd:r	acf	1.51	4BooX3	B	00	7/73	9/73	

1	2	3	4	5	6	7	8	9	10	11	12	13	14

SK 92 (43/92) *Colsterworth*   10 by 10   Revision a 1902-29 c 1947 f 1962

1	A	CPr	OPr	nd:po	-	-	6.49	4BEoo3	RB	0	7/49	7/49	
1	B	CPs		1950pc	-	ac	6.50	4BEoo3	RB	0	11/50	-	
3	B/	CPy,z,0z	OPz	1950mc	1963ch	acf	6.50	4BooX3	B	0			

SK 93 (43/93) *Grantham*   10 by 10   Revision a 1903-29 c 1938-50 f 1967

1	A	CPr	OPr	nd:po	-	-	6.49	4BEoo3	RB	10	7/49	9/49	
3	B	CPu	OPv	1955pc	-	ac	6.55	4oooX3		11	2/55	8/57	
3	B/	CPw		1955mc	1958mc	ac	6.55	4oooX3		11			
3	B//*	CPz,0z	OPz	1955ms	1968r	acf	6.55	4oooX3		11	6/68	6/68	

AA 1.3.49 on A(A), 3/-, OSPR 9/49. M on B/. The Belton Park sheet, but there is no sign of the First World War Machine Gun Corps encampment there, nor any trace of its extensive railway system.

SK 94 (43/94) *Ancaster*   10 by 10   Revision a 1903-04 c 1946-53

1	A	CPr	OPr	nd:po	-	-	6.49	4BEoo3	RB	00	7/49	7/49	
3	B	CPu	OPv	1955pc	-	ac	6.55	4oooX3		11	9/55	8/57	
3	B/	CP?w,z,0z		1955mc	1959mc	ac	6.55	4oooX3		11			1

M on B/. 1. Barkston Heath Airfield, a name possibly not present on the unrecorded CPw state.

SK 95 (43/95) *Leadenham*   10 by 10   Revision a 1903-15 c 1938-53

1	A	CPr	OPr	nd:po	-	-	6.49	4BEoo3	RB	000	7/49	7/49	
3	A	CPt		1949pc	-	-	6.49	4BEoo3	RB	000			
3	B	CPw,z	OPv	1960mc	-	ac	6.60	4oooX3		111	10/60	11/60	

M on A.

SK 96 (43/96) *Bracebridge*   10 by 10   Revision a 1904-30 c 1938-53 f 1965

1	A	CPr	OPr	nd:po	-	-	1.49	4BEoo3	RB	10	12/48	2/49	
2	B	CPt		1953pc	-	ac	6.53	4BooX3		11	10/53	-	
3	B	-	OPv	1953pc	-	ac	6.53	4BooX3		11	-	11/56	
3	B/	CPw		1953mc	1960mc	ac	6.53	4BooX3		11			1
3	B//	CPz	OPz	1953mc	1966ch	acf	6.53	4BooX3		33			

AA 1.10.48 on A(A), 3/-, OSPR 2/49. M on A, B. 1. Waddington Airfield. With a bombing range (9360-9460).

SK 97 (43/97) *Lincoln*   10 by 10   Revision a 1904-38 c 1938-51 f 1963 g 1975

1	A	CPr	OPr	1948po	-	-	6.48	4BEoo3	RB	000	7/48	9/48	
2	nc [B]	CPt		1953pc	-	ac	6.53	4BooX4		110	9/53	-	
3	B/	CPv		1953pc	1957mc	ac	6.53	4BooX4		110			
3	B//	CPz	OPz	1953mc	1965ch	acf	6.53	4BooX4		333			
3	B///*	CFz	OFz	1953^{c}ms	nd:r	acg	6.53	4BooX4		333	12/75	8/75	

AA 1.5.48 on A(A), 3/-, OSPR 9/48. M on B/.

SK 98 (43/98) *Ingham Lincolnshire*   10 by 10   Revision a 1904-07 c 1947-48 f 1970

1	A	CPr	OPr	1948po	-	-	1.47	4BE1X3	B	000	12/47	2/48	
2	B	CPt,u	OPv	1953pc	-	ac	6.53	4BooX3		111	7/53	9/57	
3	B	CPz		1953pc	-	ac	6.53	4BooX3		111			
3	B/	CPz	OPz	1953^{c}ms	nd:ch	acf	6.53	4BooX3		331			

AA 1.5.47 on A(A), 3/-, OSPR 2/48.

SK 99 (43/99) *Kirton in Lindsey*   10 by 10   Revision a 1905 c 1948-51

1	A	CPr	OPr	1947po	-	-	6.47	4BE1X3	RB	00	11/47	2/48	
2	B	CPt	OPs	1953pc	-	ac	6.53	4BooX3		21	9/53	10/53	
3	B	CPz		1953pc	-	ac	6.53	4BooX3		21			

AA 1.7.47 on A(A), 3/-, OSPR 2/48. SS: Soil, outline, 1986 on B.

SM 62 (12/62) *Bishops and Clerks* *   10 by 10   Revision a 1881-1906 c 1948

| 1 | A | CPs | OPs | 1950pc | - | ac | 6.50 | 4oEoo4 | | | 10/50 | 12/50 | |
| 2 | A | CPs,z | | 1950pc | 1952 | ac | 1.52 | 4ooo04 | | | 1/52 | - | |

1	2	3	4	5	6	7	8	9	10	11	12	13	14

SM 70 (12/70) **& Part of SM 71** *Skomer Island* 10 by 10 Revision a 1906 c 1946-48
1	A	CPr	OPr	nd:po	-	-	1.50	4BEoo4		0	4/50	5/50	
2	A	CPt		1950pc	1952rpb	-	1.50	4BooX4		0			
3	A	CP?w,z.z		1950mc	1960mc	ac	1.50	4BooX4		0			

Sheet lines are offset, 170 km E to 180 km E, 202 km N to 212 km N.

SM 72 (12/72) **& Part of SM 73** *St David's Pembrokeshire* 10 by 12 Revision a 1906 c 1948 f 1964
1	A	CPs	OPs	1950pc	-	ac	6.50	4BEoo3		0	11/50	12/50	
3	A	CPt		1950pc	-	ac	6.50	4BEoo3		0			
3	A/	CPw,x		1950mc	1958c	ac	6.50	4BooX3		0			
3	A//	CPz,0z	OPz	1950mc	1965ch	acf	6.50	4BooX3		3			

SM 80 (12/80) *Angle* 10 by 10 Revision a 1906 c 1937-50 f 1969
1	A	CPr		1948po	-	-	1.48	4BE4X3		00	2/48	-	
1	B	CPr	OPr	1948po	-	-	1.48	4BE4X3		00	-	5/48	
	C												
3	D	CPt,v	OPv	1954pc	-	ac	6.54	4BooX3		10	2/54	6/57	
3	D/	CPw,z		1954pc	1958mc	ac	6.54	4BooX3		10			
3	D//	CPz		1954cms	nd:mc	acf	6.54	4BooX3		10			

AA 1.11.47 on B(A), 3/-, OSPR 5/48. M on B, D.

SM 81 (12/81) *St Brides Bay* 10 by 10 Revision a 1905-06 c 1948
1	A	CPs	OPs	1950pc	-	ac	6.50	4BEoo4		0	10/50	11/50	
3	A/	CPu		1950pc	1956mc	ac	6.50	4BooX4		0			
3	A//	CPw,z,0z		1950mc	1959mc	ac	6.50	4BooX4		0			

SM 82 (12/82) *Newgale* 10 by 10 Revision a 1906 c 1948 f 1965
1	A	CPs	OPs	1951pc	-	ac	1.51	4BEoo4		0	1/51	1/51	
3	A/	CPw,z	OPv	1951pc	1958mc	ac	1.51	4BooX4		0			
3	A//	CPz		1951mc	1965ch	acf	1.51	4BooX4		3			
3	A///	CPz	OPz	1951mc	1965ch	acf	1.51	4BooX4		3			

M on A.

SM 83 (12/83) **& Parts of SM 73 & SM 84** *Mathry* 12 by 12 Revision a 1906 c 1948
| 1 | A | CPs | OPs | 1950pc | - | ac | 6.50 | 4BEoo3 | | | 11/50 | 12/50 | |
| 3 | A | CPu,x,z.z | | 1950pc | - | ac | 6.50 | 4BEoo3 | | | | | |

SM 90 (12/90) *Pembroke* 10 by 10 Revision a 1906-39 c 1948-50 f 1964
1	A	CPr	OPr	1948po	-	-	6.48	4BEoo3	R	1	5/48	7/48	1
3	B	CPu	OPv	1955pc	-	ac	6.55	4BooX3		1	12/55	6/57	2
3	B/	CPv,y		1955pc	1957mc	ac	6.55	4BooX3		1			
3	B//*	CPz	OPz	1955mc	1965i	acf	6.55	4oooX3		0	9/65	9/65	3

AA 1.3.48 on A(A), 3/-, OSPR 7/48. M on A, B/. SS: Soil, outline, 1973 on B//*. 1. Pembroke naval dockyard is blank, with martello towers. 2. Fort Scoveston is shown. 3. The naval dockyard area is shown.

SM 91 (12/91) *Haverfordwest** 10 by 10 Revision a 1906 c 1948 f 1970
1	A	CPr	OPr	nd:po	-	-	1.50	4BEoo3	R	3	6/50	6/50	
3	A	CPu		1950pc	-	-	1.50	4BEoo3	R	3			
3	A/	CPx,z		1950mc	1961ch	ac	1.50	4BooX3		3			
3	A//	CPz		1950cms	1961ch	acf	1.50	4BooX3		3			

SS: Soil, outline, 1973 on A//.

SM 92 (12/92) *Wolf's Castle** 10 by 10 Revision a 1906 c 1948
1	A	CPs	OPs	1951pc	-	ac	1.51	4BEoo4	R		9/51	10/51	
3	A/	CPw,z.z	OPv	1951mc	1960mc	ac	1.51	4BooX4					
3	A[/]	C0z		1951mc	1960mc	ac	1.51	4BooX4					

SM 93 (12/93) **& Part of SM 94** *Fishguard* 10 by 12 Revision a 1906 c 1937-48 f 1968
1	A	CPs	OPs	1951pc	-	ac	1.51	4BEoo3	R		4/51	7/51	
3	A	CPt,y,z		1951pc	-	ac	1.51	4BEoo3	R				
3	A/	CPz,0z	O0z	1951ms	1969ch	acf	1.51	4Booo3					

AA 1.5.51 on A(-), 3/-, OSPR 7/51.

1	2	3	4	5	6	7	8	9	10	11	12	13	14

SN 00 (22/00) *Carew**   10 by 10   Revision a 1905-06 c 1948 f 1965
1 A   CPr   OPr   1948po   -   -   6.48 4BE4X3   R   0   4/48 6/48
1 B   CPr   nd:po   -   -   6.48 4BEoo3   R   0
3 B/   CPw   1950pc 1961mc a   6.48 4BooX3   0
3 B//   CPz   OPz   1950mc 1966ch acf   6.48 4BooX3   3
AA 1.3.48 on A(A), 3/-, OSPR 6/48. M on B, B/.

SN 01 (22/01) *Llawhaden**   10 by 10   Revision a 1905-06 c 1938-48
1 A   CPr   OPs   1950pc   -   ac   6.50 4BEoo3   R   0   8/50 8/50
3 A/   CPu,x   1950pc 1956mc ac   6.50 4BooX3   0
3 A//   CPz   OPz   1950pc 1956mc ac   6.50 4BooX3   0

SN 02 (22/02) *Maenclochog**   10 by 10 Revision a 1905-06 c 1948 f 1965 g 1972
                                              j 1970 reservoir
1 A   CPs   OPs   1951pc   -   ac   1.51 4BEoo4   R   8/51 10/51
3 A/   CPw   1951mc 1960mc ac   1.51 4BooX4
3 A//   CPz   1951mc 1966ch acf   1.51 4BooX4
3 A///* CPz   OPz   1951ᶜms nd:v   acgj 1.51 4BooX4   3/73 3/73 1
M on A. 1. Llys-y-fran Reservoir (OSPR).

SN 03 (22/03) *Newport Pembrokeshire*   10 by 10 Revision a 1904-06 c 1946-48 f 1965
1 A   CPs   OPs   1951pc   -   ac   6.51 4BEoo4   RB   8/51 10/51
3 A   CPw   1951pc -   ac   6.51 4BooX4   B
3 A/   CPz   1951mc 1966ch acf   6.51 4BooX4
M on A.

SN 04 (22/04) *Dinas Head*   10 by 10 Revision a 1904-06 c 1948
1 A   CPs   OPs   1950pc   -   ac   6.50 4BEoo4   10/50 11/50
2 A   CPt   1950pc 1952   ac   6.50 4Booo4
3 A/   CPw.z   1950mc 1960mc ac   6.50 4BooX4

SN 10 (22/10) *Tenby**   10 by 10 Revision a 1905-06 c 1938-48 e 1948
1 A   CPr   OPr   1950pc   -   ace   1.50 4BEoo3   R   6/50 6/50
2 A   CPs   1950pc 1952   ace   1.52 4Booo3   R
2 A   CPu   1950pc 1952rpb ace   1.52 4BooX3   R
3 A   CPx,z   1950pc -   ace   1.52 4BooX3   R
3 A/   CPz   1950pc -   ace   1.52 4BooX3

SN 11 (22/11) *Narberth**   10 by 10 Revision a 1905-37 c 1948
1 A   CPr   OPr   nd:po   -   -   1.50 4BEoo3   R   0   5/50 6/50
3 A/   CPv,z.z   1950pc 1956mc ac   1.50 4BooX3   0

SN 12 (22/12) *Llandissilio*   10 by 10 Revision a 1905-06 c 1948
1 A   CPs   OPs   1950pc   -   ac   6.50 4BEoo4   R   12/50 1/51
3 A/   CPw,z   1950mc 1960mc ac   6.50 4BooX4
M on A.

SN 13 (22/13) *Eglwyswrw**   10 by 10 Revision a 1904-06 c 1948
1 A   CPs   OPs   1950pc   -   ac   6.50 4BEoo4   R   12/50 12/50
3 A/   CPw,z   1950mc 1960mc ac   6.50 4BooX4
M on A. SS: Soil, outline, 1976 on A/.

SN 14 (22/14) **& Part of SN 15** *Cardigan** 10 by 12 Revision a 1904-06 c 1938-48
1 A   CPs   OPs   1951pc   -   ac   1.51 4BEoo3   R   6/51 7/51
3 A   CPu   1951pc -   ac   1.51 4BEoo3   R
3 A/   CPz   OPz   1951ᶜms -   ac   1.51 4Booo3

SN 20 (22/20) *Pendine**   10 by 10 Revision a 1905 c 1948
1 A   CPr   OPr   nd:po   -   -   1.50 4BEoo3   5/50 6/50
3 A   CPw,z   1950pc -   ac   1.50 4BooX3

1	2	3	4	5	6	7	8	9	10	11	12	13	14

SN 21 (22/21) *St Clears**   10 by 10   Revision a 1904-05 c 1948
| 1 | A | CPr | OPr | nd:po | - | - | 1.50 | 4BEoo3 | R | | 5/50 | 6/50 |
| 3 | A/ | CPw,z | | 1950mc | 1959mc | ac | 1.50 | 4BooX3 | | | | |

SN 22 (22/22) *Meidrim**   10 by 10   Revision a 1905-06 c 1948
| 1 | A | CPs | OPs | 1951pc | - | ac | 1.51 | 4BEoo4 | R | | 3/51 | 5/51 |
| 3 | A/ | CPw,z | | 1951mc | 1960mc | ac | 1.51 | 4BooX4 | | | | |

SN 23 (22/23) *Boncath*   10 by 10   Revision a 1904-05 c 1948
| 1 | A | CPs | OPs | 1951pc | - | ac | 6.51 | 4BEoo4 | R | | 9/51 | 10/51 |
| 3 | A/ | CPw,z | | 1951mc | 1960mc | ac | 6.60 | 4BooX4 | | | | |

SN 24 *Llechryd**   10 by 10   Revision a 1904-05 c 1938-48
2	A	CPt	OPs	1952pc	-	ac	6.52	4Booo4	R		6/52	6/52
3	A	CPu		1952pc	-	ac	6.52	4Booo4	R			
3	A/	CPz	OPz	1952pc	-	ac	6.52	4Booo4				

SS: Soil, Land Use Capability, coloured, 1980 on A/.

SN 25 (22/25) **& Part of SN 15** *Aberporth**   12 by 10   Revision a 1904 c 1948-49
| 1 | A | CPs | OPs | 1950pc | - | ac | 6.50 | 4BEoo3 | | | 11/50 | 12/50 |
| 3 | A/ | CP?w,z.z | OPz | 1950mc | 1958mc | ac | 6.50 | 4BooX3 | | | | |

SN 30 (22/30) *Mouth of the Towy**   10 by 10   Revision a 1905-13 c 1938-48
| 1 | A | CPr | OPr | nd:po | - | - | 1.50 | 4BEoo3 | R | 0 | 5/50 | 6/50 |
| 3 | A/ | CPv,z | OPv | 1950pc | 1956mc | ac | 1.50 | 4BooX3 | | 0 | | |

SN 31 (22/31) *Llanstephan Carmarthenshire**   10 by 10   Revision a 1904-13 c 1948 f 1972
1	A	CPr	OPr	nd:po	-	-	6.50	4BEoo3	R		5/50	6/50
3	A/	CPw,z		1950mc	1960mc	ac	6.50	4BooX3				
3	A//	CPz	OPz	1950mc	1960mc	ac	6.50	4BooX3				
3	A///*	CPz	OPz	1950^{c}ms	nd:r	acf	6.50	4BooX3			6/74	7/74

SN 32 *Cynwyl Elfed*   10 by 10   Revision a 1904-05 c 1938-48
| 2 | A | CPt | OPs | 1952pc | - | ac | 6.52 | 4Booo4 | | | 6/52 | 7/52 |
| 3 | A/ | CPw,z.z | | 1952mc | 1959mc | ac | 6.52 | 4BooX4 | | | | |

AA 1.3.52 on A(A), 4/-, OSPR 7/52.

SN 33 (22/33) *Llangeler**   10 by 10   Revision a 1904-05 c 1948
| 1 | A | CPs | OPs | 1950pc | - | ac | 6.50 | 4BEoo4 | R | | 12/50 | 12/50 |
| 3 | A/ | CPw.z | | 1950mc | 1960mc | ac | 6.50 | 4BooX4 | | | | |

SN 34 *Newcastle Emlyn**   10 by 10   Revision a 1904 c 1938-48
2	A	CPt	OPs	1952pc	-	ac	6.52	4Booo4	R		9/52	10/52
3	A	CPu		1952pc	-	ac	6.52	4Booo4	R			
3	A/	CPz	OPz	1952pc	-	ac	6.52	4Booo4				

SN 35 & Part of SN 36 *New Quay*   10 by 12   Revision a 1904 c 1938-48
| 2 | A | CPt,x | OPs | 1952pc | - | ac | 1.52 | 4Booo3 | | | 7/52 | 6/52 |
| 3 | A | CPz | | 1952pc | - | ac | 1.52 | 4Booo3 | | | | |

SN 40 (22/40) *Burry Port**   10 by 10   Revision a 1905-13 c 1938-53 f 1969
1	A	CPr	OPr	nd:po	-	-	6.48	4BEoo3	R	0	1/49	2/49
1	A	CPs		1949pc	1951	-	1.51	4BEoo3	R	0		
3	B	CPu		1956pc	-	ac	6.56	4oooX3		1	4/56	-
3	B/	CPv		1956pc	1957mc	ac	6.56	4oooX3		1		
3	B//	CPz	OPz	1956^{c}ms	nd:ch	acf	6.56	4oooX3		3		

AA 1.1.49 on A(A), 3/-, OSPR 2/49.

SN 41 (22/41) *Llangendeirne**   10 by 10   Revision a 1905-13 c 1948
| 1 | A | CPr | OPr | nd:po | - | - | 1.50 | 4BEoo3 | R | | 4/50 | 5/50 |
| 3 | A/ | CPv,y,z | | 1950pc | 1958mc | ac | 1.50 | 4BooX3 | | | | |

AA 1.3.50 on A(A), 3/-, OSPR 5/50. SS: Soil, outline, 1973 on A/.

1	2	3	4	5	6	7	8	9	10	11	12	13	14

SN 42 (22/42) *Carmarthen*   10 by 10   Revision a 1904-05 c 1948 f 1965
1   A   CPr   OPr   nd:po   -   -   1.50   4BEoo3   R   4/50   5/50
3   B   CPu,x   OPv   1955pc   -   ac   6.55   4BooX3   11/55   6/57
3   B/   CPz   OPz   1955mc   1966mc   acf   6.55   4oooX3
AA 1.3.50 on A(A), 3/-, OSPR 5/50.

SN 43 (22/43) *Pencader**   10 by 10   Revision a 1904 c 1948
1   A   CPs   OPs   1951pc   -   ac   6.51   4BEoo4   R   11/51   11/51
3   A/   CPw,z,0z   1951mc   1960mc   ac   6.51   4BooX4

SN 44 *Llandyssul*   10 by 10   Revision a 1904 c 1948
2   A   CPt   OPs   1952pc   -   ac   6.52   4Booo4   R   7/52   7/52
3   A/   CPw,z   1952mc   1960mc   ac   6.52   4BooX4

SN 45 (22/45) *Llanarth*   10 by 10   Revision a 1904 c 1938-48
1   A   CPs   OPs   1951pc   -   ac   6.51   4BEoo4   R   10/51   11/51
3   A/   CPw.z   1951mc   1960mc   ac   6.51   4BooX4

SN 46 (22/46) *Aberayron*   10 by 10   Revision a 1904 c 1938-50
1   A   CPr   OPr   nd:po   -   -   6.49   4BEoo3   R   3/50   5/50
3   B   CPu,z   OPv   1955pc   -   ac   6.55   4oooX3   1/56   5/57

SN 50 (22/50) *Llanelly (North)** 10 by 10 Revision a 1904-13 c 1938-48
1   A   CPr   OPr   nd:po   -   -   1.49   4BEoo3   R   10/49   11/49
3   A/   CPw.z   1949mc   1960mc   ac   1.49   4BooX3
AA 1.9.49 on A(A), 3/-, OSPR 11/49.

SN 51 (22/51) *Cross Hands Carmarthenshire** 10 by 10 Revision a 1905-13 c 1948
1   A   CPs   OPs   1950pc   -   ac   6.50   4BEoo3   R   10/50   10/50
3   A/   CPw,z   1950mc   1959c   ac   6.50   4BooX3

SN 52 (22/52) *Llanfynydd**   10 by 10   Revision a 1904-05 c 1948
1   A   CPr   OPr   nd:po   -   -   1.50   4BEoo3   R   4/50   5/50
3   A/   CPw,z   1950mc   1959mc   ac   1.50   4BooX3

SN 53 *Abergorlech**   10 by 10   Revision a 1903-04 c 1948
2   A   CPt   OPs   1952pc   -   ac   6.52   4Booo4   8/52   9/52
3   A   CPx,z   1952pc   -   ac   6.52   4Booo4

SN 54 *Lampeter**   10 by 10   Revision a 1904 c 1938-48
2   A   CPt   OPs   1952pc   -   ac   6.52   4Booo4   R   8/52   9/52
3   A/   CPw,z   1952mc   1960mc   ac   6.52   4BooX4

SN 55 *Llanfihangel Ystrad**   10 by 10   Revision a 1904 c 1948
2   A   CPs   OPs   1951pc   -   ac   6.51   4Booo4   R   11/51   11/51
3   A/   CPw,z   1951mc   1960mc   ac   6.51   4BooX4

SN 56 (22/56) *Llanon**   10 by 10   Revision a 1904 c 1948-50
1   A   CPr   OPr   nd:po   -   -   1.49   4BEoo3   2/49   2/49
3   B   CPu   OPv   1956pc   -   ac   6.56   4oooX3   5/56   11/57
3   B/   CPw,z   1956mc   1959mc   ac   6.56   4oooX3

SN 57 (22/57) *Llanfarian*   10 by 10   Revision a 1904 c 1937-52
1   A   CPr   OPr   nd:po   -   -   6.47   4BEoo3   R   1/49   2/49
3   B   CPu   OPv   1956pc   -   ac   6.56   4oooX3   4/56   5/57
3   B/   CPw.z,0z   1956mc   1959mc   ac   6.56   4oooX3
AA 1.1.49 on A(A), 3/-, OSPR 2/49.

SN 60 (22/60) *Clydach Glamorgan** 10 by 10 Revision a 1913-42 c 1938-48
1   A   CPr   1947po   -   -   1.48   4BE4X3   R   10/47   -
1   B   OPr   1947po   -   -   1.48   4BE4X3   R   -   12/47

1	2	3	4	5	6	7	8	9	10	11	12	13	14
2	B	CPt		1947pc	1953c	-	6.53	4BooX3					
3	B		OPs	1947pc	1953c	-	6.53	4BooX3					
3	C	CPu	OPv	1956pc	-	ac	6.56	4oooX3			4/56	12/57	
3	C/	CPx,z		1956mc	1961ch	ac	6.56	4oooX3					

AA 1.7.47 on B(A), 3/-, OSPR 12/47; AA 1.11.53 on B(B), 4/-, OSPR 8/54. M on A.

SN 61 (22/61) *Ammanford*   10 by 10 Revision a 1905-13 c 1938-48 f 1975
1	A	CPs	OPs	1950pc	-	ac	6.50	4BEoo3	R		11/50	11/50	
3	A/	CPv.z		1950pc	1957c	ac	6.50	4BooX3					
3	A//*	CFz	OFz	1950^{c}ms	nd:Ch	acf	6.50	4oooX3			11/76	11/76	

M on A.

SN 62 (22/62) *Llandeilo**   10 by 10 Revision a 1904-05 b 1905 c 1948 f 1966 g 1976
1	A	CPr	OPr	nd:po	-	-	1.50	4BEoo3	R		5/50	5/50	
3	B	CPu	OPv	1956pc	-	ac	6.56	4oooX3			2/56	6/57	
3	B/	CPw		1956mc	1959mc	bc	6.56	4oooX3					
3	B//	CPz	OPz	1956mc	1967ch	bcf	6.56	4oooX3					
3	B///*	CFz	OFz	1956^{c}ms	nd:Ch	bcg	6.56	4oooX3				12/76	1

M on A. SS: Soil, Land Use Capability, coloured, 1979 on B//. 1. OSPR (outline edition) not found.

SN 63 *Llansawel*   10 by 10 Revision a 1904-05 c 1948 f 1975
| 2 | A | CPt | OPs | 1952pc | - | ac | 6.52 | 4Booo4 | | | 6/52 | 6/52 | |
| 3 | A/ | CFz | | 1952^{c}ms | nd:ch | acf | 6.52 | 4Booo4 | | | | | |

M on A.

SN 64 *Pumpsaint*   10 by 10 Revision a 1904 c 1938-48
| 2 | A | CPs,t,w | OPs | 1952pc | - | ac | 1.52 | 4Booo4 | | | 3/52 | 4/52 | |
| 3 | A | CPz,0z | | 1952pc | - | ac | 1.52 | 4Booo4 | | | | | |

M on A.

SN 65 *Tregaron*   10 by 10 Revision a 1904 c 1948
| 2 | A | CPt | OPs | 1952pc | - | ac | 6.52 | 4Booo4 | R | | 6/52 | 6/52 | |
| 3 | A/ | CPv,z,0z | OPz | 1952pc | 1956mc | ac | 6.52 | 4BooX4 | | | | | |

M on A.

SN 66 (22/66) *Bronant*   10 by 10 Revision a 1904 c 1948-49
| 1 | A | CPr | OPr | nd:po | - | - | 1.49 | 4BEoo3 | R | | 1/49 | 1/49 | |
| 3 | B | CPu.0z | OPv | 1955pc | - | ac | 6.55 | 4oooX3 | | | 11/55 | 11/57 | |

M on A.

SN 67 (22/67) *Llanilar*   10 by 10 Revision a 1904 c 1938-48
1	A	CPr	OPr	nd:po	-	-	1.48	4BEoo3	R		11/48	12/48	
1	B	CPs		1950pc	-	ac	6.50	4BEoo3	R				
2	B		OPs	1950pc	-	ac	6.50	4BooX3	R		-	2/53	
3	B/	CPw,z		1950mc	1959mc	ac	6.50	4BooX3					

M on B.

SN 68 (22/68) **& Part of SN 58** *Aberystwyth* 15 by 10 Revision a 1903-04 c 1937-48
1	A	CPr	OPr	nd:po	-	-	6.47	4BEoo3	R		2/49	4/49	
3	B	CPt	OPv	1954pc	-	ac	6.54	4BooX3			2/54	6/57	
3	B/	CPv		1954pc	1958mc	ac	6.54	4BooX3					
3	B//	CPw,z.z		1954mc	1960mc	ac	6.54	4BooX3					

AA 1.2.49 on A(A), 3/-, OSPR 4/49.

SN 69 (22/69) **& Part of SN 59** *Aberdovey* 15 by 10 Revision a 1900-04 c 1948-52
1	A	CPr	OPr	nd:po	-	-	1.48	4BEoo3	RB		1/49	1/49	
3	B	CPu	OPv	1956pc	-	ac	6.56	4oooX3			5/56	6/57	
3	B/	CPw		1956mc	1959mc	ac	6.56	4oooX3					
3	B//	CPw.x,z		1956mc	1960mc	ac	6.56	4oooX3					
3	B//	G0z		1956mc	1960mc	ac	6.56	4oooX3					1

AA 1.1.49 on A(A), 3/-, OSPR 1/49. M on B//. 1. Overprinted in green with public rights of way information.

1	2	3	4	5	6	7	8	9	10	11	12	13	14

SN 70 (22/70) *Pontardawe**   10 by 10   Revision a 1913-42 c 1938-48
1	A	CPr	OPr	1948po	-	-	1.48	4BE4X3	RB		1/48	3/48	
3	B	CPv	OP?	1956pc	-	ac	6.56	4oooX3			7/56	2/58	
3	B/	CPw,z		1956mc	1960mc	ac	6.56	4oooX3					

AA 1.4.47 on A(A), 3/-, OSPR 3/48. M on A.

SN 71 *Cwmllynfell*   10 by 10   Revision a 1903-14 c 1948 f 1976
| 3 | A | CPt,u.x,z | OPs | 1953pc | - | ac | 6.53 | 4BooX4 | | | 11/53 | 11/53 | |
| 3 | A/* | CFz | OFz | 1953^{c}ms | nd:Ch | acf | 6.53 | 4BooX4 | | | 2/77 | 2/77 | |

M on A.

SN 72 *Llangadog**   10 by 10   Revision a 1903-05 c 1948 f 1975
2	A	CPs	OPs	1952pc	-	ac	1.52	4Booo4	R		5/52	6/52	
3	A/	CPw,z		1952mc	1959mc	ac	1.52	4BooX4					
3	A//*	CFz	OFz	1952^{c}ms	nd:Ch	acf	1.52	4BooX4			12/76	1/77	

M on A. SS: Soil, coloured, 1982 on A//*.

SN 73 *Llandovery*   10 by 10   Revision a 1904-05 c 1948 f 1976
2	A	CPt	OPs	1952pc	-	ac	6.52	4Booo4	R		11/52	10/52	
3	A/	CPw.z		1952mc	1960mc	ac	6.52	4BooX4					
3	A//	CFz		1952^{c}ms	nd:ch	acf	6.52	4BooX4					

M on A.

SN 74 *Mynydd Mallaen*   10 by 10   Revision a 1904 c 1948 f 1972; j 1972 reservoir
2	A	CPt	OPs	1953pc	-	ac	6.53	4BooX4			10/53	11/53	
3	A/	CPw.z		1953mc	1959mc	ac	6.53	4BooX4					
3	A//*	CPz,0z	OPz	1953^{c}ms	nd:v	acfj	6.53	4BooX4			9/73	10/73	1

M on A. 1. Llyn Brianne Reservoir.

SN 75 *Llyn Berwyn*   10 by 10   Revision a 1904 c 1948
| 2 | A | CPt | OPs | 1954pc | - | ac | 1.54 | 4BooX4 | | | 10/53 | 11/53 | |
| 3 | A | CPv.z | | 1954pc | | ac | 1.54 | 4BooX4 | | | | | |

M on A.

SN 76 *Ystrad Meurig*   10 by 10   Revision a 1904 c 1948
| 3 | A | CPt | OPs | 1953pc | - | ac | 6.53 | 4BooX4 | | | 11/53 | 11/53 | |
| 3 | A/ | CPw,z | | 1953mc | 1959mc | ac | 6.53 | 4BooX4 | | | | | |

M on A.

SN 77 (22/77) *Devil's Bridge*   10 by 10   Revision a 1904-05 c 1948 f 1970
1	A	CPr	OPr	nd:po	-	-	1.48	4BEoo3	R		1/49	2/49	
3	B	CPu	OPv	1956pc	-	ac	6.56	4oooX4			3/56	5/57	
3	B/	CPw		1956mc	1960mc	ac	6.56	4oooX4					
3	B//	CPz,0z	OPz	1956^{c}ms	nd:ch	acf	6.56	4oooX4					

M on A.

SN 78 (22/78) *Pont-erwyd*   10 by 10   Revision a 1900-04 f 1963
1	A	CPr	OPr	nd:po	-	-	1.48	4BEoo3	B		11/48	11/48	
1	B	CPr	OP?	nd:po	-	-	1.48	4BEoo3	B			4/54	
3	B/	CPw		1950mc	1959mc	a	1.48	4BooX3					
3	B//*	CPz,0z	OPz	1950mc	1964v	af	1.48	4BooX3			9/64	10/64	

M on B.

SN 79 (22/79) *Anglers' Retreat*   10 by 10   Revision a 1900-04 c 1948
| 1 | A | CPr,u | OPr | nd:po | - | - | 1.49 | 4BEoo3 | RB | | 12/48 | 2/49 | |
| 3 | B | CPu,y,z.0z | OPv | 1955pc | - | ac | 6.55 | 4oooX3 | | | 1/56 | 5/57 | |

AA 1.12.48 on A(A), 3/-, OSPR 2/49.

SN 80 (22/80) *Vale of Neath**   10 by 10   Revision a 1913-35 c 1948
| 1 | A | CPr | OPr | nd:po | - | - | 1.49 | 4BEoo3 | R | | 10/49 | 10/49 | |
| 3 | A/ | CP?v,y,z | | 1949pc | 1957c | ac | 6.49 | 4BooX3 | | | | | |

M on A.

184

1	2	3	4	5	6	7	8	9	10	11	12	13	14

SN 81 *Pant-y-cwrt*   10 by 10   Revision a 1903-14 c 1942-48 f 1966 g 1976
3   A   CPt,z   OPs   1953pc   -   ac   6.53   4BooX4   11/53   11/53
3   A/   CPz,0z   OPz   1953mc   1967ch   acf   6.53   4BooX4
3   A//*   CFz   OFz   1953^{c}ms   nd:Ch   acg   6.53   4oooX4   1
M on A. 1. OSPR not found.

SN 82 *Trecastle**   10 by 10   Revision a 1903-04 c 1948 f 1963 g 1969 h 1976
                                                 j 1963 reservoir
3   A   CPt   OPs   1953pc   -   ac   6.53   4BooX4   11/53   12/53
3   A/
3   A//*   CPz   OPz   1953mc   1964v   acfj   6.53   4BooX4   5/64   4/64   1
3   A//*/   CPz   1953^{c}ms   nd:ch   acgj   6.53   4BooX4
3   A//*//*   CFz   OFz   1953^{c}ms   nd:Ch   achj   6.53   4BooX4   2
M on A. 1. Usk Reservoir. 2. OSPR not found.

SN 83 *Mynydd Bwlch-y-Groes*   10 by 10   Revision a 1903-04 c 1938-48 f 1975
3   A   CPt,v.z   OPs   1953pc   -   ac   6.53   4BooX4   12/53   12/53
3   A/*   CFz   OFz   1953^{c}ms   nd:Ch   acf   6.53   4BooX4   2/77   1
M on A. 1. OSPR (outline edition) not found.

SN 84 *Llanwrtyd Wells*   10 by 10   Revision a 1903-05 c 1938-48 f 1976; j 1972 reservoir
3   A   CPt,v.z   OPs   1953pc   -   ac   6.53   4BooX4   11/53   12/53
3   A/*   CFz   OFz   1953^{c}ms   nd:v   acfj   6.53   4BooX4   12/76   1
M on A. 1. Llyn Brianne Reservoir. 1. OSPR (outline edition) not found.

SN 85 *Abergwesyn*   10 by 10   Revision a 1903-04 c 1938-48
2   A   CPt,u   OPs   1953pc   -   ac   6.53   4BooX4   11/53   11/53
3   A   CPz,0z   1953pc   -   ac   6.53   4BooX4
M on A.

SN 86 *Claerwen*   10 by 10   Revision a 1903-04 c 1948-52
2   A   CPt   OPs   1953pc   -   ac   6.53   4BooX4   7/53   7/53
3   A/   CPv,z   1953pc   1956mc   ac   6.53   4BooX4
M on A.

SN 87 *Afon Elan*   10 by 10   Revision a 1901-03 c 1948
2   A   CPt   OPs   1953pc   -   ac   6.53   4BooX4   5/53   5/53
3   A/   CPv.z   1953pc   1957mc   ac   6.53   4BooX4
M on A.

SN 88 *Pumlumon-Arwystli*   10 by 10   Revision a 1900-05 c 1948-49 f 1970
                                               j 1964-65 afforestation k 1969 reservoir
2   A   CPt   OPs   1953pc   -   ac   6.53   4BooX4   7/53   7/53
3   A/   CPv.z   1953pc   1957mc   ac   6.53   4BooX4
3   A//*   CPz,0z   OPz   1953^{c}ms   nd:fl   acfjk   6.53   4BooX4   4/72   8/72   1
M on A. 1. Llyn Clywedog.

SN 89 *Staylittle*   10 by 10   Revision a 1900-01 c 1948-49; j 1981 selected
3   A   CPt   OPs   1953pc   -   ac   6.53   4BooX4   12/53   12/53
3   A/   CP?v,z,0z   1953pc   1956mc   ac   6.53   4BooX4
3   A//*   CFz   OFz   1982^{c}m-s   nd:frw   acj   6.83   4BooX4   10/82   10/82

SN 90 (22/90) *Hirwaun**   10 by 10   Revision a 1903-14 c 1947-49 f 1967
1   A   CPr   OPr   1947po   -   -   1.48   4BE2X3   RB   8/47   1
3   B   CPu   OP?   1954pc   -   ac   6.54   4oooX3   11/54   1/58
3   B/   CPw,z   1954mc   1958c   ac   6.54   4oooX3
3   B//*   CPz   OPz   1954mc   1968r   acf   6.54   4oooX3   2/68   1/68
M on A. 1. OSPR (outline edition) not found.

1	2	3	4	5	6	7	8	9	10	11	12	13	14

SN 91 (22/91) *Ystradfellte*   10 by 10   Revision a 1903-14   c 1948   f 1970   g 1976
1	A	CPr	OPr	nd:po	-	-	6.48	4BEoo3			8/49	9/49	
2	A	CPt		1949pc	1953rpb	-	6.48	4BooX3					
3	A	CPx		1949pc	-	-	6.48	4BooX3					
3	A/	CPz	OPz	1949ᶜms	nd:ch	acf	6.48	4BooX3					
3	A//		OFz	1949ᶜms	nd:ch	acg	6.48	4BooX3					

M on A.

SN 92 *Sennybridge**   10 by 10   Revision a 1903   c 1948   f 1965   g 1976
2	A	CPt	OPs	1953pc	-	ac	6.53	4BooX4			10/53	10/53	
3	A/	CP?w,x		1953mc	1959mc	ac	6.53	4BooX4					
3	A//	CPz,0z	OPz	1953mc	1965ch	acf	6.53	4BooX4					
3	A///*	CFz	OFz	1953ᶜms	nd:Ch	acg	6.53	4BooX4					1

M on A. 1. OSPR not found.

SN 93 *Llanfihangel Nant Bran**   10 by 10   Revision a 1903   c 1942-48
| 2 | A | CPt | OPs | 1953pc | - | ac | 6.53 | 4BooX4 | | | 8/53 | 8/53 | |
| 3 | A | CPw.z | | 1953mc | - | ac | 6.53 | 4BooX4 | | | | | |

M on A.

SN 94 *Garth Brecknockshire*   10 by 10   Revision a 1902-04   c 1945-48
| 2 | A | CPt | OPs | 1953pc | - | ac | 6.53 | 4BooX4 | | | 10/53 | 11/53 | 1 |
| 3 | A/ | CPw.z,0z | | 1953mc | 1958mc | ac | 6.53 | 4BooX4 | | | | | |

M on A. 1. Showing the Artillery Range (9142).

SN 95 *Beulah Brecknockshire**   10 by 10   Revision a 1902-04   c 1941-48
| 3 | A | CPt | OPs | 1953pc | - | ac | 6.53 | 4BooX4 | | | 12/53 | 12/53 | |
| 3 | A/ | CPw,z,0z | | 1953mc | 1959mc | ac | 6.53 | 4BooX4 | | | | | |

M on A.

SN 96 *Rhayader*   10 by 10   Revision a 1902-04   c 1948   f 1966
2	A	CPt	OPs	1953pc	-	ac	1.53	4BooX4	R		1/53	2/53	1
3	A/	CP?w,z		1953mc	1960mc	ac	1.53	4BooX4					
3	A//	CPz,0z	O0z	1953mc	1966ch	acf	1.53	4BooX4					

M on A. 1. The Elan Aqueduct tunnels are marked. There are some indications of the Elan Valley railway system - the junction at 962671 and an alignment next to the road.

SN 97 *Llangurig*   10 by 10   Revision a 1901-04   c 1948
| 2 | A | CPt | OPs | 1952pc | - | ac | 6.52 | 4Booo4 | R | | 11/52 | 11/52 | |
| 3 | A/ | CPw,z | | 1952mc | 1960mc | ac | 6.52 | 4BooX4 | | | | | |

M on A.

SN 98 *Llanidloes*   10 by 10   Revision a 1901   c 1938-48   f 1969; j 1969 reservoir
2	A	CPt	OPs	1952pc	-	ac	6.52	4Booo4	R		1/53	1/53	
3	A/	CPw		1952mc	1960mc	ac	6.52	4BooX4					
3	A//*	CPz	OPz	1952ms	1969v	acfj	6.52	4BooX4			1/70	1/70	1

M on A. 1. Llyn Clywedog.

SN 99 *Carno**   10 by 10   Revision a 1900-01   c 1948
| 2 | A | CPt | OPs | 1953pc | - | ac | 1.53 | 4BooX4 | | | 5/53 | 5/53 | |
| 3 | A/ | CPw.z | | 1953mc | 1961mc | ac | 1.53 | 4BooX4 | | | | | |

SO 00 (32/00) [Merthyr Tydfil]   10 by 10   Revision a 1903-15   c 1938-48   f 1964
1	A	CPr	OPr	1947po	-	-	6.47	4BE2X3	RB		6/47	10/47	
3	B	CPu	OPv	1954pc	-	ac	6.54	4oooX3			10/54	11/57	
3	B/	CP?w,y		1954pc	1958mc	ac	6.54	4oooX3					
3	B//*	CPz	OPz	1954mc	1965r	acf	6.54	4oooX3			8/65	7/65	

AA 1.4.47 on A(A), 3/-, OSPR 10/47. M on A.

1	2	3	4	5	6	7	8	9	10	11	12	13	14

SO 01 (32/01) *Ponsticill*   10 by 10   Revision a 1903-15   c 1948   f 1969   g 1976
1	A	CPr	OPr	1947po	-	-	1.48	4BE2X3	RB		8/47	11/47
3	B	CPu	OPv	1955pc	-	ac	6.55	4oooX3			5/55	11/57
3	B/	CPw,z		1955mc	1960mc	ac	6.55	4oooX3				
3	B//	CPz	OPz	1955^{c}ms	1970ch	acf	6.55	4oooX3				
3	B///*	CFz	OFz	1955^{c}ms	nd:Ch	acg	6.55	4oooX3			5/77	5/77

AA 1.5.47 on A(A), 3/-, OSPR 11/47. M on A.

SO 02 (32/02) *Brecon*   10 by 10   Revision a 1903   c 1938-48   f 1976
1	A	CPs	OPs	1951pc	-	ac	6.51	4BEoo4	R		8/51	9/51
3	A	CPt,x.z		1951pc	-	ac	6.51	4BooX4	R			
3	A/	CPz		1951ms	-	ac	6.51	4BooX4				
3	A//*	CFz	OFz	1951^{c}ms	nd:Ch	acf	6.51	4BooX4			2/77	2/77

AA 1.7.51 on A(A), 3/-, OSPR 9/51. M on A.

SO 03 (32/03) *Llandefaelog-fach*   10 by 10   Revision a 1903   c 1938-48
| 1 | A | CPs | OPs | 1951pc | - | ac | 1.51 | 4BEoo4 | RB | | 7/51 | 8/51 |
| 3 | A/ | CPw,z,0z | | 1951mc | 1959mc | ac | 1.51 | 4BooX4 | B | | | |

M on A.

SO 04 (32/04) *Aberedw*   10 by 10   Revision a 1902-03   c 1946-48
| 1 | A | CPs | OPs | 1951pc | - | ac | 1.51 | 4BEoo4 | RB | | 5/51 | 7/51 |
| 3 | A/ | CPu,z | OPz | 1951pc | 1956mc | ac | 1.51 | 4BooX4 | B | | | |

M on A.

SO 05 *Builth Wells*   10 by 10   Revision a 1902-03   c 1938-48   f 1965
2	A	CPs	OPs	1951pc	-	ac	6.51	4Booo4	RB		2/52	4/52
3	A/	CPv		1951pc	1957mc	ac	6.51	4BooX4	B			
3	A//	CPz,0z		1951mc	1966ch	acf	6.51	4BooX4	B			
3	A///	CFz		1951^{c}ms	-	acf	6.51	4BooX4	B			

AA 1.12.51 on A(A), 3/-, OSPR 4/52. M on A/.

SO 06 (32/06) *Llandrindod Wells*   10 by 10   Revision a 1902   c 1938-48
| 1 | A | CPs | OPs | 1951pc | - | ac | 6.51 | 4BEoo4 | RB | | 12/51 | 1/52 |
| 3 | A/ | CPw.y,z,0z | | 1951mc | 1960mc | ac | 6.51 | 4BooX4 | | | | |

AA 1.1.52 on A(A), 3/-, OSPR 1/52. M on A.

SO 07 (32/07) *Abbey-Cwmhir*   10 by 10   Revision a 1901-02   c 1946-48
| 1 | A | CPs,x | OPs | 1950pc | - | ac | 6.50 | 4BEoo4 | | | 10/50 | 12/50 |
| 3 | A | CPz | | 1950pc | - | ac | 6.50 | 4BEoo4 | | | | |

This sheet was almost totally unrevised; even the sheet number was not adjusted until ten years after the event.

SO 08 (32/08) *Llandinam* *   10 by 10   Revision a 1901-02   c 1938-48
| 1 | A | CPs | OPs | 1951pc | - | ac | 6.51 | 4BEoo4 | RB | | 12/51 | 1/52 |
| 3 | A/ | CPw,z | | 1951mc | 1960mc | ac | 6.51 | 4BooX4 | | | | |

AA 1.12.51 on A(A), 3/-, OSPR 1/52.

SO 09 (32/09) *Caersws*   10 by 10   Revision a 1900-01   c 1938-48
| 1 | A | CPs | OPs | 1951pc | - | ac | 1.51 | 4BEoo4 | RB | | 3/51 | 6/51 |
| 3 | A/ | CPw,z | | 1951mc | 1960mc | ac | 1.51 | 4BooX4 | | | | |

AA 1.4.51 on A(A), 3/-, OSPR 6/51. SS: Soil, outline, 1974 on A/.

SO 10 (32/10) [*Ebbw Vale*]   10 by 10   Revision a 1914-16   c 1945-49   f 1966
1	A	CPr	OPr	nd:po	-	-	6.48	4BEoo3	R		8/48	10/48
3	B	CPu	OPv	1956pc	-	ac	6.56	4oooX3			5/56	5/57
3	B/	CPx,z	OPx	1956mc	1961mc	ac	6.56	4oooX3				
3	B//*	CPz	OPz	1956mc	1967r	acf	6.56	4oooX3			6/67	6/67

AA 1.9.48 on A(A), 3/-, OSPR 10/48. M on A.

1	2	3	4	5	6	7	8	9	10	11	12	13	14

SO 11 (32/11) *Brynmawr*   10 by 10   Revision a 1899-1916 c 1938-49 f 1966 g 1975
1	A	CPs		1951pc	-	ac	6.51	4BEoo4	R		8/51	-	
1	B		OPs	1951pc	-	ac	6.51	4BEoo4	R		-	11/51	
1	B [=A/]		OPs	1952pc	-	ac	6.51	4BEoo4	R				1
3	A/	CP?v,x		1951pc	1957mc	ac	6.51	4BooX4					
3	A//*	CPz	OPz	1951mc	1967r	acf	6.51	4BooX4			11/67	10/67	
3	A//*/*	CFz	OFz	1951ᶜms	nd:Ch	acg	6.51	4BooX4			11/76	11/76	

AA 1.6.46 on B(A), 3/-, OSPR 11/51. M on A. 1. A copy in the Ordnance Survey Record Map Library has the B print code crossed out, with A/ added in manuscript. Either would appear to be incorrect, so too the continuation of the print code sequence.

SO 12 (32/12) *Llangorse**   10 by 10   Revision a 1903 c 1945-48 f 1975
1	A	CPs	OPs	1951pc	-	ac	1.51	4BEoo4	RB		5/51	5/51	
3	A/	CP?w,z.z	OPz	1951mc	1959mc	ac	1.51	4BooX4	B				
3	A//*	CFz	OFz	1951ᶜms	nd:Ch	acf	1.51	4BooX4	B		12/76		1

M on A. 1. OSPR (outline edition) not found.

SO 13 (32/13) *Talgarth Brecknockshire*   10 by 10   Revision a 1903-27 c 1938-48 f 1976
1	A	CPs	OPs	1951pc	-	ac	1.51	4BEoo4	RB		6/51	7/51	
3	A/	CPv,z	OPz	1951pc	1956mc	ac	1.51	4BooX4	B				
3	A//*	CFz	OFz	1951ᶜms	nd:Ch	acf	1.51	4BooX4	B		2/77	2/77	

M on A/.

SO 14 (32/14) *Painscastle*   10 by 10   Revision a 1902-27 c 1938-48
1	A	CPs	OPs	1951pc	-	ac	1.51	4BEoo4	RB		5/51	7/51	
3	A	CPu		1951pc	-	ac	1.51	4BEoo4	RB				
3	A/	CPw,z.z,0z	OPx	1951mc	1961mc	ac	1.51	4BooX4	B				

M on A.

SO 15 (32/15) *Llanfihangel-nant-Melan*   10 by 10   Revision a 1902 c 1948
| 1 | A | CPs | OPs | 1951pc | - | ac | 1.51 | 4BEoo4 | B | | 7/51 | 8/51 | |
| 3 | A/ | CPv,z | | 1951pc | 1958mc | ac | 1.51 | 4BooX4 | B | | | | |

M on A/.

SO 16 (32/16) *Penybont*   10 by 10   Revision a 1902 c 1948
| 1 | A | CPs | OPs | 1951pc | - | ac | 6.51 | 4Booo4 | RB | | 11/51 | 11/51 | |
| 3 | A/ | CPw,z | | 1951mc | 1959mc | ac | 6.51 | 4BooX4 | B | | | | |

M on A/.

SO 17 *Llanbister**   10 by 10   Revision a 1901-02 c 1946-48
| 2 | A | CPs | OP? | 1951pc | - | ac | 6.51 | 4Booo4 | RB | | 2/52 | 2/52 | |
| 3 | A/ | CPx,z | | 1951mc | 1961mc | ac | 6.51 | 4BooX4 | B | | | | |

SO 18 (32/18) *Dolfor*   10 by 10   Revision a 1901-02 c 1938-49
1	A	CPs	OPs	1951pc	-	ac	1.51	4BEoo4	B		6/51	6/51	
2	A		OPs	1951pc	-	ac	1.52	4Booo4	B				1
3	A/	CPw,y,z	OPv	1951mc	1958mc	ac	1.51	4BooX4	B				

AA 10.1.52 on A(A), 3/-, OSPR 2/52. 1. Unrecorded except as the base of the Administrative Areas edition.

SO 19 *Newtown Montgomeryshire*   10 by 10   Revision a 1901 c 1938-49
| 2 | A | CPs | OPs | 1951pc | - | ac | 6.51 | 4BEoo4 | R | | 1/52 | 3/52 | |
| 3 | A/ | CPw,z,0z | OPv | 1951mc | 1961mc | ac | 6.51 | 4oooX4 | | | | | |

AA 1.1.52 on A(A), 3/-, OSPR 3/52.

SO 20 (32/20) *Pontypool*   10 by 10   Revision a 1915-17 c 1944-49
| 1 | A | CPr | OPr | nd:po | - | - | 1.50 | 4BEoo3 | RB | | 3/50 | 5/50 | 1 |
| 3 | A/ | CP?v,y,z,0z | OPz | 1950pc | 1956mc | ac | 1.50 | 4BooX3 | B | | | | |

AA 1.3.50 on A(A), 3/-, OSPR 5/50. M on A. 1. The coloured edition appears in both OSPR 3/50 and 4/50.

	1	2	3	4	5	6	7	8	9	10	11	12	13	14

SO 21 (32/21) *Abergavenny (West)*   10 by 10 Revision a 1903-17 c 1938-48 f 1965 g 1976
                                                  j 1965 railway
1 A   CPr   OPr   nd:po   -   -   1.50 4BEoo3   RB   3/50   5/50
3 B   CPv,x   OPv,z   1957pc   -   ac   6.57 4oooX3   5/57   9/57
3 B/*   CPz   OPz   1957mc   1965r   acfj   6.57 4oooX3   12/65 12/65
3 B/*/* CFz   OFz   1957ᶜms nd:Ch   acg   6.57 4oooX3   5/77   5/77
AA 1.12.49/NT 26.9.49 on A(A), 3/-, OSPR 5/50. M on A.

SO 22 (32/22) *Llanthony*   10 by 10 Revision a 1891-1903 c 1948 e 1945 f 1976
1 A   CPs   OPs   1951pc   -   ace   6.51 4BEoo4   B   1/52   1/52
3 A/   CPw,x,z OPv,y,z 1951pc   1958mc ace   6.51 4BooX4   B
3 A//* CFz   OFz   1951ᶜms nd:Ch   acef   6.51 4BooX4   B   2/77   2/77
M on A.

SO 23 *Black Hill Herefordshire*   10 by 10 Revision a 1903-27 c 1945-48
2 A   CPt   OPs   1952pc   -   ac   6.52 4Booo4   9/52   10/52
3 A/   CPv,x,z OPv,z   1952pc   1957mc ac   6.52 4BooX4
M on A/.

SO 24 (32/24) *Hay-on-Wye*   10 by 10 Revision a 1902-27 c 1938-48
1 A   CPs   OPs   1951pc   -   ac   6.51 4BEoo4   RB   12/51 1/52
3 A/   CPv,y,z OPv   1951pc   1958c   ac   6.51 4BooX4   B
M on A/.

SO 25 (32/25) *Kington (West) Herefordshire* 10 by 10 Revision a 1902-27 c 1938-49 f 1966
1 A   CPs   OPs   1951pc   -   ac   6.51 4BEoo4   RB   8/51   10/51
3 A/   CPw   1960ᶜmc 1960mc ac   6.51 4oooX4
3 A//   CPz   OPz   1960ᶜmc 1966ch acf   6.51 4oooX4
M on A/.

SO 26 *New Radnor*   10 by 10 Revision a 1902-27 c 1949
2 A   CPs   OPs   1951pc   -   ac   6.51 4Booo4   B   1/52   4/52
3 A/   CPw,y,z OPv,z,0z 1951mc 1959mc ac   6.51 4BooX4   B
AA 1.11.51 on A(A), 3/-, OSPR 4/52.

SO 27 *Knighton Radnorshire*   10 by 10 Revision a 1901-26 c 1938-49
2 A   CPt   OPs   1952pc   -   ac   6.52 4Booo4   R   6/52   8/52 1
3 A   OPs   1952pc   -   ac   6.52 4BooX4   R
3 A/   CP?v,y,z,0z   1952pc   1957mc ac   6.52 4BooX4
AA 1.2.52 on A(-), 4/-, OSPR 8/52. 1. Showing the Knighton Tunnel on the Elan Aqueduct.

SO 28 *Newcastle Shropshire*   10 by 10 Revision a 1901-24 c 1938-49
2 A   CPt   OPs   1952pc   -   ac   6.52 4Booo4   7/52   7/52
3 A/   CPv,y,z   1952pc   1957mc ac   6.52 4BooX4

SO 29 *Montgomery**   10 by 10 Revision a 1901-24 c 1938-49
2 A   CPs   OPs   1952pc   -   ac   1.52 4Booo4   R   3/52   4/52
3 A/   CPv,y,z OPz   1952pc   1956mc ac   1.52 4BooX4

SO 30 (32/30) *Usk*   10 by 10 Revision a 1916-18 c 1938-49 f 1965 g 1974
                                                  j 1965 reservoir k 1965 railway
1 A   CPr   OPr   nd:po   -   -   1.50 4BEoo3   R   2/50   2/50
3 B   CPu   OPv   1956pc   -   ac   6.56 4oooX3   4/56   5/57
3 B/   CPx   OPx   1956mc   1961mc ac   6.56 4oooX3   1
3 B//* CPz   OPz   1956mc   1966rv acfjk 6.56 4oooX3   5/66   8/66
3 B///* CFz   OFz   1956ᶜms nd:r   acg   6.56 4oooX3   6/75   5/75 2
M on A. 1. There is a large blank space at 3401 where full topography had been present earlier. 2. This print code should read B//*/*.

1	2	3	4	5	6	7	8	9	10	11	12	13	14

SO 31 (32/31) *Abergavenny (East)*   10 by 10 Revision a 1916-18 c 1944-45 f 1968 g 1975
                                                   j 1965 building
1   A   CPr   OPr   nd:po   -   -   6.49   4BEoo3   R   2/50   3/50
3   B   CPv   OPv   1956pc   -   ac   6.56   4oooX3   2/57   6/57
3   B/   CPw,y   OPv,z   1956mc   1959mc   ac   6.56   4oooX3   2/57   6/57
3   B//*   CPz   OPz   1956ms   1968r   acf   6.56   4oooX3   11/68 9/68
3   B//*/* CFz   OFz   1956ᶜms   nd:br   acgj   6.56   4oooX3   2/77   2/77
AA 1.9.49/NT 26.9.49 on A(A), 3/-, OSPR 3/50. M on A.

SO 32 (32/32) *Longtown*   10 by 10 Revision a 1903-18 c 1945 f 1976
1   A   CPs   OPs   1951pc   -   ac   6.51   4BEoo4   RB   11/51 11/51
3   A/   CPw,y,z   OPv   1951mc   1959mc   ac   6.51   4BooX4   B
3   A//*   CFz   OFz   1951ᶜms   nd:r   acf   6.51   4oooX4   B   1
M on A. 1. OSPR not found.

SO 33 *Vowchurch*   10 by 10 Revision a 1903-16 c 1945 e 1945 f 1976
2   A   CPs   OPs   1952pc   -   ae   1.52   4Booo4   R   5/52   6/52
3   A/   CP?w,y,z OPz   1952mc   1960mc   ae   1.52   4BooX4
3   A//   CFz   1952ᶜms   nd:ch   acf   1.52   4oooX4

SO 34 *Staunton on Wye*   10 by 10 Revision a 1902-03 c 1945
2   A   CPs   OPs   1952pc   -   ac   1.52   4Booo4   RB   3/52   4/52   1
3   A/   CPv,y,z.z OPv   1952pc   1957mc   ac   1.52   4BooX4   B
SS: Soil, outline, 1972 on A/. 1. Probably the last new sheet to be published showing bridle roads.

SO 35 *Kington (East) Herefordshire* 10 by 10 Revision a 1902-27 c 1945-49
2   A   CPs   OPs   1952pc   -   ac   1.52   4Booo4   R   1   2/52   4/52
3   A/   CPw,z   OPv   1952mc   1959mc   ac   1.52   4BooX4   1

SO 36 *Presteigne*   10 by 10 Revision a 1902-27 c 1938-49
2   A   CPt   OPs   1952pc   -   ac   6.52   4Booo4   R   1   7/52   10/52
3   A/   CPw,x,z,0z OPv   1952pc   1958mc   ac   6.52   4BooX4   1
AA 1.4.52 on A(A), 4/-, OSPR 10/52.

SO 37 *Bucknell Shropshire*   10 by 10 Revision a 1901-26 c 1949
2   A   CPt   OPs   1952pc   -   ac   6.52   4Booo4   R   8/52   10/52
3   A/   CPw,y,z   1952mc   1958mc   ac   6.52   4BooX4
AA 1.6.52 on A(A), 4/-, OSPR 10/52.

SO 38 *Bishops's Castle*   10 by 10 Revision a 1901-24 c 1938-49
2   A   CPt,u   OPs   1952pc   -   ac   6.52   4Booo4   R   6/52   7/52
3   A/   CPw.z,0z OPz.0z   1952mc   1961mc   ac   6.52   4BooX4

SO 39 *Lydham*   10 by 10 Revision a 1901-24 c 1949
2   A   CPt   OPs   1952pc   -   ac   6.52   4Booo4   6/52   8/52
3   A/   CPv,w,y,z.z OPv   1952mc   1957mc   ac   6.52   4BooX4

SO 40 (32/40) *Raglan*   10 by 10 Revision a 1916-18 c 1938-49 f 1973 g 1976
1   A   CPr   OPr   nd:po   -   -   1.49   4BEoo3   RB   2/49   2/49
1   B   CPs   OPv   1951pc   -   ac   1.51   4BEoo3   R   6/51   -
3   B/   CPw,z   1951mc   1960mc   ac   1.51   4BooX3
3   B//*   CPz   OPz   1951ᶜms   nd:r   acf   1.51   4BooX3   10/73 10/73
3   B//*/ CFz   1951ᶜms   nd:ch   acg   1.51   4BooX3   3/77
M on B.

SO 41 (32/41) *Rockfield Monmouthshire* 10 by 10 Revision a 1903-18 c 1943-45 f 1972
1   A   CPr   OPr   nd:po   -   -   1.49   4BEoo3   RB   1/49   1/49
3   B   CPv,x,z   OPv   1956pc   -   ac   6.56   4oooX3   1/57   1/57   1
3   B/*   CPz   OPz   1956ᶜms   nd:r   acf   6.56   4oooX3   11/72 3/73
M on B.

1	2	3	4	5	6	7	8	9	10	11	12	13	14

SO 42 (32/42) *Garway**　　　　　　　　10 by 10　Revision a 1903-18 c 1944-45
1　A　　CPr　　　OPr　　　nd:po　　-　　-　　1.50　4BEoo3　R　　　5/50　6/50
3　B　　CPv,y,z　OPv,x,z　1956pc　-　　ac　6.56　4oooX3　　　　1/57　1/57

SO 43 (32/43) *Thruxton Herefordshire*　10 by 10　Revision a 1903-28 c 1945
1　A　　CPr　　　OPr　　　nd:po　　-　　-　　1.50　4BEoo3　R　　0　6/50　7/50
3　B　　CPv　　　OPv　　　1957pc　-　　ac　6.57　4oooX3　　　0　3/57　5/57
3　B/　 CPx,z.z　　　　　　1957mc　1960mc　ac　6.57　4oooX3　　0
AA 1.3.50 on A(A), 3/-, OSPR 7/50.

SO 44 (32/44) *Credenhill*　　　　　　　　10 by 10　Revision a 1902-28 c 1945
1　A　　CPr　　　OPr　　　nd:po　　-　　-　　1.50　4BEoo3　RB　　7/50　7/50
3　B　　CPv　　　OPv　　　1956pc　-　　ac　6.56　4oooX3　　　　1/57　2/57
3　B/　 CPw,z　　　　　　　1956mc　1960mc　ac　6.56　4oooX3
AA 1.2.50 on A(A), 3/-, OSPR 8/50.

SO 45 (32/45) *Leominster (West)*　　　　10 by 10　Revision a 1902-27 c 1945-49
1　A　　CPr　　　OPr　　　nd:po　　-　　-　　6.49　4BEoo3　RB　0　1/50　2/50
3　B　　CPv,y,z　OPv　　　1957pc　-　　ac　6.57　4oooX3　　　1　7/57　8/57
AA 1.9.49 on A(A), 3/-, OSPR 2/50.

SO 46 (32/46) *Bircher*　　　　　　　　　10 by 10　Revision a 1902-27 c 1938-49
1　A　　CPr　　　OPr　　　nd:po　　-　　-　　1.50　4BEoo3　RB　0　4/50　5/50
3　B　　CPv　　　OPv　　　1956pc　-　　ac　6.56　4oooX3　　　1　1/57　2/57
3　B/　 CPw,z　　OPx　　　1956mc　1961mc　ac　6.56　4oooX3　　1
M on B.

SO 47 (32/47) *Bromfield Shropshire*　　10 by 10　Revision a 1901-24 c 1949
1　A　　CPr　　　OPr　　　nd:po　　-　　-　　1.50　4BEoo3　RB　　1/50　3/50
3　B　　CPv,w.z　OPv　　　1957pc　-　　ac　6.57　4oooX3　　　　3/57　5/57
M on B.

SO 48 (32/48) *Craven Arms*　　　　　　10 by 10　Revision a 1901-02 c 1949 f 1966
1　A　　CPr　　　OPr　　　nd:po　　-　　-　　1.50　4BEoo3　RB　　2/50　3/50
3　B　　CPv,x　　OPv　　　1957pc　-　　ac　6.57　4oooX3　　　　3/57　-
3　B/　 CPz,0z　　OPz,0z　1957mc　1966ch　acf　6.57　4oooX3
GS: Solid and Drift, 1969 on B/.

SO 49 (32/49) *Church Stretton*　　　　10 by 10　Revision a 1901-25 c 1930-49 f 1970
1　A　　CPr　　　OPr　　　nd:po　　-　　-　　1.50　4BEoo4　RB　　4/50　5/50　1
3　B　　CPv　　　OPv　　　1956pc　-　　ac　6.56　4oooX4　　　　11/56　11/56
3　B/　 CPw.y,z　OPv　　　1956mc　1961ch　ac　6.56　4oooX4
3　B//　CPz　　　O0z　　　1956^{c}ms　nd:ch　acf　6.56　4oooX4
3　B//　G0z　　　　　　　　1956^{c}ms　nd:ch　acf　6.56　4oooX4　　　　6/77　-　2
AA 1.3.50 on A(A), 3/-, OSPR 5/50. GS: Solid and Drift, 1968 on B/. 1. Showing an outline of the Church Stretton bypass. 2. Overprinted in green with public rights of way information.

SO 50 (32/50) *St Briavels*　　　　　　　10 by 10　Revision a 1918-20 c 1938-49
1　A　　CPr　　　OPr　　　nd:po　　-　　-　　1.49　4BEoo3　RB　　3/49　4/49
1　B　　CPs　　　　　　　　1951pc　-　　ac　1.51　4BEoo3　R　　　7/51　-
3　B/　 CPv　　　　　　　　1951pc　1957mc　ac　1.51　4BooX3
3　B//　CPw.z　　OPv　　　1951mc　1959mc　ac　1.51　4BooX3

SO 51 (32/51) *Monmouth*　　　　　　　10 by 10　Revision a 1900-21 c 1943-44 f 1968; j 1967 railway
1　A　　CPr　　　OPr　　　nd:po　　-　　-　　6.49　4BEoo3　RB　　8/49　9/49
2　B　　CPt　　　　　　　　1949pc　1953c　-　　6.49　4BooX3　R
3　B/　 CPv　　　　　　　　1949pc　1958c　a　　6.49　4BooX3
3　B//　CP?x,y,z　　　　　　1949mc　1961mc　a　　6.49　4BooX3
3　B///　CPz　　　OPz　　　1949ms　1968ch　acfj　6.49　4BooX3

191

1	2	3	4	5	6	7	8	9	10	11	12	13	14

SO 52 (32/52) *Ross-on-Wye (West)*   10 by 10   Revision a 1903-27 c 1943-48 f 1960 g 1972
1	A	CPr	OPr	nd:po	-	-	1.50	4BEoo3	RB		3/50	5/50	
3	B	CPv		1956pc	-	ac	6.56	4oooX3			1/57	-	
3	B/*	CPx,y,z	OPx,y	1956mc	1961r	acf	6.56	4oooX3			5/62	3/62	
3	B/*/*	CPz	OPz	1956ᶜms	nd:r	acg	6.56	4oooX3			9/73	10/73	

SS: Soil, outline, 1971 on B/*.

SO 53 (32/53) *Hereford (South)*   10 by 10   Revision a 1902-28 c 1945 f 1970; j 1965 building
1	A	CPr	OPr	nd:po	-	-	1.50	4BEoo3	R		3/50	4/50	
3	A	CPu		1950pc	-	-	1.50	4BEoo3	R				
3	A/	CPw,y,z		1950mc	1961mc	ac	1.50	4BooX3					
3	A//*	CPz,0z	OPz	1950ᶜms	nd:b	acfj	1.50	4BooX3			10/71	1/72	

AA 1.11.49/NT 23.1.48 on A(A), 3/-, OSPR 4/50. SS: Soil, outline, 1971 on A/.

SO 54 (32/54) *Hereford (North)*   10 by 10   Revision a 1902-28 c 1945 f 1968
1	A	CPr	OPr	nd:po	-	-	1.50	4BEoo3	RBC		1/50	2/50	
3	B	CPv,x,z	OPv	1957pc	-	ac	6.57	4oooX3	C		2/57	2/57	
3	B/	CPz	OPz	1957ms	1970ch	acf	6.57	4oooX3	C				

AA 1.9.49 on A(A), 3/-, OSPR 2/50. M on B.

SO 55 (32/55) *Leominster (East)*   10 by 10   Revision a 1902-27 c 1938-48 d 1938-49 e 1948
1	A	CPr	OPr	1950pc	-	ace	1.50	4BEoo3	RB		6/50	7/50	
3	B	CPv	OPv	1956pc	-	ad	6.56	4oooX3			11/56	11/56	
3	B/	CPw,z		1956mc	1961mc	ad	6.56	4oooX3					

AA 1.3.50 on A(A), 3/-, OSPR 7/50.

SO 56 (32/56) *Tenbury Wells*   10 by 10   Revision a 1901-27 c 1930-49
1	A	CPr	OPr	nd:po	-	-	1.50	4BEoo3	RBC		1/50	1/50	
3	B	CPv	OPv	1956pc	-	ac	6.56	4oooX3	C		10/56	10/56	
3	B/	CPw,z		1956mc	1960mc	ac	6.56	4oooX3	C				

M on B.

SO 57 (32/57) *Ludlow*   10 by 10   Revision a 1901-24 c 1949
1	A	CPr	OPr	nd:po	-	-	1.50	4BEoo3	RB		11/49	1/50	
3	A	CPt		1949pc	-	-	1.50	4BEoo3	RB				
3	B	CPv,x,z.z	OPv,z.z	1956pc	-	ac	6.56	4oooX3			11/56	11/56	

AA 17.8.49 on A(A), 3/-, OSPR 1/50; AA 1.1.54 on B(B). M on B.

SO 58 (32/58) *Brown Clee Hill*   10 by 10   Revision a 1901-02 c 1949
| 1 | A | CPr | OPr | nd:po | - | - | 1.50 | 4BEoo3 | B | | 3/50 | 4/50 | |
| 3 | B | CPv.x,z,0z | OPv,z | 1956pc | - | ac | 6.56 | 4oooX3 | | | 10/56 | 11/56 | |

SO 59 (32/59) *Wenlock Edge (North)*   10 by 10   Revision a 1901-25 c 1930-49
1	A	CPr	OPr	nd:po	-	-	1.50	4BEoo3	RB		3/50	4/50	
3	B	CPv	OPv	1956pc	-	ac	6.56	4oooX3			8/56	8/56	
3	B/	CPw,x,y,z.z	OPv,z.z	1956mc	1959mc	ac	6.56	4oooX3					
3	B/	G0z		1956mc	1959mc	ac	6.56	4oooX3			7/77	-	1

GS: Solid and Drift, 1969 on B/. 1. Overprinted in green with public rights of way information.

SO 60 (32/60) *Lydney**   10 by 10   Revision a 1920 c 1943-49
1	A	CPr	OPr	nd:po	-	-	1.49	4BEoo3	R		9/48	9/48	
1	B	CPr		nd:po	-	-	1.49	4BEoo3	R				
3	B	CPu		1950pc	-	-	1.49	4BEoo3	R				
3	C	CPx.z	OPx	1961mc	-	ac	6.61	4oooX3			2/61	2/61	

SO 61 (32/61) *Cinderford*   10 by 10   Revision a 1901-27 c 1948-49 f 1969
1	A	CPr	OPr	nd:po	-	-	1.49	4BEoo3	R		1/49	1/49	
3	B	CPv	OP?	1957pc	-	ac	6.57	4oooX3			1/58	1/58	
3	B/	CP?w,y,z.z		1957mc	1960mc	ac	6.57	4oooX3					
3	B//	CPz	OPz	1957ms	1969ch	acf	6.57	4oooX3					

SS: Soil, coloured, 1971 on B//.

1	2	3	4	5	6	7	8	9	10	11	12	13	14

SO 62 (32/62) *Ross-on-Wye (East)**   10 by 10   Revision a 1903-27 c 1943-49 f 1960
1	A	CPr	OPr	nd:po	-	-	1.50	4BEoo4	RB		1/50	1/50	
3	B	CPv	OPv	1957pc	-	ac	6.57	4oooX4			4/57	5/57	
3	B/*	CPx,y,z	OPx	1957mc	1961r	acf	6.57	4oooX4			11/61	11/61	

SO 63 (32/63) *Much Marcle*   10 by 10   Revision a 1901-27 c 1945-49
1	A	CPr	OPr	nd:po	-	-	1.50	4BEoo3	RB		2/50	3/50	
3	B	CPv	OPv	1957pc	-	ac	6.57	4oooX3			2/57	2/57	
3	B/	CPw,z		1957mc	1960mc	ac	6.57	4oooX3					

SO 64 (32/64) *Castle Frome*   10 by 10   Revision a 1902-27 c 1948-49 f 1965
1	A	CPr	OPr	nd:po	-	-	6.49	4BEoo3	RBC		2/50	3/50	
3	B	CPv	OPv	1956pc	-	ac	6.56	4oooX3	C		2/57	2/57	
3	B/	CP?w,y		1956mc	1960mc	ac	6.56	4oooX3	C				
3	B//*	CPz	OPz	1956mc	1966r	acf	6.56	4oooX3	C		7/66	9/66	

SO 65 (32/65) *Bromyard*   10 by 10   Revision a 1902-26 c 1938-49
1	A	CPr	OPr	nd:po	-	-	1.50	4BEoo3	RB		3/50	5/50	
3	B	CPv	OPv	1956pc	-	ac	6.56	4oooX3			1/57	2/57	
3	B/	CPw,z		1956mc	1960mc	ac	6.56	4oooX3					

M on B.

SO 66 (32/66) *Newnham**   10 by 10   Revision a 1901-26 c 1949
| 1 | A | CPr | OPr | nd:po | - | - | 1.50 | 4BEoo4 | B | | 4/50 | 5/50 | |
| 3 | B | CPv.y,z | OPv | 1956pc | - | ac | 6.56 | 4oooX4 | | | 10/56 | 11/56 | |

SO 67 (32/67) *Cleobury Mortimer**   10 by 10   Revision a 1901-25 c 1949
1	A	CPr	OPr	nd:po	-	-	1.50	4BEoo4	RB		2/50	2/50	
3	B	CPv	OPv	1956pc	-	ac	6.56	4oooX4			10/56	11/56	
3	B/	CPw.z	OPv.z	1956mc	1959mc	ac	6.56	4oooX4					

M on the 1958 index, but not recorded.

SO 68 (32/68) *Burwarton**   10 by 10   Revision a 1901-02 c 1945-49 d 1949
| 1 | A | CPr | OPr | 1950pc | - | ac | 6.50 | 4BEoo4 | B | 0 | 7/50 | 7/50 | |
| 3 | B | CPv,w,z | OPv,x,z | 1956pc | - | ad | 6.56 | 4oooX4 | | 1 | 8/56 | 10/57 | |

SO 69 (32/69) *Much Wenlock*   10 by 10   Revision a 1901-25 c 1938-49 f 1965
1	A	CPr	OPr	nd:po	-	-	1.50	4BEoo3	RB		3/50	4/50	
3	B	CPv	OPv	1956pc	-	ac	6.56	4oooX3			10/56	11/56	
3	B/	CPw	OPv	1956mc	1960mc	ac	6.56	4oooX3					
3	B//	CPz		1956mc	1966ch	acf	6.56	4oooX3					

AA 1.3.50 on A(A), 3/-, OSPR 4/50.

SO 70 (32/70) *Frampton on Severn**   10 by 10   Revision a 1920-24 c 1938-49
1	A	CPr	OPr	1948po	-	-	6.48	4BEoo3	RB	0	5/48	6/48	
1	B	CPs		1951pc	-	ac	1.51	4BEoo3	RB	0	10/51	-	1
3	B/	CPv,x,z	OPx	1951pc	1958mc	ac	1.51	4BooX3	B	0			
3	B//*	CPz,Fz	OPz,Fz	1951^{c}ms	nd:r	ac	1.51	4oooX3	B	0	8/72	8/72	

1. Bridle roads were only partially deleted: one survived in 7909.

SO 71 (32/71) *Huntley Gloucestershire**   10 by 10   Revision a 1920-21 c 1936-49 f 1973
1	A	CPr	OPr	1947po	-	-	6.47	4BE2X3	RB	0	7/47	10/47	
2	B	CPs	OPv	1952pc	-	ac	1.52	4Booo3	R	0	4/52	5/57	
3	B	CPx,z	OPz	1952pc	-	ac	1.52	4Booo3	R	0			
3	B/	CPz		1952pc	-	ac	1.52	4Booo3		0			
3	B//*	CPz	OPz	1952^{c}ms	nd:r	acf	1.52	4Booo3		0	12/73	3/74	

AA 1.5.47 on A(A), 3/-, OSPR 10/47.

1	2	3	4	5	6	7	8	9	10	11	12	13	14

SO 72 (32/72) *Newent*   10 by 10   Revision a 1903-21 c 1949 f 1960
1	A	CPr		1947po	-	-	6.47	4BE2X3	RB		10/47	-
3	B		OPr	1947po	-	-	6.47	4BE2X3	RB		-	12/47
2	C	CPs,u	OPv	1952pc	-	ac	1.52	4Booo3	R		5/52	5/57
3	C/	CP?x,y,z		1952mc	1961ch	acf	1.52	4BooX3				

AA 1.6.47/NT 17.11.47 on B(A), 3/-, OSPR 12/47.

SO 73 (32/73) *Ledbury*   10 by 10   Revision a 1903-26 c 1938-49 f 1960
1	A	CPr	OPr	nd:po	-	-	1.50	4BEoo3	RB	1/50	1/50
3	B	CPt	OPv	1954pc	-	ac	6.54	4BooX4		3/54	10/57
3	B/										
3	B//*	CPx	OPx	1954mc	1961r	acf	6.54	4BooX4		12/61	12/61
3	B//*/	CPz,0z		1954mc	1961r	acf	6.54	4BooX4			

SO 74 (32/74) *Malvern*   10 by 10   Revision a 1902-26 c 1938-49
1	A	CPr	OPr	nd:po	-	-	6.49	4BEoo3	RB	1/50	2/50
3	B	CPt		1954pc	-	ac	6.54	4BooX3		4/54	-
3	B/										
3	B//	CP?v,x,z.0z		1954pc	1957mc	ac	6.54	4BooX3			

AA 29.9.49 on A(A), 3/-, OSPR 2/50. SS: Soil, Land Use Capability, outline, 1976 on B//.

SO 75 (32/75) *Broadwas*   10 by 10   Revision a 1902-26 c 1949
| 1 | A | CPr | OPr | nd:po | - | - | 1.50 | 4BEoo3 | RB | 12/49 | 1/50 |
| 3 | B | CPw,z | OPx | 1961mc | - | ac | 6.61 | 4oooX3 | | 1/61 | 2/61 |

M on B.

SO 76 (32/76) *Witley**   10 by 10   Revision a 1882-1926 c 1938-49
1	A	CPr	OPr	nd:po	-	-	1.50	4BEoo3	B	11/49	1/50
3	A	CPu	OPv	1949pc	-	-	1.50	4BEoo3	B		
3	B	CPw,x,z	OPv,z	1960mc	-	ac	6.60	4oooX3		11/60	12/60

SO 77 (32/77) *Bewdley**   10 by 10   Revision a 1901-26 c 1938-50 f 1968
1	A	CPr	OPr	nd:po	-	-	1.50	4BEoo3	RB	12/49	1/50
3	A	CPt		1949pc	-	-	1.50	4BEoo3	RB		
3	B	CPw.x,z	OPv	1958pc	-	ac	6.58	4oooX3		11/58	10/58
3	B/	CPz	OPz	1958ms	1969ch	acf	6.58	4oooX3			

AA 20.9.49/NT 16.9.49 on A(A), 3/-, OSPR 1/50. M on A.

SO 78 (32/78) *Highley*   10 by 10   Revision a 1901-25 c 1948-50 f 1966; j 1965 reservoir
1	A	CPr	OPr	nd:po	-	-	1.49	4BEoo3	RB	2/49	3/49	
2	B	CPs	OPv	1951pc	-	ac	6.51	4Booo3	R	3/52	5/57	
3	B/	CPw,x		1951pc	1958mc	ac	6.51	4BooX3				
3	B//*	CPz,0z	OPz	1951mc	1967v	acfj	6.51	4BooX3		12/66	1/67	1

AA 1.2.49 on A(A), 3/-, OSPR 3/49. M on B. 1. Chelmarsh Reservoir.

SO 79 (32/79) *Bridgnorth*   10 by 10   Revision a 1901-25 c 1938-49 f 1965
1	A	CPr	OPr	nd:po	-	-	6.49	4BEoo3	RB	4/49	5/49
2	B	CPs		1952pc	-	ac	6.52	4Booo3	R	6/52	-
3	B/	CPv,x	OPv	1952pc	1956mc	ac	6.52	4BooX3		-	6/57
3	B//*	CPz,0z	OPz	1952mc	1965r	acf	6.52	4BooX3		9/65	10/65

AA 1.4.49 on A(A), 3/-, OSPR 5/49. M on B.

SO 80 (32/80) *Stroud*   10 by 10   Revision a 1920-21 c 1936-50
1	A	CPr	OPr	nd:po	-	-	1.49	4BEoo3	RB	10/48	12/48
2	B	CPs		1952pc	-	ac	1.52	4Booo3	R	4/52	-
3	B/	CPv,x,z	OPx	1952pc	1956mc	ac	1.52	4BooX3			

AA 1.11.48/NT 9.10.48 on A(A), 3/-, OSPR 12/48.

1	2	3	4	5	6	7	8	9	10	11	12	13	14

SO 81 (32/81) *Gloucester*   10 by 10   Revision a 1920-21 c 1936-49 f 1960 g 1966 h 1971

1	A	CPr		1947po	-	-	1.48	4BE2X3	RB	1	10/47	-	
1	B		OPr	1947po	-	-	1.48	4BE2X3	RB	1	-	12/47	
2	C	CPt	OPv	1952pc	-	ac	6.52	4Booo3	R	1	8/52	11/57	1
3	C/	CPw		1952mc	1958c	ac	6.52	4BooX3		1			
3	C//	CPx	OPx	1952mc	1961ch	acf	6.52	4BooX3		1			
3	C///*	CPz	OPz	1952mc	1967r	acg	6.52	4BooX3		1	6/67	6/67	
3	C///*/*	CPz	OPz	1952^{c}ms	nd:r	ach	6.52	4BooX3		1	5/72	8/72	

AA 1.6.47 on B(A), 3/-, OSPR 12/47. M on A, C. 1. The airfield at 8816 is only partially open space. When shown on edition C as "Airfield" the open space is enlarged, and then reduced again on edition C//.

SO 82 (32/82) *Norton Gloucestershire*   10 by 10   Revision a 1921 c 1936-49 f 1972

1	A	CPr	OPr	1947po	-	-	1.47	4BE2X3	RB	0	7/47	12/47	
2	B	CPs		1951pc	-	ac	6.51	4Booo3	R	0	4/52	-	1
3	B/	CPv		1951pc	1958mc	ac	6.51	4BooX3		0			
3	B//	CP?w,y,z		1951mc	1960mc	ac	6.51	4BooX3		0			
3	B///*	CPz	OPz	1951^{c}ms	nd:r	acf	6.51	4BooX3		0	2/73	4/73	

AA 1.2.47 on A(A), 3/-, OSPR 12/47. SS: Soil, outline, 1973 on B//. 1. Showing RAF Innsworth.

SO 83 (32/83) *Tewkesbury*   10 by 10   Revision a 1921-26 c 1938-49 f 1963

1	A	CPr	OPr	1948po	-	-	1.48	4BEoo3	RB		7/48	7/48	
3	B	CPt		1954pc	-	ac	6.54	4BooX3			4/54	-	
3	B/		OPv	1954pc	1956mc	ac	6.54	4BooX3			-	2/57	
3	B//	CPw		1954mc	1961mc	ac	6.54	4BooX3					
3	B///*	CPy,z,0z	OPy	1954mc	1963r	acf	6.54	4BooX3			8/63	8/63	

SO 84 (32/84) *Upton upon Severn*   10 by 10   Revision a 1903-26 c 1938-50 f 1960 g 1964

1	A	CPr	OPr	nd:po	-	-	1.50	4BEoo3	RB	0	12/49	1/50	
3	A		OPs	1949pc	-	-	1.50	4BEoo3	RB	0			
3	B	CPx	OPx	1961mc	-	acf	6.61	4oooX4		3	5/61	5/61	1
3	B/*	CPz	OPz	1961mc	1965r	acg	6.61	4oooX4		3	8/65	9/65	

1. With the M5 under construction, shown as a single carriageway.

SO 85 (32/85) *Worcester*   10 by 10   Revision a 1902-26 c 1938-50 f 1961 g 1965 h 1968
                                      j 1967 building

1	A	CPr	OPr,s	nd:po	-	-	1.49	4BEoo3	RB		8/48	10/48	
2	B	CPt,u	OPv	1952pc	-	ac	6.52	4Booo3	R		8/52	5/57	
3	B/*	CPx	OPx	1952mc	1962r	acf	6.52	4BooX3			7/62	7/62	1
3	B/*/*	CPz	OPz	1952mc	1966r	acg	6.52	4BooX3			1/66	1/66	
3	B/*/*/*	CPz,0z	OPz	1952ms	1969br	achj	6.52	4BooX3			1/70	2/70	

AA 1.6.48 on A(A), 3/-, OSPR 10/48. M on B. 1. Showing the M5 under construction as a dual carriageway.

SO 86 (32/86) *Droitwich (West)* *   10 by 10   Revision a 1925-26 c 1938-49 f 1965

1	A	CPr	OPr	nd:po	-	-	1.49	4BEoo3	RB		2/49	2/49	
2	B	CPs	OPs	1951pc	-	ac	6.51	4Booo3	R		2/52	-	
3	B/		OPv	1951pc	1958mc	ac	6.51	4BooX3					
3	B//	CPx		1951mc	1961ch	ac	6.51	4BooX3					
3	B///*	CPz	OPz	1951mc	1966r	acf	6.51	4BooX3			1/66	1/66	

AA 1.7.51 on B(A), 3/-, OSPR 4/52. M on B.

SO 87 (32/87) *Kidderminster*   10 by 10   Revision a 1921-26 c 1938-49 f 1970; j 1965 building

1	A	CPr	OPr	nd:po	-	-	6.48	4BEoo3	RB		9/48	11/48	
1	A	CPs		1948pc	1950	-	6.48	4BEoo3	RB				
1	B	CPs		1951pc	-	ac	6.51	4BEoo3	RB		11/51	-	
3	B	CPz	OPv,z	1951pc	-	ac	6.51	4BooX3	RB		-	1/57	
3	B/	CPz	OPz	1951pc	-	ac	6.51	4BooX3	B				
3	B//*	CPz,0z	OPz	1951^{c}ms	nd:b	acfj	6.51	4BooX3	B		1/72	5/72	

AA 1.9.48/NT 6.9.48 on A(A), 3/-, OSPR 11/48. M on B. SS: Soil, outline, 1974 on B//*.

1	2	3	4	5	6	7	8	9	10	11	12	13	14

SO 88 (32/88) *Kinver*　　　　　　　　10 by 10　Revision a 1901-25 c 1937-49
1　A　　　CPr　　OPr　　nd:po　　-　　　-　　1.49　4BEoo3　RBC　　　11/48　12/48
1　B　　　CPs　　OPs　　1951pc　-　　　ac　6.51　4BEoo4　RBC　　　8/51　10/51
3　B　　　CPu　　OPv　　1951pc　-　　　ac　6.51　4BEoo4　RBC
3　B/　　CPx,z　　　　1951mc　1961ch　ac　6.51　4BooX4　BC
AA 1.11.48/NT 17.9.48 on A(A), 3/-, OSPR 12/48; AA 1.9.51/NT 17.9.48 on B(B), 3/-, OSPR 10/51. M on B.

SO 89 (32/89) *Wombourn**　　　　　10 by 10　Revision a 1901-24 c 1937-49
1　A　　　CPr　　OPr　　1947po　-　　　-　　1.47　4BE2X3　RB　00　1/47　5/47
2　B　　　CPt　　OPs　　1952pc　-　　　ac　6.52　4Booo3　　　11　9/52　3/55
3　B/　　CPv,x,z　　　1952pc　1956mc　ac　6.52　4BooX3　　　11
AA 1.9.46/NT 1.1.47 on A(A), 3/-, OSPR 5/47; AA 1.9.46/NT 1.7.47 on A(B), 3/-. M on B.

SO 90 (32/90) *Sapperton*　　　　　　10 by 10　Revision a 1920-21 c 1938-56 f 1969
1　A　　　CPr　　OPr　　nd:po　　-　　　-　　1.49　4BEoo3　RB　0　12/49　1/50
2　B [1]　CPt　　　　　1949pc　1953rpb　-　　6.53　4BooX3　B　0
3　B [2]　CPw.z　OPv,z　1959mc　-　　　ac　6.59　4oooX3　　　1　9/59　10/59　1
3　B/　　CPz　　OPz　　1959ms　1970ch　acf　6.59　4oooX3　　　3
1. Aston Down Airfield.

SO 91 (32/91) *Birdlip*　　　　　　　10 by 10　Revision a 1920-21 c 1938-49 f 1970
1　A　　　CPr　　OPr　　1948po　-　　　-　　6.48　4BEoo3　RB　　6/48　7/48
1　A　　　　　　OPr　　nd:po　　-　　　-　　6.48　4BEoo3　RB
1　B　　　CPs　　　　　1950pc　-　　　ac　6.50　4BEoo3　RB　　12/50　-
3　B　　　CPu,x,y　OPv　1950pc　-　　　ac　6.50　4BEoo3　RB　　-　　5/57
3　B/　　CPz　　　　　1950ᶜms　nd:ch　acf　6.50　4Booo3　B
AA 1.5.46/NT 1.6.48 on A(A), 3/-, OSPR 7/48.

SO 92 (32/92) *Cheltenham**　　　　　10 by 10　Revision a 1920-21 c 1938 d 1938-49 f 1971
1　A　　　CPr　　　　　nd:po　　-　　　-　　1.49　4BEoo3　RB　0　9/48　-
1　B　　　　　　OPr　　nd:po　　-　　　-　　1.49　4BEoo3　RB　0　-　　11/48
2　C　　　CPs　　OPs　　1952pc　-　　　ac　1.52　4Booo3　R　　2　5/52　1/54　1
3　C/　　CPv　　　　　1952pc　1957mc　ac　1.52　4BooX3　　　2
3　C//　　CPw.z　OPv　　1952mc　1961mc　ac　1.52　4BooX3　　　2
3　C///*　CPz　　OPz　　1952ᶜms　nd:r　adf　1.52　4BooX3　　　2　3/72　8/72
AA 1.9.48 on B(A), 3/-, OSPR 11/48. M on C. 1. Another reference to the outline edition is in OSPR 9/54.

SO 93 (32/93) *Bredon*　　　　　　　10 by 10　Revision a 1921 c 1938-49 f 1960 g 1963 h 1973
1　A　　　CPr　　OPr　　nd:po　　-　　　-　　1.50　4BEoo3　RB　　12/49　1/50
2　B　　　CPs　　　　　1952pc　-　　　ac　1.52　4Booo3　R　　　5/52　-
3　B/　　CPv,w　OPv　　1952pc　1958mc　ac　1.52　4BooX3
3　B//*　CPx　　OPx　　1952mc　1961r　acf　1.52　4BooX3　　　　4/62　4/62　1
3　B//*/　CPy,z　　　　1952mc　1963ch　acg　1.52　4BooX3
3　B//*//*　CPz,0z　OPz　1952ᶜms　nd:r　ach　1.52　4BooX3　　　　9/73　10/73
1. With the M5 under construction.

SO 94 (32/94) *Pershore**　　　　　10 by 10　Revision a 1903-26 c 1938-50 f 1964
1　A　　　CPr　　OPr　　nd:po　　-　　　-　　6.49　4BEoo3　RB　00　10/49　11/49
2　B　　　CPs　　OPs　　1952pc　-　　　ac　6.52　4Booo4　R　　30　6/52　6/52　1
3　B　　　CPu　　　　　1952pc　-　　　ac　6.52　4Booo4　R　　30
3　B/　　CPw　　　　　1952mc　1960mc　ac　6.52　4BooX4　　　10
3　B//*　CPz　　OPz　　1952mc　1965r　acf　6.52　4BooX4　　　10　5/65　3/65
1. Pershore (Airfield). Altered to upper case without brackets on edition B/.

SO 95 (32/95) *Upton Snodsbury*　　10 by 10　Revision a 1902-26 c 1949-50 f 1964 g 1970
1　A　　　CPr　　OPr　　nd:po　　-　　　-　　6.49　4BEoo3　RB　0　10/49　11/49
3　A　　　　　　OPv　　1949pc　-　　　-　　6.49　4BEoo3　RB　0
3　B　　　CPx　　OPx　　1961mc　-　　　ac　6.61　4oooX3　　　1　4/61　3/61
3　B/*　　CPz　　OPz　　1961mc　1965r　acf　6.61　4oooX3　　　1　6/65　5/65
3　B/*/　CPz,0z　　　　1961ᶜms　nd:ch　acg　6.61　4oooX3　　　3　　　　　　1
M on A. 1. Pershore Airfield.

196

1	2	3	4	5	6	7	8	9	10	11	12	13	14

SO 96 (32/96) *Droitwich (East)*   10 by 10   Revision a 1902-26 c 1938-49 f 1963

1	A	CPr	OPr	nd:po	-	-	1.49	4BEoo3	RB			3/49	5/49	
2	B	CPs,u		1951pc	-	ac	6.51	4Booo3	R			2/52	-	
3	B		OPs	1951pc	-	ac	6.51	4Booo3	R					
3	B/	CPw		1951mc	1960mc	ac	6.51	4BooX3						
3	B//*	CPy,z	OPy	1951mc	1964r	acf	6.51	4BooX3				1/64	2/64	

AA 1.3.49 on A(A), 3/-, OSPR 5/49. M on B.

SO 97 (32/97) *Bromsgrove*   10 by 10   Revision a 1913-26 c 1937-49 f 1963 g 1965 h 1967

1	A	CPr	OPr	1947po	-	-	1.47	4BE2X3	RB			1/47	6/47	
2	B	CPs,u	OPv	1952pc	-	ac	1.52	4Booo3	R			5/52	6/57	
3	B/	CPw	OPv	1952mc	1961mc	ac	1.52	4BooX3						
3	B//*	CPz	OPz	1952mc	1964r	acf	1.52	4BooX3				4/64	4/64	
3	B//*/*	CPz	OPz	1952mc	1966r	acg	1.52	4BooX3				7/66	8/66	
3	B//*/*/*	CPz,0z	OPz	1952mc	1968r	ach	1.52	4BooX3				2/68	1/68	

AA 1.9.46/NT 1.1.47 on A(A), 3/-, OSPR 6/47; AA 1.9.46/NT 1.7.47 on A(B), 3/-, OSPR - . M on B.

SO 98 (32/98) *[Halesowen]*   10 by 10   Revision a 1913-38 c 1937-49 f 1966

1	A	CPr	OPr	1947po	-	-	1.47	4BE2X3	RB			3/47	7/47	
2	B	CPt	OPs	1953pc	-	ac	6.53	4BooX4				10/53	10/53	
3	B/	CPv		1953pc	1957mc	ac	6.53	4BooX4						
3	B//	CPx	OPx	1953mc	1961ch	ac	6.53	4BooX4						
3	B///*	CPz	OPz	1953mc	1967r	acf	6.53	4BooX4				2/67	2/67	

AA 1.9.46 on A(A), 3/-, OSPR 7/47. M on A.

SO 99 (32/99) *Wolverhampton (South)* *   10 by 10   Revision a 1912-14 c 1938-49 f 1971

1	A	CPr	OPr	1947po	-	-	6.47	4BE2X3	R			6/47	10/47	
1	B	CPs		1951pc	-	ac	1.51	4BEoo3	R			8/51	-	
2	B	CPs,t	OPs	1951pc	1952rpb	ac	1.52	4BooX3	R					
3	B/	CPw,z		1951mc	1960mc	ac	1.52	4BooX3						
3	B//*	CPz	OPz	1951^{c}ms	nd:r	acf	1.52	4BooX3				11/72	2/73	

AA 1.6.47 on A(A), 3/-, OSPR 10/47. M on B, B/.

SP 00 (42/00) *Cirencester*   10 by 10   Revision a 1920 c 1938-56

| 1 | A | CPr | OPr | nd:po | - | - | 6.49 | 4BEoo3 | RB | | | 12/49 | 1/50 | |
| 3 | B | CPw,y,z,0z | OPv | 1959mc | - | ac | 6.59 | 4oooX4 | | | | 9/59 | 7/59 | |

M on B.

SP 01 (42/01) *Andoversford*   10 by 10   Revision a 1920-21 c 1949-50 f 1976

1	A	CPr	OPr	nd:po	-	-	1.49	4BEoo3	RB	0		8/48	9/48	
3	B	CPv.x,z	OP?	1957pc	-	ac	6.57	4oooX3		3		3/58	3/58	1
3	B/*	CFz	OFz	1957^{c}ms	nd:r	acf	6.57	4oooX3		3				2

1. Showing a Polish Girls School south-west of Stowell Park (0811). 2. OSPR not found.

SP 02 (42/02) *Winchcombe*   10 by 10   Revision a 1920-21 c 1949-50

| 1 | A | CPr | OPr | nd:po | - | - | 6.48 | 4BEoo3 | RB | | | 9/48 | 9/48 | |
| 3 | B | CPw,x,z,0z | OPv | 1961mc | - | ac | 6.61 | 4oooX3 | | | | 12/60 | 12/60 | |

SP 03 (42/03) *Stanway Gloucestershire*   10 by 10   Revision a 1921 c 1938-50

1	A	CPr	OPr	nd:po	-	-	6.49	4BEoo3	RB			11/49	11/49	
1	B	CPs	OP?	1951pc	-	ac	1.51	4BEoo3	RB			1/52	8/51	
3	B/	CPw,y,z,0z		1951pc	1958mc	ac	1.51	4BooX3	B					

SP 04 (42/04) *Evesham*   10 by 10   Revision a 1921-26 c 1938-50

1	A	CPr	OPr	nd:po	-	-	6.49	4BEoo3	RB			5/49	7/49	
2	B	CPs		1952pc	-	ac	1.52	4Booo3	R			5/52	-	
3	B		OPv	1952pc	-	ac	1.52	4Booo3	R			-	6/57	
3	B/	CPx.z	OPx	1952mc	1961ch	ac	1.52	4BooX3						

AA 1.4.49 on A(A), 3/-, OSPR 7/49.

1	2	3	4	5	6	7	8	9	10	11	12	13	14

SP 05 (42/05) *Alcester*　　　　　　10 by 10　Revision a 1903-26 c 1950
1　A　　CPr　　　OPr　　　nd:po　　　-　　　　-　　　6.49　4BEoo3　RB　　　　7/49　7/49　1
1　B　　CPs　　　OPs　　　1951pc　　-　　　　ac　　1.51　4BEoo3　RB　　　　8/51　8/51
3　B/　　CPw,y,z　　　　　1951pc　　1958mc　ac　　1.51　4BooX3　B
M on B. SS: Soil, outline, 1974 on B/. 1. The south to east railway curve opened in 1942 is added.

SP 06 (42/06) *Redditch*　　　　　　10 by 10　Revision a 1902-38 c 1938-50 f 1974
　　　　　　　　　　　　　　　　　　　　　　j 1982 selected k SUSI
1　A　　CPr　　　OPr　　　nd:po　　　-　　　　-　　　1.49　4BEoo3　RB　　　　2/49　3/49
2　B　　CPs,u　　OPv　　　1952pc　　-　　　　ac　　1.52　4Booo3　R　　　　 4/52　1/57
3　B/　　CP?w,x,z.z　　　　1952mc　　1960mc　ac　　1.52　4BooX3
3　B//*　CFz　　　OFz　　　1952ᶜms　　nd:r　　acf　 1.52　4BooX3　　　　　　6/75　7/75
3　B//*/*　CFz　　OFz　　　1983ᶜm-s　 nd:br　acfjk　6.84　4BooX3　　　　　　1/84　1/84
AA 1.11.48 on A(A), 3/-, OSPR 3/49. M on B.

SP 07 (42/07) *Alvechurch*　　　　　10 by 10　Revision a 1913-26 c 1938-50 f 1975
1　15046　CPr　　　　　　　1946po　　-　　　　-　　　1.47　4BE2X3　R　　　　11/46　-
1　A　　　　　　　OPr　　　1946po　　-　　　　-　　　1.47　4BE2X3　R　　　　-　　2/47
2　B　　CPs,t　　OPt　　　1952pc　　-　　　　ac　　1.52　4Booo3　R　　　　 4/52　6/54
3　B　　CPw,z　　　　　　　1952pc　　1958mc　ac　　1.52　4BooX3
3　B/　　CPz　　　OPz　　　1952pc　　1958mc　ac　　1.52　4BooX3
3　B//*　CFz　　　OFz　　　1952ᶜms　　nd:r　　acf　 1.52　4BooX3　　　　　　11/76 11/76
AA 1.9.46 on A(A), 3/-, OSPR 2/47; AA 1.11.53/NT 1.11.53 on B(B), 4/-, OSPR 6/54. M on B.

SP 08 (42/08) *Birmingham (Central)*　10 by 10　Revision a 1913-14 c 1937-50 f 1969 g 1974
　　　　　　　　　　　　　　　　　　　　　　j 1965 building
1　20047　CPr　　　　　　　1947po　　-　　　　-　　　1.47　4BE2X3　R　　　　1/47　-　　1
1　A　　　　　　　OPr　　　1947po　　-　　　　-　　　1.47　4BE2X3　R　　　　-　　5/47
2　B　　CPt　　　　　　　　1952pc　　-　　　　ac　　6.52　4Booo3　R　　　　 8/52　11/53
2　B　　CPu　　　OPs　　　1952pc　　-　　　　ac　　6.52　4BooX3　R
3　B/　　CPw.y,z　　　　　1952mc　　1959v　　ac　　6.52　4BooX3
3　B//*　CPz　　　OPz　　　1952ms　　1970br　acfj　6.52　4BooX3　　　　　　7/70　8/70
3　B//*/*　CFz　　OFz　　　1952ᶜms　　nd:r　　acgj　6.52　4BooX3　　　　　　6/75　8/75
AA 1.10.46 on A(A), 3/-, OSPR 5/47; AA 1.11.53 on B(B), 4/-, OSPR 6/54. M on 20047. 1. A special printing (?ca 1950) in black and orange with little beside sheet numbers in the bottom margin, (no code, CPz, 1947po, 2BEoo3) is in the Department of Geography, University of Birmingham.

SP 09 (42/09) *Walsall*　　　　　　10 by 10　Revision a 1912-38 c 1950 f 1972
1　A　　CPr　　　OPr　　　1947po　　-　　　　-　　　1.47　4BE2X3　R　　　　4/47　7/47
1　B　　CPs　　　　　　　　1951pc　　-　　　　ac　　6.51　4BEoo3　R　　　　12/51　-
2　B　　　　　　　OPs　　　1951pc　　-　　　　ac　　1.52　4Booo3　R　　　　 -　　2/54
3　B/　　CPw,x,z　　　　　1951mc　　1958c　　ac　　6.51　4BooX3
3　B//*　CPz　　　OPz　　　1951ᶜms　　nd:r　　acf　 6.51　4BooX3　　　　　　2/73　4/73
AA 1.3.47 on A(A), 3/-, OSPR 7/47; AA 1.11.53 on B(B), 4/-, OSPR 1/55. M on B, B/.

SP 10 (42/10) *Fairford*　　　　　　10 by 10　Revision a 1919-20 b 1919-24 c 1956 e no date
1　A　　CPr　　　OPr　　　nd:po　　　-　　　　-　　　1.49　4BEoo3　RB　0　　12/48 12/48
1　B　　CPr　　　OPr　　　1948pc　　1950c　　ae　　6.50　4BEoo3　RB　0
3　B　　CPu　　　　　　　　1950pc　　-　　　　ae　　6.50　4BEoo3　RB　0
3　C　　CPw,y,z,0z　OPv　　1959mc　　-　　　　bc　　6.59　4oooX3　　　　2　10/59 10/59

SP 11 (42/11) *Northleach*　　　　　10 by 10　Revision a 1919-21 c 1950 f 1960
1　A　　CPr　　　OPr　　　nd:po　　　-　　　　-　　　6.49　4BEoo3　B　　0　　12/49 1/50
3　A　　CPu　　　　　　　　1949pc　　-　　　　-　　　6.49　4BEoo3　B　　0
3　B　　CPx,y,0z　OPx　　　1961mc　　-　　　　acf　 6.61　4oooX4　　　　2　11/61 11/61

SP 12 (42/12) *Stow-on-the-Wold*　　10 by 10　Revision a 1919-21 c 1950
1　A　　CPr　　　OPr　　　nd:po　　　-　　　　-　　　6.49　4BEoo3　RB　　　11/49 11/49
　　B
3　C　　CPv,x,z　OPv,x　　 1956pc　　-　　　　ac　　6.56　4oooX3　　　　　　10/56 8/57
SS: Soil, both coloured and outline, 1978 on C.

1	2	3	4	5	6	7	8	9	10	11	12	13	14

SP 13 (42/13) *Chipping Campden*   10 by 10   Revision a 1900-21 c 1949-50
1	A	CPr	OPr	nd:po	-	-	6.49	4BEoo4	RB		10/49	11/49	
2	B	CPs	OPv	1952pc	-	ac	1.52	4Booo4	R		4/52	8/57	
3	B/	CPw,y,z.z,0z		1952mc	1960mc	ac	1.52	4BooX4					

SP 14 (42/14) *Mickleton Gloucestershire*   10 by 10   Revision a 1921-22 c 1950
1	A	CPr	OPr	nd:po	-	-	6.49	4BEoo3	RB	00	7/49	7/49	
2	A	CPs		1949pc	1951	-	6.51	4Booo3	RB	00			
3	A/	CPv		1949pc	1957mc	ac	6.51	4BooX3	RB	00			
3	B	CPw,y,z	OPv	1958mc	-	ac	6.58	4oooX3		33	1/59	1/59	1

1. Showing the railway layout in the Royal Engineer's Depot at Long Marston.

SP 15 (42/15) *Stratford-upon-Avon (West)*   10 by 10   Revision a 1903-23 c 1938-50
1	A	CPr		nd:po	-	-	6.49	4BEoo3	RB	0	6/49	-	
1	B		OPr	nd:po	-	-	6.49	4BEoo3	RB	0	-	7/49	
2	C	CPs,u	OPs	1951pc	-	ac	6.51	4Booo3	R	0	1/52	1/52	
3	C/	CPw,z,0z	OPz	1951mc	1960mc	ac	6.51	4BooX3		0			

AA 1.5.49 on B(A), 3/-, OSPR 7/49. M on C. The south to east railway curve opened in 1960 is never shown.

SP 16 (42/16) *Henley-in-Arden*   10 by 10   Revision a 1903-23 c 1950 f 1965
1	A	CPr	OPr	nd:po	-	-	1.49	4BEoo3	RB	0	3/49	4/49	
2	B	CPt		1949pc	1953c	-	6.53	4BooX3	B	0			
3	C	CPw,y	OPv	1959mc	-	ac	6.59	4oooX3		0	5/59	7/59	
3	C/	CPz,0z	OPz	1959mc	1966ch	acf	6.59	4oooX3		0			

M on B.

SP 17 (42/17) *Solihull*   10 by 10   Revision a 1902-23 c 1938-50 f 1969
    j 1959-65 building k 1980 selected
1	10047	CPr		1947po	-	-	1.47	4BE2X3	R	0	1/47	-	
1	A		OPr	1947po	-	-	1.47	4BE2X3	R	0	-	5/47	
1	B	CPs	OPs	1951pc	-	ac	1.51	4BEoo3	R	1	12/51	1/54	
3	B	CPt		1951pc	-	ac	1.51	4BooX3	R	1			
3	B/	CPw,x,z		1951mc	1960mc	ac	1.51	4BooX3		1			
3	B//*	CPz	OPz	1951^{c}ms	nd:b	acfj	1.51	4BooX3		1	2/71	5/71	
3	B//*/*	CFz	OFz	1981^{c}ms	md:r	acfjk	1.51	4BooX3		1	10/81	11/81	

AA 1.9.46/NT 1.1.47 on A(A), 3/-, OSPR 5/47; AA 1.9.46/NT 1.7.47 on A(B), 3/-, OSPR -; AA 1.8.51/NT 4.5.50 on B(C), 3/-, OSPR 10/51. M on B.

SP 18 (42/18) *Birmingham (East)*   10 by 10   Revision a 1902-14 c 1938-50 f 1972 g 1977
1	A	CPr	OPr	1947po	-	-	1.47	4BE2X3	R	0	3/47	6/47	
1	B	CPs		1951pc	-	ac	6.51	4BEoo3	R	3	12/51	-	
3	B	CPu	OPs	1951pc	-	ac	6.51	4Booo3	R	3	-	1/55	
3	B/	CP?w,y,z		1951mc	1960c	ac	6.56	4BooX3		3			
3	B//*	CPz	OPz	1951^{c}ms	nd:r	acf	6.56	4BooX3		3	1/73	12/73	
3	B//*/*	CFz	OFz	1951^{c}ms	nd:r	acg	6.51	4BooX3		3	3/78	3/78	1

AA 1.10.46 on A(A), 3/-, OSPR 6/47; AA 1.11.53 on B(B), 4/-, OSPR 1/55; M on B, B/. 1. Birmingham Airport.

SP 19 (42/19) *Sutton Coldfield**   10 by 10   Revision a 1912-23 c 1937-50 f 1965 g 1971
1	A	CPr	OPr	1947po	-	-	1.47	4BE2X3	RB	1	2/47	5/47	
1	A	CPs		1947po	-	-	1.47	4BEoo3	RB	1			
2	B	CPs	OPt	1952pc	-	ac	1.52	4Booo3	R	1	4/52	6/54	1
3	B/	CPv		1952pc	1956mc	ac	1.52	4BooX3		1			
3	B//	CPx		1952mc	1960mc	ac	1.52	4BooX3		1			
3	B///	CPz	OPz	1952mc	1966ch	acf	1.52	4BooX3		3			
3	B////*	CPz	OPz	1952^{c}ms	nd:r	acg	1.52	4BooX3		3	1/73	4/73	

AA 1.1.46 on A(A), 3/-, OSPR 5/47; AA 1.11.53 on B(B), 4/-, OSPR 6/54. M on A. 1. The outline price state has "Price 2/6 net. Paper flat." blocked out.

1	2	3	4	5	6	7	8	9	10	11	12	13	14

SP 20 (42/20) *Carterton*   10 by 10   Revision a 1910-19 c 1956 e 1947 f 1971

1	A	CPr	OPr	nd:po	-	-	6.49	4BEoo3	RB	00	5/49	6/49	
1	B	CPs		1949pc	1951c	ae	1.51	4BEoo3	RB	00			
3	B/	CPv		1949pc	1956c	ae	1.51	4BooX3	B	00			
3	B//	CPv		1949pc	1958c	ae	1.51	4BooX3	B	00			
3	C	CPw,x,z	OPv	1959mc	-	ac	6.59	4oooX3		13	10/59	10/59	1
3	C/*	CPz	OPz	1959ᶜms	nd:r	acf	6.59	4oooX3		13	1/73	3/73	

1. Brize Norton Airfield.

SP 21 (42/21) *Burford Oxfordshire*   10 by 10   Revision a 1919 c 1950 f 1970

1	A	CPr	OPr	nd:po	-	-	6.49	4BEoo3	RB	0	6/49	7/49	
1	B	CPs		1951pc	-	ac	1.51	4BEoo3	RB	0	4/51	-	
3	B/	CPw,y.z	OPv	1951pc	1958c	ac	1.51	4BooX3	B	1			1
3	B//	CPz,0z	OPz	1951ᶜms	nd:ch	acf	1.51	4BooX3	B	3			2

1. Little Rissington Aerodrome. 2. Little Rissington Airfield.

SP 22 (42/22) *Adlestrop*   10 by 10   Revision a 1919 c 1938-50

1	A	CPr	OPr	nd:po	-	-	6.49	4BEoo3	RB		6/49	7/49	
1	B	CPs	OPv	1951pc	-	ac	1.51	4BEoo3	RB		8/51	2/57	
3	B	CPx,y,z	OPx	1951pc	-	ac	1.51	4BEoo3	RB				

One of the least revised of sheets, the original National Grid reference system remaining in use throughout, and bench marks, railway company names and bridle roads never deleted.

SP 23 (42/23) *Moreton-in-Marsh*   10 by 10   Revision a 1919-21 c 1950 f 1966

1	A	CPr	OPr	nd:po	-	-	6.49	4BEoo3	RB	0	6/49	6/49	
1	B	CPs	OPv	1951pc	-	ac	6.51	4BEoo3	RB	0	1/52	1/57	
3	B	CPx,z		1951pc	-	ac	6.51	4BEoo3	RB	0			
3	B/	CPz,0z	O0z	1951mc	1967ch	acf	6.51	4BooX3	B	0			

SP 24 (42/24) *Shipston on Stour*   10 by 10   Revision a 1900-22 c 1949-50

1	A	CPr	OPr	nd:po	-	-	1.49	4BEoo3	RB		6/49	7/49	
1	A	CPs		1949pc	1951	-	1.51	4BEoo3	RB				
3	B	CPx,y,z,0z	OPx	1961mc	-	ac	6.61	4oooX3			7/61	3/61	

The alignment of the Stratford & Moreton tramway is on the grey plate.

SP 25 (42/25) *Stratford-upon-Avon (East)*   10 by 10   Revision a 1900-23 c 1938-50

1	A	CPr	OPr	nd:po	-	-	6.49	4BEoo3	RBC	00	5/49	7/49	
2	B	CPs	OPs	1951pc	-	ac	6.51	4Booo3	RC	00	1/52	-	
3	B	CPx.z		1951pc	-	ac	6.51	4BooX3	RC	00			
3	B/	CPz,0z	OPz	1951pc	-	ac	6.51	4BooX3	C	00			

AA 1.3.49 on A(A), 3/-, OSPR 7/49. M on B. The alignment of the Stratford & Moreton tramway is on the grey plate.

SP 26 (42/26) *Warwick**   10 by 10   Revision a 1923 c 1938-50 f 1970

1	A	CPr	OPr	nd:po	-	-	6.49	4BEoo3	RB		4/49	5/49	1
1	B	CPr		nd:po	-	-	6.49	4BEoo3	RB				
2	C	CPs		1951pc	-	ac	6.51	4Booo3	R		1/52	-	
3	D	CPt		1951pc	1953c	ac	6.53	4BooX3					
3	D/	CPw,y,z		1951mc	1958c	ac	6.53	4BooX3					
3	D//*	CPz	OPz	1951ᶜms	nd:r	acf	6.53	4BooX3			5/71	6/71	

AA 1.2.49 on A(A), 3/-, OSPR 5/49. M on D. 1. The Grand Union Canal is solid blue through Hatton Locks.

SP 27 (42/27) *Kenilworth**   10 by 10   Revision a 1902-25 c 1937-50 f 1965 g 1970

1	A	CPr		1947po	-	-	1.48	4BE2X3	RB	0	8/47	11/47	
1	B	CPs	OPs	1951pc	-	ac	6.51	4BEoo3	RB	0	8/51	9/51	
3	B/	CPv		1951pc	1957mc	ac	6.51	4BooX3	B	0			
3	B//	CP?w,x		1951mc	1960mc	ac	6.51	4BooX3	B	0			
3	B///	CPz	OPz	1951mc	1966ch	acf	6.51	4BooX3	B	3			
3	B////*	CPz	OPz	1951ᶜms	nd:b	acg	6.51	4BooX3	B	3	1/72	5/72	

AA 4.9.47 on A(A), 3/-, OSPR 11/47. M on B.

1	2	3	4	5	6	7	8	9	10	11	12	13	14

SP 28 (42/28) *Meriden** 10 by 10 Revision a 1902-37 c 1938-49 f 1962 g 1964 h 1972

1	A	CPr	OPr	1947po	-		1.48	4BE2X3	R		8/47	11/47	
1	B	CPs	OPv	1951pc	-	ac	1.51	4BEoo3	R		12/51	1/57	
3	B	CPt		1951pc	-	ac	1.51	4BEoo3	R				
3	B/	CPw		1951mc	1959mc	ac	1.51	4BooX3					
3	B//	CPy		1951mc	1963ch	acf	1.51	4BooX3					
3	B///	CPz	OPz	1951mc	1965ch	acg	1.51	4BooX3					
3	B////*	CPz	OPz	1951^{c}ms	nd:r	ach	1.51	4BooX3			12/72	6/73	

AA 4.9.47 on A(A), 3/-, OSPR 11/47. M on B.

SP 29 (42/29) [Kingsbury] 10 by 10 Revision a 1921-23 c 1938-50

1	A	CPr	OPr	nd:po	-	-	1.49	4BEoo3	RB		8/48	9/48	
1	B	CPs	OPs	1951pc	-	ac	1.51	4BEoo3	RB		8/51	9/51	
3	B		OPs	1951pc	-	ac	1.51	4BEoo3	RB				
3	B/	CPv,x,z		1951pc	1957c	ac	1.51	4BooX3					

M on B.

SP 30 (42/30) *Witney (South)* 10 by 10 Revision a 1910-19 c 1938-56

1	A	CPr	OPr,s	nd:po	-	-	1.49	4BEoo4	RB	0	5/49	6/49	
2	B	CPt,u		1949pc	1953c	-	6.53	4BooX4	B	0			
3	C	CPw,x,y,z	OPv,z	1960mc	-	ac	6.60	4ooX4		1	1/60	2/60	

SS: Soil, coloured, 1983 on C.

SP 31 (42/31) *Witney (North)* 10 by 10 Revision a 1919 c 1938-50

1	A	CPr	OPr	nd:po	-	-	6.49	4BEoo3	RB	0	7/49	7/49	
2	B	CPs,u	OPv	1952pc	-	ac	1.52	4Booo3	R	2	4/52	1/57	
3	B	CPy,z	OPz	1952pc	-	ac	1.52	4Booo3	R	2			
3	B/	C0z		1952pc	-	ac	1.52	4Booo3	R	2			1

1. Note the continuing absence of an adjoining sheet index, and the presence of railway company names.

SP 32 (42/32) *Chipping Norton* 10 by 10 Revision a 1919 c 1950

1	A	CPr	OPr	nd:po	-	-	1.49	4BEoo3	RB	00	6/49	7/49	
3	A	CPu		1949pc	-	-	1.49	4BEoo3	RB	00			
3	B	CPw,x,z	OPv	1958pc	-	ac	6.58	4ooX3		32	1/59	12/58	

SP 33 (42/33) *Hook Norton* 10 by 10 Revision a 1919-20 c 1949-50

1	A	CPr	OPr	nd:po	-	-	6.49	4BEoo3	RB		5/49	6/49	
3	A	CPu		1949pc	-	-	6.49	4BEoo3	RB				
3	B	CPw,x.y,z	OPv	1959mc	-	ac	6.59	4ooX3			10/59	10/59	

SP 34 (42/34) *Edge Hill** 10 by 10 Revision a 1904-20 c 1949

1	A	CPr	OPr	nd:po	-	-	1.49	4BEoo3	B	0	5/49	6/49	
2	A	CPu		1949pc	-	-	1.49	4BooX3	B	0			
3	A/	CPw		1949pc	1958mc	a	1.49	4BooX3	B	0			
3	B	CPw.y,z	OPv.z	1959mc	-	ac	6.59	4ooX4		3	8/59	11/59	

The munitions depot at Kineton is not shown, nor its railway system.

SP 35 (42/35) *Kineton Warwickshire* 10 by 10 Revision a 1900-23 c 1938-50

1	A	CPr	OPr	nd:po	-	-	6.49	4BEoo3	RB	0	6/49	7/49	
2	B	CPs	OPs	1952pc	-	ac	1.52	4Booo3	R	1	6/52	2/53	
3	B/	CPu		1952pc	1956mc	ac	1.52	4BooX3		1			
3	B//	CPw.z,0z		1952mc	1961mc	ac	1.52	4BooX3		1			1

The munitions depot at Kineton is not shown, nor its railway system. 1. Gaydon Airfield.

SP 36 (42/36) *Royal Leamington Spa** 10 by 10 Revision a 1903-23 c 1938-50

1	A	CPr	OPr	nd:po	-	-	1.49	4BEoo3	RB		2/49	3/49	
2	B	CPs,t	OPs	1951pc	-	ac	6.51	4Booo3	R		12/51	1/56	
3	B/	CPw,x,z		1951pc	1958mc	ac	6.51	4BooX3					
3	B//		OPz	1951pc	1958mc	ac	6.51	4BooX3					

AA 1.11.48 on A(A), 3/-, OSPR 3/49. M on B. SS: Soil, outline, 1973 on B//.

1	2	3	4	5	6	7	8	9	10	11	12	13	14

SP 37 (42/37) *Coventry (South)* *   10 by 10   Revision a 1903-23 c 1938-50 f 1966 g 1971
                                                         j 1966 airfield

1	A	CPr	OPr	1947po	-	-	6.47	4BE2X3	RB	0	7/47	11/47	
2	B	CPs	OPv	1952pc	-	ac	1.52	4Booo3	R	0	4/52	1/57	
3	B/	CPw,x,z		1952pc	1958mc	ac	1.52	4BooX3		0			
3	B//*	CPz	OPz	1952mc	1967r	acf	1.52	4BooX3		0			
3	B//*/*	CPz	OPz	1952^{c}ms	nd:ar	acgj	1.52	4BooX3		3	5/72		1

AA 4.9.47 on A(A), 3/-, OSPR 11/47. M on A. 1. Coventry Civic Airport. OSPR (outline edition) not found.

SP 38 (42/38) *Coventry (North)* *   10 by 10   Revision a 1922-23 c 1938-50

1	A	CPr	OPr	1947po		-	6.47	4BE1X3	RB	0	8/47	11/47
1	B	CPs		1951pc		ac	1.51	4BEoo3	RB	0	8/51	-
3	B/	CPv,x,z		1951pc	1957c	ac	1.51	4BooX3	B	0		

AA 4.9.47 on A(A), 3/-, OSPR 11/47. M on A.

SP 39 (42/39) [Nuneaton]   10 by 10   Revision a 1922-27 c 1938-50 f 1966
                                                        j 1959 motor proving ground

1	A	CPr	OPr	1947po	-	-	1.48	4BE2X3	RB	0	12/47	2/48	
2	B	CPs	OPs	1952pc	-	ac	1.52	4Booo3	R	0	4/52	6/52	
3	B/	CPv,x		1952pc	1957c	ac	1.52	4BooX3		0			
3	B//*	CPz	OPz	1952mc	1967mr	acfj	1.52	4BooX3		3	2/67	2/67	1

AA 1.9.47 on A(A), 3/-, OSPR 2/48. M on B. 1. The airfield is now in use as a Motor Proving Ground.

SP 40 (42/40) *Oxford (West)*   10 by 10   Revision a 1910-19 c 1938-47 d 1938-56
                                                        f 1961 g 1968 j 1965 reservoir

1	A	CPr		1948po	-	-	6.48	4BEoo3	RB	01	7/48	-	
1	A		OPr	nd:po	-	-	6.48	4BEoo3	RB	01	-	9/48	
1	B	CPs		1948pc	1951c	ac	1.51	4BEoo3	RB	01			
2	C	CPt		1948pc	1953c	ac	6.53	4BooX3	B	01			
3	C		OPs	1948pc	1953c	ac	6.53	4BooX3	B	01			
3	C/	CP?		1948pc	1957mc	ac	6.53	4BooX3	B	01			M
3	D	CPw	OPv	1960mc	-	ad	6.60	4oooX3		10	1/60	2/60	1
3	D/*	CPx	OPx	1961mc	1961r	adf	6.60	4oooX3		10	12/61	12/61	2
3	D/*/*	CPz	OPz	1960ms	1968rv	adgj	6.60	4oooX3		10	9/68	9/68	3

AA 1.4.48 on A(A), 3/-, OSPR 9/48. M on C/. 1. Northern dispersals of Abingdon Airfield are shown north of the road. 2. Showing the Oxford Western Bypass under construction. 3. Farmoor Reservoir.

SP 41 (42/41) *Woodstock Oxfordshire*   10 by 10   Revision a 1919 c 1936-50 f 1961

1	A	CPr	OPr	1947po	-	-	6.47	4BE2X3	RB	0	7/47	11/47
2	B	CPt,u	OPv	1952pc	-	ac	6.52	4Booo3	R	2	10/52	10/57
3	B/	CPw		1952mc	1960mc	ac	6.52	4BooX3		2		
3	B//	CPx,z		1952mc	1963ch	acf	6.52	4BooX3		2		

AA 1.5.47 on A(A), 3/-, OSPR 11/47.

SP 42 (42/42) *Barton*   10 by 10   Revision a 1919-20 c 1949-50

1	A	CPr	OPr	nd:po	-	-	6.49	4BEoo3	RB	00	6/49	7/49
2	A	CPt		1949pc	1952	-	1.52	4Booo3	RB	00		
3	A/	CPw		1949pc	1958mc	a	1.52	4BooX3	B	00		
3	B	CPw,x,z	OPv	1959mc	-	ac	6.59	4oooX4		30	5/59	5/59

SP 43 (42/43) *Bloxham*   10 by 10   Revision a 1919-20 c 1949

1	A	CPr	OPr	nd:po	-	-	1.49	4BEoo3	RB	0	6/49	7/49
3	A	CPu		1949pc	-	-	1.49	4Booo3	RB	0		
3	B	CPv	OPv	1957pc	-	ac	6.57	4oooX3		3	1/58	11/57
3	B/	CPy,z	OPy	1957pc	-	ac	6.57	4oooX3		3		

SP 44 (42/44) *Banbury (North)*   10 by 10   Revision a 1920-21 c 1938-49

1	A	CPr	OPr	nd:po	-	-	6.49	4BEoo3	RB	0	7/49	7/49
2	B	CPs,u	OPv	1951pc	-	ac	6.51	4Booo3	R	0	2/52	10/57
3	B/	CPw,z.z	OPv	1951mc	1958mc	ac	6.51	4BooX3		0		

202

1	2	3	4	5	6	7	8	9	10	11	12	13	14

SP 45 (42/45) *Priors Marston**   10 by 10 Revision a 1899-1923 c 1949-50
| 1 | A | CPr | OPr | nd:po | - | - | 1.49 | 4BEoo3 | RB | 0 | 5/49 | 6/49 | |
| 3 | B | CPv,z.z,0z | OPv | 1957pc | - | ac | 6.57 | 4oooX3 | | 1 | 5/57 | 8/57 | |

SP 46 (42/46) *Southam Warwickshire*   10 by 10 Revision a 1899-1923 c 1938-50 f 1972
j 1971 reservoir
1	A	CPr	OPr	nd:po	-	-	1.49	4BEoo3	RB		5/49	6/49	
2	B	CPs,u	OPv	1951pc	-	ac	6.51	4Booo4	R		1/52	2/57	
3	B/	CPw.z		1951mc	1960mc	ac	6.51	4BooX4					
3	B//*	CPz	OPz	1951^{c}ms	nd:v	acfj	6.51	4BooX4			9/73	12/73	1

1. Draycote Water.

SP 47 (42/47) *Rugby (West)*   10 by 10 Revision a 1903-23 c 1938-50 f 1973; j 1971 reservoir
1	A	CPr	OPr	1948po	-	-	1.48	4BE4X3	RB	0	2/48	7/48	
1	B	CPs		1951pc	-	ac	6.51	4BEoo3	RB	0	11/51	-	
2	B		OPs	1951pc	-	ac	6.51	4BooX3	RB	0	-	2/53	
3	B/	CPv		1951pc	1957mc	ac	6.51	4BooX3	B	0			
3	B//*	CPx,z	OPx	1951mc	1961r	ac	6.51	4BooX3	B	0	2/61		1
3	B//*/*	CPz	OPz	1951^{c}ms	nd:rv	acfj	6.51	4BooX3	B	0	1/74	5/74	2

AA 1.1.48 on A(A), 3/-, OSPR 7/48. M on A. SS: Soil, coloured, 1977 on B//*/*. 1. OSPR (outline edition) not found. 2. Draycote Water reservoir.

SP 48 (42/48) [Wolvey]   10 by 10 Revision a 1901-23 c 1938-50
1	A	CPr	OPr	1948po	-	-	6.48	4BEoo3	RBC	00	6/48	8/48	
2	A	CPu		1948pc	-	-	6.48	4BooX3	RBC	00			
3	B	CPw,z	OPv,z	1960mc	-	ac	6.60	4oooX3	C	20	5/60	6/60	

AA 1.4.48 on A(A), 3/-, OSPR 8/48. M on A.

SP 49 (42/49) [Hinckley]   10 by 10 Revision a 1901-27 c 1938-50
1	A	CPr	OPr	nd:po	-	-	6.48	4BEoo3	RB		7/48	9/48	
2	B	CPt	OPs	1952pc	-	ac	6.52	4Booo3	R		7/52	7/53	
3	B/	CPw.z		1952mc	1960mc	ac	6.52	4BooX3					

AA 1.5.48 on A(A), 3/-, OSPR 9/48. M on B.

SP 50 (42/50) *Oxford (East)*   10 by 10 Revision a 1910-32 c 1936-39 d 1936-56
e 1947 f 1963 g 1968; j 1965 building
1	A	CPr	OPr	1947po	-	-	1.48	4BE1X3	RB		9/47	12/47	
2	B	CPt		1952pc	-	ace	6.52	4Booo4	R		7/52	-	
3	B/	CPv		1952pc	1957mc	ace	6.52	4BooX4					
3	B//	CPw		1952mc	1959mc	ace	6.52	4BooX4					
3	C	CPw	OPv	1960mc	-	ad	6.60	4oooX4			5/60	6/60	
3	C/*	CPz	OPz	1960mc	1964r	adf	6.60	4oooX4			6/64	6/64	
3	C/*/*	CPz	OPz	1960ms	1968br	adgj	6.60	4oooX4			12/68	12/68	

AA 1.3.47 on A(A), 3/-, OSPR 12/47. M on B/.

SP 51 (42/51) *Islip Oxfordshire*   10 by 10 Revision a 1911-19 c 1936-50
1	A	CPr	OPr	1947po	-	-	1.48	4BE2X3	RB	0	11/47	2/48	
1	B	CPs		1951pc	-	ac	6.51	4BEoo3	R	0	12/51	-	
2	C	CPt	OPs	1951pc	1953c	ac	6.53	4BooX3		0	-	12/55	
3	C/	CPv		1951pc	1957mc	ac	6.53	4BooX3		0			
3	C//	CPx,z,0z		1951mc	1961ch	ac	6.53	4BooX3		0			

AA 1.9.47 on A(A), 3/-, OSPR 2/48. The ordnance depot at Graven Hill and its railway system are not shown.

SP 52 (42/52) *Bicester*   10 by 10 Revision a 1919-20 c 1938-50
1	A	CPr	OPr	nd:po	-	-	6.49	4BEoo3	RB	011	6/49	6/49	
2	B	CPs		1951pc	-	ac	6.51	4Booo3	R	011	3/52	-	
3	C	CPt	OPv	1952pc	1954c	ac	6.53	4BooX3		011	-	8/57	
3	C/	CPw.y,z	OPv	1952mc	1960c	ac	6.53	4BooX3		011			

The ordnance depot at Graven Hill and its railway system are not shown.

1	2	3	4	5	6	7	8	9	10	11	12	13	14

SP 53 (42/53) *Brackley*   10 by 10   Revision a 1899-1925 c 1949-50

1	A	CPr	OPr	nd:po	-	-	6.49	4BEoo3	RB	00	6/49	7/49	
2	B	CPt,u		1949pc	1953c	-	6.53	4BooX3	B	00			
3	C	CPw.y,z	OPx	1961mc	-	ac	6.61	4oooX3		32	12/60	1/61	

SP 54 (42/54) *Culworth*   10 by 10   Revision a 1899-1923 c 1949-50

1	A	CPr	OPr	nd:po	-	-	1.49	4BEoo3	RB	0	6/49	7/49	
3	A	CPu		1949pc	-	-	1.49	4BEoo3	RB	0			
3	B	CPw,z	OPv	1961mc	-	ac	6.61	4oooX3		3	1/61	1/61	

SP 55 (42/55) *Woodford Halse*   10 by 10   Revision a 1899-1923 c 1938-50

1	A	CPr	OPr	nd:po	-	-	1.49	4BEoo3	RB	0	5/49	6/49	
2	B	CPs		1951pc	-	ac	6.51	4Booo3	R	0	12/51	-	
3	B/	CPv.x,z	OPv	1951pc	1956mc	ac	6.51	4BooX3		0	-	2/57	

M on B/.

SP 56 (42/56) *Daventry**   10 by 10   Revision a 1899-1923 c 1938-50 f 1967 g 1975

1	A	CPr	OPr	nd:po	-	-	6.49	4BEoo3	RB		5/49	6/49	
1	B	CPs		1951pc	-	ac	6.51	4BEoo3	RB		11/51	-	1
3	B	CPu	OPv	1951pc	-	ac	6.51	4BEoo3	RB		-	2/57	
3	B/	CPw	OPz	1951mc	1960mc	ac	6.51	4BooX3	B				
3	B//*	CPz	OPz	1951mc	1968r	acf	6.51	4BooX3	B		3/68	2/68	
3	B//*/*	CFz	OFz	1951^{c}ms	nd:r	acg	6.51	4BooX3	B		4/76	12/75	

M on B/. 1. Showing the BBC Station at Borough Hill (5862).

SP 57 (42/57) *Rugby (East)*   10 by 10   Revision a 1899-1923 c 1938-50 f 1965 g 1972
                                         j 1965 radio station

1	A	CPr	OPr	nd:po	-	-	1.49	4BEoo3	RB		2/49	3/49	
2	B	CPs,u	OPs,v	1951pc	-	ac	6.51	4Booo3	R		1/52	2/57	
3	B/	CPw		1951mc	1960mc	ac	6.51	4BooX3					
3	B//*	CPz	OPz	1951mc	1966dr	acfj	6.51	4BooX3			10/66	10/66	
3	B//*/*	CPz	OPz	1951^{c}ms	nd:r	acg	6.51	4BooX3			3/73	4/73	

AA 1.11.48 on A(A), 3/-, OSPR 3/49. M on A.

SP 58 (42/58) *[Lutterworth]*   10 by 10   Revision a 1899-1923 c 1950 f 1964; j 1959 airfield

1	A	CPr	OPr	nd:po	-	-	1.49	4BEoo3	RB	00	5/49	6/49	
2	B	CPs	OPs	1951pc	-	ac	6.51	4Booo3	R	00	2/52	11/53	
3	B/	CPv		1951pc	1956mc	ac	6.51	4BooX3		00			
3	B//	CPw,y		1951mc	1958mc	ac	6.51	4BooX3		00			
3	B///*	CPz	OPz	1951mc	1965ar	acfj	6.51	4BooX3		33	6/65	6/65	1

M on B//. 1. Bruntingthorpe Airfield.

SP 59 (42/59) *[Narborough]*   10 by 10   Revision a 1901-28 c 1938-50 f 1963 g 1965

1	A	CPr	OPr	1947po	-	-	1.48	4BE2X3	RB		10/47	12/47	1
2	B	CPs		1952pc	-	ac	1.52	4Booo3	R		4/52	-	
3	B	CPu		1952pc	-	ac	1.52	4Booo3	R				
3	B/	CPw		1952pc	1958mc	ac	1.52	4BooX3					
3	B//*	CPy	OPy	1952mc	1963r	acf	1.52	4BooX3			11/63	11/63	2
3	B//*/*	CPz	OPz	1952mc	1965r	acg	1.52	4BooX3			12/65	12/65	

AA 1.6.47 on A(A), 3/-, OSPR 12/47. M on B/. 1. The alignment of the temporary connection between the L&NW and GC Railways at Whetstone is shown (5598). 2. Showing the M1 under construction.

SP 60 (42/60) *Tiddington Oxfordshire*   10 by 10   Revision a 1919 c 1938 d 1938-56 f 1964 g 1975

1	A	CPr	OPr	1948po	-	-	6.48	4BEoo3	RB	1	5/48	6/48	
2	B	CPt,u	OPs	1948pc	1953c	-	6.53	4BooX3	B	1			
3	B/	CPw		1948mc	1959c	ac	6.53	4BooX3	B	1			
3	C	CPw	OPv	1960mc	-	ad	6.60	4oooX3		1	8/60	6/60	
3	C/*	CPz	OPz	1960mc	1965r	adf	6.60	4oooX3		1	1/65	2/65	
3	C/*/*	CFz	OFz	1960^{c}ms	nd:r	adg	6.60	4oooX3		1	11/76	11/76	

M on B, C. SS: Soil, coloured, 1983 on C/*.

1	2	3	4	5	6	7	8	9	10	11	12	13	14

SP 61 (42/61) *Brill*   10 by 10   Revision a 1898-1919 c 1945-50 f 1963
1	A	CPr	OPr	nd:po	-	-	6.48	4BEoo3	RB	10	8/48	9/48	
1	B	CPs		1951pc	-	ac	1.51	4BEoo3	RB	10	4/51	-	
2	B	CPu	OPs	1951pc	-	ac	1.51	4Booo3	RB	10	-	4/52	
3	B/	CPy,z		1951mc	1964ch	acf	1.51	4BooX3	B	30			

The ordnance depot at Upper Arncott and its railway system are not shown.

SP 62 (42/62) *Marsh Gibbon*   10 by 10   Revision a 1898-1920 c 1938-50
1	A	CPr	OPr	nd:po	-	-	6.49	4BEoo3	RB	0	6/49	6/49	
1	B	CPs		1951pc	-	ac	6.51	4BEoo3	RB	0	11/51	-	
3	B	CPu,x,z	OPs,z	1951pc	-	ac	6.51	4BEoo3	RB	0	-	1/54	
3	B/	CPz,0z	OPz,0z	1951pc	-	ac	6.51	4BEoo3	B	0			

A sheet where the revised form of the National Grid was never implemented.

SP 63 (42/63) *Buckingham (West)*   10 by 10   Revision a 1899-1923 c 1938-50
1	A	CPr	OPr	nd:po	-	-	6.49	4BEoo3	RB	00	9/49	9/49	
3	B	CPu	OPv	1955pc	-	ac	6.55	4oooX3		11	11/55	8/57	
3	B/	CPw.z		1955mc	1960mc	ac	6.55	4oooX3		11			

SP 64 (42/64) *Towcester*   10 by 10   Revision a 1899-1923 c 1950
1	A	CPr	OPr	nd:po	-	-	1.49	4BEoo3	RB	0	5/49	6/49	
3	B	CPu	OPv	1955pc	-	ac	6.55	4oooX3		3	1/56	8/57	1
3	B/	CPw.z		1955mc	1959mc	ac	6.55	4oooX3		3			

1. Showing the full runway layout of Silverstone Airfield, taken over as a motor racing circuit.

SP 65 (42/65) *Weedon Bec*   10 by 10   Revision a 1899-1923 c 1937-50 f 1969
                                       j 1966 industrial
1	A	CPr	OPr	1948po	-	-	6.48	4BEoo3	RB		6/48	7/48	
3	B	CPu	OPv	1955pc	-	ac	6.55	4oooX3			1/56	2/57	
3	B/	CPw		1955mc	1960mc	ac	6.55	4oooX3					
3	B//*	CPz	OPz	1955^{c}ms	nd:ir	acfj	6.55	4oooX3			3/71	5/71	

M on B.

SP 66 (42/66) *Long Buckby*   10 by 10   Revision a 1899-1938 c 1950 f 1965
1	A	CPr	OPr	nd:po	-	-	1.49	4BEoo3	RB		10/48	11/48	
3	B	CPu		1954pc	-	ac	6.54	4oooX3			12/54	-	
3	B/	CPw		1954pc	1958mc	ac	6.54	4oooX3					
3	B//*	CPz	OPz	1954mc	1966r	acf	6.54	4oooX3			2/66	2/66	

M on B/. SS: Soil, Land Use Capability, both coloured, both outline, 1978 on B//*. A line of trees marks a boundary in 6463: Ash, Lime, Ash, Ash, Limes, Oak, Ash, present on Northamptonshire six-inch sheet 43NE since the Edition of 1926.

SP 67 (42/67) *West Haddon*   10 by 10   Revision a 1899-1923 c 1950
1	A	CPr	OPr	nd:po	-	-	1.49	4BEoo3	RB		5/49	6/49	
3	B	CPu	OPv	1955pc	-	ac	6.55	4oooX3			1/55	8/57	
3	B/	CPw,z		1955pc	1958mc	ac	6.55	4oooX3					

M on B/.

SP 68 (42/68) *Husbands Bosworth*   10 by 10   Revision a 1899-1928 c 1938-50 f 1968
1	A	CPr	OPr	nd:po	-	-	6.49	4BEoo3	RB	00	6/49	6/49	
3	B	CPu	OPv	1954pc	-	ac	6.54	4oooX3		11	12/54	8/57	
3	B/	CPw,z		1954pc	1958mc	ac	6.54	4oooX3		11			
3	B//	CPz	OPz	1954ms	1969ch	acf	6.54	4oooX3		11			

M on B/.

SP 69 (42/69) *Wigston Magna*   10 by 10   Revision a 1902-28 c 1938-50
1	A	CPr	OPr	1947po	-	-	1.48	4BE2X3	RB		11/47	2/48	
	B												
3	C	CPu	OPv	1955pc	-	ac	6.55	4oooX3			12/54	2/57	
3	C/	CPw,z		1955pc	1958mc	ac	6.55	4oooX3					

AA 1.8.47 on A(A), 3/-, OSPR 2/48. M on C/.

1	2	3	4	5	6	7	8	9	10	11	12	13	14

SP 70 (42/70) *Thame* 10 by 10 Revision a 1919 c 1938 d 1938-56

1	A	CPr	OPr	nd:po	-	-	1.48	4BEoo3	RB	0	2/49	10/48	
1	B	CPr		nd:po	-	-	1.48	4BEoo3	RB	0			
1	C	CPs		1948pc	1950c	ac	6.50	4BEoo3	RB	0			
2	C	CPt		1948pc	1950c	ac	6.50	4BooX3	RB	0			
3	D	CPw,z	OPv	1960mc	-	ad	6.60	4oooX3		2	2/60	4/60	1

M on C. 1. Aylesbury & Thame Airport.

SP 71 (42/71) *Waddesdon* 10 by 10 Revision a 1876-1923 b 1898-1923 c 1938 d 1938-50

1	A	CPr	OPr	nd:po	-	-	6.48	4BEoo3	RB	0	12/48	2/49	
1	B	CPs		1950pc	-	ac	6.50	4BEoo3	RB	0	12/50	-	
	C												
3	D	CPu,y,z.z	OPv	1955pc	-	bd	6.55	4oooX3		1	9/55	2/57	

AA 1.9.48 on A(A), 3/-, OSPR 2/49.

SP 72 (42/72) *Winslow** 10 by 10 Revision a 1898-1924 c 1942-50 d 1950 f 1965

1	A	CPr	OPr	nd:po	-	-	1.49	4BEoo3	RB	0	5/49	6/49	
1	B	CPs	OPs	1951pc	-	ac	1.51	4BEoo3	RB	0	11/51	11/51	
3	C	CPu	OPv	1955pc	-	ad	6.55	4oooX3		1	9/55	2/57	
3	C/	CP?x,y		1955mc	1961ch	ad	6.55	4oooX3		1			
3	C//	CPz	OPz	1955mc	1966ch	adf	6.55	4oooX3		3			

SP 73 (42/73) *Buckingham (East)* 10 by 10 Revision a 1899-1924 c 1938-50 f 1968
  j 1966 reservoir

1	A	CPr	OPr	nd:po	-	-	1.49	4BEoo3	RB	0	5/49	6/49	
3	B	CPu,z	OPv	1956pc	-	ac	6.56	4oooX3		1	2/56	2/57	
3	B/*	CPz	OPz	1956ms	1969v	acfj	6.56	4oooX3		1	4/69	8/69	1

AA 1.1.49 on A(A), 3/-, OSPR 6/49. 1. Foxcote Reservoir.

SP 74 (42/74) *Stony Stratford* 10 by 10 Revision a 1898-1924 c 1938-50 d 1945-53
  f 1968 g 1975; j 1984 selected

1	A	CPr	OPr	nd:po	-	-	1.49	4BEoo3	RB		5/49	7/49	
3	B	CPu	OPv	1955pc	-	ac	6.55	4oooX3			12/55	9/57	
3	B/	CPw,z		1955mc	1960mc	ac	6.55	4oooX3					
3	B//*	CPz	OPz	1955ms	1968r	acf	6.55	4oooX3			8/68	8/68	
3	B//*/*	CFz	OFz	1955^{c}ms	nd:r	acg	6.55	4oooX3			4/76	4/76	
3	B//*/*/*	CFz	OFz	1985^{c}m-s	nd:r	adgj	6.85	4oooX3			6/85	6/85	

AA 1.1.49 on A(A), 3/-, OSPR 7/49.

SP 75 (42/75) *Northampton (South)* 10 by 10 Revision a 1899-1924 c 1937-50 f 1960-61

1	A	CPr	OPr	1947po	-	-	6.47	4BE2X3	RB		4/47	7/47	
3	B	CPu	OPv	1956pc	-	ac	6.56	4oooX4			4/56	2/57	
3	B/	CPw		1956mc	1960mc	ac	6.56	4oooX4					
3	B//*	CPx,z	OPx	1956mc	1961r	acf	6.56	4oooX4			12/61	12/61	

AA 1.2.47 on A(A), 3/-, OSPR 7/47. M on B.

SP 76 (42/76) *Northampton (North)** 10 by 10 Revision a 1899-1938 c 1937-50 f 1963
  j 1963 reservoir

1	A	CPr	OPr	1947po	-	-	6.47	4BE2X3	RB		7/47	10/47	
3	B	CPu,v	OPv	1955pc	-	ac	6.55	4oooX3			3/55	2/57	
3	B/*	CPz	OPz	1955mc	1964rv	acfj	6.55	4oooX3			8/64	11/64	1

AA 1.5.47 on A(A), 3/-, OSPR 10/47. M on B. 1. Pitsford Reservoir.

SP 77 (42/77) *Brixworth* 10 by 10 Revision a 1899-1924 c 1950 f 1966; j 1963 reservoir

1	A	CPr	OPr	nd:po	-	-	1.49	4BEoo3	RB	0	5/49	5/49	
3	B	CPu	OPv	1954pc	-	ac	6.54	4oooX3		1	11/54	8/57	
3	B/	CPw		1954mc	1960mc	ac	6.54	4oooX3		1			1
3	B//*	CPz	OPz	1954mc	1967v	acfj	6.54	4oooX3		1	3/67	3/67	2

M on B. 1. Harrington Airfield. 2. Pitsford Reservoir.

1	2	3	4	5	6	7	8	9	10	11	12	13	14

SP 78 (42/78) *Market Harborough* 10 by 10 Revision a 1899-1928 c 1938-50 f 1965
1 A   CPr   OPr   nd:po   -   -   6.49 4BEoo3 RB 0 6/49   7/49
3 B   CPu   OPv   1954pc -   ac   6.54 4oooX3   1 11/54 9/57
3 B/   CPw   1954mc 1958mc   ac   6.54 4oooX3   1
3 B// CPz   1954mc 1966ch   acf   6.54 4oooX3   3
AA 1.2.49 on A(A), 3/-, OSPR 7/49. M on B/.

SP 79 (42/79) *Hallaton* 10 by 10 Revision a 1902-28 c 1950
1 A   CPr   OPr   nd:po   -   -   6.49 4BEoo3 RB   5/49   6/49
3 B   CPu   OPv   1954pc -   ac   6.54 4oooX3   12/54 2/57
3 B/   CPw.z   1954mc 1960mc   ac   6.54 4oooX3
M on B.

SP 80 (42/80) *Wendover* 10 by 10 Revision a 1897-1919 c 1938 d 1938-56
                                         f 1962 g 1968; j 1966 building
1 A   CPr   OPr   nd:po   -   -   1.49 4BEoo3 RB   12/48 1/49
2 B   CPt   1948pc 1950c   ac   1.52 4BooX3 RB
3 B/   CPv   1948pc 1956c   ac   1.52 4BooX3 B
3 B//   CPw   1948mc 1959c   ac   1.52 4BooX3 B
3 C   CPw   OPv   1960mc -   ad   6.60 4oooX3   7/60   8/60
3 C/   CPy,z   1960mc 1963ch   adf   6.60 4oooX3
3 C//* CPz,0z   OPz   1960ms 1969b   adgj   6.60 4oooX3   11/69 1/70

SP 81 (42/81) *Aylesbury* 10 by 10 Revision a 1897-1923 c 1938-50 f 1962 g 1969
                                         j 1968 building
1 A   CPr   OPr   nd:po   -   -   1.49 4BEoo3 RB 0 1/49   2/49
1 B   CPr   nd:po   -   -   1.49 4BEoo3 RB 0
3 C   CPu   OPv   1955pc -   ac   6.55 4oooX3   0 1/56   6/56
3 C/   CPw   1955mc 1960mc   ac   6.55 4oooX3   0
3 C//   CPy,z   1955mc 1963ch   acf   6.55 4oooX3   0
3 C///* CPz,0z   OPz   1955ms 1970br   acgj   6.55 4oooX3   0 5/70   8/70
AA 1.9.48/NT 14.7.47 on A(A), 3/-, OSPR 2/49.

SP 82 (42/82) *Stewkley* 10 by 10 Revision a 1898-1924 c 1938-50
1 A   CPr   OPr   nd:po   -   -   1.48 4BEoo3 RB 0 3/49   4/49
3 B   CPu   OPv   1955pc -   ac   6.55 4oooX3   1 10/55 2/57
3 B/   CP?w,x.z   1955mc 1960mc   ac   6.55 4oooX3   1

SP 83 (42/83) *Bletchley Buckinghamshire* 10 by 10 Revision a 1923-24 c 1938-50; j 1984 selected k SUSI
From B/* renamed *Milton Keynes & Bletchley*
1 A   CPr   OPr   nd:po   -   -   6.48 4BEoo3 RB   5/49   7/49
3 B   CPu.x,z OPv   1956pc   -   ac   6.56 4oooX3   2/56   2/57
3 B/* CFz   OFz   1985^{c}m-s   nd:brl   acjk 6.85 4oooX3   9/85   9/85
AA 1.1.49 on A(A), 3/-, OSPR 7/49.

SP 84 (42/84) *Newport Pagnell** 10 by 10 Revision a 1898-1938 c 1938-50 f 1963
1 A   CPr   OPr   nd:po   -   -   6.48 4BEoo3 RB   5/49   6/49
3 B   CPu   OPv   1955pc -   ac   6.55 4oooX3   1/56   9/57
3 B/   CPw   1955mc 1960mc   ac   6.55 4oooX3
3 B//* CPz   OPz   1955mc 1964r   acf   6.55 4oooX3   6/64   9/64
AA 1.1.49 on A(A), 3/-, OSPR 6/49.

SP 85 (42/85) *Olney** 10 by 10 Revision a 1899-1924 c 1950
1 A   CPr   OPr   nd:po   -   -   6.48 4BEoo3 RB   6/48   7/48
3 B   CPu   OPv   1955pc -   ac   6.55 4oooX4   11/55 2/57
3 B/   CPw,z.z   1955mc 1959mc   ac   6.55 4oooX4
M on B.

1	2	3	4	5	6	7	8	9	10	11	12	13	14

SP 86 (42/86) *Wellingborough (West)*   10 by 10   Revision a 1899-1924   c 1938-50
j 1983 selected k SUSI

1	A	CPr	OPr	nd:po	-	-	1.49	4BEoo3	RB	00	10/48	12/48	
	B												
3	C	CPu,x,z	OPv	1955pc	-	ac	6.55	4oooX3		13	1/55	8/57	1
3	C/*	CFz	OFz	1983ᶜm-s	nd:ru	acjk	6.84	4oooX3		13	6/84	6/84	

AA 1.7.48 on A(A), 3/-, OSPR 12/48. M on C. 1. Possibly a unique reference to an aquadrome.

SP 87 (42/87) *Kettering**   10 by 10   Revision a 1899-1924   c 1938-50

1	A	CPr	OPr	nd:po	-	-	1.49	4BEoo3	RB		11/48	12/48	
3	B	CPu	OPv	1954pc	-	ac	6.54	4oooX3			12/54	8/57	
3	B/	CPw,z		1954pc	1958mc	ac	6.54	4oooX3					

AA 1.9.48 on A(A), 3/-, OSPR 12/48. M on B/.

SP 88 (42/88) *Corby (West)*   10 by 10   Revision a 1899-1924   c 1938-50   f 1965   g 1968
j 1965 airfield k 1966 building

1	A	CPr	OPr	nd:po	-	-	1.49	4BEoo3	RB	0	4/49	5/49	
3	B	CPu	OPv	1954pc	-	ac	6.54	4oooX3		1	12/54	8/57	
3	B/	CPw		1954mc	1958mc	ac	6.54	4oooX3		1			
3	B//*	CPz	OPz	1954mc	1965r	acfj	6.54	4oooX3		3	9/65	10/65	
3	B//*/*	CPz	OPz	1954ᶜms	nd:br	acgjk	6.54	4oooX3		3	1/71	3/71	

AA 1.2.49 on A(A), 3/-, OSPR 5/49. M on B/.

SP 89 (42/89) *Uppingham*   10 by 10   Revision a 1899-1928   c 1938-50   f 1968; j 1965 building

1	A	CPr		nd:po	-	-	6.48	4BEoo3	RB		5/49	-	
1	B		OPr	nd:po	-	-	6.48	4BEoo3	RB		-	6/49	
	C												
3	D	CPu	OPv	1955pc	-	ac	6.55	4oooX3			1/55	2/57	
3	D/	CPw,z		1955mc	1961mc	ac	6.55	4oooX3					
3	D//*	CPz	OPz	1955ms	1969b	acfj	6.55	4oooX3			10/69	11/69	

AA 1.2.49 on B(A), 3/-, OSPR 6/49.

SP 90 (42/90) *Chesham*   10 by 10   Revision a 1897-1937   c 1938-47   d 1938-56
e 1948 f 1963

1	A	CPr	OPr	nd:po	-	-	1.49	4BEoo3	RB	0	9/48	10/48	
1	B												
1	C	CPs		1951pc	-	ace	1.51	4BEoo3	RB	0	12/51	-	
3	C	CPu	OPv	1951pc	-	ace	1.51	4BEoo3	RB	0	-	10/57	
3	D	CPw	OPv	1960mc	-	ad	6.60	4oooX3		1	4/60	5/60	1
3	D/	CPy,z,0z	OPz	1960mc	1963ch	adf	6.60	4oooX3		1			

AA 1.6.48/NT 6.9.48 on A(A), 3/-, OSPR 10/48. 1. Bovingdon Airfield.

SP 91 (42/91) *Tring*   10 by 10   Revision a 1922-23 b 1897-1923 c 1938 d 1937-51
e 1948

1	10046	CPr		1946po	-	-	1.47	4BE2X3	RB	0	12/46	-	
1	A		OPr	1947po	-	-	1.47	4BE2X3	RB	0	-	3/47	
1	B	CPs		1950pc	-	ace	6.50	4BEoo3	RB	0	10/50	-	
3	B	CPt		1950pc	-	ace	6.50	4BEoo3	RB	0			
3	C	CPu	OPv	1956pc	-	bd	6.56	4oooX3		1	5/56	2/57	
3	C/	CPw.z,0z		1956mc	1960mc	bd	6.56	4oooX3		1			

AA 1.9.46 on A(A), 3/-, OSPR 3/47.

SP 92 (42/92) *Leighton Buzzard*   10 by 10   Revision a 1900-24   c 1937-51

1	10047	CPr		1947po	-	-	1.47	4BE2X3	RB		1/47	-	1
1	A		OPr	1947po	-	-	1.47	4BE2X3	RB		-	4/47	
3	B	CPu	OPv	1955pc	-	ac	6.55	4oooX3			9/55	2/57	2
3	B/	CP?w,y,z.z		1955mc	1960mc	ac	6.55	4oooX3					

AA 1.10.46 on A(A), 3/-, OSPR 4/47. M on A, B. 1. The narrow gauge railway built for transporting sand is shown as standard gauge. 2. The industrial line is now shown as a narrow gauge tramway.

1	2	3	4	5	6	7	8	9	10	11	12	13	14

SP 93 (42/93) *Woburn*   10 by 10   Revision a 1898-1924 c 1938-51 f 1963
1 10047 CPr   1947po   -   -   1.47 4BE2X3 RB   12/46 -
1 A   OPr   1946po   -   -   1.47 4BE2X3 RB   -   4/47
3 B   CPu   OPv   1955pc   -   ac   6.55 4oooX3   8/55   2/57
3 B/   CPw   1955mc 1960mc ac   6.55 4oooX3
3 B//*   CPy,z   OPy   1955mc 1964r   acf   6.55 4oooX3   12/63 1/64
AA 1.9.46 on A(A), 3/-, OSPR 4/47.

SP 94 (42/94) *Cranfield**   10 by 10   Revision a 1898-1924 c 1938-51 f 1963
1 A   CPr   OPr   nd:po   -   -   6.48 4BEoo3 B   0   3/49   4/49
2 A   CPt   1949pc   1953rpb -   6.48 4BooX3 B   0
3 B   CPu   OPv   1956pc   -   ac   6.56 4oooX3   1   5/56   2/57
3 B/
3 B//   CPy,z   OPz   1956mc 1963ch acf   6.56 4oooX3   1   1
1. Cranfield Airfield.

SP 95 (42/95) *Turvey*   10 by 10   Revision a 1899-1924 c 1950 d 1948-51
1 A   CPr   OPr   nd:po   -   -   6.48 4BEoo3 RB   0   3/49   4/49
1 B   CPs   1951pc   -   ac   6.51 4BEoo3 RB   0   1/52   -
3 C   CPu   OPv   1956pc   -   ad   6.56 4oooX3   1   2/56   2/57
3 C/   CPw,z   1956mc 1960mc ad   6.56 4oooX3   1
M on C, C/.

SP 96 (42/96) *Wellingborough (East)*   10 by 10   Revision a 1899-1924 c 1938-50
1 A   CPr   OPr   1948po   -   -   6.48 4BEoo3 RB   00   6/48   8/48
3 B   CPu   OPv   1955pc   -   ac   6.55 4oooX3   10   2/55   8/57
3 B/   CP?w,z   1955mc 1960mc ac   6.55 4oooX3   10
AA 1.4.48 on A(A), 3/-, OSPR 8/48. M on B, B/.

SP 97 (42/97) *Irthlingborough**   10 by 10   Revision a 1899-1924 c 1938-50
1 A   CPr   OPr   1948po   -   -   6.48 4BEoo3 RB   6/48   7/48
   B
3 C   CPu   1955pc   -   ac   6.55 4oooX3   1/55   -
3 C/   CPv,x,z   1955pc   1958mc ac   6.55 4oooX3
AA 1.4.48 on A(A), 3/-, OSPR 7/48. M on C/.

SP 98 (42/98) *Corby (East)*   10 by 10   Revision a 1899-1924 c 1938-50 d 1937-50
1 A   CPr   OPr   nd:po   -   -   6.48 4BEoo3 B   00   9/48   11/48
1 B   CPs   1951pc   -   ac   6.51 4BEoo4 B   00   1/52   -
3 C   CPu   OPv   1955pc   -   ad   6.55 4oooX4   11   1/55   1/57
3 C/   CPw,z   1955mc 1960mc ad   6.55 4oooX4   11
AA 1.6.48/NT 14.9.48 on A(A), 3/-, OSPR 11/48. M on C, C/.

SP 99 (42/99) *Harringworth*   10 by 10   Revision a 1899-1938 c 1937-50
1 A   CPr   OPr   nd:po   -   -   6.49 4BEoo3 RB   00   6/49   7/49
   B
3 C   CPu   OPv   1955pc   -   ac   6.55 4oooX3   11   2/55   2/57
3 C/   CPx,z   1955pc   -   ac   6.55 4oooX3   11
AA 1.4.49 on A(A), 3/-, OSPR 7/49.

SR 99 (11/99) **& Part of SR 89** *Castlemartin*   12 by 10   Revision a 1906 c 1948-50
1 A   CPr   OPr   1948po   -   -   6.48 4BEoó3   6/48   7/48
3 B   CPu   OPv   1956pc   -   ac   6.56 4oooX3   5/56   7/56
3 B/   CP?w,y,z.z   1956mc 1960mc ac   6.56 4oooX3
AA 1.6.48 on A(A), 3/-, OSPR 7/48. M on A, B/.

SS 09 (21/09) *Manorbier*   10 by 10   Revision a 1904-05 c 1948
1 A   CPr   OPr   1948po   -   -   1.48 4BE4X3 R   3/48   4/48
1 B   CPr   nd:po   -   -   1.48 4BEoo3 R
3 B/   CPw,z   1949pc   1960mc ac   1.48 4BooX3
AA 1.8.46/NT 1.12.47 on A(A), 3/-, OSPR 4/48. M on B, B/.

1	2	3	4	5	6	7	8	9	10	11	12	13	14

SS 14 (21/14) [Lundy Island] 10 by 10 Revision a 1903
1 A   CPs   OPs   1950pc   -   a   6.50   4BEoo4   10/50 12/50 1
1 A   CPs   1950pc   1951   a   1.51   4BEoo4
Superseded in 1960 by Sheet 856 in the [Second Series] Provisional Edition, whereon Lundy is inset. 1. A proof copy with a "r" price code, and no print code, otherwise of the same specification, is recorded.

SS 19 (21/19) *Caldy Island** 10 by 10 Revision a 1905-06 c 1948 f 1968
1 A   CPr   OPr   nd:po   -   -   1.50   4BEoo3   R   5/50   5/50
2 A   CPt,u   1950pc   1952   -   1.52   4Booo3   R
3 A/* CPz   OPz   1950ms   1969r acf   1.52   4Booo3   8/69   10/69

SS 20 (21/20) **& Part of SS 10** [Bude] 15 by 10 Revision a 1905-32 c 1938 d 1938-58 f 1963
1 A   CPs   OPs   1950pc   -   ac   6.50   4BEoo3   R   8/50   9/50
3 A   CPx   1950pc   -   ac   6.50   4BEoo3   R
3 B   CPz   OPz   1965mc   -   adf   6.65   4oooX3   3/65   6/65
AA 1.5.50 on A(A), 3/-, OSPR 9/50.

SS 21 (21/21) **& Part of SS 11** [Kilkhampton] 12 by 10 Revision a 1904-05 c 1958 f 1963
1 A   CPs   OPs   1950pc   -   a   6.50   4BEoo3   9/50   9/50
3 B   CPt   1950pc   1953c   a   6.53   4BooX3
3 C   CPz   OPz   1965mc   -   acf 6.65   4oooX3   1/65   2/65

SS 22 (21/22) [Hartland Point] 10 by 10 Revision a 1904 c 1958 f 1960
1 A   CPs   OPs   1950pc   -   a   6.50   4BEoo3   8/50   9/50
3 A   CPt   1950pc   -   a   6.50   4BEoo3
3 B   CPy,z OPy   1963mc   -   acf 6.63   4oooX3   9/63   9/63
M on A.

SS 30 (21/30) [Holsworthy] 10 by 10 Revision a 1905 c 1938 d 1938-58 f 1963
1 A   CPr   OPs   1950pc   -   ac   1.50   4BEoo3   R   7/50   8/50
3 B   CPz   OPz   1964mc   -   adf   6.64   4oooX3   6/64   7/64

SS 31 (21/31) [Bradworthy] 10 by 10 Revision a 1904-05 c 1958 f 1960
1 A   CPr   OPs   1950pc   -   a   6.50   4BEoo3   8/50   8/50
3 A   CPu   1950pc   -   a   6.50   4BEoo3
3 B   CPz   OPz   1964mc   -   acf 6.64   4oooX3   8/64   9/64

SS 32 (21/32) [Clovelly] 10 by 10 Revision a 1904 c 1958 f 1963
1 A   CPr   OPs   1950pc   -   a   6.50   4BEoo3   8/50   8/50
3 B   CPz   OPz   1964mc   -   acf 6.64   4oooX3   6/64   5/64
M on A.

SS 40 (21/40) [Black Torrington] 10 by 10 Revision a 1905 c 1958 f 1961
1 A   CPr   OPr   nd:po   -   -   6.49   4BEoo3   R   3/50   4/50
3 B   CPx,z OPx   1962mc   -   acf 6.62   4oooX3   12/62 12/62

SS 41 (21/41) [Great Torrington (West)] 10 by 10 Revision a 1903-32 c 1938-57 f 1961
1 A   CPr   OPr   nd:po   -   -   1.50   4BEoo3   RB 3/50   4/50
3 B   CPy   OPy   1963mc   -   acf 6.63   4oooX3   11/63 11/63

SS 42 (21/42) [Bideford] 10 by 10 Revision a 1903-32 c 1938 d 1958 f 1961-62
1 A   CPs   OPs   1950pc   -   ac   6.50   4BEoo3   R   8/50   9/50
3 B   CPy,z OPy   1963mc   -   adf   6.63   4oooX3   6/63   6/63
AA 1.6.50/NT 31.7.50 on A(A), 3/-, OSPR 9/50. M on A.

SS 43 (21/43) [Braunton] 10 by 10 Revision a 1903-32 c 1938 d 1938-58
1 A   CPr   OPs   1950pc   -   ac   6.50   4BEoo3   R   0   7/50   9/50
3 A   CPt   1950pc   -   ac   6.50   4BEoo3   R   0
3 B   CPx,z OPx   1961mc   -   ad   6.61   4oooX3   1   5/61   5/61   1
AA 1.6.50/NT 2.6.50 on A(A), 3/-, OSPR 9/50. M on A. 1. Chivenor Airfield.

1	2	3	4	5	6	7	8	9	10	11	12	13	14

SS 44 (21/44) [Woolacombe] 10 by 10 No revision dates listed
1 A   CPr   OPs   nd:po   -   -   1.50 4BEoo3   R   8/50   10/50
3 A   CPu   1950pc   -   -   1.50 4BEoo3   R
M on A. Superseded in 1960 by Sheet 856 in the [Second Series] Provisional Edition.

SS 48 (21/48) **& Part of SS 38** *Rhossili* 12 by 10 Revision a 1913-14
1 A   CPr   OPr   nd:po   -   -   6.48 4BEoo3   7/48   8/48
1 B   CPr   nd:po   -   -   6.48 4BEoo3
3 B   CPt   1948pc   1949c   -   6.48 4BEoo3
3 B/   CP?w,y,z.z   1948mc   1960mc   a   6.48 4BooX3

SS 49 (21/49) **& Part of SS 39** *Llangennith** 12 by 10 Revision a 1913 c 1938-53
1 A   CPr   OPr   1948po   -   -   6.48 4BE4X3   4/48   6/48
3 B   CPv   OPv   1956pc   -   ac   6.56 4oooX3   7/56   9/57
3 B/   CP?x,z.z   1956mc   1961mc   ac   6.56 4oooX3
AA 1.3.48 on A(A), 3/-, OSPR 6/48.

SS 50 (21/50) [Hatherleigh] 10 by 10 Revision a 1904 c 1957 f 1961
1 A   CPr   OPr   nd:po   -   -   1.49 4BEoo3   R   12/49 1/50
3 B   CPy,z   OPy   1963mc   -   acf   6.63 4oooX3   11/63 11/63

SS 51 (21/51) [Great Torrington (East)] 10 by 10 Revision a 1904 c 1957-58 f 1961
1 A   CPr   OPr   nd:po   -   -   1.50 4BEoo3   R   3/50   4/50
3 B   CPy   OPy   1963mc   -   acf   6.63 4oooX3   6/63   6/63

SS 52 (21/52) [Newton Tracey] 10 by 10 Revision a 1903-04 c 1938-58 f 1961
1 A   CPr   OPr   nd:po   -   -   1.50 4BEoo3   R   3/50   4/50
3 B   CPx,y   OPx   1962mc   -   acf   6.62 4oooX4   10/62 12/62

SS 53 (21/53) [Barnstaple] 10 by 10 Revision a 1903-32 c 1938-58
1 A   CPr   OPr   nd:po   -   -   1.49 4BEoo3   R   0   5/50   6/50
2 A   CPt   1950pc   1952rpb -   1.49 4BooX3   R   0
3 B   CPx,z   OPx   1961mc   -   ac   6.61 4oooX3   1   6/61   5/61
AA 1.5.50 on A(A), 3/-, OSPR 6/50. M on A.

SS 54 (21/54) [Ilfracombe] 10 by 10 No revision dates listed
1 A   CPr   OPr   nd:po   -   -   1.50 4BEoo3   R   5/50   7/50
3 A   CPu   1950pc   -   -   1.50 4BEoo3   R
AA 1.3.50/NT 10.5.50 on A(A), 3/-, OSPR 7/50. M on A. Superseded in 1960 by Sheet 856 in the [Second Series] Provisional Edition.

SS 58 (21/58) **& Part of SS 68** *Oxwich Bay* 15 by 10 Revision a 1913-14 c 1947-48
1 A   CPr   OPr   1947po   -   -   1.48 4BE2X3   R   8/47   11/47
1 B   CPr   nd:po   -   -   1.48 4BEoo3   R
3 B   OPs   1947pc   1949c   -   1.48 4BEoo3   R
3 B/   CP?w.z   1947pc   1958mc   ac   1.48 4BooX3
AA 1.6.47 on A(A), 3/-, OSPR 11/47; AA 1.11.53 on B(B), 4/-, OSPR 2/55. M on B.

SS 59 (21/59) *Llanelly (South)** 10 by 10 Revision a 1913-36
1 A   CPr   1947po   -   -   6.48 4BE4X3   R   0   11/47 -
1 B   OPr   1947po   -   -   6.48 4BE4X3   R   0   -   1/48
1 C   CPr   nd:po   -   -   6.48 4BEoo3   R   0
3 C   OPt   1949pc   -   -   6.48 4BEoo3   R   0
3 C/   CPw,z,0z   1949pc   1958mc   a   6.48 4BooX3   0
AA 1.4.47 on B(A), 3/-, OSPR 1/48; AA 1.11.53 on C(B), 4/-, OSPR 8/54. M on C.

SS 60 (21/60) *Winkleigh* 10 by 10 Revision a 1904 c 1957 f 1961
1 A   CPr   OPr   nd:po   -   -   1.49 4BEoo3   0   12/49 1/50
3 B   CPz   OPz   1964mc   -   acf   6.64 4oooX3   3   4/64   4/64

211

1	2	3	4	5	6	7	8	9	10	11	12	13	14

SS 61 (21/61) *Chulmleigh**   10 by 10   Revision a 1903-04 c 1957 f 1961
| 1 | A | CPr | OPr | nd:po | - | - | 6.49 | 4BEoo3 | R | 0 | 3/50 | 5/50 |
| 3 | B | CPx,z | OPx | 1963mc | - | acf | 6.63 | 4oooX3 | | 3 | 1/63 | 1/63 |

SS: Soil, Land Use Capability, coloured, 1980 on B.

SS 62 (21/62) *Chittlehampton**   10 by 10   Revision a 1903 c 1957 f 1961
| 1 | A | CPr | OPr | nd:po | - | - | 6.48 | 4BEoo3 | R | | 3/50 | 4/50 |
| 3 | B | CPy,z,0z | OPy | 1963mc | - | acf | 6.63 | 4oooX3 | | | 11/63 | 11/63 |

SS 63 (21/63) *Brayford*   10 by 10   Revision a 1903 c 1957 f 1960
| 1 | A | CPr | OPr | nd:po | - | - | 1.49 | 4BEoo3 | R | | 3/50 | 4/50 |
| 3 | B | CPy,z | OPy | 1963mc | - | acf | 6.63 | 4oooX3 | | | 9/63 | 9/63 |

M on A. SS: Soil, Land Use Capability, coloured, 1981 on B.

SS 64 (21/64) *Parracombe*   10 by 10   Revision a 1903-05 c 1938-58
1	A	CPr	OPr	nd:po	-	-	1.50	4BEoo3	B		3/50	5/50
3	A	CPu		1950pc	-	-	1.50	4BEoo3	B			
3	B	CPx,z,0z	OPx	1962mc	-	ac	6.62	4oooX3			8/62	8/62

M on A.

SS 69 (21/69) *Swansea*   10 by 10   Revision a 1913-36 c 1935-48 f 1959 g 1969 h 1974
1	A	CPr	OPr	1947po	-	-	1.48	4BE2X3	R		8/47	11/47
1	B	CPs		1951pc	-	ac	6.51	4BEoo4	R		1/52	-
3	B		OPs	1951pc	-	ac	1.52	4Booo4	R		-	8/54
3	B/	CP?w,x		1951mc	1958c	acf	1.52	4BooX4				
3	B//	CPz		1951^{c}ms	nd:ch	acg	1.52	4BooX4				
3	B///*	CFz	OFz	1951^{c}ms	nd:r	ach	1.52	4BooX4			7/74	6/74

AA 1.6.47 on A(A), 3/-, OSPR 11/47; AA 1.11.53 on B(B), 4/-, OSPR 8/54. M on B.

SS 70 (21/70) *Copplestone*   10 by 10   Revision a 1903 c 1938-57 f 1961
| 1 | A | CPr | OPr | nd:po | - | - | 6.49 | 4BEoo3 | R | | 1/50 | 1/50 |
| 3 | B | CPz | OPz | 1964mc | - | acf | 6.64 | 4oooX3 | | | 5/64 | 5/64 |

SS 71 (21/71) *Chawleigh*   10 by 10   Revision a 1903-04 c 1957 f 1960
| 1 | A | CPr | OPr | nd:po | - | - | 6.49 | 4BEoo3 | | | 3/50 | 4/50 |
| 3 | B | CPy,z | OPy | 1963mc | - | acf | 6.63 | 4oooX3 | | | 9/63 | 9/63 |

SS 72 (21/72) *South Molton**   10 by 10   Revision a 1903 c 1957 f 1961
| 1 | A | CPr | OPr | nd:po | - | - | 1.49 | 4BEoo3 | R | | 4/50 | 5/50 |
| 3 | B | CPy,z,0z | OPy | 1963mc | - | acf | 6.63 | 4oooX3 | | | 9/63 | 9/63 |

SS 73 (21/73) *Simonsbath**   10 by 10   Revision a 1902-03 c 1957 f 1961
1	A	CPr	OPr	nd:po	-	-	6.49	4BEoo3			11/49	12/49
2	A	CPt		1949pc	1953rpb	-	6.49	4BooX3				
3	A/	CPx		1949mc	1960mc	a	6.49	4BooX3				
3	B	CPy,z	OPy,z	1963mc	-	acf	6.63	4oooX3			11/63	11/63

M on A.

SS 74 (21/74) **& Part of SS 75** *Lynton**   10 by 12   Revision a 1902-03 c 1938 d 1957-58 f 1961 g 1973
1	A	CPr	OPr	nd:po	-	-	6.49	4BEoo3	B		11/49	12/49
1	A	CPs		1949pc	1951	-	1.51	4BEoo3	B			
3	A/	CPw		1949pc	1960mc	ac	1.51	4BooX3				
3	B	CPy,z,z+	OPy	1963mc	-	adf	6.63	4BooX3			8/63	8/63
3	B/	CPz,0z	OPz	1963^{c}ms	nd:ch	adg	6.63	4BooX3				

M on A. SS: Soil, coloured, 1983 on B/.

1	2	3	4	5	6	7	8	9	10	11	12	13	14

SS 78 (21/78) *Port Talbot**   10 by 10   Revision a 1913-43 c 1938-55 f 1965 g 1968
j 1965 industrial k 1965 railway

1	A	CPr	OPr	nd:po	-	-	1.48	4BEoo3	R		8/48	9/48	
3	B	CPu	OPv	1956pc	-	ac	6.56	4oooX3			6/56	6/57	
3	B/	CPw		1956mc	1960	ac	6.56	4oooX3					
3	B/*	CPz	OPz	1956mc	1966ir	acfjk	6.56	4oooX3			5/66	5/66	1
3	B/*/*	CPz	OPz	1956ms	1969r	acgjk	6.56	4oooX3			3/69	3/69	

AA 1.8.48 on A(A), 3/-, OSPR 9/48. M on A. 1. The print code is illogical. A motorway is under construction.

SS 79 (21/79) *Neath*   10 by 10   Revision a 1913-36 c 1938-48 e 1948 f 1966 g 1970
j 1964 building k 1964 afforestation

1	A	CPr	OPr	1948po	-	-	6.48	4BE4X3	R		2/48	3/48	
1	B	CPr		1950pc	-	ace	6.50	4BEoo3	R				
3	B	CPu	OPt	1950pc	-	ace	6.50	4BEoo3	R				
3	B/*	CPx,z	OPx	1950mc	1961r	ace	6.50	4BooX3			6/61	6/61	
3	B/*/*	CPz	OPz	1950mc	1968r	acef	6.50	4BooX3			1/68	3/68	
3	B/*/*/*	CPz	OPz	1950^{c}ms	nd:bf	acegjk	6.50	4BooX3			5/71	6/71	

AA 1.4.47 on A(A), 3/-, OSPR 3/48; AA 1.11.53 on B(B), 4/-, OSPR 8/54. M on B, B/*.

SS 80 (21/80) *Crediton**   10 by 10   Revision a 1903-04 c 1938-58 f 1961

| 1 | A | CPr | OPr | nd:po | - | - | 1.48 | 4BEoo3 | | | 2/49 | 3/49 | |
| 3 | B | CPx,y,z | OPx | 1962mc | - | acf | 6.62 | 4oooX4 | | | 9/62 | 9/62 | |

SS 81 (21/81) *Witheridge*   10 by 10   Revision a 1903-04 c 1957-58 f 1961

| 1 | A | CPr | OPr | nd:po | - | - | 6.49 | 4BEoo3 | | | 3/50 | 4/50 | |
| 3 | B | CPx,y,z | OPx | 1962mc | - | acf | 6.62 | 4oooX3 | | | 5/62 | 1/62 | |

SS 82 (21/82) *Anstey Devonshire**   10 by 10   Revision a 1902-03 c 1958 f 1961

1	A	CPr	OPr	nd:po	-	-	6.48	4BEoo3	RB		11/49	12/49	
3	A/	CP?v,w		1949pc	1957mc	a	6.48	4BooX3	B				
3	B	CPx,z	OPx,z	1962mc	-	acf	6.62	4oooX3			5/62	2/62	

SS 83 (21/83) *Exford**   10 by 10   Revision a 1902-03 c 1958 f 1961

1	A	CPr	OPr	nd:po	-	-	1.49	4BEoo3	B		11/49	12/49	
2	A	CPu		1949pc	-	-	1.52	4Booo3	B				
3	B	CPx,z,0z	OPx	1962mc	-	acf	6.62	4oooX4			4/62	1/62	

M on A.

SS 84 (21/84) *Porlock*   10 by 10   Revision a 1902-28 c 1958 f 1961

1	A	CPr	OPr	nd:po	-	-	1.49	4BEoo3	B		11/49	12/49	
2	B	CPt		1949pc	1953c	-	6.53	4BooX3	B				
3	C	CPx,z,0z	OPx	1962mc	-	acf	6.62	4oooX4			6/62	4/62	

M on A.

SS 87 (21/87) **& Part of SS 77** *Porthcawl**   12 by 10   Revision a 1913-41

| 1 | A | CPr | OPr | nd:po | - | - | 6.48 | 4BEoo3 | R | 2 | 6/49 | 7/49 | |
| 3 | A/ | CPw,z.z | | 1949mc | 1959mc | a | 6.48 | 4BooX3 | | 2 | | | |

AA 1.6.49 on A(A), 3/-, OSPR 7/49. M on A/.

SS 88 (21/88) *Pyle**   10 by 10   No revision dates listed

1	A	CPr	OPr	nd:po	-	-	6.48	4BEoo3	R	2	8/49	9/49	
3	A	CPu.x		1949pc	-	-	6.48	4BEoo3	R	2			
3	A/	CPz	OPz	1949ms	-	-	6.48	4BooX3		2			

AA 1.8.49 on A(A), 3/-, OSPR 9/49. M on A.

SS 89 (21/89) *Maesteg*   10 by 10   Revision a 1914-42 c 1945-48 f 1969

1	A	CPr	OPr	nd:po	-	-	6.48	4BEoo3	RB		10/49	11/49	
2	A	CPs		1949pc	1952	-	1.52	4Booo3	RB				
3	A/	CP?w,y,z		1949pc	1958mc	ac	1.52	4BooX3	RB				
3	A//	CPz	OPz	1949^{c}ms	nd:ch	acf	1.52	4BooX3	B				

AA 1.8.49 on A(A), 3/-, OSPR 11/49. M on A.

1	2	3	4	5	6	7	8	9	10	11	12	13	14

SS 90 (21/90) *Silverton**   10 by 10   Revision a 1903-04 c 1957-58 f 1961 g 1965
1	A	CPr	OPr	nd:po	-	-	1.49	4BEoo3	R		5/49	5/49	
3	A	CPu		1949pc	-	-	1.49	4BEoo3	R				
3	B	CPx	OPx	1963mc	-	acf	6.63	4oooX3			1/63	1/63	
3	B/	CPz,0z	OPz	1963mc	1966ch	acg	6.63	4oooX3					

AA 1.4.49/NT 8.4.49 on A(A), 3/-, OSPR 5/49.

SS 91 (21/91) *Tiverton Devonshire**   10 by 10   Revision a 1903-33 c 1938-58 f 1961
| 1 | A | CPr | OPr | nd:po | - | - | 6.49 | 4BEoo3 | R | | 4/50 | 5/50 | |
| 3 | B | CPx,y,z | OPx | 1962mc | - | acf | 6.62 | 4oooX3 | | | 5/62 | 2/62 | |

AA 1.1.50 on A(A), 3/-, OSPR 5/50.

SS 92 (21/92) *Dulverton**   10 by 10   Revision a 1902-03 c 1957-58 f 1961 g 1966
1	A	CPr	OPr	nd:po	-	-	1.50	4BEoo3	RB		1/50	1/50	
3	A/	CPw		1950mc	1959mc	a	1.50	4BooX3	B				
3	B	CPx,y,z	OPx	1962mc	-	acf	6.62	4oooX4			5/62	2/62	
3	B/	CPz,0z	OPz	1962mc	1966ch	acg	6.62	4oooX4					

SS 93 (21/93) *Brendon Hills (West)*   10 by 10   Revision a 1902 c 1958 f 1961; j 1981 selected
1	A	CPr	OPr	nd:po	-	-	6.49	4BEoo3	B		11/49	12/49	
3	A/	CPv		1949pc	1957mc	a	6.49	4BooX3	B				
3	B	CPx,z,0z	OPx,z	1962mc	-	acf	6.62	4oooX4			3/62	1/62	
3	B/*	CFz	OFz	1981^{c}ms	nd:v	acfj	6.62	4oooX4			12/81	12/81	1

M on A. 1. Wimbleball Lake Reservoir.

SS 94 (21/94) *Minehead*   10 by 10   Revision a 1902-36 b 1902-28 c 1938 d 1936-58 f 1961
1	A	CPr	OPr	nd:po	-	-	1.49	4BEoo3	RB		12/49	2/50	
3	A	CPt		1949pc	-	-	1.49	4BEoo3	RB				
3	A/	CPw		1949mc	1959mc	ac	1.49	4BooX3	B				
3	B	CPy,z	OPy,z	1963mc	-	bdf	6.63	4oooX4			4/63	4/63	

AA 1.8.49/NT 22.3.49 on A(A), 3/-, OSPR 2/50. M on A.

SS 96 (21/96) *Llantwit Major**   10 by 10   Revision a 1914 c 1938-48 f 1970
1	A	CPr	OPr	1948po	-	-	1.48	4BEoo3	R	0	7/48	8/48	
1	B	CPr		nd:po	-	-	1.48	4BEoo3	R	0			
3	B	CPx		nd:po	-	-	1.48	4BEoo3	R	0			
3	B/	CPz		nd:po	-	-	1.48	4BEoo3		0			
3	B//*	CPz	OPz	1948^{c}ms	nd:b	acf	1.48	4BooX3		3	6/71	7/71	

SS 97 (21/97) *Bridgend (South) Glamorgan**   10 by 10   Revision a 1914 c 1938-48 f 1967
1	A	CPr		nd:po	-	-	1.49	4BEoo3	R	0	10/48	-	
1	B	CPr	OPr	nd:po	-	-	1.49	4BEoo3	R	0	-	12/48	
	C												
	D												
3	D/	CP?w,y		1950mc	1959mc	ac	6.50	4BooX3		0			
3	D//*	CPz	OPz	1950mc	1968r	acf	6.50	4BooX3		0	1/68	1/68	

AA 1.10.48 on B(A), 3/-, OSPR 12/48.

SS 98 (21/98) *Pencoed**   10 by 10   Revision a 1914-42 c 1938-48
1	A	CPr	OPr	1948po	-	-	1.48	4BE4X3	RB		10/48	1/48	
1	B	CPs		1950pc	-	ac	6.50	4BEoo3	R				
2	B		OPs	1950pc	-	ac	6.50	4BooX3	R		-	3/53	
3	B/	CP?w,x,z		1950mc	1959mc	ac	6.50	4BooX3					

AA 1.6.46 on A(A), 3/-, OSPR 1/48.

SS 99 (21/99) *Rhondda**   10 by 10   Revision a 1913-40 c 1948
| 1 | A | CPr | OPr | 1947po | - | - | 1.48 | 4BE2X3 | RB | | 9/47 | 11/47 | 1 |
| 3 | B | CPu | OPv | 1955pc | - | ac | 6.55 | 4oooX3 | | | 1/56 | 6/57 | |

214

1	2	3	4	5	6	7	8	9	10	11	12	13	14
3	B/		CPw,y,z	1955mc	1960mc	ac	6.55	4oooX3					

AA 1.6.46 on A(A), 3/-, OSPR 11/47. 1. No fewer than eleven place names are in the large Times Roman alphabet apparently reserved for village names.

ST 00 (31/00) *Cullompton*   10 by 10   Revision a 1903-04 c 1957-58 f 1961 g 1970
1	A	CPr	OPr	nd:po	-	-	1.48	4BEoo3	R		2/49	3/49
	A/											
3	A//	CPw		1949mc	1959mc	a	1.48	4BooX3				
3	B	CPx,z	OPx	1962mc	-	acf	6.62	4oooX3			12/62	12/62
3	B/*	CPz	OPz	1962^{c}ms	-	acg	6.62	4oooX3			5/71	6/71

ST 01 (31/01) *Willand*   10 by 10   Revision a 1903-29 c 1957-58
1	A	CPr	OPr	nd:po	-	-	1.48	4BEoo4	R		10/49	11/49
3	A	CPu		1949pc	-	-	1.48	4BEoo4	R			
3	B	CPx,z	OPx	1961mc	-	ac	6.61	4oooX4			6/61	6/61

ST 02 (31/02) *Wiveliscombe*   10 by 10   Revision a 1902-30 c 1957-58 f 1961
1	A	CPr	OPr	nd:po	-	-	6.48	4BEoo3	RB		10/49	11/49
3	B	CPx,z	OPx,z	1961mc	-	acf	6.61	4oooX4			9/61	9/61

M on B.

ST 03 (31/03) *Brendon Hills (East)*   10 by 10   Revision a 1903 c 1938 d 1938-61 f 1961
1	A	CPr	OPr	nd:po	-	-	6.48	4BEoo3	B		11/49	12/49
3	A/	CPw		1949mc	1960mc	ac	6.48	4BooX3	B			
3	B	CPx,y,z	OPx,z	1962mc	-	adf	6.62	4oooX4			3/62	1/62

ST 04 (31/04) *Watchet*   10 by 10   Revision a 1903 c 1938 d 1938-58 f 1961
1	A	CPr	OPr	nd:po	-	-	6.49	4BEoo3	R		12/49	12/49
3	A/	CPw		1949pc	1960mc	ac	6.49	4BooX3				
3	B	CPx,y,z	OPx,y	1962mc	-	adf	6.62	4oooX3			1/62	1/62

ST 06 (31/06) *Rhoose*   10 by 10   Revision a 1914-43 c 1938-50 f 1964
1	A	CPr	OPr	nd:po	-	-	6.48	4BEoo3	R	00	8/48	9/48	
1	B	CPs		1951pc	-	ac	1.51	4BEoo3	R	00	1/52	-	
2	B		OPs	1951pc	-	ac	1.51	4Booo3	R	00	-	8/51	
3	B/	CPw		1951mc	1959c	ac	1.51	4BooX3		03			1
3	B//	CPz	OPz	1951mc	1966ch	acf	1.51	4BooX3		33			2

AA 1.7.48 on A(A), 3/-, OSPR 9/48; AA 1.5.52 on B(B), 4/-, OSPR 10/52. M on B. 1. Cardiff (Rhoose) Airport. 2. Glamorgan (Rhoose) Airport.

ST 07 (31/07) *Bonvilston*   10 by 10   Revision a 1914-42 c 1938-48 f 1966
1	A	CPr	OPr	1948po	-	-	1.48	4BEoo3	R		7/48	8/48
1	B	CPs	OP?	1951pc	-	ac	1.51	4BEoo3	R		6/51	8/51
3	B/	CP?v,y		1951pc	1957mc	ac	1.51	4BooX3				
3	B//*	CPz	OPz	1951mc	1967r	acf	1.51	4BooX3			8/67	9/67

M on B.

ST 08 (31/08) *Pontypridd*   10 by 10   Revision a 1914-43 c 1948 f 1970
1	A	CPr	OPr	1947po	-	-	1.48	4BE4X3	R		10/47	12/47
1	B	CPr		nd:po	-	-	1.48	4BEoo3	R			
3	B/	CPw,y		1950pc	1958mc	ac	1.48	4oooX3				
3	B//	CPz	OPz	1950^{c}ms	nd:ch	acf	1.48	4oooX3				

AA 1.3.47 on A(A), 3/-, OSPR 12/47. M on B.

ST 09 (31/09) *Mountain Ash*   10 by 10   Revision a 1914-15 c 1938-48
1	A	CPr	OPr	1947po	-	-	1.48	4BE2X3	RBC		8/47	11/47
2	A	CPt		1947pc	-	-	1.48	4BooX3	RBC			
3	B	CPv	OPv	1956pc	-	ac	6.56	4oooX3	C		7/56	8/57
3	B/	CPw,z		1956mc	1958mc	ac	6.56	4oooX3	C			

AA 1.6.47/NT 1.9.47 on A(A), 3/-, OSPR 11/47.

1	2	3	4	5	6	7	8	9	10	11	12	13	14

ST 10 (31/10) *Honiton* 10 by 10 Revision a 1903 c 1938-57 f 1961 g 1967
1	A	CPr	OPr	nd:po	-	-	6.48	4BEoo3	R	00	8/49	9/49
3	A	CPu		1949pc	-	-	6.48	4BEoo3	R	00		
3	B	CPx,z	OPx	1962mc	-	acf	6.62	4oooX3		33	9/62	9/62
3	B/*	CPz	OPz	1962ms	1968r	acg	6.62	4oooX3		33	3/68	2/68

AA 1.8.49 on A(A), 3/-, OSPR 9/49. SS: Soil, Land Use Capability, outline, 1971 on B/*.

ST 11 (31/11) *Blackdown Hills (West)* 10 by 10 Revision a 1903-29 c 1957 f 1961
1	A	CPr	OPr	nd:po	-	-	1.49	4BEoo3	R	0	10/49	11/49
3	A	CPt		1949pc	-	-	1.49	4BEoo3	R	0		
3	B	CPx,z	OPx	1961mc	-	acf	6.61	4oooX4		3	10/61	10/61

ST 12 (31/12) *Wellington Somerset* 10 by 10 Revision a 1902-30 c 1938-57 f 1961
1	A	CPr	OPr	nd:po	-	-	6.49	4BEoo3	R		10/49	11/49
3	A	CPu		1949pc	-	-	6.49	4BEoo3	R			
3	B	CPy,z	OPy	1963mc	-	acf	6.63	4oooX4			5/63	5/63

ST 13 (31/13) *Quantock Forest* 10 by 10 Revision a 1902-29 c 1957 f 1961
1	A	CPr	OPr	nd:po	-	-	6.49	4BEoo3	R		11/49	12/49
2	A	CPu		1949pc	-	-	6.49	4BooX3	R			
3	A/	CPw		1949mc	1959mc	a	6.49	4BooX3				
3	B	CPx,z	OPx	1962mc	-	acf	6.62	4oooX4			1/62	1/62

ST 14 (31/14) *Quantoxhead* 10 by 10 Revision a 1902-28 c 1936-57 f 1961
1	A	CPr	OPr	nd:po	-	-	6.49	4BEoo3			11/49	12/49
1	B	CPs		1949pc	1951c	a	1.51	4BEoo3				
3	C	CPx,z,0z	OPx	1962mc	-	acf	6.62	4oooX4			6/62	7/62

ST 16 (31/16) *Barry* 10 by 10 Revision a 1915-43 c 1938-50 f 1966
1	A	CPr	OPr	1948po	-	-	1.48	4BE2X3	R		12/47	3/48
3	B	CPu	OPv	1956pc	-	ac	6.56	4oooX3			4/56	5/57
3	B/	CP?w,y		1956mc	1960mc	ac	6.56	4oooX3				
3	B//	CPz	OPz	1956mc	1967ch	acf	6.56	4oooX3				

AA 1.6.46 on A(A), 3/-, OSPR 3/48. M on A.

ST 17 (31/17) *Cardiff (South West)* 10 by 10 Revision a 1915-43 c 1938-53 f 1968 g 1972
j 1964 building
1	A	CPr	OPr	1947po	-	-	1.48	4BE2X3	R		9/47	11/47
2	A	CPt	OPt	1947pc	1953rpb	-	1.48	4BooX3	R			
3	B	CPv		1956pc	-	ac	6.56	4oooX3			6/56	-
3	B/	CPv.x,z		1956pc	1958mc	ac	6.56	4oooX3				
3	B//*	CPz	OPz	1956ms	1970b	acfj	6.56	4oooX3			2/70	3/70
3	B//*/*	CPz	OPz	1956[c]ms	nd:r	acgj	6.56	4oooX3			5/73	6/73

AA 1.7.47 on A(A), 3/-, OSPR 11/47; AA 1.11.53 on A(B), 4/-, OSPR 6/54. M on A.

ST 18 (31/18) *Caerphilly* 10 by 10 Revision a 1915-43 c 1938-48 f 1960 g 1970
1	A	CPr	OPr	1947po	-	-	1.48	4BE4X3	R		12/47	1/48
1	B	CPr		nd:po	-	-	1.48	4BEoo3	R			
3	B		OPs	1947pc	1949c	-	1.48	4BEoo3	R			
3	B/											
3	B//	CPx,z		1947mc	1961ch	af	1.48	4BooX3				
3	B///*	CPz	OPz	1947[c]ms	nd:b	acg	1.48	4BooX3			11/71	1/72

AA 1.3.47 on A(A), 3/-, OSPR 1/48; AA 1.11.53 on B(B), 4/-, OSPR 11/54. M on B, B//.

ST 19 (31/19) *Gelligaer* 10 by 10 Revision a 1914-16 c 1938-48
1	A	CPr		1948po	-	-	6.48	4BE4X3	R		4/48	-
1	B [1]		OPr	1948po	-	-	6.48	4BE4X3	R		-	5/48
1	B [2]	CPs		1951pc	-	ac	6.51	4BEoo3	R		12/51	-
3	B/	CPw,z.z		1951mc	1960mc	ac	6.51	4BooX3				

AA 1.4.47 on B(A), 3/-, OSPR 5/48.

1	2	3	4	5	6	7	8	9	10	11	12	13	14

ST 20 (31/20) *Yarcombe*   10 by 10   Revision a 1901-29 c 1938-57
1	A	CPr	OPr	nd:po	-	-	1.49	4BEoo3	RB		9/49	10/49	
1	A	CPs		1949pc	1951	-	1.51	4BEoo3	RB				
3	B	CPw,x,z,0z	OPv	1958mc	-	ac	6.58	4oooX3			1/59	12/58	

ST 21 (31/21) *Blackdown Hills (East)*   10 by 10   Revision a 1901-29 c 1957
1	A	CPr	OPr	nd:po	-	-	6.48	4BEoo3	B	0	10/49	11/49	
2	B	CPt		1949pc	1953rpb	-	6.48	4BooX3	B	0			
3	C	CPw,x,z	OPv	1959mc	-	ac	6.59	4oooX3		0	4/59	5/59	

ST 22 (31/22) *Taunton*   10 by 10   Revision a 1902-29 b 1901-29 c 1938 d 1938-57
                              f 1966 g 1976
1	A	CPr	OPr	nd:po	-	-	1.49	4BEoo3	RB		10/49	12/49	
1	B	CPs		1949pc	1950c	ac	6.49	4BEoo3	RB				
3	C	CPw,x,y	OPv	1959mc	-	ad	6.59	4oooX3			4/59	7/59	
3	C/	CPz	OPz	1959mc	1966ch	adf	6.59	4oooX3					
3	C//*	CFz	OFz	1959^{c}ms	nd:r	adg	6.59	4oooX3					1

AA 1.10.49 on A(A), 3/-, OSPR 12/49. 1. OSPR not found.

ST 23 (31/23) *Bridgwater (West)*   10 by 10   Revision a 1902-29 c 1938-55 f 1967; j 1964 reservoir
1	A	CPr	OPr	nd:po	-	-	6.49	4BEoo3			11/49	1/50	
3	B	CPw,x	OPv	1959mc	-	ac	6.59	4oooX3			6/59	9/59	
3	B/*	CPz,0z	OPz	1959mc	1967rv	acfj	6.59	4oooX3			11/67	11/67	1

AA 1.11.49 on A(A), 3/-, OSPR 1/50. 1. Hawkridge Reservoir.

ST 24 (31/24) *Stert Point*   10 by 10   Revision a 1902-29 c 1938-56 f 1966
                                 j 1961 power station
1	A	CPr	OPr	nd:po	-	-	1.50	4BEoo3			11/49	12/49	
			OP?								-	8/57	
3	B	CPw,x	OPv	1959mc	-	ac	6.59	4oooX3			9/59	10/59	
3	B/*	CPz,0z	OPz,0z	1959mc	1967p	acfj	6.59	4oooX3			1/67	2/67	1

1. Hinkley Point Nuclear Power Station.

ST 26 (31/26) *The Holmes*   10 by 10   Revision a 1898-1902 c 1949-56
| 1 | A | CPr | OPr | 1947po | - | - | 1.47 | 4BE2X3 | | | 1/47 | 5/47 | |
| 3 | B | CPw,z | OPv | 1958mc | - | ac | 6.58 | 4oooX3 | | | 1/59 | 1/59 | |

AA 1.12.46 on A(A), 3/-, OSPR 5/47.

ST 27 (31/27) *Cardiff (East)*   10 by 10   Revision a 1915-43 c 1947-53 f 1972; j 1970 industrial
1	A	CPr	OPr	1947po	-	-	1.47	4BE2X3	R	0	1/47	4/47	
3	A		OPt	1947pc	-	-	1.47	4BEoo3	R	0			
3	B	CPu	OPv	1956pc	-	ac	6.56	4oooX3		3	5/56	3/58	1
3	B/	CPw,z		1956mc	1960mc	ac	6.56	4oooX3		3			
3	B//*	CPz	OPz	1956^{c}ms	nd:ir	acfj	6.56	4oooX3		0	2/73	6/73	

AA 1.12.46 on A(A), 3/-, OSPR 4/47; AA 1.11.53 on A(B), 4/-, OSPR 8/54. M on A. 1. Cardiff (Pengam Moors) Airport (Ministry of Civil Aviation).

ST 28 (31/28) *Rogerstone*   10 by 10   Revision a 1915-47 c 1938-49 f 1969 g 1973
                                j 1964 building
1	A	CPr		1947po	-	-	1.48	4BE4X3	R		11/47	-	
1	B [1]		OPr	1947po	-	-	1.48	4BE4X3	R		-	1/48	
1	B [2]	CPr		1950pc	-	ac	6.50	4BEoo3	R				
3	B [2]		OPs	1950pc	-	ac	6.50	4BEoo3	R				
3	B/	CPw.z		1950mc	1961ch	ac	6.50	4BooX3					
3	B//*	CPz	OPz	1959ms	1970br	acfj	6.50	4BooX3			1/71	3/71	
3	B//*/*	CPz	OPz	1959^{c}ms	nd:r	acgj	6.50	4BooX3			6/74	6/74	

AA 1.3.47 on B(A), 3/-, OSPR 1/48; AA 1.11.53 on B(B), 4/-, OSPR 11/54. M on B [2], B/.

1	2	3	4	5	6	7	8	9	10	11	12	13	14

ST 29 (31/29) *Risca*   10 by 10   Revision a 1916-17 c 1938-49 f 1967
1	A	CPr	OPr	1947po	-	-	1.48	4BE4X3	R		11/47	1/48	
1	B	CPs		1950pc	-	ac	6.50	4BEoo3	R				
3	B/	CPx		1950mc	-	ac	6.50	4BooX3					
3	B//	CPz	OPz	1950mc	1968ch	acf	6.50	4BooX3					

AA 1.7.47 on A(A), 3/-, OSPR 1/48.

ST 30 (31/30) *Chard*   10 by 10   Revision a 1901-38 c 1957
1	A	CPr	OPr	nd:po	-	-	1.49	4BEoo3	RB		9/49	10/49	
2	B	CPt		1949pc	1953c	-	6.53	4BooX3	B				
3	C	CPw,x,z,0z	OPv,x	1958mc	-	ac	6.58	4oooX3			11/58	1/59	

ST 31 (31/31) *Ilminster*   10 by 10   Revision a 1901-28 c 1957
| 1 | A | CPr | OPr | nd:po | - | - | 6.49 | 4BEoo3 | RB | 0 | 10/49 | 11/49 | |
| 3 | B | CPw,x.z,0z | OPv | 1958mc | - | ac | 6.58 | 4oooX3 | | 0 | 1/59 | 1/59 | |

ST 32 (31/32) *Curry*   10 by 10   Revision a 1901-38 c 1957
1	A	CPr	OPr	nd:po	-	-	6.48	4BEoo4	R		10/49	11/49	
1	A	CPs		1949pc	1951	-	1.51	4BEoo4	R				
3	B	CPw,x,z,0z	OPv	1958mc	-	ac	6.58	4oooX4			1/59	2/59	

ST 33 (31/33) *Bridgwater (East)*   10 by 10   Revision a 1902-29 c 1938-56 f 1961 g 1966 h 1977
1	A	CPr	OPr	nd:po	-	-	6.49	4BEoo3	R	0	11/49	1/50	
3	B	CPw	OPv	1960mc	-	ac	6.60	4oooX3		1	12/59	12/59	
3	B/	CPy		1960mc	1963ch	acf	6.60	4oooX3		1			
3	B//	CPz	OPz	1960mc	1966ch	acg	6.60	4oooX3		1			
3	B///*	CFz	OFz	1960^{c}ms	nd:r	ach	6.60	4oooX3		1	4/78	4/78	

AA 1.11.49/NT 17.1.48 on A(A), 3/-, OSPR 1/50.

ST 34 (31/34) *Burnham-on-Sea*   10 by 10   Revision a 1902-29 c 1938-56 f 1965 g 1978
                                                          j 1981 selected k SUSI
1	A	CPr	OPr	nd:po	-	-	1.50	4BEoo3	R		11/49	1/50	
3	B	CPw	OPv	1959mc	-	ac	6.59	4oooX3			10/59	10/59	
3	B/	CPz	OPz	1959mc	1966ch	acf	6.59	4oooX3					
3	B//*	CFz	OFz	1979^{c}ms	nd:r	acg	6.59	4oooX3			4/80	4/80	
3	B//*/*	CFz	OFz	1981^{c}m-s	nd:b	acgjk	6.59	4oooX3			1/82	1/82	

AA 1.10.49 on A(A), 3/-, OSPR 1/50.

ST 35 (31/35) **& Part of ST 25** *Brent Knoll* 15 by 10 Revision a 1902-30 c 1938-56 f 1975
1	A	CPr	OPr	nd:po	-	-	6.49	4BEoo3	R	0	12/49	1/50	
3	B	CPw,x,z.z	OPv.z	1960mc	-	ac	6.60	4oooX3		2	12/59	1/60	
3	B/*	CFz	OFz	1960^{c}ms	nd:r	acf	6.60	4oooX3		2	5/76	5/76	1

AA 1.8.49 on A(A), 3/-, OSPR 1/50. 1. Both coloured and outline editions were listed again in OSPR 9/76.

ST 36 (31/36) *Weston-super-Mare*   10 by 10   Revision a 1902-30 c 1938-39 d 1938-56 f 1975
1	A	CPr	OPr	1948po	-	-	1.48	4BE2X3	RB	0	2/48	3/48	
3	B	CPu		1956pc	-	ac	6.55	4oooX3	B	0	1/56	-	
3	C	CPw,x,z.z	OPv,z	1959mc	-	ad	6.59	4oooX3		3	10/59	9/59	1
3	C/*	CFz	OFz	1959^{c}ms	nd:r	adf	6.59	4oooX3		3	6/76	6/76	

AA 1.9.47/NT 13.2.48 on A(A), 3/-, OSPR 3/48. 1. Weston-super-Mare Aerodrome.

ST 37 (31/37) *Mouth of the Severn*   10 by 10   Revision a 1900-30 c 1938-49
1	A	CPr	OPr	1947po	-	-	6.47	4BE2X3			6/47	8/47	
1	B	CPs		1951pc	-	ac	1.51	4BEoo3			3/51	-	
3	B		OPs	1951pc	-	ac	1.51	4BEoo3			-	11/54	
3	B/	CP?w,y,z		1951mc	1960mc	ac	1.51	4BooX3					

AA 1.7.47 on A(A), 3/-, OSPR 8/47; AA 1.1.54 on B(B), 4/-, OSPR 11/54.

1	2	3	4	5	6	7	8	9	10	11	12	13	14

ST 38 (31/38) *Newport Monmouthshire*   10 by 10   Revision a 1916-37   c 1936-50   f 1964   g 1967
j 1964 industrial

1	A	CPr		1947po	-	-	1.48	4BE2X3	R		10/47	-	
1	B		OPr	1947po	-	-	1.48	4BE2X3	R		-	11/47	
3	C	CPu	OPv	1956pc	-	ac	6.56	4oooX3			4/56	6/57	
3	C/	CPw		1956mc	1960mc	ac	6.56	4oooX3					
3	C//*	CPz	OPz	1956mc	1965i	acfj	6.56	4oooX3			6/65	8/65	
3	C//*/*	CPz	OPz	1956ms	1968r	acgj	6.56	4oooX3			5/68	4/68	

AA 1.7.47 on B(A), 3/-, OSPR 11/47. M on B.

ST 39 (31/39) *Caerleon*   10 by 10   Revision a 1917-18   c 1938-49   f 1966   g 1970   h 1974
j 1966 building

1	A	CPr	OPr	1947po	-	-	1.48	4BE2X3	R		8/47	11/47	
1	B	CPs		1951pc	-	ac	1.51	4BEoo3	R		4/51	-	
3	B/	CP?v,x		1951pc	1956mc	ac	1.51	4BooX3					
3	B//	CPz	OPz	1951mc	1967ch	acf	1.51	4BooX3					
3	B///*	CPz	OPz	1951^{c}ms	nd:br	acgj	1.51	4BooX3			8/71	9/71	
3	B///*/*	CFz	OFz	1951^{c}ms	nd:r	achj	1.51	4BooX3			1/75	11/74	

AA 1.6.47 on A(A), 3/-, OSPR 11/47.

ST 40 (31/40) *Crewkerne*   10 by 10   Revision a 1901-28   c 1938-57

1	A	CPr	OPr	nd:po	-	-	6.48	4BEoo3	RB		11/49	12/49	
2	A	CPt		1949pc	1953rpb	-	6.48	4BooX3	RB				
3	B	CPw,x,z	OPv,x	1958mc	-	ac	6.58	4oooX3			1/59	3/59	

ST 41 (31/41) *South Petherton*   10 by 10   Revision a 1901-28   c 1938-57

1	A	CPr	OPr	nd:po	-	-	6.48	4BEoo3	RB		10/49	11/49	
2	A	CPt		1949pc	1952	-	1.52	4Booo3	RB				
3	B	CPw	OPv	1958mc	-	ac	6.58	4oooX3			2/59	3/59	
3	B/	CPw.z,0z		1958mc	1961mc	ac	6.58	4oooX3					

ST 42 (31/42) *Langport*   10 by 10   Revision a 1901-28   c 1957

| 1 | A | CPr | OPr | nd:po | - | - | 1.49 | 4BEoo3 | R | | 11/49 | 12/49 | |
| 3 | B | CPw,z | OPv | 1959mc | - | ac | 6.59 | 4oooX4 | | | 4/59 | 7/59 | |

ST 43 (31/43) *Street*   10 by 10   Revision a 1902-29   c 1938-56   f 1971

1	A	CPr	OPr	nd:po	-	-	1.49	4BEoo3	R		12/49	1/50	
3	B	CPw	OPv	1959mc	-	ac	6.59	4oooX3			8/59	7/59	
3	B/	CPz		1959mc	-	ac	6.59	4oooX3					
3	B//*	CPz,0z	OPz	1959^{c}ms	nd:r	acf	6.59	4oooX3			5/72	8/72	

ST 44 (31/44) *Wedmore*   10 by 10   Revision a 1902-29   c 1938-56

1	A	CPr	OPr	nd:po	-	-	1.50	4BEoo3	R		12/49	1/50	
1	A	CPs		1949pc	1951	-	1.51	4BEoo3	R				
3	B	CPw,x,z	OPv	1959mc	-	ac	6.59	4oooX3			4/59	5/59	

ST 45 (31/45) *Cheddar*   10 by 10   Revision a 1902-29   c 1956   f 1967

1	A	CPr	OPr	nd:po	-	-	6.49	4BEoo3	RB		11/49	1/50	
1	A	CPs		1949pc	1951	-	6.51	4BEoo3	RB				
3	B	CPw,x,z	OPv	1959mc	-	ac	6.59	4oooX4			6/59	9/59	
3	B/	CPz,0z	OPz	1959ms	1968ch	acf	6.59	4oooX4					

GS: Solid and Drift, 1969 on B. Another edition, 1983.

ST 46 (31/46) *Congresbury*   10 by 10   Revision a 1902-30   c 1949-56   f 1966   g 1974

1	A	CPr	OPr	1947po	-	-	6.47	4BE2X3	R	0	7/47	10/47	
1	B	CPr		nd:po	-	-	6.47	4BEoo3	R	0			
3	C	CPw,x	OPv	1959mc	-	ac	6.59	4oooX4		3	9/59	7/59	
3	C/	CPz	OPz	1959mc	1966ch	acf	6.59	4oooX4		3			
3	C//*	CFz	OFz	1959^{c}ms	nd:r	acg	6.59	4oooX4		3	4/76	11/75	1

AA 1.2.47 on A(A), 3/-, OSPR 10/47. 1. The outline edition also appears in OSPR 4/76.

219

1	2	3	4	5	6	7	8	9	10	11	12	13	14

ST 47 (31/47) *Portishead*   10 by 10   Revision a 1899-1930 c 1938-50 f 1974

1	20046	CPr		1946po		-	1.47	4BE2X3	R		9/46	-	
1	2546		OPr	1946po		-	1.47	4BE2X3	R		-	12/46	
3	A		OPt	1946pc		-	1.47	4BEoo3	R				1
3	B	CPv	OPv	1956pc		ac	6.56	4oooX3			7/56	5/57	
3	B/	CPx,z.z	OPx	1956mc	1961ch	ac	6.56	4oooX3					
3	B//*	CFz	OFz	1956^{c}ms	nd:r	acf	6.56	4oooX3			9/75	9/75	

AA 1.7.46 on 2546(2546), 3/-, OSPR 12/46; AA 1.1.54 on A(A), 4/-, OSPR 8/54. M on 20046. GS: *Clevedon - Portishead*, Solid and Drift, 1968 on B/. 1. Used for the Administrative Areas map, so far not recorded as an outline edition.

ST 48 (31/48) *Rogiet*   10 by 10   Revision a 1899-1919 c 1945-49 f 1968

1	A	CPr		1947po		-	1.48	4BE4X3	R		11/47	-
1	B		OPr	1947po		-	1.48	4BE4X3	R		-	1/48
1	C	CPs		1950pc		ac	6.50	4BEoo3	R			
3	C	CPz		1950pc		ac	6.50	4Booo3	R			
3	C/*	CPz	OPz	1950ms	1968ry	acf	6.50	4BooX3			8/68	9/68

AA 1.4.47 on B(A), 3/-, OSPR 1/48. M on C.

ST 49 (31/49) *Caerwent*   10 by 10   Revision a 1917-19 c 1949 f 1966

1	A	CPr	OPr	nd:po		-	6.49	4BEoo3			2/49	4/49
3	B	CPu	OPv	1956pc		ac	6.56	4oooX3			4/56	5/57
3	B/	CPw		1956mc	1959mc	ac	6.56	4oooX3				
3	B//*	CPz	OPz	1956mc	1966r	acf	6.56	4oooX3			11/66	12/66

ST 50 (31/50) *Melbury*   10 by 10   Revision a 1901-28 c 1957

1	A	CPr	OPr	nd:po		-	6.49	4BEoo3	RB		10/49	11/49
2	B	CPt		1949pc	1953c	-	6.53	4BooX3	B			
3	C	CPw,z	OPv	1958mc		ac	6.58	4oooX3			11/58	1/59

ST 51 (31/51) *Yeovil*   10 by 10   Revision a 1901-28 c 1938-57 f 1973

1	A	CPr	OPr	nd:po		-	6.48	4BEoo3	RB	1	11/49	1/50	
3	B	CPw	OPv	1958mc		ac	6.58	4oooX3		1	1/59	12/58	1
3	B/	CPw.x,y,z		1958mc	1960mc	ac	6.58	4oooX3		1			
3	B//*	CFz	OFz	1958^{c}ms	nd:r	acf	6.58	4oooX3		1	4/75	6/74	

AA 1.10.49/NT 12.8.49 on A(A), 3/-, OSPR 1/50. 1. Yeovil Aerodrome.

ST 52 (31/52) *Ilchester*   10 by 10   Revision a 1901-28 c 1956; j 1981 selected

1	A	CPr	OPr	nd:po		-	6.49	4BEoo3	RB	0	10/49	11/49	
1	B	CPs		1949pc	1951c	a	1.51	4BEoo3	RB	0			
3	C	CPw,z,0z	OPv	1959mc		ac	6.59	4oooX3		1	3/59	4/59	1
3	C/*	CFz	OFz	1981^{c}ms	nd:r	acj	6.59	4oooX3		1	12/81	12/81	

1. Yeovilton Airfield.

ST 53 (31/53) *Glastonbury*   10 by 10   Revision a 1928-29 c 1938 d 1938-56

1	A	CPr	OPr	nd:po		-	1.49	4BEoo3	RB		1/50	1/50
3	A/	CPv		1949pc	1957mc	ac	1.49	4BooX3	B			
3	B	CPw,z.z,0z	OPv	1960mc		ad	6.60	4oooX4			10/60	6/60

ST 54 (31/54) *Wells Somerset*   10 by 10   Revision a 1902-29 c 1938-56

1	A	CPr	OPr	nd:po		-	6.49	4BEoo3	RB		12/49	12/49
2	A	CPs		1949pc	1952	-	1.52	4Booo3	RB			
3	B	CPw,y,z	OPv	1960mc		ac	6.60	4oooX3			10/60	11/60

ST 55 (31/55) *Compton Martin*   10 by 10   Revision a 1902-29 c 1956-57

1	A	CPr	OPr	nd:po		-	6.49	4BEoo3			11/49	1/50
1	A	CPs		1949pc	1951	-	1.51	4BEoo3				
3	B	CPx,y,z,0z	OPx,z	1961mc		ac	6.61	4oooX3			2/61	2/61

1	2	3	4	5	6	7	8	9	10	11	12	13	14

ST 56 (31/56) *Chew Magna*   10 by 10   Revision a 1902-30 c 1938-49 d 1938-56 f 1969
j 1966 building

1	10046	CPr		1946po	-	-	1.47	4BE2X3	R	10	11/46	-	1
1	A		OPr	1946po	-	-	1.47	4BE2X3	R	10	-	2/47	
1	B	CPs		1951pc	-	ac	6.51	4BEoo3	R	10	8/51	-	
3	B		OPs	1951pc	-	ac	6.51	4BEoo3	R	10	-	8/54	
3	C	CPx,z	OP?,z	1961mc	-	ad	6.61	4oooX4		33	7/61	3/61	2
3	C/*	CPz,0z	OPz	1961^{c}ms	nd:br	adfj	6.61	4oooX4		33	4/71	6/71	3

AA 1.9.46 on A(A), 3/-, OSPR 2/47; AA 1.1.54 on B(B), 4/-, OSPR 8/54. M on A, B. 1. Bristol Airport. 2. Bristol (Whitchurch) Airport, Bristol (Lulsgate Bottom) Aerodrome. 3. The print code has a short bar and the star beneath the letter. The imprint has "Director General of the Ordance Survey" (sic).

ST 57 (31/57) *Bristol (West)*   10 by 10   Revision a 1902-36 c 1938-49 f 1961 g 1972

1	20046	CPr		1946po	-	-	1.47	4BE2X3	RB	0	11/46	-	
1	A		OPr	1946po	-	-	1.47	4BE2X3	RB	0	-	2/47	
1	B	CPs		1951pc	-	ac	1.51	4BEoo3	R	0	10/51	-	
2	B		OPs	1951pc	-	ac	1.51	4BooX3	R	0			
3	B/	CPw		1951mc	1959c	ac	1.51	4BooX3		1			
3	B//	CPx,z		1951mc	1962ch	acf	1.51	4BooX3		1			
3	B///*	CPz	OPz	1951^{c}ms	nd:r	acg	1.51	4BooX3		1	3/73	3/73	

AA 1.9.46/NT 1.9.46 on A(A), 3/-, OSPR 2/47; AA 1.1.54/NT 1.9.46 on B(B), 4/-, OSPR 11/54. M on A, B//.

ST 58 (31/58) *Aust*   10 by 10   Revision a 1912-36 c 1938-49 f 1965 g 1970
j 1950 tide lines k 1964 industrial l 1970 airfield

1	A	CPr	OPr	1947po	-	-	6.47	4BE2X3	R	0	4/47	8/47	
1	B												
1	C	CPs		1951pc	-	ac	1.51	4BEoo3	R	0	12/51	-	
3	C	CPt	OPs	1951pc	-	ac	1.51	4BEoo3	R	0	-	8/54	
3	C/	CPw,z		1951mc	1959mc	ac	1.51	4BooX3		1			
3	C//*	CPz	OPz	1951mc	1966ir	acfjk	1.51	4oooX3		2	1/66	1/66	1
3	C//*/*	CPz,0z	OPz	1951^{c}ms	nd:r	acgjkl	1.51	4oooX3		3	11/71	1/72	

AA 30.1.47 on A(A), 3/-, OSPR 8/47; AA 1.1.54 on C(B), 4/-, OSPR 8/54. M on C, C/. 1. With the M4 under construction. Filton Airfield.

ST 59 (31/59) *Chepstow*   10 by 10   Revision a 1900-24 b 1918-20 c 1949 f 1968

1	A	CPr	OPr	nd:po	-	-	6.48	4BEoo3	R		10/48	11/48	1
1	B	CPr		nd:po	-	-	6.48	4BEoo3	R				
3	B	CPu		1949pc	-	-	6.48	4BEoo3	R				
3	B/	CPw		1949mc	1959mc	a	6.48	4BooX3					
3	C	CPw,x,y	OPv	1960mc	-	bc	6.60	4oooX3			4/60	4/60	
3	C/*	CPz,0z	OPz	1960ms	1968r	bcf	6.60	4oooX3			12/68	12/68	

AA 1.10.48 on A(A), 3/-, OSPR 11/48. 1. Showing the shipworkers' camp at Beachley, and its railway system.

ST 60 (31/60) *Cerne Abbas*   10 by 10   Revision a 1900-01 c 1956-57

1	A	CPr	OPr	nd:po	-	-	6.48	4BEoo3	B		4/49	4/49	
1	B	CPs		1949pc	1950c	a	6.50	4BEoo3	B				
3	C	CPw,y,z	OPv	1958mc	-	ac	6.58	4oooX3			11/58	1/59	

ST 61 (31/61) *Sherborne Dorset*   10 by 10   Revision a 1927 c 1938-57

1	A	CPr	OPr	nd:po	-	-	6.49	4BEoo3	RB		11/49	1/50	
3	A	CPu		1949pc	-	-	6.49	4BEoo3	RB				
3	B	CPw,x,y,z.z,0z	OPv,z	1958mc	-	ac	6.58	4oooX3			11/58	1/59	

ST 62 (31/62) *Charlton Horethorne*   10 by 10   Revision a 1901-28 b 1901-29 c 1956

1	A	CPr	OPr	nd:po	-	-	6.48	4BEoo3	RB		11/49	1/50	
1	B	CPs		1949pc	1951c	a	1.51	4BEoo3	RB				
3	C	CPv,y,z,0z	OP?	1958pc	-	bc	6.58	4oooX4			5/58	5/58	

ST 63 (31/63) *Bruton*   10 by 10   Revision a 1901-29 c 1956

1	A	CPr	OPr	nd:po	-	-	6.48	4BEoo3	RB		1/50	1/50	
1	B	CPs		1949pc	1951c	a	1.51	4BEoo3	RB				
3	C	CPv,w,z.z,0z	OPv	1958pc	-	ac	6.58	4oooX3			3/58	7/58	

1	2	3	4	5	6	7	8	9	10	11	12	13	14

ST 64 (31/64) *Shepton Mallet*   10 by 10   Revision a 1902-29 c 1938-56
1	A	CPr	OPr	nd:po	-	-	6.49	4BEoo3	R		11/49	1/50
3	A	CPu		1949pc	-	-	6.49	4BEoo3	R			
3	B	CPv,x,z,0z	OP?	1958pc	-	ac	6.58	4oooX3			3/58	3/58

ST 65 (31/65) *Radstock*   10 by 10   Revision a 1902-29 c 1938-56
1	A	CPr	OPr	nd:po	-	-	1.50	4BEoo3	R		12/49	2/50
2	A	CPs		1949pc	-	-	1.52	4Booo3	R			
3	B	CPv,y,z.z,0z	OPv	1958pc	-	ac	6.58	4oooX3			3/58	7/58

AA 1.10.49 on A(A), 3/-, OSPR 2/50.

ST 66 (31/66) *Keynsham*   10 by 10   Revision a 1882-1930 c 1947-49 d 1938-56 f 1967
1	10046	CPr		1946po	-	-	1.47	4BE2X3	R		11/46	-
1	A		OPr	1946po	-	-	1.47	4BE2X3	R		-	1/47
1	B	CPs		1951pc	-	ac	1.51	4BEoo3	R		8/51	-
3	B		OPs	1951pc	-	ac	1.51	4BEoo3	R		-	8/54
3	C	CPv	OP?	1958pc	-	ad	6.58	4oooX3			3/58	3/58
3	C/	CPx,z		1958mc	1961ch	ad	6.58	4oooX3				
3	C//*	CPz	OPz	1958mc	1968r	adf	6.58	4oooX3			1/68	1/68

AA 1.8.46 on A(A), 3/-, OSPR 1/47; AA 1.1.54 on B(B), 4/-, OSPR 8/54. M on A.

ST 67 (31/67) *Bristol (East)*   10 by 10   Revision a 1902-36 c 1938-49 f 1967 g 1971 h 1977
1	20046	CPr		1946po	-	-	1.47	4BE2X3	R		10/46	-
1	A		OPr	1946po	-	-	1.47	4BE2X3	R		-	3/47
1	B	CPs		1951pc	-	ac	6.51	4BEoo3	R		1/52	-
3	B	CPt	OPs	1951pc	-	ac	6.51	4BEoo3	R		-	8/54
3	B/	CPw		1951mc	1959c	ac	6.51	4BooX3				
3	B//	CPw,y,z		1951mc	1960mc	ac	6.51	4BooX3				
3	B///*	CPz	OPz	1951ms	1968r	acf	6.51	4BooX3			5/68	5/68
3	B///*/*	CPz	OPz	1951^{c}ms	nd:r	acg	6.51	4BooX3			8/72	8/72
3	B///*/*/*	CFz	OFz	1951^{c}ms	nd:r	ach	6.51	4BooX3			7/78	7/78

AA 1.8.46 on A(A), 3/-, OSPR 3/47; AA (B) not found; AA 1.1.54/NT 1.7.47 on B(C), 4/-, OSPR 8/54. M on A.

ST 68 (31/68) *Tytherington Gloucestershire*   10 by 10 Revision a 1916-36 c 1938-49 f 1962 g 1967 h 1972
1	A	CPr	OPr	1947po	-	-	1.48	4BE2X3	RB	1	10/47	12/47
1	B	CPr		nd:po	-	-	1.48	4BEoo3	RB	1		M
1	C	CPs		1951pc	-	ac	1.51	4BEoo3	R	1	12/51	-
3	C	CPt		1951pc	-	ac	1.51	4BEoo3	R	1		
3	C/	CPw		1951mc	1960mc	ac	1.51	4BooX3		1		
3	C//*	CPy	OPy	1951mc	1963r	acf	1.51	4BooX3		1	6/63	6/63
3	C//*/*	CPz	OPz	1951mc	1968r	acg	1.51	4BooX3		1	1/68	1/68
3	C//*/*/*	CPz	OPz	1951^{c}ms	nd:r	ach	1.51	4BooX3		1	9/72	12/72

AA 1.8.47 on A(A), 3/-, OSPR 12/47. M on B.

ST 69 (31/69) *Berkeley*   10 by 10   Revision a 1919-20 c 1946-49 f 1964 g 1972 h 1976
j 1970 reservoir
1	A	CPr	OPr	nd:po	-	-	1.49	4BEoo3	R		2/49	3/49
1	B	CPs	OPs	1951pc	-	ac	6.51	4BEoo3	R		6/51	3/55
3	B/	CPv		1951pc	1958mc	ac	6.51	4BooX3				
3	B//	CPx		1951pc	1958mc	ac	6.51	4BooX3				
3	B///	CPz	OPz	1951mc	1965ch	acf	6.51	4BooX3				
3	B////*	CPz	OPz	1951^{c}ms	nd:rv	acgj	6.51	4BooX3			3/73	3/73
3	B////*/	CFz	OFz	1951^{c}ms	nd:ch	achj	6.51	4BooX3			6/77	6/77

ST 70 (31/70) *Bulbarrow Hill*   10 by 10   Revision a 1900-27 c 1956-57
1	A	CPr	OPr	nd:po	-	-	1.49	4BEoo3	B		5/49	5/49
1	A	CPr		1949pc	1950	-	1.49	4BEoo3	B			
3	B	CPw,x,z	OPv	1958pc	-	ac	6.58	4oooX4			10/58	1/59

M on A, B.

1	2	3	4	5	6	7	8	9	10	11	12	13	14

ST 71 (31/71) *Sturminster Newton* 10 by 10 Revision a 1900-27 c 1938-57

1	A		CPr	OPr	nd:po	-	-	6.48	4BEoo3	RB	0	11/49	1/50
3	A		CPt		1949pc	-	-	6.48	4BEoo3	RB	0		
3	B		CPw,x.z,0z	OPv	1958pc	-	ac	6.58	4oooX3		3	10/58	10/58

M on A, B.

ST 72 (31/72) *Wincanton* 10 by 10 Revision a 1900-29 c 1938-56; j 1980 selected

1	A		CPr	OPr	nd:po	-	-	6.49	4BEoo3	RB	0	11/49	1/50
3	B		CPv	OP?	1957pc	-	ac	6.57	4oooX3		3	2/58	2/58
3	B		CPz		1957mc	-	ac	6.57	4oooX3		3		
3	B/*		CFz	OFz	1981^{c}ms	nd:r	acj	6.57	4oooX3		3	4/81	4/81

M on A.

ST 73 (31/73) *Stourhead* 10 by 10 Revision a 1900-29 c 1956

1	A		CPr	OPr	nd:po	-	-	1.49	4BEoo3	RB		7/49	8/49
2	B		CPt		1949pc	1953c	-	6.53	4BooX3	B			
3	C		CPv,z,x,0z	OP?,z	1958pc	-	ac	6.58	4oooX3			3/58	3/58

M on A, with legends, A.

ST 74 (31/74) *Frome* 10 by 10 Revision a 1902-39 c 1938-55

1	A		CPr	OPr	nd:po	-	-	1.50	4BEoo3	RB		9/49	9/49
3	A		CPt		1949pc	-	-	1.50	4BEoo3	RB			
3	B		CPv,w,z.z	OP?	1957pc	-	ac	6.57	4oooX3			1/58	1/58

M on A.

ST 75 (31/75) *Wellow Somerset* 10 by 10 Revision a 1902-29 b 1899-29 c 1936-38 d 1938-56

1	A		CPr	OPr	1947po	-	-	1.48	4BE2X3	RB		9/47	12/47	
?	B		CP?		1947pc	1951c	ac	1.51	4B?o?3	RB				M
3	C		CPu		1955pc	-	bc	6.55	4oooX3	B		8/55	-	
3	D		CPv	OP?	1958pc	-	bd	6.58	4oooX3			3/58	3/58	
3	D/		CPw,z,0z		1958mc	1960mc	bd	6.58	4oooX3					

AA 1.7.46 on A(A), 3/-, OSPR 12/47. M on A, with legends, B.

ST 76 (31/76) *Bath* 10 by 10 Revision a 1902-39 c 1936-56 f 1965

1	A		CPr	OPr	1947po	-	-	6.47	4BE2X3	RB	0	9/47	12/47
1	B		CPr		nd:po	-	-	6.47	4BEoo3	RB	0		
2	C		CPt		1948pc	1953c	-	6.53	4BooX3	B	0		
3	D		CPv,x	OPv	1958pc	-	ac	6.58	4oooX3		3	5/58	9/58
3	D/*		CPz	OPz	1958mc	1965r	acf	6.58	4oooX3		3	9/65	9/65
3	D/*/		CPz	OPz	1958mc	1965r	acf	6.58	4oooX3		3		

AA 1.4.47/NT 19.11.47 on A(A), 3/-, OSPR 12/47. M on A, with legends, B.

ST 77 (31/77) *Marshfield Gloucestershire* 10 by 10 Revision a 1902-19 c 1938-49 f 1968 g 1971
                                                              j 1966 airfield

1	A		CPr		1947po	-	-	1.48	4BE2X3	B	00	10/47	-	
1	B [1]			OPr	1947po	-	-	1.48	4BE2X3	B	00	-	1/48	
1	B [2]		CPs		1951pc	-	ac	1.51	4BEoo3	B	00	1/52	-	
3	B/		CPw,x		1951mc	1958mc	ac	1.51	4BooX3	B	10			
3	B//*		CPz	OPz	1951ms	1969r	acfj	1.51	4BooX3	B	33	1/70	5/70	1
3	B//*/*		CPz	OPz	1951^{c}ms	nd:r	acgj	1.51	4BooX3	B	33	12/72	4/73	

AA 1.5.46 on B(A), 3/-, OSPR 1/48. M on B [2]. 1. Colerne Airfield.

ST 78 (31/78) *Chipping Sodbury* 10 by 10 Revision a 1919-20 c 1949-53 f 1971

1	A		CPr	OPr	nd:po	-	-	6.49	4BEoo3	RB		4/49	4/49
2	A		CPt		1949pc	1952	-	1.52	4Booo3	RB			
3	B		CPv,x.z	OPv	1957pc	-	ac	6.57	4oooX3			1/58	-
3	B/*		CPz	OPz	1957^{c}ms	nd:r	acf	6.57	4oooX3			10/72	12/72

M on A.

1	2	3	4	5	6	7	8	9	10	11	12	13	14

ST 79 (31/79) *Dursley*　　　　　　10 by 10　Revision a 1919-21 c 1949 f 1973
1	A	CPr	OPr	nd:po	-	-	1.49	4BEoo3	R		11/48	12/48	
1	B	CPs		1951pc	-	ac	1.51	4BEoo3	R		7/51	-	
3	C	CPt,v.x,y,z	OPv	1951pc	1954c	ac	6.54	4BooX3			-	6/57	
3	C/*	CPz	OPz	1951^{c}ms	nd:r	acf	6.54	4BooX3			12/73	1/74	

ST 80 (31/80) *Blandford Forum*　　　10 by 10　Revision a 1900-28 c 1938-56
1	A	CPr	OPr	1948po	-	-	6.48	4BE2X3	RB		3/48	4/48	
1	B	CPr		nd:po	-	-	6.48	4BEoo3	RB				
3	B	CPu		1949pc	-	-	6.48	4BooX3	RB				
3	C	CPw,x,z	OPv	1958mc	-	ac	6.58	4oooX3			11/58	1/59	

M on B, C.

ST 81 (31/81) *Shillingstone*　　　　10 by 10　Revision a 1900-28 c 1938-56
| 1 | A | CPr,t | OPr | nd:po | - | - | 6.48 | 4BEoo3 | RB | | 11/49 | 12/49 | |
| 3 | B | CPw,x,z.z | OPv | 1958pc | - | ac | 6.58 | 4oooX3 | | | 9/58 | 10/58 | |

M on A, B.

ST 82 (31/82) *Shaftesbury*　　　　　10 by 10　Revision a 1900-28 c 1938-56; j 1983 selected
1	A	CPr	OPr	nd:po	-	-	1.49	4BEoo3	RB		7/49	7/49	
	B												
3	C	CPv	OP?	1957pc	-	ac	6.57	4oooX3			3/58	5/58	
3	C/	CPw,z	OPz	1957mc	1960mc	ac	6.57	4oooX3					
3	C//*	CFz	OFz	1984^{c}m-s	nd:r	acj	6.85	4oooX3			9/84	9/84	

M on A, with legends, A, ?B, C/.

ST 83 (31/83) *Mere Wiltshire*　　　　10 by 10　Revision a 1900-23 c 1956
1	A	CPr	OPr	nd:po	-	-	1.49	4BEoo3	B		7/49	8/49	
1	A	CPr		1949pc	1950	-	6.50	4BEoo3	B				
3	B	CPv,w.z	OPv	1957pc	-	ac	6.57	4oooX3			2/58	2/58	

M on A, ?with legends (unrecorded), A, B.

ST 84 (31/84) *Warminster*　　　　　10 by 10　Revision a 1922-29 c 1936-42 d 1936-56 e 1947
　　　　　　　　　　　　　　　　　　　　　　　j 1983 selected
1	A	CPr	OPr	nd:po	-	-	6.49	4BEoo3	RB		8/49	8/49	
1	B	CPs		1949pc	1950c	ace	6.50	4BEoo3	RB				
3	C	CPv,x,z.z	OP?	1958pc	-	ad	6.58	4oooX3			3/58	3/58	
3	C/*	CFz	OFz	1983^{c}m-s	nd:r	adj	6.85	4oooX3			6/84	6/84	

M on ?A, ?with legends (not recorded), B, C.

ST 85 (31/85) *Trowbridge*　　　　　10 by 10　Revision a 1902-36 c 1937-39 d 1937-56 e 1947
1	A	CPr	OPr	nd:po	-	-	1.49	4BEoo3	RB		8/49	9/49	1
1	B	CPs		1949pc	1950c	ace	6.50	4BEoo3	RB				
3	C	CPv,x,z	OP?,z	1958pc	-	ad	6.58	4oooX3			3/58	3/58	

M on A, ?with legends (not recorded), B. 1. The eastern railway curve is not shown: it is on edition C.

ST 86 (31/86) *Bradford-on-Avon*　　　10 by 10　Revision a 1919-36 c 1938-56
1	A	CPr	OPr	1948po	-	-	6.48	4BEoo3	RB		7/48	8/48	
?	B	CPs		1948pc	1951c	-	?	4B?oo3	RB				M
2	B	CPt		1948pc	1951c	-	1.52	4Booo3	RB				
3	C	CPt		1948pc	1953c	-	6.53	4BooX3	B				
3	D	CPw.x,z	OPv	1958pc	-	ac	6.58	4oooX3			6/58	8/58	

M on A, with legends, B.

ST 87 (31/87) *Castle Combe*　　　　10 by 10　Revision a 1919-25 c 1938-49 f 1972
1	A	CPr	OPr	1948po	-	-	1.48	4BE2X3	RB	00	3/48	4/48	
	B												
2	C	CPt	OPv	1952pc	-	ac	6.52	4Booo3	R	12	7/52	9/57	
3	C/	CPw,z		1952mc	1960mc	ac	6.52	4BooX3		12			1
3	C//*	CPz,0z	OPz	1952^{c}ms	nd:r	acf	6.52	4BooX3		12	2/73	3/73	

M on C. 1. Colerne Airfield.

1	2	3	4	5	6	7	8	9	10	11	12	13	14

ST 88 (31/88) *Sherston*   10 by 10   Revision a 1919-21 c 1938-49 f 1974
1	A	CPr	OPr	nd:po	-	-	6.49	4BEoo3	RB	0	7/49	8/49	
1	B	CPs		1951pc	-	ac	1.51	4BEoo3	RB	0	4/51	-	
3	B/	CPw,y,z		1951mc	1958mc	ac	1.51	4BooX3	B	0			
3	B//*	CFz	OFz	1951ᶜms	nd:r	acf	1.51	4BooX3	B	0	4/75	11/74	

M on B.

ST 89 (31/89) *Nailsworth*   10 by 10   Revision a 1919-21 c 1936-49
1	A	CPr	OPr	nd:po	-	-	1.49	4BEoo3	B		2/49	2/49	
2	B	CPs		1951pc	-	ac	6.51	4Booo4			3/52	-	
3	B/	CPv	OPv	1951pc	1956mc	ac	6.51	4BooX4			-	5/57	
3	B//	CPw,z,0z		1951mc	1961mc	ac	6.51	4BooX4					

M on B.

ST 90 (31/90) *Tarrant*   10 by 10   Revision a 1900-28 c 1936-38 d 1936-57
1	A	CPr	OPr	1948po	-	-	1.48	4BE2X3	RB	0	3/48	5/48	
1	B	CPr		nd:po	-	-	1.48	4BEoo3	RB	0			
3	B/	CPw		1948pc	1958mc	ac	1.48	4BooX3	B	0			
3	C	CPx,y,z	OPx	1961mc	-	ad	6.61	4oooX3		1	5/61	5/61	1

M on B. 1. Tarrant Rushton Airfield.

ST 91 (31/91) *Tollard Royal*   10 by 10   Revision a 1900-28 c 1956-57
1	A	CPr	OPr	nd:po	-	-	6.48	4BEoo3	B		11/48	11/48	
1	B	CPr		nd:po	-	-	6.48	4BEoo3	B				
3	C	CPx,z	OPx	1961mc	-	ac	6.61	4oooX3			6/61	9/61	

M on A, with legends, B.

ST 92 (31/92) *Tisbury*   10 by 10   Revision a 1923-24 c 1941 d 1956-57 e 1947
1	A	CPr	OPr	nd:po	-	-	1.48	4BEoo3	RB		7/48	9/48	
1	B	CPs		1950pc	-	ace	6.50	4BEoo3	RB		12/50	-	
3	C	CPw,z.z	OPv	1958pc	-	ad	6.58	4oooX3			9/58	10/58	

M on A, ?with legends (unrecorded), B, C.

ST 93 (31/93) *Hindon*   10 by 10   Revision a 1923 c 1939-42 d 1956-57 e no date
1	A	CPr	OPr	nd:po	-	-	1.48	4BEoo3	RB		10/48	10/48	
1	B	CPs		1951pc	-	ace	1.51	4BEoo3	RB		6/51	-	
3	C	CPw,y,z	OPv	1958pc	-	ad	6.58	4oooX3			10/58	10/58	

M on B, C.

ST 94 (31/94) *Heytesbury*   10 by 10   Revision a 1922-23 c 1938-42 d 1938-57
1	A	CPr	OPr	nd:po	-	-	1.48	4BEoo3	RB		10/48	10/48	
?	A	CP?		1947pc	1952	-	1.52	4B?o?3	RB				M
3	B	CPu	OPv	1955pc	-	ac	6.55	4oooX3	B		7/55	9/57	
3	C	CPw,y,z	OPv	1958pc	-	ad	6.58	4oooX3			9/58	10/58	

M on A, with legends, A.

ST 95 (31/95) *Cheverell*   10 by 10   Revision a 1922 b 1922-23 c 1938-57 e 1947
1	A	CPr	OPr	nd:po	-	-	1.48	4BEoo3	RB	0	9/48	10/48	
3	B	CPt	OPs	1953pc	-	ae	6.53	4BooX3	RB	0	12/53	-	
3	C	CPw,x,y,z	OPv	1958pc	-	bc	6.58	4BooX3		3	9/58	11/58	

AA 1.9.48 on A(A), 3/-, OSPR 10/48. M on A, with legends, B, C.

ST 96 (31/96) *Melksham*   10 by 10   Revision a 1922-36 c 1938-56 f 1961 g 1971 h 1974
1	A	CPr	OPr	nd:po	-	-	1.49	4BEoo3	RB		2/49	2/49	
3	B	CPt		1949pc	1953c	-	6.53	4BooX3	B				
3	C	CPx,z	OPx	1961mc	-	acf	6.61	4oooX4			11/61	11/61	
3	C/*	CPz	OPz	1961ᶜms	nd:r	acg	6.61	4oooX4			2/72	4/72	
3	C/*/	CFz	OFz	1961ᶜms	nd:r	ach	6.61	4oooX4			5/75	5/75	

AA 1.1.49/NT 4.1.49 on A(A), 3/-, OSPR 2/49. M on A, ?with legends (unrecorded), A, B.

1	2	3	4	5	6	7	8	9	10	11	12	13	14

ST 97 (31/97) *Chippenham Wiltshire*   10 by 10   Revision a 1921-22 c 1938-56 f 1973 g 1976
1	A	CPr	OPr	nd:po	-	-	1.49	4BEoo3	RB	0	2/49	5/49	
2	A	CPt		1949pc	1952	-	1.49	4Booo3	RB	0			
3	B	CPw,y,z	OPv	1959mc	-	ac	6.59	4oooX4		1	6/59	7/59	
3	B/*	CPz	OPz	1959ᶜms	nd:r	acf	6.59	4oooX4		1	11/73	12/73	
3	B/*/*	CFz	OFz	1959ᶜms	nd:r	acg	6.59	4oooX4		1			1

AA 1.2.49 on A(A), 3/-, OSPR 5/49. M on A. 1. OSPR not found.

ST 98 (31/98) *Malmesbury*   10 by 10   Revision a 1919-22 c 1938-56 f 1975
1	A	CPr	OPr	nd:po	-	-	1.49	4BEoo3	RB	0	3/49	4/49	
2	B	CPt		1949pc	1953c	-	6.53	4BooX3	B	0			
3	C	CPw,y,z	OPv	1959mc	-	ac	6.59	4oooX4		1	10/59	10/59	1
3	C/*	CFz	OFz	1959ᶜms	nd:r	acf	6.59	4oooX4		1	7/76	7/76	

M on A. 1. Hullavington Airfield.

ST 99 (31/99) *Kemble*   10 by 10   Revision a 1919-21 c 1938-49 d 1938-56 e 1947
1	A	CPr	OPr	nd:po	-	-	6.49	4BEoo3	RB	000	7/49	7/49	
1	B	CPs		1951pc	-	ace	1.51	4BEoo3	RB	000	3/51	-	
3	C	CPw,y.z,0z	OPv	1959mc	-	ad	6.59	4oooX3		112	6/59	9/59	

SU 00 (41/00) *Verwood*   10 by 10   Revision a 1900-36 c 1938-57
1	A	CPr	OPr	1948po	-	-	6.47	4BE2X3	RB		1/48	3/48	
1	B	CPr		nd:po	-	-	6.47	4BEoo3	RB				
3	B	CPt		1948pc	-	-	6.47	4BEoo3	RB				
3	C	CPx,z	OPx	1961mc	-	ac	6.61	4oooX3			9/61	9/61	

AA 1.10.47 on A(A), 3/-, OSPR 3/48. M on B.

SU 01 (41/01) *Cranborne*   10 by 10   Revision a 1900-24 c 1938-57
1	A	CPr	OPr	1948po	-	-	6.47	4BE2X3	R		2/48	5/48	
1	B	CPr		nd:po	-	-	6.47	4BEoo3	R				
3	C	CPx,z	OPx	1961mc	-	ac	6.61	4oooX3			7/61	6/61	

M on A, B.

SU 02 (41/02) *Bishopstone Wiltshire*   10 by 10   Revision a 1900-24 c 1938-57
1	A	CPr	OPr	1948po	-	-	6.48	4BE2X3	B		4/48	5/48	
1	B	CPr		nd:po	-	-	1.49	4BEoo3	B				
3	C	CPw,y,z	OPv	1958pc	-	ac	6.58	4oooX4			8/58	10/58	

M on A, B.

SU 03 (41/03) *Wilton Wiltshire*   10 by 10   Revision a 1923-24 c 1938-42 d 1938-57 e no date
j 1985 selected
1	A	CPr	OPr	nd:po	-	-	1.49	4BEoo3	RB		12/48	1/49	
1	B	CPs		1951pc	-	ace	1.51	4BEoo3	RB				1
2	B		OPs	1951pc	-	ace	1.52	4Booo3	RB		-	6/52	
3	C	CPw,x,z	OPv	1958pc	-	ad	6.58	4oooX3			10/58	12/58	
3	C/*	CFz	OFz	1986ᶜm-s	nd:rs	adj	6.86	4oooX3			6/86	6/86	2

M on B, with legends, B. SS: Soil, Land Use Capability, outline, 1976 on C. 1. There was a B printing apparently unique to GSGS 4627 (qv) that preceded this edition. 2. One of the final batch of three revised reprints in this series (with TF 14, TL 64).

SU 04 (41/04) *Shrewton*   10 by 10   Revision a 1922-39 c 1956-57
1	A	CPr	OPr	nd:po	-	-	1.49	4BEoo3	B		10/48	11/48	
2	B	CPt		1948pc	1953c	-	6.53	4BooX3	B				
3	C	CPw,y,z	OPv	1958pc	-	ac	6.58	4oooX3			9/58	11/58	

M on A, with legends, B. One of the Salisbury Plain sheets, showing West Down (later Westdown) Camp, Greenland Camp, Rollestone Camp, and several other unnamed military sites.

SU 05 (41/05) *Urchfont*   10 by 10   Revision a 1922-39 c 1938-57
1	A	CPr	OPr	nd:po	-	-	1.48	4BEoo3	RB		10/48	11/48	
3	A	CPt		1948pc	-	-	6.48	4BooX3	RB				
3	B	CPw,x,z	OPv	1958pc	-	ac	6.58	4oooX3			9/58	11/58	

M on A, with legends, A, B.

1	2	3	4	5	6	7	8	9	10	11	12	13	14

SU 06 (41/06) *Devizes*   10 by 10   Revision a 1921-23 c 1938-56 f 1961

1	A	CPr	OPr	nd:po	-	-	1.49	4BEoo3	RB		5/49	6/49
3	B	CPx,y,z,0z	OPx	1961mc	-	acf	6.61	4oooX3			10/61	9/61

AA 1.5.49 on A(A), 3/-, OSPR 6/49. M on A, with legends, A.

SU 07 (41/07) *Lyneham Wiltshire*   10 by 10   Revision a 1921-36 c 1938-56 f 1970
                                                                            j 1965-70 airfield k 1965-70 building

1	A	CPr	OPr	nd:po	-	-	1.49	4BEoo3	B	000	8/48	10/48	
2	A	CPt		1948pc	1952	-	1.49	4Booo3	B	000			
3	B	CPw,z	OPv	1959mc	-	ac	6.59	4oooX3		121	6/59	9/59	1
3	B/*	CPz,0z	OPz	1959ᶜms	nd:ab	acfjk	6.59	4oooX3		321	11/71	1/72	

AA 1.9.48/NT 2.4.48 on A(A), 3/-, OSPR 10/48. M on A. 1. Lyneham Airfield.

SU 08 (41/08) *Wootton Bassett*   10 by 10   Revision a 1921-22 c 1936-38 d 1936-56 e 1947
                                                                            f 1972

1	A	CPr	OPr	nd:po	-	-	6.48	4BEoo3	RBC		9/48	10/48
1	B	CPs		1948pc	1951c	ace	1.51	4BEoo3	RBC			
3	C	CPw,y,z	OPv	1959mc	-	ad	6.59	4oooX3	C		9/59	9/59
3	C/*	CPz,0z	OPz	1959ᶜms	nd:r	adf	6.59	4oooX3	C		3/73	7/73

M on B.

SU 09 (41/09) *South Cerney*   10 by 10   Revision a 1920-38 c 1956

1	A	CPr	OPr	nd:po	-	-	1.49	4BEoo3	RB	00	9/48	9/48	
1	A	CPs		1948pc	1951	-	1.51	4BEoo3	RB	00			
3	B	CPw,z	OPv	1959mc	-	ac	6.59	4oooX3		31	8/59	7/59	1

M on A, B. 1. South Cerney Airfield.

SU 10 (41/10) *Ringwood*   10 by 10   Revision a 1900-08 c 1938-57 f 1961 g 1965 h 1971

1	A	CPr	OPr	1948po	-	-	6.47	4BE2X3	R	0	1/48	3/48
1	B	CPr		nd:po	-	-	6.47	4BEoo3	R	0		
2	C	CPt		1949pc	1953c	-	6.47	4BooX3	R	0		
3	D	CPx	OPx	1961mc	-	acf	6.61	4oooX3		3	10/61	9/61
3	D/	CPz	OPz	1961mc	1966ch	acg	6.61	4oooX3		3		
3	D//*	CPz	OPz	1961ᶜms	nd:r	ach	6.61	4oooX3		3	5/72	8/72

AA 1.2.48/NT 15.1.48 on A(A), 3/-, OSPR 3/48. M on B.

SU 11 (41/11) *Fordingbridge*   10 by 10   Revision a 1900-24 c 1938-57 f 1961

1	A	CPr	OPr	1948po	-	-	1.48	4BE2X3	RB		1/48	4/48
1	B	CPr		nd:po	-	-	1.48	4BEoo3	RB			
3	B	CPu		1949pc	-	-	1.48	4BEoo3	RB			
3	C	CPx,y,z	OPx	1961mc	-	acf	6.61	4oooX3			7/61	10/61

AA 1.10.47 on A(A), 3/-, OSPR 4/48. M on A, B, C.

SU 12 (41/12) *Salisbury (South)*   10 by 10   Revision a 1900-24 c 1938-57 f 1965 g 1972
                                                                               j 1980 selected

1	A	CPr	OPr	1948po	-	-	6.48	4BE2X3	RB		5/48	6/48
1	B	CPr		nd:po	-	-	6.48	4BEoo3	RB			
3	C	CPw,x	OPv	1958pc	-	ac	6.58	4oooX3			9/58	10/58
3	C/*	CPz	OPz	1958mc	1966r	acf	6.58	4oooX3			6/66	7/66
3	C/*/*	CPz	OPz	1958ᶜms	nd:r	acg	6.58	4oooX3			7/73	10/73
3	C/*/*/*	CFz	OFz	1980ᶜms	nd:r	acgj	6.58	4oooX3			2/81	3/81

AA 1.3.48 on A(A), 3/-, OSPR 6/48. M on A, B, C.

SU 13 (41/13) *Salisbury (North)*   10 by 10   Revision a 1923-24 c 1938-57; j 1980 selected

1	A	CPr	OPr	nd:po	-	-	1.48	4BEoo3	RB	10	1/49	3/49	1
1	B	CPr		nd:po	-	-	1.48	4BEoo3	RB	10			
3	C	CPw	OPv	1958pc	-	ac	6.58	4oooX4		10	10/58	1/59	
3	C/	CPx,z,z-		1958mc	1961ch	ac	6.58	4oooX4		11			2
3	C//*	CFz	OFz	1981ᶜms	nd:r	acj	6.58	4oooX4		11	5/81	5/81	

AA 1.12.48/NT 22.12.48 on A(A), 3/-, OSPR 3/49. M on A, with legends, B. 1. Showing South Camp, also the railway connection to Porton Camp. 2. Old Sarum Airfield, Boscombe Down Airfield.

227

1	2	3	4	5	6	7	8	9	10	11	12	13	14

SU 14 (41/14) *Amesbury*   10 by 10   Revision a 1922-23 c 1957 f 1972

1	A	CPr	OPr	nd:po	-	-	1.48	4BEoo3	RB	10	2/49	2/49	
3	A	CPt		1948pc	-	-	1.48	4BooX3	RB	10			
3	B	CPw,x,z	OPv	1958pc	-	ac	6.58	4oooX4		10	11/58	1/59	
3	B/*	CPz	OPz	1958^{c}ms	nd:r	acf	6.58	4oooX4		11	3/73	7/73	

M on A, with legends, A. One of the Salisbury Plain sheets, showing Larkhill Camp, Durrington Camp, Charters Camp (later Carter Barracks), Sling Camp, Bulford Camp (later Barracks), Barracks (near Netheravon).

SU 15 (41/15) *Upavon*   10 by 10   Revision a 1922-23 c 1939-42 d 1956-57 e 1947

1	A	CPr	OPr	nd:po	-	-	1.48	4BEoo3	RB	0	12/48	1/49	
1	B	CPs	OP?	1951pc	-	ace	1.51	4BEoo3	RB	0	1/52	8/51	
3	C	CPw,x,z.z	OPv	1958pc	-	ad	6.58	4oooX3		0	10/58	10/58	

M on A, with legends, B.

SU 16 (41/16) *Marlborough*   10 by 10   Revision a 1922 c 1938 d 1938-56

1	A	CPr	OPr	nd:po	-	-	1.49	4BEoo3	RB		5/49	5/49	
2	A	CPt		1949pc	-	-	1.49	4Booo3	RB				
3	A/	CPw		1949pc	1959mc	ac	1.49	4BooX3	B				
3	B	CPx,z,0z	OPx	1961mc	-	ad	6.61	4oooX4			5/61	5/61	

M on A, with legends, A, A/.

SU 17 (41/17) *Marlborough Downs*   10 by 10   Revision a 1922 b 1921-43 c 1938 d 1938-56 e 1947 f 1970

1	A	CPr	OPr	1948po	-	-	6.48	4BEoo3	RB	0	7/48	7/48	1
1	B	CPs		1951pc	-	ace	1.51	4BEoo3	RB	0	3/51	-	
2	B	CPt		1951pc	1953rpb	ace	1.51	4BooX3	RB	0			
3	C	CPw,z	OPv	1960mc	-	bd	6.60	4oooX3		1	12/59	12/59	2
3	C/	CPz,0z	OPz	1960^{c}ms	nd:ch	bdf	6.60	4oooX3		3			

M on B. 1. Chiseldon Camp (1877). 2. Wroughton Airfield.

SU 18 (41/18) *Swindon*   10 by 10   Revision a 1921-43 c 1936-43 d 1938-56 e 1947 f 1966 g 1973 h 1977; j 1965 industrial

1	A	CPr	OPr	1948po	-	-	6.48	4BEoo3	RBC	0	7/48	8/48	
1	B	CPs	OPs	1951pc	-	ace	1.51	4BEoo3	RBC	0	1/52	9/51	
3	C	CPw,z	OP?	1959mc	-	ad	6.59	4oooX3	C	0	9/59	9/59	
3	C/*	CPz	OPz	1959mc	1967r	adf	6.59	4oooX3	C	0	2/67	2/67	
3	C/*/*	CPz	OPz	1959^{c}ms	nd:ir	adgj	6.59	4oooX3	C	0	10/73	12/73	
3	C/*/*/*	CFz	OFz	1959^{c}ms	nd:r	adhj	6.59	4oooX3	C	0	3/78	3/78	

AA 1.7.48 on A(A), 3/-, OSPR 8/48. M on B.

SU 19 (41/19) *Cricklade*   10 by 10   Revision a 1919-22 c 1956 f 1965

1	A	CPr	OPr	nd:po	-	-	1.49	4BEoo3	RB	00	10/48	10/48	
3	B	CPt		1948pc	1950c	-	1.49	4BEoo3	RB	00			
3	C	CPw	OPv	1959mc	-	ac	6.59	4oooX3		13	10/59	10/59	1
3	C/	CPz	OPz	1959mc	1965ch	acf	6.59	4oooX3		33			

M on A, C. 1. Fairford Airfield.

SU 20 (41/20) *Burley Hampshire**   10 by 10   Revision a 1907 c 1957 e 1947 f 1961 g 1972

1	A	CPr	OPr	1948po	-	-	1.48	4BE2X3	R		1/48	3/48	
3	B	CPv		1956pc	-	ae	6.56	4oooX3					
3	B/		OPv	1956pc	1960mc	ae	6.56	4oooX3					
3	C	CPx,z	OPx	1961mc	-	acf	6.61	4oooX4			11/61	11/61	
3	C/*	CPz	OPz	1961^{c}ms	nd:r	acg	6.61	4oooX4			1/73	3/73	

AA 1.10.47 on A(A), 3/-, OSPR 3/48. M on A.

SU 21 (41/21) *[Cadnam]*   10 by 10   Revision a 1909-27 c 1956-57 f 1961

1	A	CPr	OPr	1948po	-	-	1.48	4BE2X3	0		1/48	5/48	
1	B	CPr		nd:po	-	-	1.48	4BEoo3	0				
3	B	CPu		1948pc	-	-	1.48	4BooX3	0				
3	B/	CPw		1948mc	1959mc	a	1.48	4BooX3	0				
3	C	CPx,y,z	OPx	1961mc	-	acf	6.61	4oooX4	3		11/61	11/61	

AA 1.3.48/NT 24.2.48 on A(A), 3/-, OSPR 5/48. M on A, B.

1	2	3	4	5	6	7	8	9	10	11	12	13	14

SU 22 (41/22) *Whiteparish*   10 by 10   Revision a 1900-24 c 1957

1	A	CPr		1948po	-	-	1.48	4BE2X3	RBC		2/48	-	
1	B		OPr	1948po	-	-	1.48	4BE2X3	RBC		-	6/48	
1	C	CPr	OPs	nd:po	-	-	1.48	4BEoo3	RBC				
3	D	CPw,z	OPv	1958pc	-	ac	6.58	4oooX3	C		9/58	11/58	

AA 1.4.48/NT 12.4.48 on B(A), 3/-, OSPR 6/48; AA 1.6.51/NT 12.4.48 on C(A), 3/-, OSPR 8/51. M on A,C, D.

SU 23 (41/23) *Winterslow*   10 by 10   Revision a 1908-24 c 1956-57

1	A	CPr	OPr	1948po	-	-	6.48	4BE2X3	RB	0	4/48	5/48	1
1	B	CPr		nd:po	-	-	6.48	4BEoo3	RB	0			
3	B	CPv		1949pc	-	a	6.48	4BooX3	RB	0			
3	C	CPw,y,z.z	OPv	1958pc	-	ac	6.58	4oooX4		1	10/58	10/58	

M on A, B. 1. Showing Porton Camp, and its railway connection.

SU 24 (41/24) *Tidworth*   10 by 10   Revision a 1908-23 c 1956-57

1	A	CPr	OPr	nd:po	-	-	1.48	4BEoo3	RB	0	7/48	8/48	1
3	A	CPt		1948pc	-	-	1.48	4BEoo3	RB	0			
3	B	CPw,y,z	OPv	1958pc	-	ac	6.58	4oooX3		3	9/58	11/58	2

M on A, with legends, A, B. 1. One of the Salisbury Plain sheets, showing Tidworth Barracks (with railway shown as GWR), Perham Down Camp, Parkhouse Camp and part of Ludgershall. 2. Thruxton Aerodrome.

SU 25 (41/25) *Collingbourne*   10 by 10   Revision a 1922-23 c 1956-57

1	A	CPr	OPr	nd:po	-	-	1.48	4BEoo3	RB		8/48	9/48	
1	B	CPr		nd:po	-	-	1.48	4BEoo3	RB				
1	B	CP?		1948pc	1949c	-		4BEoo3	RB				M
3	C	CPw,y.z	OPv	1958pc	-	ac	6.58	4oooX3			9/58	11/58	1

M on A, with legends, B, C. 1. Windmillhill Camp.

SU 26 (41/26) *Savernake Forest*   10 by 10   Revision a 1909-22 c 1956 f 1961

1	A	CPr	OPr	nd:po	-	-	6.49	4BEoo3	RB		5/49	6/49	
3	A/	CPw		1949mc	1959mc	a	6.49	4BooX3	B				
3	B	CPx,z,0z	OPx	1961mc	-	acf	6.61	4oooX4			11/61	11/61	

M on A, with legends, A.

SU 27 (41/27) *Aldbourne*   10 by 10   Revision a 1921-22 c 1938 d 1938-56 e 1947 f 1972

1	A	CPr	OPr	nd:po	-	-	6.48	4BEoo3	RB		10/48	10/48	
1	B	CPs		1948pc	1950c	ace	6.50	4BEoo3	RB				
3	C	CPw,z	OPv	1960mc	-	ad	6.60	4oooX3			1/60	2/60	
3	C/*	CPz,0z	OPz	1960^{c}ms	nd:r	adf	6.60	4oooX3			3/73	6/73	

M on B.

SU 28 (41/28) *Shrivenham*   10 by 10   Revision a 1910-22 c 1938-43 d 1938-56
                                                  e 1947 f 1974

1	A	CPr	OPr	nd:po	-	-	1.49	4BEoo3	RB		10/48	10/48	
1	B	CPs	OPv	1950pc	-	ace	6.50	4BEoo3	RB		12/50	1/57	
3	C	CPw,x,z	OPv	1960mc	-	ad	6.60	4oooX4			1/60	1/60	
3	C/*	CFz	OFz	1960^{c}ms	nd:r	adf	6.60	4oooX4			3/75	1/75	

M on B, C.

SU 29 (41/29) *Faringdon*   10 by 10   Revision a 1910-32 c 1942 d 1955-56 e 1947

1	A	CPr	OPr	nd:po	-	-	1.49	4BEoo3	RB		10/48	10/48	
1	B	CPs		1951pc	-	ace	1.51	4BEoo3	RB		3/51	-	
3	C	CPw,y,z	OPv	1959mc	-	ad	6.59	4oooX4			6/59	9/59	

M on B.

SU 30 (41/30) *Beaulieu**   10 by 10   Revision a 1907-34 c 1938-57 f 1960 g 1963

1	A	CPr	OPr	1947po	-	-	6.47	4BE2X3	R	0	3/47	6/47	
1	B	CPr		nd:po	-	-	6.47	4BEoo3	R	0			
3	B/	CP?v,w		1948pc	1957mc	a	6.47	4BooX3		0			
3	C	CPx	OPx	1961mc	-	acf	6.61	4oooX3		3	8/61	7/61	
3	C/*	CPz	OPz	1961mc	1964r	acg	6.61	4oooX3		3	11/64	11/64	

AA 1.1.47 on A(A), 3/-, OSPR 6/47. M on B.

1	2	3	4	5	6	7	8	9	10	11	12	13	14

SU 31 (41/31) *Totton*   10 by 10   Revision a 1907-34 c 1938-47 d 1938-57
                                              e no date f 1960

1	10047	CPr		1947po	-	-	1.47	4BE2X3	R		12/46	-	1
1	A		OPr	1946po	-	-	1.47	4BE2X3	R		-	3/47	
1	B	CPs	OPs	1951pc	-	ace	1.51	4BEoo3	R		7/51	1/56	
3	B/	CPv		1951pc	1958mc	ace	1.51	4BooX3					
3	C	CPx,z	OPx	1962mc	-	adf	6.62	4oooX3			3/62	3/62	

AA 1.9.46 on A(A), 3/-, OSPR 3/47. M on A, B. 1. Showing the Ordnance Survey Office at Maybush.

SU 32 (41/32) *Romsey*   10 by 10   Revision a 1908 c 1938-42 d 1938-56 f 1964

1	A	CPr	OPr	1947po	-	-	1.47	4BE2X3	RB		2/47	5/47	
1	B	CPr		nd:po	-	-	1.47	4BEoo3	RB				
3	B/	CPv	OPv	1949pc	1956mc	ac	1.47	4BooX3	B				
3	C	CPv	OP?	1958pc	-	ad	6.58	4oooX3			3/58	5/58	
3	C/	CPz	OPz	1958mc	1964ch	adf	6.58	4oooX3					

AA 1.12.46 on A(A), 3/-, OSPR 5/47. M on A, B, C.

SU 33 (41/33) *Stockbridge*   10 by 10   Revision a 1908 c 1956 f 1970

1	A	CPr	OP?	1948po	-	-	6.48	4BEoo3	RB	00	6/48	7/48	
1	B	CPr	OPr	nd:po	-	-	6.48	4BEoo3	RB	00			
3	B	CPu		1949pc	-	-	6.48	4BEoo3	RB	00			
3	C	CPw,z	OPv,z	1958pc	-	ac	6.58	4oooX3		11	8/58	10/58	1
3	C/	CPz	OPz	1958^{c}ms	nd:ch	acf	6.58	4oooX3		22			2

M on A, with legends, B; B, with layers. 1. Middle Wallop Aerodrome. 2. Middle Wallop Airfield.

SU 34 (41/34) *Andover*   10 by 10   Revision a 1908-23 c 1938 d 1938-56 e 1947 f 1971
                                              j 1971 airfield

1	A	CPr	OPr	1948po	-	-	1.48	4BEoo3	RB	0	7/48	9/48	
1	B	CPr		1950pc	-	ace	6.50	4BEoo3	RB	0			
3	C	CPv,x,z	OP?	1958pc	-	ad	6.58	4oooX3		1	5/58	5/58	
3	C/*	CPz,0z	OPz	1958^{c}ms	nd:ar	adfj	6.58	4oooX3		2	6/72	8/72	1

AA 1.3.48 on A(A), 3/-, OSPR 9/48. M on A, with legends, B. 1. Andover Airfield.

SU 35 (41/35) *Hurstbourne Tarrant*   10 by 10   Revision a 1909-23 c 1938-56

1	A	CPr	OPr	nd:po	-	-	6.48	4BEoo3	B		9/48	9/48	
3	A	CPt		1948pc	-	-	6.48	4BEoo3	B				
3	B	CPv,w,z	OP?	1958pc	-	ac	6.58	4oooX3			3/58	3/58	

M on A, with legends, A.

SU 36 (41/36) *Hungerford Berkshire*   10 by 10   Revision a 1909-10 b 1902-23 c 1956

1	A	CPr	OPr	nd:po	-	-	1.49	4BEoo3	RB		12/48	1/49	
3	A/	CPv		1948pc	1956mc	a	1.49	4BooX3	B				
3	B	CPx.z,0z	OPx	1961mc	-	bc	6.61	4oooX4			2/61	2/61	

M on A.

SU 37 (41/37) *Lambourn*   10 by 10   Revision a 1909-22 c 1956 f 1973

1	A	CPr	OPr	nd:po	-	-	6.49	4Booo4	RB	0	5/49	5/49	
2	A	CPs,t		1949pc	1952	-	6.49	4Booo4	RB	0			
3	B	CPw,z	OPv	1960mc	-	ac	6.60	4oooX4		1	1/60	1/60	
3	B/*	CPz,0z	OPz	1960^{c}ms	nd:r	acf	6.60	4oooX4		1	12/73	2/74	

M on A, B.

SU 38 (41/38) *Lambourn Downs*   10 by 10   Revision a 1909-10 c 1938 d 1938-56

1	A	CPr	OPr	nd:po	-	-	1.49	4BEoo3	BC	3	5/49	5/49	
1	B	CPr,u		nd:po	-	-	1.49	4BEoo3	BC	0			
3	B/	CPv		1949pc	1957mc	ac	1.49	4BooX3	BC	0			
3	C	CPw,y,z	OPv	1959mc	-	ad	6.59	4oooX4		1	10/59	10/59	

M on B.

1	2	3	4	5	6	7	8	9	10	11	12	13	14

SU 39 (41/39) *Stanford in the Vale*   10 by 10   Revision a 1910-19 c 1956

1	A	CPr	OPr	nd:po	-	-	6.49	4BEoo3	RBC	23	5/49	5/49	1
1	B	CPr		nd:po	-	-	6.49	4BEoo3	RBC	20			
3	C	CPw.x,z,0z	OPv	1959mc	-	ac	6.59	4oooX4		21	10/59	10/59	

M on B. 1. The outline edition appears in both OSPR 5/49 and 1/50.

SU 40 (41/40) *Southampton Water*   10 by 10   Revision a 1907-32 c 1938-48 d 1938-57
                                                 e 1947-48 f 1960 g 1964 h 1971
                                                 j 1969 building k 1969 industrial l 1969 afforestation

1	10046	CPr		1946po	-	-	1.47	4BE2X3	R	00	12/46	-	1
1	A		OPr	1946po	-	-	1.47	4BE2X3	R	00	-	4/47	
1	B	CPs		1951pc	-	ace	1.51	4BEoo3	R	00	4/51	-	
3	B		OPs	1951pc	-	ace	1.51	4BEoo3	R	00	-	3/55	
3	C	CPx	OPx	1962mc	-	adf	6.62	4oooX3		11	4/62	5/62	2
3	C/*	CPz	OPz	1962mc	1964r	adg	6.62	4oooX3		11	12/64	12/64	
3	C/*/*	CPz	OPz	1962^{c}ms	nd:bfi	adhjkl	6.62	4oooX3		11	1/72	5/72	

AA 1.9.46 on A(A), 3/-, OSPR 4/47. M on A. 1. Showing the Royal Victoria Hospital with its railway connection. 2. Southampton Water Aerodrome (on Southampton Water). At 4707 is Aerodrome (Air Service Training), at 4807 Naval Training School (Mercury). Calshot Castle is restored. There are towers along the coast (4700, 4802).

SU 41 (41/41) *Southampton**   10 by 10   Revision a 1908-34 c 1938-46 d 1938-57

1	15047	CPr		1947po	-	-	1.47	4BE2X3	R	1	1/47	-	1
1	A		OPr	1947po	-	-	1.47	4BE2X3	R	1	-	4/47	
2	A	CPt		1947pc	1952rpb	-	1.47	4BooX3	R	1			
3	A/	CPv		1947pc	1957c	ac	1.47	4BooX3		1			2
3	A//	CPw		1947mc	1958c	ac	1.47	4BooX3		1			
3	B	CPx,y,z	OPx	1961mc	-	ad	6.61	4oooX3		1	4/61	5/61	

AA 1.9.46 on A(A), 3/-, OSPR 4/47. M on A, A/. 1. Showing the Ordnance Survey Office, London Road. 2. Southampton (Eastleigh) Airport.

SU 42 (41/42) *Winchester (South)*   10 by 10   Revision a 1908-31 c 1938-57 f 1968

1	A	CPr	OPr	1947po	-	-	6.47	4BE2X3	RB		3/47	6/47	
1	B	CPr		nd:po	-	-	6.47	4BEoo3	RB				
3	C	CPv,x,z	OPv	1958pc	-	ac	6.58	4oooX3			5/58	9/58	
3	C/*	CPz	OPz	1958ms	1969r	acf	6.58	4oooX3			10/69	10/69	

AA 1.12.46 on A(A), 3/-, OSPR 6/47. M on A, B.

SU 43 (41/43) *Winchester (North)*   10 by 10   Revision a 1908-31 c 1938-57 f 1970

1	A	CPr	OPr	1948po	-	-	1.48	4BE2X3	RB	1	5/48	6/48	
1	B	CPr		nd:po	-	-	1.48	4BEoo3	RB	1			
2	B	CPt		1948pc	49c/52rpb	-	1.52	4BooX3	RB	1			
3	C	CPw,y,z	OPv	1958pc	-	ac	6.58	4oooX3		2	6/58	8/58	
3	C/	CPz	OPz	1958^{c}ms	nd:ch	acf	6.58	4oooX3		2			

AA 1.2.48 on A(A), 3/-, OSPR 6/48. M on A, B, C.

SU 44 (41/44) *Whitchurch Hampshire*   10 by 10   Revision a 1908-09 c 1938-56 f 1965

1	A	CPr	OPr	nd:po	-	-	6.48	4BEoo3	R		8/48	10/48	
3	A	CPu		1948pc	-	-	6.48	4Booo3	R				
3	B	CPv,x	OPv	1958pc	-	ac	6.58	4oooX3			5/58	7/58	
3	B/*	CPz,0z	OPz	1958mc	1966r	acf	6.58	4oooX3			6/66	7/66	

AA 1.3.48 on A(A), 3/-, OSPR 10/48. M on A, with legends, A, B, B/*.

SU 45 (41/45) *Litchfield*   10 by 10   Revision a 1909 c 1938-56 f 1976

1	A	CPr	OPr	nd:po	-	-	6.48	4BEoo3	R		8/48	8/48	
3	B	CPv,w,z	OP?	1958pc	-	ac	6.58	4oooX3			3/58	3/58	
3	B/*	CFz	OFz	1958^{c}ms	-	acf	6.58	4oooX3					1

M on A, with legends, A; A, with layers and legends. 1. OSPR not found.

SU 46 (41/46) *Newbury*   10 by 10   Revision a 1909-38 c 1956 f 1966

| 1 | A | CPr | OPr | nd:po | - | - | 6.49 | 4BEoo3 | RB | 0 | 2/49 | 4/49 | |

1	2	3	4	5	6	7	8	9	10	11	12	13	14
1	B	CPr		nd:po	-	-	6.49	4BEoo3	RB	0			
3	C	CPx	OPx	1961mc	-	ac	6.61	4oooX3		1	7/61	5/61	
3	C/*	CPz,0z	OPz	1961mc	1966r	acf	6.61	4oooX3		1	8/66	9/66	

AA 1.7.46/NT 3.2.49 on A(A), 3/-, OSPR 4/49. M on A, with legends, B; A. with layers and legends.

SU 47 (41/47) *Chieveley*   10 by 10   Revision a 1909-32 c 1956 f 1973

1	A	CPr	OPr	nd:po	-	-	1.49	4BEoo3	RB	0	2/49	2/49	
3	A	CPt		1949pc	-	-	1.49	4BEoo3	RB	0			
3	B	CPw,y,z	OPv	1960mc	-	ac	6.60	4oooX3		1	6/60	4/60	1
3	B/*	CPz	OPz	1960ᶜms	nd:r	acf	6.60	4oooX3		1	10/73	11/73	

M on A, B. 1. Welford Airfield.

SU 48 (41/48) *Harwell*   10 by 10   Revision a 1909-32 c 1938-40 d 1938-56
e 1947 f 1965 g 1973

1	A	CPr	OPr	nd:po	-	-	1.49	4BEoo3	RB	1	2/49	3/49	
1	B	CPs		1949pc	1950c	ace	6.50	4BEoo3	RB	1			
3	C	CPw,x	OPv	1960mc	-	ad	6.60	4oooX3		1	2/60	2/60	
3	C/	CPz	OPz	1960mc	1966ch	adf	6.60	4oooX3		1			
3	C//*	CPz	OPz	1960ᶜms	nd:r	adg	6.60	4oooX3		1	3/74	6/74	

M on B, C.

SU 49 (41/49) *Abingdon*   10 by 10   Revision a 1910-32 c 1938-56 f 1976

1	A	CPr	OPr	nd:po	-	-	1.49	4BEoo3	RB	12	12/48	11/48	1
2	B	CPt		1948pc	1953c	-	6.53	4BooX3	B	12			
3	C	CPw,y,z	OPv	1959mc	-	ac	6.59	4oooX3		12	9/59	10/59	2
3	C/*	CFz	OFz	1959ᶜms	nd:r	acf	6.59	4oooX3		12	3/77	3/77	

M on A, C. 1. Showing the Milton and Steventon complexes. 2. Abingdon Airfield.

SU 50 (41/50) *Fareham*   10 by 10   Revision a 1907-31 c 1938-42 d 1938-57
e 1947 f 1960 g 1965

1	15046	CPr		1946po	-	-	1.47	4BE2X3	R	33	12/46	-	1
1	A		OPr	1946po	-	-	1.47	4BE2X3	R	33	-	2/47	
3	B	CPv	OPv	1956pc	-	ace	6.56	4oooX3		11	1/57	9/57	2
3	C	CPx	OPx	1961mc	-	adf	6.61	4oooX3		13	10/61	9/61	3
3	C/	CPz	OPz	1961mc	1965ch	adg	6.61	4oooX3		33			

AA 1.9.46 on A(A), 3/-, OSPR 2/47. M on A. 1. The following fortifications are shown: Fort Wallington, Fort Fareham, Fort Elson (later named, 5903), Fort Brockhurst, Fort Rowner, Fort Grange. There are blank areas at 5203, 5903. 2. Grid North is incorrectly shown as west of True North, later corrected. 3. Lee-on-Solent Airfield.

SU 51 (41/51) *Bishops Waltham*   10 by 10   Revision a 1907-11 c 1938-57 f 1960

1	A	CPr	OPr	1947po	-	-	6.47	4BE2X3	RB		1/48	7/47	
1	B	CPr		nd:po			6.47	4BEoo3	RB				M
3	B	CPt		1949pc	-	-	6.47	4BEoo3	RB				
3	B/	CPw		1949mc	1958mc	a	6.47	4BooX3					
3	B//	CPw		1949mc	1960mc	a	6.47	4BooX3					
3	C	CPx,z	OPx	1961mc	-	acf	6.61	4oooX3			10/61	10/61	

AA 1.12.46/NT 1.6.47 on A(A), 3/-, OSPR 7/47. M on B.

SU 52 (41/52) *Cheriton Hampshire*   10 by 10   Revision a 1908-40 c 1938-57 f 1961

1	A	CPr	OPr	1947po	-	-	6.47	4BE2X3	B		2/47	5/47	
1	B	CPr		nd:po	-	-	6.47	4BEoo3	B				
3	C	CPv.z	OP?	1958pc	-	ac	6.58	4oooX3			3/58	3/58	
3	C	CPx,z		1958pc	-	acf	6.58	4oooX3					

AA 1.12.46 on A(A), 3/-, OSPR 5/47. M on B.

SU 53 (41/53) *New Alresford*   10 by 10   Revision a 1908-31 c 1938-57 f 1962

1	A	CPr	OPr	1948po	-	-	6.48	4BE4X3	RB		4/48	5/48	
3	A	CPu		1948pc	-	-	6.48	4BooX3	RB				
3	B	CPv	OP?	1958pc	-	ac	6.58	4oooX3			3/58	5/58	
3	B/	CPy,z	OPz	1958mc	1964ch	acf	6.58	4oooX3					

M on A, B.

1	2	3	4	5	6	7	8	9	10	11	12	13	14

SU 54 (41/54) *Micheldever*   10 by 10   Revision a 1908-30 c 1938-56 f 1972

1	A	CPr	OPr	nd:po	-	-	1.49	4BEoo3	R		12/48	1/49	
3	B	CPv,z	OPv	1958pc	-	ac	6.58	4oooX3			5/58	8/58	
3	B/*	CPz,0z	OPz	1958ᶜms	nd:r	acf	6.58	4oooX3			1/73	2/73	

M on A, with legends, A.

SU 55 (41/55) *Kingsclere*   10 by 10   Revision a 1909-30 c 1938-56

| 1 | A | CPr | OPr | nd:po | - | - | 1.49 | 4BEoo3 | RB | | 8/48 | 9/48 | |
| 3 | B | CPw.x,z,0z | OPv | 1958pc | - | ac | 6.58 | 4oooX4 | | | 10/58 | 10/58 | |

M on A, with legends, A; A, with layers and legends.

SU 56 (41/56) *Thatcham*   10 by 10   Revision a 1909-32 c 1938-40 d 1938-56 e 1947

1	A	CPr	OPr	nd:po	-	-	1.49	4BEoo3	RB	03	10/48	11/48	1
1	B	CPs		1948pc	1950c	ace	6.50	4BEoo3	RB	03			
3	B/	CPw		1948pc	1959c	ace	1.52	4BooX3		10			2
3	C	CPw.z	OPv	1961mc	-	ad	6.61	4oooX3		10	11/60	11/60	

AA 1.7.48 on A(A), 3/-, OSPR 11/48. M on A, with legends, B, C; A, with layers and legends. 1. The airport at Aldermaston is shown (unnamed) on this edition. 2. Greenham Common Airfield.

SU 57 (41/57) *Hermitage Berkshire*   10 by 10   Revision a 1910-32 b 1909-32 c 1938 d 1938-56 e 1947 f 1972

1	A	CPr	OPr	nd:po	-	-	1.49	4BEoo3	RB	1	9/48	10/48	
1	B	CPs		1948pc	1950c	ace	6.50	4BEoo3	RB	1			
3	C	CPw,x,z	OPv	1960mc	-	bd	6.60	4oooX3		3	2/60	2/60	
3	C/*	CPz,0z	OPz	1960ᶜms	nd:r	bdf	6.60	4oooX3		3	9/73	9/73	

M on B.

SU 58 (41/58) *Blewbury*   10 by 10   Revision a 1910-32 c 1956

1	A	CPr	OPr	1948po	-	-	6.48	4BEoo3	RB	1	7/48	8/48	
1	A	CPs		1948pc	1951	-	1.51	4BEoo3	RB	1			
3	A/	CPw		1948pc	1958c	a	1.51	4BooX3	B	1			
3	B	CPw,y,z.z	OPv	1960mc	-	ac	6.60	4oooX3		1	7/60	4/60	

M on A, B.

SU 59 (41/59) *Dorchester Oxfordshire*   10 by 10   Revision a 1910-23 c 1938-56 e 1947

1	A	CPr	OPr	nd:po	-	-	1.49	4BEoo3	RB	10	9/48	9/48	1
2	A	CPs		1948pc	1952	-	1.52	4Booo3	RB	10			
	A/												
3	A//	CPw		1948pc	1958mc	ae	1.52	4BooX3	B	10			
3	B	CPw.z,0z	OPv	1960mc	-	ac	6.60	4oooX3		31	7/60	8/60	

M on A, B. 1. The Milton complex is shown, with sidings from the GWR.

SU 60 (41/60) *Portsmouth (North)*   10 by 10   Revision a 1907-31 c 1938-42 d 1938-57 e 1947-48 f 1962 g 1978; j 1978 tide lines

1	15046	CPr		1946po	-	-	1.47	4BE2X3	RB	1	12/46	-	1
1	A		OPr	1946po	-	-	1.47	4BE2X3	RB	1	-	4/47	2
2	A		OPs	1946pc	-	-	1.52	4Booo3	RB	1			
3	B	CPt		1954pc	-	ace	6.54	4oooX3	B	1	8/54	-	
3	B/	CPv		1954pc	1957mc	ace	6.54	4oooX3	B	1			
3	B//		OPv	1954pc	1960mc	ace	6.54	4oooX3	B	1			
3	B//	CPx		1954mc	1960mc	ace	6.54	4oooX3	B	1			
3	C	CPz	OPz	1964mc	-	adf	6.64	4oooX3		1	11/64	1/65	3
3	C/	CPz	OPz	1964mc	-	adf	6.64	4oooX3		1			
3	C//*	CFz	OFz	1979ᶜms	nd:r	adgj	6.64	4oooX3		1	2/80	2/80	

AA 1.9.46 on A(A), 3/-, OSPR 4/47; AA 1.11.53 on A(B), 4/-, OSPR 6/54. M on A. 1. The naval dockyard, and areas on Gosport (6101) are blank. Some fortifications are shown: Fort Nelson, Fort Southwick, Fort Widley, Fort Purbrook, Farlington Redoubt, Moat. 2. Not recorded, and specification unconfirmed. 3. Portsmouth City Airport. The naval dockyard area is shown in full, though there are still blank spaces on the Gosport side.

1	2	3	4	5	6	7	8	9	10	11	12	13	14

SU 61 (41/61) *Hambledon Hampshire*   10 by 10   Revision a 1907-30 b 1904 c 1938-47 d 1938-57 e 1947 f 1962

1	A	CPr	OPr	1947po	-	-	6.47	4BE2X3	RB		9/47	11/47	
1	B	CPs		1950pc	-	ace	6.50	4BEoo3	RB		12/50	-	
3	B		OPs	1950pc	-	ace	6.50	4BEoo3	B		-	8/54	
3	C	CPy,z,0z	OPy	1963mc	-	bdf	6.63	4oooX3			11/63	11/63	

AA 1.9.46 on A(A), 3/-, OSPR 11/47; AA 1.11.53 on B(B), 4/-, OSPR 8/54. M on A.

SU 62 (41/62) *West Meon*   10 by 10   Revision a 1908 c 1956-57

1	A	CPr	OPr	nd:po	-	-	1.49	4BEoo3	RB		11/48	11/48	
1	B	CPr		nd:po	-	-	1.49	4BEoo3	RB				
1	B	CP		1948pc	1949c	-	1.49	4BEoo3	RB				M
3	B/	CPv		1948pc	1956c	a	1.49	4BooX3	B				
3	C	CPw,x,z	OPv	1959mc	-	ac	6.59	4oooX4			4/59	7/59	

M on B.

SU 63 (41/63) *Four Marks*   10 by 10   Revision a 1908-39 c 1938 d 1938-46 f 1961-63

1	A	CPr	OPr	nd:po	-	-	1.49	4BEoo3	RB		12/48	1/49	
1	B	CPr		nd:po	-	-	1.49	4BEoo3	RB				
1	B	CP		1948pc	1949c	-	1.49	4BEoo3	RB				M
3	B/	CPv		1948pc	1956c	ac	1.49	4BooX3	B				
3	C	CPz,0z	OPz	1964mc	-	adf	6.64	4oooX3			6/64	6/64	

AA 1.8.48 on A(A), 3/-, OSPR 1/49. M on B, C.

SU 64 (41/64) *Lasham*   10 by 10   Revision a 1908-30 c 1938-57 f 1969 g 1973

1	A	CPr	OPr	nd:po	-	-	1.49	4BEoo3	B	0	1/49	2/49	1
3	B	CPv,z	OPv	1958pc	-	ac	6.58	4oooX3		1	6/58	11/58	
3	B/	CPz	OPz	1958ms	1970ch	acf	6.58	4oooX3		3			
3	B//*	CPz,0z	OPz	1958^{c}ms	nd:r	acg	6.58	4oooX3		3	11/73	12/73	

AA 1.9.48 on A(A), 3/-, OSPR 2/49. M on A, with legends, A. 1. Bridle roads appear as bridle paths (6446).

SU 65 (41/65) *Basingstoke*   10 by 10   Revision a 1909-30 c 1938-56 f 1965 g 1972 h 1977 j 1981 selected k SUSI

1	A	CPr	OPr	nd:po	-	-	1.49	4BEoo3	RB		10/48	10/48	1
3	B	CPt		1948pc	1953c	-	6.53	4BooX3	B				
3	C	CPw,x	OPv	1958pc	-	ac	6.58	4oooX3			8/58	10/58	
3	C/	CPz	OPz	1958mc	1967ch	acf	6.58	4oooX3					
3	C//*	CPz	OPz	1958^{c}ms	nd:r	acg	6.58	4oooX3			1/73	2/73	
3	C//*/*	CFz	OFz	1958^{c}ms	nd:r	ach	6.58	4oooX3					2
3	C//*/*/*	CFz	OFz	1981^{c}m-s	nd:br	achjk	6.58	4oooX3			1/82	1/82	3

M on A, with legends, A, C. 1. Showing the Central Ordnance Depot at Bramley, also Park Prewett Mental Hospital with its railway connection. 2. OSPR not found. 3. The areas of building development are shaded grey.

SU 66 (41/66) *Mortimer*   10 by 10   Revision a 1908-09 c 1936 d 1936-56 e 1947 f 1974

1	A	CPr	OPr	1947po	-	-	1.47	4BE2X3	RB	0	10/47	11/47	1
1	B	CPs		1951pc	-	ace	1.51	4BEoo3	R	3	6/51	-	
3	B/	CPw		1951pc	1958mc	ace	1.52	4BooX3		3			
3	B//	CPw		1951mc	1960mc	ace	1.52	4BooX3		1			
3	C	CPx,z	OPx	1961mc	-	ad	6.61	4oooX4		0	7/61	6/61	
3	C/*	CFz	OFz	1961^{c}ms	-	adf	6.61	4oooX4		0	5/75	5/75	

AA 1.9.47 on A(A), 3/-, OSPR 11/47. M on A, B, C. 1. Burghfield Royal Ordnance Factory is not shown. This is a classic example of a sheet with uncoloured second class roads, with none at all coloured in the early states.

SU 67 (41/67) *Reading (West)*   10 by 10   Revision a 1909-32 c 1938 d 1938-56 f 1963 g 1974

1	A	CPr	OPr	1947po	-	-	6.47	4BE1X3	RB		10/47	12/47	
1	A	CPs		1947pc	1951	-	6.51	4BEoo3	RB				
3	A/	CPv		1951pc	1957mc	ac	6.51	4BooX3	B				
3	A//	CPw		1951mc	1959mc	ac	6.51	4BooX3	B				
3	A///	CPw		1951mc	1961ch	ac	6.51	4BooX3	B				1
3	B	CPz	OPz	1964mc	-	adf	6.64	4oooX3			6/64	6/64	
3	B/*	CFz	OFz	1964^{c}ms	nd:r	adg	6.64	4oooX3			12/74	9/74	

AA 1.5.47/NT 17.11.47 on A(A), 3/-, OSPR 12/47. M on A. 1. NB This state still has uncoloured minor roads.

234

1	2	3	4	5	6	7	8	9	10	11	12	13	14

SU 68 (41/68) *Wallingford*   10 by 10   Revision a 1898-1918 c 1950 d 1950-56

1	A	CPr	OPr	nd:po	-	-	1.49	4BEoo3	RB		10/48	11/48	
1	B	CPr		nd:po	-	-	1.49	4BEoo3	RB				
2	B	CPs,t		1948pc	1949c	-	1.52	4Booo3	RB				
3	B/	CPv		1948pc	1957c	ac	1.52	4BooX3	B				
3	C	CPw.x,z	OPv	1960mc	-	ad	6.60	4oooX3			4/60	4/60	

M on A, with legends, B, C.

SU 69 (41/69) *Watlington Oxfordshire*   10 by 10   Revision a 1910-19 c 1938-56 f 1970; j 1970 airfield

1	A	CPr	OPr	nd:po	-	-	1.49	4BEoo3	B	00	3/49	4/49	
3	A	CPt		1949pc	1953rpb	-	6.52	4BooX3	B	00			
3	B	CPw,x,z	OPv	1959mc	-	ac	6.59	4oooX3		11	9/59	10/59	1
3	B/*	CPz,0z	OPz	1959^{c}ms	nd:ar	acfj	6.59	4oooX3		33	11/71	1/72	2

M on A, B. 1. Benson Airfield. 2. Chalgrove Airfield.

SU 70 (41/70) *Havant*   10 by 10   Revision a 1909-37 c 1938-57 f 1965 g 1971

1	A	CPr	OPr	1947po	-	-	1.48	4BE2X3	RB	1	9/47	11/47	
1	B	CPr		nd:po	-	-	1.48	4BEoo3	RB	1			
2	B		OPs	1948pc	-	-	1.48	4BooX3	RB	1			
3	C	CPw,x	OPv	1959mc	-	ac	6.59	4oooX4		1	4/59	4/59	1
3	C/*	CPz	OPz	1959mc	1966ar	acf	6.59	4oooX4		3	7/66	10/66	
3	C/*/*	CPz,0z	OPz	1959^{c}ms	nd:r	acg	6.59	4oooX4		3	7/72	12/72	

AA 1.9.46 on A(A), 3/-, OSPR 11/47; AA 1.11.53 on B(B), 4/-, OSPR 6/54. M on B, C. 1. Thorney Island Airfield.

SU 71 (41/71) *Harting*   10 by 10   Revision a 1908-31 c 1938-57

1	A	CPr		1948po	-	-	1.48	4BE2X3	RB		2/48	-	
1	B		OPr	1948po	-	-	1.48	4BE2X3	RB		-	6/48	
1	C	CPr		nd:po	-	-	1.48	4BEoo3	RB				
3	D	CPv	OPs	1948pc	1953c	-	6.53	4BooX3	B				
3	E	CPw,x,z,0z	OP?,0z	1958mc	-	ac	6.58	4oooX4			1/59	3/59	

AA 1.4.48 on B(A), 3/-, OSPR 6/48; AA 1.11.53 on D(B), 4/-, OSPR 1/55. M on C, E.

SU 72 (41/72) *Petersfield*   10 by 10   Revision a 1908-31 c 1938 d 1938-57

1	A	CPr	OPr	nd:po	-	-	6.48	4BEoo3	RB		8/48	9/48	
1	B	CPr		nd:po	-	-	6.48	4BEoo3	RB				
2	B	CPt		1948pc	49c/52rpb	-	6.48	4BooX3	RB				
3	B/	CPv		1948pc	1957mc	ac	6.48	4BooX3					
3	C	CPw,x,z	OPv	1958mc	-	ad	6.58	4oooX3			2/59	3/59	

AA 1.4.48 on A(A), 3/-, OSPR 9/48. M on B.

SU 73 (41/73) *Alton Hampshire*   10 by 10   Revision a 1908-09 c 1938-39 d 1956-57
          f 1963 g 1972

1	A	CPr	OPr	nd:po	-	-	1.49	4BEoo3	RB		2/49	3/49	1
1	B	CPr		nd:po	-	-	1.49	4BEoo3	RB				
3	B/												
3	B//	CPx		1949mc	1961mc	ac	1.49	4BooX3	B				
3	C	CPz		1964mc	-	adf	6.64	4oooX3			7/64	11/64	
3	C/*	CPz,0z	OPz	1964^{c}ms	nd:r	adg	6.64	4oooX3			5/73	6/73	

AA 1.8.48/NT 15.1.49 on A(A), 3/-, OSPR 3/49. M on B, C. 1. Showing the military railway.

SU 74 (41/74) *Froyle*   10 by 10   Revision a 1908-30 c 1938-56 f 1962 g 1972

1	A	CPr	OPr	nd:po	-	-	1.49	4BEoo3	RB	0	10/48	12/48	
3	A	CPt		1948pc	-	-	1.49	4BEoo3	RB	0			
3	B	CPw	OPv	1958pc	-	ac	6.58	4oooX3		1	6/58	10/58	1
3	B/	CPy,z	OPz	1958mc	1963ch	acf	6.58	4oooX3		1			2
3	B//*	CPz,0z	OPz	1958^{c}ms	nd:r	acg	6.58	4oooX3		1	3/73	5/73	

AA 1.9.48 on A(A), 3/-, OSPR 12/48. M on A, with legends, A, B, B/. 1. Odiham Aerodrome. 2. Odiham Airfield.

1	2	3	4	5	6	7	8	9	10	11	12	13	14

SU 75 (41/75) *Hartley Wintney* 10 by 10 Revision a 1909-30 c 1938-56 f 1972
1	A	CPr	OPr	1948po	-	-	6.48	4BEoo3	RB	0	7/48	8/48	
1	A	CPs		1948pc	1951	-	1.51	4BEoo3	RB	0			
3	B	CPv,x,z.z	OPv	1958pc	-	ac	6.58	4oooX3		3	6/58	8/58	
3	B/*	CPz,0z	OPz	1958^{c}ms	nd:r	acf	6.58	4oooX3		3	5/73	6/73	

AA 1.4.48 on A(A), 3/-, OSPR 8/48. M on A, with legends, A, B.

SU 76 (41/76) *Arborfield* 10 by 10 Revision a 1909-32 c 1938-47 d 1938-56 f 1974
1	A	CPr	OPr	1948po	-	-	1.48	4BE4X3			12/47	1/48	1
1	B	CPs		1950pc	-	ac	6.50	4BEoo3			12/50	-	
3	B	CPt		1950pc	-	ac	6.50	4Booo3					
3	C	CPw,x,z	OPv	1958mc	-	ad	6.58	4oooX3			11/58	10/58	
3	C/*	CFz	OFz	1958^{c}ms	nd:r	adf	6.58	4oooX3			11/75	8/75	

AA 1.11.47 on A(A), 3/-, OSPR 1/48. M on A, B, C. 1. Army Remount Depot, later Arborfield Garrison.

SU 77 (41/77) *Reading (East)* 10 by 10 Revision a 1910-33 c 1938 d 1938-56
e 1947 f 1961 g 1971
1	A	CPr	OPr	1947po	-	-	6.47	4BE2X3	R	1	11/47	1/48	1
1	B	CPs		1951pc	-	ace	1.51	4BEoo3	R	1	2/51	-	
2	B	CPt		1951pc	-	ace	1.52	4Booo3	R	1			
3	B/	CPw		1951mc	1959c	ace	1.52	4oooX3		1			
3	B//	CP?w,z		1951mc	1960c	ace	1.52	4oooX3		1			
3	C	CPz	OPz	1964mc	-	adf	6.64	4oooX3		1	6/64	6/64	
3	C/*	CPz	OPz	1964^{c}ms	nd:r	adg	6.64	4oooX3		1	1/73	4/73	2

AA 1.10.47 on A(A), 3/-, OSPR 1/48. M on A, B. 1. Reading Aerodrome. 2. With the M4 under construction.

SU 78 (41/78) *Henley-on-Thames* 10 by 10 Revision a 1897-1933 c 1938 d 1938-45
1	A	CPr	OPr	nd:po	-	-	6.48	4BEoo3	RB		8/48	9/48	
2	A	CPt		1948pc	1952rpb	-	6.48	4BooX3	RB				
3	A/	CPv		1948pc	1956mc	ac	6.48	4BooX3	RB				
3	B	CPw,x,z	OPv	1960mc	-	ad	6.60	4oooX3			6/60	8/60	

AA 1.4.48 on A(A), 3/-, OSPR 9/48. M on A, with legends, A.

SU 79 (41/79) *Stokenchurch* 10 by 10 Revision a 1918-19 c 1938 d 1938-56 f 1970 g 1977
1	A	CPr	OPr	nd:po	-	-	1.49	4BEoo3	RB		1/49	2/49	
2	A	CPu		1949pc	1952	-	1.49	4BooX3	RB				
3	A/		OPv	1949pc	1957c	ac	1.49	4BooX3	B				
3	B	CPw,y,z	OPv	1960mc	-	ad	6.60	4oooX3			4/60	4/60	
3	B/*	CPz	OPz	1960^{c}ms	nd:r	adf	6.60	4oooX3			7/71	7/71	
3	B/*/*	CFz	OFz	1960^{c}ms	nd:r	adg	6.60	4oooX3			7/78	7/78	

M on A.

SU 80 (41/80) *Chichester* 10 by 10 Revision a 1909-33 c 1938-57
1	A	CPr	OPr	nd:po	-	-	6.48	4BEoo3	RB	0	11/48	1/49	
1	B	CPr		nd:po	-	-	6.48	4BEoo3	RB	0			
3	B	CPt		1948pc	1949c	-	6.48	4BEoo3	RB	0			
3	C	CPw,x,z.z	OPv	1958mc	-	ac	6.58	4oooX3		2	1/59	1/59	1

AA 1.9.48/NT 18.11.48 on A(A), 3/-, OSPR 1/49. M on B. 1. Showing Goodwood motor circuit, on the site of West Hampnett airfield.

SU 81 (41/81) *Cocking* 10 by 10 Revision a 1910 c 1957 e 1947
1	A	CPr	OPr	1948po	-	-	1.48	4BEoo3	RB		5/48	6/48	
1	B	CPs		1948pc	1951c	ae	1.51	4BEoo3	RB				
3	C	CPw,x,z,0z	OPv	1958mc	-	ac	6.58	4oooX4			12/58	1/59	

M on B.

SU 82 (41/82) *Midhurst* 10 by 10 Revision a 1908-30 c 1938-57
1	A	CPr	OPr	1948po	-	-	6.48	4BEoo3	R		5/48	5/48	
1	B	CPr		nd:po	-	-	6.48	4BEoo3	R				
2	B	CPt		1948pc	1949c	-	6.48	4Booo3	R				
3	C	CPw,x,z	OPv	1958mc	-	ac	6.58	4oooX3			4/59	5/59	

M on B.

1	2	3	4	5	6	7	8	9	10	11	12	13	14

SU 83 (41/83) *Hindhead* 10 by 10 Revision a 1908-13 c 1938 d 1938,57 e 1947 f 1963

1	A	CPr	OPr	1948po	-	-	6.48	4BEoo3	RB		7/48	8/48	1
1	B	CPr		nd:po	-	-	6.48	4BEoo3	RB				
1	B B/	CPs		1948pc	1951c	ace	6.51	4BEoo3	RB				
3	B//	CPw		1948mc	1959c	ace	1.52	4BooX3	B				
3	B///	CPw.x		1948mc	1961ch	ace	1.52	4BooX3	B				
3	C	CPz,0z	OPz,0z	1964mc	-	adf	6.64	4oooX3			6/64	9/64	

AA 1.4.48/NT 15.6.48 on A(A), 3/-, OSPR 8/48. M on A, with legends, B, C. 1. Showing the military railway.

SU 84 (41/84) *Farnham Surrey* 10 by 10 Revision a 1909-34 b 1909-35 c 1938 d 1938-56
f 1964 g 1974

1	A	CPr	OPr	nd:po	-	-	1.49	4BEoo3	RB		9/48	10/48	
1	B	CPr		nd:po	-	-	1.49	4BEoo3	RB				
2	B	CPt		1948pc	49c/52rpb	-	1.52	4BooX3	RB				
3	C	CPv,x	OPv	1958pc	-	bd	6.58	4oooX3			5/58	7/58	
3	C/	CPz	OPz	1958mc	1965ch	bdf	6.58	4oooX3					
3	C//*	CFz	OFz	1958ᶜms	nd:r	bdg	6.58	4oooX3			11/74	8/74	

AA 1.7.48/NT 6.9.48 on A(A), 3/-, OSPR 10/48. M on A, with legends, B, C, C/; A, with layers and legends.

SU 85 (41/85) *Farnborough Hampshire* 10 by 10 Revision a 1909-35 c 1938-39 d 1938-56 f 1972
j 1969 airfield k 1969 gas installation

1	A	CPr	OPr	nd:po	-	-	1.49	4BEoo3	RB	00	12/48	1/49	1
3	A		OPs	1948pc	1950	-	1.52	4Booo3	RB	00			
3	B	CPw	OPv	1958pc	-	ad	6.58	4oooX3		30	9/58	11/58	2
3	B/	CPx,z		1958mc	1961ch	ad	6.58	4oooX3		30			
3	B//*	CPz	OPz	1958ᶜms	nd:air	adfjk	6.58	4oooX3		33	3/73	4/73	3

AA 1.7.48 on A(A), 3/-, OSPR 1/49; AA 1.10.53 on A(B), 4/-, OSPR 4/54. M on A, with legends, A, B, B/; A, with layers and legends. 1. Showing details of the Aldershot army properties. 2. Blackbushe Airport. 3. The actual revision notice wording is: "Royal Aircraft and National Gas Turbine establishments revised 1969".

SU 86 (41/86) *Wokingham* 10 by 10 Revision a 1907-32 c 1937-39 d 1937-56 e 1947
f 1961 g 1966 h 1972

1	A	CPr	OPr	nd:po	-	-	1.49	4BEoo3	RB		12/48	1/49	1
1	B	CPs		1951pc	-	ace	6.51	4BEoo3	RB		11/51	-	
2	B	CPt	OPv	1951pc	-	ace	1.52	4Booo3	RB		-	9/57	
3	C	CPv	OPv	1958pc	-	ad	6.58	4oooX3			5/58	7/58	
3	C/	CPx		1958pc	-	ad	6.58	4oooX3					
3	C//	CPz	OPz	1958mc	1967ch	adg	6.58	4oooX3					
3	C///*	CPz	OPz	1958ᶜms	nd:r	adh	6.58	4oooX3			3/73	4/73	

AA 1.11.48/NT 19.11.48 on A(A), 3/-, OSPR 1/49. M on A, with legends, A, B, C, C/. 1. The locations marked include Royal Military College, Sandhurst, RAF Staff College, Bracknell, Ambarrow (Ministry of Supply).

SU 87 (41/87) *White Waltham* 10 by 10 Revision a 1910-38 c 1956 f 1963 g 1972

1	A	CPr	OPr	1948po	-	-	1.48	4BEoo3	RB	1	7/48	8/48	
1	B	CPr	OPs	nd:po	-	-	1.48	4BEoo3	RB	1			
3	B	CPt		1949pc	-	-	1.48	4BooX3	RB	1			
3	C	CPx	OPx	1961mc	-	ac	6.61	4oooX3		1	3/61	3/61	1
3	C/	CPz	OPz	1961mc	1964ch	acf	6.61	4oooX3		1			
3	C//*	CPz	OPz	1961ᶜms	nd:r	acg	6.61	4oooX3		1	3/73	4/73	

AA 1.4.48 on A(A), 3/-, OSPR 8/48; AA 1.7.51 on B(A), 3/-, OSPR 9/51. M on A, with legends, B, C. 1. Showing the M4 under construction.

SU 88 (41/88) *Marlow* 10 by 10 Revision a 1919-32 c 1938-56 f 1963 g 1972

1	A	CPr	OPr	nd:po	-	-	1.49	4BEoo3	RB		10/48	12/48	
1	B	CPr		nd:po	-	-	1.49	4BEoo3	RB				
2	C	CPt,u		1948pc	1953c	-	1.49	4BooX3	RB				
3	D	CPw	OPv	1960mc	-	ac	6.60	4oooX3			6/60	8/60	
3	D/	CPy,z		1960mc	1963ch	acf	6.60	4oooX3					
3	D//*	CPz	OPz	1960ᶜms	nd:r	acg	6.60	4oooX3			10/73	12/73	

AA 1.7.48/NT 2.11.48 on A(A), 3/-, OSPR 12/48. M on B, with legends, C. SS: Soil, outline, 1973 on D//*; Land Use Capability, outline, 1976 on D/*.

237

1	2	3	4	5	6	7	8	9	10	11	12	13	14

SU 89 (41/89) *High Wycombe*   10 by 10   Revision a 1919-24 c 1938 d 1938-56 e 1947-48
f 1969 g 1978; j 1966 building

1	A	CPr		nd:po	-	-	6.49	4BEoo3	RB	0	4/49	-	
1	B		OPr	nd:po	-	-	6.49	4BEoo3	RB	0	-	5/49	
1	C	CPs		1949pc	1950c	ace	6.50	4BEoo3	RB	0			
3	C/	CPw		1949mc	1958c	ace	6.50	4BooX3	B	0			
3	D	CPw,y.z	OPv	1961mc	-	ad	6.61	4oooX3		1	11/60	11/60	
3	D/*	CPz	OPz	1961ms	1969br	adfj	6.61	4oooX3		1	10/69	1/70	
3	D/*/*	CFz	OFz	1961ᶜms	nd:r	adgj	6.61	4oooX3		1	11/78	11/78	

AA 1.3.49/NT 3.2.49 on B(A), 3/-, OSPR 5/49. M on B, with legends, C.

SU 90 (41/90) **& Part of SZ 99** *Bognor Regis*   10 by 12   Revision a 1910-34 c 1938-40 d 1938-57
e 1946-47 f 1965 g 1972

1	A	CPr	OPr	nd:po	-	-	1.49	4BEoo3	RB	00	12/48	1/49	
1	B	CPs	OP?	1951pc	-	ace	6.51	4BEoo3	RB	00	11/51	2/58	
3	C	CPw	OPv	1958mc	-	ad	6.58	4oooX3		11	1/59	3/59	1
3	C/	CPw		1958mc	1961ch	ad	6.58	4oooX3		11			
3	C//*	CPz	OPz	1958mc	1966r	adf	6.58	4oooX3		31	7/66	8/66	
3	C//*/*	CPz	OPz	1958ᶜms	nd:r	adg	6.58	4oooX3		31	12/72	2/73	

AA 1.9.48 on A(A), 3/-, OSPR 1/49. M on A, C. 1. Tangmere Airfield, Ford Airfield.

SU 91 (41/91) *Sutton Sussex*   10 by 10   Revision a 1909-10 c 1945-47 d 1945-57 e 1947

1	A	CPr	OPr	nd:po	-	-	6.48	4BEoo3	RB		8/48	9/48	
1	B	CPs		1950pc	-	ace	6.50	4BEoo3	RB				
2	C	CPt		1950pc	1953c	ace	1.51	4BooX3	RB				
3	D	CPw,x,z	OPv	1958mc	-	ad	6.58	4oooX3			1/59	12/58	

M on B.

SU 92 (41/92) *Petworth*   10 by 10   Revision a 1909-10 c 1938-57

1	A	CPr	OPr	1948po	-	-	6.48	4BEoo3	RB		5/48	5/48	
1	B	CPr		nd:po	-	-	1.48	4BEoo3	RB				
3	B	CPt		1948pc	1949c	-	1.48	4BEoo3	RB				
3	C	CPw,x,z	OPv	1958mc	-	ac	6.58	4oooX3			1/59	4/59	

M on B.

SU 93 (41/93) *Haslemere*   10 by 10   Revision a 1913 c 1938 d 1954-57 e 1947 f 1963

1	A	CPr	OPr	1948po	-	-	1.48	4BE4X3	RB		4/48	6/48	
1	B	CPr		nd:po	-	-	1.48	4BEoo3	RB				
1	C	CPs		1948pc	1950c	ace	6.50	4BEoo3	RB				
	C/												
3	C//	CPw.x		1948mc	1960mc	ace	1.52	4BooX3	B				
3	D	CPz,0z	OPz	1965mc	-	adf	6.65	4oooX3			1/65	2/65	

AA 1.3.48/NT 12.1.48 on A(A), 3/-, OSPR 6/48. M on A, C.

SU 94 (41/94) *Godalming*   10 by 10   Revision a 1913-34 c 1938 d 1938-56 e 1947-48

1	A	CPr	OPr	nd:po	-	-	1.49	4BEoo3	RB		9/48	11/48	
1	B	CPs		1950pc	-	ace	6.50	4BEoo3	RB		1/51	-	
2	B		OPs	1950pc	-	ace	6.50	4Booo3	RB				1
3	B	CPt		1950pc	-	ace	6.50	4Booo3	RB				
3	B/	CPu		1950pc	1956mc	ace	6.50	4BooX3	B				
3	B//												
3	B///	CPw		1950mc	1960mc	ace	6.50	4BooX3	B				
3	C	CPx.z.0z	OPx	1961mc	-	ad	6.61	4oooX4			6/61	5/61	

AA 1.7.48/NT 15.9.48 on A(A), 3/-, OSPR 11/48; AA 1.5.52/NT 15.9.48 on B(B), 4/-, OSPR 1/53. M on A, with legends, B, C. 1. Recorded so far only as the base to the Administrative Areas edition.

SU 95 (41/95) *Pirbright*   10 by 10   Revision a 1912-35 c 1938 d 1938-56 e 1947-48

1	A	CPr	OPr	nd:po	-	-	1.49	4BEoo3	RB		11/48	1/49	1
1	B	CPr		1948pc	1950c	ace	6.50	4BEoo3	RB				
2	B	CPt		1948pc	1950c	ace	1.52	4BooX3	RB				
3	B/												

238

1	2	3	4	5	6	7	8	9	10	11	12	13	14
3	B//	CPw		1948mc	1960mc	ace	1.52	4BooX3	B				
3	C	CPx,z	OPx	1961mc	-	ad	6.61	4oooX4			5/61	5/61	

AA 1.7.48 on A(A), 3/-, OSPR 1/49. M on A, B, C. 1. The military sites depicted include Inkerman Barracks and encampments at Bisley, Pirbright, Deepcut and Blackdown, with the military railway closed beyond Pirbright.

SU 96 (41/96) *Chobham Common*   10 by 10   Revision a 1909-34 c 1937-38 d 1937-56 e 1947-48
                                                    f 1961 g 1968 h 1972

1	A	CPr	OPr	nd:po	-	-	1.49	4BEoo3	RB		11/48	1/49	
1	B	CPs		1951pc	-	ace	1.51	4BEoo3	RB		4/51	-	
2	B		OPs	1951pc	-	ace	1.52	4Booo3	RB		-	3/54	
3	B/	CPu		1951pc	1956mc	ace	1.51	4BooX3	B				
3	B//	CPw		1951pc	1958mc	ace	1.51	4BooX3	B				
3	C	CPx,z	OPx	1961mc	-	adf	6.61	4oooX3			9/61	9/61	
3	C/	CPz	OPz	1961ms	1968ch	adg	6.61	4oooX3					
3	C//*	CPz	OPz	1961ᶜms	nd:r	adh	6.61	4oooX3			2/73	4/73	

AA 1.8.48 on A(A), 3/-, OSPR 1/49. M on A, with legends, B, B/, C.

SU 97 (41/97) *Windsor*   10 by 10   Revision a 1931-34 c 1938 d 1955-56 e 1948
                                              f 1961 g 1963 h 1968

1	A	CPr	OPr	1948po	-	-	6.48	4BEoo3	RB		5/48	7/48	
1	B	CPs		1950pc	-	ace	6.50	4BEoo3	R		11/50	-	
3	B	CPu		1950pc	-	ace	1.52	4Booo3	R				
3	B/	CPv		1950pc	1957mc	ace	1.52	4BooX3					
3	B//	CPw		1950mc	1960mc	ace	1.52	4BooX3					
3	B///	CPx,y,z	OPy	1950mc	1961ch	acef	1.52	4BooX3					1
3	C	CPz	OPz	1965mc	-	adg	6.65	4oooX4			3/65	5/65	
3	C/*	CPz	OPz	1965ms	1969r	adh	6.65	4oooX4			3/69	6/69	

AA 1.3.48/NT 10.10.47 on A(A), 3/-, OSPR 7/48. M on A, with legends, B, B///. 1. Showing the M4 under contruction.

SU 98 (41/98) *Burnham Beeches*   10 by 10   Revision a 1923-33 c 1938 d 1938-56 e 1947-48
                                                       f 1961 g 1972

1	A	CPr		1948po	-	-	6.48	4BEoo3	RB		7/48	-	
1	A		OPr	nd:po	-	-	6.48	4BEoo3	RB		-	9/48	
1	B	CPr		1950pc	-	ace	6.50	4BEoo3	RB				
3	B	CPu	OPs	1950pc	-	ace	6.48	4BEoo3	RB				
3	C	CPw	OPv	1960mc	-	ad	6.60	4oooX3			6/60	6/60	
3	C/*	CPx,z	OPx	1960mc	1962r	adf	6.60	4oooX3			12/62	8/62	
3	C/*/*	CPz	OPz	1960ᶜms	nd:r	adg	6.60	4oooX3			5/73	7/73	

AA 1.3.48/NT 15.7.48 on A(A), 3/-, OSPR 9/48; AA 1.10.53/NT 12.4.54 on B(B), 4/-, OSPR 6/54. M on B, ?C (unrecorded), C/*.

SU 99 (41/99) *Beaconsfield*   10 by 10   Revision a 1897-1923 c 1938-50 d 1938-56
                                                    e 1947-48 f 1960

1	A	CPr	OPr	nd:po	-	-	1.49	4BEoo3	RB		10/48	11/48	
1	B	CPs		1951pc	-	ace	6.51	4BEoo3	RB		12/51	-	
2	B		OPs	1951pc	-	ace	6.51	4BEoo3	RB		-	7/52	
3	B	CPu		1951pc	-	ace	1.52	4BooX3	RB				
3	B/	CPw		1951pc	1958mc	ace	1.52	4BooX3	RB				
3	C	CPx,z,0z	OPx	1961mc	-	adf	6.61	4oooX3			5/61	5/61	

M on B, C.

SV 80 (00/80) [*St Agnes*]   10 by 10   Revision a 1906 c 1931 d 1958

| 1 | A | CPs | OPs | 1951pc | - | ac | 1.51 | 4BEoo4 | | | 2/51 | 3/51 | |
| 3 | B | CPw | OPv | 1960mc | - | ad | 6.60 | 4oooX4 | | | 12/59 | 1/60 | |

Heights are in feet above the assumed Mean Sea Level at Hugh Town, St Mary's (but edition A gives Newlyn).

SV 81 (00/81) [*Tresco*]   10 by 10   Revision a 1906 c 1931 d 1958

| 1 | A | CPs | OPs | 1951pc | - | ac | 1.51 | 4BEoo4 | | | 1/51 | 1/51 | |
| 3 | B | CPw,x | OPv | 1960mc | - | ad | 6.60 | 4oooX4 | | | 12/59 | 2/60 | |

Heights are in feet above the assumed Mean Sea Level at Hugh Town, St Mary's (but edition A gives Newlyn).

1	2	3	4	5	6	7	8	9	10	11	12	13	14

SV 91 (00/91) **& Part of SV 90** [St Mary's]   10 by 12   Revision a 1906 c 1931 d 1958
1	A	CPs	OPs	1950pc	-	ac	6.50	4BEoo3	0		12/50	1/51	
3	A	CPu		1950pc	-	ac	6.50	4BEoo3	0				
3	B	CPw	OPv	1960mc	-	ad	6.60	4oooX3	0		12/59	2/60	

Heights are in feet above the assumed Mean Sea Level at Hugh Town, St Mary's (but editions A give Newlyn).

Isles of Scilly   20 by 15   Revision a 1906 c 1958
	A	CPz		1964mc	-	ac	1964	4oo7X5	3		5/64	-	
	A/	CPz		1964ms	1969ch	ac	1964	4oo7X5	3				
	A//	CPz		1964^{c}ms	1974ch	ac	1964	4oo7X5	3				

The sheet co-ordinates are 79 km E to 99 km E, 4 km N to 19 km N. Heights are in feet above the assumed Mean Sea Level at Hugh Town, St Mary's. This map superseded the previous three sheets, as well as the map of the islands at two inches to the mile originating as part of the one-inch Fifth (Relief) Edition, which had been in print since 1933.

SW 32 (10/32) *Land's End*   10 by 10   Revision a 1906-07 c 1938 d 1938-58
1	A	CPs	OPs	1950pc	-	ac	6.50	4BEoo4	0		12/50	1/51	
2	A	CPt		1950pc	1952	ac	6.50	4Booo4	0				
3	B	CPw,x.z	OPv	1960mc	-	ad	6.60	4oooX4	1		1/60	2/60	1

1. Land's End (St Just) Aerodrome.

SW 33 (10/33) *St Just*   10 by 10   Revision a 1906 c 1938-58
1	A	CPs	OPs	1951pc	-	a	1.51	4BEoo4			1/51	1/51	
2	A	CPt		1951pc	1952	a	1.52	4Booo4					
3	B	CPw,z	OPv	1959mc	-	ac	6.59	4oooX4			11/59	12/59	

SW 42 (10/42) *Newlyn*   10 by 10   Revision a 1906-08 c 1938 d 1938-58 f 1963
1	A	CPs	OPs	1951pc	-	ac	1.51	4BEoo4			2/51	4/51	
2	A	CPt		1951pc	1953rpb	ac	1.51	4BooX4					
3	B	CPw	OPv	1960mc	-	ad	6.60	4oooX4			12/59	2/60	1
3	B/	CPy,z	OPz	1960mc	1963ch	adf	6.60	4oooX4					

AA 1.2.51 on A(A), 3/-, OSPR 4/51. 1. The Ordnance Survey Tidal Observatory is named.

SW 43 (10/43) **& Part of SW 44** *Penzance*   10 by 12   Revision a 1906 c 1942 d 1938-58
1	A	CPs	OPs	1951pc	-	ac	1.51	4BEoo3	R		2/51	4/51	
2	A	CPt		1951pc	1952	ac	1.51	4Booo3	R				
3	B	CPw.z,z+	OPv	1960mc	-	ad	6.60	4oooX4			2/60	4/60	

AA 1.2.51 on A(-), 3/-, OSPR 4/51.

SW 52 (10/52) *Prah Sands*   10 by 10   Revision a 1907 c 1938-58
1	A	CPs	OPs	1950pc	-	a	6.50	4BEoo4			10/50	12/50	
2	A	CPt		1950pc	1952rpb	a	6.50	4BooX4					
3	B	CPw,z.z	OPv	1959mc	-	ac	6.59	4oooX4			11/59	12/59	

SW 53 (10/53) *Hayle*   10 by 10   Revision a 1906-09 c 1938 d 1938-58
1	A	CPs	OPs	1951pc	-	ac	1.51	4BEoo4	R		1/51	2/51	
3	A	CPt		1951pc	-	ac	1.51	4BEoo4	R				
3	B	CPw,x.z	OPv	1960mc	-	ad	6.60	4oooX4			1/60	2/60	

AA 1.1.51 on A(A), 3/-, OSPR 2/51. SS: Soil, Land Use Capability, coloured, 1978 on B.

SW 54 (10/54) *St Ives Cornwall*   10 by 10   Revision a 1906 c 1938 d 1938-58
| 1 | A | CPs | OPs | 1950pc | - | ac | 6.50 | 4BEoo4 | | | 11/50 | 12/50 | |
| 3 | B | CPw,x.z.z | OPv | 1960mc | - | ad | 6.60 | 4oooX4 | | | 12/59 | 1/60 | |

M on A.

SW 61 (10/61) *Mullion*   10 by 10   Revision a 1906 c 1958
1	A	CPs	OPs	1950pc	-	a	6.50	4BEoo4	0		10/50	11/50	
3	A	CPu		1950pc	-	a	6.50	4BEoo4	0				
3	B	CPw,x,z	OPv	1960mc	-	ac	6.60	4oooX4	3		12/59	12/59	

M on A, B.

1	2	3	4	5	6	7	8	9	10	11	12	13	14

SW 62 (10/62) *Helston* 10 by 10 Revision a 1906-07 c 1938-46 d 1938-58 f 1966
1	A		CPs	OPs	1951pc	-	ac	1.51	4BEoo4	R	0	1/51	2/51	
3	A		CPu		1951pc	-	ac	1.51	4BEoo4	R	0			
3	B		CPw,x,z	OPv	1960mc	-	ad	6.60	4oooX4		1	4/60	2/60	1
3	B/		CPz,0z	OPz	1960mc	1967ch	adf	6.60	4oooX4		3			

AA 1.1.51 on A(A), 3/-, OSPR 2/51. 1. Culdrose Airfield.

SW 63 (10/63) *Camborne (South)* 10 by 10 Revision a 1906-07 c 1938 d 1938-58
1	A		CPs	OPs	1951pc	-	ac	1.51	4BEoo4	R		4/51	6/51
3	A		CPt		1951pc	-	ac	1.51	4BEoo4	R			
3	B		CPw,x,z	OPv	1960mc	-	ad	6.60	4oooX4			1/60	2/60

AA 1.4.51 on A(A), 3/-, OSPR 6/51.

SW 64 (10/64) *Camborne (North)* 10 by 10 Revision a 1906 c 1938 d 1938-58
1	A		CPs	OPs	1951pc	-	ac	1.51	4BEoo4	R	0	2/51	3/51
2	A		CPt		1951pc	1952	ac	1.51	4Booo4	R	0		
3	B		CPw,y,z.z	OPv	1960mc	-	ad	6.60	4oooX4		0	6/60	5/60

AA 1.2.51/NT 26.1.51 on A(A), 3/-, OSPR 3/51.

SW 71 (10/71) **& Part of SW 81** *Lizard* 12 by 10 Revision a 1906 c 1958
1	A		CPs	OPs	1951pc	-	a	1.51	4BEoo3			1/51	2/51
2	A		CPs		1951pc	1951	a	1.52	4Booo3				
3	B		CPw,z.z,z+	OPv	1960mc	-	ac	6.60	4oooX3			2/60	2/60

M on A.

SW 72 (10/72) **& Part of SW 82** *Helford River* 15 by 10 Revision a 1906 c 1958 f 1970
1	A		CPs	OPs	1951pc	-	a	1.51	4BEoo3			3/51	3/51
2	A		CPt		1951pc	1952	a	1.51	4Booo3				
3	B		CPw,z	OPv	1960mc	-	ac	6.60	4oooX3			2/60	2/60
3	B/		CPz		1960^{c}ms	nd:ch	acf	6.60	4oooX3				

M on A.

SW 73 (10/73) *Penryn* 10 by 10 Revision a 1906 c 1933-38 d 1933-58 f 1970
j 1968 reservoir
1	A		CPs	OPs	1951pc	-	ac	1.51	4BEoo4	R		4/51	6/51	
3	A		CPt		1951pc	-	ac	1.51	4BEoo4	R				
3	B		CPw,x,z	OPv	1960mc	-	ad	6.60	4oooX4			1/60	2/60	
3	B/*		CPz	OPz	1960^{c}ms	nd:v	adfj	6.60	4oooX4			6/71	7/71	1

AA 1.4.51 on A(A), 3/-, OSPR 6/51. 1. Stithians Reservoir.

SW 74 (10/74) *Redruth (East)* 10 by 10 Revision a 1906 c 1933-38 d 1933-58
1	A		CPs	OPs	1951pc	-	ac	1.51	4BEoo4	R		4/51	6/51
3	A		CPu		1951pc	-	ac	1.51	4BEoo4	R			
3	B		CPw,y,z.z	OPv	1960mc	-	ad	6.60	4oooX4			1/60	1/60

AA 1.4.51 on A(A), 3/-, OSPR 6/51.

SW 75 (10/75) **& Parts of SW 65 & SW 76** 12 by 12 Revision a 1906-33 c 1938 d 1938-58 f 1960
Perranporth
1	A		CPs	OPs	1951pc	-	ac	1.51	4BEoo4	R	0	3/51	3/51
3	A		CPt		1951pc	-	ac	1.51	4Baoo3	R	0		
3	A/		CPv		1951pc	1956mc	ac	1.51	4BooX3		0		
3	A//		CPw		1951pc	1958mc	ac	1.51	4BooX3		0		
3	B		CPz	OPz	1965mc	-	adf	6.65	4oooX3		3	4/65	6/65

AA 16.2.51/NT 20.11.50 on A(A), 3/-, OSPR 3/51.

SW 83 (10/83) **& Part of SW 93** *Falmouth* 15 by 10 Revision a 1906 b 1906-33 c 1933-38 d 1938-58
1	A		CPs	OPs	1951pc	-	ac	1.51	4BEoo3	R		2/51	4/51
2	A		CPt		1951pc	1952	ac	1.52	4Booo3	R			
3	B		CPw.x,z	OPv	1960mc	-	bd	6.60	4oooX3			5/60	8/60

AA 1.2.51/NT 16.2.51 on A(A), 3/-, OSPR 4/51.

1	2	3	4	5	6	7	8	9	10	11	12	13	14

SW 84 (10/84) *Truro*   10 by 10   Revision a 1906 c 1933-38 d 1933-58 f 1972
1	A	CPs	OPs	1951pc	-	ac	1.51	4BEoo4	R		4/51	5/51
3	A	CPt		1951pc	-	ac	1.51	4BEoo4	R			
3	A/	CPw		1951pc	1958mc	ac	1.51	4BooX4				
3	B	CPw,z	OPv	1960mc	-	ad	6.60	4oooX4			11/60	4/60
3	B/*	CPz	OPz	1960ᶜms	nd:r	adf	6.60	4oooX4			3/73	7/73

AA 1.2.51 on A(A), 3/-, OSPR 5/51.

SW 85 (10/85) *Summercourt*   10 by 10   Revision a 1905-06 b 1906 c 1958 f 1962 g 1975
| 1 | A | CPs | OPs | 1951pc | - | a | 1.51 | 4BEoo4 | R | | 2/51 | 3/51 |
| 3 | B | CPz | OPz | 1964mc | - | bcf | 6.64 | 4oooX4 | | | 10/64 | 10/64 |
| 3 | B/* | CFz | OFz | 1964ᶜms | nd:r | bcg | 6.64 | 4oooX4 | | | | 12/75 | 1

1. OSPR (coloured edition) not found.

SW 86 (10/86) **& Part of SW 76** *Newquay* 12 by 10 Revision a 1905-33 c 1938 d 1938-62 f 1962 g 1972
1	A	CPs	OPs	1951pc	-	ac	1.51	4BEoo3	R	00	2/51	4/51
3	A	CPu		1951pc	-	ac	1.51	4BEoo3	R	00		
3	B	CPz,z+	OPz	1964mc	-	adf	6.64	4oooX3		33	6/64	8/64
3	B/	CPz-	OPz-	1964ᶜms	nd:ch	adg	6.64	4oooX3		33		

AA 1.1.51/NT 22.2.51 on A(A), 3/-, OSPR 4/51.

SW 87 (10/87) *Trevose Head*   10 by 10   Revision a 1905-06 c 1938 d 1938-58 f 1963
1	A	CPs	OPs	1950pc	-	ac	6.50	4BEoo4		0	10/50	12/50
3	A	CPt		1950pc	-	ac	6.50	4BEoo4		0		
3	B	CPz	OPz	1964mc	-	adf	6.64	4oooX4		3	8/64	9/64

SW 94 (10/94) **& Part of SW 93** *Tregony**   10 by 12 Revision a 1906-33 c 1938 d 1938-58
1	A	CPs		1951pc	-	ac	6.51	4BEoo3	R		3/51	-
1	B		OPs	1951pc	-	ac	6.51	4BEoo3			-	6/51
3	A	CPt		1951pc	-	ac	6.51	4BEoo3				
3	C	CPw.z	OPv	1960mc	-	ad	6.60	4oooX3			12/59	2/60

AA 1.3.51/NT 3.3.50 on B(A), 3/-, OSPR 6/51.

SW 95 (10/95) *Nanpean*   10 by 10   Revision a 1905-33 c 1938 d 1958 f 1963
| 1 | A | CPs | OPs | 1951pc | - | ac | 1.51 | 4BEoo4 | R | | 3/51 | 6/51 |
| 3 | B | CPz,0z | OPz | 1965mc | - | adf | 6.65 | 4oooX4 | | | 1/65 | 2/65 |

AA 1.2.51 on A(A), 3/-, OSPR 6/51.

SW 96 (10/96) *St Columb Major*   10 by 10   Revision a 1905-06 c 1938 d 1938-58 f 1963 g 1978
1	A	CPs	OPs	1951pc	-	ac	1.51	4BEoo4	R		1/51	1/51
3	A	CPu		1951pc	-	ac	1.51	4BEoo4	R			
3	B	CPz	OPz	1964mc	-	adf	6.64	4oooX4			6/64	6/64
3	B/*	CFz	OFz	1964ᶜms	nd:r	adg	6.64	4oooX4			10/80	10/80

SW 97 (10/97) **& Part of SW 98** *Padstow* 10 by 12 Revision a 1905-06 c 1938 d 1938-58 c 1963
1	A	CPs	OPs	1951pc	-	ac	1.51	4BEoo3	R		4/51	5/51
2	A	CPt		1951pc	1952	ac	1.51	4Booo3				
3	A/	CPw		1951mc	1959mc	ac	1.51	4BooX3				
3	B	CPz	OPz	1964mc	-	adf	6.64	4oooX3			11/64	11/64

SX 04 (20/04) **& Part of SX 03** *Mevagissey* 10 by 12 Revision a 1906-34 b 1905-33 c 1938 d 1938-58
1	A	CPs	OPs	1950pc	-	ac	1.50	4BEoo3			11/50	12/50
3	A	CPu		1950pc	-	ac	1.50	4BEoo3				
3	B	CPw,x,z.z	OPv	1960mc	-	bd	6.60	4oooX3			7/60	5/60

AA 1.10.50/NT 2.10.50 on A(A), 3/-, OSPR 12/50.

SX 05 (20/05) *St Austell*   10 by 10   Revision a 1905-33 c 1938 d 1938-58 f 1963
1	A	CPs	OPs	1950pc	-	ac	6.50	4BEoo4	R		12/50	2/51
3	A/	CP?w,z		1950mc	1960mc	ac	6.50	4BooX4				
3	B	CPz	OPz	1965mc	-	adf	6.65	4oooX4			6/65	8/65

AA 1.12.50 on A(A), 3/-, OSPR 2/51.

1	2	3	4	5	6	7	8	9	10	11	12	13	14
SX 06 (20/06) *Bodmin*						10 by 10	Revision a 1905-06 c 1938 d 1958 f 1963						
1	A	CPs	OPs	1950pc	-	ac	6.50	4BEoo4	R		11/50	11/50	
3	B	CPz	OPz	1965mc	-	adf	6.65	4oooX4			2/65	3/65	
SX 07 (20/07) *St Tudy*						10 by 10	Revision a 1905-06 c 1958 f 1963						
1	A	CPs	OPs	1950pc	-	a	6.50	4BEoo4	R		12/50	1/51	
3	A/	CPw		1950mc	1958mc	a	6.50	4BooX4					
3	B	CPz	OPz	1964mc	-	acf	6.64	4oooX4			11/64	1/65	
SX 08 (20/08) **& Part of SX 09** *Tintagel*						10 by 12	Revision a 1905 c 1957-58 f 1963						
1	A	CPs	OPs	1950pc	-	a	6.50	4BEoo3	R	0	10/50	11/50	
3	B	CPz	OPz	1965mc	-	acf	6.65	4oooX3		1	1/65	2/65	
SX 15 (20/15) **& Part of SX 14** *Fowey*						10 by 12	Revision a 1905-33 c 1938 d 1958 f 1963						
1	A	CPs	OPs	1950pc	-	ac	6.50	4BEoo3	R		11/50	11/50	
	A/												
	A//												
3	A///	CPx		1950mc	1961ch	ac	6.50	4oooX3					
3	B	CPz	OPz	1965mc	-	adf	6.65	4oooX3			2/65	3/65	
SX 16 (20/16) *St Neot*						10 by 10	Revision a 1905-06 c 1940 d 1958 f 1963						
1	A	CPs	OPs	1950pc	-	ac	6.50	4BEoo3	R		11/50	12/50	
3	A/	CPv		1950pc	1956mc	ac	6.50	4BooX3					
3	B	CPz	OPz	1965mc	-	adf	6.65	4oooX3			1/65	1/65	
SX 17 (20/17) *Bodmin Moor (West)*						10 by 10	Revision a 1905 c 1958 f 1964						
1	A	CPs	OPs	1950pc	-	-	6.50	4BEoo3			10/50	11/50	
3	B	CPz	OPz	1964mc	-	acf	6.64	4oooX3			11/64	12/64	

M on A.

SX 18 (20/18) *Camelford*						10 by 10	Revision a 1905 c 1958 f 1963 g 1973						
1	A	CPs	OPs	1950pc	-	a	6.50	4BEoo3	R	0	10/50	10/50	
3	B	CPz	OPz	1965mc	-	acf	6.65	4oooX3		3	1/65	1/65	
3	B/	CPz	OPz	1965[c]ms	nd:ch	acg	6.65	4oooX3		3			

M on A. SS: Soil, Land Use Capability, outline, 1976 on B/.

SX 19 (20/19) *Crackington Haven*						10 by 10	Revision a 1905 c 1957-58 f 1963						
1	A	CPs	OPs	1950pc	-	a	6.50	4BEoo3			11/50	11/50	
3	A	CPv		1950pc	-	a	6.50	4BooX3					
3	B	CPz	OPz	1964mc	-	acf	6.64	4oooX3			11/64	12/64	
SX 25 (20/25) *Looe**						10 by 10	Revision a 1905 c 1938-48 d 1938-58 f 1963						
1	A	CPr	OPr	1947po	-	-	6.47	4BE2X3	R		6/47		1
1	B	CPr		1950pc	-	ac	6.50	4BEoo3	R				
3	B	CPu	OPv	1950pc	-	ac	6.50	4BEoo3	R		-	11/57	
3	B/	CPw.z		1950mc	1958mc	ac	6.50	4BooX3					
3	C	CPz	OPz	1965mc	-	adf	6.65	4oooX3			3/65	4/65	
3	C/	CPz	OPz	1965mc	-	adf	6.65	4oooX3					

AA 1.5.46/NT 1.7.47 on A(A), 3/-, OSPR not found. 1. OSPR (outline edition) not found.

SX 26 (20/26) *Liskeard**						10 by 10	Revision a 1905 b 1904 c 1929-38 d 1929-58						
							e 1955 f 1963 g 1978						
1	A	CPr		1947po	-	-	1.48	4BE2X3	R		9/47	-	
1	B		OPr	1947po	-	-	1.48	4BE2X3	R		-	11/47	
3	C	CPu	OPv	1955pc	-	ace	6.55	4oooX3			11/55	6/57	
3	D	CPz	OPz	1965mc	-	bdf	6.65	4oooX3			1/65	1/65	
3	D/*	CFz	OFz	1965[c]ms	nd:r	bdg	6.65	4oooX3			11/78	11/78	

AA 1.5.46 on B(A), 3/-, OSPR 11/47.

1	2	3	4	5	6	7	8	9	10	11	12	13	14

SX 27 (20/27) *Bodmin Moor (East)*   10 by 10   Revision a 1905 c 1958 f 1963
| 1 | A | | CPs | OPs | 1950pc | | a | 6.50 | 4BEoo3 | | 8/50 | 9/50 | |
| 3 | B | | CPy,z| OPy | 1963mc | | acf | 6.63 | 4oooX3 | | 10/63 | 10/63 | |

M on A.

SX 28 (20/28) [*Egloskerry*]   10 by 10   Revision a 1905 c 1938 d 1938-58 f 1963
| 1 | A | | CPs | OPs | 1950pc | | ac | 6.50 | 4BEoo3 | R | 9/50 | 9/50 | |
| 3 | B | | CPz | OPz | 1965mc | | adf | 6.65 | 4oooX3 | | 1/65 | 1/65 | |

M on A.

SX 29 (20/29) [*Week St Mary*]   10 by 10   Revision a 1905-32 c 1952-58 f 1963
| 1 | A | | CPs | OPs | 1950pc | | a | 6.50 | 4BEoo3 | | 9/50 | 10/50 | |
| 3 | B | | CPz | OPz | 1964mc | | acf | 6.64 | 4oooX3 | | 11/64 | 11/64 | |

SX 35 (20/35) *St Germans**   10 by 10   Revision a 1905 c 1958 f 1962 g 1968
1	15046		CPr		1946po		-	6.46	4BE2X3	R	8/46	-	1
1	2546			OPr	1946po		-	6.46	4BE2X3	R	-	11/46	2
3	B		CPz	OPz	1965mc		acf	6.65	4oooX3		3/65	3/65	
3	B/		CPz	OPz	1965ms	1968ch	acg	6.65	4oooX3				

AA 1.7.46 on 2546(2546), 3/-, OSPR 11/46. M on A. 1. Showing the Tregantle military railway system. 2. The outline and AA editions appear also in OSPR 10/47.

SX 36 (20/36) *Callington*   10 by 10   Revision a 1905 c 1958 f 1960-62
1	10046		CPr		1946po		-	6.46	4BE2X3	R	10/46	-	
1	A			OPr	1946po		-	6.46	4BE2X3	R	-	2/47	
3	B		CPz	OPz	1965mc		acf	6.65	4oooX3		1/65	1/65	

AA 1.9.46 on A(A), 3/-, OSPR 2/47. M on A.

SX 37 (20/37) *Kit Hill*   10 by 10   Revision a 1905 c 1958 f 1960
1	A		CPr	OPr	1947po		-	1.47	4BE2X3	R	2/47	5/47	
1	B		CPr		nd:po		-	1.47	4BEoo3	R			
3	C		CPz,0z	OPz	1964mc		acf	6.64	4oooX4		6/64	7/64	
3	C/		CFz		1964^{c}ms		acf	6.64	4oooX4				

AA 1.5.46 on A(A), 3/-, OSPR 5/47.

SX 38 (20/38) [*Launceston*]   10 by 10   Revision a 1905-32 c 1938 d 1958 f 1963
1	A		CPs	OPs	1950pc		ac	6.50	4BEoo3	R	9/50	10/50	
3	A		CPt		1950pc		ac	6.50	4BEoo3	R			
3	B		CPz	OPz	1964mc		adf	6.64	4oooX3		8/64	8/64	

SX 39 (20/39) [*North Tamerton*]   10 by 10   Revision a 1905 c 1958 f 1960
| 1 | A | | CPs | OPs | 1950pc | | a | 6.50 | 4BEoo3 | R | 9/50 | 9/50 | |
| 3 | B | | CPx,y| OP?,z | 1961mc | | acf | 6.61 | 4oooX3 | | 8/61 | 8/61 | |

SX 45 (20/45) **& Part of SX 44** *Plymouth*   10 by 12   Revision a 1953-54 c 1957 f 1964 g 1966
1	20047		CPr		1947po		-	6.46	4BE2X3	R	12/46	-	
1	A			OPr	1947po		-	6.46	4BE2X3	R	-	5/47	
2	A		CPt		1946pc	1952	-	1.52	4Booo3	R			
3	A			OPs	1946pc	1952	-	1.52	4Booo3	R			
3	A		CRv	ORu	1956pc		a	1956	5oo5X5		7/56	7/56	1
3	B		CRw	ORu	1960mc		ac	1960	5oo6X5		12/59	1/60	
3	B/*		CRz	ORz	1960mc	1964r	acf	1960	5oo6X5		10/64	11/64	
3	B/*/		CRz	ORz	1960mc	1967ch	acg	1960	5oo6X5				2

AA 1.9.46 on A(A), 3/-, OSPR 5/47; AA 1.11.53 on A(B), 4/-, OSPR 2/55. M on A, Provisional Edition. Naval sites shown include HMS Raleigh, HMS Fisgard, Crownhill Fort, Manadon (RN Engineering College), Breakwater Fort, Staddon Fort, Stamford Fort, Woodland Fort, Agaton Fort, Ernesettle Battery, Whitesand Bay Battery and various barracks. 1. A proof (no print code, CRu 1955pc a 1955 5oo5X5) is recorded. 2. Devonport naval dockyard detail is supplied for the first time.

1	2	3	4	5	6	7	8	9	10	11	12	13	14

SX 46 (20/46) *Bere Alston*   10 by 10   Revision a 1954 c 1957

1	10046	CPr		1946po	-	-	6.46	4BE2X3	R	0	11/46	-	
1	A		OPr	1946po	-	-	6.46	4BE2X3	R	0	-	2/47	
3	A	CPt	OPs	1946pc	-	-	6.46	4BEoo3	R	0			
3	A	CRv	ORu	1956pc	-	a	1956	5oo5X5		1	7/56	7/56	1
3	B	CRw,y,z.z	ORu,z	1959mc	-	ac	1959	5oo6X5		1	9/59	9/59	2

AA 1.9.46 on A(A), 3/-, OSPR 2/47; AA 1.11.53/NT 3.2.54 on A(B), 4/-, OSPR 1/55. M on A, Provisional Edition. 1. A proof (state CRu, 1956pc) is recorded. 2. A proof (state CRw 1959mc) is recorded.

SX 47 (20/47) *Tavistock*   10 by 10   Revision a 1954-55 c 1958

1	A	CPr	OPr	1947po	-	-	6.47	4BE2X3	R		3/47	6/47	
1	B	CPr		nd:po	-	-	6.47	4BEoo3	R				
3	A	CRv	ORu	1956pc	-	a	1956	5oo5X5			7/56	7/56	
3	B	CRw,x.z	OR?,z	1959mc	-	ac	1959	5oo6X5			9/59	9/59	

AA 1.5.46 on A(A), 3/-, OSPR 6/47. SS: Soil, Land Use Capability, coloured, 1977 on B, Provisional Edition.

SX 48 (20/48) [*Lewtrenchard*]   10 by 10   Revision a 1904-05 c 1958 f 1961

| 1 | A | CPr | OPr | nd:po | - | - | 6.49 | 4BEoo3 | R | | 3/50 | 4/50 | |
| 3 | B | CPx,y,z | OPx | 1962mc | - | acf | 6.62 | 4oooX3 | | | 10/62 | 10/62 | |

SX 49 (20/49) [*Germansweek*]   10 by 10   Revision a 1904-05 c 1958 f 1961

| 1 | A | CPr | OPr | nd:po | - | - | 1.50 | 4BEoo3 | RB | | 3/50 | 4/50 | |
| 3 | B | CPy,z | OPy | 1963mc | - | acf | 6.63 | 4oooX3 | | | 11/63 | 11/63 | |

SX 54 (20/54) *Newton Ferrers*   10 by 10   Revision a 1949-51 c 1957

1	10046	CPr		1946po	-	-	1.47	4BE2X3	B		12/46	-	
1	A		OPr	1946po	-	-	1.47	4BE2X3	B		-	2/47	
3	A	CRv	ORu	1956pc	-	a	1956	5oo5X5			7/56	7/56	
3	B	CRw,y,z.z	ORu	1959mc	-	ac	1959	5oo6X5			8/59	12/59	

AA 1.4.46 on A(A), 3/-, OSPR 2/47. M on A, Provisional Edition.

SX 55 (20/55) *Plympton*   10 by 10   Revision a 1954 c 1957 f 1963 g 1972

1	20046	CPr		1946po	-	-	6.46	4BE2X3	RB		8/46	-	
1	A		OPr	1946po	-	-	6.46	4BE2X3	RB		-	1/47	
3	A		OPt	1946pc	-	-	6.46	4BEoo3	R				
3	A	CRv	ORu	1956pc	-	a	1956	5oo5X5			7/56	7/56	
3	B	CRw	ORu	1959mc	-	ac	1959	5oo6X5			10/59	9/59	
3	B/	CRy,z		1959mc	1963ch	acf	1959	5oo6X5					
3	B//*	CRz	ORz	1959^{c}ms	nd:r	acg	1959	5oo6X5			10/72	12/72	

AA 1.6.46 on A(A), 3/-, OSPR 1/47; AA 1.11.53 on A(B), 4/-, OSPR 8/54. M on A, Provisional Edition.

SX 56 (20/56) *Yelverton Devonshire*   10 by 10   Revision a 1950-52 c 1957

1	10046	CPr		1946po	-	-	6.46	4BE2X3	R	00	11/46	-	
1	A		OPr	1946po	-	-	6.46	4BE2X3	R	00	-	2/47	
3	A	CPt	OPs	1946pc	-	-	6.46	4BEoo3	R	00			
3	A	CRv	ORu	1956pc	-	a	1956	5oo5X5		31	7/56	7/56	1
3	B	CRw	ORu	1960mc	-	ac	1960	5oo6X5		31	12/59	9/59	
3	B	CRx,z	ORz	1959mc	-	ac	1959	5oo6X5		31			

AA 1.9.46 on A(A), 3/-, OSPR 2/47; AA 1.11.53 on A(B), 4/-, OSPR 1/55. M on A, Provisional Edition. 1. Plymouth Airport.

SX 57 (20/57) *Princetown*   10 by 10   Revision a 1950-52 c 1957 f 1971

1	A	CPr	OPr	1947po	-	-	6.47	4BE2X3	R		4/47	7/47	
1	B	CPr		nd:po	-	-	6.47	4BEoo3	R				
3	A	CRv	ORu	1956pc	-	a	1956	5oo5X5			7/56	7/56	
3	B	CRw,x,z	ORu	1959mc	-	ac	1959	5oo6X5			6/59	9/59	
3	B/	CRz	ORz	1959^{c}ms	nd:ch	acf	1959	5oo6X5					

AA 1.5.46 on A(A), 3/-, OSPR 7/47. M on B, Provisional Edition, with and without legends, B, Regular Edition.

1	2	3	4	5	6	7	8	9	10	11	12	13	14

SX 58 (20/58) [Lydford]　　　　　　　　10 by 10　Revision a 1904-05 c 1958 f 1961
1　A　　CPr　　OPr　　nd:po　　-　　-　　1.48　4BEoo3　R　　　3/49　4/49
1　B　　CPs　　　　　1949pc　1950c　a　6.50　4BEoo3　R
3　B　　　　　　OPs　1949pc　1950c　a　6.50　4BooX3　R
3　C　　CPy,z　OPy　1963mc　-　　acf　6.63　4oooX3　　　　11/63　11/63
M on A, with legends, B.

SX 59 (20/59) [Okehampton]　　　　　　10 by 10　Revision a 1904-32 c 1938 d 1957-58 f 1961
1　A　　CPr　　OPr　　nd:po　　-　　-　　1.48　4BEoo3　R　　　3/49　4/49
3　A/　 CP?x　　　　 1961　　　　　ac　1.48　4BooX3　　　　　　　　　　　M
3　B　　CPy　　OPy　1963mc　-　　adf　6.63　4oooX3　　　　5/63　5/63
M on A, ?with legends (unrecorded), A, A/.

SX 64 (20/64) **& Part of SX 63** *Bigbury*　　10 by 12　Revision a 1953-54 c 1957
1　7547　CPr　　　　 1947po　　-　　-　　1.47　4BE1X3　　　　1/47　-
1　A　　　　　　OPr　1947po　　-　　-　　1.47　4BE1X3　　　　-　　5/47
3　A　　CPt　　　　　1947pc　　-　　-　　1.47　4BEoo3
3　A　　CRv　　ORu　1956pc　　-　　a　　1956　5oo5X5　　　　7/56　7/56
3　B　　CRw,x.z ORu　1960mc　　-　　ac　1960　5oo6X5　　　　12/59　1/60
AA 1.4.46/NT 13.12.46 on A(A), 3/-, OSPR 5/47. M on A, Provisional Edition.

SX 65 (20/65) *Ivybridge*　　　　　　　　10 by 10　Revision a 1953-54 c 1957
1　A　　CPr　　OPr　　1947po　　-　　-　　6.47　4BE2X3　R　　5/47　7/47
1　B　　CPr　　　　　nd:po　　　-　　-　　6.47　4BEoo3　R
3　A　　CRv　　ORu　1956pc　　-　　a　　1956　5oo5X5　　　　7/56　7/56
3　B　　CRw,x,z.z ORu 1959mc　　-　　ac　1959　5oo6X5　　　　10/59　10/59
AA 1.4.46 on A(A), 3/-, OSPR 7/47. M on B, Provisional Edition, B, Regular Edition. SS: Soil, Land Use Capability, outline, 1976 on B, Regular Edition.

SX 66 (20/66) *Brent Moor*　　　　　　　10 by 10　Revision a 1954 c 1957
1　A　　CPr　　OPr　　1947po　　-　　-　　1.47　4BE2X3　R　　4/47　7/47
1　B　　CPr　　　　　nd:po　　　-　　-　　1.47　4BEoo3　R
3　A　　CRv　　ORu　1956pc　　-　　a　　1956　5oo5X5　　　　7/56　7/56
3　B　　CRw,x,z.z ORu 1959mc　　-　　ac　1959　5oo6X5　　　　8/59　12/59
AA 1.4.46 on A(A), 3/-, OSPR 7/47. M on B, Provisional Edition, B, Regular Edition.

SX 67 (20/67) *Two Bridges*　　　　　　　10 by 10　Revision a 1954 c 1957
1　A　　CPr　　OPr　　1947po　　-　　-　　1.47　4BE2X3　　　　1/47　4/47
1　B　　CPr　　　　　nd:po　　　-　　-　　1.47　4BEoo3
3　A　　CRv　　ORu　1956pc　　-　　a　　1956　5oo5X5　　　　7/56　7/56
3　B　　CRw,x,z.z ORu 1959mc　　-　　ac　1959　5oo6X5　　　　8/59　12/59
AA 1.5.46 on A(A), 3/-, OSPR 4/47. M on A, B, Provisional Edition, B, Regular Edition.

SX 68 (20/68) [Gidleigh]　　　　　　　　10 by 10　Revision a 1904 c 1957-58 f 1961
1　A　　CPr　　OPr　　1948po　　-　　-　　1.48　4BE2X3　　　　1/48　4/48
1　B　　CPr　　　　　nd:po　　　-　　-　　1.48　4BEoo3
3　B/　 CPw　　　　　1948pc　1958mc a　1.48　4BooX3
3　C　　CPx　　OP?　1963mc　　-　　acf　6.63　4oooX3　　　　3/63　3/63
AA 1.5.46 on A(A), 3/-, OSPR 4/48. M on A, B, B/.

SX 69 (20/69) *Sticklepath Devonshire*　　10 by 10　Revision a 1904 c 1957-58 f 1961
1　A　　CPr　　OPr　　nd:po　　　-　　-　　6.48　4BEoo3　R　　5/49　5/49
3　A/　 CPw　　　　　1949mc　1958mc a　6.48　4BooX3
3　B　　CPy,z　OPy　1963mc　　-　　acf　6.63　4oooX3　　　　9/63　9/63
M on A, with legends, A.

SX 73 (20/73) **& Part of SX 63** *Salcombe**　12 by 10　Revision a 1904-05 c 1936-53 d 1936-57
1　A　　CPr　　OPr　　1947po　　-　　-　　1.48　4BE2X3　　　　12/47　2/48
3　B　　CPu　　OPs　1955pc　　-　　ac　6.55　4oooX3　　　　　5/55　1/56
3　B/　 CPw　　　　　1955pc　1958c　ac　6.55　4oooX3
3　C　　CPw,x　OPv　1959mc　　-　　ad　6.59　4oooX3　　　　12/59　10/59
AA 1.4.46/NT 17.1.48 on A(A), 3/-, OSPR 2/48. M on A.

1	2	3	4	5	6	7	8	9	10	11	12	13	14

SX 74 (20/74) *Kingsbridge*   10 by 10   Revision a 1904-05 c 1936-53 d 1936-57
1 A   CPr   OPr   1948po   -   -   6.48   4BE2X3   R   3/48   4/48
3 B   CPu   OPv   1955pc   -   ac   6.55   4oooX3   4/55   6/57
3 B/   CPv   1955pc   1957mc   ac   6.55   4oooX3
3 C   CPw,y,z OPv   1959mc   -   ad   6.59   4oooX3   11/59 12/59
AA 1.4.46 on A(A), 3/-, OSPR 4/48.

SX 75 (20/75) *Harbertonford*   10 by 10   Revision a 1904-05 c 1957
1 A   CPr   OPr   1948po   -   -   6.48   4BE2X3   R   3/48   4/48
1 B   CPr   nd:po   -   -   6.48   4BEoo3   R
3 C   CPw,x,z OPv   1959mc   -   ac   6.59   4oooX3   9/59   9/59
AA 1.1.48 on A(A), 3/-, OSPR 4/48.

SX 76 (20/76) *Buckfastleigh*   10 by 10   Revision a 1904-32 c 1938 d 1938-57
1 A   CPr   1948po   -   -   6.47   4BE2X3   R   10/48 -
1 B   OPr   1948po   -   -   6.47   4BE2X3   R   -   2/48
1 C   CPr   nd:po   -   -   6.47   4BEoo3   R
3 C/   CPv   1947pc   1958mc   ac   6.47   4BooX3
3 D   CPw,x,z OPv   1959mc   -   ad   6.59   4oooX3   11/59 12/59
AA 1.4.46 on B(A), 3/-, OSPR 2/48.

SX 77 (20/77) *Widecombe in the Moor*   10 by 10   Revision a 1904 c 1938 d 1938-57 f 1979
1 A   CPr   OPr   1948po   -   -   6.48   4BE2X3   4/48   4/48
1 B   CPr   nd:po   -   -   6.48   4BEoo3
3 B/   CPv   1949pc   1957mc   ac   6.48   4BooX3
3 C   CPw,z,0z OPv   1959mc   -   ad   6.59   4oooX3   11/59 11/59
3 C/* CFz   OFz   1979^{c}ms nd:r   adf 6.59   4oooX3   4/80   4/80
AA 1.3.48/NT 27.2.48 on A(A), 3/-, OSPR 4/48. M on A, ?, C.

SX 78 (20/78) [*Moretonhampstead*]   10 by 10   Revision a 1904 c 1936-39 d 1936-58 f 1961
1 A   CPr   OPr   1948po   -   -   1.48   4BEoo3   R   7/48   8/48
1 B   CPs   1951pc   -   ac   1.51   4BEoo4   R   6/51   -
3 B/   CPw,x   1951mc   1959mc   ac   1.51   4BooX4
3 C   CPz   OPz   1964mc   -   adf 6.64   4oooX4   9/64   9/64
M on A, with legends, B/.

SX 79 (20/79) *Crockernwell*   10 by 10   Revision a 1904 c 1938 d 1957 f 1961
1 A   CPr   OPr   nd:po   -   -   1.49   4BEoo3   R   12/49 1/50
3 A/   CPw   1949mc   1958mc   ac   1.49   4BooX3
3 B   CPy,z   OPy   1964mc   -   adf 6.64   4oooX3   12/63 1/64

SX 83 (20/83) [*Start Point*]   10 by 10   Revision a 1905 c 1957
1 A   CPr   OPr   nd:po   -   -   6.48   4BEoo3   7/48   8/48
1 B   CPr   nd:po   -   -   6.48   4BEoo3
3 C   CPw,x,z OP?   1959mc   -   ac   6.59   4oooX3   9/59   9/59
M on B.

SX 84 (20/84) [*Slapton*]   10 by 10   Revision a 1904-05 c 1957
1 A   CPr   OPr   nd:po   -   -   1.48   4BEoo3   12/48 1/49
1 B   CPs   1948pc   1950c   a   6.50   4BEoo3
3 C   CPw   OPv   1959mc   -   ac   6.59   4oooX3   10/59 10/59
3 C/   CPw,y,z   1959mc   1960mc   ac   6.59   4oooX3
AA 1.12.48 on A(A), 3/-, OSPR 1/49.

SX 85 (20/85) [*Dartmouth*]   10 by 10   Revision a 1904-33 c 1938 d 1938-57
1 A   CPr   OPr   nd:po   -   -   1.48   4BEoo3   R   11/48 12/48 1
1 A   CPr   1948pc   1950   -   6.50   4BEoo3   R
3 A/   CPw   1948pc   1959c   ac   6.50   4BooX3
3 B   CPw,x,z OPv   1960mc   -   ad   6.60   4oooX4   1/60   2/60
AA 1.11.48 on A(A), 3/-, OSPR 12/48. 1. Showing Britannia Royal Naval College.

1	2	3	4	5	6	7	8	9	10	11	12	13	14

SX 86 (20/86) [Paignton]   10 by 10   Revision a 1904-37 c 1957
1 A   CPr   OPr   nd:po   -   -   1.48 4BEoo3 R   1/49   2/49
3 A/   CPv   1948pc   1958mc   a   1.48 4BooX3
3 B   CPw,x OPv   1960mc   -   ac   6.60 4oooX3   12/59 2/60
AA 1.1.49 on A(A), 3/-, OSPR 2/49. One of the first sheets to be superseded by Second Series mapping in 1965.

SX 87 (20/87) [Newton Abbot]   10 by 10   Revision a 1904-33 c 1938-58 f 1961 g 1964
1 A   CPr   OPr   nd:po   -   -   1.49 4BEoo3 R   1/49   2/49
3 A   CPt   1948pc   -   -   1.49 4BEoo3 R
3 A/   CPw   1948mc   1958mc   a   1.49 4BooX3
3 A// CPw   1948mc   1960mc   a   1.49 4BooX3
3 B   CPy   OPy   1963mc   -   acf   6.63 4oooX4   3/63   2/63
3 B/   CPz   1963mc   1965ch   acg   6.63 4oooX4
AA 1.1.49/NT 28.11.47 on A(A), 3/-, OSPR 2/49.

SX 88 (20/88) *Christow*   10 by 10   Revision a 1904-32 c 1938 d 1938-58 f 1961
1 A   CPr   OPr   nd:po   -   -   1.48 4BEoo3 R   12/48 1/49
3 A/   CPw   1948pc   1958mc   ac   1.48 4BooX3
3 B   CPy,z OPy   1963mc   -   adf   6.63 4oooX3   6/63   6/63

SX 89 (20/89) *Tedburn St Mary*   10 by 10   Revision a 1903-32 c 1936-38 d 1936-57 f 1961
1 A   CPr   OPr   nd:po   -   -   1.48 4BEoo3 R   1/49   1/49
1 B   CPs   1950pc   -   ac   6.50 4BEoo3 R   12/50 -
2 B   OPs   1950pc   -   ac   6.50 4BEoo3 R
3 B/   CPw   1950mc   1959mc   ac   1.52 4BooX3
3 C   CPy,z OPy   1963mc   -   adf   6.63 4oooX3   11/63 11/63
AA 1.9.48 on A(A), 3/-, OSPR 1/49.

SX 95 (20/95) **& Part of SX 94** [Brixham] 10 by 12   Revision a 1904 c 1957
1 A   CPr   OPr   nd:po   -   -   1.49 4BEoo3 R   2/49   4/49
3 B   CPw,x,z OPv   1959mc   -   ac   6.59 4oooX3   10/59 10/59
AA 1.2.49 on A(A), 3/-, OSPR 4/49.

SX 96 (20/96) [Torquay]   10 by 10   Revision a 1904-36 c 1938-57
1 A   CPr   OPr   nd:po   -   -   1.48 4BEoo3 R   2/49   4/49
3 B   CPw   OPv   1959mc   -   ac   6.59 4oooX3   11/59 12/59
AA 1.2.49 on A(A), 3/-, OSPR 4/49. One of the first sheets to be superseded by Second Series mapping in 1965.

SX 97 (20/97) [Teignmouth]   10 by 10   Revision a 1904-33 c 1938 d 1938-58 f 1961 g 1964
1 A   CPr   OPr   nd:po   -   -   1.48 4BEoo3 R   11/48 11/48
3 A/   CPw   1948mc   1960mc   ac   1.48 4BooX3
3 B   CPx   OPx   1963mc   -   adf   6.63 4oooX3   2/63   12/62
3 B/   CPz   OPz   1963mc   1965ch   adg   6.63 4oooX3
AA 1.10.48 on A(A), 3/-, OSPR 11/48.

SX 98 (20/98) [Topsham]   10 by 10   Revision a 1903-33 c 1938-58 f 1961
1 A   CPr   OPr   nd:po   -   -   6.48 4BEoo3 R   12/48 2/49
3 A   CPu   1948pc   -   -   6.48 4BEoo3 R
3 A/   CPw   1948mc   1960mc   a   6.48 4BooX3
3 B   CPy,z OPy   1963mc   -   acf   6.63 4oooX4   8/63   9/63
AA 1.10.48/NT 29.10.48 on A(A), 3/-, OSPR 2/49. M on A.

SX 99 (20/99) *Exeter*   10 by 10   Revision a 1903-33 c 1938 d 1938-57 f 1961 g 1966
1 A   CPr   OPr   nd:po   -   -   1.48 4BEoo3 R   0 12/48 1/49
3 A   CPt   1948pc   -   -   6.48 4BooX3   0
3 A/   CPw   1948mc   1959mc   ac   6.48 4BooX3   0
3 B   CPy   OPy   1963mc   -   adf   6.63 4oooX4   3 6/63   6/63 1
3 B/   CPz   OPz   1963mc   1967ch   adg   6.63 4oooX4   3
AA 1.9.48/NT 25.11.48 on A(A), 3/-, OSPR 1/49. 1. Exeter Airport.

1	2	3	4	5	6	7	8	9	10	11	12	13	14

SY 08 (30/08) **& Part of SY 07** [Exmouth]   10 by 12   Revision a 1903-33 c 1938 d 1938-57 f 1961
1	A	CPr	OPr	nd:po	-	-	1.48	4BEoo3	R		11/48	1/49	
2	A	CPt		1948pc	1952	-	1.48	4Booo3	R				
	A/												
3	A//	CP?		1948mc	1960mc	ac	?	4BooX3					M
3	B	CPy	OPy	1964mc	-	adf	6.64	4oooX4			2/64	2/64	

AA 15.4.48 on A(A), 3/-, OSPR 1/49. M on A//.

SY 09 (30/09) *Whimple**   10 by 10   Revision a 1903-33 c 1938 d 1938-57 f 1961
1	A	CPr	OPr	nd:po	-	-	1.48	4BEoo3	R	0	12/48	2/49
3	A/	CPw		1948pc	1958mc	ac	1.48	4BooX3		0		
3	B	CPx,z	OPx	1961mc	-	adf	6.61	4oooX4		3	10/61	9/61

AA 1.12.48/NT 3.12.48 on A(A), 3/-, OSPR 2/49.

SY 18 (30/18) [Sidmouth]   10 by 10   Revision a 1903-33 c 1938 d 1938-57 f 1961
1	A	CPr	OPr	nd:po	-	-	1.48	4BEoo3	R		3/49	5/49
3	A	CPs,t		1949pc	-	-	1.48	4BEoo3	R			
3	A/	CPx		1949pc	1960mc	ac	1.48	4BooX3				
3	B	CPy,z	OPy	1963mc	-	adf	6.63	4oooX4			6/63	7/63

AA 4.1.49/NT 28.2.49 on A(A), 3/-, OSPR 5/49.

SY 19 (30/19) *Sidbury Devonshire**   10 by 10   Revision a 1903-33 b 1903-38 c 1938 d 1957 f 1961
1	A	CPr	OPr	nd:po	-	-	1.49	4BEoo3	R		9/49	11/49
1	B	CPs		1949pc	1950c	ac	6.50	4BEoo3	R			
3	C	CPx,z	OPx	1963mc	-	bdf	6.63	4oooX3			1/63	1/63

AA 7.8.49 on A(A), 3/-, OSPR 11/49.

SY 28 (30/28) *Beer*   10 by 10   Revision a 1906-33 c 1938-57
| 1 | A | CPr | OPr | nd:po | - | - | 1.48 | 4BEoo3 | | | 2/49 | 3/49 |
| 3 | B | CPw,y,z | OPv | 1958mc | - | ac | 6.58 | 4oooX3 | | | 12/58 | 1/59 |

SY 29 (30/29) *Seaton Devonshire*   10 by 10   Revision a 1903-33 c 1938-57
| 1 | A | CPr | OPr | nd:po | - | - | 1.49 | 4BEoo4 | RB | | 9/49 | 10/49 |
| 3 | B | CPw,y,z | OPv | 1958mc | - | ac | 6.58 | 4oooX3 | | | 12/58 | 2/59 |

Note the frame on this sheet: type 4 on the A printing, type 3 on the B.

SY 39 (30/39) **& Part of SY 38** *Lyme Regis* 10 by 12 Revision a 1901-28 c 1938-57
1	A	CPr	OPr	nd:po	-	-	1.49	4BEoo3	RB		10/49	11/49
3	A	CPt		1948pc	-	-	1.49	4BEoo3	RB			
3	B	CPw,x,z	OPv	1959mc	-	ac	6.59	4oooX3			3/59	7/59

SY 49 (30/49) **& Part of SY 48** *Bridport*   10 by 12 Revision a 1901-28 c 1938-57
1	A	CPr	OPr	nd:po	-	-	6.48	4BEoo3	RB		10/49	11/49
3	A	CPt		1949pc	-	-	6.48	4Booo3	RB			
3	B	CPw,z	OPv	1959mc	-	ac	6.59	4oooX3			4/59	5/59	1

AA 1.6.49 on A(A), 3/-, OSPR 11/49. 1. There is an experimental B printing (1958pc, CPw, MV 6.58), with the grey plate in green. The other colours are not present.

SY 58 (30/58) *Abbotsbury*   10 by 10   Revision a 1901-28 c 1957
1	A	CPr	OPr	nd:po	-	-	1.48	4BEoo3	RB		10/48	10/48
	B											
3	C	CPw,z	OPv	1958mc	-	ac	6.58	4oooX3			1/59	12/58

M on A.

SY 59 (30/59) *Maiden Newton*   10 by 10   Revision a 1901-28 c 1957
1	A	CPr	OPr	nd:po	-	-	1.49	4BEoo3	RB		9/49	10/49
1	B	CPs		1948pc	1950c	a	6.50	4BEoo3	RB			
3	C	CPw,x,z.z	OPv	1958pc	-	ac	6.58	4oooX3			1/59	12/58

1	2	3	4	5	6	7	8	9	10	11	12	13	14

SY 67 (30/67) **& Parts of SY 66 & SY 77** 12 by 12 Revision a 1901-39 c 1938-39 d 1938-57 e 1930
[Weymouth]
1	A		CPr	OPr	1948po	-	-	6.48	4BEoo3	R		6/48	7/48	
3	A		CP		1948pc	1953rpb	ace	6.48	4BooX3	R				M
3	B		CPv	OPv	1956pc	-	ace	6.56	4oooX3			3/57	3/57	
3	C		CPw	OPv	1958pc	-	ad	6.58	4oooX3			12/58	2/59	
3	C/		CPx		1958mc	1961ch	ad	6.58	4oooX3					

AA 1.3.48 on A(A), 3/-, OSPR 7/48. M on A.

SY 68 (30/68) [Melcombe Regis] 10 by 10 Revision a 1901-27 c 1938 d 1938-57
1	A		CPr	OPr	1948po	-	-	1.48	4BEoo3	RB		6/48	8/48
1	B		CPr		1948pc	1950c	ac	6.50	4BEoo3	RB			
3	B/		CPv		1948pc	1957c	ac	6.50	4BooX3	B			
3	C		CPw,x,z	OP?	1958pc	-	ad	6.58	4oooX3			10/58	1/59

AA 1.4.48/NT 8.6.48 on A(A), 3/-, OSPR 8/48. M on B.

SY 69 (30/69) *Dorchester Dorset* 10 by 10 Revision a 1900-28 c 1938 d 1938-57 f 1969
1	A		CPr	OPr	nd:po	-	-	1.48	4BEoo3	RB		10/48	11/48
1	B		CPs		1951pc	-	ac	1.51	4BEoo3	RB		3/51	-
2	B			OPs	1951pc	-	ac	1.51	4BooX3	RB			
3	C		CPw,z	OPv	1958pc	-	ad	6.58	4oooX3			10/58	12/58
3	C/		CPz	OPz	1958ms	1970ch	adf	6.58	4oooX3				

AA 1.8.48 on A(A), 3/-, OSPR 11/48. M on C.

SY 78 (30/78) [Broadmayne] 10 by 10 Revision a 1900-28 c 1938-57
1	A		CPr	OPr	1948po	-	-	1.48	4BEoo3	RB		6/48	8/48	
1	B		CPr		nd:po	-	-	1.48	4BEoo3	RB				M
2	B		CPt		1949pc	-	-	1.48	4Booo3	RB				
3	C		CPw,x,z	OPv	1958pc	-	ac	6.58	4oooX3			9/58	10/58	

AA 1.4.48 on A(A), 3/-, OSPR 8/48. M on B.

SY 79 (30/79) *Puddletown* 10 by 10 Revision a 1900-28 c 1956
1	A		CPr	OPr	nd:po	-	-	1.48	4BEoo3	B		12/48	1/49
1	B		CPr		nd:po	-	-	1.48	4BEoo3	B			
3	C		CPw,x,z	OPv	1958pc	-	ac	6.58	4oooX3			10/58	1/59
3	C/		CPz	OPz	1958pc	-	ac	6.58	4oooX3				

AA 1.9.48/NT 1.9.48 on A(A), 3/-, OSPR 1/49. M on B.

SY 88 (30/88) **& Part of SY 87** *Lulworth Cove* 10 by 12 Revision a 1900-27 c 1956 f 1964
1	A		CPr	OPr	1948po	-	-	1.48	4BEoo3	RB		7/48	8/48	1
3	A		CPu		1948pc	-	-	1.48	4BooX3	RB				
3	B		CPw,x	OPv	1958mc	-	ac	6.58	4oooX3			11/58	1/59	
3	B/		CPz	OPz	1958mc	1964ch	acf	6.58	4oooX3					

M on A. 1. Showing Bovington Camp, with its railway connections, also Lulworth Camp.

SY 89 (30/89) *Bere Regis* 10 by 10 Revision a 1900 c 1956
1	A		CPr	OPr	1948po	-	-	6.47	4BE2X3	B		3/48	4/48
1	B		CPr		nd:po	-	-	6.47	4BEoo3	B			
3	C		CPw,x,z.z	OPv	1958pc	-	ac	6.58	4oooX3			10/58	10/58

AA 1.11.47/NT 26.11.47 on A(A), 3/-, OSPR 4/48. M on B.

SY 97 (30/97) **& Part of SZ 07** *Swanage* 15 by 10 Revision a 1900-26 c 1938 d 1938-57 f 1961
1	A		CPr	OPr	1948po	-	-	1.48	4BE2X3	RB		1/48	6/48
1	B		CPr		nd:po	-	-	1.48	4BEoo3	RB			
3	B/		CPv		1948pc	1956mc	ac	1.52	4BooX3	B			
3	C		CPx,z	OPx	1962mc	-	adf	6.62	4oooX3			1/62	2/62

AA 1.10.47 on A(A), 3/-, OSPR 6/48. M on B.

1	2	3	4	5	6	7	8	9	10	11	12	13	14

SY 98 (30/98) [Wareham]   10 by 10   Revision a 1900-37 c 1938 d 1938-57 e 1947 f 1961
1	A	CPr	OPr	1948po	-	-	6.48	4BE2X3	RB		2/48	4/48	1
1	A	CPs		1948pc	1951	-	6.51	4BEoo3	RB				
3	B	CPu	OPv	1956pc	-	ace	6.56	4oooX3			4/56	6/57	
3	B/												
3	B//	CPw		1956mc	1959mc	ace	6.56	4oooX3					
3	C	CPx,z	OPx	1961mc	-	adf	6.61	4oooX3			10/61	9/61	

AA 1.10.47/NT 16.2.48 on A(A), 3/-, OSPR 4/48. M on A. 1. A proof copy (CPr, 1947po, MV 6.47, 4BE1X3) dated 11 October 1947 is recorded.

SY 99 (30/99) [Hamworthy]   10 by 10   Revision a 1900-26 c 1956-57 f 1962
1	A	CPr	OPr	1948po	-	-	1.48	4BE2X3	RB	1/48	3/48
1	B	CPr		nd:po	-	-	1.48	4BEoo3	RB		
2	B	CPt		1948pc	1952	-	1.48	4Booo3	RB		
3	C	CPx	OPx	1962mc	-	acf	6.62	4oooX3		4/62	5/62

AA 1.10.47 on A(A), 3/-, OSPR 3/48. M on B.

SZ 08 (40/08) *Poole Harbour*   10 by 10   Revision a 1900-26 c 1938 d 1938-57 f 1961
1	A	CPr	OPr	1947po	-	-	1.47	4BE2X3	B		3/47	5/47
1	B	CPr		nd:po	-	-	1.47	4BEoo3	B			
3	B	CPt		1949pc	-	-	1.47	4BEoo3	B			
3	B/	CPv		1949pc	1957mc	ac	1.47	4BooX3	B			
3	C	CPx,z	OPx	1961mc	-	adf	6.61	4oooX3			10/61	10/61

AA 1.12.46 on A(A), 3/-, OSPR 5/47. M on B.

SZ 09 (40/09) *Bournemouth (West)*   10 by 10   Revision a 1900-36 c 1938-57 f 1961
1	no code	CPz		1944po	-	-	6.44	3BE1X1	R			1
1	1500/9/45 Cr	CPr		1945po	-	-	6.45	4BE1X3	R	11/45	-	2
1	500/9/45 Cr		OPr	1945po	-	-	6.45	4BE1X3	R	-	11/45	
1	5200/2/46 Cr	CPr		1945po	-	-	6.45	4BE1X3	R			3
1	no code(7046)		APr	1945po	-	-	6.45	4BE1X3	R	-	4/46	4
3	B	CPx,z	OPx	1962mc	-	acf	6.62	4oooX4		5/62	6/62	

M on [A]. For pre-publication printings see page 18. 1. A three colour printing issued for consultation. The sheet was remade to four colour specification for publication, 33/89 (SJ 89) becoming the prototype three colour sheet. 2. The first sheet published, in four colours. The price is situated bottom left below the legend (also 1946). The abbreviation of "Magnetic Variation" was "Mag.Var^m", soon altered to "Mag.Var". 3. "W" is added to the MV value in the top margin. 4. AA 1.10.45 on no code(7046), 3/-, OSPR 4/46. This issue is entered into the list as the only independent Administrative Areas edition to be printed, all others being overprinted on outline editions.

SZ 19 (40/19) *Bournemouth (East)*   10 by 10   Revision a 1907-31 c 1938 d 1938-59
                                                    e 1947 f 1962 g 1970
1	A	CPr	OPr	1947po	-	-	1.47	4BE2X3	R	31	1/47	5/47	
1	B	CPr		1950pc	-	ace	6.50	4BEoo3	R	31			
3	B/	CPw	OPv	1950pc	1958mc	ace	6.50	4BooX3		31			
3	C	CPx,z	OPx	1962mc	-	adf	6.62	4oooX4		33	8/62	8/62	1
3	C/*	CPz,0z	OPz	1962^cms	nd:r	adg	6.62	4oooX4		33	7/71	8/71	

AA 1.9.46 on A(A), 3/-, OSPR 5/47. M on B. 1. Bournemouth (Hurn) Airport, with enlarged layout.

SZ 29 (40/29) *New Milton*   10 by 10   Revision a 1907-36 c 1938-57 f 1961
1	A	CPr	OPr	1947po	-	-	1.47	4BE2X3	RB	3	2/47	6/47
1	B	CPr		nd:po	-	-	1.47	4BEoo3	RB	3		
2	C	CPt		1953pc	-	ace	6.53	4BooX3	B	3		
3	D	CPx,z	OPx	1961mc	-	acf	6.61	4oooX4		3	7/61	10/61

AA 1.12.46 on A(A), 3/-, OSPR 6/47. M on B.

SZ 38 (40/38) **& Part of SZ 28** *Totland*   12 by 10   Revision a 1905 b 1906-07 c 1938 d 1938-57
1	A	CPr	OPr	1948po	-	-	6.47	4BE2X3	R	1/48	4/48
1	B	CPr		nd:po	-	-	6.47	4BEoo3	R		
3	B/	CPw		1949mc	1960mc	ac	6.47	4BooX3			
3	C	CPx,y,z.z	OPx	1961mc	-	bd	6.61	4oooX3		5/61	8/61

AA 1.10.47/NT 6.2.48 on A(A), 3/-, OSPR 4/48. M on B.

1	2	3	4	5	6	7	8	9	10	11	12	13	14

SZ 39 (40/39) *Lymington** 10 by 10 Revision a 1907-39 c 1942-57 f 1960
1 A   CPr   OPr   1947po   -   -   6.47 4BE2X3   R   10/47 12/47
2 A   CPs   1947pc   1951   -   1.52 4Booo3   R
3 B   CPx,z   OPx   1961mc   -   acf   6.61 4oooX3   8/61   8/61
AA 1.6.47 on A(A), 3/-, OSPR 12/47. M on A.

SZ 47 (40/47) *Chale* 10 by 10 Revision a 1906-07 c 1938-57
1 A   CPr   OPr   nd:po   -   -   6.48 4BEoo3   B   8/48   9/48
1 B   CPr   nd:po   -   -   6.48 4BEoo3   B
3 C   CPx,z   OPx   1961mc   -   ac   6.61 4oooX3   3/61   5/61
M on B.

SZ 48 (40/48) *Newport (West) IOW** 10 by 10 Revision a 1906-07 c 1938 d 1938-57
1 A   CPr   OPr   nd:po   -   -   1.48 4BEoo3   RB   8/48   10/48
3 A/   CP?v,w   1948mc   1957mc ac   1.48 4BooX3   B
3 B   CPx,y,z OPx   1961mc   -   ad   6.61 4oooX3   4/61   3/61
AA 1.4.48 on A(A), 3/-, OSPR 10/48. M on A.

SZ 49 (40/49) *Cowes* 10 by 10 Revision a 1907 b 1907-31 c 1938 d 1938-57
1 A   CPr   OPr   1947po   -   -   6.47 4BE2X3   R   1   12/47 2/48   1
1 B   CPr   nd:po   -   -   6.47 4BEoo3   R   1   2
3 B/   CPw   1949mc   1959mc ac   6.47 4BooX3   1   3
3 C   CPw.z   OPv   1961mc   -   bd   6.61 4oooX3   1   11/60 11/60
AA 1.6.47 on A(A), 3/-, OSPR 2/48. M on B. 1. Parkhurst Prison is blank. 2. Parkhurst Prison is shown. 3. Cowes (Somerton) Aerodrome.

SZ 57 (40/57) *Ventnor* 10 by 10 Revision a 1906-07 c 1938-42 d 1938-57
1 A   CPr   OPr   nd:po   -   -   1.48 4BEoo3   RB   8/48   9/48
3 B   CPu   OPv   1955pc   -   ac   6.55 4oooX3   B   8/55   1/57
3 C   CPx,z   OPx   1961mc   -   ad   6.61 4oooX3   6/61   5/61
AA 1.4.48/NT 20.7.48 on A(A), 3/-, OSPR 9/48. M on A.

SZ 58 (40/58) *Newport (East) IOW* 10 by 10 Revision a 1906-07 c 1938-42 d 1938-57 e 1947
1 A   CPr   OPr   nd:po   -   -   1.48 4BEoo3   RB   0   9/48   -
1 B   OPr   nd:po   -   -   1.48 4BEoo3   RB   0   -   10/48
3 C   CPv   1956pc   -   ace   6.56 4oooX3   0   11/56 -
3 C/   OPv   1956pc   1960mc ace   6.56 4oooX3   0   -   2/60
3 D   CPx,z   OPv   1961mc   -   ad   6.61 4oooX3   1   1/61   2/61   1
AA 1.5.48/NT 9.9.48 on B(A), 3/-, OSPR 10/48. M on A. 1. Sandown Airport.

SZ 59 (40/59) *Ryde** 10 by 10 Revision a 1906-31 c 1938-57
1 A   CPr   OPr   1947po   -   -   6.47 4BE2X3   RB   2/47   5/47   1
1 B   CPr   nd:po   -   -   6.47 4BEoo3   RB
3 B   CPt   1949pc   -   -   6.47 4BEoo3   RB
3 C   CPx,z   OPx   1961mc   -   ac   6.61 4oooX3   6/61   5/61
AA 1.12.46 on A(A), 3/-, OSPR 5/47. M on B. 1. Showing Fort Gomer, Browndown Battery, Stokes Bay Lines.

SZ 68 (40/68) *Bembridge* 10 by 10 Revision a 1907 c 1938-57
1 A   CPr   OPr   nd:po   -   -   1.48 4oEoo3   RB   0   8/48   9/48   1
2 B   CPt   1948pc   1953c   -   6.53 4oooX3   B   0
3 C   CPx.z   OPx   1961mc   -   ac   6.61 4oooX3   1   7/61   5/61   2
AA 1.4.48/NT 21.7.48 on A(A), 3/-, OSPR 9/48. M on A. 1. Bembridge Fort, St Helens Fort (offshore). There are apparently no bench marks marked in the north east of the Isle of Wight. 2. Bembridge Airport.

SZ 69 (40/69) *Portsmouth (South)* 10 by 10 Revision a 1907-31 c 1938-42 d 1938-57 f 1963
1 20046 CPr   1946po   -   -   1.47 4BE2X3   1   10/46 -   1
1 A   CPr   OPr   1946po   -   -   1.47 4BE2X3   1   -   2/47
1 B   CPr   OPr   nd:po   -   -   1.47 4oEoo3   0
3 C   CPt   OPs   1949pc   1953c   -   6.53 4oooX3   0   2
3 C/   CPv   1949pc   1957mc ac   6.53 4oooX3   0
3 C//   CP?w.x   1949mc   1959mc ac   6.53 4oooX3   0

1	2	3	4	5	6	7	8	9	10	11	12	13	14
3	D	CPz	OPz	1964mc	-	adf	6.64	4oooX4		1	6/64	7/64	
3	D/	CPz	OPz	1964mc	-	adf	6.64	4oooX4		1			

AA 1.9.46 on A(A), 3/-, OSPR 2/47; AA 1.11.53 on C(B), 4/-. M on B. 1. There are apparently no bench marks marked in the north east of the Isle of Wight. Showing Fort Gilkicker, Fort Monckton, HMS Hornet, Fort Blockhouse, Vernon Shore Establishment, Southsea Castle, Lumps Fort, Fort Cumberland and various barracks. 2. Showing Spitbank Fort, Horse Sand Fort, No Man's Fort, Warner Light in Spithead.

SZ 79 (40/79) *South Hayling*   10 by 10   Revision a 1909-33 c 1938-57

1	A	CPr	OPr	1947po	-	-	6.47	4BE2X3	R		4/47	7/47	
1	B	CPr		nd:po	-	-	6.47	4BEoo3	R				
3	B		OPs	1949pc	-	-	6.47	4BEoo3	R		-	9/55	
3	C	CPw,x,z.z	OPv	1958mc	-	ac	6.58	4oooX4			1/59	2/59	

AA 1.9.46 on A(A), 3/-, OSPR 7/47. M on B.

SZ 89 (40/89) & Part of SZ 99 *Selsey Bill*   12 by 10 Revision a 1909-40 c 1938-57

1	A	CPr	OPr	1948po	-	-	6.48	4BEoo3			8/48	10/48	
1	A	CPs		1948pc	1951	-	6.51	4BEoo3					
3	B	CPw,x.z,z+	OPv	1959mc	-	ac	6.59	4oooX3			3/59	5/59	

AA 1.5.48 on A(A), 3/-, OSPR 10/48. M on A.

TA 00 (54/00) *Brigg**   10 by 10   Revision a 1905-06 c 1938-51 f 1960

| 1 | A | CPr | OPr | nd:po | - | - | 1.49 | 4BEoo3 | RB | 00 | 10/48 | 11/48 | |
| 3 | B | CPx,z | OPx | 1962mc | - | acf | 6.62 | 4oooX4 | | 11 | 5/62 | 3/62 | 1 |

AA 1.7.48 on A(A), 3/-, OSPR 11/48. 1. Caistor Airfield (Disused).

TA 01 (54/01) *Elsham**   10 by 10   Revision a 1905-30 c 1946-53

1	A	CPr	OPr	1947po	-	-	1.47	4BE1X3	RB	00	5/47	7/47	
2	B	CPt	OPv	1953pc	-	ac	6.53	4BooX3		11	9/53	6/57	
3	B	CPx,z		1953pc	-	ac	6.53	4BooX3		11			

AA 1.12.46 on A(A), 3/-, OSPR 7/47. M on B.

TA 02 (54/02) *Kingston-upon-Hull (South West)* 10 by 10 Revision a 1906-30 c 1938-52 f 1960 g 1965

1	20046	CPr		1946po	-	-	6.46	4BE1X3	R		6/46	-	
1	2546		OPr	1946po	-	-	6.46	4BE1X3	R		-	9/46	
2	B	CPt		1953pc	-	ac	6.53	4BooX3			10/53	-	
3	B/	CPv		1953pc	1957c	ac	6.53	4BooX3					
3	B//	CPx		1953mc	1961ch	acf	6.53	4BooX3					
3	B///	CPz	OPz	1953mc	1966ch	acg	6.53	4oooX3					

AA 1.4.46 on 2546(2546), 3/-, OSPR 9/46. M on A.

TA 03 (54/03) *Kingston-upon-Hull (North West)* 10 by 10 Revision a 1925-26 c 1938-52 f 1975
                                                  j 1983 selected k SUSI

1	27546	CPr		1946po	-	-	6.46	4BE1X3	RB		6/46	-	1
1	2546		OPr	1946po	-	-	6.46	4BE1X3	RB		-	9/46	
3	B	CPt	OPv	1953pc	-	ac	6.53	4BooX4			12/53	10/56	
3	B/	CPw,z		1953mc	1959mc	ac	6.53	4BooX4					
3	B//*	CFz	OFz	1953ᶜms	nd:r	acf	6.53	4BooX4			4/76	4/76	
3	B//*/*	CFz	OFz	1983ᶜm-s	nd:br	acfjk	6.85	4BooX4			7/84	7/84	

AA 1.4.46 on 2546(2546), 3/-, OSPR 9/46. M on A. 1. Beverley has a "Training Course for Race Horses" with the BR symbol, though there is no "BR" noted.

TA 04 (54/04) *Beverley (North)*   10 by 10   Revision a 1908-26 c 1946-52; j 1982 selected

1	10046	CPr		1946po	-	-	6.46	4BE1X3	RB	0	11/46	-	
1	A		OPr	1946po	-	-	6.46	4BE1X3	RB	0	-	2/47	
2	B	CPt	OPv	1953pc	-	ac	6.53	4BooX3		3	7/53	6/57	1
3	B/	CPw,z		1953mc	1959c	ac	6.53	4BooX3		1			2
3	B//*	CFz	OFz	1983ᶜm-s	nd:rs	acj	6.84	4BooX3		1	11/83	11/83	

AA 1.9.46 on A(A), 3/-, OSPR 2/47. M on A, B. 1. Leconfield Aerodrome. 2. Leconfield Airfield.

1	2	3	4	5	6	7	8	9	10	11	12	13	14

TA 05 (54/05) *Great Driffield*   10 by 10 Revision a 1908-26 c 1946-53 f 1970; j 1982 selected
1   A   CPr   OPr   1947po   -   -   6.47 4BE1X3   RB   00   10/47 1/48
2   B   CPt   OPv   1953pc   -   ac   6.53 4BooX4   11   9/53   6/57
3   B/   CPw,z   1953mc   1960mc   ac   6.53 4BooX4   11   1
3   B//   CPz   OPz   1953ᶜms   nd:ch   acf   6.53 4BooX4   13
3   B///*   CFz   OFz   1982ᶜm-s nd:r   acfj 6.53 4BooX4   13   4/82   4/82
AA 1.9.47 on A(A), 3/-, OSPR 1/48. M on A. 1. Driffield Airfield.

TA 06 (54/06) *Kilham Yorkshire**   10 by 10 Revision a 1909-26 c 1945-52
1   15046   CPr   1946po   -   -   6.46 4BE1X3   RB   0   6/46   -
1   2546   OPr   1946po   -   -   6.46 4BE1X3   RB   0   -   10/46
2   B   CPt   OPv   1953pc   -   ac   6.53 4BooX3   1   9/53   6/57
3   B/   CPw,z.z   1953mc   1959mc   ac   6.53 4BooX3   1
AA 1.4.46 on 2546(2546), 3/-, OSPR 10/46. M on A.

TA 07 (54/07) *Hunmanby**   10 by 10 Revision a 1909-26 c 1938-50
1   A   CPr   OPr   1947po   -   -   1.47 4BE1X3   RB   10/47 12/47
3   B   CPt,u,z.z,0z OPv   1954pc   -   ac   6.54 4BooX3   1/54   11/56
AA 1.12.46 on A(A), 3/-, OSPR 12/47. M on A.

TA 08 (54/08) **& Part of TA 18** *Scarborough** 15 by 10 Revision a 1909-39 b 1909-27 c 1938 d 1938-50
1   A   CPr   OPr   1948po   -   -   6.48 4BE2X3   RB   4/48   6/48
1   B   CPs   1951pc   -   ac   6.51 4BEoo3   RB   11/51 -
3   C   CPt   1954pc   -   bd   6.54 4BooX3   3/54   -
3   C/   CPv,x.z,z+   1954pc   1957mc   bd   6.54 4oooX3
AA 1.1.48 on A(A), 3/-, OSPR 6/48. M on B, C.

TA 10 (54/10) *Caistor*   10 by 10 Revision a 1905-06 c 1938-53 f 1976
1   A   CPr   OPr   1947po   -   -   6.47 4BE1X3   B   11/47 2/48
2   B   CPt   OPv   1953pc   -   ac   6.53 4BooX3   7/53   6/57
3   B/   CPw,x,z   1953mc   1959mc   ac   6.53 4BooX3
3   B//*   CFz   OFz   1953ᶜms   nd:r   acf   6.53 4BooX3   1
AA 1.9.47 on A(A), 3/-, OSPR 2/48. 1. OSPR not found.

TA 11 (54/11) *Immingham*   10 by 10 Revision a 1905-30 c 1946-52 f 1965
1   A   CPr   1947po   -   -   6.47 4BE1X3   RB   00   10/47 -
1   B   OPr   1947po   -   -   6.47 4BE1X3   RB   00   -   1/48
2   C   CPt   OPv   1953pc   -   ac   6.53 4BooX3   01   9/53   11/56
3   C/
3   C//   CPx   1953mc   1961ch   ac   6.53 4BooX3   01
3   C///   CPz   OPz   1953mc   1967ch   acf   6.53 4BooX3   33
AA 1.6.47 on B(-), 3/-, OSPR 1/48. M on B.

TA 12 (54/12) *Kingston-upon-Hull (South East)* 10 by 10 Revision a 1906-30 c 1938-53
1   15046   CPr   1946po   -   -   6.46 4BE1X3   RB   0   6/46   -
1   2546   OPr   1946po   -   -   6.46 4BE1X3   RB   0   -   9/46
3   B   CPt   1953pc   -   ac   6.53 4BooX3   1   11/53 -
3   B/
3   B//   CPx,z   1953mc   1961ch   ac   6.53 4BooX3   1
AA 1.3.46 on 2546(2546), 3/-, OSPR 10/46. M on A.

TA 13 (54/13) *Kingston-upon-Hull (North East)* 10 by 10 Revision a 1925-26 c 1938-52
                                                      j 1983 selected k SUSI
1   27546   CPr   1946po   -   -   6.46 4BE1X3   RB   6/46   -
1   2546   OPr   1946po   -   -   6.46 4BE1X3   RB   -   9/46
2   B   CPt   OPv   1953pc   -   ac   6.53 4BooX4   10/53 6/57
3   B   CPv,y,z,0z   1953pc   -   ac   6.53 4BooX4
3   B/*   CFz   OFz   1984ᶜm-s nd:br   acjk 6.85 4BooX4   9/84   9/84
AA 1.4.46 on 2546(2546), 3/-, OSPR 9/46. M on A.

254

1	2	3	4	5	6	7	8	9	10	11	12	13	14

TA 14 (54/14) *Brandesburton** 10 by 10 Revision a 1908-26 c 1938-53 f 1965
1	A	CPr		1948po	-	-	6.48	4BEoo3	RB	0	5/48	-	
1	B		OPr	1948po	-	-	6.48	4BEoo3	RB	0	-	6/48	
1	B	CPr		nd:po	-	-	6.48	4BEoo3	RB	0			
2	C	CPt	OPv	1953pc	-	ac	6.53	4BooX3		1	9/53	6/57	
3	C/	CPw		1953pc	1958mc	ac	6.53	4BooX3		1			
3	C//	CPz	OPz	1953mc	1965ch	acf	6.53	4BooX3		3			

AA 1.3.48 on B(A), np, OSPR 6/48. M on B, C. SS: Soil, coloured, 1983 on C//.

TA 15 (54/15) & Part of TA 25 *Beeford** 12 by 10 Revision a 1908-26 c 1945-53 f 1965
1	A	CPr	OPr	1947po	-	-	6.47	4BE1X3	B	0	12/47	3/48	
1	B	CPr		nd:po	-	-	6.47	4BEoo3	B	0			
2	C	CPt	OPv	1953pc	-	ac	6.53	4BooX3		1	11/53	3/58	
3	C/	CPw		1953mc	1960mc	ac	6.53	4BooX3		1			
3	C//	CPz,0z		1953mc	1966ch	acf	6.53	4BooX3		3			

AA 1.9.47 on A(A), 3/-, OSPR 3/48. M on B.

TA 16 (54/16) *Bridlington** 10 by 10 Revision a 1909-26 c 1938-53
1	A	CPr	OPr	1948po	-	-	1.48	4BE1X3	RB	0	1/48	3/48	
2	B	CPt	OPv	1953pc	-	ac	6.53	4BooX3		1	9/53	11/56	
3	B/	CPw,z.z		1953mc	1959mc	ac	6.53	4BooX3		1			1

AA 1.10.47 on A(A), 3/-, OSPR 3/48. M on A. 1. Carnaby Airfield.

TA 17 (54/17) *Reighton** 10 by 10 Revision a 1909-26 c 1938-50
1	A	CPr	OPr	1948po	-	-	6.48	4BEoo3	R		5/48	6/48	
3	B	CPt		1954pc	-	ac	6.54	4BooX3			2/54	-	
3	B/	CPv.z		1954pc	1957mc	ac	6.54	4BooX3					1

AA 1.3.48 on A(A), 3/-, OSPR 6/48. M on A. 1. Showing Filey holiday camp, with its railway access.

TA 20 (54/20) *Grimsby (South)** 10 by 10 Revision a 1905-32 c 1938-51 f 1963 g 1966
1	15046	CPr		1946po	-	-	6.46	4BE2X3	RB	0	8/46	-	
1	2546		OPr	1946po	-	-	6.46	4BE2X3	RB	0	-	11/46	
3	B	CPt	OPv	1953pc	-	ac	6.53	4BooX3	R	1	11/53	6/57	
3	B/	CPy		1953mc	1964ch	acf	6.53	4BooX3		1			
3	B//	CPz	OPz	1953mc	1967ch	acg	6.53	4BooX3		3			

AA 1.6.46 on 2546(2546), 3/-, OSPR 11/46. M on A.

TA 21 (54/21) *Grimsby (North)** 10 by 10 Revision a 1906-32 c 1938-52
1	12546	CPr		1946po	-	-	6.46	4BE1X3	RB		6/46	-	
1	2546		OPr	1946po	-	-	6.46	4BE1X3	RB		-	10/46	
2	B	CPt		1953pc	-	-	6.53	4BooX3	R		2/53	-	
3	B/	CPw		1953pc	1958mc	ac	6.53	4BooX3					
3	B//	CPw,z,0z		1953mc	1960mc	ac	6.53	4BooX3					

AA 1.4.46 on 2546(2546), 3/-, OSPR 10/46. M on A.

TA 22 (54/22) *Keyingham** 10 by 10 Revision a 1906-26 c 1938-52
1	10046	CPr		1946po	-	-	1.47	4BE1X3	RB		9/46	-	
1	A		OPr	1946po	-	-	1.47	4BE1X3	RB		-	1/47	
2	B	CPt	OPv	1953pc	-	ac	6.53	4BooX3			9/53	6/57	
3	B/	CPw,z,0z		1953mc	1959mc	ac	6.53	4BooX3					

AA 1.7.46 on A(A), 3/-, OSPR 1/47. M on A.

TA 23 (54/23) & Part of TA 33 *Aldbrough* 12 by 10 Revision a 1925-26 c 1946-52
1	10046	CPr		1946po	-	-	6.46	4BE2X3	B		9/46	-	
1	2547		OPr	1947po	-	-	6.46	4BE2X3	B		-	1/47	
2	B	CPt	OPv	1953pc	-	ac	6.53	4BooX3			9/53	12/57	
3	B	CPw,z.z+		1953pc	-	ac	6.53	4BooX3					

AA 1.6.46 on 2547(2547), 3/-, OSPR 1/47. M on 10046, B.

TA 24 (54/24) *Hornsea** 10 by 10 Revision a 1925-26 c 1938-52
| 1 | A | CPr | OPr | 1947po | - | - | 1.48 | 4BE1X3 | RB | | 9/47 | 12/47 | |

255

1	2	3	4	5	6	7	8	9	10	11	12	13	14

2   B   CPt   OPv   1953pc   -   ac   6.53   4BooX3   9/53   6/57
3   B   CPz   1953pc   -   ac   6.53   4BooX3
AA 1.6.47 on A(A), 3/-, OSPR 12/47. M on A, B.

TA 27 (54/27) & Part of TA 26 *Flamborough Head** 10 by 10 Revision a 1926 c 1938-50
1   A   CPr   OPr   1948po   -   6.47   4BE1X3   12/47 2/48
1   B   CPr   nd:po   -   6.47   4BEoo3
3   C   CPt   OPv   1953pc   -   ac   6.53   4BooX3   12/53 6/57
3   C/   CP?w,y,z.z   1953mc   1960mc   ac   6.53   4BooX3
Sheet lines are offset, at 520 km E to 530 km E, 468 km N to 478 km N. AA 1.8.47 on A(A), 3/-, OSPR 2/48. M on A.

TA 30 (54/30) & Part of TA 40 *Cleethorpes** 15 by 10 Revision a 1905-32 c 1946-53
1   A   CPr   OPr   1947po   -   -   1.47   4BE2X3   R   00   6/47   10/47
3   B   CPt,u,x.z   OP?,z 1953pc   -   ac   6.53   4BooX4   10   12/53 2/58
AA 1.3.47 on A(A), 3/-, OSPR 10/47. M on A.

TA 31 (54/31) & Part of TA 41 *Spurnhead** 15 by 10 Revision a 1905-27 c 1946-52
1   A   CPr   OPr   1947po   -   6.47   4BE2X3   R   6/47   9/47 1
2   B   CPt   OP?   1953pc   -   ac   6.53   4BooX3   4/53   2/58
3   B/   CP?w,x.z   1953mc   1959mc   ac   6.53   4BooX3
AA 1.4.47 on A(A), 3/-, OSPR 9/47. M on A. 1. Showing the Kilnsea to Spurnhead railway system.

TA 32 (54/32) & Parts of TA 33 & TA 42 *Withernsea** 12 by 12 Revision a 1906-25 c 1938-52
1   A   CPr   OPr   1947po   -   -   1.48   4BE2X3   R   12/47 2/48
2   B   CPt   OPv   1953pc   -   ac   6.53   4BooX3   10/53 6/57
3   B   CPz.z   1953pc   -   ac   6.53   4BooX3
AA 1.9.47 on A(A), 3/-, OSPR 2/48. M on A, B.

TF 00 (53/00) *Stamford*   10 by 10 Revision a 1899-1929 c 1938-50 f 1960 g 1965
1   A   CPr   OPr   nd:po   -   -   1.49   4BEoo3   RB   00   11/48 1/49
3   B   CPu   OPv   1955pc   -   ac   6.55   4oooX3   11   4/55   10/56
3   B/   CPw   1955mc   1958mc   ac   6.55   4oooX3   11
3   B//*   CPz   OPz   1955mc   1964r   acf   6.55   4oooX3   11   12/64 11/64
3   B//*/   CPz,0z   OPz   1955mc   1966ch   acg   6.55   4oooX3   22   1
AA 1.8.48 on A(A), 3/-, OSPR 1/49. 1. Wittering Airfield.

TF 01 (53/01) *Essendine*   10 by 10 Revision a 1902-29 c 1938-50
1   A   CPr   OPr   nd:po   -   -   6.49   4BEoo3   RB   7/49   7/49   1
3   B   CPu   OPv   1955pc   -   ac   6.55   4oooX3   1/55   9/57
3   B/   CPw,z   1955pc   1958mc   ac   6.55   4oooX3
1. Showing the alignment of the Edenham Railway.

TF 02 (53/02) *Edenham**   10 by 10 Revision a 1903-29 c 1938-50
1   A   CPr   OPr   nd:po   -   -   1.49   4BEoo3   RB   0   7/49   7/49   1
3   B   CPu,v,z   OPv   1954pc   -   ac   6.54   4oooX3   1   12/54 8/57
1. Showing the alignment of the Edenham Railway.

TF 03 (53/03) *Osbournby**   10 by 10 Revision a 1903 c 1947-50
1   A   CPr   OPr   nd:po   -   -   6.49   4BEoo3   B   0   7/49   8/49
3   B   CPu   OPv   1955pc   -   ac   6.55   4oooX3   1   3/55   8/57
3   B/   CPw,z   1955mc   1960mc   ac   6.55   4oooX3   1   1
1. Folkingham Airfield.

TF 04 (53/04) *Sleaford**   10 by 10 Revision a 1903-04 c 1938-51 f 1976
1   A   CPr   OPr   nd:po   -   -   6.49   4BEoo3   RB   0   6/49   7/49   1
3   B   CPu   OPv   1955pc   -   ac   6.55   4oooX3   1   8/55   10/56
3   B/   CPw,z.z   1955mc   1959mc   ac   6.55   4oooX3   1   2
3   B//*   CFz   OFz   1955^{c}ms nd:r   acf   6.55   4oooX3   1   3
AA 1.3.49 on A(A), 3/-, OSPR 7/49. SS: Soil, Land Use Capability, both coloured, both outline, 1978 on B/. 1. The Cranwell railway is shown, not the airfield. 1. Cranwell Airfield. 3. OSPR not found.

1	2	3	4	5	6	7	8	9	10	11	12	13	14

TF 05 (53/05) *Ruskington*   10 by 10   Revision a 1904 c 1947-53
1	A	CPr	OPr	nd:po	-	-	1.49	4BEoo3	RB	000	6/49	7/49	
2	B	CPt	OPv	1953pc	-	ac	6.53	4BooX3		111	9/53	8/57	
3	B/	CPw,z		1953pc	1958mc	ac	6.53	4BooX3		111			1

1. Cranwell Aerodrome.

TF 06 (53/06) *Potter Hanworth*   10 by 10   Revision a 1904-30 c 1946-51
1	A	CPr	OPr	nd:po	-	-	6.48	4BEoo3	RB	00	9/48	10/48	
2	B	CPt		1953pc	-	ac	6.53	4BooX3		11	9/53	-	
3	B/	CPv,z		1953pc	1957mc	ac	6.53	4BooX3		11			

TF 07 (53/07) *Reepham Lincolnshire**   10 by 10   Revision a 1904-38 c 1946-53
1	A	CPr	OPr	nd:po	-	-	1.49	4BEoo3	RB	00	11/48	11/48	
2	B	CPt	OP?	1953pc	-	ac	6.53	4BooX3		11	9/53	10/53	
3	B	CPz	OPv	1953pc	-	ac	6.53	4BooX3		11			

TF 08 (53/08) *Faldingworth**   10 by 10   Revision a 1904-05 c 1951
| 1 | A | CPr | OPr | nd:po | - | - | 6.49 | 4BEoo3 | RB | 00 | 3/49 | 4/49 | |
| 3 | B | CPw,x,z.z | OPv | 1960mc | - | ac | 6.60 | 4oooX3 | | 10 | 1/60 | 2/60 | |

TF 09 (53/09) *South Kelsey**   10 by 10   Revision a 1905 c 1951
1	A	CPr	OPr	1948po	-	-	1.48	4BE2X3	RB		3/48	5/48	
1	B	CPr		nd:po	-	-	1.48	4BEoo3	RB				
3	C	CPw	OPv	1958pc	-	ac	6.58	4oooX3			9/58	10/58	
3	C/	CPw,z		1958mc	1961ch	ac	6.58	4oooX3					

TF 10 (53/10) *Peterborough (North West)*   10 by 10   Revision a 1899-1925 c 1938-50 f 1975
1	A	CPr	OPr	1948po	-	-	6.48	4BEoo3	RB	0	5/48	7/48	
3	B	CPu	OPv	1955pc	-	ac	6.55	4oooX3		2	5/55	8/57	
3	B/	CP?w,x,z		1955mc	1960mc	ac	6.55	4oooX3		2			
3	B//*	CFz		1955^{c}ms	nd:r	acf	6.55	4oooX3		2			1

AA 1.3.48 on A(A), 3/-, OSPR 7/48. 1. OSPR not found.

TF 11 (53/11) *Baston**   10 by 10   Revision a 1899-1929 c 1938-50
| 1 | A | CPr | OPr | nd:po | - | - | 1.49 | 4BEoo3 | R | | 7/49 | 8/49 | |
| 3 | B | CPu.x,z | OPv | 1955pc | - | ac | 6.55 | 4oooX3 | | | 4/55 | 8/57 | |

M on A.

TF 12 (53/12) *Dowsby**   10 by 10   Revision a 1903-29 c 1938-50
1	A	CPr	OPr	nd:po	-	-	1.49	4BEoo3	R		8/49	8/49	
3	B	CPu	OPv	1955pc	-	ac	6.55	4oooX3			1/55	8/57	
3	B/	CP?w,z		1955mc	1960mc	ac	6.55	4oooX3					

TF 13 (53/13) *Billingborough*   10 by 10   Revision a 1903-29 c 1938-50
| 1 | A | CPr | OPr | nd:po | - | - | 1.49 | 4BEoo3 | RB | | 9/49 | 9/49 | |
| 3 | B | CPu,z | OPv | 1955pc | - | ac | 6.55 | 4oooX3 | | | 1/55 | 8/57 | |

TF 14 (53/14) *Heckington*   10 by 10   Revision a 1903-04 c 1946-51; j 1985 selected
1	A	CPr	OPr	nd:po	-	-	1.49	4BEoo3	RB		7/49	8/49	
3	B	CPu,z	OPv	1955pc	-	ac	6.55	4oooX3			8/55	8/57	
3	B/*	CFz	OFz	1986^{c}m-s	-	acj	6.86	4oooX3			6/86	6/86	1

1. One of the final batch of three revised reprints in this series (with SU 03, TL 64).

TF 15 (53/15) *Billinghay*   10 by 10   Revision a 1903-04 c 1951-53
| 1 | A | CPr | OPr | nd:po | - | - | 1.49 | 4BEoo3 | R | | 5/49 | 5/49 | |
| 3 | B | CPw,x,z | OPv | 1959mc | - | ac | 6.59 | 4oooX3 | | | 8/59 | 11/59 | |

TF 16 (53/16) *Woodhall Spa**   10 by 10   Revision a 1904 c 1938-53
1	A	CPr	OPr	nd:po	-	-	1.49	4BEoo3	RB	0	5/49	4/49	
2	B	CPt	OPs	1953pc	-	ac	6.53	4BooX3		1	9/53	7/55	
3	B/	CP?w,z		1953mc	1959mc	ac	6.53	4BooX3		1			

SS: Soil, outline, 1973 on B/; Land Use Capability, outline, 1974 on B/; Soil Drainage, 1974 on B/.

1	2	3	4	5	6	7	8	9	10	11	12	13	14

TF 17 (53/17) *Wragby Lincolnshire**   10 by 10   Revision a 1904-05 c 1946-53
1 A   CPr   OPr   nd:po   -   -   6.48   4BEoo3   RB   0   3/49   4/49
2 B   CPt,u   OPv   1953pc   -   ac   6.53   4BooX3   1   9/53   7/57
3 B   CPz   1953pc   -   ac   6.53   4BooX3   1

TF 18 (53/18) *Market Rasen**   10 by 10   Revision a 1905-48 c 1950-53
1 A   CPr   OPr   nd:po   -   -   6.48   4BEoo3   RB   00   12/48 1/49
2 B   CPt   OPv   1953pc   -   ac   6.53   4BooX3   11   9/53   8/57
3 B/   CP?w.y,z   1953mc   1960mc   ac   6.53   4BooX3   11   1
1. Ludford Magna Airfield.

TF 19 (53/19) *Thoresway**   10 by 10   Revision a 1905-07 c 1951
1 A   CPr   OPr   nd:po   -   -   6.48   4BEoo3   RB   0   11/48 12/48
1 A   CPs   1948pc   1951   -   1.51   4BEoo3   RB   0
3 A   CPw   1948mc   1960mc   a   1.51   4BooX3   B   1   1
3 B   CPw.z,0z OPv   1960mc   -   ac   6.60   4oooX4   1   11/60 5/60
1. Binbrook Airfield. A/ would seem a more accurate print code for this sheet.

TF 20 (53/20) *Peterborough (North East)** 10 by 10 Revision a 1899-1925 c 1938-50
1 A   CPr   OPr   1948po   -   -   1.48   4BE2X3   RB   3/48   3/48
1 A   CPr   OPr   1948po   -   -   1.48   4BE2X3   RB   1
1 B   CPr   nd:po   -   -   1.48   4BEoo3   RB
3 C   CPu,z   OPv   1955pc   -   ac   6.55   4oooX3   8/55   10/56
AA 1.11.47 on A(A), 3/-, OSPR 7/48. 1. The magnetic variation value of 11°04'W is altered to 11°00'W.

TF 21 (53/21) *Cowbit**   10 by 10   Revision a 1900-29 c 1938-50
1 A   CPr   OPr   nd:po   -   -   6.48   4BEoo3   R   8/49   9/49
3 B   CPu,z   OPv   1955pc   -   ac   6.55   4oooX3   4/55   10/56
AA 1.3.49 on A(A), 3/-, OSPR 9/49.

TF 22 (53/22) *Spalding*   10 by 10   Revision a 1903-29 c 1938-50 f 1961
1 A   CPr   OPr   nd:po   -   -   1.49   4BEoo3   R   9/49   9/49
3 B   CPu   OPv   1955pc   -   ac   6.55   4oooX3   5/55   8/57
3 B/
3 B//   CP?x,z   1955mc   1962ch   acf   6.55   4oooX3
AA 1.3.49 on A(A), 3/-, OSPR 9/49.

TF 23 (53/23) *Donington Lincolnshire**   10 by 10   Revision a 1903-29 c 1938-50
1 A   CPr   OPr   nd:po   -   -   1.49   4BEoo3   RB   10/49 11/49
3 B   CPu,y,z   OPv   1955pc   -   ac   6.55   4oooX3   2/55   10/56

TF 24 (53/24) *Swineshead Lincolnshire**   10 by 10   Revision a 1903-04 c 1938-51
1 A   CPr   OPr   nd:po   -   -   1.49   4BEoo3   R   0   5/49   5/49
3 B   CPu   OPv   1955pc   -   ac   6.55   4oooX3   1   6/55   8/57
3 B/   CP?w.z   1955mc   1960mc   ac   6.55   4oooX3   1

TF 25 (53/25) *Coningsby**   10 by 10   Revision a 1903-04 c 1946-53
1 A   CPr   OPr   nd:po   -   -   1.49   4BEoo3   R   0   4/49   4/49
2 B   CPt   OPv   1953pc   -   ac   6.53   4BooX3   1   10/53 8/57
3 B/   CP?w,z   1953mc   1960mc   ac   6.53   4BooX3   1   1
1. Coningsby Airfield.

TF 26 (53/26) *Horncastle**   10 by 10   Revision a 1904 c 1946-53
1 A   CPr   OPr   nd:po   -   -   1.49   4BEoo3   RB   0   5/49   4/49
2 B   CPt   OPs   1953pc   -   ac   6.53   4BooX3   1   9/53   9/55
3 B/   CP?w,z   1953mc   1959mc   ac   6.53   4BooX3   1

TF 27 (53/27) *West Ashby**   10 by 10   Revision a 1904-05 c 1946-51
1 A   CPr   OPr   nd:po   -   -   1.48   4BEoo3   B   2/49   3/49
3 B   CPw.y,z   OPx   1961mc   -   ac   6.61   4oooX3   12/60 2/61

1	2	3	4	5	6	7	8	9	10	11	12	13	14

TF 28 (53/28) *Donington on Bain**   10 by 10   Revision a 1905 c 1938-53
1 A   CPr   OPr   nd:po   -   -   1.49   4BEoo3   RB   0   12/48   1/49
3 B   CPt   OPs   1953pc   -   ac   6.53   4BooX3   1   12/53   12/53
3 B/   CPw,z   1953mc   1960mc   ac   6.53   4BooX3   1   1

M on A, B. SS: Soil, Land Use Capability, coloured, 1978 on B/. 1. Ludford Magna Airfield.

TF 29 (53/29) *Ludborough**   10 by 10   Revision a 1905-07 c 1937-51
1 A   CPr   OPr   nd:po   -   -   1.49   4BEoo3   RB   00   12/48   12/48
  B
2 C   CPt   OPs   1953pc   -   ac   6.53   4BooX3   11   9/53   3/54
3 C/   CPw.z   1953mc   1960mc   ac   6.53   4BooX3   11

TF 30 (53/30) *Parson Drove**   10 by 10   Revision a 1899-1928 c 1938-50
1 A   CPr   OPr   nd:po   -   -   6.49   4BEoo3   R   8/49   8/49
3 B   CPu   OPv   1955pc   -   ac   6.55   4oooX3   5/55   8/57
3 B/   CP?w,z   1955mc   1960mc   ac   6.55   4oooX3

TF 31 (53/31) *Gedney Hill**   10 by 10   Revision a 1900-29 c 1938-50
1 A   CPr   OPr   nd:po   -   -   6.48   4BEoo3   R   9/49   11/49
2 A   CPt   1949pc   1953rpb   -   6.48   4BooX3   R
3 B   CPu,w   1955pc   -   ac   6.55   4oooX3   6/55   -
3 B/   CPz   1955pc   -   ac   6.55   4oooX3

TF 32 (53/32) *Holbeach**   10 by 10   Revision a 1903-29 c 1938-50
1 A   CPr   OPr   nd:po   -   -   1.49   4BEoo3   RB   8/49   9/49
1 B   CPs   OPv   1951pc   -   ac   6.51   4BEoo3   RB   1/52   8/57   1
3 B/   CPx,z   1951mc   1961ch   ac   6.51   4BooX3   B

1. The "LMS & LNER" of edition A is altered to "M & GNR" here. By this time it was "BR".

TF 33 (53/33) *Kirton Lincolnshire**   10 by 10   Revision a 1903-29 c 1938-55
1 A   CPr   OPr   nd:po   -   -   6.49   4BEoo3   RB   10/49   11/49
3 B   CPu,y,z   OPv,z   1956pc   -   ac   6.56   4oooX3   2/56   10/56

TF 34 (53/34) *Boston*   10 by 10   Revision a 1903-04 c 1938-51
1 A   CPr   OPr   nd:po   -   -   1.49   4BEoo3   R   5/49   7/49
3 B   CPu   OPv   1955pc   -   ac   6.55   4oooX3   9/55   10/56
3 B/   CP?w,z   OPz   1955mc   1960mc   ac   6.55   4oooX3
AA 1.1.49 on A(A), 3/-, OSPR 7/49.

TF 35 (53/35) *Stickney**   10 by 10   Revision a 1904 c 1946-51
1 A   CPr   OPr   nd:po   -   -   1.49   4BEoo3   R   5/49   5/49
2 A   CPt   1949pc   1952   -   1.52   4Booo3   R
3 B   CPw,z   OPv   1958mc   -   ac   6.58   4oooX3   11/58   12/58

TF 36 (53/36) *Old Bolingbroke*   10 by 10   Revision a 1904 c 1946-53
1 A   CPr   OPr   nd:po   -   -   1.49   4BEoo3   B   0   5/49   5/49
2 B   CPt,u   OPs   1953pc   -   ac   6.53   4BooX3   1   10/53   1/56
3 B   CPz   1953pc   -   ac   6.53   4BooX3   1

TF 37 (53/37) *Tetford*   10 by 10   Revision a 1904-05 c 1946-51
1 A   CPr   OPr   nd:po   -   -   6.48   4BEoo3   B   5/49   5/49
3 B   CPw,z,0z   OPv   1960mc   -   ac   6.60   4oooX4   11/60   11/60

TF 38 (53/38) *Louth**   10 by 10   Revision a 1905 c 1947-53
1 A   CPr   OPr   nd:po   -   -   1.49   4BEoo3   RB   0   12/48   1/49
2 B   CPt   OPv   1953pc   -   ac   6.53   4BooX3   1   10/53   6/57
3 B/   CPv,x,z,0z   1953pc   1957mc   ac   6.53   4BooX3   1
AA 1.10.48 on A(A), 3/-, OSPR 1/49.

1	2	3	4	5	6	7	8	9	10	11	12	13	14

TF 39 (53/39) *Covenham**   10 by 10 Revision a 1905 c 1946-51 f 1972; j 1972 reservoir
1 A   CPr   OPr   nd:po   -   -   6.48 4BEoo3   RB   11/48 12/48
2 B   CPt   OPv   1953pc   -   ac   6.53 4BooX3   8/53   6/57
3 B/   CPv,z   1953pc   1958mc   ac   6.53 4BooX3
3 B//*   CPz   OPz   1953ᶜms nd:rv   acfj 6.53 4BooX3   5/73   9/73   1
1. Covenham Reservoir.

TF 40 (53/40) *Wisbech (South)*   10 by 10 Revision a 1900-25 c 1938-50; j 1980 selected
1 A   CPr   OPr   nd:po   -   -   6.49 4BEoo3   R   10/49 11/49 1
3 B   CPu   OPs   1955pc   -   ac   6.55 4oooX3   11/55 10/55
3 B/   CP?w,z   1955mc   1960mc   ac   6.55 4oooX3
3 B//*   CFz   OFz   1981ᶜms nd:r   acj   6.55 4oooX3   3/81   3/81
AA 1.6.49/NT 6.7.49 on A(A), 3/-, OSPR 11/49. M on A. 1. The Wisbech & Outwell tramway is shown as a black line down the centre of the road.

TF 41 (53/41) *Wisbech (North)**   10 by 10 Revision a 1900-29 c 1938-51
1 A   CPr   OPr   nd:po   -   -   6.49 4BEoo3   R   0   9/49   11/49
3 B   CPu   OPv   1955pc   -   ac   6.55 4oooX3   1   9/55   11/56
3 B/   CPx,z   1955mc   1961ch   ac   6.55 4oooX3   1
AA 1.7.49/NT 17.7.49 on A(A), 3/-, OSPR 11/49. M on A.

TF 42 (53/42) *Sutton Bridge**   10 by 10 Revision a 1903-29 c 1938-52
1 A   CPr   OPr   nd:po   -   -   1.50 4BEoo3   RB   0   10/49 11/49
3 B   CPu,x,z   OPv   1955pc   -   ac   6.55 4oooX3   1   8/55   10/56

TF 43 (53/43) *Gat Sand**   10 by 10 Revision a 1903-29 c 1946-52
1 A   CPr   OPr   nd:po   -   -   1.49 4BEoo3   7/49   7/49
3 B   CPu   OPv   1955pc   -   ac   6.55 4oooX3   6/55   6/57
3 B/   CPx,z.z   OPx   1955mc   1961ch   ac   6.55 4oooX3

TF 44 (53/44) **& Part of TF 54** *Leverton**   15 by 10 Revision a 1903-04 c 1946-51
1 A   CPr   OPr   nd:po   -   -   1.49 4BEoo3   7/49   8/49
3 B   CPu,x,z.z OPv   1955pc   -   ac   6.55 4oooX3   8/55   11/57

TF 45 (53/45) *Friskney**   10 by 10 Revision a 1904 c 1946-51
1 A   CPr   OPr   nd:po   -   -   6.48 4BEoo3   RB   11/48 12/48
2 B   CPt   1953pc   -   ac   6.53 4BooX3   8/53   -
3 B/   CP?w,z   OPz   1953pc   1958mc   ac   6.53 4BooX3

TF 46 (53/46) *Firsby*   10 by 10 Revision a 1904-05 c 1946-53 f 1965
1 A   CPr   OPr   nd:po   -   -   6.48 4BEoo3   RB   0   11/48 12/48
2 B   CPt   OPv   1953pc   -   ac   6.53 4BooX3   1   10/53 10/56
3 B/   CPw   1953mc   1959mc   ac   6.53 4BooX3   1
3 B//   CPz   OPz   1953mc   1966ch   acf   6.53 4BooX3   3
M on A, B.

TF 47 (53/47) *Alford Lincolnshire*   10 by 10 Revision a 1904-05 c 1946-53
1 A   CPr   OPr   nd:po   -   -   6.48 4BEoo3   RB   11/48 12/48
2 B   CPt   OPv   1953pc   -   ac   6.53 4BooX3   10/53 6/57
3 B/   CPw,z   1953mc   1960mc   ac   6.53 4BooX3

TF 48 (53/48) **& Part of TF 58** *Mablethorpe** 15 by 10 Revision a 1905 c 1938-53
1 A   CPr   OPr   nd:po   -   -   6.48 4BEoo3   RB   0   12/48 1/49
2 B   CPt,x   OPv   1953pc   -   ac   6.53 4BooX4   1   11/53 8/57
3 B   CPz+,z+   1953pc   -   ac   6.53 4BooX4   1   1
AA 1.9.48 on A(A), 3/-, OSPR 1/49. 1. On the second printing the heading "Provisional Edition" is moved right.

TF 49 (53/49) *North Somercotes**   10 by 10 Revision a 1905 c 1946-51
1 A   CPr   OPr   nd:po   -   -   6.48 4BEoo3   R   11/48 12/48
2 B   CPt   OPs   1953pc   -   ac   6.53 4BooX3   6/53   12/53
3 B/   CP?w,z.0z OPz   1953mc   1959mc   ac   6.53 4BooX3

1	2	3	4	5	6	7	8	9	10	11	12	13	14

TF 50 (53/50) *Outwell**   10 by 10 Revision a 1900-27 c 1950 f 1963
1   A   CPr   OPr   nd:po   -   -   6.49   4BEoo3   R   8/49   9/49
    B
3   C   CPu   OPv   1955pc   -   ac   6.55   4oooX3   4/55   6/57
3   C/
3   C//   CPz   OPz   1955mc   1964ch acf 6.55   4oooX3
M on A. The Wisbech & Outwell tramway is in part shown as a black line down the centre of the road.

TF 51 (53/51) *St John's Highway**   10 by 10 Revision a 1900-27 c 1938-50
1   A   CPr   OPr   nd:po   -   -   6.49   4BEoo3   R   10/49 11/49
1   B   CPs   1951pc   -   ac   6.51   4BEoo3   R   1/52   -
2   B   OPv   1951pc   -   ac   1.52   4Booo3   R   -   6/57
3   B/   CPz   1951pc   -   ac   1.52   4BooX3
M on B, B/ (listed as B).

TF 52 (53/52) *Terrington Norfolk**   10 by 10 Revision a 1903-29 c 1938-50
1   A   CPr   OPr   nd:po   -   -   6.49   4BEoo3   R   7/49   8/49
3   B   CPu,y,z OPv   1955pc   -   ac   6.55   4oooX3   8/55   6/57
M on A.

TF 55 (53/55) *Wainfleet Sand**   10 by 10 Revision a 1904 c 1938-51
1   A   CPr   OPr   nd:po   -   -   6.48   4BEoo3   RB   11/48 11/48
2   B   CPt   1953pc   -   ac   6.53   4BooX3   4/53   -
3   B/   CPv,z.z OPz   1953pc   1956mc ac   6.53   4BooX3

TF 56 (53/56) *Skegness**   10 by 10 Revision a 1904-05 c 1938-51
1   A   CPr   OPr   nd:po   -   -   6.48   4BEoo3   R   0   2/49   2/49
2   B   CPt   OPs   1953pc   -   ac   6.53   4BooX3   1   9/53   4/56   1
3   B/   CPw,z.z OPv   1953mc   1960mc ac   6.53   4BooX3   1
AA 1.10.48 on A(A), 3/-, OSPR 2/49. 1. There is a railway triangle at 559629 - for turning steam locomotives?

TF 57 (53/57) *Chapel St Leonards**   10 by 10 Revision a 1905 c 1938-51
1   A   CPr   OPr   nd:po   -   -   6.48   4BEoo3   RB   12/48 2/49
3   B   CPt   OPv   1953pc   -   ac   6.53   4BooX3   11/53 10/56
3   B/   CP?w,z.z   1953mc   1960mc ac   6.53   4BooX3
AA 1.9.48 on A(A), 3/-, OSPR 2/49.

TF 60 (53/60) *Downham Market**   10 by 10 Revision a 1900-27 c 1938-50
1   A   CPr   OPr   nd:po   -   -   1.50   4BEoo3   R   0   10/49 11/49
3   B   CPu   OPv   1955pc   -   ac   6.55   4oooX3   1   7/55   8/57
3   B/   CPy,z   1955mc   1963ch ac   6.55   4oooX3   1
M on A.

TF 61 (53/61) *King's Lynn (South)**   10 by 10 Revision a 1904-27 c 1938-50 f 1961 g 1969
1   A   CPr   OPr   nd:po   -   -   6.49   4BEoo3   RB   9/49   9/49
3   B   CPu   OPv   1955pc   -   ac   6.55   4oooX3   9/55   10/56
3   B/   CPw   1955mc   1961mc ac   6.55   4oooX3
3   B//   CPx   1955mc   1962ch acf 6.55   4oooX3
3   B///* CPz   OPz   1955ms   1970r acg 6.55   4oooX3   5/70   8/70
AA 1.5.49 on A(-), 3/-, OSPR 9/49. M on A.

TF 62 (53/62) *King's Lynn (North)**   10 by 10 Revision a 1904-27 c 1938-52 f 1967
1   A   CPr   OPr   nd:po   -   -   1.50   4BEoo3   R   10/49 11/49
3   B   CPu   OPv   1955pc   -   ac   6.55   4oooX3   10/55 8/57
3   B/   CP?w,z   1955mc   1960mc ac   6.55   4oooX3
3   B//* CPz   OPz   1955ms   1968r acf 6.55   4oooX3   5/68   4/68
AA 1.6.49 on A(A), 3/-, OSPR 11/49. M on A.

261

1	2	3	4	5	6	7	8	9	10	11	12	13	14

TF 63 (53/63) *Snettisham**   10 by 10   Revision a 1904-27 c 1938-52 f 1961
1 A   CPr   OPr   nd:po   -   -   6.49   4BEoo3   R   8/49   9/49
3 B   CPu   OPv   1955pc   -   ac   6.55   4oooX3   8/55   10/56
3 B/   CPz   OPz   1955mc   1965ch   acf   6.55   4oooX3
M on A.

TF 70 (53/70) *Beechamwell**   10 by 10   Revision a 1904-26 c 1938-50 f 1968
1 A   CPr   OPr   nd:po   -   -   1.48   4BEoo3   R   0   3/49   4/49
3 B   CPu,y   OPv   1955pc   -   ac   6.55   4oooX3   1   5/55   10/56
3 B/   CPz,0z   OPz,0z 1955ms   1969ch   acf   6.55   4oooX3   3
M on A.

TF 71 (53/71) *Narborough*   10 by 10   Revision a 1904-26 c 1938-50
1 A   CPr   OPr   nd:po   -   -   1.48   4BEoo3   RB   0   3/49   4/49
3 B   CPu   OPv   1955pc   -   a   6.55   4oooX3   1   10/55 6/57
3 B/   CPw,z   1955mc   1959mc   ac   6.55   4oooX3   1
M on A.

TF 72 (53/72) *Hillington**   10 by 10   Revision a 1904-27 c 1938-50 f 1966
1 A   CPr   OPr   nd:po   -   -   1.48   4BEoo3   RB   0   3/49   4/49
3 B   CPu   OPv   1955pc   -   ac   6.55   4oooX3   1   6/55   10/56
3 B/   CPw   1955mc   1960mc   ac   6.55   4oooX3   1
3 B//   CPz   1955mc   1966ch   acf   6.55   4oooX3   3
M on A.

TF 73 (53/73) *Docking**   10 by 10   Revision a 1904-40 c 1950
1 A   CPr   OPr   nd:po   -   -   1.48   4BEoo3   RB   0   3/49   4/49
3 B   CPu,z   OPv   1955pc   -   ac   6.55   4oooX3   1   3/55   11/56
M on A.

TF 74 (53/74) **& Part of TF 64** *Hunstanton**   15 by 10   Revision a 1904-27 c 1928-50
1 A   CPr   OPr   nd:po   -   -   1.48   4BEoo3   3/49   5/49
3 B   CPu   OP?   1955pc   -   ac   6.55   4oooX3   9/55   2/58
3 B/   CP?w,x,z.z,0z   1955mc   1960mc   ac   6.55   4oooX3
AA 1.1.49/NT 9.3.49 on A(A), 3/-, OSPR 5/49. M on A.

TF 80 (53/80) *Swaffham*   10 by 10   Revision a 1904-26 c 1938-50 f 1965
1 A   CPr   OPr   nd:po   -   -   1.48   4BEoo3   R   00 2/49   2/49
3 B   CPu   OPv   1955pc   -   ac   6.55   4oooX3   12 1/56   6/57
3 B/   CPz   1955mc   1966ch   acf   6.55   4oooX3   32
M on A.

TF 81 (53/81) *Castle Acre*   10 by 10   Revision a 1904-27 c 1950; j 1983 selected
1 A   CPr   OPr   nd:po   -   -   1.48   4BEoo3   R   2/49   2/49
3 B   CPu,y,z OPv   1955pc   -   ac   6.55   4oooX3   7/55   6/57
3 B/*   CFz   OFz   1983^{c}m-s nd:r acj   6.84   4oooX3   1/84   1/84
M on A.

TF 82 (53/82) *Helhoughton**   10 by 10   Revision a 1904-39 c 1950 f 1965
1 A   CPr   OPr   nd:po   -   -   1.48   4BEoo3   R   000 2/49   3/49
3 B   CPu   OPv   1955pc   -   ac   6.55   4oooX3   111 9/55   8/57
3 B/   CPw   1955mc   1959mc   ac   6.55   4oooX3   111   1
3 B//   CPz,0z   OPz,0z 1955mc   1966ch   acf   6.55   4oooX3   330
M on A. 1. West Raynham Airfield.

TF 83 (53/83) *Creake*   10 by 10   Revision a 1904-39 c 1950 f 1966 g 1969
1 A   CPr   OPr   nd:po   -   -   1.48   4BEoo3   R   00 1/49   2/49
3 B   CPu   OPs   1955pc   -   ac   6.55   4oooX3   11 3/55   6/57
3 B/   CPw   1955mc   1958mc   ac   6.55   4oooX3   11
3 B//*   CPz   OPz   1955mc   1967r   acf   6.55   4oooX3   11 6/67   6/67
3 B//*/ CPz,0z   1955mc   1970ch   acg   6.55   4oooX3   33   1
M on A. 1. Sculthorpe Aerodrome.

262

1	2	3	4	5	6	7	8	9	10	11	12	13	14

TF 84 (53/84) *Burnham Market*   10 by 10   Revision a 1904-27 c 1939-50 f 1965
1	A	CPr	OPr	nd:po	-	-	1.49	4BEoo3	R		2/49	2/49	
3	B	CPu	OPv	1955pc	-	ac	6.55	4oooX3			6/55	6/57	
3	B/	CPx		1955mc	1961mc	ac	6.55	4oooX3					
3	B//	CPz	OPz	1955mc	1966ch	acf	6.55	4oooX3					

M on A.

TF 90 (53/90) *Shipdham*   10 by 10   Revision a 1904-26 c 1938-50
1	A	CPr	OPr	nd:po	-	-	1.48	4BEoo3	RB	00	2/49	2/49	
3	B	CPu	OPv	1955pc	-	ac	6.55	4oooX3		11	11/55	6/57	
3	B/	CPw,z		1955mc	1960mc	ac	6.55	4oooX3		11			1

M on A. 1. Watton Airfield.

TF 91 (53/91) *East Dereham*   10 by 10   Revision a 1904-27 c 1938-50 f 1965 g 1976
                                                                    j 1981 selected
1	A	CPr	OPr	nd:po	-	-	1.48	4BEoo3	RB	00	3/49	3/49	
3	B	CPu,y	OPv	1955pc	-	ac	6.55	4oooX3		11	6/55	8/57	
3	B/	CPz	OPz	1955mc	1966ch	acf	6.55	4oooX3		32			
3	B//*	CFz	OFz	1955^{c}ms	nd:r	acg	6.55	4oooX3		32	3/78	3/78	
3	B//*/*	CFz	OFz	1981^{c}ms	md:r	acgj	6.55	4oooX3		32	12/81	12/81	

M on A.

TF 92 (53/92) *Fakenham (South) Norfolk**   10 by 10   Revision a 1904 c 1939-50
1	A	CPr	OPr	nd:po	-	-	1.48	4BEoo3	RB		2/49	3/49	
3	B	CPu	OPv	1955pc	-	ac	6.55	4oooX3			10/55	6/57	
3	B/	CPx,z,0z		1955mc	1961ch	ac	6.55	4oooX3					

M on A.

TF 93 (53/93) *Walsingham*   10 by 10   Revision a 1904-26 c 1939-50 f 1965
1	A	CPr	OPr	nd:po	-	-	1.48	4BEoo3	RB	00	3/49	4/49	
3	B	CPu	OPv	1955pc	-	ac	6.55	4oooX3		11	5/55	6/57	
3	B/	CPw		1955mc	1960mc	ac	6.55	4oooX3		11			
3	B//	CPz,0z		1955mc	1966ch	acf	6.55	4oooX3		33			

M on A.

TF 94 (53/94) *Wells-next-the-Sea*   10 by 10   Revision a 1904-27 c 1938-50 f 1966
1	A	CPr	OPr	nd:po	-	-	1.48	4BEoo3	RB	0	2/49	3/49	
3	B	CPu	OPv	1955pc	-	ac	6.55	4oooX3		1	5/55	6/57	
3	B/	CP?w,z		1955mc	1960mc	ac	6.55	4oooX3		1			
3	B//	CPz		1955mc	1966ch	acf	6.55	4oooX3		3			

M on A, B/.

TG 00 (63/00) *Hingham**   10 by 10   Revision a 1905-26 c 1938-50 f 1965
1	A	CPr		1948po	-	-	1.48	4BE2X3	RB		1/48	-	
1	B		OPr	1948po	-	-	1.48	4BE2X3	RB		-	3/48	
1	C	CPr		nd:po	-	-	1.48	4BEoo3	RB				
3	D	CPu,z	OPv	1955pc	-	ac	6.55	4oooX3			7/55	10/56	
3	D/	CPz	OPz	1955mc	-	acf	6.55	4oooX3					

AA 1.10.47 on B(A), 3/-, OSPR 3/48. M on C.

TG 01 (63/01) *Hockering*   10 by 10   Revision a 1904-26 c 1938-50; j 1981 selected
1	A	CPr	OPr	1948po	-	-	1.48	4BE2X3	R	00	2/48	5/48	
1	B	CPr		nd:po	-	-	1.48	4BEoo3	R	00			
3	C	CPu	OPv	1955pc	-	ac	6.55	4oooX3		11	10/55	10/56	
3	C/	CPw,z		1955mc	1960mc	ac	6.55	4oooX3		11			
3	C//*	CFz	OFz	1981^{c}ms	nd:r	acj	6.55	4oooX3		11	1/82	1/82	

M on B.

1	2	3	4	5	6	7	8	9	10	11	12	13	14

TG 02 (63/02) *Foulsham**   10 by 10   Revision a 1904-05 c 1938-50 f 1965
1	A	CPr	OPr	1948po	-	-	1.48	4BE2X3	R	0	1/48	4/48	
1	B	CPr		nd:po	-	-	1.48	4BEoo3	R	0			
1	B	CPs		1948pc	1951c	a	1.51	4BEoo3	R	0			
3	C	CPu	OPv	1955pc	-	ac	6.55	4oooX3		1	6/55	6/56	
3	C/	CPx		1955mc	1961ch	ac	6.55	4oooX3		1			
3	C//	CPz,0z	OPz,0z	1955mc	1966ch	acf	6.55	4oooX3		3			

AA 1.10.47 on A(A), 3/-, OSPR 4/48. M on B.

TG 03 (63/03) *Holt Norfolk*   10 by 10   Revision a 1904-26 c 1950
| 1 | A | CPr | OPr | nd:po | - | - | 6.48 | 4BEoo3 | R | | 3/49 | 4/49 | |
| 3 | B | CPu,x,z,0z | OPv | 1955pc | - | ac | 6.55 | 4oooX3 | | | 5/55 | 6/57 | |

M on A.

TG 04 (63/04) *Blakeney Norfolk*   10 by 10   Revision a 1904-26 c 1950 f 1966
1	A	CPr	OPr	nd:po	-	-	6.48	4BEoo3	RB	0	2/49	3/49	
3	A	CPt		1949pc	-	-	6.48	4Booo3	RB	0			
3	B	CPu,z	OPv	1955pc	-	ac	6.55	4oooX3		1	10/55	8/57	
3	B/	CPz	OPz	1955mc	1967ch	acf	6.55	4oooX3		1			

M on A.

TG 10 (63/10) *Wymondham Norfolk**   10 by 10   Revision a 1905-26 c 1938-53 f 1965 g 1970
1	15046	CPr		1946po	-	-	6.46	4BE2X3	R	0	9/46	-	
1	2547		OPr	1947po	-	-	6.46	4BE2X3	R	0	-	1/47	
3	B	CPv		1956pc	-	ac	6.56	4oooX3		1	8/56	-	
3	B/	CP?w,y		1956mc	1959mc	ac	6.56	4oooX3		1			
3	B//*	CPz	OPz	1956mc	1965r	acf	6.56	4oooX3		1	10/65	11/65	
3	B//*/	CPz		1956^{c}ms	nd:ch	acg	6.56	4oooX3		3			

AA 1.6.46 on 2547(2547), 3/-, OSPR 1/47. M on A.

TG 11 (63/11) *Attlebridge**   10 by 10   Revision a 1905-27 c 1938-53 f 1972
1	10046	CPr		1946po	-	-	1.47	4BE2X3	R	0	10/46	-	
1	A		OPr	1946po	-	-	1.47	4BE2X3	R	0	-	1/47	
3	B	CPv	OPv	1956pc	-	ac	6.56	4oooX3		1	10/56	10/56	
3	B/	CP?w,y,z		1956mc	1960mc	ac	6.56	4oooX3		1			
3	B//*	CPz	OPz	1956^{c}ms	nd:r	acf	6.56	4oooX3		1	7/73	9/73	

AA 1.8.46 on A(A), 3/-, OSPR 1/47. M on A. SS: Soil, Land Use Capability, coloured, 1980 on B//*.

TG 12 (63/12) *Aylsham*   10 by 10   Revision a 1905-26 b 1904-26 c 1938-52 f 1970
1	A	CPr	OPr	1948po	-	-	1.48	4BE2X3	R	00	2/48	4/48	
1	B	CPs		1950pc	-	a	6.50	4BEoo3	R	00	10/50	-	
3	C	CPv,x,z	OPy	1956pc	-	bc	6.56	4oooX3		11	7/56	-	
3	C/	CPz,0z	OPz,0z	1956^{c}ms	nd:ch	bcf	6.56	4oooX3		33			

AA 1.11.47/NT 22.8.47 on A(A), 3/-, OSPR 4/48. M on B.

TG 13 (63/13) *Barningham Norfolk*   10 by 10   Revision a 1904-27 c 1946-52
1	A	CPr	OPr	nd:po	-	-	1.49	4BEoo3	RB	0	3/49	4/49	
3	B	CPu	OPv	1956pc	-	ac	6.56	4oooX3		2	4/56	8/57	
3	B/	CPw,z,0z		1956mc	1960mc	ac	6.56	4oooX3		2			

M on A.

TG 14 (63/14) **& Part of TG 24** *Sheringham* 15 by 10   Revision a 1926-27 c 1938-52
1	A	CPr	OPr	nd:po	-	-	6.48	4BEoo3	R		3/49	5/49	
3	A	CPu		1949pc	-	-	6.48	4BEoo3	R				
3	B	CPv	OP?	1956pc	-	ac	6.56	4oooX3			10/56	2/58	
3	B/	CP?w,x,z		1956mc	1960mc	ac	6.56	4oooX3					
3	B/	CPz+,0z		1956^{c}ms	nd:mc	ac	6.56	4oooX3					1

AA 1.12.48/NT 17.2.49 on A(A), 3/-, OSPR 5/49. M on A. 1. A change in imprint is unusual, unless it implies a change in content, in which case B// would be a more appropriate print code.

264

1	2	3	4	5	6	7	8	9	10	11	12	13	14

TG 20 (63/20) *Norwich (South)**   10 by 10   Revision a 1905-26 c 1938-52
1	30046	CPr		1946po	-	-	6.46	4BE2X3	RC		9/46	-	1
1	2547		OPr	1947po	-	-	6.46	4BE2X3	RC		-	1/47	
3	B	CPv	OPv	1956pc	-	ac	6.56	4oooX3			1/57	2/57	2
3	B/	CPw,z		1956mc	1958mc	ac	6.56	4oooX3					

AA 1.6.46 on 2547(2547), 3/-, OSPR 1/47. M on A. 1. There is no canal, but the River Yare is solid blue through Norwich, as on the one-inch New Popular Edition Sheet 126. 2. The River Yare is stippled blue. Two proof copies of edition B are recorded: a proof of the black plate only (showing the price of the coloured map), printed back to back with TL 29; also a two coloured proof with black and grey plates (coloured green) only.

TG 21 (63/21) *Norwich (North)**   10 by 10   Revision a 1905-27 c 1938-54 f 1964 g 1969 h 1973
                                                    j 1967-69 airfield k 1967-69 building
1	15046	CPr		1946po	-	-	6.46	4BE2X3	R	00	9/46	-	
1	2546		OPr	1946po	-	-	6.46	4BE2X3	R	00	-	12/46	
3	B	CPv	OPv	1956pc	-	ac	6.56	4oooX3		11	2/57	2/57	
3	B/	CPw,z		1956mc	1958mc	ac	6.56	4oooX3		11			
3	B//	CPz	OPz	1956mc	1965ch	acf	6.56	4oooX3		13			
3	B///*	CPz	OPz	1954ᶜms	nd:ab	acgjk	6.56	4oooX3		32	1/71	2/71	1
3	B///*/*	CPz	OPz	1956ᶜms	nd:r	achjk	6.56	4oooX3		32	11/73	2/74	

AA 1.7.46 on 2546(2546), 3/-, OSPR 12/46. M on A. 1. Norwich Airport.

TG 22 (63/22) *North Walsham (South)*   10 by 10   Revision a 1903-26 c 1938 d 1938-53 f 1970
                                                      j 1982 selected
1	A	CPr	OPr	1948po	-	-	1.48	4BE2X3	RB	0	1/48	4/48	
1	B	CPs		1950pc	-	ac	6.50	4BEoo3	RB	0	11/50	-	
3	C	CPv		1956pc	-	ad	6.56	4oooX3		1	7/56	-	
3	C/	CP?w,z		1956mc	1960mc	ad	6.56	4oooX3		1			1
3	C//	CPz	OPz	1956ᶜms	nd:ch	adf	6.56	4oooX3		3			
3	C///*	CFz	OFz	1982ᶜm-s	nd:r	adfj	6.56	4oooX3		3	3/83	3/83	

AA 1.10.47 on A(A), 3/-, OSPR 4/48. M on B. 1. Coltishall Airfield.

TG 23 (63/23) **& Part of TG 24** *North Walsham (North)* 10 by 12 Revision a 1905-27 c 1938-52
1	A	CPr		nd:po	-	-	1.49	4BEoo3	RB		4/49	-	
1	B		OPr	nd:po	-	-	1.49	4BEoo3	RB		-	5/49	
3	C	CPv	OPv	1956pc	-	ac	6.56	4oooX3			8/56	12/57	1
3	C/	CP?w,y,z.z		1956mc	1960mc	ac	6.56	4oooX3					

AA 1.12.48/NT 10.3.49 on B(A), 3/-, OSPR 5/49. M on B. 1. An experimental version without print code is recorded, with the grey plate in green. The four colour squares are present in the margin.

TG 30 (63/30) *Brundall*   10 by 10   Revision a 1905-26 c 1946-52
1	A	CPr	OPr	1947po	-	-	1.48	4BE2X3	R		12/47	3/48	
3	B	CPv		1956pc	-	ac	6.56	4oooX3			7/56	-	
3	B/	CPw,y,z.z,0z		1956mc	1960mc	ac	6.56	4oooX3					

AA 1.9.47 on A(A), 3/-, OSPR 3/48. M on A.

TG 31 (63/31) *Horning**   10 by 10   Revision a 1905-26 c 1938-54
1	A	CPr	OPr	1947po	-	-	1.48	4BE2X3	R	0	11/47	2/48	
3	B	CPv	OPv	1956pc	-	ac	6.56	4oooX3		1	10/56	10/56	
3	B/	CPw,y,z		1956mc	1960mc	ac	6.56	4oooX3		1			

AA 1.9.47 on A(A), 3/-, OSPR 2/48. M on A. SS: Soil, Land Use Capability, outline, 1977 on B/.

TG 32 (63/32) *Stalham*   10 by 10   Revision a 1905-38 c 1938-54 f 1969
1	A	CPr	OPr	1948po	-	-	1.48	4BE2X3	R	0	2/48	5/48	
3	B	CPv,x,z	OPv	1956pc	-	ac	6.56	4oooX3		1	10/56	12/57	
3	B/	CPz	OPz	1956ms	1970ch	acf	6.56	4oooX3		1			

AA 1.10.47 on A(A), 3/-, OSPR 5/48. M on A.

TG 33 (63/33) *Mundesley**   10 by 10   Revision a 1905-26 c 1938-52 f 1970
1	A	CPr	OPr	nd:po	-	-	1.48	4BEoo3	R		3/49	4/49	
3	B	CPv,w	OPz	1956pc	-	ac	6.56	4oooX3			10/56	-	
3	B/	CPz	OPz	1956ᶜms	nd:ch	acf	6.56	4oooX3					

M on A.

265

1	2	3	4	5	6	7	8	9	10	11	12	13	14

TG 40 (63/40) *Burgh Castle**   10 by 10   Revision a 1903-26 c 1942-52
1 15046   CPr   1946po   -   -   1.47   4BE1X3   R   6/46 -
1 2546   OPr   1946po   -   -   1.47   4BE1X3   R   -   9/46
3 B   CPv   1956pc   -   ac   6.56   4oooX3   7/56 -
3 B   CPx,z   1956mc   -   ac   6.56   4oooX3
AA 1.4.46 on 2546(2546), 3/-, OSPR 9/46. M on A.

TG 41 (63/41) **& Part of TG 51** *Caistor-on-Sea** 15 by 10 Revision a 1905-38 c 1938-52 f 1973
1 15046   CPr   1946po   -   -   1.47   4BE2X3   R   00 8/46 -
1 2547   OPr   1947po   -   -   1.47   4BE2X3   R   00 -   1/47
3 B   CPv   OP?   1956pc   -   ac   6.56   4oooX3   10 10/56 2/58
3 B/   CP?w,y,z   1956mc   1960mc ac   6.56   4oooX3   10
3 B//*   CPz   OPz   1956ᶜms   nd:r   acf   6.56   4oooX3   10 1/74 3/74
AA 1.6.46 on 2547(2547), 3/-, OSPR 1/47. M on A.

TG 42 (63/42) *Hickling Norfolk**   10 by 10   Revision a 1905-38 c 1938-53 f 1969
1 A   CPr   OPr   nd:po   -   -   1.48   4BEoo3   10/48 10/48
? B   CPs   1948pc   1952c -   1.52   4Booo3   M
3 C   CPv,z   OPv   1956pc   -   ac   6.56   4oooX3   6/56 6/57
3 C/   CPz   OPz   1956ms   1970ch acf   6.56   4oooX3
M on B.

TG 50 (63/50) *Great Yarmouth*   10 by 10   Revision a 1926 c 1946-54 f 1970
1 15046   CPr   1946po   -   -   1.47   4BE1X3   R   8/46 -
1 2546   OPr   1946po   -   -   1.47   4BE1X3   R   -   11/46
3 B   CPv.y,z OPy   1956pc   -   ac   6.56   4oooX3   8/56 -
3 B/   CPz   OPz   1956ᶜms   nd:ch acf   6.56   4oooX3
AA 1.3.46 on 2546(2546), 3/-, OSPR 11/46. M on A.

TL 00 (52/00) *Hemel Hempstead*   10 by 10   Revision a 1922-37 b 1912-39 c 1938-50 d 1938-55
                                                    e 1948 f 1964 g 1975
1 A   CPr   OPr   1948po   -   -   6.48   4BEoo3   RB   00 5/48 7/48
   B
2 C   CPs   OPs   1951pc   -   ace   1.52   4Booo3   RB   31 12/51 -
3 D   CPv   OPv   1956pc   -   bd   6.56   4oooX3   11 11/56 1/57
3 D/*   CPx   OPx   1956mc   1961r bd   6.56   4oooX3   11 2/61   1
3 D/*/   CPz   OPz   1956mc   1965ch bdf   6.56   4oooX3   33
3 D/*//* CFz   1956ᶜms   nd:r   bdg   6.56   4oooX3   33 9/76
AA 1.3.48/NT 6.4.48 on A(A), 3/-, OSPR 7/48; AA 1.3.52/NT 4.5.50 on C(A) 4/-, OSPR 8/52. 1. Bovingdon
Airfield. OSPR (outline edition) not found.

TL 01 (52/01) *Markyate*   10 by 10   Revision a 1922-39 c 1937-51 f 1962
1 15046   CPr   1946po   -   -   1.47   4BE1X3   B   10/46 -
1 A   OPr   1946po   -   -   1.47   4BE1X3   B   -   1/47
3 B   CPu   OPv   1955pc   -   ac   6.55   4oooX3   1/56 1/57
3 B/   CPw   1955mc   1960mc ac   6.55   4oooX3
3 B//*   CPy,z   OPy   1955mc   1963r acf   6.55   4oooX3   11/63 12/63
AA 1.9.46/NT 1.9.46 on A(A), 3/-, OSPR 1/47.

TL 02 (52/02) *Luton*   10 by 10   Revision a 1899-24 c 1937-51 f 1962 g 1972 h 1978
                                                    j 1966 building
1 10046   CPr   1946po   -   -   1.47   4BE2X3   RB   11/46 -
1 A   OPr   1946po   -   -   1.47   4BE2X3   RB   -   2/47
2 B   CPt   1946pc   1953c -   1.53   4BooX3   RB
3 C   CPv   OPv   1956pc   -   ac   6.56   4oooX3   10/56 6/60
3 C/*   CPy   OPy   1956mc   1963r acf   6.56   4oooX3   3/63 4/63
3 C/*/*   CPz   OPz   1956ᶜms   nd:br acgj 6.56   4oooX3   5/72 8/72
3 C/*/*/* CFz   OFz   1979ᶜms   nd:r   achj 6.56   4oooX3   10/80 10/80
AA 1.9.46/NT 1.9.46 on A(A), 3/-, OSPR 2/47. M on A.

1	2	3	4	5	6	7	8	9	10	11	12	13	14

TL 03 (52/03) *Ampthill*   10 by 10   Revision a 1899-1924 c 1938-51 f 1964

1	10047	CPr		1947po	-	-	1.47	4BE2X3	RB		1/47	-	
1	A		OPr	1947po	-	-	1.47	4BE2X3	RB		-	4/47	
3	B	CPu	OPv	1956pc	-	ac	6.56	4ooX3			4/56	9/57	
3	B/	CPw		1956mc	1960mc	ac	6.56	4ooX3					
3	B//*	CPz	OPz	1956mc	1965r	acf	6.56	4ooX3			1/65	2/65	

AA 1.9.46 on A(A), 3/-, OSPR 4/47. M on A.

TL 04 (52/04) *Bedford (South)**   10 by 10   Revision a 1900-24 c 1937-53 f 1970

1	A	CPr	OPr	1948po	-	-	1.48	4BEoo3	RB		7/48	8/48	
2	A	CPt		1948pc	1952rpb	-	1.48	4BooX3	RB				
3	B	CPu	OPs	1956pc	-	ac	6.56	4ooX3			3/56	4/56	
3	B/	CPw,z		1956mc	1960mc	ac	6.56	4ooX3					
3	B//	CPz	OPz	1956^{c}ms	nd:ch	acf	6.56	4ooX3					

AA 1.5.48 on A(A), 3/-, OSPR 8/48.

TL 05 (52/05) *Bedford (North)*   10 by 10   Revision a 1899-1924 c 1938-51 f 1971
                                                j 1966,69 airfield

1	A	CPr	OPr	1948po	-	-	1.48	4BEoo3	RB	000	7/48	9/48	
3	A	CPt		1948pc	-	-	1.48	4BEoo3	RB	000			
3	B	CPu,y,z	OPs	1956pc	-	ac	6.56	4ooX3		110	4/56	4/56	
3	B/*	CPz	OPz	1956^{c}ms	nd:a	acfj	6.56	4ooX3		330	1/72	4/72	1

AA 1.5.48 on A(A), 3/-, OSPR 9/48. 1. Twinwood Airfield.

TL 06 (52/06) *Riseley*   10 by 10   Revision a 1899-1924 c 1950 f 1968

1	A	CPr	OPr	nd:po	-	-	1.49	4BEoo3	B	00	10/48	12/48	
	B												
3	C	CPu,x	OPv	1955pc	-	ac	6.55	4ooX3		11	1/55	1/57	
3	C/	CPz	OPz	1955ms	1969ch	acf	6.55	4ooX3		33			

AA 1.8.48 on A(A), 3/-, OSPR 12/48. M on C.

TL 07 (52/07) *Bythorn**   10 by 10   Revision a 1899-1924 c 1938-50 f 1969

1	A	CPr	OPr	1948po	-	-	6.48	4BE4X3	RB	00	3/48	5/48	
3	B	CPu,z	OPv	1955pc	-	ac	6.55	4ooX3		11	3/55	1/57	
3	B/	CPz	OPz	1955ms	1970ch	acf	6.55	4ooX3		31			

AA 1.1.48 on A(A), 3/-, OSPR 5/48. M on B.

TL 08 (52/08) *Oundle*   10 by 10   Revision a 1899-1924 c 1938-50 f 1965

1	A	CPr	OPr	nd:po	-	-	6.49	4BEoo3	RB	0	6/49	7/49	
3	B	CPu,y	OPv	1955pc	-	ac	6.55	4ooX3		1	1/55	9/57	
3	B/	CPz	OPz	1955mc	1966ch	acf	6.55	4ooX3		3			

M on B.

TL 09 (52/09) *Wansford Northamptonshire*   10 by 10   Revision a 1899-1924 c 1938-50 f 1964 g 1966

1	A	CPr	OPr	nd:po	-	-	6.48	4BEoo3	RB	0	9/48	10/48	
	B												
3	C	CPu	OPv	1955pc	-	ac	6.55	4ooX3		1	3/55	1/57	
3	C/	CPw		1955mc	1960mc	ac	6.55	4ooX3		1			
3	C//	CPz	OPz	1955mc	1964ch	acf	6.55	4ooX3		1			
3	C///*	CPz	OPz	1955mc	1967r	acg	6.55	4ooX3		1	10/67	10/67	

TL 10 (52/10) *St Albans*   10 by 10   Revision a 1912-39 c 1937-55 f 1960

1	A	CPr	OPr	1948po	-	-	6.48	4BE2X3	RB	10	4/48	5/48	1
1	B	CPr		nd:po	-	-	6.48	4BEoo3	RB	10			
2	C	CPs		1948pc	1951c	-	1.52	4Booo3	RB	10			
3	D	CPv	OPv	1957pc	-	ac	6.57	4ooX3		33	1/58	11/57	2
3	D/*	CPx,z,0z	OP?,z	1957mc	1961r	acf	6.57	4ooX3		33	5/61	5/61	

AA 1.1.48 on A(A), 3/-, OSPR 5/48. M on B. 1. Showing the Middlesex, and another (1706), County Mental Hospitals with their railway connection. 2. Radlett Aerodrome.

1	2	3	4	5	6	7	8	9	10	11	12	13	14

TL 11 (52/11) *Harpenden*   10 by 10   Revision a 1922-39 c 1937-51

1	10047	CPr		1947po	-	-	1.47	4BE2X3	RB		12/46	-	
1	A		OPr	1946po	-	-	1.47	4BE2X3	RB		-	4/47	
1	B	CPr		nd:po	-	-	1.47	4BEoo3	RB				
3	C	CPu	OPv	1956pc	-	ac	6.56	4oooX3			5/56	1/57	
3	C/	CPw,x,z		1956mc	1959mc	ac	6.56	4oooX3					

AA 1.9.46 on A(A), 3/-, OSPR 4/47.

TL 12 (52/12) *Hitchin*   10 by 10   Revision a 1921-22 c 1938-51 d 1937-51 f 1970
j 1966 building k 1970 airfield

1	10046	CPr		1946po	-	-	1.47	4BE2X3	RB	1	11/46	-	
1	A		OPr	1946po	-	-	1.47	4BE2X3	RB	1	-	4/47	
3	B	CPt		1954pc	-	ac	6.54	4BooX3	B	1	1/54	-	1
3	C	CPu	OPv	1956pc	-	ad	6.56	4oooX3		1	4/56	1/57	
3	C/	CPw,y,z		1956mc	1959mc	ad	6.56	4oooX3		1			
3	C//*	CPz	OPz	1956^{c}ms	nd:b	adfjk	6.56	4oooX3		3	3/71	5/71	

AA 1.9.46 on A(A), 3/-, OSPR 4/47. 1. Luton Airport.

TL 13 (52/13) *Shefford*   10 by 10   Revision a 1899-1939 c 1937-51

1	10047	CPr		1947po	-	-	1.47	4BE2X3	RB	0	1/47	-	
1	A		OPr	1947po	-	-	1.47	4BE2X3	RB	0	-	4/47	
3	B	CPu	OPv	1956pc	-	ac	6.56	4oooX3		1	4/56	8/57	
3	B/	CPx,z.z		1956mc	1961ch	ac	6.56	4oooX3		1			1

AA 1.9.46 on A(A), 3/-, OSPR 4/47. 1. Henlow Airfield.

TL 14 (52/14) *Biggleswade*   10 by 10   Revision a 1900-24 c 1938-53 f 1962

1	A	CPr	OPr	nd:po	-	-	1.49	4BEoo3	RB	1	4/49	5/49	
3	B	CPu	OPv	1956pc	-	ac	6.56	4oooX3		1	2/56	1/57	
3	B/*	CPy,z	OPy	1956mc	1963r	acf	6.56	4oooX3		1	7/63	7/63	

TL 15 (52/15) *St Neots (South)* *   10 by 10   Revision a 1900-24 c 1938-53 f 1964 g 1972

1	A	CPr	OPr	nd:po	-	-	6.48	4BEoo3	RB	0	5/49	5/49	
2	A	CPt		1949pc	1952rpb	-	6.48	4BooX3	RB	0			
3	B	CPu	OPv	1956pc	-	ac	6.56	4oooX3		1	2/56	8/57	
3	B/												
3	B//	CPz	OPz	1956mc	1964ch	acf	6.56	4oooX3		1			
3	B///*	CPz	OPz	1956^{c}ms	nd:r	acg	6.56	4oooX3		1	9/73	10/73	

TL 16 (52/16) *St Neots (North)*   10 by 10   Revision a 1900-24 c 1938-50 f 1963 g 1969 h 1972
j 1966 airfield k 1966 reservoir

1	A	CPr	OPr	nd:po	-	-	6.48	4BEoo3	RB	00	4/49	4/49	
1	B	CPs	OPv	1951pc	-	ac	6.51	4Booo4	R	00	1/52	8/57	
3	B/	CPz	OPz	1951mc	1964ch	acf	6.51	4BooX4		10			
3	B//*	CPz	OPz	1951ms	1970arv	acgjk	6.51	4BooX4		32	2/70	3/70	1
3	B//*/*	CPz	OPz	1951^{c}ms	nd:r	achjk	6.51	4BooX4		32	5/73	6/73	

1. Little Staughton Airfield, Grafham Water.

TL 17 (52/17) *Alconbury* *   10 by 10   Revision a 1894-1924 c 1950 f 1968

1	A	CPr	OPr	nd:po	-	-	1.49	4BEoo3	RB	00	6/49	7/49	
1	B	CPs	OPv	1951pc	-	ac	6.51	4BEoo3	RB	00	9/51	8/57	
3	B/	CPx		1951mc	1961ch	ac	6.51	4BooX3		11			
3	B//	CPz	OPz	1951ms	1969ch	acf	6.51	4BooX3		31			

TL 18 (52/18) *Sawtry* *   10 by 10   Revision a 1899-1924 c 1950 f 1961

1	A	CPr	OPr	nd:po	-	-	1.49	4BEoo3	RB	00	6/49	7/49	
3	B	CPu	OPv	1954pc	-	ac	6.54	4oooX3		11	11/54	1/57	
3	B/	CPx,z		1954mc	1963ch	acf	6.54	4oooX3		11			

TL 19 (52/19) *Peterborough (South West)*   10 by 10   Revision a 1899-25 c 1938-50 f 1960 g 1971 h 1978

| 1 | A | CPr | OPr | 1948po | - | - | 1.48 | 4BE4X3 | RB | 0 | 2/48 | 4/48 | |
| 3 | B | CPu | OPv | 1955pc | - | ac | 6.55 | 4oooX4 | | 2 | 1/55 | 8/57 | |

1	2	3	4	5	6	7	8	9	10	11	12	13	14
3	B/	CPx,z		1955mc	1961ch	acf	6.55	4ooX4		2			
3	B//	CPz	OPz	1955^{c}ms	nd:ch	acg	6.55	4ooX4		2			
3	B///*	CFz	OFz	1979^{c}ms	nd:rv	ach	6.55	4ooX4		2	11/80	11/80	

AA 1.10.47 on A(A), 3/-, OSPR 4/48.

TL 20 (52/20) *Hatfield** 10 by 10 Revision a 1922-38 c 1937-55 f 1961 g 1966
j 1961 airfield k 1966 Greater London

1	A	CPr	OPr	1947po	-	-	6.47	4BE2X3	RB	1	5/47	8/47	
1	B	CPr		nd:po	-	-	1.50	4BEoo3	RB	1			
2	C	CPt		1950pc	1953c	-	1.50	4BooX3	RB	1			
3	D	CPv	OPv	1957pc	-	ac	6.57	4ooX3		1	3/57	8/57	
3	D/	CPw		1957mc	1960mc	ac	6.57	4ooX3		1			
3	D//	CPx		1957mc	1963ch	acf	6.57	4ooX3		1			
3	D///*	CPz	OPz	1957mc	1966ar	acgjk	6.57	4ooX3		3	10/66	10/66	1

AA 15.4.47/NT 1.7.47 on A(A), 3/-, OSPR 8/47. M on B. 1. Hatfield Aerodrome.

TL 21 (52/21) *Welwyn* 10 by 10 Revision a 1922-39 c 1937-49 d 1937-51
e 1948 f 1961 g 1968 h 1978; j 1966 building

1	A	CPr	OPr	nd:po	-	-	1.49	4BEoo3	RB	0	8/49	10/49	
1	B	CPs		1951pc	-	ace	6.51	4BEoo3	R	1	12/51	-	
3	C	CPv	OPv	1956pc	-	ad	6.56	4ooX3		1	7/56	8/57	1
3	C/												
3	C//*	CPx	OPx	1956mc	1962r	adf	6.56	4ooX3		1	8/62	12/62	
3	C//*/*	CPz	OPz	1956ms	1969br	adgj	6.56	4ooX3		1	11/69	11/69	
3	C//*/*/*	CFz	OFz	1956^{c}ms	nd:r	adhj	6.56	4ooX3		1	4/80	4/80	

AA 1.1.49 on A(A), 3/-, OSPR 10/49. 1. Panshanger Aerodrome.

TL 22 (52/22) *Stevenage* 10 by 10 Revision a 1922-39 c 1946-51 f 1964 g 1968 h 1976
j 1966 building

1	A	CPr	OPr	nd:po	-	-	6.48	4BEoo3	RB		5/49	6/49	
3	B	CPu	OPv	1956pc	-	ac	6.56	4ooX3			5/56	8/57	
3	B/	CPx		1956mc	1961ch	ac	6.56	4ooX3					
3	B//*	CPz	OPz	1956mc	1965r	acf	6.56	4ooX3			3/65	4/65	
3	B//*/*	CPz	OPz	1956ms	1969br	acgj	6.56	4ooX3			6/69	8/69	
3	B//*/*/*	CFz	OFz	1956^{c}ms	nd:r	achj	6.56	4ooX3			2/77	2/77	

AA 1.1.49 on A(A), 3/-, OSPR 6/49.

TL 23 (52/23) *Baldock* 10 by 10 Revision a 1901-37 c 1938-51 f 1968; j 1966 building

1	A	CPr	OPr	nd:po	-	-	1.49	4BEoo3	RB		3/49	5/49	
3	A	CPt		1949pc	-	-	1.49	4BEoo3	RB				
3	B	CPu	OPv	1956pc	-	ac	6.56	4ooX3			5/56	8/57	
3	B/	CPx,z		1956mc	1961ch	ac	6.56	4ooX3					
3	B//*	CPz,0z	OPz	1956ms	1970br	acfj	6.56	4ooX3			2/70	3/70	

AA 1.1.49 on A(A), 3/-, OSPR 5/49.

TL 24 (52/24) *Potton* 10 by 10 Revision a 1900-24 c 1937-53 f 1964

1	A	CPr	OPr	nd:po	-	-	1.49	4BEoo3	RB	0	4/49	4/49	
3	B	CPt		1949pc	1954c	-	6.53	4BooX3	B	0			
3	C	CPu	OPv	1956pc	-	ac	6.56	4ooX3		1	2/56	8/57	
3	C/	CPx		1956mc	1961ch	ac	6.56	4ooX3		1			
3	C//*	CPz,0z	OPz	1956mc	1965r	acf	6.56	4ooX3		1	6/65	2/65	

TL 25 (52/25) *Gransden* 10 by 10 Revision a 1900-39 c 1947-53

| 1 | A | CPr | OPr | nd:po | - | - | 1.49 | 4BEoo3 | RB | 0 | 5/49 | 5/49 | |
| 3 | B | CPu,y,z | OPv | 1956pc | - | ac | 6.56 | 4ooX3 | | 1 | 2/56 | 8/57 | |

TL 26 (52/26) *Papworth Everard* 10 by 10 Revision a 1900-38 c 1950 f 1965

1	A	CPr	OPr	nd:po	-	-	6.48	4BEoo3	RB	0	7/49	8/49	
3	B	CPu	OPv	1955pc	-	ac	6.55	4ooX3		1	1/55	8/57	
3	B/	CPw,x		1955mc	1960mc	ac	6.55	4ooX3		1			1
3	B//	CPz		1955mc	1966ch	acf	6.55	4ooX3		3			

1. Graveley Airfield.

1	2	3	4	5	6	7	8	9	10	11	12	13	14

TL 27 (52/27) *Huntingdon** 10 by 10 Revision a 1900-24 c 1938-50 f 1968 g 1975
j 1966 building

1	A	CPr	OPr	nd:po	-	-	1.49	4BEoo3	RB	000	7/49	7/49	
3	B	CPu	OPv	1955pc	-	ac	6.55	4oooX3		111	1/55	10/56	
3	B/	CPw,y	OPz	1955mc	1960mc	ac	6.55	4oooX3		111			1
3	B//*	CPz	OPz	1955ms	1969b	acfj	6.55	4oooX3		111	8/69	8/69	
3	B//*/*	CFz	OFz	1955^{c}ms	nd:r	acgj	6.55	4oooX3		111	1/76	1/76	

M on A, B. 1. Alconbury Airfield 06, Wyton Airfield 84, Warboys Airfield 99.

TL 28 (52/28) *Ramsey Huntingdonshire** 10 by 10 Revision a 1924 c 1938-50

1	A	CPr	OPr	nd:po	-	-	1.49	4BEoo3	RB	0	7/49	7/49	
3	B	CPu	OPv	1954pc	-	ac	6.54	4oooX3		1	1/55	10/56	
3	B/	CPw,x,z.z,0z		1954pc	1958mc	ac	6.54	4oooX3		1			

TL 29 (52/29) *Whittlesey* 10 by 10 Revision a 1899-1925 c 1938-50 f 1971

1	A	CPr	OPr	1948po	-	-	1.48	4BE4X3	R		3/48	7/48	
2	A	CPt		1948pc	1952	-	1.48	4Booo3	R				
3	B	CPu	OPv	1955pc	-	ac	6.55	4oooX3			2/55	10/56	1
3	B/	CPw,z		1955mc	1960mc	ac	6.55	4oooX3					
3	B//	CPz	OPz	1955^{c}ms	nd:ch	acf	6.55	4oooX3					

AA 1.1.48 on A(A), 3/-, OSPR 7/48. 1. A proof of the black plate (1955pc, CPr, MV 6.55) is recorded, with manuscript date 19 July 1955. It is back to back with a grey printing of the black plate of TG 20.

TL 30 (52/30) *Cheshunt* 10 by 10 Revision a 1921-35 b 1912-35 c 1938 d 1938-54
f 1966; j 1966 Greater London

1	A	CPr	OPr	1947po	-	-	1.48	4BE4X3	RB		11/47	1/48	
1	B	CPr		nd:po	-	-	1.48	4BEoo3	RB				
2	B	CPs		1947pc	1951	-	1.52	4Booo3	RB				
3	B/	CPv		1947pc	1956mc	ac	1.52	4BooX3	B				
3	B//	CPw	OPv	1947pc	1958mc	ac	1.52	4BooX3	B				
3	C	CPw.x	OPv	1960mc	-	bd	6.60	4oooX4			4/60	5/60	
3	C/	CPz	OPz	1960mc	1967ch	bdfj	6.60	4oooX4					

AA 1.4.47 on A(A), 3/-, OSPR 1/48. M on B.

TL 31 (52/31) *Hertford* 10 by 10 Revision a 1915-39 c 1938-51 f 1976

1	A	CPr	OPr	nd:po	-	-	1.49	4BEoo3	RB		3/49	5/49	
2	A	CPs		1949pc	1951	-	6.51	4Booo3	RB				
3	B	CPv,z.z	OPz	1956pc	-	ac	6.56	4oooX3			8/56	-	
3	B/*	CFz	OFz	1956^{c}ms	nd:r	acf	6.56	4oooX3					1

AA 1.12.48 on A(A), 3/-, OSPR 5/49. 1. OSPR not found.

TL 32 (52/32) *Buntingford* 10 by 10 Revision a 1919-38 c 1946-51

1	A	CPr	OPr	nd:po	-	-	1.49	4BEoo3	RB		3/49	4/49	
3	A	CPt		1949pc	-	-	1.49	4BEoo3	RB				
3	B	CPv,x,z	OPv	1956pc	-	ac	6.56	4oooX3			8/56	10/56	
3	B/	CPz	OPz	1956pc	-	ac	6.56	4oooX3					

TL 33 (52/33) *Barkway** 10 by 10 Revision a 1922-37 b 1901-38 c 1937-47 d 1938-53
f 1970

1	A	CPr	OPr	nd:po	-	-	6.48	4BEoo3	RB		5/49	5/49	
1	B	CPs		1948pc	1950c	ac	6.50	4BEoo3	RB				
3	C	CPv	OPv	1956pc	-	bd	6.56	4oooX3			7/56	9/57	
3	C/	CPw,x		1956mc	1959mc	bd	6.56	4oooX3					
3	C//	CPz	OPz	1956^{c}ms	nd:ch	bdf	6.56	4oooX3					

TL 34 (52/34) *Royston Hertfordshire** 10 by 10 Revision a 1896-1938 c 1937-53

1	A	CPr	OPr	nd:po	-	-	1.48	4BEoo3	RB	00	5/49	6/49	
2	A	CPt		1949pc	1952	-	1.48	4Booo3	RB	00			
3	B	CPv	OPv	1956pc	-	ac	6.56	4oooX3		11	7/56	8/57	
3	B/	CPw,z		1956mc	1960mc	ac	6.56	4oooX3		11			1

M on A. SS: Soil, outline, 1985 on B/. 1. Bassingbourn Airfield.

1	2	3	4	5	6	7	8	9	10	11	12	13	14

TL 35 (52/35) *Eversden**   10 by 10   Revision a 1900-24 b 1900-37 c 1937-47 d 1938-47
e 1938-52 (psr); f 1965

1	A	CPr	OPr	nd:po	-	-	6.48	4BEoo3	RB	0	7/49	9/49	
1	B	CPs	OPs	1951pc	-	ac	1.51	4BEoo3	RB	0	9/51	10/51	
3	C	CPv,x	OPv	1956pc	-	be	6.56	4oooX3		1	7/56	1/57	
3	C/	CPz	OPz	1956mc	1966ch	bef	6.56	4oooX3		3			

M on B, C.

TL 36 (52/36) *Swavesey**   10 by 10   Revision a 1900-24 c 1937-50 f 1969

1	A	CPr	OPr	nd:po	-	-	1.49	4BEoo3	RB	0	8/49	8/49	
3	B	CPs		1949pc	1953c	-	6.53	4BooX3	B	0			
3	C	CPu.x,z	OPv	1955pc	-	ac	6.55	4oooX3		0	9/55	8/57	
3	C/	CPz	OPz	1955ms	1970ch	acf	6.55	4oooX3		0			

TL 37 (52/37) *St Ives Huntingdonshire**   10 by 10   Revision a 1924 c 1937-50

1	A	CPr	OPr	nd:po	-	-	1.49	4BEoo3	RB	0	7/49	7/49	
3	A	CPt		1949pc	-	-	1.49	4BEoo3	RB	0			
3	B	CPu,x,z,0z	OPv	1956pc	-	ac	6.56	4oooX3		1	3/56	10/56	

TL 38 (52/38) *Chatteris**   10 by 10   Revision a 1924-25 c 1938-50

1	A	CPr	OPr	nd:po	-	-	6.48	4BEoo3	R		9/49	10/49	
3	B	CPu	OPs	1955pc	-	ac	6.55	4oooX3			9/55	12/55	
3	B/	CP?w,y,z		1955mc	1958mc	ac	6.55	4oooX3					

TL 39 (52/39) *Benwick*   10 by 10   Revision a 1924-25 c 1938-50

1	A	CPr	OPr	nd:po	-	-	6.49	4BEoo3	R		7/49	9/49	
3	B	CPu	OPv	1955pc	-	ac	6.55	4oooX3			9/55	8/57	
3	B	CPz		1955mc		ac	6.55	4oooX3					

AA 1.4.49 on A(A), 3/-, OSPR 9/49.

TL 40 (52/40) *Epping*   10 by 10   Revision a 1915-35 c 1936-48 d 1936-55
e 1946-47 f 1961 g 1965 h 1976

1	A	CPr	OPr	nd:po	-	-	6.48	4BEoo3	RB	0	9/48	9/48	
1	B B/	CPs	OPs	1951pc	-	ace	1.51	4BEoo3	RB	0	11/51	1/52	
3	B//	CPv		1951pc	1958mc	ace	1.52	4BooX3	B	0			
3	C		OPv	1959mc	-	ad	6.59	4oooX3		1		4/60	1
3	C/	CPx		1959mc	1961ch	adf	6.59	4oooX3		1			2
3	C//	CPz	OPz	1959mc	1966ch	adg	6.59	4oooX3		3			
3	C///*	CFz	OFz	1959ᶜms	nd:r	adh	6.59	4oooX3		3			3

AA 1.1.52 on B(A), 3/-, OSPR 1/52. M on B. 1. OSPR (coloured edition) not found. 2. North Weald Airfield. 3. OSPR not found.

TL 41 (52/41) *Harlow*   10 by 10   Revision a 1915-22 c 1938-47 d 1938-55
e 1946-47 f 1962 g 1977

1	A	CPr	OPr	nd:po	-	-	6.49	4BEoo3	RB	00	5/49	7/49	
1	B	CPs		1951pc	-	ace	1.51	4BEoo3	RB	30	6/51	-	
2	B		OPs	1951pc	-	ace	1.51	4Booo3	RB	30	-	1/52	
3	C	CPw	OPv	1958mc	-	ad	6.58	4oooX3		21	1/59	2/59	
3	C/	CPw		1958mc	1961mc	ad	6.58	4oooX3		21			
3	C//*	CPz	OPz	1958mc	1965r	adf	6.58	4oooX3		21	1/65	2/65	
3	C//*/*	CFz	OFz	1958ᶜms	nd:r	adg	6.58	4oooX3		21	3/78	3/78	

AA 1.4.49 on A(A), 3/-, OSPR 7/49; AA 1.6.51 on B(A), 3/-, OSPR 8/51. M on B.

TL 42 (52/42) *Bishop's Stortford**   10 by 10   Revision a 1915-38 c 1938-51

1	A	CPr	OPr	nd:po	-	-	6.48	4BEoo3	RB		12/48	2/49	
3	B	CPv	OPv	1956pc	-	ac	6.56	4oooX3			8/56	8/56	
3	B/	CPx		1956mc	1961ch	ac	6.56	4oooX3					
3	B//	CPz		1956mc	1961ch	ac	6.56	4oooX3					

AA 1.11.48 on A(A), 3/-, OSPR 2/49. M on A.

1	2	3	4	5	6	7	8	9	10	11	12	13	14

TL 43 (52/43) *Great Chishill*   10 by 10   Revision a 1896-22 b 1896-38 c 1946-48 d 1938-52

1	A	CPr	OPr	nd:po	-	-	1.49	4BEoo3	B	0	5/49	5/49	
1	B	CPs		1951pc	-	ac	1.51	4BEoo3	B	0	10/51	-	
3	C	CPv	OPv	1956pc	-	bd	6.56	4oooX3		1	7/56	10/57	
3	C/	CPw,z		1956mc	1960mc	bd	6.56	4oooX3		1			

M on B.

TL 44 (52/44) *Whittlesford*   10 by 10   Revision a 1896-1938 c 1938-52 f 1965 g 1972
                                              j 1983 selected

1	A	CPr	OPr	nd:po	-	-	6.49	4BEoo3	R	00	7/49	9/49	
3	B	CPu		1949pc	1953c	-	6.53	4BooX3		00			
3	C	CPv		1956pc	-	ac	6.56	4oooX3		11	10/56	-	
3	C/	CPx		1956mc	1961ch	ac	6.56	4oooX3		11			1
3	C//*	CPz	OPz	1956mc	1966ar	acf	6.56	4oooX3		11	10/66	10/66	
3	C//*/*	CPz	OPz	1956^{c}ms	nd:r	acg	6.56	4oooX3		11	4/73	6/73	
3	C//*/*/*	CFz	OFz	1984^{c}m-s	nd:r	acgj	6.85	4oooX3		11	9/84	9/84	

M on A. 1. Duxford Airfield.

TL 45 (52/45) *Cambridge*   10 by 10   Revision a 1901-38 c 1938-52 f 1968
                                           j 1966 building k 1980 selected

1	A	CPr	OPr	1947po	-	-	1.47	4BE2X3	RB	0	3/48	7/47	
1	B	CPr		nd:po	-	-	1.47	4BEoo3	RB	0			
3	B	CPu		1948pc	-	-	1.47	4BEoo3	RB	0			
3	C	CPv	OPv	1956pc	-	ac	6.56	4oooX3		2	10/56	11/56	1
3	C/	CPw.z		1956mc	1960mc	ac	6.56	4oooX3		2			
3	C//*	CPz	OPz	1956ms	1969br	acfj	6.56	4oooX3		2	10/69	10/69	
3	C//*/*	CFz	OFz	1981^{c}ms	nd:rsv	acfjk	6.56	4oooX3		2	7/81	7/81	

AA 1.12.46 on A(A), 3/-, OSPR 7/47. M on B. 1. Cambridge (Teversham) Aerodrome.

TL 46 (52/46) *Histon*   10 by 10   Revision a 1901-25 c 1938-50 f 1970
                                         j 1966,70 airfield k 1980 selected

1	A	CPr	OPr	1947po	-	-	6.47	4BE2X3	RB	00	8/47	11/47	
2	A	CPt		1947pc	1951	-	6.51	4BooX3	RB	00			
3	B	CPu	OPv	1955pc	-	ac	6.55	4oooX3		11	1/56	8/57	
3	B/	CPw		1955mc	1961mc	ac	6.55	4oooX3		11			1
3	B//*	CPz	OPz	1955^{c}ms	nd:ar	acfj	6.55	4oooX3		33	4/72	8/72	2
3	B//*/*	CFz	OFz	1981^{c}ms	nd:rv	acfjk	6.55	4oooX3		33	5/81	5/81	

AA 1.1.47 on A(A), 3/-, OSPR 11/47. M on A, B. 1. Oakington Airfield, Waterbeach Airfield. 2. Oakington Airfield, revised 1970, Waterbeach Airfield, revised 1966.

TL 47 (52/47) *Haddenham Cambridgeshire**   10 by 10   Revision a 1900-24 c 1937-50

1	A	CPr	OPr	nd:po	-	-	6.48	4BEoo3	R	0	7/49	8/49	
3	B	CPu	OPv	1955pc	-	ac	6.55	4oooX3		1	9/55	10/56	
3	B/	CPw,z,0w		1955mc	1960mc	ac	6.55	4oooX3		1			1

1. Mepal Airfield. Remarkably the price of 3/- erroneously remains even on the 0w printing.

TL 48 (52/48) *Mepal**   10 by 10   Revision a 1924-25 c 1938-50 f 1969

1	A	CPr	OPr	nd:po	-	-	1.49	4BEoo3	R	0	9/49	10/49	
3	B	CPu	OPv	1955pc	-	ac	6.55	4oooX3		1	8/55	10/56	
3	B/	CPw		1955mc	1960mc	ac	6.55	4oooX3		1			
3	B//	CPz	OPz	1955ms	1970ch	acf	6.55	4oooX3		2			

TL 49 (52/49) *March*   10 by 10   Revision a 1900-25 c 1938-50

| 1 | A | CPr | OPr | nd:po | - | - | 6.49 | 4BEoo3 | R | | 8/49 | 9/49 | |
| 3 | B | CPu,x,z | OPv | 1955pc | - | ac | 6.55 | 4oooX3 | | | 6/55 | 10/56 | |

M on A.

TL 50 (52/50) *Chipping Ongar*   10 by 10   Revision a 1915-18 c 1938 d 1938-55 e 1947

1	A	CPr	OPr	nd:po	-	-	1.49	4BEoo3	RB	0	4/49	4/49	
1	B	CPs		1949pc	1950c	ace	6.50	4BEoo3	RB	0			
3	C	CPw,x,z,0z	OPv	1959mc	-	ad	6.59	4oooX3		3	5/59	6/59	

M on B.

1	2	3	4	5	6	7	8	9	10	11	12	13	14

TL 51 (52/51) *Hatfield Heath*   10 by 10   Revision a 1915-18 c 1938-52 f 1966; j 1982 selected
1 A   CPr   OPr   nd:po   -   -   1.49 4BEoo3   B   0   2/49   2/49
2 A   CPt   1949pc   1952rpb   -   1.49 4BooX3   B   0
3 B   CPw,y   OPv   1959mc   -   ac   6.59 4oooX4   1   4/59   6/59
3 B/   CPz   OPz   1959mc   1967ch   acf   6.59 4oooX4   3
3 B//*   CFz   OFz   1982ᶜm-s   nd:r   acfj 6.83 4oooX4   3   10/82 10/82
M on A.

TL 52 (52/52) *Stansted Mountfitchet**   10 by 10   Revision a 1915-19 c 1946-47 d 1938-51
                                                        e 1947 f 1968; j 1982 selected
1 A   CPr   nd:po   -   -   1.49 4BEoo3   RB   00 2/49   -
1 B   OPr   nd:po   -   -   1.49 4BEoo3   RB   00 -   4/49
1 C   CPs   1951pc   -   ace   1.51 4BEoo3   RB   30 6/51   -
3 D   CPv   OPv   1956pc   -   ad   6.56 4oooX3   31 1/57   1/57   1
3 D/   CP?w,y   1956mc   1960mc   ad   6.56 4oooX3   11   2
3 D//   CPz   OPz   1956ms   1969ch   adf   6.56 4oooX3   31   3
3 D///*   CFz   OFz   1982ᶜm-s   nd:ry   adfj 6.83 4oooX3   31 3/83   3/83
AA 1.4.49/NT 4.2.49 on B(A), 3/-, OSPR 4/49. M on C. 1. Stansted Aerodrome. 2. Stansted Mountfitchet Airfield. 3. Stansted Airport.

TL 53 (52/53) *Saffron Walden*   10 by 10   Revision a 1916-19 c 1938-52; j 1982 selected
1 A   CPr   OPr   nd:po   -   -   1.49 4BEoo3   RB   0   7/49   8/49
2 A   CPt   1949pc   1952rpb   -   1.49 4BooX3   RB   0
3 B   CPv,x,z OPv   1956pc   -   ac   6.56 4oooX3   1   10/56 11/56
3 B/*   CFz   OFz   1983ᶜm-s   nd:r   acj   6.83 4oooX3   1   12/82 12/82
M on A.

TL 54 (52/54) *Linton Cambridgeshire*   10 by 10   Revision a 1901-19 b 1901-38 c 1938-49 d 1938-52
                                                        f 1965 g 1959; j 1965 airfield
1 A   CPr   OPr   nd:po   -   -   1.49 4BEoo3   RB   0   8/49   9/49
1 B   CPs   1951pc   -   ac   1.51 4BEoo3   RB   0   1/52   -
3 C   CPv   OPv   1956pc   -   bd   6.56 4oooX3   1   10/56 10/56
3 C/   CPx   1956mc   1961ch   bd   6.56 4oooX3   1
3 C//*   CPz   OPz   1956mc   1965ar   bdfj 6.56 4oooX3   2   12/65 1/66
3 C//*/* CPz   OPz   1956ms   1969r   bdgj 6.56 4oooX3   2   1/70   2/70
M on B. SS: Soil, outline, 1985 on C//*/*.

TL 55 (52/55) *Fulbourn*   10 by 10   Revision a 1901-25 b 1901-38 c 1938 d 1938-52
1 A   CPr   OPr   nd:po   -   -   1.49 4BEoo3   R   7/49   8/49
1 B   CPs   1949pc   1950c   ac   6.50 4BEoo3   R
3 C   CPv,z,0z OPz   1956pc   -   bd   6.56 4oooX3   7/56   -
M on B.

TL 56 (52/56) *Burwell Cambridgeshire**   10 by 10   Revision a 1901-25 c 1938 d 1938-50
1 A   CPr   OPr   nd:po   -   -   1.49 4BEoo3   R   9/49   10/49
1 B   CPs   1949pc   1950c   ac   6.50 4BEoo3   R   M
3 C   CPu   OPv   1955pc   -   ad   6.55 4oooX3   9/55   8/57
3 C/   CP?w,x,z   1955mc   1960mc   ad   6.55 4oooX3
M on B.

TL 57 (52/57) *Soham**   10 by 10   Revision a 1900-25 c 1938-50
1 A   CPr   OPr   nd:po   -   -   6.48 4BEoo3   R   0   9/49   11/49
1 B   CPs   1951pc   -   ac   6.51 4BEoo3   R   0   10/51 -
3 C   CPu   OPv   1955pc   -   ac   6.55 4oooX3   1   5/55   8/57
3 C/   CPw,z   1955mc   1960mc   ac   6.55 4oooX3   1
M on B.

1	2	3	4	5	6	7	8	9	10	11	12	13	14

TL 58 (52/58) *Ely (North)*   10 by 10   Revision   a 1900-25   c 1950
1	A	CPr	OPr	nd:po	-	-	6.49	4BEoo3	RB		10/49	11/49	
1	B	CPs		1951pc	-	ac	6.51	4BEoo3	RB		10/51	-	
2	B		OPv	1951pc	-	ac	1.52	4Booo3	RB		-	8/57	
3	B	CPy		1951pc	-	ac	1.52	4Booo3	RB				
3	B/	CPz	OPz	1951pc	-	ac	6.51	4Booo3	B				

M on B.

TL 59 (52/59) *Welney**   10 by 10   Revision   a 1900-26   c 1950   f 1965
1	A	CPr	OPr	nd:po	-	-	6.49	4BEoo3	R		8/49	9/49	
1	B	CPs		1949pc	1950c	a	6.50	4BEoo3	R				M
3	C	CPu,w	OPv	1955pc	-	ac	6.55	4oooX3			10/55	8/57	
3	C/	CPz	OPz	1955mc	1966ch	acf	6.55	4oooX3					

M on B.

TL 60 (52/60) *Chelmsford (West)*   10 by 10   Revision   a 1915-19   c 1938-55   f 1961
1	A	CPr	OPr	nd:po	-	-	1.49	4BEoo3	RB		10/48	12/48	
1	B	CPr	OPs	nd:po	-	-	6.49	4BEoo3	RB				
2	C		OPs	1948pc	1952c	-	1.52	4Booo3	RB				
3	D	CPw	OPv	1958mc	-	ac	6.58	4oooX3			2/59	3/59	
3	D/	CPx,z		1958mc	1962ch	acf	6.58	4oooX3					

AA 1.8.48 on A(A), 3/-, OSPR 12/48; AA 1.1.52 on C(B), 3/-, OSPR 3/52. M on B.

TL 61 (52/61) *Great Waltham**   10 by 10   Revision   a 1915-19   c 1938-46   d 1938-52   e 1947
1	A	CPr	OPr	nd:po	-	-	1.49	4BEoo3	B		8/49	9/49	
1	B	CPs	OPv	1950pc	-	ace	6.50	4BEoo3	B		1/51	10/56	
3	C	CPw,y,0z	OPv	1959mc	-	ad	6.59	4oooX3			3/59	4/59	

M on B.

TL 62 (52/62) *Great Dunmow*   10 by 10   Revision   a 1915-19   c 1946-53
1	A	CPr,u	OPr	nd:po	-	-	6.49	4BEoo3	RB	00	8/49	9/49	
3	B	CPw	OPv	1958pc	-	ac	6.58	4oooX3		10	9/58	10/58	
3	B	CPw,z,0z		1958mc	-	ac	6.58	4oooX3		10			

M on A.

TL 63 (52/63) *Thaxted*   10 by 10   Revision   a 1916-19   c 1946-52
1	A	CPr	OPr	nd:po	-	-	6.49	4BEoo3	B	0	8/49	9/49	
1	A	CPs		1949pc	1951	-	1.51	4BEoo3	B	0			
3	B	CPv	OPv	1956pc	-	ac	6.56	4oooX3		1	1/57	5/57	
3	B/	CPw,z		1956mc	1960mc	ac	6.56	4oooX3		1			

M on A.

TL 64 (52/64) *Haverhill*   10 by 10   Revision   a 1901-24   c 1938   d 1946-52   f 1969
                                                               j 1985 selected   k SUSI
1	A	CPr	OPr	nd:po	-	-	6.49	4BEoo3	RB	00	8/49	9/49	
1	B	CPs		1949pc	1950c	ac	6.50	4BEoo3	RB	00			
3	C	CPv	OPv	1957pc	-	ad	6.57	4oooX3		11	2/57	2/57	
3	C/	CPw,y		1957mc	1960mc	ad	6.57	4oooX3		11			
3	C//	CPz	OPz	1957ms	1970ch	adf	6.57	4oooX3		11			
3	C///*	CFz	OFz	1985^{c}m-s	nd:brsy	adfjk	6.86	4oooX3		11	6/86	6/86	1

M on B. 1. One of the final batch of three revised reprints in this series (with SU 03, TF 14).

TL 65 (52/65) *Dullingham**   10 by 10   Revision   a 1901-25   c 1938-50   d 1938-52
1	A	CPr	OPr	nd:po	-	-	6.49	4BEoo3	RB	0	8/49	9/49	
1	B	CPs		1951pc	-	ac	6.51	4BEoo3	RB	0	11/51	-	
3	C	CPv,x,z	OPv	1956pc	-	ad	6.56	4oooX3		1	8/56	8/57	

M on B.

1	2	3	4	5	6	7	8	9	10	11	12	13	14

TL 66 (52/66) *Newmarket**   10 by 10   Revision a 1924-25 c 1938-50 f 1976

1	A	CPr	OPr	nd:po	-	-	6.49	4BEoo3	RB	00	8/49	9/49	
3	B	CPu	OPt	1955pc	-	ac	6.55	4oooX3		22	10/55	10/55	
3	B/	CPx,z		1955mc	1961ch	ac	6.55	4oooX3		22			
3	B//*	CFz	OFz	1955^{c}ms	nd:r	acf	6.55	4oooX3		22	10/77	10/77	

AA 1.4.49 on A(A), 3/-, OSPR 9/49. M on A.

TL 67 (52/67) *Isleham**   10 by 10   Revision a 1901-25 c 1938-50

1	A	CPr	OPr	nd:po	-	-	6.49	4BEoo3	R	0	8/49	9/49	
1	B	CPs		1951pc	-	ac	6.51	4BEoo3	R	0	11/51	-	
3	C	CPu	OPv	1955pc	-	ac	6.55	4oooX3		1	3/55	8/57	
3	C/	CPw,z		1955mc	1960mc	ac	6.55	4oooX3		1			1
3	C//	CPz,0z	OPz	1955ms	1969ch	ac	6.55	4oooX3		3			

M on A. 1. Mildenhall Airfield.

TL 68 (52/68) *Shippea Hill*   10 by 10   Revision a 1900-25 c 1950 f 1964

1	A	CPr	OPr	nd:po	-	-	6.49	4BEoo3	R	0	8/49	9/49
3	B	CPu	OPv	1955pc	-	ac	6.55	4oooX3		1	9/55	8/57
3	B/	CPw		1955mc	1960mc	ac	6.55	4oooX3		1		
3	B//	CPz,0z	OPz	1955mc	1965ch	acf	6.55	4oooX3		1		

M on A.

TL 69 (52/69) *Southery*   10 by 10   Revision a 1900-26 c 1950 f 1965
                                                    j 1964 cut-off channel

1	A	CPr	OPr	nd:po	-	-	6.49	4BEoo3	R		9/49	11/49
3	B	CPu	OPv	1955pc	-	ac	6.55	4oooX3			10/55	10/56
3	B/	CPw		1955mc	1960mc	ac	6.55	4oooX3				
3	B//*	CPz	OPz	1955mc	1966or	acfj	6.55	4oooX3			7/66	9/66

M on A.

TL 70 (52/70) *Chelmsford (East)*   10 by 10   Revision a 1919 c 1938-54 f 1968 g 1972
                                                    j 1962-67 building

1	A	CPr		nd:po	-	-	1.49	4BEoo3	RB		2/49	-
1	B		OPr	nd:po	-	-	1.49	4BEoo3	RB		-	5/49
3	A	CPt		1949pc	-	-	1.49	4BEoo3	RB			
3	C	CPw,x	OPv	1958mc	-	ac	6.58	4oooX4			1/59	2/59
3	C/*	CPz	OPz	1958ms	1969br	acfj	6.58	4oooX4			10/69	9/69
3	C/*/*	CPz	OPz	1958^{c}ms	nd:r	acgj	6.58	4oooX4			2/73	4/73

AA 1.10.48 on B(A), 3/-, OSPR 5/49. M on A.

TL 71 (52/71) *Little Waltham**   10 by 10   Revision a 1915-20 c 1938-54 f 1968 g 1972
                                                    j 1966 reservoir

1	A	CPr	OPr	nd:po	-	-	6.49	4BEoo3	RB	0	8/49	9/49	
2	A	CPt		1949pc	1953rpb	-	6.49	4BooX3	RB	0			
3	B	CPv,x	OP?	1958pc	-	ac	6.58	4oooX3		3	3/58	3/58	
3	B/*	CPz	OPz	1958ms	1969rv	acfj	6.58	4oooX3		3	5/69	5/69	1
3	B/*/*	CPz,0z	OPz	1958^{c}ms	nd:r	acgj	6.58	4oooX3		3	7/73	9/73	

M on A. SS: Soil, Land Use Capability, coloured, 1980 on B/*/*. 1. Leighs Reservoirs.

TL 72 (52/72) *Braintree**   10 by 10   Revision a 1915-21 c 1938-54

1	A	CPr	OPr	nd:po	-	-	1.49	4BEoo3	RB	0	8/49	9/49
3	A	CPu		1949pc	-	-	1.49	4BEoo3	RB	0		
3	B	CPw,xz,0z	OPv	1958pc	-	ac	6.58	4oooX3		1	1/59	2/59

AA 1.4.49 on A(A), 3/-, OSPR 9/49. M on A.

TL 73 (52/73) *Hedingham*   10 by 10   Revision a 1919-24 c 1938-54

1	A	CPr	OPr	nd:po	-	-	1.49	4BEoo3	RB	00	8/49	9/49
2	A	CPs		1949pc	1952	-	1.52	4Booo3	RB	00		
3	B	CPw,z	OPv	1958pc	-	ac	6.58	4oooX3		30	8/58	10/58

M on A.

275

1	2	3	4	5	6	7	8	9	10	11	12	13	14

TL 74 (52/74) *Clare Suffolk**   10 by 10   Revision a 1902-24 c 1946-54
1 A   CPr   OPr   nd:po   -   -   6.49 4BEoo3 RB 0   8/49 9/49
3 A   CPu   1949pc   -   -   6.49 4BEoo3 RB 0
3 B   CPv   OPv   1957pc   -   ac   6.57 4oooX3   0   1/58 11/57
3 B/ CPw,z   1957mc   1960mc   ac   6.57 4oooX3   0
M on A.

TL 75 (52/75) *Wickhambrook*   10 by 10   Revision a 1901-25 c 1946-53
1 A   CPr   OPr   nd:po   -   -   6.49 4BEoo3 B   00 9/49 9/49
3 B   CPv,w,z OP?   1957pc   -   ac   6.57 4oooX3   30 1/58 1/58
M on A.

TL 76 (52/76) *Barrow Suffolk*   10 by 10   Revision a 1901-25 c 1938-50
1 A   CPr   OPr   nd:po   -   -   1.49 4BEoo3 RB   9/49 9/49
? B   CP?   1949pc   1952c   -   1.52 4BooX3 RB   M
3 C   CPu   OPv   1955pc   -   ac   6.55 4oooX3   10/55 10/56
3 C/ CPw,z   1955mc   1960c   ac   6.55 4oooX3
M on A, B, C/.

TL 77 (52/77) *Mildenhall*   10 by 10   Revision a 1903-26 c 1938-50 f 1962 g 1967
                                                  j 1964 cut-off channel
1 A   CPr   OPr   nd:po   -   -   1.49 4BEoo3 B   0   9/49 9/49
2 B   CPs   OPs   1952pc   -   ac   1.52 4Booo3   0   4/52 6/52
3 B/   CPy   1952mc   1964ch   acf 1.52 4BooX3   0
3 B//* CPz   OPz   1952mc   1968o   acgj 1.52 4BooX3   0   1/68 1/68
M on B.

TL 78 (52/78) *Brandon Suffolk**   10 by 10   Revision a 1903-26 c 1938-50 f 1967
1 A   CPr   OPr   nd:po   -   -   6.49 4BEoo3 R   00 8/49   9/49
2 A   CPt   1949pc   1953rpb -   1.52 4BooX3 R   00
3 B   CPu,x OPv   1955pc   -   ac   6.55 4oooX3   11 10/55 10/56
3 B/ CPz   OPz   1955ms   1968ch   acf 6.55 4oooX3   11
M on A.

TL 79 (52/79) *Methwold**   10 by 10   Revision a 1903-26 c 1950 f 1964
                                                  j 1964 cut-off channel
1 A   CPr   OPr   nd:po   -   -   1.49 4BEoo3 R   00 10/49 11/49
3 B   CPu   OPv   1955pc   -   ac   6.55 4oooX3   11 10/55 8/57
3 B/   CPw   1955mc   1960mc   ac   6.55 4oooX3   11
3 B//* CPz   OPz   1955mc   1965ao   acfj 6.55 4oooX3   33 8/65   9/65
M on A.

TL 80 (52/80) *Maldon*   10 by 10   Revision a 1919-20 c 1938-54
1 A   CPr   OPr   nd:po   -   -   6.48 4BEoo3 RB   11/48 1/49
1 B   CPr   nd:po   -   -   6.48 4BEoo3 RB
3 C   CPv   OPv   1958pc   -   ac   6.58 4oooX3   5/58   7/58
3 C/ CP?w,y,z,0z   1958mc   1960mc   ac   6.58 4oooX3
AA 1.9.48 on A(A), 3/-, OSPR 1/49. M on B.

TL 81 (52/81) *Witham**   10 by 10   Revision a 1919-20 c 1938-54 f 1963 g 1967
1 A   CPr   OPr   nd:po   -   -   6.49 4BEoo3 RB   8/49 9/49
2 A   CPs   1949pc   1953rpb -   1.52 4BooX3 RB
3 B   CPv   OP?   1958pc   -   ac   6.58 4oooX3   3/58 5/58
3 B/   CPz   OPz   1958mc   1964ch   acf 6.58 4oooX3
3 B//* CPz   OPz   1958ms   1968r   acg 6.58 4oooX3   4/68 3/68
AA 1.4.49 on A(A), 3/-, OSPR 9/49. M on A. GS: Solid and Drift, 1972 on B//*. Another edition, 1981.

1	2	3	4	5	6	7	8	9	10	11	12	13	14

TL 82 (52/82) *Coggeshall*   10 by 10   Revision a 1920 c 1938 d 1938-54 e 1948
1	A		CPr	OPr	nd:po	-	-	1.49	4BEoo3	R	00	8/49	9/49
1	B		CPs		1949pc	1950c	ace	6.50	4BEoo3	R	00		
2	B			OPv	1949pc	1950c	ace	1.52	4Booo3	R	00	-	8/57
3	C		CPw,x,z,0z	OPv	1958pc	-	ad	6.58	4oooX3		33	10/58	10/58

M on A.

TL 83 (52/83) *Halstead Essex**   10 by 10   Revision a 1919-24 c 1938-54
1	A		CPr	OPr	nd:po	-	-	1.49	4BEoo3	R		9/49	9/49
3	B		CPu		1955pc	-	ac	6.55	4oooX3			6/55	-
3	B/		CP?v,y,z		1955pc	1957mc	ac	6.55	4oooX3				

M on A.

TL 84 (52/84) *Sudbury Suffolk**   10 by 10   Revision a 1902-24 c 1938-54
1	A		CPr	OPr	nd:po	-	-	6.49	4BEoo3	RB	0	8/49	8/49
3	B		CPu	OPv	1955pc	-	ac	6.55	4oooX3		1	7/55	10/56
3	B/		CP?w,x,z,0z		1955mc	1960mc	ac	6.55	4oooX3		1		

M on A.

TL 85 (52/85) *Hartest*   10 by 10   Revision a 1902-24 c 1950-53
1	A		CPr	OPr	nd:po	-	-	1.49	4BEoo3	RB	0	8/49	9/49
3	B		CPu	OPv	1955pc	-	ac	6.55	4oooX3		1	6/55	10/56
3	B/		CPw,z		1955mc	1961mc	ac	6.55	4oooX3		1		

M on A.

TL 86 (52/86) *Bury St Edmunds*   10 by 10   Revision a 1901-25 c 1938-50 f 1965 g 1974
    j 1980 selected
1	A		CPr	OPr	nd:po	-	-	6.49	4BEoo3	RB	0	9/49	11/49
3	B		CPu	OPv	1955pc	-	ac	6.55	4oooX3		1	11/55	8/57
3	B/		CP?w,y		1955mc	1960mc	ac	6.55	4oooX3		1		
3	B//		CPz	OPz	1955mc	1966ch	acf	6.55	4oooX3		3		
3	B///*		CFz	OFz	1955^{c}ms	nd:r	acg	6.55	4oooX3		3	12/75	5/75
3	B///*/*		CFz	OFz	1981^{c}ms	nd:r	acgj	6.55	4oooX3		3	2/81	2/81

AA 1.4.49 on A(A), 3/-, OSPR 11/49. M on A.

TL 87 (52/87) *Barnham Suffolk**   10 by 10   Revision a 1903-26 c 1938-50 f 1969
1	A		CPr	OPr	nd:po	-	-	6.49	4BEoo3	RB	0	9/49	9/49
2	A		CPt		1949pc	1953rpb	-	6.49	4BooX3	RB	0		
3	B		CPu,y	OPv	1955pc	-	ac	6.55	4oooX3		1	11/55	8/57
3	B/		CPz	OPz	1955ms	1970ch	acf	6.55	4oooX3		3			1

M on A. 1. Honington Airfield.

TL 88 (52/88) *Thetford**   10 by 10   Revision a 1903-26 c 1941 d 1938-50 e 1945
1	A		CPr	OPr	nd:po	-	-	1.49	4BEoo3	R	0	8/49	9/49
1	B		CPs		1949pc	1951c	ace	1.51	4BEoo3	R	0		
3	C		CPu,x,z	OPv	1955pc	-	ad	6.55	4oooX3		2	11/55	8/57

M on A, B, C.

TL 89 (52/89) *Mundford**   10 by 10   Revision a 1904-26 c 1941 d 1950 e 1945
1	A		CPr	OPr	nd:po	-	-	1.49	4BEoo3		00	8/49	9/49
1	B		CPs		1951pc	-	ace	1.51	4BEoo3		10	8/51	-
3	C		CPu,z	OPv	1955pc	-	ad	6.55	4oooX3		22	1/56	8/57

M on A, B, C.

TL 90 (52/90) *Lower Mayland**   10 by 10   Revision a 1920-21 c 1938-53
1	A		CPr		1948po	-	-	1.48	4BEoo3	RB		7/48	-
1	A			OPr	nd:po	-	-	1.48	4BEoo3	RB		-	9/48
1	B		CPr		nd:po	-	-	1.48	4BEoo3	RB			
3	C		CPv	OP?	1958pc	-	ac	6.58	4oooX3			3/58	3/58
3	C/		CPx.z		1958mc	1961mc	ac	6.58	4oooX3				

AA 1.4.48 on A(A), 3/-, OSPR 9/48. M on B.

1	2	3	4	5	6	7	8	9	10	11	12	13	14

TL 91 (52/91) *Tollesbury*   10 by 10   Revision a 1920-21 c 1938 d 1938-53
                                                    e 1947-48 f 1965
1	A	CPr	OPr	nd:po	-	-	1.49	4BEoo3	R	0	10/49	11/49	
1	B	CPs		1949pc	1950c	ace	6.50	4BEoo3	R	0			
3	C	CPu	OPv	1955pc	-	ad	6.55	4oooX3		1	1/56	10/56	
3	C/	CPx		1955mc	1961mc	ad	6.55	4oooX3		1			
3	C//	CPz,0z	OPz	1955mc	1966ch	adf	6.55	4oooX3		3			1

M on A. 1. Abberton Reservoir.

TL 92 (52/92) *[Colchester (West)]*   10 by 10   Revision a 1920-21 c 1938-53
1	A	CPr	OPr	nd:po	-	-	1.49	4BEoo3	RB	00	9/49	9/49	1
3	B	CPu	OPv	1955pc	-	ac	6.55	4oooX3		11	7/55	8/57	
3	B/	CP?w,y		1955mc	1958mc	ac	6.55	4oooX3		11			

AA 1.4.49 on A(A), 3/-, OSPR 9/49. M on A. 1. Colchester Camp.

TL 93 (52/93) *Nayland*   10 by 10   Revision a 1902-24 c 1938-54 f 1965
1	A	CPr	OPr	nd:po	-	-	1.49	4BEoo3	RB	0	7/49	8/49	
3	B	CPu,x	OPv	1955pc	-	ac	6.55	4oooX3		1	6/55	8/57	
3	B/	CPz	OPz	1955mc	1966ch	acf	6.55	4oooX3		3			

M on A.

TL 94 (52/94) *Lavenham*   10 by 10   Revision a 1902-24 c 1938-54
1	A	CPr	OPr	nd:po	-	-	6.49	4BEoo3	RB	0	8/49	9/49	
3	B	CPu	OPv	1955pc	-	ac	6.55	4oooX3		1	3/55	8/57	
3	B/	CP?w.y,z,0z		1955mc	1960mc	ac	6.55	4oooX3		1			

M on A.

TL 95 (52/95) *Hitcham Suffolk*   10 by 10   Revision a 1902-25 c 1950-53 f 1965
1	A	CPr	OPr	nd:po	-	-	1.49	4BEoo3	R	00	8/49	9/49	
3	B	CPu	OPs	1955pc	-	ac	6.55	4oooX4		11	3/55	7/55	
3	B/	CPx		1955mc	1961ch	ac	6.55	4oooX4		11			1
3	B//	CPz,0z	OPz	1955mc	1966ch	acf	6.55	4oooX4		33			

M on A. 1. Rattlesden Airfield.

TL 96 (52/96) *Woolpit*   10 by 10   Revision a 1903 c 1950 f 1969
1	A	CPr	OPr	nd:po	-	-	6.49	4BEoo3	R	0	8/49	9/49	
3	B	CPt		1949pc	1953c	-	6.53	4BooX3		0			
3	C	CPu,x	OPs	1955pc	-	ac	6.55	4oooX3		1	1/56	1/56	
3	C/	CPz	OPz	1955^{c}ms	nd:ch	acf	6.55	4oooX3		1			

M on A.

TL 97 (52/97) *Ixworth*   10 by 10   Revision a 1903 c 1950 f 1966
1	A	CPr	OPr	nd:po	-	-	6.49	4BEoo3	B	000	9/49	9/49	
2	A	CPs		1949pc	1952	-	1.52	4Booo3	B	000			
3	B	CPu,y	OPv	1955pc	-	ac	6.55	4oooX3		111	11/55	10/56	
3	B/	CPz	OPz	1955mc	1967ch	acf	6.55	4oooX3		333			

M on A.

TL 98 (52/98) *East Harling*   10 by 10   Revision a 1903-26 c 1950
1	A	CPr	OPr	nd:po	-	-	6.49	4BEoo3	R	00	9/49	9/49	
3	A	CPt		1949pc	-	-	1.52	4Booo3	R	00			
3	B	CPu,y,z	OPv	1955pc	-	ac	6.55	4BooX3		21	11/55	8/57	

M on A.

TL 99 (52/99) *Caston*   10 by 10   Revision a 1904-26 c 1950 f 1970
1	A	CPr	OPr	nd:po	-	-	1.49	4BEoo3	R	00	8/49	9/49	
3	B	CPu.z	OPv	1955pc	-	ac	6.55	4oooX3		12	11/55	8/57	
3	B/	CPz	OPz	1955^{c}ms	nd:ch	acf	6.55	4oooX3		32			1

M on A, B. SS: Soil, outline, 1986 on B/. 1. Watton Airfield.

1	2	3	4	5	6	7	8	9	10	11	12	13	14

TM 00 (62/00) *Bradwell-on-Sea* 10 by 10 Revision a 1900-21 c 1938 d 1938-53 e 1947
1 A   CPr   OPr   nd:po   -   -   1.48 4BEoo3   0   8/48   9/48
1 B   CPs   1948pc   1950c   ace   6.50 4BEoo3   0
3 C   CPv,x,z   OP?   1957pc   -   ad   6.57 4oooX3   3   1/58   2/58
M on B.

TM 01 (62/01) *Brightlingsea* 10 by 10 Revision a 1920-21 c 1938-53
1 A   CPr   OPr   nd:po   -   -   6.48 4BEoo3   R   10/48 11/48 1
1 B   CPr   nd:po   -   -   6.48 4BEoo3   R
3 C   CPu,x,z   OPv   1955pc   -   ac   6.55 4oooX3   1/56   6/57
M on B, C. 1. Showing Martello Towers.

TM 02 (62/02) [*Colchester (East)*] 10 by 10 Revision a 1920-21 c 1938-53
1 15046   CPr   1946po   -   -   6.46 3BE1X1   RB   0   9/46   -
1 2547   OPr   1947po   -   -   6.46 3BE1X1   RB   0   -   1/47
3 B   CPu   OPv   1955pc   -   ac   6.55 3oooX2   1   7/55   8/57
3 B/   CPw,z   1955mc   1961ch ac   6.55 3oooX2   1
AA 1.6.46 on 2547(2547), 3/-, OSPR 1/47. M on A.

TM 03 (62/03) *Stratford St Mary** 10 by 10 Revision a 1902-24 c 1938-53 f 1967
1 A   CPr   1948po   -   -   1.48 3BE1X1   R   00 12/47 -
1 B   OPr   1948po   -   -   1.48 3BE1X1   R   00 -   3/48
1 C   CPr   nd:po   -   -   1.48 3BEoo1   R   00
3 D   CPu   OPv   1955pc   -   ac   6.55 3oooX2   11 7/55   11/56
3 D/   CPw   1955mc   1961mc ac   6.55 3oooX2   11
3 D//   CPz   OPz   1955ms   1968ch acf   6.55 3oooX2   11
AA 1.10.47/NT 8.1.48 on B(A), 3/-, OSPR 3/48. M on A, C.

TM 04 (62/04) *Hadleigh Suffolk* 10 by 10 Revision a 1902-25 c 1938-54
1 A   CPr   OPr   1947po   -   -   6.46 3BE1X1   RB   5/47   8/47
3 B   CPu   OPv   1955pc   -   ac   6.55 3oooX2   4/55   6/57
3 B/   CP?w,y,z   1955pc   1958mc ac   6.55 3oooX2
AA 1.12.46 on A(A), 3/-, OSPR 8/47. M on A.

TM 05 (62/05) *Stowmarket* 10 by 10 Revision a 1902-25 c 1938-53
1 A   CPr   OPr   1947po   -   -   1.47 3BE1X1   RB   0   5/47   7/47
3 B   CPu   OPv   1955pc   -   ac   6.55 3oooX2   1   3/55   6/57
3 B/   CP?w,z   1955mc   1959c   ac   6.55 3oooX2   1   1
AA 1.9.46 on A(A), 3/-, OSPR 7/47. M on A. 1. Wattisham Airfield.

TM 06 (62/06) *Finningham** 10 by 10 Revision a 1903-25 c 1938-50; j 1981 selected
1 10046   CPr   1946po   -   -   6.46 3BE1X1   RB   0   11/46 -
1 A   OPr   1946po   -   -   6.46 3BE1X1   RB   0   -   3/47
3 B   CPu,y,z   OPv   1955pc   -   ac   6.55 3oooX2   1   11/55 5/57
3 B/*   CFz   OFz   1981^{c}ms nd:ry   acj   6.55 3oooX2   1   2/82   3/82
AA 1.8.46 on A(A), 3/-, OSPR 3/47. M on A.

TM 07 (62/07) *Botesdale** 10 by 10 Revision a 1903-25 c 1938-50
1 A   CPr   OPr   1947po   -   -   1.47 3BE1X1   R   7/47   10/47
3 B   CPu   OPv   1955pc   -   ac   6.55 3oooX2   7/55   6/57
3 B/   CPw.z   1955mc   1961mc ac   6.55 3oooX2
AA 1.1.47 on A(-), 3/-, OSPR 10/47. M on A.

TM 08 (62/08) *Kenninghall** 10 by 10 Revision a 1903-26 c 1938-50 f 1965
1 A   CPr   OPr   1947po   -   -   6.47 3BE1X1   RB   00 10/47 2/48
3 B   CPu   OPv   1955pc   -   ac   6.55 3oooX2   11 9/55   11/56
3 B/   CPx   1955mc   1961ch ac   6.55 3oooX2   11
3 B//   CPz   OPz   1955mc   1965ch acf   6.55 3oooX2   33
AA 1.8.47 on A(A), 3/-, OSPR 2/48. M on A.

1	2	3	4	5	6	7	8	9	10	11	12	13	14

TM 09 (62/09) *Attleborough Norfolk** 10 by 10 Revision a 1904-26 c 1938-50
1　A　　　CPr　　OPr　　1947po　　-　　　-　　　6.47　3BE1X1　　RB　　000　2/47　6/47
3　B　　　CPu　　OPv　　1955pc　　-　　　ac　　6.55　3oooX2　　　　　111　8/55　6/57
3　B/　　 CPx,z　　　　　1955mc　1961ch　ac　　6.55　3oooX2　　　　　111
AA 1.6.46 on A(-), 3/-, OSPR 6/47. M on A.

TM 11 (62/11) **& Part of TM 21** *Clacton-on-Sea** 15 by 10 Revision a 1921-22 c 1938-54
1　A　　　CPr　　OPr　　nd:po　　-　　　-　　　1.49　4BEoo3　　R　　　　10/48　12/48　1
1　B　　　CPr　　OPr　　nd:po　　-　　　-　　　1.49　4BEoo3　　R　　　　　　　　　　M
3　C　　　CPu　　　　　1955pc　　-　　　ac　　6.55　4oooX3　　　　　　　9/55　-
AA 1.8.48 on A(A), 3/-, OSPR 12/48. M on B. 1. Showing Martello Towers.

TM 12 (62/12) *Weeley** 10 by 10 Revision a 1921-22 c 1938-54
1　10046　CPr　　　　　1946po　　-　　　-　　　6.46　3BE1X1　　R　　　　9/46　-　　1
1　2546　　　　　OPr　　1946po　　-　　　-　　　6.46　3BE1X1　　R　　　　-　　11/46
3　B　　　CPu　　OPv　　1955pc　　-　　　ac　　6.55　3oooX2　　　　　　　7/55　5/57
3　B/　　 CP?w,y.z,0z　　1955mc　1960mc　ac　　6.55　3oooX2
AA 1.6.46 on 2546(2546), 3/-, OSPR 11/46. M on A. SS: Soil, coloured, 1980 on B/; Land Use Capability, coloured, 1981 on B/. 1. Showing the unbuilt Mistley-Weeley railway (at 1329-1327, 1426, 1524).

TM 13 (62/13) *Holbrook Suffolk** 10 by 10 Revision a 1902-25 c 1938 d 1938-54 e 1948
1　A　　　CPr　　OPr　　1947po　　-　　　-　　　6.47　3BE1X1　　R　　　　9/47　12/47　1
3　B　　　CPt　　　　　1954pc　　-　　　ace　　6.54　3BooX2　　　　　　 5/54　-
3　C　　　CPu　　OPv　　1955pc　　-　　　ad　　6.55　3oooX2　　　　　　6/55　11/56
3　C/　　 CPx,z　　　　 1955mc　　-　　　ad　　6.55　3oooX2
AA 1.7.47 on A(A), 3/-, OSPR 12/47. M on A. 1. Showing the unbuilt Mistley-Weeley railway (at 1220-1321).

TM 14 (62/14) *Ipswich** 10 by 10 Revision a 1924-25 c 1950-53
1　A　　　CPr　　OPr　　1947po　　-　　　-　　　6.47　3BE1X1　　RC　　1　　8/47　10/47
3　B　　　CPu,x,y,z OPv　1955pc　　-　　　ac　　6.55　3oooX2　　C　　 1　　7/55　6/57
AA 1.6.47 on A(A), 3/-, OSPR 10/47. M on A.

TM 15 (62/15) *Coddenham** 10 by 10 Revision a 1903-25 c 1938-53 f 1976
1　A　　　CPr　　OPr　　1947po　　-　　　-　　　6.47　3BE1X1　　RB　　　　8/47　12/47
3　B　　　CPu　　OPv　　1955pc　　-　　　ac　　6.55　3oooX2　　　　　　　6/55　11/56
3　B/　　 CPx,z　　　　 1955mc　1961ch　ac　　6.55　3oooX2
3　B//*　 CFz　　OFz　　1955^{c}ms　nd:r　acf　6.55　3oooX2
AA 1.3.47 on A(A), 3/-, OSPR 12/47. M on A.

TM 16 (62/16) *Debenham** 10 by 10 Revision a 1903-25 c 1946-52
1　10046　CPr　　　　　1946po　　-　　　-　　　1.47　3BE1X1　　RB　　0　　11/46　-
1　A　　　　　　OPr　　1946po　　-　　　-　　　1.47　3BE1X1　　RB　　0　　-　　3/47
3　B　　　CPu,x,z OPv　　1956pc　　-　　　ac　　6.56　3oooX2　　　　　1　　5/56　6/57
AA 1.9.46 on A(A), 3/-, OSPR 3/47. M on A.

TM 17 (62/17) *Eye Suffolk* 10 by 10 Revision a 1903-25 c 1938-52 d 1938-53 f 1966
　　　　　　　　　　　　　　　　　　　j 1980 selected
1　A　　　CPr　　OPr　　1947po　　-　　　-　　　6.47　3BE1X1　　R　　　00　　9/47　11/47
2　B　　　CPs　　　　　1947pc　　1952c　　　　1.52　3Booo2　　R　　　00　　　　　　M
3　C　　　CPu　　OPv　　1956pc　　-　　　ac　　6.56　3oooX2　　　　　 11　　4/56　11/56
3　C/　　 CP?w,z　　　　1956mc　1960mc　ad　　6.56　3oooX2　　　　　 11
3　C//　　CPz　　OPz　　1956mc　1966ch　adf　6.56　3oooX2　　　　　 31
3　C///*　CFz　　OFz　　1980^{c}ms　nd:r　adfj　6.56　3oooX2　　　　31　　2/81　2/81
AA 1.6.47 on A(A), 3/-, OSPR 11/47. M on B.

TM 18 (62/18) *Diss (North)** 10 by 10 Revision a 1903-26 c 1938-53
1　10046　CPr　　　　　1946po　　-　　　-　　　6.46　3BE1X1　　RB　　000　11/46　-
1　A　　　　　　OPr　　1946po　　-　　　-　　　6.46　3BE1X1　　RB　　000　-　　2/47
3　B　　　CPu　　OPv　　1956pc　　-　　　ac　　6.56　3oooX2　　　　　111　5/56　6/57
3　B/　　 CPx,z　　　　 1956mc　1961mc　ac　　6.56　3oooX2　　　　　111　　　　　1
AA 1.9.46 on A(A), 3/-, OSPR 2/47. M on A. 1. Tibenham Airfield.

1	2	3	4	5	6	7	8	9	10	11	12	13	14

TM 19 (62/19) *Ashwellthorpe**   10 by 10   Revision a 1904-26 c 1946-53
1   A   CPr   OPr   1947po   -   -   6.47   3BE1X1   RB   0   10/47   1/48
2   A   CPt   1947pc   1953rpb -   1.52   3BooX2   RB   0
3   B   CPu,z   OPv   1956pc   -   ac   6.56   3oooX2   1   4/56   11/56
AA 1.8.47 on A(A), 3/-, OSPR 1/48. M on A.

TM 22 (62/22) *Walton on the Naze**   10 by 10   Revision a 1921-22 1938-54
1   15046   CPr   1946po   -   -   6.46   3BE1X1   R   7/46   -
1   2546   OPr   1946po   -   -   6.46   3BE1X1   R   -   5/47
3   B   CPu   OPv   1955pc   -   ac   6.55   3oooX2   5/55   6/57
3   B/   CPw.z,0z   1955mc   1960mc   ac   6.55   3oooX2
AA 1.4.46 on 2546(2546), 3/-, OSPR 5/47. M on A.

TM 23 (62/23) *Harwich**   10 by 10   Revision a 1922-25 c 1938-54 f 1964
1   20046   CPr   1946po   -   -   6.46   3BE1X1   RB   0   7/46   -
1   2546   OPr   1946po   -   -   6.46   3BE1X1   RB   0   -   10/46
3   B   CPu,x   OPv   1955pc   -   ac   6.55   3oooX2   0   10/55 6/57
3   B/   CPz   OPz   1955mc   1965ch   acf 6.55   3oooX2   0
AA 1.6.46 on 2546(2546), 3/-, OSPR 10/46. M on A.

TM 24 (62/24) *Woodbridge**   10 by 10   Revision a 1902-25 c 1938-53
1   A   CPr   OPr   1947po   -   -   6.47   3BE1X1   R   0   9/47   12/47
3   B   CPu   OPv   1955pc   -   ac   6.55   3oooX2   1   6/55   11/56
3   B/   CP?w,z   1955mc   1960mc   ac   6.55   3oooX2   1   1
AA 1.7.47/NT 26.11.47 on A(A), 3/-, OSPR 12/47. M on A. 1. Martlesham Heath Airfield.

TM 25 (62/25) *Ufford Suffolk**   10 by 10   Revision a 1903-25 c 1938-53
1   10047   CPr   1947po   -   -   1.47   3BE1X1   R   0   1/47   -
1   A   OPr   1947po   -   -   1.47   3BE1X1   R   0   -   4/47
3   B   CPu   OPv   1955pc   -   ac   6.55   3oooX2   1   5/55   5/57
3   B/   CPw   1955mc   1960mc   ac   6.55   3oooX2   1
3   B//   CPz   OPz   1955mc   1965ch   acf 6.55   3oooX2   3
AA 1.9.46 on A(A), 3/-, OSPR 4/47. M on A.

TM 26 (62/26) *Framlingham**   10 by 10   Revision a 1903-25 c 1945-52
1   A   CPr   1947po   -   -   6.47   3BE1X1   R   9/47   -
1   B   OPr   1947po   -   -   6.47   3BE1X1   R   -   12/47
3   C   CPv   1956pc   -   ac   6.56   3oooX2   7/56   -
3   C/   CPx,z   1956mc   1961ch   ac   6.56   3oooX2
AA 1.6.47 on B(A), 3/-, OSPR 12/47. M on B.

TM 27 (62/27) *Stradbroke*   10 by 10   Revision a 1903-25 c 1946-53
1   15046   CPr   1946po   -   -   6.46   3BE1X1   R   0   7/46   -   1
1   2547   OPr   1947po   -   -   6.46   3BE1X1   R   0   -   1/47
3   B   CPv,y,z OPz   1956pc   -   ac   6.56   3oooX2   1   8/56   -   2
AA 1.4.46 on 2547(2547), 3/-, OSPR 1/47. M on A. 1. The alignment of the Mid-Suffolk Light Ry beyond Laxfield is not shown: the railway is described as under construction. 2. The alignment of the Mid-Suffolk Light Ry beyond Laxfield is shown.

TM 28 (62/28) *Harleston Norfolk**   10 by 10   Revision a 1903-26 c 1946-53
1   15046   CPr   1946po   -   -   6.46   3BE1X1   RB   01 7/46   -
1   2546   OPr   1946po   -   -   6.46   3BE1X1   RB   01 -   11/46
3   B   CPv,y,z OPy   1956pc   -   ac   6.56   3oooX2   11 7/56   -
AA 1.4.46 on 2546(2546), 3/-, OSPR 12/46. M on A. SS: Soil, Land Use Capability, coloured, 1978 on B.

TM 29 (62/29) *Hempnall*   10 by 10   Revision a 1904-26 c 1946-53
1   15046   CPr   1946po   -   -   6.46   3BE1X1   RB   1   6/46   -
1   2546   OPr   1946po   -   -   6.46   3BE1X1   RB   1   -   5/47
3   B   CPv,y,z   1956pc   -   ac   6.56   3oooX2   1   7/56   -
AA 1.4.46 on 2546(2546), 3/-, OSPR 5/47. M on A.

1	2	3	4	5	6	7	8	9	10	11	12	13	14

TM 33 (62/33) *Felixstowe (East)**   10 by 10   Revision a 1924-25 c 1938-53
1	15046	CPr		1946po	-	-	6.46	3BE1X1			4/46	-	
1	2546		OPr	1946po	-	-	6.46	3BE1X1			-	4/46	1
3	B	CPu,x.z	OPv	1955pc	-	ac	6.55	3oooX2			6/55	5/57	

AA 1.3.46 on 2546(2546), 3/-, OSPR 7/46. M on A. 1. The outline edition appears also in OSPR 7/46.

TM 34 (62/34) *Hollesley**   10 by 10   Revision a 1902-25 c 1953 f 1969; j 1970 airfield
1	A	CPr	OPr	1947po	-	-	1.47	3BE1X1		0	6/47	8/47	
2	B	CPs		1949pc			1.52	3Booo2		0			M
3	C	CPu	OPv	1955pc	-	ac	6.55	3oooX2		1	8/55	6/57	
3	C/	CPw,y		1955mc	1960mc	ac	6.55	3oooX2		1			1
3	C//*	CPz	OPz	1955^{c}ms	nd:a	acfj	6.55	3oooX2		3	12/70	2/71	

AA 1.1.47 on A(A), 3/-, OSPR 8/47. M on B. 1. Woodbridge Airfield.

TM 35 (62/35) *Wickham Market*   10 by 10   Revision a 1902-25 c 1945-53; j 1982 selected
1	A	CPr		1947po	-	-	1.48	3BE1X1	R	00	11/47	-	
1	B		OPr	1947po	-	-	1.48	3BE1X1	R	00	-	2/48	
3	C	CPu	OPv	1955pc	-	ac	6.55	3oooX2		11	6/55	11/56	
3	C/	CP?w,z		1955mc	1960mc	ac	6.55	3oooX2		11			
3	C//*	CFz	OFz	1983^{c}m-s	nd:ry	acj	6.55	3oooX2		11	2/83	2/83	

AA 1.8.47 on B(A), 3/-, OSPR 2/48. M on A. 1. Bentwater Airfield.

TM 36 (62/36) *Saxmundham*   10 by 10   Revision a 1903-25 c 1945-53
1	A	CPr	OPr	1947po	-	-	1.48	3BE1X1	RB	0	11/47	2/48	
3	B	CPv		1956pc	-	ac	6.56	3oooX2		1	7/56	-	
3	B/	CPw,z		1956mc	1960mc	ac	6.56	3oooX2		1			

AA 1.9.47 on A(A), 3/-, OSPR 2/48. M on A.

TM 37 (62/37) *Halesworth*   10 by 10   Revision a 1903-26 c 1946-53
1	15046	CPr		1946po	-	-	1.47	3BE1X1	R	00	10/46	-	
1	2547		OPr	1947po	-	-	1.47	3BE1X1	R	00	-	1/47	
3	B	CPv		1956pc	-	ac	6.56	3oooX2		11	7/56	-	
3	B/	CP?x,z		1956mc	1961ch	ac	6.56	3oooX2		11			

AA 1.7.46 on 2547(2547), 3/-, OSPR 1/47. M on A.

TM 38 (62/38) *Bungay*   10 by 10   Revision a 1925-26 c 1946-53
1	15046	CPr		1946po	-	-	6.46	3BE1X1	RB	00	10/46	-	
1	A		OPr	1946po	-	-	6.46	3BE1X1	RB	00	-	3/47	
3	B	CPu	OPv	1956pc	-	ac	6.56	3oooX2		11	5/56	6/57	
3	B/	CPx,z		1956mc	1961mc	ac	6.56	3oooX2		11			

AA 1.8.46 on A(A), 3/-, OSPR 3/47. M on A.

TM 39 (62/39) *Loddon**   10 by 10   Revision a 1905-26 c 1946-51 f 1977
1	15046	CPr		1946po	-	-	6.46	3BE1X1	RB	0	6/46	-	
1	2546		OPr	1946po	-	-	6.46	3BE1X1	RB	0	-	10/46	
3	B	CPv,x,z	OPv	1956pc	-	ac	6.56	3oooX2		1	7/56	6/57	
3	B/*	CFz	OFz	1956^{c}ms	nd:r	acf	6.56	3oooX2		1	7/78	7/78	

AA 1.4.46 on 2546(2546), 3/-, OSPR 10/46. M on A.

TM 44 (62/44) *Orford Ness**   10 by 10   Revision a 1902-25 c 1938-53
1	5046	CPr		1946po	-	-	6.46	3BE1X1			6/46	-	
1	1046		OPr	1946po	-	-	6.46	3BE1X1			-	9/46	
1	B	CPr		nd:po	-	-	6.46	3BEoo1					
3	C	CPu,x.z	OPv	1955pc	-	ac	6.55	3oooX2			5/55	11/56	

AA 1.4.46 on 1046(1046) 3/-, OSPR 9/46. M on B.

TM 45 (62/45) *Aldeburgh**   10 by 10   Revision a 1925-26 c 1938-55
1	15046	CPr		1946po	-	-	6.46	3BE1X1	R		7/46	-	
1	2546		OPr	1946po	-	-	6.46	3BE1X1	R		-	12/46	
3	B	CPu	OPv	1955pc	-	ac	6.55	3oooX2			6/55	5/57	

1	2	3	4	5	6	7	8	9	10	11	12	13	14
3	B/	CP?w,x,z		1955mc	1960mc	ac	6.55	3oooX2					
3	B//	CPz.z	OPz	1955mc	1960mc	ac	6.55	3oooX2					

AA 1.5.46 on 2546(2546), 3/-, OSPR 12/46. M on A.

TM 46 (62/46) *Leiston* 10 by 10 Revision a 1903-26 c 1945-53 f 1968
                                       j 1966 power station

1	15046	CPr		1946po	-	-	6.46	3BE1X1	R	1	8/46	-	
1	2546		OPr	1946po	-	-	6.46	3BE1X1	R	1	-	11/46	
3	B	CPv		1956pc	-	ac	6.56	3oooX2		1	7/56	-	
3	B/	CPw,z		1956mc	1960mc	ac	6.56	3oooX2		1			
3	B//*	CPz	OPz	1956ms	1968p	acfj	6.56	3oooX2		1	11/68	8/68	1

AA 1.5.46 on 2546(2546), 3/-, OSPR 11/46. M on A. 1. Sizewell Power Station.

TM 47 (62/47) **& Part of TM 57** *Southwold* 15 by 10 Revision a 1903-25 c 1946-53 f 1965

1	20046	CPr		1946po	-	-	6.46	3BE1X1	R	0	8/46	-	
1	2546		OPr	1946po	-	-	6.46	3BE1X1	R	0	-	11/46	
3	B	CPv,x	OP?	1956pc	-	ac	6.56	3oooX2		1	10/56	1/58	
3	B/	CPz	OPz	1956mc	1966ch	acf	6.56	3oooX2		1			

AA 1.4.46 on 2546(2546), 3/-, OSPR 11/46. M on A.

TM 48 (62/48) **& Part of TM 58** *Beccles (South)** 15 by 10 Revision a 1925-26 c 1938-53 f 1962

1	A	CPr	OPr	1948po	-	-	6.47	3BE1X1	RB	00	12/47	3/48	
3	B	CPv	OPv	1956pc	-	ac	6.56	3oooX4		11	10/56	9/57	
3	B/												
3	B//	CPy,z	OPz	1956mc	1963ch	acf	6.56	3oooX4		11			

AA 1.7.47 on A(A), 3/-, OSPR 3/48. M on A.

TM 49 (62/49) *Beccles (North)** 10 by 10 Revision a 1903-26 c 1938-53

1	20046	CPr		1946po	-	-	1.46	3BE1X1	R		4/46	-	
1	2546		OPr	1946po	-	-	1.46	3BE1X1	R		-	4/46	1
3	B	CPu,z	OPv	1956pc	-	ac	6.56	3oooX2			5/56	8/57	

AA 1.3.46 on 2546(2546), 3/-, OSPR 9/46. M on A. SS: Soil, Land Use Capability, Soil Drainage, outline, 1970 on B. 1. The outline edition appears also in OSPR 9/46.

TM 59 (62/59) *Lowestoft** 10 by 10 Revision a 1926 c 1938 d 1938-54 e 1945

1	A	CPr	OPr	1947po	-	-	1.47	3BE1X1	R		5/47	7/47	
2	B	CPs		1947pc	1951	ace	1.52	3Booo2	R				M
3	C	CPv	OPv	1956pc	-	ad	6.56	3oooX2			10/56	11/56	
3	C/	CPw,z		1956mc	1960mc	ad	6.56	3oooX2					

AA 1.12.46 on A(A), 3/-, OSPR 7/47. M on B.

TQ 00 (51/00) *Littlehampton* 10 by 10 Revision a 1909-43 c 1937-46 d 1937-57 e 1946

1	A	CPr	OPr	1948po	-	-	6.48	4BE2X3	RB	0	4/48	5/48	1
1	B	CPr		nd:po	-	-	1.48	4BEoo3	RB	0			
1	C	CPs		1951pc	-	ace	1.51	4BEoo3	RB	0	8/51	-	
2	C		OPs	1951pc	-	ace	1.51	4BooX3	RB	0	-	2/53	
3	D	CPw.x,z	OPv	1959mc	-	ad	6.59	4oooX3		1	3/59	4/59	

AA 1.1.48/NT 12.3.48 on A(A), 3/-, OSPR 5/48. M on B. 1. The original railway alignment at Ford is evident.

TQ 01 (51/01) *Storrington* 10 by 10 Revision a 1909-39 c 1938-57

1	A	CPr	OPr	nd:po	-	-	6.48	4BEoo3	RB		8/48	9/48	
1	B	CPr		nd:po	-	-	6.48	4BEoo3	RB				
3	B	CPt		1948pc	1950c	-	6.48	4BEoo3	RB				
3	C	CPw,x,z.z,0z	OPv	1959mc	-	ac	6.59	4oooX3			4/59	9/59	

M on B.

TQ 02 (51/02) *Billingshurst* 10 by 10 Revision a 1909-10 c 1938-43 d 1938-57 e 1945-46

1	A	CPr	OPr	nd:po	-	-	6.48	4BEoo3	RB		9/48	9/48	
1	B	CPs	OPv	1951pc	-	ace	1.51	4BEoo3	RB		8/51	1/57	
3	C	CPw,z,0z	OPv	1959mc	-	ad	6.59	4oooX3			4/59	5/59	

M on A, B.

1	2	3	4	5	6	7	8	9	10	11	12	13	14

TQ 03 (51/03) *Cranleigh*   10 by 10   Revision a 1909-37 b 1907-37 c 1954-57 f 1963

1	A	CPr	OPr	1948po	-	-	6.48	4BEoo3	RB	0	5/48	5/48	
2	B	CPt		1948pc	1952c	-	6.52	4Booo3	RB	0			
3	B/												
3	B//	CPw		1948mc	1958mc	a	6.48	4BooX3	B	0			
3	B///	CPw.z		1948mc	1960mc	a	6.48	4BooX3	B	0			
3	C	CPz,0z	OPz	1965mc	-	bcf	6.65	4oooX3		3	4/65	4/65	

M on A, B.

TQ 04 (51/04) *Shere*   10 by 10 Revision a 1913-34 c 1938-55

1	A	CPr	OPr	1948po	-	-	6.48	4BE4X3	RB		4/48	6/48	
1	B	CPr		nd:po	-	-	6.48	4BEoo3	RB				
1	C	CPr		nd:po	-	-	6.48	4BEoo3	RB				
2	C	CPt		1948pc	1953c	-	6.48	4BooX3	RB				
3	D	CPv		1956pc	-	ac	6.56	4oooX3			10/56	-	
3	D/	CPw,z,0z		1956mc	1961ch	ac	6.56	4oooX3					

AA 1.2.48/NT 31.3.48 on A(A), 3/-, OSPR 6/48. M on C.

TQ 05 (51/05) *Woking*   10 by 10 Revision a 1913-35 c 1938-55

1	A	CPr	OPr	1948po	-	-	6.48	4BE2X3	RB	0	4/48	5/48	
1	B	CPr		1950pc	-	-	6.48	4BEoo3	RB	0			
2	C	CPt		1950pc	1953c	-	6.48	4BooX3	R	0			
3	D	CPv	OPv	1956pc	-	ac	6.56	4oooX3		1	1/57	2/57	
3	D/	CPw,z,0z		1956mc	1960mc	ac	6.56	4oooX3		1			

AA 1.2.48 on A(A), 3/-, OSPR 5/48. M on B.

TQ 06 (51/06) *Chertsey*   10 by 10 Revision a 1912-34 c 1937-40 d 1938-55
                                                    e 1947-48 f 1963 g 1969; j 1963 industrial

1	A	CPr	OPr	1948po	-	-	1.48	4BE2X3	R	00	4/48	6/48	1
1	B	CPr		nd:po	-	-	1.48	4BEoo3	R	00			
1	C	CPs		1951pc	-	ace	1.51	4BEoo3	R	00	6/51	-	
2	C		OPs	1951pc	-	ace	1.52	4Booo3	R	00			
2	D	CPt		1951pc	1953c	ace	1.52	4BooX3	R	00			
3	E	CPv	OPv	1957pc	-	ad	6.57	4oooX3		12	2/57	5/57	2
3	E/	CPw		1957pc	-	ad	6.57	4oooX3		12			
3	E//*	CPz	OPz	1957pc	1965ri	adfj	6.57	4oooX3		32	5/65	7/65	
3	E//*/	CPz	OPz	1957ms	1970ch	adgj	6.57	4oooX3		32			

AA 1.1.48 on A(A), 3/-, OSPR 6/48. M on B, C. 1. Brooklands race track is shown on this sheet: later the airfield layout was shown here as well. 2. Fair Oaks Aerodrome.

TQ 07 (51/07) *Staines*   10 by 10 Revision a 1912-35 b 1897-1935 c 1938 d 1938-55
                                                    e 1948 f 1965 g 1972
                                                    j 1965 Greater London k 1971 reservoir

1	A	CPr	OPr	1948po	-	-	6.48	4BEoo3	R	11	6/48	8/48	1
1	B	CPs		1951pc	-	ace	1.51	4BEoo3	R	31	6/51	-	2
2	B		OPs	1951pc	-	ace	1.51	4Booo3	R	31			
3	B	CPt		1951pc	-	ace	6.51	4Booo3	R	31			
3	C	CPv	OPv	1956pc	-	bd	6.56	4oooX3		31	1/57	2/57	
3	C/	CPw	OPz	1956mc	1960mc	bd	6.56	4oooX3		31			
3	C//*	CPz	OPz	1956mc	1966r	bdfj	6.56	4oooX3		31	5/66	3/66	
3	C//*/												
3	C/*//*	CPz	OPz	1956^{c}ms	nd:rv	bdgjk	6.56	4oooX3		31	11/73	3/74	3

AA 1.4.48/NT 1.6.48 on A(A), 3/-, OSPR 8/48; AA 1.3.52/NT 1.6.48 on B(-), 4/-, OSPR 6/52. SAAEGL 1.4.65 on C/(a), 8/6, OSPR 5/65. M on B. 1. The airport is the original Heathrow. 2. London Airport. 3. Heathrow Airport - London. Showing the alignment of General Roy's Base. Wraysbury Reservoir.

TQ 08 (51/08) *Uxbridge*   10 by 10 Revision a 1897-1935 b 1897-1938 c 1938 d 1938-49
                                                    e 1938-55 (psr); f 1947-48 g 1972

1	A	CPr	OPr	1947po	-	-	1.48	4BE2X3	RB	00	9/47	11/47	
1	B	CPr		1950pc	-	acf	1.50	4BEoo3	RB	30			1
1	C	CPr		1950pc	-	acf	1.51	4BEoo3	RB	30			M

284

1	2	3	4	5	6	7	8	9	10	11	12	13	14
1	D	CPs	OPs	1951pc	-	adf	6.51	4BEoo3	R	30	12/51	6/53	
3	E	CPv	OPv	1956pc	-	be	6.56	4oooX3		11	10/56	10/56	2
3	E/	CPw.z	OPz	1956mc	1960mc	be	6.56	4oooX3		11			3
3	E//*	CPz	OPz	1956^{c}ms	nd:r	beg	6.56	4oooX3		11	5/73	6/73	

AA 1.6.47/NT 4.10.47 on A(A), 3/-, OSPR 11/47. SAAEGL 1.4.65 on E/(a), 8/6, OSPR 3/65. M on C. 1. Northolt Airport. 2. Northolt Airport is stripped of its layout, and merely noted as Airfield. Uxbridge RAF Station (0683) is now noted, and its buildings are in black. 3. Northolt Airfield. Denham Aerodrome.

TQ 09 (51/09) *Rickmansworth*   10 by 10   Revision a 1912-35 b 1912-38 c 1934-49 d 1938-55 e 1947-48 f 1970 g 1976

1	A	CPr	OPr	1948po	-	-	1.48	4BE2X3	RB	1	12/47	3/48	
1	B	CPr		nd:po	-	-	1.48	4BEoo3	RB	1			
2	C	CPt		1952pc	-	ace	6.52	4Booo3	R	1	10/52	-	
3	D	CPv	OPv	1956pc	-	bd	6.56	4oooX3		1	11/56	11/56	
3	D/	CPw	OPz	1956mc	1960mc	bd	6.56	4oooX3		1			
3	D//	CPz,0z	OPz	1956^{c}ms	nd:ch	bdf	6.56	4oooX3		1			
3	D///*	CFz	OFz	1956^{c}ms	nd:r	bdg	6.56	4oooX3		1			1

AA 1.10.47 on A(A), 3/-, OSPR 3/48. SAAEGL 1.4.65 on D/(a), 8/6, OSPR 3/65. M on B. 1. OSPR not found.

TQ 10 (51/10) *Worthing*   10 by 10   Revision a 1909-44 c 1938-57

1	A	CPr	OPr	1948po	-	-	6.48	4BE2X3	RB	0	4/48	5/48	
2	A	CPt		1948pc	1952rpb	-	6.51	4BooX3	RB	0			
3	B	CPw,x,z.z	OPv	1958mc	-	ac	6.58	4oooX4		0	2/59	4/59	

AA 1.2.48/NT 17.3.48 on A(A), 3/-, OSPR 5/48. M on A.

TQ 11 (51/11) *Steyning*   10 by 10   Revision a 1909-39 c 1938-57 f 1972

1	A	CPr	OPr	1948po	-	-	6.48	4BEoo3	RB		5/48	6/48	
1	B	CPr		nd:po	-	-	6.48	4BEoo3	RB				
2	B	CPu		1948pc	1949c	-	6.48	4Booo3	RB				
3	C	CPw,x,z	OPv	1958mc	-	ac	6.58	4oooX3			1/59	2/59	
3	C/*	CPz,0z	OPz	1958^{c}ms	nd:r	acf	6.58	4oooX3			4/73	6/73	

M on B.

TQ 12 (51/12) *Southwater*   10 by 10   Revision a 1909-33 c 1938 d 1938-57 e 1945-46 f 1966

1	A	CPr	OPr	1948po	-	-	1.48	4BEoo3	RB		5/48	7/48	
1	B	CPs		1950pc	-	ace	6.50	4BEoo3	RB		12/50	-	
2	B		OPs	1950pc	-	ace	1.52	4Booo3	RB		-	5/52	
3	C	CPw,x	OPv	1958mc	-	ad	6.58	4oooX4			1/59	2/59	
3	C/*	CPz,0z	OPz	1958mc	1967r	adf	6.58	4oooX4			1/67	2/67	

AA 1.3.48 on A(A), 3/-, OSPR 7/48. M on B.

TQ 13 (51/13) *Horsham*   10 by 10   Revision a 1895-1932 c 1937-38 d 1937-56 e 1945 f 1963 g 1967 h 1975

1	A	CPr	OPr	1948po	-	-	6.48	4BEoo3	RB		6/48	7/48	
1	B	CPr		nd:po	-	-	6.48	4BEoo3	RB				
1	C	CPs	OPv	1948pc	1950c	ace	6.50	4BEoo3	RB		-	1/57	
3	C/	CPw,x		1948mc	1959c	ace	6.50	4BooX3	B				
3	D	CPz	OPz	1965mc	-	adf	6.65	4oooX3			1/65	1/65	
3	D/*	CPz	OPz	1965ms	1968r	adg	6.65	4oooX3			5/68	3/68	
3	D/*/*	CFz	OFz	1965^{c}ms	nd:r	adh	6.65	4oooX3			4/76	12/75	1

AA 1.4.48 on A(A), 3/-, OSPR 7/48. **M on** C. 1. The outline edition appears in both OSPR 12/75 and 4/76.

TQ 14 (51/14) *Dorking*   10 by 10   Revision a 1911-34 c 1938-55

1	A	CPr	OPr	**1948po**	-	-	6.48	4BEoo3	RB		7/48	8/48	
1	B	CPr		**1949pc**	1950	-	6.48	4BEoo3	RB				
3	B	CPt		**1949pc**	1950	-	6.48	4BEoo3	RB				
3	C	CPv	OPv	1956pc	-	ac	6.56	4oooX3			11/56	1/57	1
3	C/	CPw,x,z		1956mc	1959mc	ac	6.56	4oooX3					

AA 1.4.48/NT 27.9.47 on A(A), 3/-, OSPR 8/48. M on B. 1. An experimental printing is recorded, with the grey plate in green. The colour blocks are present, but otherwise it appears to be a normal C printing.

1	2	3	4	5	6	7	8	9	10	11	12	13	14

TQ 15 (51/15) *Leatherhead*   10 by 10   Revision a 1912-34 c 1938-55

1	A	CPr	OPr	1948po	-	-	1.48	4BE2X3	RB		12/47	3/48	
1	B	CPr		nd:po	-	-	1.48	4BEoo3	RB				
1	C	CPr		nd:po	-	-	1.48	4BEoo3	RB				
1	C/D	CPr		nd:po	1950	-	1.48	4BEoo3	RB				
3	E	CPt	OPv	1949pc	1953c	-	6.53	4BooX3	B				
3	F	CPv	OPv	1956pc	-	ac	6.56	4oooX3			2/57	2/57	
3	F/	CPw,x,z		1956mc	1959mc	ac	6.56	4oooX3					

AA 1.10.46/NT 8.1.48 on A(A), 3/-, OSPR 3/48. M on C.

TQ 16 (51/16) *Esher*   10 by 10   Revision a 1932 c 1938 d 1938-55 e 1947 f 1973
j 1963 reservoir

1	A	CPr		1947po	-	-	6.47	4BE2X3	RB		10/47	-	1
1	B		OPr	1947po	-	-	6.47	4BE2X3	RB		-	1/48	
3	C	CPt		1954pc	-	ace	6.54	4oooX4			6/54	-	2
3	D	CPv	OPv	1956pc	-	ad	6.56	4oooX4			1/57	2/57	
3	D/	CPx,y,z	OPz	1956mc	1961ch	ad	6.56	4oooX4					
3	D//*	CPz	OPz	1956ᶜms	nd:rv	adfj	6.56	4oooX4			11/73	6/74	3

AA 1.10.46/NT 5.11.47 on B(A), 3/-, OSPR 1/48. SAAEGL 1.4.65 on D/(a), no price, OSPR 5/65. M on B, D/.
1. Two Ordnance Survey Offices are marked. Hospital railways are marked (also on TQ 26). Cardinal Wolsey's waterworks for Hampton Court are marked (also on TQ 17, 26, 27). 2. The Ordnance Survey Office labels are deleted. 3. Queen Elizabeth II Reservoir.

TQ 17 (51/17) *Hounslow*   10 by 10   Revision a 1910-35 b 1910-40 c 1938 d 1938-50
e 1938-55 (psr); f 1947-48 g 1967

1	A	CPr	OPr	1947po	-	-	6.47	4BE2X3	R	110	9/47	11/47	1
1	B	CPs		1950pc	-	acf	6.50	4BEoo3	R	210	12/50	-	
3	C	CPt	OPs	1954pc	-	bdf	6.54	4oooX4		21	8/54	11/55	
3	D	CPv,w	OPv	1957pc	-	be	6.57	4oooX4		113	3/57	8/57	
3	D/*	CPx.z	OPx,z	1957mc	1961r	be	6.57	4oooX4		113	4/61	5/61	
3	D/*/*	CPz	OPz	1957ms	1968r	beg	6.57	4oooX4		113	4/68	5/68	

AA 1.3.47/NT 2.10.47 on A(A), 3/-, OSPR 11/47. SAAEGL 1.4.65 on D/*(a) (OPz), 8/6, OSPR 5/65. M on A, B. 1. Heston Airport, London Air Park. Also note Kneller Hall, Twickenham, Kempton Park (also on TQ 16), the RASC depot at Feltham, with its railway connection.

TQ 18 (51/18) *Harrow*   10 by 10   Revision a 1912-36 b 1912-35 c 1938-49 d 1938-55
e 1947

1	A	CPr	OPr	1947po	-	-	6.47	4BE2X3	R	1	8/47	11/47	
1	B	CPr		nd:po	-	-	6.47	4BEoo3	R	3			
1	C	CPs	OPs	1951pc	-	ace	6.51	4BEoo3	RC	3	11/51	6/55	1
3	D	CPv	OPv	1956pc	-	bd	6.56	4oooX3		1	11/56	11/56	
3	D/	CPw,x,z.z	OPz	1956mc	1960mc	bd	6.56	4oooX3		1			2

AA 1.3.47 on A(A), 3/-, OSPR 11/47. SAAEGL 1.4.65 on D/(a), 8/6, OSPR 5/65. M on B, C. 1. A section of the canal is in solid blue. 2. Northolt Airfield.

TQ 19 (51/19) *Watford*   10 by 10   Revision a 1912-35 c 1938-55 f 1960 g 1969
j 1968 building k 1968 reservoir l 1969 Greater London

1	A	CPr	OPr	1947po	-	-	6.47	4BE2X3	RB	0	7/47	11/47	1
1	B	CPr		nd:po	-	-	6.47	4BEoo3	RB	0			
3	B	CPt		1950pc	-	-	6.47	4BEoo3	RB	0			
3	C	CPv	OPv	1956pc	-	ac	6.56	4oooX3		3	11/56	11/56	2
3	C/	CPx	OPz	1956mc	1961ch	acf	6.56	4oooX3		3			
3	C//*	CPz	OPz	1956ᶜms	nd:brv	acgjkl	6.56	4oooX3		3	2/71	6/71	3

AA 1.3.47 on A(A), 3/-, OSPR 11/47. SAAEGL 1.4.65 on C/(a), 8/6, OSPR 5/65. M on B. 1. The unbuilt Northern Line continuation beyond Edgware is on the grey plate. 2. Elstree Aerodrome. 3. Hillfield Reservoir is built on the site of Elstree Aerodrome, part of which is still depicted.

1	2	3	4	5	6	7	8	9	10	11	12	13	14

TQ 20 (51/20) [Hove]　　　　　　　　10 by 10　Revision a 1909-40 c 1938-57
1　15046　CPr　　　　1946po　-　-　1.47　4BE2X3　RB　0　11/46　-
1　A　　　　OPr　1946po　-　-　1.47　4BE2X3　RB　0　-　2/47
2　A　CPt　OPs　1946pc　1952　-　1.51　4Booo3　RB　0
3　B　CPw,x　OPv　1958mc　-　ac　6.58　4oooX3　　2　11/58　1/59　1

AA 1.7.46 on A(A), 3/-, OSPR 2/47; AA 1.12.53 on A(B), 4/-, OSPR 11/54. M on A. 1. Brighton Hove & Worthing Municipal Airport.

TQ 21 (51/21) [Hurstpierpoint]　　　　10 by 10　Revision a 1909-38 c 1937-57
1　A　CPr　OPr　1947po　-　-　1.47　4BE2X3　RB　　1/47　5/47
1　B　CPr　　　nd:po　-　-　1.47　4BEoo3　RB
3　B　CPt　　　1949pc　-　-　1.51　4BEoo3　B
3　C　CPw,x　OPv　1958mc　-　ac　6.58　4oooX3　　　1/59　2/59

AA 1.9.46/NT 1.1.47 on A(A), 3/-, OSPR 5/47. M on B. One of the first sheets to be superseded by Second Series mapping in December 1965.

TQ 22 (51/22) *Cowfold*　　　　　　　10 by 10　Revision a 1909-38 c 1938-56 f 1963
1　A　CPr　OPr　1948po　-　-　1.48　4BEoo3　B　　5/48　7/48
1　B [1]　CPs　　　nd:po　-　-　　　4BEoo3　B　　　　　M
3　B [2]　CPw,x　OPv　1958mc　-　ac　6.58　4oooX3　　　11/58　12/58
3　B/　CPz　OPz　1958mc　1964ch　acf　6.58　4oooX3

AA 1.3.48 on A(A), 3/-, OSPR 7/48. M on B.

TQ 23 (51/23) *Crawley*　　　　　　　10 by 10　Revision a 1909-32 b 1909-37 c 1938-46 d 1938-56
　　　　　　　　　　　　　　　　　　　　　f 1961 g 1963 h 1976
1　A　CPr　OPr　1948po　-　-　1.48　4BEoo3　RB　0　8/48　7/48
1　B　CPr　　　nd:po　-　-　1.48　4BEoo3　RB　0
2　B　CPt　　　1948pc　1953c　-　1.51　4BooX3　RB　0
3　B/
3　B//　CPw　　　1948mc　1960mc　ac　1.51　4BooX3　B　0
3　B///*　CPx,z　OPx　1948mc　1961r　acf　1.51　4BooX3　B　3　10/61　12/61
3　C　CPz,0z　OPz　1965mc　-　bdg　6.65　4oooX3　　3　7/65　7/65　1
3　C/*　CFz　OFz　1965ᶜms　nd:r　bdh　6.65　4oooX3　　3　　　　　2

M on B. 1. This was the last fully revised edition until NS 27 in 1970. 2. OSPR not found.

TQ 24 (51/24) *Horley Surrey*　　　　10 by 10　Revision a 1910-34 c 1938-48 d 1938-55 e 1945-48
1　A　CPr　OPr　1948po　-　-　6.48　4BEoo3　RB　10　7/48　9/48　1
1　B　CPs　　　1950pc　-　ace　6.50　4BEoo3　RB　12
3　B　CPu　　　1950pc　-　ace　1.51　4BEoo3　RB　12
3　C　CPv　OPv　1957pc　-　ad　6.57　4oooX3　　32　2/57　2/57
3　C/*　CPw.y,z　OPv　1957mc　1960r　ad　6.57　4oooX3　　32　11/60　11/60　2

AA 1.4.48 on A(A), 3/-, OSPR 9/48. M on B. 1. Gatwick Airport. 2. Gatwick Airport is now much enlarged.

TQ 25 (51/25) *Burgh Heath*　　　　　10 by 10　Revision a 1932-34 c 1938 d 1938-55 e 1948 f 1966
1　A　CPr　OPr　1947po　-　-　6.47　4BE2X3　RB　　7/47　10/47
1　B　CPr　　　nd:po　-　-　6.47　4BEoo3　RB
1　C　CPs　　　1947pc　1950c　ace　6.50　4BEoo3　RB
2　C　CPt,u　OPs,v　1947pc　1952c　ace　6.50　4Booo3　R　　　8/54
3　D　CPv　OPv　1956pc　-　ad　6.56　4oooX3　　　10/56　11/56
3　D/　CPx　OPz　1956mc　1960mc　ad　6.56　4oooX3
3　D//　CPz,0z　OPz　1956mc　1967ch　adf　6.56　4oooX3

AA 1.10.46/NT 12.8.47 on A(A), 3/-, OSPR 10/47. SAAEGL 1.4.65 on D/(a), 8/6, OSPR 5/65. M on C.

TQ 26 (51/26) *Merton Surrey*　　　　10 by 10　Revision a 1932-40 c 1938-55
1　A　CPr　OPr　1947po　-　-　6.47　4BE2X3　RB　0　6/47　10/47
1　B　CPr　　　nd:po　-　-　6.47　4BEoo3　RB　0
2　C　CPt,u　　　1950pc　1953c　-　6.53　4BooX3　B　0
3　D　CPv,w　OPv　1956pc　-　ac　6.56　4oooX3　　3　2/57　2/57
3　D/　CPw,z.z　OPz　1956mc　1961ch　ac　6.56　4oooX3　　3

AA 3.7.47/NT 21.7.47 on A(A), 3/-, OSPR 10/47. SAAEGL 1.4.65 on D/(a), 8/6, OSPR 5/65. M on B, C.

1	2	3	4	5	6	7	8	9	10	11	12	13	14

TQ 27 (51/27) [Wandsworth] 10 by 10 Revision a 1912-34 c 1938-50 d 1938-55
                                          e 1947-48 f 1961 g 1965; j 1965 Greater London

1	A	CPr	OPr	1947po	-	-	6.47	4BE2X3	RB		8/47	12/47	1
1	B	CPr		nd:po	-	-	6.47	4BEoo3	RB				
2	C	CPs	OPs	1952pc	-	ace	1.52	4Booo3	R		3/52	-	
3	D	CPv,w	OPv	1956pc	-	ad	6.56	4oooX3			1/57	1/57	
3	D/*	CPx	OPx,z	1956mc	1961r	adf	6.56	4oooX3			7/61	7/61	
3	D/*/*	CPz	OPz	1956mc	1966r	adgj	6.56	4oooX3			6/66	6/66	

AA 1.3.47/NT 24.9.47 on A(A), 3/-, OSPR 12/47; AA 1.12.53/NT 24.9.47 on C(B), 4/-, OSPR 2/55. SAAEGL 1.4.65 on D/*(a) (OPz), 8/6, OSPR 5/65. M on B, C. 1. The West London Extension Railway is labelled LNER.

TQ 28 (51/28) [Paddington] 10 by 10 Revision a 1912-36 c 1938-55

1	A	CPr	OPr	1947po	-	-	6.47	4oE2X3	RC	0	7/47	11/47	1
1	B	CPr		nd:po	-	-	6.47	4oEoo3	RC	0			M
2	B	CPs		1948pc	1952	-	1.52	4oooo3	RC	0			
3	B	CPt		1948pc	1952	-	1.52	4oooX3	RC	0			
3	C	CPv,w,y	OPv,z	1957pc	-	ac	6.57	4oooX3		1	1/58	11/57	

AA 1.3.47/NT 21.8.47 on A(A), 3/-, OSPR 11/47. SAAEGL 1.4.65 on C(a) (OPz), 8/6, OSPR 4/65. M on B. 1. LPTB labels are on the West London line, and on Willesden to Hampstead line.

TQ 29 (51/29) *Barnet* 10 by 10 Revision a 1935-36 c 1938-55 f 1968
                                    j 1967 airfield k 1967 railway l 1968 Greater London

1	A	CPr	OPr	1947po	-	-	6.47	4BE2X3	R	1	6/47	11/47	
1	B	CPr		nd:po	-	-	6.47	4BEoo3	R	1			M
2	B	CPs	OP?	1949pc	1951	-	6.51	4Booo3	R	1		4/54	
3	C	CPv	OPv	1956pc	-	ac	6.56	4oooX3		1	10/56	11/56	
3	C/	CPw,x,z		1956mc	1958mc	ac	6.56	4oooX3		1			
3	C//*	CPz	OPz	1956ms	1969r	acfjkl	6.56	4oooX3		3	10/69	2/70	

AA 1.6.47 on A(A), 3/-, OSPR 11/47. SAAEGL 1.4.65 on C/:L1476(a) (TQ 29 & Part of TL 20, 10 by 12 km), 8/6, OSPR 4/65. M on B.

TQ 30 (51/30) [Brighton] 10 by 10 Revision a 1909-40 c 1937-57

1	20046	CPr		1946po	-	-	1.47	4BE2X3	RB		11/46	-	
1	A		OPr	1946po	-	-	1.47	4BE2X3	RB		-	2/47	
1	B	CPr		nd:po	-	-	1.47	4BEoo3	RB				
3	B	CPu	OPs	1949pc	-	-	1.47	4BEoo3	RB				
3	C	CPw,x	OPv	1959mc	-	ac	6.59	4oooX3			5/59	5/59	

AA 1.8.46 on A(A), 3/-, OSPR 2/47; AA 1.12.53 on B(B), 4/-, OSPR 1/55. M on B.

TQ 31 (51/31) [Burgess Hill] 10 by 10 Revision a 1908-38 c 1938-57

1	A	CPr	OPr	1947po	-	-	1.47	4BE2X3	RB		1/47	4/47	
1	B	CPr		nd:po	-	-	1.47	4BEoo3	RB				
2	B	CPt	OPs	1949pc	-	-	1.52	4Booo3	RB				
3	C	CPw.x	OPv	1959mc	-	ac	6.59	4oooX3			8/59	7/59	

AA 1.9.46 on A(A), 3/-, OSPR 4/47; AA 1.12.53 on B(B), 4/-, OSPR 2/55. M on B. One of the first sheets to be superseded by Second Series mapping in December 1965.

TQ 32 (51/32) *Haywards Heath* 10 by 10 Revision a 1909-37 c 1937-57

1	A	CPr	OPr	1948po	-	-	6.48	4BEoo3	RB		6/48	8/48	
1	B	CPr		nd:po	-	-	6.48	4BEoo3	RB				
3	B	CPu		1948pc	1949c	-	6.48	4BooX3	RB				
3	C	CPw,y,z	OPv	1959mc	-	ac	6.59	4oooX3			8/59	7/59	

AA 1.3.48 on A(A), 3/-, OSPR 8/48. M on B, C.

TQ 33 (51/33) *East Grinstead* 10 by 10 Revision a 1904 c 1938 d 1938-57 f 1963

1	A	CPr	OPr	1948po	-	-	1.48	4BEoo3	RB		5/48	5/48	
1	B	CPr		nd:po		-	?	4BEoo3	RB				M
2	B	CPt		1948pc	49c/52rpb	-	1.48	4BooX3	RB				
3	B		OPs	1948pc	49c/52rpb	-	1.48	4BooX3	RB		-	4/54	

288

1	2	3	4	5	6	7	8	9	10	11	12	13	14
3	B/	CPv		1948pc	1957c	ac	1.48	4BooX3	B				
3	B//	CP?w,x.y		1948mc	1960c	ac	1.48	4BooX3	B				
3	C	CPz	OPz	1965mc	-	adf	6.65	4oooX3				3/65	4/65

AA 1.6.53/NT 3.2.54 on B(A), 4/-, OSPR 4/54. M on B.

TQ 34 (51/34) *Lingfield*   10 by 10   Revision a 1907-34 c 1938-55

1	A	CPr	OPr	1948po	-	-	6.48	4BEoo3	RB	1	6/48	6/48	
1	B	CPr		nd:po	-	-	6.48	4BEoo3	RB	2			
2	B	CPt		1949pc	1952	-	1.52	4Booo3	RB	2			
3	C	CPv	OPv	1956pc	-	ac	6.56	4oooX3		2	1/57	2/57	1
3	C/	CPw,y,z		1956mc	1960mc	ac	6.56	4oooX3		2			

M on B. 1. Redhill Aerodrome.

TQ 35 (51/35) *Caterham*   10 by 10   Revision a 1907-34 b 1907-36 c 1934-44 d 1938-55
                                                      e 1946-48 f 1976

1	A	CPr	OPr	1947po	-	-	1.48	4BE2X3	RB	1	12/47	3/48	
1	B	CPr		nd:po	-	-	1.48	4BEoo3	RB	1			
1	C	CPs		1951pc	-	ace	1.51	4BEoo3	B	1	6/51	-	
3	C	CPu		1951pc	-	ace	1.51	4BEoo3	B	1			
3	D	CPv.x,z	OPv,z	1957pc	-	bd	6.57	4oooX3		3	1/58	11/57	1
3	D/*	CFz	OFz	1957^{c}ms	nd:r	bdf	6.57	4oooX3		3	5/77	5/77	

AA 1.10.47/NT 17.12.47 on A(A), 3/-, OSPR 3/48. SAAEGL 1.4.65 on D(a) (OPz), 8/6, OSPR 5/65. M on C. 1. Kenley Aerodrome.

TQ 36 (51/36) *Croydon*   10 by 10   Revision a 1908-39 c 1938-55 f 1964
                                                      j 1965 Greater London

1	A	CPr	OPr	1947po	-	-	6.47	4BE2X3	RB	1	8/47	11/47	1
1	B	CPr		nd:po	-	-	6.47	4BEoo3	RB	1			M
2	B	CPt	OPs	1948pc	1952c	-	6.47	4Booo3	RB	1			
3	C	CPv,x	OPv,z	1957pc	-	ac	6.57	4oooX4		3	9/57	9/57	
3	C/	CPz	OPz	1957mc	1965ch	acfj	6.57	4oooX4		3			

AA 3.7.47/NT 3.7.47 on A(A), 3/-, OSPR 11/47; AA 1.12.53/NT 3.7.47 on B(B), 4/-, OSPR 11/54. SAAEGL 1.4.54 on C(a) (OPz), 8/6, OSPR 5/65. M on B, C. 1. Croydon Airport.

TQ 37 (51/37) *Lewisham*   10 by 10   Revision a 1913-31 c 1938-50 d 1938-55
                                                      e 1946-47 f 1961

1	A	CPr		1947po	-	-	6.47	4oE2X3	R		8/47	-	
1	B		OPr	1947po	-	-	6.47	4oE2X3	R		-	11/47	
1	C	CPr,s		nd:po	-	-	6.48	4oEoo3	R				
2	D	CPs		1952pc	-	ace	1.52	4oooo3	R		3/52	-	
2	D		OPs	1952pc	-	ace	1.52	4oooX3	R		-	12/53	
3	D/	CPw		1952pc	1958c	ace	1.52	4oooX3					
3	E	CPx	OPx,z	1961mc	-	adf	6.61	4oooX3			5/61	7/61	
3	E/	CPz	OPz	1961mc	-	adf	6.61	4oooX3					

AA 1.3.47 on B(A), 3/-, OSPR 11/47; AA 1.12.53 on D(B), 4/-, OSPR 1/55. SAAEGL 1.4.65 on E(a) (OPz), 8/6, OSPR 5/65. M on C, D.

TQ 38 (51/38) *Hackney*   10 by 10   Revision a 1910-35 b 1919-35 c 1938-46 d 1938-55
                                                       f 1966; j 1966 Greater London

1	A	CPr	OPr	1947po	-	-	1.48	4oE2X3	RC		7/47	11/47	
1	B	CPr		nd:po	-	-	1.48	4oEoo3	RC				M
2	B	CPs	OPs	1948pc	1952	-	1.52	4oooo3	RC				
3	B/	CPw		1948pc	1958mc	ac	1.52	4oooX3	C				
3	C	CPw,y	OPv	1960mc	-	bd	6.60	4oooX3			5/60	4/60	
3	C/*	CPz	OPz	1960mc	1967r	bdfj	6.60	4oooX3			3/67	3/67	

AA 1.6.47/NT 1.9.47 on A(A), 3/-, OSPR 11/47; AA 1.12.53/NT 1.9.47 on B(B), 4/-, OSPR 11/54. SAAEGL 1.4.65 on C(a) (OPz), 8/6, OSPR 5/65. M on B.

1	2	3	4	5	6	7	8	9	10	11	12	13	14

TQ 39 (51/39) *Edmonton*   10 by 10   Revision a 1910-40 c 1938-55 f 1971
1	A	CPr	OPr	1947po	-	-	1.48	4BE4X3	RB		11/47	12/47	
1	B	CPr	OPs	nd:po	-	-	1.48	4BEoo3	RB				
2	B	CPs		1947pc	1951c	-	6.51	4Booo3	RB				
3	C	CPw,x.z	OPv,z	1960mc	-	ac	6.60	4oooX4			1/60	2/60	
3	C/*	CPz	OPz	1960ᶜms	nd:r	acf	6.60	4oooX4			3/72	12/72	

AA 1.7.47 on A(A), 3/-, OSPR 12/47. SAAEGL 1.4.65 on C:L1475(a) (OPz) (TQ 39 & Part of TL 30, 10 by 12 km), 8/6 OSPR 3/65. M on B.

TQ 40 (51/40) **& Part of TV 49** *Newhaven*   10 by 12   Revision a 1908-40 c 1937 d 1937-57 f 1968
1	A	CPr	OPr	1948po	-	-	6.48	4BE2X3	RB		4/48	6/48	
1	B	CPr		nd:po	-	-	6.48	4BEoo3	RB				
1	B	CPs		1948pc	1951	-	6.51	4BEoo3	RB				
3	B/		OPv	1948pc	1956mc	ac	6.51	4BooX3	RB				
3	C	CPw,y,z	OPv	1959mc	-	ad	6.59	4oooX3			8/59	12/59	
3	C/	CPz,z+,0z	OPz	1959ms	-	adf	6.59	4oooX3					

AA 1.1.48 on A(A), 3/-, OSPR 6/48. M on B.

TQ 41 (51/41) *Lewes*   10 by 10   Revision a 1908-30 c 1937-57
1	A	CPr	OPr	1948po	-	-	1.48	4BE4X3	RB		4/48	5/48	
1	B	CPr		nd:po		-	1.48	4BEoo3	RB				M
3	B	CPt		1949pc	-	-	1.48	4BEoo3	RB				
3	C	CPw,x,z	OPv	1959mc	-	ac	6.59	4oooX3			4/59	5/59	

AA 1.2.48 on A(A), 3/-, OSPR 5/48. M on B.

TQ 42 (51/42) *Uckfield*   10 by 10   Revision a 1908-30 c 1957 f 1965
1	A	CPr	OPr	1948po	-	-	6.48	4BEoo3	RB		7/48	8/48	
1	B	CPr		nd:po	-	-	6.50	4BEoo3	RB				M
3	C	CPt	OPs	1948pc	1953c	-	6.53	4BooX3	B				
3	D	CPw,x	OPv	1959mc	-	ac	6.59	4oooX3			2/59	3/59	
3	D/	CPz	OPz	1959mc	1966ch	acf	6.59	4oooX3					

M on B.

TQ 43 (51/43) *Forest Row*   10 by 10   Revision a 1907-38 c 1946 d 1946-57 f 1963
1	A	CPr	OPr	nd:po	-	-	6.48	4BEoo3	RB		10/48	10/48	
1	B	CPr		nd:po	-	ac	6.50	4BEoo3	RB				M
1	B	CPs		1948pc	1951c	ac	6.51	4BEoo3	RB				
3	B		OPs	1948pc	1951c	ac	6.51	4BooX3	RB		-	4/54	
3	B/												
3	B//	CP?w,x		1948mc	1959c	ac	6.51	4BooX3	B				
3	C	CPz	OPz	1965mc	-	adf	6.65	4oooX3			1/65	2/65	

AA 1.6.53/NT 11.2.54 on B(A), 4/-, OSPR 4/54. M on B.

TQ 44 (51/44) *Edenbridge*   10 by 10   Revision a 1907-37 c 1937-55 f 1973; j 1971 reservoir
1	A	CPr	OPr	nd:po	-	-	6.48	4BEoo3	RB		10/48	10/48	
1	B	CPs		nd:po	-	-		4BEoo3	RB				M
1	B	CPs		1948pc	1949c	-	6.48	4BEoo3	RB				
3	C	CPv	OPv	1957pc	-	ac	6.57	4oooX3			5/57	8/57	
3	C/	CPw,x,z		1957mc	1960mc	ac	6.57	4oooX3					
3	C//*	CPz,0z	OPz	1957ᶜms	nd:rv	acfj	6.57	4oooX3			12/73	11/74	1

M on B. 1. Bough Beech Reservoir.

TQ 45 (51/45) *Westerham*   10 by 10   Revision a 1907-39 c 1938-55
1	A	CPr	OPr	1947po	-	-	1.48	4BE2X3	RB	0	11/47	2/48	
1	B	CPr		nd:po	-	-	1.48	4BEoo3	RB	0			
3	B	CPt		1949pc	-	-	1.48	4BEoo3	RB	0			
3	C	CPv	OPv	1957pc	-	ac	6.57	4oooX3		1	1/58	11/57	
3	C/	CPw,y,z	OPv,z	1957mc	1960mc	ac	6.57	4oooX3		1			

AA 1.9.47/NT 15.12.47 on A(A), 3/-, OSPR 2/48. SAAEGL 1.4.65 on C/(a) (OPz), 8/6, OSPR 5/65. M on B.

290

1	2	3	4	5	6	7	8	9	10	11	12	13	14

TQ 46 (51/46) *Orpington*	10 by 10	Revision a 1906-40 b 1932-39 c 1938 d 1938-55
1	A		CPr	OPr	1947po	-	-	6.47	4BE2X3	RB	0	6/47	11/47	
1	B		CPr		nd:po	-	-	6.47	4BEoo3	RB	0			M
2	B		CPt		1949pc	1952	-	1.52	4Booo3	RB	0			
3	B/		CPv		1949pc	1956mc	ac	1.52	4BooX3	B	0			
3	C		CPv.x,z	OPv,z	1957pc	-	bd	6.57	4oooX3		0	9/57	9/57	

AA 1.2.47/NT 21.7.47 on A(A), 3/-, OSPR 11/47. SAAEGL 1.4.65 on C:L1478(a) (OPz) (TQ 46 & Part of TQ 56, 12 by 10 km), 8/6, OSPR 5/65. M on B.

TQ 47 (51/47) [*Woolwich*]	10 by 10	Revision a 1913-31 c 1938 d 1938-55 f 1966
j 1966 Greater London
1	A		CPr	OPr	1947po	-	-	6.47	4BE2X3	R		6/47	9/47	1
1	B		CPr		nd:po	-	-	1.48	4BEoo3	R				
2	B		CPt,u	OPs	1948pc	1952	-	1.52	4Booo3	R				
3	B/		CPw		1948mc	1958c	ac	1.52	4BooX3	R				
3	C		CPx	OPx,z	1961mc	-	ad	6.61	4oooX3			7/61	2/61	
3	C/		CPz	OPz	1961mc	1966ch	adfj	6.61	4oooX3					

AA 1.3.47 on A(A), 3/-, OSPR 9/47; AA 1.12.53 on B(B), 4/-, OSPR 8/54. SAAEGL 1.4.65 on C(a) (OPz), 8/6, OSPR 5/65. M on B. The Woolwich Arsenal area is blank. Showing the Royal Military Academy.

TQ 48 (51/48) [*Barking*]	10 by 10	Revision a 1907-20 c 1938-49 d 1938-55 e 1946
1	A		CPr		1947po	-	-	1.48	4BE4X3	RC	0	12/47	-	
1	A			OPr	1948po	-	-	1.48	4BE4X3	RC	0	-	4/48	
1	B		CPs		1951pc	-	ace	6.51	4BEoo3	RC	0	12/51	-	
2	B		CPt	OPs	1951pc	1953rpb	ace	6.51	4BooX3	RC	0	-	8/54	
3	C		CPw,y	OPv,z	1959mc	-	ad	6.59	4oooX3		3	6/59	7/59	

AA 1.10.47/NT 15.11.47 on A(A), 3/-, OSPR 4/48; AA 1.12.53 on B(B), 4/-, OSPR 8/54. SAAEGL 1.4.65 on C(a) (OPz), 8/6, OSPR 4/65. M on A. The Woolwich Arsenal area is blank.

TQ 49 (51/49) *Loughton**	10 by 10	Revision a 1914-38 c 1954-55
1	A		CPr	OPr	1947po	-	-	6.47	4BE2X3	RB	00	10/47	11/47	
1	B		CPr		nd:po	-	-	6.47	4BEoo3	RB	00			
2	B		CPt	OPs	1950pc	-	-	6.47	4Booo3	RB	00			
3	C		CPw,x,z,0z	OPv,z	1959mc	-	ac	6.59	4oooX3		32	5/59	12/59	1

AA 1.10.47 on A(A), 3/-, OSPR 11/47. SAAEGL 1.4.65 on C(a) (OPz), 8/6, OSPR 3/65. M on B. 1. There is another entry for the coloured map in OSPR 2/60.

TQ 50 (51/50) *Polegate*	10 by 10	Revision a 1908-39 c 1937-57 f 1972
1	A		CPr	OPr	1947po	-	-	6.47	4BE2X3	RB		11/47	2/48	
1	B		CPr		nd:po	-	-	6.47	4BEoo3	RB				
3	B		CPt		1949pc	-	-	6.47	4BEoo3	RB				
3	C		CPw	OPv	1959mc	-	ac	6.59	4oooX4			5/59	5/59	
3	C/		CPx,z		1959mc	1961ch	ac	6.59	4oooX4					
3	C//		CPz,0z	OPz	1959^{c}ms	nd:ch	acf	6.59	4oooX4					1

AA 1.1.47/NT 15.12.47 on A(A), 3/-, OSPR 2/48. M on B. 1. Arlington Reservoir.

TQ 51 (51/51) *East Hoathly*	10 by 10	Revision a 1908-29 c 1937 d 1937-57 e 1946
1	A		CPr	OPr	1948po	-	-	1.48	4BE2X3	RB		4/48	5/48	1
1	B		CPs		1951pc	-	ace	1.51	4BEoo3	RB		10/51	-	
2	B			OPs	1951pc	-	ace	1.52	4BooX3	RB		-	2/53	
3	C		CPw,x,z,0z	OPv	1959mc	-	ad	6.59	4oooX4			8/59	9/59	

M on B. 1. The East Sussex County Mental Hospital at Hellingly has a private railway.

TQ 52 (51/52) *Mayfield*	10 by 10	Revision a 1908-29 b 1908-30 c 1946 d 1957
1	A		CPr	OPr	1948po	-	-	6.48	4BE2X3	RB		4/48	5/48	
1	B		CPs		1951pc	-	ac	1.51	4BEoo3	RB		5/51	-	
3	C		CPw,y,z.z,0z	OPv	1959mc	-	bd	6.59	4oooX4			9/59	7/59	

M on A.

1	2	3	4	5	6	7	8	9	10	11	12	13	14

TQ 53 (51/53) *Royal Tunbridge Wells*   10 by 10   Revision a 1907-30 c 1954-57 e 1946 f 1963

1	A	CPr		1948po	-	-	6.48	4BEoo3	RB		5/48	-	
1	B		OPr	1948po	-	-	6.48	4BEoo3	RB		-	6/48	
1	C	CPr		nd:po	-	-	6.48	4BEoo3	RB				
1	D	CPs		1951pc	-	ae	1.51	4BEoo3	RB		6/51	-	
	D/												
3	D//	CPw		1951pc	1958mc	ae	1.51	4BooX3	B				
3	D///	CPx		1951mc	1960mc	ae	1.51	4BooX3	B				
3	E	CPz	OPz	1965mc	-	acf	6.65	4oooX3			2/65	2/65	

AA 1.3.48 on B(A), 3/-, OSPR 6/48. M on C.

TQ 54 (51/54) *Tonbridge*   10 by 10   Revision a 1907-37 b 1907-38 c 1938-46 d 1936-55
                                                      e 1946 f 1975

1	A	CPr	OPr	nd:po	-	-	6.48	4BEoo3	RB		9/48	10/48	
1	B	CPr		1948pc	1950c	ace	6.48	4BEoo3	RB				
3	B	CPu		1948pc	1950c	ace	6.48	4BEoo3	RB				
3	C	CPv	OPv	1957pc	-	bd	6.57	4oooX3			6/57	8/57	
3	C/	CPw.z		1957mc	1961ch	bd	6.57	4oooX3					
3	C//*	CFz	OFz	1957^{c}ms	nd:r	bdf	6.57	4oooX3			4/76	4/76	

AA 1.6.48/NT 16.8.48 on A(A), 3/-, OSPR 10/48. M on B.

TQ 55 (51/55) *Sevenoaks*   10 by 10   Revision a 1907 b 1907-37 c 1936-40 d 1938-55
                                                      e 1946 f 1965

1	A	CPr		1948po	-	-	6.48	4BEoo3	RB		7/48	-	
1	B		OPr	nd:po	-	-	6.48	4BEoo3	RB		-	10/48	
1	C	CPr		nd:po	-	-	6.48	4BEoo3	RB				
2	D	CPt	OPs	1952pc	-	ace	6.52	4Booo3	R		8/52	9/52	
2	D	CPu		1952pc	-	ace	6.52	4BooX3	R				
3	E	CPv,w	OPv	1957pc	-	bd	6.57	4oooX3			9/57	9/57	
3	E/*	CPz	OPz	1957mc	1965r	bdf	6.57	4oooX3			12/65	10/65	

AA 1.9.48/NT 15.7.48 on B(A), 3/-, OSPR 10/48. M on D.

TQ 56 (51/56) *Swanley**   10 by 10   Revision a 1907-36 b 1930-39 c 1936-39 d 1936-55
                                                      e 1946 f 1967

1	A	CPr	OPr	1947po	-	-	6.47	4BE2X3	R		11/47	2/48	
1	B	CPr		1947po	-	-	6.47	4BE2X3	R				1
1	C	CPr		nd:po	-	-	6.47	4BEoo3	R				
1	D												
1	E	CPs		1950pc	-	ace	6.50	4BEoo3	R		12/50	-	
2	E	CPu	OPs	1950pc	-	ace	1.51	4BooX3	R		-	2/53	
3	F	CPv	OPv	1957pc	-	bd	6.57	4oooX3			5/57	6/57	
3	F/	CPw.y		1957mc	1960mc	bd	6.57	4oooX3					
3	F//*	CPz	OPz	1957ms	1968r	bdf	6.57	4oooX3			6/68	2/68	

AA 1.9.47 on A(A), 3/-, OSPR 2/48. M on E. 1. One of only three coloured edition B sheets with legends.

TQ 57 (51/57) [*Dartford*]   10 by 10   Revision a 1907-31 c 1938-46 d 1938-55
                                                      e 1946 f 1963

1	A	CPr	OPr	1948po	-	-	6.48	4BE4X3	R		4/48	6/48	
1	B	CPs		1950pc	-	ace	6.50	4BEoo3	R				M
2	C	CPt		1950pc	1953c	ace	1.50	4BooX3	R				
3	C/	CPw		1950mc	1958c	ace	6.50	4BooX3					
3	D	CPw	OPv	1960mc	-	ad	6.60	4oooX3			5/60	6/60	
3	D/*	CPz	OPz	1960mc	1964r	adf	6.60	4oooX3			11/64	12/64	

AA 1.2.48/NT 17.3.48 on A(A), 3/-, OSPR 6/48. SAAEGL 1.4.65 on D/*(a), 8/6, OSPR 5/65. M on B.

TQ 58 (51/58) [*Hornchurch*]   10 by 10   Revision a 1914-15 b 1914-31 c 1938 d 1938-55

| 1 | A | CPr | OPr | nd:po | - | - | 6.48 | 4BEoo3 | RB | 1 | 8/48 | 10/48 | |
| 2 | A | CPs,u | OPs | 1948pc | 1951 | - | 1.52 | 4BEoo3 | RB | 1 | | | |

1	2	3	4	5	6	7	8	9	10	11	12	13	14
3	A/	CPv		1948pc	1958mc	ac	1.52	4BooX3	B	1			
3	B	CPw,x,z	OPv	1959mc	-	bd	6.59	4oooX4		2	10/59	10/59	

AA 1.6.48 on A(A), 3/-, OSPR 10/48. SAAEGL 1.4.65 on B:L1477(a) (OPz) (TQ 58 & Part of TQ 68, 12 by 10 km), 8/6, OSPR 4/65. M on A.

TQ 59 (51/59) *Harold Hill**   10 by 10   Revision a 1914-15 c 1938-55 f 1967

1	A	CPr	OPr	1948po	-	-	6.48	4BEoo3	RB		5/48	7/48	
1	B	CPr		nd:po	-	-	6.48	4BEoo3	RB				
2	B		OPs	1949pc	-	-	1.52	4Booo3	RB				
2	B	CPt	OPs	1949pc	-	-	1.52	4BooX3	RB				
3	C	CPw.x	OPv,z	1959mc	-	ac	6.59	4oooX4			9/59	8/59	
3	C/*	CPz	OPz	1959mc	1968r	acf	6.59	4oooX4			2/68	1/68	

AA 1.3.48 on A(A), 3/-, OSPR 7/48; AA 1.1.53 on B(B), 4/-, OSPR 1/54. SAAEGL 1.4.65 on C(a) (OPz), 8/6, OSPR 3/65. M on B. SS: Soil, Land Use Capability, Soil Drainage, outline, 1971 on C/*.

TQ 60 (51/60) *Pevensey*   10 by 10   Revision a 1908-40 c 1937-50 d 1937-57

1	A	CPr		1947po	-	-	6.47	4BE2X3	RB		10/47	-	1
1	B		OPr	1947po	-	-	6.47	4BE2X3	RB		-	1/48	
3	B	CPu	OPv	1955pc	-	ac	6.55	4oooX3	B		6/55	9/57	
3	C	CPw,y,z	OPv	1959mc	-	ad	6.59	4oooX4			6/59	8/59	

AA 1.1.47 on B(A), 3/-, OSPR 1/48. M on A. 1. There are Martello Towers at 6401, 6402, 6503, 6805.

TQ 61 (51/61) *Herstmonceux*   10 by 10   Revision a 1908-29 c 1937-38 d 1937-57 e 1946

1	A	CPr	OPr	1948po	-	-	1.48	4BE2X3	B		4/48	4/48	
1	B	CPs		nd:po	-	-	6.50	4BEoo3	B				M
2	C	CPs	OPs	1951pc	-	ace	6.51	4Booo3	B		11/51	11/51	
3	D	CPw,x,z,0z	OPv	1959mc	-	ad	6.59	4oooX3			8/59	9/59	1

M on B. 1. Showing the Royal Greenwich Observatory at Herstmonceux.

TQ 62 (51/62) *Burwash*   10 by 10   Revision a 1908-39 c 1947-57

| 1 | A | CPr | OPr | 1948po | - | - | 1.48 | 4BEoo3 | RB | | 6/48 | 6/48 | |
| 3 | B | CPw,x,z,0z | OPv | 1959mc | - | ac | 6.59 | 4oooX3 | | | 4/59 | 4/59 | |

M on A.

TQ 63 (51/63) *Wadhurst*   10 by 10   Revision a 1908-39 c 1954-57 f 1963

1	A	CPr	OPr	1948po	-	-	6.48	4BEoo3	RB		5/48	6/48	
1	B	CPr		nd:po	-	-	6.48	4BEoo3	RB				
1	B	CPs		1948pc	1951	-	1.51	4BEoo3	RB				
3	C	CPz	OPz	1965mc	-	acf	6.65	4oooX3			2/65	3/65	

AA 1.3.48 on A(A), 3/-, OSPR 6/48. M on B.

TQ 64 (51/64) *Paddock Wood*   10 by 10   Revision a 1907-39 c 1936-55

1	A	CPr	OPr	1948po	-	-	6.48	4BE4X3	R		4/48	6/48	
1	B	CPr		nd:po	-	-	6.48	4BEoo3	R				
3	C	CPv,x,z	OPv	1957pc	-	ac	6.57	4oooX3			3/57	5/57	

AA 1.2.48 on A(A), 3/-, OSPR 6/48. M on B.

TQ 65 (51/65) *Wrotham Heath*   10 by 10   Revision a 1907-38 c 1936-55

1	A	CPr	OPr	1948po	-	-	6.48	4BE2X3	RB	0	4/48	5/48	
1	B	CPr		nd:po	-	-	6.48	4BEoo3	RB	0			
1	B	CPr		1949pc	1950	-	6.48	4BEoo3	RB	0			
3	B	CPu		1949pc	1950	-	6.48	4BEoo3	RB	0			
3	C	CPv.x,z	OPv	1957pc	-	ac	6.57	4oooX3		0	1/58	11/57	

AA 1.2.48 on A(A), 3/-, OSPR 5/48. M on B.

TQ 66 (51/66) *Meopham**   10 by 10   Revision a 1907-37 b 1932-39 c 1938-39 d 1936-55 e 1946

1	A	CPr	OPr	1948po	-	-	1.48	4BE2X3	RB		12/47	2/48	
1	B	CPs		1947pc	1950c	ace	6.50	4BEoo3	RB				
3	C	CPv,w,y,z	OPv	1957pc	-	bd	6.57	4oooX3			5/57	6/57	

AA 1.10.47/NT 2.1.48 on A(A), 3/-, OSPR 2/48. M on A.

1	2	3	4	5	6	7	8	9	10	11	12	13	14

TQ 67 (51/67) *Gravesend* 10 by 10 Revision a 1915-40 c 1938-55 f 1970

1	A	CPr	OPr	1948po	-	-	6.48	4BE2X3	RB	1	2/48	6/48	
1	B	CPr		nd:po	-	-	6.48	4BEoo3	RB	1			
2	B	CPt	OPv	1948pc	48c/52rpb	-	6.48	4BooX3	RB	1			
3	C	CPv.x	OP?	1958pc	-	ac	6.57	4ooX3		2	3/58	3/58	
3	C/*	CPz	OPz	1958^{c}ms	nd:bipr	acf	6.57	4ooX3		2	2/72	10/72	

AA 1.1.48 on A(A), 3/-, OSPR 6/48. M on B.

TQ 68 (51/68) *Laindon* 10 by 10 Revision a 1915-20 c 1936-40 d 1938-55
  e no date f 1972

1	A	CPr		1948po	-	-	1.48	4BE2X3	R		2/48	-	
1	B [1]		OPr	1948po	-	-	1.48	4BE2X3	R		-	4/48	
1	B [2]	CPs	OP?	1951pc	-	ace	1.51	4BEoo3	R		6/51	8/51	
3	C	CPw,x,z	OPv	1958mc	-	ad	6.58	4ooX4			2/59		1
3	C/*	CPz	OPz	1958^{c}ms	nd:r	adf	6.58	4ooX4			11/73	1/74	

AA 1.11.47 on B(A), 3/-, OSPR 4/48. M on B [2]. 1. OSPR (outline edition) not found.

TQ 69 (51/69) *Billericay* 10 by 10 Revision a 1915-20 c 1938-55 f 1967

1	A	CPr	OPr	1948po	-	-	6.48	4BE2X3	RB		4/48	6/48	
1	B	CPr		nd:po	-	-	6.48	4BEoo3	RB				
3	B	CPu		1949pc	-	-	6.48	4BEoo3	RB				
3	C	CPw,y	OPv	1959mc	-	ac	6.59	4ooX3			9/59	7/59	
3	C/*	CPz	OPz	1959mc	1968r	acf	6.59	4ooX3			3/68	3/68	

AA 1.2.48 on A(A), 3/-, OSPR 6/48. M on B.

TQ 70 (51/70) *Bexhill* 10 by 10 Revision a 1925-40 c 1937-50 d 1937-57
  e 1946 f 1960

1	A	CPr	OPr	1947po	-	-	1.47	4BE2X3	R	0	8/47	11/47	
3	B	CPu	OPv	1955pc	-	ace	6.55	4ooX3		0	11/55	1/57	
3	B/												
3	B//	CP?w,x.x		1955mc	1960mc	ace	6.55	4ooX3		0			
3	C	CPz	OPz	1965mc	-	adf	6.65	4ooX3		1	4/65	6/65	

AA 1.1.47 on A(A), 3/-, OSPR 11/47. M on A.

TQ 71 (51/71) *Battle Sussex* 10 by 10 Revision a 1927-28 c 1938 d 1937-57 e 1946 f 1964

1	A	CPr	OPr	1947po	-	-	6.47	4BE2X3	RB		11/47	2/48	
1	B	CPr		nd:po	-	-	6.47	4BEoo3	RB				
2	B		OPs	1949pc	-	ace	6.47	4BooX3	RB		-	7/53	
3	B/	CPw		1949mc	1958mc	ace	6.47	4BooX3	B				
3	B//	CPw,z		1949mc	1960mc	ace	6.47	4ooX3	B				
3	C	CPz,0z	OPz	1965mc	-	adf	6.65	4ooX3			4/65	4/65	

AA 1.9.47/NT 24.12.47 on A(A), 3/-, OSPR 2/48. M on B.

TQ 72 (51/72) *Robertsbridge* 10 by 10 Revision a 1907-29 c 1947 d 1937-57 e 1946 f 1963

1	A	CPr	OPr	1948po	-	-	6.48	4BEoo3	RB		5/48	5/48	
1	B	CPs		1951pc	-	ace	1.51	4BEoo3	RB		6/51	-	
2	B		OPs	1951pc	-	ace	1.51	4BooX3	RB		-	11/53	
3	B/	CPw,x		1951mc	1960c	ace	1.51	4BooX3	B				
3	C	CPz	OPz	1965mc	-	adf	6.65	4ooX3			2/65	3/65	

M on B.

TQ 73 (51/73) *Hawkhurst* 10 by 10 Revision a 1906-37 c 1938 d 1938-57 f 1963

1	A	CPr	OPr	1948po	-	-	1.48	4BEoo3	RB		5/48	5/48	
1	B	CPr		nd:po	-	-	1.48	4BEoo3	RB				
2	B	CPt		1949pc	1952rpb	-	1.48	4BooX3	RB				
3	B/	CPw,x		1949mc	1960mc	ac	1.48	4BooX3	B				
3	C	CPz	OPz	1965mc	-	adf	6.65	4ooX3			3/65	2/65	

M on B.

1	2	3	4	5	6	7	8	9	10	11	12	13	14

TQ 74 (51/74) *Marden Kent*   10 by 10   Revision a 1906-38 c 1955 f 1968

1	A	CPr	OPr	1948po	-	-	6.48	4BEoo3	R		5/48	5/48	
3	A	CPu		1948pc	-	-	6.48	4BEoo3	R				
3	A/	CPw		1948mc	1959mc	a	6.48	4BooX3					
3	B	CPx,y	OPx	1961mc	-	ac	6.61	4oooX3			4/61	5/61	
3	B/	CPz	OPz	1961ms	1969ch	acf	6.61	4oooX3					

M on A.

TQ 75 (51/75) *Maidstone*   10 by 10   Revision a 1907-39 c 1936-39 d 1936-55 f 1962

1	A	CPr	OPr	nd:po	-	-	1.49	4BEoo3	RB		12/48	2/49	
2	A	CPt	OPs	1948pc	1952rpb	-	1.49	4BooX3	RB				
3	A/												
3	A//	CPw		1948pc	1958mc	ac	1.49	4BooX3	B				
3	B	CPw	OPv	1960mc	-	ad	6.60	4oooX3			10/60	9/60	
3	B/*	CPy,z	OPy	1960mc	1963r	adf	6.60	4oooX3			4/63	4/63	

AA 1.10.48/NT 11.12.48 on A(A), 3/-, OSPR 2/49. M on A.

TQ 76 (51/76) *Chatham**   10 by 10   Revision a 1906-33 b 1906-39 c 1938-48 d 1938-55
                                                                                                          e 1946 f 1961 g 1966

1	A	CPr	OPr	1947po	-	-	6.47	4BE2X3	RB	1	8/47	12/47	1
1	B												
1	C	CPs	OP?	1951pc	-	ace	1.51	4BEoo3	RB	1	6/51	8/51	
	C/												
3	C//	CPw		1951pc	1958c	ace	1.51	4BooX3	B	1			
3	D	CPw	OPv	1960mc	-	bd	6.60	4oooX4		1	5/60	6/60	2
3	D/	CPx		1960mc	1962ch	bdf	6.60	4oooX4		1			3
3	D//*	CPz	OPz	1960mc	1967r	bdg	6.60	4oooX4		1	11/67	10/67	

AA 1.4.47 on A(A), 3/-, OSPR 12/47. M on A, C, D/. 1. Showing Fort Borstal, Fort Bridgewoods, Fort Horsted, Fort Luton, an unnamed fort (7866), School of Military Engineering, and barracks. The Chatham Dockyard area is blank. 2. A greater area of Chatham Dockyard is blank. 3. Showing the M2 under construction.

TQ 77 (51/77) *Cooling*   10 by 10   Revision a 1906-31 c 1938 d 1938-55 e 1946

1	A	CPr	OPr	1947po	-	-	6.47	4BE2X3	RB	6/47	10/47	1
3	B	CPt		1954pc	-	ace	6.54	4BooX3		3/54	-	2
3	C	CPw,y	OPv	1960mc	-	ad	6.60	4oooX3		2/60	4/60	3
3	C	CPz		1961mc	-	ad	6.60	4oooX3				

AA 1.1.47 on A(A), 3/-, OSPR 10/47. M on A, B. 1. Only a small area of Chatham Dockyard is blank. 2. The full detail in the dockyard is shown. 3. The entire area inside the Medway loop is blank.

TQ 78 (51/78) *Basildon*   10 by 10   Revision a 1906-32 c 1937-40 d 1937-55
                                                                                     e 1946-48 f 1966

1	A	CPr	OPr	1948po	-	-	1.48	4BE2X3	R		2/48	3/48
1	B	CPs	OPs	1951pc	-	ace	6.51	4BEoo3	R		11/51	11/51
3	B/	CPv		1951pc	1957mc	ace	1.52	4BooX3				
3	C	CPw,x	OPv	1959mc	-	ad	6.59	4oooX3			8/59	4/60
3	C/*	CPz	OPz	1959mc	1967r	adf	6.59	4oooX3			5/67	5/67

AA 1.11.47 on A(A), 3/-, OSPR 3/48. M on B.

TQ 79 (51/79) *Wickford*   10 by 10   Revision a 1919-20 c 1938-54 f 1962 g 1968
                                                                                     j 1967 building

1	A	CPr	OPr	1947po	-	-	6.47	4BE2X3	RB	12/47	2/48
2	A	CPs		1947pc	-	-	1.52	4Booo3	RB		
3	A	CPv		1947pc	1952	-	1.52	4Booo3	RB		
3	B	CPv	OP?	1957pc	-	ac	6.57	4oooX3		2/58	2/58
3	B/	CP?y,z		1957mc	1963ch	acf	6.57	4oooX3			
3	B//*	CPz	OPz	1957ms	1970br	acgj	6.57	4oooX3		3/70	5/70

AA 1.10.47 on A(A), 3/-, OSPR 2/48. M on A.

1	2	3	4	5	6	7	8	9	10	11	12	13	14

TQ 81 (51/81) **& Part of TQ 80** *Hastings* 10 by 12 Revision a 1927-40 c 1938 d 1938-57 e 1955
j 1981 selected k SUSI
1	A	CPr	OPr	1947po	-	-	1.48	4BE2X3	RB		12/47	3/48	
3	B	CPu	OPv	1956pc	-	ace	6.56	4oooX3			5/56	6/57	
3	B/	CPw		1956pc	1958mc	ace	6.56	4oooX3					
3	C	CPx,z,z+,0z	OPx	1961mc	-	ad	6.61	4oooX3			5/61	5/61	
3	C/*	CFz	OFz	1981^{c}m-s	nd:b	adjk	6.61	4oooX3			6/82	6/82	

AA 1.9.47/NT 17.1.48 on A(A), 3/-, OSPR 3/48. M on A. GS: *Hastings - Rye,* Solid and Drift, 1977.

TQ 82 (51/82) *Northiam* 10 by 10 Revision a 1906-37 c 1938 d 1938-57
1	A	CPr	OPr	1948po	-	-	6.48	4BEoo3	RB		5/48	5/48	
	A/												
3	A//	CPw		1948mc	1959mc	ac	6.48	4BooX3	B				
3	B	CPw,z	OPv	1960mc	-	ad	6.60	4oooX4			11/60	10/60	

M on A.

TQ 83 (51/83) *Tenterden* 10 by 10 Revision a 1907 b 1906-32 c 1937-38 d 1937-57
1	A	CPr	OPr	1948po	-	-	6.48	4BEoo3	RB		5/48	5/48	
1	B	CPr		nd:po	-	-	6.48	4BEoo3	RB				
3	B/	CPw		1948mc	1960mc	ac	1.52	4BooX3	B				
3	C	CPx,y,z	OPx	1961mc	-	bd	6.61	4oooX3			7/61	6/61	

M on B.

TQ 84 (51/84) *Headcorn* 10 by 10 Revision a 1906-39 c 1936-55
1	A	CPr	OPr	1948po	-	-	6.48	4BEoo3	RB		6/48	6/48	
3	A	CPt		1948pc	-	-	6.52	4Booo3	RB				
3	B	CPw,x,z,0z		1958pc	-	ac	6.58	4oooX3			7/58		1

M on A. 1. OSPR (outline edition) not found.

TQ 85 (51/85) *Harrietsham* 10 by 10 Revision a 1906-07 c 1938-48 d 1938-55
f 1961 g 1966
1	A	CPr	OPr	1947po	-	-	6.47	4BE2X3	RB	0	8/47	11/47	
1	B	CPs		1951pc	-	ac	6.51	4BEoo3	RB	0	10/51	-	
3	C	CPw	OPv	1958pc	-	ad	6.58	4oooX3		1	11/58	2/59	
3	C/*	CPx	OPx	1958mc	1962r	adf	6.58	4oooX3		1	5/62	5/62	
3	C/*/*	CPz	OPz	1958mc	1966r	adg	6.58	4oooX3		1	1/67	1/67	

AA 1.6.47/NT 1.10.47 on A(A), 3/-, OSPR 11/47. M on A, B.

TQ 86 (51/86) *Rainham Kent* 10 by 10 Revision a 1906-32 b 1906-40 c 1931-40 d 1938-55
e 1946-47 f 1964
1	A	CPr	OPr	1947po	-	-	6.47	4BE2X3	RB		5/47	8/47	
1	B	CPs	OPv	1951pc	-	ace	1.51	4BEoo3	RB		5/51	1/57	
3	C	CPw.x	OPv	1958pc	-	bd	6.58	4oooX3			8/58	11/58	
3	C/*	CPz	OPz	1958mc	1965r	bdf	6.58	4oooX3			6/65	4/65	

AA 1.2.47 on A(A), 3/-, OSPR 8/47. M on A, B. SS: Soil, Land Use Capability, outline, 1976 on C/*.

TQ 87 (51/87) *Isle of Grain** 10 by 10 Revision a 1906-40 c 1938-55
1	A	CPr	OPr	1947po	-	-	6.47	4BE2X3	R		6/47	8/47	1
1	B	CPr		nd:po	-	-	6.47	4BEoo3	R				
3	C	CPw,x.z	OPv	1958pc	-	ac	6.58	4oooX3			1/59	12/58	

AA 1.4.47 on A(A), 3/-, OSPR 8/47. M on B. 1. Isle of Grain forts: Dummy Battery, Wing Battery.

TQ 88 (51/88) *Southend-on-Sea* 10 by 10 Revision a 1906-32 c 1938-54
1	20046	CPr		1946po	-	-	1.47	4BE2X3	R	1	9/46	-	
1	A		OPr	1946po	-	-	1.47	4BE2X3	R	1	-	1/47	
3	B	CPv.x,z	OPv	1958pc	-	ac	6.58	4oooX3		3	8/58	10/58	1

AA 1.7.46 on A(A), 3/-, OSPR 1/47. M on A. 1. Southend Municipal Airport.

1	2	3	4	5	6	7	8	9	10	11	12	13	14

TQ 89 (51/89) *Hockley Essex* 10 by 10 Revision a 1919-20 c 1938 d 1938-54 e 1947 f 1966
1	A		CPr	OPr	1947po	-	-	6.47	4BE2X3	R	6/47	8/47	
1	B		CPs		1950pc	-	ace	6.50	4BEoo3	R			M
2	B		CPu		1950pc	-	ace	1.52	4Booo3	R			
3	C		CPw,x	OPv	1958pc	-	ad	6.58	4oooX3		11/58	12/58	
3	C/		CPz	OPz	1958mc	1967ch	adf	6.58	4oooX3				

AA 1.3.47/NT 1.7.47 on A(A), 3/-, OSPR 8/47. M on B.

TQ 91 (51/91) *Winchelsea* 10 by 10 Revision a 1906-37 c 1938-50 d 1938-57
1	nc [A]	CPr		1948po	-	-	1.48	4BE2X3	RB	3/48	-	1
1	A		OPr	1948po	-	-	1.48	4BE2X3	RB	-	5/48	
1	nc [A]	CPr		nd:po	-	-	1.48	4BEoo3	RB			
2	B	CPu		1949pc	-	-	1.48	4BooX3	R			
3	B/	CPw		1949pc	1958c	ac	1.48	4BooX3				
3	C	CPw,z	OPv	1960mc	-	ad	6.60	4oooX4		10/60	11/60	

M on A, B. 1. Martello towers at 9119, 9418.

TQ 92 (51/92) *Rye* 10 by 10 Revision a 1906-40 c 1946-57
1	A	CPr	OPr	1948po	-	-	6.48	4BE2X3	RB	4/48	5/48	
1	B	CPr		nd:po	-	-		4BEoo3	RB			M
3	B	CPt		1949pc	-	-	6.49	4BooX3	RB			
3	C	CPx,z	OPx	1961mc	-	ac	6.61	4oooX3		5/61	5/61	

M on B.

TQ 93 (51/93) *Woodchurch Kent* 10 by 10 Revision a 1906-31 c 1936-40 d 1936-55 e 1946
1	A	CPr	OPr	nd:po	-	-	6.48	4BEoo3	RB	8/48	9/48	
1	B	CPs		1951pc	-	ace	1.51	4BEoo3	RB	3/51	-	
2	B		OPv	1951pc	-	ace	1.52	4Booo3	RB	-	1/57	
3	C	CPx,z	OPx	1961mc	-	ad	6.61	4oooX4		6/61	5/61	

M on B.

TQ 94 (51/94) *Charing* 10 by 10 Revision a 1906-40 c 1936-55
| 1 | A | CPr | OPr | nd:po | - | - | 6.48 | 4BEoo3 | RB | 10/48 | 10/48 | |
| 3 | B | CPw,x,z.z | OPv | 1958pc | - | ac | 6.58 | 4oooX3 | | 11/58 | 1/59 | |

AA 1.6.48 on A(A), 3/-, OSPR 10/48. M on A.

TQ 95 (51/95) *Doddington Kent* 10 by 10 Revision a 1906-39 c 1938-55 f 1966
1	A	CPr	OPr	1948po	-	-	6.48	4BEoo3	RB	7/48	9/48	
1	B	CPs		nd:po	-	-	6.48	4BEoo3	RB			M
3	C	CPt		1949pc	1954c	-	6.54	4BooX3	B			
3	D	CPw	OPv	1958pc	-	ac	6.58	4oooX3		9/58	10/58	
3	D/*	CPz	OPz	1958mc	1967r	acf	6.58	4BooX3		11/67	10/67	

AA 1.4.48 on A(A), 3/-, OSPR 9/48. M on B.

TQ 96 (51/96) *Sittingbourne* 10 by 10 Revision a 1906-40 c 1938-55 f 1961
1	A	CPr	OPr	nd:po	-	-	6.48	4BEoo3	RB	8/48	9/48	
1	B	CPs		nd:po	-	-	6.48	4BEoo3	RB			M
2	B	CPt		1948pc	1949c	-	6.48	4BooX3	RB			
3	B/		OPv	1948pc	1956c	a	6.48	4BooX3	B	-	2/57	
3	C	CPw	OPv	1958pc	-	ac	6.58	4oooX3		11/58	12/58	
3	C/	CPx,z		1958mc	1962ch	acf	6.58	4oooX3				

AA 1.3.48 on A(A), 3/-, OSPR 9/48. M on B.

TQ 97 (51/97) *Sheerness** 10 by 10 Revision a 1906-46 c 1938-55 f 1961
1	A	CPr	OPr	1948po	-	-	6.48	4BE2X3	R	4/48	5/48	1
1	B	CPr		nd:po	-	-	6.48	4BEoo3	R			
3	C	CPw	OPv	1958mc	-	ac	6.58	4oooX4		1/59	3/59	
3	C/	CPx,z		1958mc	1963ch	acf	6.58	4oooX4				

AA 1.2.48 on A(A), 3/-, OSPR 5/48. M on B. 1. The dockyard at Sheerness is blank, but showing fortifications.

1	2	3	4	5	6	7	8	9	10	11	12	13	14

TQ 98 (51/98) *Shoeburyness* 10 by 10 Revision a 1910-21 c 1938 d 1938-54 e 1947
1	10046	CPr		1946po	-	-	1.47	4BE1X3	R		10/46	-	1
1	A		OPr	1946po	-	-	1.47	4BE1X3	R		-	1/47	
3	B	CPu	OPv	1956pc	-	ace	6.56	4ooX3			5/56	5/57	
3	C	CPw	OPv	1958pc	-	ad	6.58	4ooX3			9/58	10/58	
3	C/	CPw,z		1958mc	1960mc	ad	6.58	4ooX3					

AA 1.7.46 on A(A), 3/-, OSPR 1/47. M on A. 1. Showing the Shoeburyness military complex and railway.

TQ 99 (51/99) *Burnham-on-Crouch* 10 by 10 Revision a 1919-21 c 1938-47 d 1938-53
1	A	CPr	OPr	1947po	-	-	1.47	4BE2X3	R		2/47	5/47	
3	B	CPt		1954pc	-	ac	6.54	4BooX3			1/54	-	
3	C	CPv,x,z	OP?	1958pc	-	ad	6.58	4ooX3			3/58	3/58	

AA 1.12.46 on A(A), 3/-, OSPR 5/47. M on A. SS: Soil, Land Use Capability, outline, 1976 on C.

TR 01 (61/01) *Dungeness** 10 by 10 Revision a 1906-40 c 1938-46 d 1938-57 f 1968
                                                j 1967 power station
1	A	CPr	OPr	1948po	-	-	1.48	4BEoo3	RB		5/48	5/48	1
1	B	CPr		nd:po	-	-	1.48	4BEoo3	RB				
3	B/	CPv,w		1949pc	1957mc	ac	1.48	4BooX3					
3	C	CPx	OPx	1961mc	-	ad	6.61	4ooX3			2/61	2/61	
3	C/*	CPz	OPz	1961ms	1969p	adfj	6.61	4ooX3			9/69	2/70	
3	C/*/	CFz		1961ms	nd:p	adfj	6.61	4ooX3					

M on B. 1. Dungeness Coastguard Station (Lloyd's Signal Station). With MV altered to 11°00'.

TR 02 (61/02) **& Part of TR 12** *New Romney** 12 by 10 Revision a 1906-40 c 1937-46 d 1937-57 e 1946-47
1	A	CPr	OPr	1948po	-	-	1.48	4BEoo3	RB	0	6/48		1
1	B	CPs		1951pc	-	ace	1.51	4BEoo3	RB	0	10/51	-	
2	B		OPs	1951pc	-	ace	1.51	4BEoo3	RB	0	-	5/56	
3	B/	CPv		1951pc	1957mc	ace	1.51	4BooX3	B	0			
3	C	CPw,x,z	OPv	1960mc	-	ad	6.60	4ooX3		3	6/60	6/60	2

M on A, with legends, ?B, C. 1. Lydd Camp, Martello Towers (1029). 1. OSPR (outline edition) not found. 2. Lydd (Ferryfield) Airport.

TR 03 (61/03) *Aldington Kent* 10 by 10 Revision a 1906-07 c 1937-40 d 1937-57 e 1946
1	A	CPr	OPr	nd:po	-	-	6.48	4BEoo3	RB		8/48	9/48	
2	B	CPt,u	OPv	1952pc	-	ace	1.51	4BooX3	RB		1/53	10/56	
3	C	CPw,z,0z	OPx	1961mc	-	ad	6.61	4ooX3			12/60	2/61	

M on A, with legends.

TR 04 (61/04) *Ashford Kent* 10 by 10 Revision a 1906-40 c 1938-55
1	A	CPr	OPr	nd:po	-	-	6.48	4BEoo3	RB		9/48	9/48	1
3	A	CPu		1948pc	-	-	6.48	4BEoo3	RB				
3	B	CPw	OPv	1958pc	-	ac	6.58	4ooX3			11/58	12/58	
3	B/	CPx,z,0z		1958mc	1961ch	ac	6.58	4ooX3					

AA 1.5.48 on A(A), 3/-, OSPR 11/48. M on A, with legends, A, B/. SS: Soil, Land Use Capability, outline, 1973 on B/. 1. A20 Ashford bypass shown as under construction.

TR 05 (61/05) *Selling** 10 by 10 Revision a 1906 c 1937-55 f 1963
1	A	CPr	OPr	1948po	-	-	6.48	4BEoo3	RB		6/48	7/48	
2	B	CPt		1948pc	1953c	-	6.53	4BooX3	RB				
3	C	CPw	OPv	1958pc	-	ac	6.58	4ooX4			8/58	11/58	
3	C/*	CPy,z	OPz	1958mc	1964r	acf	6.58	4ooX4			1/64	2/64	

M on A.

TR 06 (61/06) *Faversham* 10 by 10 Revision a 1903-32 c 1938 d 1938-55 e 1947
1	A	CPr	OPr	1948po	-	-	6.48	4BE2X3	RB		4/48	5/48	
	B												
2	C	CPs		1949pc	1952c	ace	1.52	4Booo3	RB				M
2	D	CPt		1949pc	1953c	ace	6.53	4BooX3	B				
3	E	CPw,x,z	OPv	1958pc	-	ad	6.58	4ooX3			6/58		1

AA 1.1.48 on A(A), 3/-, OSPR 5/48. M on C. 1. OSPR (outline edition) not found.

298

1	2	3	4	5	6	7	8	9	10	11	12	13	14

TR 07 (61/07) *Leysdown-on-Sea**   10 by 10   Revision a 1906-31 c 1931-55

1	A	CPr	OPr	1948po		-	6.47	4BE2X3	R		1/48	4/48	
2	B	CPs		1948pc	1951c	-	?	4Booo3	R				M
3	C	CPv	OP?	1958pc		ac	6.58	4oooX3			3/58	3/58	
3	C/	CPx,z		1958pc		ac	6.58	4oooX3					

AA 1.5.47 on A(A), 3/-, OSPR 4/48. M on B.

TR 09 (61/09) **& Part of TR 08** *Foulness*   10 by 12   Revision a 1910-20 c 1950-53

1	A	CPr	OPr	1947po		-	6.47	4BE2X3			5/47	8/47	
1	B	CPs		1950pc	-	a	6.50	4BEoo3			1/51	-	
3	B	CPu		1950pc	-	a	6.50	4BooX3					
3	C	CPv,z.z	OP?	1957pc	-	ac	6.57	4oooX3			3/58	5/58	1

AA 1.2.47 on A(A), 3/-, OSPR 8/47. M on B. 1. Showing barracks, Observation Tower, Observation Box.

TR 13 (61/13) *Hythe Kent*   10 by 10   Revision a 1906-40 c 1938-57 f 1970

1	A	CPr	OPr	1948po		-	6.48	4BEoo3	RB	10	7/48	9/48	1
	B												
2	C	CPs,t	OPv	1949pc	1953c	-	6.48	4oooX3	RB	10			2
3	D	CPw,z	OPv	1960mc		ac	6.60	4oooX4		10	8/60	11/60	3
3	D/	CPz,0z	OPz	1960ᶜms	nd:ch	acf	6.60	4oooX4		30			4

AA 1.3.48 on A(A), 3/-, OSPR 9/48. M on A, with legends, C. 1. Showing Martello Towers, Shorncliffe Camp, and barracks. 2. Lympne Airport. 3. With Dymchurch Redoubt, Army School of Education. 4. Ashford Airport. The earlier printing has the ?job number 807/4/5/74 in the bottom right hand corner.

TR 14 (61/14) *Lyminge**   10 by 10   Revision a 1906-40 c 1937-50 d 1937-57

1	A	CPr	OPr	nd:po		-	1.48	4BEoo3	RB		8/48	9/48	
3	B	CPt		1949pc		-	6.48	4BEoo3	B				
3	B/		OPv	1949pc	1956mc	ac	6.48	4BooX3	B		-	2/57	
3	C	CPx,z	OPx	1961mc		ad	6.61	4oooX3			6/61	5/61	

M on A, with legends, B/.

TR 15 (61/15) *Canterbury*   10 by 10   Revision a 1906-37 c 1938-57 f 1963

1	A	CPr	OPr	1948po		-	6.48	4BEoo3	RB		7/48	8/48	
2	B	CPs,u		1949pc		-	1.52	4Booo3	RB				
3	C	CPx	OPx	1961mc		ac	6.61	4oooX3			7/61	6/61	
3	C/	CPz	OPz	1961mc	1964ch	acf	6.61	4oooX3					

AA 1.1.48 on A(A), 3/-, OSPR 8/48. M on B.

TR 16 (61/16) *Whitstable**   10 by 10   Revision a 1906-39 c 1938 d 1938-57

1	A	CPr	OPr	1948po		-	6.48	4BE2X3	RB		4/48	6/48	
3	B	CPu		1955pc		ac	6.55	4oooX3			1/55	-	
3	C	CPw,z	OPv	1960mc		ad	6.60	4oooX3			7/60	7/60	

AA 1.2.48 on A(A), 3/-, OSPR 6/48. M on A.

TR 23 (61/23) *Folkestone**   10 by 10   Revision a 1905-39 c 1937-46 d 1937-57

1	15046	CPr		1946po		-	1.47	4BE2X3	RB		12/46	-	
1	A		OPr	1946po		-	1.47	4BE2X3	RB		-	4/47	
3	A/	CPw		1946pc	1959c	ac	1.47	4BooX3	B				
3	B	CPw,z	OPv	1960mc		ad	6.60	4oooX4			8/60	11/60	

AA 1.9.46 on A(A), 3/-, OSPR 4/47. M on A.

TR 24 (61/24) *Lydden**   10 by 10   Revision a 1905-40 c 1937-50 d 1937-57

1	A	CPr	OPr	1947po		-	1.47	4BE2X3	RB		3/47	6/47	
1	B	CPr		nd:po		-	1.47	4BEoo3	RB				
3	B/	CPw		1950mc	1959mc	ac	1.47	4BooX3	B				
3	C	CPw,y,z	OPv	1961mc		ad	6.61	4oooX3			11/60	1/61	

AA 1.11.46 on A(A), 3/-, OSPR 6/47. M on B.

299

1	2	3	4	5	6	7	8	9	10	11	12	13	14

TR 25 (61/25) *Aylesham*   10 by 10 Revision a 1905-39 c 1938 d 1938-57
1	A	CPr	OPr	nd:po	-	-	6.48	4BEoo3	RB		9/48	9/48	
1	B	CPr		nd:po	-	-	6.48	4BEoo3	RB				
3	B/	CPv		1948pc	1956c	ac	1.51	4BooX3	B				
3	B//	CPw		1948mc	1958c	ac	1.51	4BooX3	B				
3	C	CPw,z	OPv	1960mc	-	ad	6.60	4oooX3			11/60	9/60	

M on B.

TR 26 (61/26) **& Part of TR 27** *St Nicholas at Wade* 10 by 12 Revision a 1905-39 c 1938-57
1	A	CPr	OPr	1948po	-	-	6.48	4BE2X3	R		6/48	6/48	
1	B	CPr		nd:po	-	-	6.48	4BEoo3	R				
3	C	CPw,x,z.z	OPv	1960mc	-	ac	6.60	4oooX3			6/60	6/60	

AA 1.2.48 on A(A), 3/-, OSPR 6/48. M on B.

TR 34 (61/34) **& Part of TR 33** *Dover*   10 by 12 Revision a 1905-38 c 1937-38 d 1937-57 f 1968
1	15046	CPr		1946po	-	-	6.46	4BE2X3	RB		8/46	-	1
1	2546		OPr	1946po	-	-	6.46	4BE2X3	RB		-	12/46	
	A/												
3	A//	CPw		1946mc	1959c	ac	6.46	4BooX3	B				
3	B	CPw,y	OPv	1960mc	-	ad	6.60	4oooX3			8/60	11/60	
3	B/	CPz	OPz	1960ms	1968ch	adf	6.60	4oooX3					

AA 1.6.46 on 2546(2546), 3/-, OSPR 12/46. M on A. 1. Dover Castle 21 and the Citadel (prison) 3040 are both left blank: they are restored on edition B.

TR 35 (61/35) *Deal*   10 by 10 Revision a 1905-39 c 1937-57
1	A	CPr	OPr	1948po	-	-	1.48	4BE2X3	RB		4/48	6/48	
1	A	CPs		1948pc	1951	-	6.51	4BEoo3	RB				
3	B	CPx,z	OPx	1961mc	-	ac	6.61	4oooX3			3/61	3/61	

AA 1.1.48 on A(A), 3/-, OSPR 6/48. M on A. SS: Soil, outline, 1972 on B; Land Use Capability, outline, 1973 on B.

TR 36 (61/36) **& Part of TR 37 & TR 46** *Margate* 12 by 12 Revision a 1905-40 c 1938-57
1	A	CPr	OPr	1948po	-	-	1.48	4BEoo3	RB	01	6/48	7/48	1
1	A	CPs		1948pc	1951	-	6.51	4BEoo3	RB	01			
3	A/	CPw		1948pc	1958c	ac	1.52	4BooX3	B	01			
3	A//	CPw		1948mc	1960mc	ac	1.52	4BooX3	B	11			2
3	B	CPx,z,0z	OPx	1961mc	-	ac	6.61	4oooX4		11	12/61	7/61	3

AA 1.3.48 on A(A), 3/-, OSPR 7/48. M on A. 1. Manston Airfield unnamed, no features. The Richborough area is blank. Ramsgate railway tunnel is marked, also the alignment of the Ramsgate to Margate railway. 2. Manston Airfield. 3. Ramsgate Municipal Airport. There is no sign of Manston's railway alignment.

TV 59 (50/59) **& Parts of TV 49 & TV 69** *Eastbourne* 15 by 10 Revision a 1925-39 c 1938-50 d 1938-57
1	A	CPr	OPr	1947po	-	-	1.48	4BE2X3	B		12/47	3/48	1
1	B	CPr		nd:po	-	-	1.48	4BEoo3	B				
3	B/	CPv		1950pc	1957mc	ac	1.48	4BooX3	B				
3	C	CPw,y.z	OPv	1959mc	-	ad	6.59	4oooX3			7/59	6/59	

AA 1.2.47/NT 22.1.48 on A(A), 3/-, OSPR 3/48. M on B. 1. Showing martello tower (5098).

Appendices

Isle of Man

Isle of Man mapping at the 1:25,000 scale is in black outline, photo-reduced from the six-inch map and prepared by the Ordnance Survey for the Government of the Isle of Man. With National Grid at 1 km intervals. The copies recorded are in the Ordnance Survey Record Map Library.

Sheet number	Sheet size	Publication date	Revision a 1955-56 b 1956 c 1967-72
NX 30	10 by 10	1974[c]	bc
NX 40	10 by 10	1974[c]	bc
SC 16 & Part of SC 17	10 by 12	1974[c]	bc
SC 26	10 by 10	1974[c]	bc
SC 27	10 by 10	1974[c]	ac
SC 28	10 by 10	1974[c]	bc
SC 37 & Part of SC 36	10 by 12	1974[c]	bc
SC 38	10 by 10	1974[c]	bc
SC 39	10 by 10	1974[c]	bc
SC 47	10 by 10	1974[c]	bc
SC 48	10 by 10	1974[c]	ac
SC 49	10 by 10	1974[c]	bc

1 2 3   4   5   6   7 8 9   10 11   12   13   14

GSGS 4627, later M821, coloured edition

See page 58 for an explanation of headings and layout. Most of these sheets are in the British Library.

G 2	NH 74	[A]	nd:po	-	-	6.54	4BEoo4	R		1EG	1952p	5000/11/52 ASE RCE
G 2	NH 75	[A]	nd:po	-	-	6.54	4BEoo4	R	0	1EG	1952p	5000/11/52 ASE RCE
G 2	NH 84	[A]	nd:po	-	-	6.54	4BEoo4			1EG	1952p	5000/11/52 ASE RCE
G 2	NH 85	[A]	nd:po	-	-	6.54	4BEoo4	R	0	1EG	1952p	5000/10/52 ASE RCE
G 2	NH 94	[A]	1950pc	-	ac	6.54	4BEoo4			1EG	1952p	5000/11/52 ASE RCE
G 2	NH 95+	A	nd:po	-	-	6.54	4BEoo3	R		1EG	1952p	5000/11/52 ASE RCE
G 2	NJ 05	[A]	1950pc	-	ac	6.54	4BEoo4	R		1EG	1952p	5000/12/52 ASE RCE
G 2	NJ 06+	[A]	nd:po	-	-	6.54	4BEoo3	R	0	1EG	1952p	5000/12/52 ASE RCE
G 2	NJ 15	[A]	1950pc	-	a	6.54	4BEoo4		0	1EG	1952p	5000/12/52 ASE RCE
G 2	NJ 16+	[A]	1950pc	-	ac	6.54	4BEoo3	R	00	1EG	1952p	5000/11/52 ASE RCE
G 2	NJ 24	[A]	1950pc	-	ac	6.54	4BEoo4	R		1EG	1952p	5000/11/52 ASE RCE
G 2	NJ 25	[A]	1950pc	-	ac	6.54	4BEoo4	R	0	1EG	1952p	5000/11/52 ASE RCE
G 2	NJ 26+	[A]	1950pc	-	ac	6.54	4BEoo3	R	000	1EG	1952p	5000/12/52 ASE RCE
G 2	NJ 81	[A]	nd:po	-	-	6.54	4BEoo3	R	0	1EG	1952p	2500/12/52 ASE RCE
G 3	NJ 81	A	1949pc	-	-	1.49	4BooX3	R	0	1EG	1952b	5000/10/55 OS
G 2	NJ 92+	[A]	1951pc	-	ac	6.54	4BEoo3	R		1EG	1952p	2500/1/53 ASE RCE
M 3	NJ 92+	B	nd	-	ad	6.57	4oooX3			E2G	nd:p	3000/6/62 ASE RCE
G 2	NJ 96	[A]	1950pc	-	ac	6.54	4BEoo4			1EG	1952p	2500/12/52 ASE RCE
M 3	NJ 96	B	nd	-	ad	6.57	4oooX4			E2G	nd:p	3000/6/62 ASE RCE
G 2	NO 41	B	nd:po	-	-	6.54	4BEoo3	R		1EG	1952p	2500/12/52 ASE RCE
M 3	NO 41	C/	nd	-	ac	6.56	4oooX3			E2G	nd:p	5000/6/62/ASE RCE
G 2	NO 42+	[A]	1947po	-	-	6.54	4BEoo3	R	0	1EG	1952p	2500/12/52 ASE RCE
M 3	NO 42+	B/	nd	-	ac	6.56	4oooX3		1	E2G	nd:p	5000/6/62/ASE RCE
G 2	NO 53+	A	nd:po	-	-	6.54	4BEoo3	R	0	1EG	1952p	2500/12/52 ASE RCE

Also used for GSGS Misc.1584 *"Barry-Buddon"* : *Location of ranges and tr[ainin]g facilities No.40 WETC,* First Edition-GSGS (500/8/52 SPC RE), overprinted in blue.

G 3	NO 53+	B	1957pc	-	ac	6.57	4oooX3		3	E2G	1957b	5000/11/57/5539/OS

Also used for GSGS Misc.1584 *Location of range "Barry Buddon" and trg facilities No.40 WETC,* Edition 2-GSGS, no date [1957] (1000/10/57/5668/OS); also *Barry-Buddon Trg Area,* Edition 3-GSGS, 1957 (1000/12/67/3687/R); also *Barry-Buddon Training Area : Manoeuvre Area Boundary,* Edition 4-GSGS, 1959 (500/2/59/5909/OS), all overprinted in blue.

G 3	NS 26	[B]	1955pc	-	ac	6.55	4oooX3			E1G	1956b	5000/11/56
G 1	NS 27	A:s	nd:po	-	-	6.48	4BEoo3	R		1EG	1950p	5000/11/50 SPC RE
G 2	NS 28	A:s	1949pc	-	-	1.52	4Booo3	R		2EG	1952p	5000/11/51
M 3	NS 33+	C	1957pc	-	ad	6.57	4oooX3		0	E1G	1961b	2500/3/61/6235/OS
M 3	NS 34	B	1958pc	-	ac	6.58	4oooX3			E1G	1960b	2500/2/60 13 FSS
M 3	NS 34	B/	1958mc	1960mc	ac	6.58	4oooX3			E1G	nd:b	3000/4/63/ASE RCE
G 3	NS 35	A	1949pc	-	a	1.49	4BooX3	RB		E1G	1957b	5000/4/57
G 3	NS 36	A	1949pc	-	a	6.48	4BooX3	RB		E1G	1957b	5000/2/57
G 1	NS 37	A:s	nd:po	-	-	6.48	4BEoo3	R		1EG	1950p	5000/11/50 SPC RE
G 3	NS 38	A	1949pc	-	ac	6.49	4BooX3	R		E1G	1957b	5000/11/57/5284/OS
M 3	NS 43	C	1957pc	-	ad	6.57	4oooX3			E1G	1960b	2500/2/60 13 FSS
M 3	NS 44	B	1957pc	-	ac	6.57	4oooX3			E1G	1960b	2500 2 60 13 FSS
G 3	NS 45	B	1950pc	-	ac	6.50	4BooX3	R		E1G	1957b	5000/4/57
G 1	NS 46	B:s	nd:po	-	-	6.48	4BEoo3	R	00	1EG	1950p	5000/11/50 SPC RE
G 1	NS 47	C:s	1951pc	-	ac	1.51	4BEoo3	R		1EG	1951p	5000/2/51 SPC RE
G 3	NS 48	A	1949pc	-	a	1.50	4BooX3	R		E1G	1957b	5000/11/57/5284/OS
M 3	NS 53	B	1957pc	-	ac	6.57	4oooX4			E1G	1960b	2500/2/60 13 FSS
G 3	NS 54	A	1950pc	-	ac	1.50	4BooX3			E1G	1957b	5000/2/57
G 1	NS 55	B:s	nd:po	-	-	1.51	4BEoo3	R		1EG	1950p	5000/12/50 SPC RE
G 1	NS 56	B:s	nd:po	-	-	1.51	4BEoo3	R	0	1EG	1951p	5000/2/51 SPC RE
G 1	NS 57	A:s	1947po	-	-	1.51	4BEoo3	R		1EG	1951p	5000/2/51 SPC RE

1	2	3	4	5	6	7	8	9	10	11	12	13	14
G	3	NS 58	B	1949pc	1950c	a	6.50	4BooX3	R		E1G	1957b	5000/11/57/5284/OS
G	3	NS 64	B	1950pc	1951c	a	1.51	4BooX3	R		E1G	1957b	5000/2/57
G	1	NS 65	A:s	1947po	-	-	1.51	4BEoo3	R		1EG	1950p	5000/12/50 SPC RE
G	1	NS 66	B:s	1947pc	1950c	ac	1.51	4BEoo3	R		1EG	1951p	5000/2/51 SPC RE
G	1	NS 67	B:s	nd:po	-	-	1.51	4BEoo3	R		1EG	1950p	5000/12/50 SPC RE
G	3	NS 68	A	1949pc	-	ac	6.49	4BooX3			E1G	1957b	5000/11/57/5284/OS
G	3	NS 74	B	1956pc	-	ac	6.56	4oooX3			E1G	1956b	5000/1/57
G	1	NS 75	B:s	nd:po	-	-	1.51	4BEoo3	R		1EG	1950p	5000/2/51 SPC RE
G	1	NS 76	C:s	nd:po	-	-	1.51	4BEoo3	R		1EG	1950p	5000/2/51 SPC RE
G	2	NS 77	[B]	nd:po	-	-	6.54	4BEoo3	R		1EG	1952p	2500/11/52 ASE RCE
G	3	NS 77	B	1950pc	-	-	1.48	4BooX3	R		E1G	1952b	5000/9/55 OS
G	3	NS 78	B	1956pc	-	ac	6.56	4oooX3			E1G	1956b	5000/11/56
G	3	NS 84	B	1956pc	-	ac	6.56	4oooX4			E1G	1956b	5000/12/56
G	2	NS 86	[A]	nd:po	-	-	6.54	4BEoo3	R		1EG	1952p	2500/11/52 ASE RCE
G	3	NS 86	A	1949pc	-	-	1.50	4BooX3	R		E1G	1952b	5000/9/55 OS
G	2	NS 87	A	nd:po	-	-	6.54	4BEoo3	R		1EG	1952p	2500/10/52 ASE RCE
M	3	NS 87	B	1956pc	-	ac	6.56	4oooX3			E2G	1962b	3000/10/62/6394/OS
G	3	NS 88	B	1956pc	-	ac	6.56	4oooX3			E1G	1957b	5000/2/57
G	3	NS 89	B	1956pc	-	ac	6.56	4oooX3			E1G	1956b	5000/12/56
G	2	NS 96	[A]	nd:po	-	-	6.54	4BEoo3	R		1EG	1952p	2500/10/52 ASE RCE
G	3	NS 96	A	1949pc	-	-	1.50	4BooX3	R		E1G	1952b	5000/9/55 OS
G	2	NS 97	[A]	nd:po	-	-	6.54	4BEoo3	R		1EG	1952p	2500/11/52 ASE RCE
M	3	NS 97	B/	1956mc	1961mc	ac	6.56	4oooX3L			E2G	1962b	3000/6/62/6394/OS
G	1	NS 98	B:s	nd:po	-	-	1.51	4BEoo3	R	0	1EG	1951p	5000/12/50 SPC RE
G	2	NT 06	B	nd:po	-	-	6.54	4BEoo3	R		1EG	1952p	2500/10/52 ASE RCE
G	3	NT 06	[B]	1949pc	-	-	1.49	4BooX3	R		E1G	1952b	5000/9/55 OS
G	1	NT 07	B:s	nd:po	-	-	1.51	4BEoo3	R		1EG	1951p	5000/2/51 SPC RE
G	1	NT 08	A:s	nd:po	-	-	1.51	4BEoo3	R		1EG	1951p	5000/5/51 SPC RE
G	3	NT 09	B	1956pc	-	ac	6.56	4oooX3			E1G	1956b	5000/1/57

NT 14 appears as issued on the 1958 index, but has not been recorded.

1	2	3	4	5	6	7	8	9	10	11	12	13	14
G	3	NT 15	B	1955pc	-	ac	6.55	4oooX3			E1G	1956b	5000/12/55 OS
G	1	NT 16	[A]:r	1947po	-	-	6.50	4BEoo3	R	0	WOE	1950p	4503/14 RE/Sep 50/3000
G	1	NT 16	A	1947po			6.47	4BEoo3	R	0	1EG	1950p	5000/9/51
G	1	NT 17	B:s	nd:po	-	-	1.51	4BEoo3	R	0	1EG	1951p	5000/5/51 SPC RE
G	1	NT 18	B:s	1951pc	-	ac	1.51	4BEoo3	R	0	1EG	1951p	5000/6/51 SPC RE
G	3	NT 19	B	1956pc	-	ac	6.56	4oooX3L			E1G	1956b	5000/1/57
G	3	NT 25	B	1955pc	-	ac	6.55	4oooX3L			E1G	1956b	5000/2/56 OS
G	3	NT 26	C	1955pc	-	ac	6.55	4oooX3			E1G	1955b	7000/10/55 OS
G	1	NT 27	A:s	1947po	-	-	1.51	4BEoo3	R		1EG	1951p	5000/5/51 SPC RE
G	1	NT 28	B:s	nd:po	-	-	1.51	4BEoo3	R		1EG	1951p	5000/5/51 SPC RE
M	3	NT 28	B	1956pc	-	ac	6.56	4oooX3L			E2G	nd:p	2000/4/63 ASE RCE
G	3	NT 35	B	1955pc	-	ac	6.55	4oooX3			E1G	1956b	5000/12/55 OS
G	2	NT 36	[A]	1947po	-	-	6.54	4BEoo3	R		1EG	1952p	2500/11/52 ASE RCE
G	3	NT 36	B	1955pc	-	ac	6.55	4oooX3			E2G	1955b	5000/11/55 OS
G	2	NT 37	[A]	1947po	-	-	6.54	4BEoo3	R		1EG	1952p	2500/12/52 ASE RCE
M	3	NT 37	C/	1955mc	1960mc	ac	6.55	4oooX3			E2G	1962b	5000/7/62/6394/OS
M	3	NT 40	A:u	1954pc	-	ace	6.54	4oooX4			E1G	1957b	2500/10/57/333 2SPC(Air)
G	2	NT 41	A:s	1950pc	-	-	1.52	4Booo4	B		2EG	1952p	5000/11/51
G	2	NT 46	A	1948pc	1952	-	6.54	4Booo3	R		1EG	1952p	2500/10/52 ASE RCE
M	3	NT 46	B/	1955mc	1960mc	ac	6.55	4oooX3			E2G	1962b	5000/6/62/6394/OS
G	2	NT 47	A	nd:po	-	-	6.54	4BEoo3	R	0	1EG	1952p	2500/12/52 ASE RCE
M	3	NT 47	B	nd	-	ac	6.55	4oooX3		2	E2G	nd:p	5000/6/62/ASE RCE
G	2	NT 48	A	1948pc	-	-	6.54	4Booo3		0	1EG	1952p	2500/3/53 ASE RCE
M	3	NT 48	B	nd	-	ac	6.55	4oooX3		1	E2G	nd:p	5000/6/62/ASE RCE
G	3	NT 50	A	1954pc	-	ac	6.54	4oooX4			E1G	1955b	7000/5/55 OS
G	2	NT 51	A:s	1950pc	-	-	1.52	4Booo3	R		2EG	1952p	5000/11/51
M	3	NT 52	B/	1954mc	1959mc	ac	6.54	4oooX4			E1G	1960b	2500/2/61/6235/OS

NT 55 appears as an additional sheet on an official copy of the 1958 index, but has not been recorded.

1	2	3	4	5	6	7	8	9	10	11	12	13	14
G	2	NT 56	A	nd:po	-	-	6.54	4BEoo3	B		1EG	1952p	2500/12/52 ASE RCE
M	3	NT 56	B/	nd	-	ac	6.54	4oooX3			E2G	nd:p	5000/6/62/ASE RCE

1	2	3	4	5	6	7	8	9	10	11	12	13	14
G	2	NT 57	A	1948pc	-	-	6.54	4Booo3	R	0	1EG	1952p	2500/3/53 ASE RCE
M	3	NT 57	B/	nd	-	ac	6.54	4oooX3		1	E2G	nd:p	5000/6/62/ASE RCE
G	2	NT 58+	A	nd:po	-	-	6.54	4BEoo3	R	0	1EG	1952p	2500/10/52 ASE RCE
M	3	NT 58+	B/	nd	-	ac	6.54	4oooX3		1	E2G	nd:p	5000/6/62/ASE RCE
G	3	NT 60	A	1954pc	-	ac	6.54	4oooX4			E1G	1955b	7000/5/55 OS
G	3	NT 61	B	1954pc	-	ac	6.54	4oooX4			E1G	1955b	7000/5/55 OS
M	3	NT 62	B/	1954mc	1960mc	ac	6.54	4oooX4			E1G	1961b	2500/2/61/6235/OS
M	3	NT 62	B/	1954mc	1960mc	ac	6.54	4oooX4			E1G	1961b	no code, 13 FSS RE

NT 65 appears as an additional sheet on an official copy of the 1958 index, but has not been recorded.

1	2	3	4	5	6	7	8	9	10	11	12	13	14
G	2	NT 66	[A]	nd:po	-	-	6.54	4BEoo3	B		1EG	1952p	2500/12/52 ASE RCE
M	3	NT 66	B/	1954mc	1961mc	ac	6.54	4oooX3			E2G	nd:p	5000/4/63 ASE RCE
G	2	NT 67	A	nd:po	-	-	6.54	4BEoo3	RB		1EG	1952p	2500/10/52 ASE RCE
G	3	NT 67	B	1954pc	-	ac	6.54	4oooX3			E2G	1955b	5000/10/55 OS
G	3	NT 70	A	1954pc	-	ac	6.54	4oooX4			E1G	1955b	7000/5/55 OS
G	3	NT 71	A	1954pc	-	ac	6.54	4oooX4			E1G	1955b	7000/6/55 OS
G	2	NT 76	[A]	1950pc	1951c	ac	6.54	4BEoo3	RB		1EG	1952p	2500/12/52 ASE RCE
G	3	NT 76	C	1954pc	-	ad	6.54	4oooX3			E2G	1955b	5000/10/55 OS
G	2	NT 77	A	nd:po	-	-	6.54	4BEoo3	R		1EG	1952p	2500/3/53 ASE RCE
M	3	NT 77	B/	1954mc	1959mc	ac	6.54	4BooX3			E2G	nd:p	3000/4/63 ASE RCE
G	3	NT 80	A	1954pc	-	ac	6.54	4oooX4			E1G	1955b	7000/6/55 OS
G	3	NT 81	A	1954pc	-	ac	6.54	4oooX4			E1G	1955b	7000/3/55 OS
G	2	NT 86+	[A]	1950pc	1951c	a	6.54	4BEoo3	R		1EG	1952p	2500/12/52 ASE RCE
M	3	NT 86+	C/	1954mc	1960mc	ac	6.54	4oooX3			E2G	1962b	3000/8/62/6394/OS
G	3	NT 90	A	1954pc	-	ac	6.54	4oooX4			E1G	1955b	7000/5/55 OS
G	2	NT 96	[A]	nd:po	-	-	6.54	4BEoo3	R		1EG	1952p	2500/12/52 ASE RCE
M	3	NT 96	B/	1954pc	1958mc	ac	6.54	4oooX3			E2G	1962b	3000/6/62/6394/OS
G	2	NU 20	[A]	nd:po	-	-	6.54	4BEoo3	RB	0	1EG	1952p	2500/3/53 ASE RCE
G	3	NU 20	B	1954pc		ac	6.54	4oooX3		1	E2G	1955b	5000/11/55 OS
G	1	NX 64	A:r	nd:po	-	-	1.50	4BEoo3			1EG	1951p	5000/11/51
G	1	NX 74	A:r	nd:po	-	-	1.50	4BEoo3			1EG	1951p	5000/11/51
G	3	NX 74	B	1954pc	-	ac	6.54	4BooX3			E2G	1955b	5000/12/55 OS
M	3	NY 34	A	nd:mo	-	a	1.50	4BooX3	RB		E1G	1958b	2500/12/58/5756/OS
M	3	NY 35	C/	1954pc	1958mc	ad	6.54	4oooX3		11	E1G	1959b	2500/2/59/5756/OS
M	3	NY 36	C	1954pc	-	ac	6.54	4oooX3			E1G	1958b	2500/1/59/5756/OS
M	3	NY 36	C/	1954mc	1959mc	ac	6.54	4oooX3			E2G	1961b	4000/10/61/6335/OS
G	3	NY 39	A	1954pc	-	ac	6.54	4oooX4			E1G	1955b	7000/4/55 OS
M	3	NY 44	B/	1953mc	1958mc	ac	1.53	4BooX3			E1G	1959b	2500/2/59/5756/OS
M	3	NY 45	B/	1954pc	1958mc	ac	6.54	4BooX3			E1G	1958b	2500/11/58/5756/OS
M	3	NY 46	[C]	1954pc	-	ad	6.54	4BooX3		11	E1G	1958b	2500/1/59/5756/OS
M	3	NY 46	C/	1954mc	1960mc	ad	6.54	4BooX3		11	E2G	1961b	3000/10/61/6335/OS
G	3	NY 49	A	1954pc	-	ac	6.54	4oooX4			E1G	1955b	7000/5/55 OS
M	3	NY 59	A:u	1954pc	-	ac	6.54	4oooX4			E1G	1957b	2500/10/57/333 13 FSS RE
M	3	NY 59	A:u	1954pc	-	ac	6.54	4oooX4			E1G	1957b	2500/10/57/333 2SPC(Air)
G	2	NY 71	B:s	1947pc	-	-	1.48	4Booo3	RB		1EG	1952p	6000/6/52 SPC RE
M	3	NY 71	C/	1953pc	1957mc	ac	1.53	4BooX3			E2G	nd:b	5000/4/63 ASE RCE
G	2	NY 72	A:s	1950pc	-	-	1.50	4Booo3	B		1EG	1952p	6000/6/52 SPC RE
M	3	NY 72	B	nd	-	ac	6.58	4oooX3			E2G	nd:p	2000/6/62/ASE RCE
M	3	NY 72	B	1958pc	-	ac	6.58	4oooX3			E2G	nd:p	3000/5/64/653/SPC RE
G	2	NY 81	B:t	1952pc	-	ac	1.52	4Booo3	R		1EG	1952p	6000/6/52 SPC RE
M	2	NY 81	B	1952pc	-	ac	6.52	4Booo3	R		E1G	nd:p	3000/5/64/653/SPC RE
G	2	NY 82	A:t	1950pc	-	-	1.50	4Booo3	B		1EG	1952p	6000/6/52 SPC RE
M	3	NY 82	B	1960mc	-	ac	6.60	4oooX3			E2G	nd:p	5000/4/63/ASE RCE
G	2	NY 88	A:s	1950pc	-	-	1.52	4Booo3	RB		2EG	1952p	6000/11/51
G	2	NY 89	A:s	1950pc	-	-	1.52	4Booo4	B		2EG	1952p	6000/11/51
G	1	NY 91	[A]:r	1947po	-	-	1.50	4BEoo3	RB		WOE	1950p	4503/14 RE/Sep 50/3000
G	1	NY 92	A:r	nd:po	-	-	1.50	4BEoo3	RB		WOE	1950p	4503/14 RE/Sep 50/3000
M	3	NY 93	A	1950pc	-	a	1.50	4BooX3	R		E1G	1958b	5000/6/58/5711/OS

NY 96 appears as an additional sheet on an official copy of the 1958 index, but has not been recorded.

1	2	3	4	5	6	7	8	9	10	11	12	13	14
G	2	NY 97	A:s	1950pc	-	-	1.52	4Booo3	RB		2EG	1952p	6000/11/51
G	2	NY 98	A:s	1950pc	-	-	1.52	4Booo3	RB		2EG	1952p	6000/11/51

1	2	3	4	5	6	7	8	9	10	11	12	13	14
G	3	NY 99	A	1950pc	-	-	1.50	4BooX3	B		1EG	1952b	5000/10/52
G	1	NZ 00	A:r	1948po	-	-	1.50	4BEoo3	B		WOE	1950p	4503/14 RE/Sep 50/5000
G	3	NZ 00	B	1952pc	-	ac	1.52	4BooX3			2EG	1955b	5000/6/55 OS
M	3	NZ 00	B/	1952mc	1959c	ac	1.52	4BooX3			E3G	1961b	2000/11/61/6335/OS
G	1	NZ 01	B:r	nd:po	-	-	6.50	4BEoo3	RB		WOE	1950p	4503/14 RE/Nov 50/3000
G	1	NZ 02	A:r	nd:po	-	-	1.50	4BEoo3	RB		WOE	1950p	4503/14 RE/Sep 50/3000
G	3	NZ 03	B/	1952pc	1956mc	ac	6.52	4BooX3			E1G	1957b	3000/5/57
G	3	NZ 05	B	1954pc	-	ac	6.54	4BooX3			E1G	1956b	5000/5/56
M	3	NZ 05	B/	1954mc	1959mc	ac	6.54	4BooX3			E2G	1961b	2000/10/61/6335/OS
G	2	NZ 06	A	1948po	-	-	6.54	4BEoo3	RB	0	1EG	1952p	2500/3/53 ASE RCE
G	3	NZ 06	B	1954pc	-	ac	6.54	4BooX3		1	E2G	1955b	5000/8/55 OS
G	2	NZ 07	B	1950pc	-	ac	6.54	4BEoo3	B	0	1EG	1952p	2500/3/53 ASE RCE
G	3	NZ 07	B	1950pc	-	ac	6.50	4BooX3	B	0	E1G	1952b	5000/12/55 OS
G	1	NZ 10	A:r	1948po	-	-	6.50	4BEoo3	RB		WOE	1950p	4503/14 RE/Oct 50/5000
G	1	NZ 11	[A]:r	1946po	-	-	1.50	4BEoo3	RB		WOE	1950p	4503/14 RE/Sep 50/3000
G	2	NZ 11	[A]:r	1946po	-	-	6.52	4BEoo3	RB		2EG	nd:p	8000/10/52 SPC RE
G	1	NZ 12	A:r	nd:po	-	-	1.50	4BEoo3	RB		WOE	1950p	4503/14 RE/Nov 50/3000
G	1	NZ 12	nc:r	nd:po	-	-	1.50	4BEoo3	RB		WOE	1950p	P11/460/9000/5/52 + last
M	3	NZ 13	C/:v	1953pc	1957mc	ac	1.53	4BooX3			E1G	1957b	2500/10/57/333 2 SPC(Air)
G	3	NZ 14	A	1949pc	-	ac	1.49	4BooX3	RB		E1G	1956b	5000/1/57
G	3	NZ 15	[B]	1954pc	-	ac	6.54	4oooX3			E1G	1956b	5000/6/56
G	3	NZ 16	C	1951pc	-	ac	1.51	4BooX3	RB		1EG	1952b	5000/11/52
M	3	NZ 16	C[//]	1960^{c}mc	1960mc	ac	6.51	4BooX3	B		E2G	nd:b	2000/4/63 ASE RCE
G	3	NZ 17	B	1948pc	1950c	-	6.48	4BooX3	R	3	E1G	1954b	5500/11/54 SPC RE

Used for GSGS Misc.1662/3 Staff College Entrance Examination 1955, Tactics B, Map B.

1	2	3	4	5	6	7	8	9	10	11	12	13	14
G	3	NZ 17	B	1948pc	1950c	-	6.48	4BooX3	R	3	E1G	1956b	5000/12/55 OS
M	3	NZ 17	B/	1948mc	1959c	ac	6.48	4BooX3		3	E2G	1961b	2000/11/61/6355/OS
G	2	NZ 20	C:s	1949pc	-	-	1.52	4Booo3	RB	00	2EG	1952p	8000/6/52
G	2	NZ 21	B:s	1950pc	-	-	1.52	4Booo3	R		2EG	1952p	9000/6/52
G	2	NZ 22	A:s	1948pc	-	-	1.52	4Booo3	RB		2EG	1952p	9000/6/52
G	3	NZ 24	B	1953pc	-	ac	6.53	4BooX3			E1G	1956b	5000/6/56
M	3	NZ 24	B/	1953mc	1961ch	ac	6.53	4BooX3			E1G	1962b	2000/3/62/6394/OS
G	3	NZ 25	A	1947pc	-	-	6.47	4BooX3	RB		1EG	1952p	5000/12/52
M	3	NZ 25	[?B//]	nd	-	ac	6.54	4BooX3			E2G	nd:b	5000/6/62/ASE RCE
G	1	NZ 26	A:s	1947po	-	-	6.47	4BEoo3	R		1EG	1950p	6000/12/50 SPC RE
G	3	NZ 27	B	1954pc	-	ac	6.54	4BooX3		0	E1G	1956b	5000/2/56 OS
M	3	NZ 32	C	1952pc	-	ad	6.52	4BooX3			E1G	1958b	5000/6/58/5711/OS
M	3	NZ 32	C/	1952mc	1960mc	ad	6.52	4BooX3			E2G	1961b	2000/10/61/6335/OS
M	3	NZ 33	B	1953pc	-	ac	1.53	4BooX3	R		E1G	1958b	5000/9/58/5711/OS
M	3	NZ 33	B	nd	-	ac	1.53	4BooX3	R		E1G	nd:b	2000/9/61 ASE RCE
G	3	NZ 34	C	1953pc	-	bd	1.53	4BooX3			E1G	1956b	5000/2/56 OS
M	3	NZ 34	C//	1953mc	1959mc	bd	1.53	4BooX3			E2G	1961b	2000/10/61/6335/OS
G	3	NZ 35+	B	1951pc	-	ac	6.51	4BooX3	RB	0	1EG	1953b	5000/3/53 OS
G	1	NZ 36+	B:r	nd:po	-	-	1.47	4BEoo3	RB		1EG	1950p	6000/10/50 SPC RE
G	3	NZ 37+	B	1954pc	-	ac	6.54	4BooX3			E1G	1956b	5000/1/56 OS
M	3	NZ 37+	B/	1954pc	1957c	ac	6.54	4BooX3			E2G	1962b	2000/1/62/6335/OS
G	1	NZ 41	A:r	1947po	-	-	1.48	4BEoo3	RB	0	1EG	1950p	6000/11/50 SPC RE
G	1	NZ 42	A:s	1947po	-	-	6.47	4BEoo3	RB	0	1EG	1950p	6000/12/50 SPC RE
G	3	NZ 43+	B	1953pc	-	ac	1.53	4BooX3	R		1EG	1953b	5000/4/53 OS
M	3	NZ 43+	B/	1953pc	1957c	ac	1.53	4BooX3			E1G	1962b	2000/12/62/6394/OS
G	2	NZ 44	[A]	1947po	-	-	6.54	4BEoo3	RB		1EG	1952p	2500/12/52 ASE RCE
G	3	NZ 44	B	1953pc	-	ac	1.53	4BooX3	R		2EG	1954b	5000/5/54 OS
G	1	NZ 51	B:r	1947po	-	-	1.48	4BEoo3	RB		1EG	1950p	6000/11/50 SPC RE
G	1	NZ 52	A:r	1947po	-	-	6.47	4BEoo3	R	0	1EG	1950p	6000/9/50 SPC RE

The Newlyn legend is inadvertently blocked out with the HBM, and then reinserted below it.

1	2	3	4	5	6	7	8	9	10	11	12	13	14
G	2	NZ 61	[A]	1947po	-	-	6.54	4BEoo3	RB		1EG	1952p	2500/12/52 ASE RCE
G	3	NZ 61	[B]	1954pc	-	ac	6.54	4BooX4			E2G	1955b	5000/10/55 OS
G	3	NZ 62	A	1946pc	-	-	6.46	4BooX3	RB	0	1EG	1952b	5000/11/52
M	3	NZ 62	[B]	1954pc	-	ac	6.54	4BooX3		0	E2G	nd:b	5000/4/63 ASE RCE

1 2 3	4	5	6	7	8 9	10	11	12	13	14
G 2 NZ 71+	[A]	1947po	-	-	6.54 4BEoo3	RB		1EG	1952p	2500/12/52 ASE RCE
G 3 NZ 71+	B	1954pc	-	ac	6.54 4BooX3			2EG	1954b	5000/9/54 OS
G 1 NZ 80	A:s	1948po	-	-	1.51 4BEoo3	RB		1EG	1951p	5000/1/51
G 3 NZ 80	B	1954pc	-	ac	6.54 4BooX3			E2G	1955b	5000/10/55 OS
G 2 NZ 81	[A]	1947po	-	-	6.54 4BEoo3	RB		1EG	1952p	2500/11/52 ASE RCE
G 3 NZ 81	B	1954pc	-	ac	6.54 4BooX3			2EG	1954b	5000/10/54 OS
G 1 NZ 90+	A:s	1947po	-	-	1.51 4BEoo3	RB		1EG	1951p	5000/1/51
G 3 NZ 90+	B	1954pc	-	ac	6.54 4BooX3			2EG	1954b	5000/5/54 OS
G 1 SD 17	A:s	1946po	-	-	6.46 3BEoo1	R	00	1EG	1950p	6000/12/50 SPC RE
M 3 SD 18+	B/	1952pc	1956mc	ac	6.52 3BooX2			E1G	1959b	2500/1/59/5756/OS
M 3 SD 18+	B/	nd	-	ac	6.52 3BooX2			E1G	1959b	4000/9/61 ASE RCE
G 1 SD 20+	[A]:s	1946po	-	-	1.51 3BEoo1	R	0	1EG	1951p	6000/5/51
M 3 SD 20+	B	nd	-	ac	6.53 3BooX2		1	E2G	nd:p	2000/6/62 ASE RCE
G 1 SD 26+	A:r	1946po	-	-	6.46 3BEoo1	R		1EG	1950p	6000/9/50 SPC RE
G 1 SD 27	A:s	1946po	-	-	6.46 3BEoo1	R		1EG	1950p	5000/11/50 SPC RE
M 3 SD 28	B	1953pc	-	ac	1.53 3BooX2	R		E1G	1958b	2500/1/59/5756/OS
M 3 SD 28	B	nd	-	ac	1.53 3BooX2	R		E1G	1958b	4000/9/61 ASE RCE
G 1 SD 30	[A]:s	1946po	-	-	1.51 3BEoo1	R	0	1EG	1951p	5000/5/51
G 3 SD 31+	B	1953pc	-	ac	6.53 3BooX4		100	E1G	1956b	5000/3/56 OS
M 3 SD 31+	B/	1953pc	1958mc	ac	6.53 3BooX4		100	E2G	1962b	2000/6/62/6335/OS
G 3 SD 32+	B	1954pc	-	ac	6.54 3oooX2			E1G	1956b	5000/3/56 OS
G 3 SD 40	B	1953pc	-	ac	1.53 3BooX2	R		E1G	1956b	5000/2/56 OS
M 3 SD 40	B	1953pc	-	ac	1.53 3BooX2	R		E1G	1962b	5000/6/62/6394/OS
G 3 SD 41	B	1953pc	-	ac	6.53 3BooX2	R	1	E1G	1956b	5000/1/56 OS
G 3 SD 42	B	1955pc	-	ac	6.55 4oooX4		1	E1G	1956b	5000/2/56 OS
M 3 SD 46	C	1952mc	-	ac	6.52 4BooX3	R	0	E1G	1961b	2500/1/61/6235/OS
M 3 SD 47	B/	1952mc	1961mc	ac	6.52 4BooX3			E1G	1962b	2500/1/62/6235/OS
G 3 SD 50	B	1953pc	-	ac	6.53 3BooX2			E1G	1956b	5000/12/55 OS
G 3 SD 51	B	1950pc	-	ac	6.50 3BooX2	R		E1G	1956b	5000/7/56
G 3 SD 52	C	1955pc	-	ac	6.55 3oooX2			E1G	1956b	5000/1/56 OS
G 3 SD 53	B	1955pc	-	ac	6.55 3oooX2			E1G	1956b	5000/4/56
G 3 SD 60	B	1953pc	-	ac	1.53 3BooX4	R		E1G	1956b	5000/12/55 OS
G 3 SD 61	C	1954pc	-	ac	6.54 3oooX2			E1G	1956b	5000/1/56 OS
G 3 SD 62	B	1955pc	-	ac	6.55 3oooX2			E1G	1956b	5000/1/56 OS
G 3 SD 63	B	1954pc	-	ac	6.54 3oooX2		1	E1G	1956b	5000/4/56
G 1 SD 70	B:r	nd:po	-	-	1.47 3BEoo1	R		1EG	1951p	6000/6/51
G 3 SD 71	B	1950pc	-	ace	1.47 3BooX2	R		E1G	1956b	5000/7/56
G 3 SD 72	B	1954pc	-	ac	6.54 3oooX2			E1G	1956b	5000/1/56 OS
G 1 SD 80	B:r	1950pc	-	ac	6.50 3BEoo1	RB		1EG	1951p	6000/7/51
M 3 SD 80	B/	1950mc	1960mc	ac	6.50 3BooX2	B		E1G	1962b	2000/5/62/6394/OS
G 1 SD 81	D	1950pc	-	-	1.52 3BooX2	R		E1G	1956b	5000/12/56
G 3 SD 82	B	1954pc	-	ac	6.54 3oooX2			E1G	1956b	5000/6/56
G 1 SD 90	C:s	1951pc	-	ac	1.51 3BEoo1	R		1EG	1950p	9000/8/51
G 3 SD 91	C	1954pc	-	ac	6.54 3oooX2			E1G	1956b	5000/7/56
G 3 SD 92	B	1954pc	-	ac	6.54 3oooX2			E1G	1956b	5000/5/56
M 3 SD 92	B/	1954mc	1960mc	ac	6.54 3oooX2			E2G	1962b	2000/1/62/6335/OS
G 3 SE 00	C	1955pc	-	ac	6.55 4oooX3			E1G	1956b	5000/1/56 OS
G 3 SE 01	C	1955pc	-	ac	6.55 4oooX3			E1G	1956b	5000/1/56 OS
M 3 SE 01	C/	1955pc	1958mc	ac	6.55 4oooX3			E2G	1961b	2000/10/61/6335/OS
G 2 SE 02	B:s	1951pc	-	ac	1.52 4Booo3	RB		2EG	1952p	6000/2/52
G 1 SE 03	B:s	1950pc	-	ac	6.50 4BEoo3	RB		1EG	1951p	6000/6/51
G 3 SE 04	C	1952pc	-	ac	6.52 4BooX4	R		E1G	1956b	5000/1/57
G 3 SE 05	E	1957pc	-	ad	6.57 4oooX3			E1G	1957b	5000/11/57/5284/OS
M 3 SE 05	E/	1957pc	1958mc	ad	6.57 4oooX3			E2G	1961b	2000/10/61/6335/OS
G 1 SE 09	A:r	1948po	-	-	6.50 4BEoo3	RB		WOE	1950p	4503/14 RE/Sep 50/5000
G 3 SE 09	A	1948pc	-	-	6.48 4BooX3	RB		1EG	1950b	5000/10/54 OS
G 3 SE 10	C	1955pc	-	ac	6.55 4oooX3			E1G	1956b	5000/12/55 OS
M 3 SE 10	C/	1955mc	1960mc	ac	6.55 4oooX3			E2G	1961b	2000/10/61/6335/OS
G 1 SE 11	B:r	nd:po	-	-	6.47 4BEoo3	R	0	1EG	1951p	6000/6/51

1 2 3	4	5	6	7	8 9	10	11	12	13	14
G 1 SE 12	A:s	1946po	-	-	1.47 4BEoo3	R		1EG	1951p	6000/6/51

The print code letter is faint, sometimes invisible.

G 1 SE 13	C:s	1947pc	1950c	ac	6.47 4BEoo3	R		1EG	1951p	6000/7/51

Also overprinted in red as GSGS Misc.488/29: *Zone 1 (Part of Red Area 12)* (250/11/50).

G 3 SE 14	C	1952pc	-	ac	6.52 4BooX3	R		E1G	1956b	5000/1/57
M 3 SE 14	C/	1952mc	1959mc	ac	6.52 4BooX3			E2G	1961b	2000/10/61/6335/OS
G 3 SE 15	B	1952pc	-	ac	6.52 4BooX4			E1G	1956b	5000/1/57
M 3 SE 15	B/	1952mc	1960mc	ac	6.52 4BooX4			E2G	1961b	2000/11/61/6335/OS
G 1 SE 19	A:r	1948po	-	-	6.50 4BEoo3	RB		WOE	1950p	4503/14 RE/Sep 50/5000
G 3 SE 19	B	1952pc	-	ac	6.52 4BooX3	R		2EG	1955b	5000/10/55 OS
M 3 SE 19	B/	1952mc	1958mc	ac	6.52 4BooX3			E3G	1961b	4000/11/61/6335/OS
G 3 SE 20	D	1955pc	-	ac	6.55 4oooX3			E1G	1956b	5000/2/56 OS
M 3 SE 20	D/	1955mc	1958mc	ac	6.55 4oooX3			E2G	1961b	2000/10/61/6335/OS
G 1 SE 21	B:r	nd:po	-	-	1.47 4BEoo3	RB		1EG	1951p	6000/7/51
G 1 SE 22	[A]:s	1946po	-	-	1.47 4BEoo3	R		1EG	1951p	6000/7/51
G 1 SE 23	B:r	nd:po	-	-	1.47 4BEoo3	RB		1EG	1951p	6000/8/51

Also overprinted in red in GSGS Misc.[488]: *Yorkshire – West Riding* : *Detail Map of Red Area 13,* 1950.

G 3 SE 24	C	1952pc	-	ac	6.52 4BooX3	R	3	E1G	1956b	5000/1/57
G 3 SE 25	B	1952pc	-	ac	6.52 4BooX3	R		E1G	1956b	5000/1/57
M 3 SE 25	B	1952pc	-	ac	6.52 4BooX3	R		E1G	1961b	no code, 135 SER (TA)

The colour tabs are on this printing.

G 1 SE 27	A	1948po	-	-	6.50 4BEooo3	RB		WOE	1950p	4503/14 RE/Sep 50/5000
G 2 SE 27	A	1948po	-	-	6.52 4BEooo3	RB		2EG	1952p	9000/11/52 SPC RE
G 1 SE 29	A:r	1948po	-	-	6.50 4BEoo3	RB	1	WOE	1950p	4503/14 RE/Oct 50/5000
G 3 SE 29	B	1952pc	-	ac	6.52 4BooX3	R	1	2EG	1953b	5000/1/53
G 3 SE 30	C	1955pc	-	bd	6.55 4oooX3			E1G	1956b	5000/1/56 OS
M 3 SE 30	C/	1955mc	1960mc	bd	6.55 4oooX3			E2G	1961b	2000/11/61/6335/OS
G 1 SE 31	A:s	1947po	-	-	6.47 4BEoo3	RB		1EG	1951p	5000/8/51
G 1 SE 32	B:r	nd:po	-	-	6.47 4BEoo3	RB		1EG	1951p	5000/8/51
G 1 SE 33	B:r	nd:po	-	-	1.47 4BEoo3	RB		1EG	1951p	5000/8/51
G 3 SE 34	B	1951pc	-	ac	1.51 4BooX3	RB		E1G	1956b	5000/7/56
G 3 SE 35	B	1952pc	-	ac	6.52 4BooX4	R		E1G	1956b	5000/1/57
M 3 SE 35	B/	1952pc	1957mc	ac	6.52 4BooX4			E2G	1961b	2000/11/61/6335/OS
G 1 SE 36	B	nd:po	-	-	6.50 4BEoo3	RB		WOE	1950p	4501/14 RE/Sep 50/5000
G 1 SE 36	B	nd:po	-	-	6.50 4BEoo3	RB		1EG	1950p	9000/3/52 SPC RE
G 2 SE 36	[B]:r	nd:po	-	-	6.52 4BEoo3	RB		2EG	1952p	9000/11/52 SPC
G 1 SE 37	A	1948po	-	-	6.50 4BEoo3	RB	00	WOE	1950p	4503/14 RE/Sep 50/5000
G 3 SE 37	C	1953pc	-	ac	1.53 4BooX3	R	11	2EG	1954b	5000/5/54 OS
G 3 SE 40	B/	1952pc	1956mc	ac	6.52 4BooX3		11	E1G	1956b	5000/1/57
G 3 SE 41	B	1953pc	-	ac	1.53 4BooX3	R		E1G	1956b	5000/1/56 OS
M 3 SE 41	B/	1953pc	1958mc	ac	1.53 4oooX3			E2G	1961b	2000/10/61/6335/OS
G 3 SE 42	B	1954pc	-	ac	6.54 4BooX3			E1G	1956b	5000/2/56 OS
M 3 SE 42	B/	1954mc	1958c	ac	6.54 4BooX3			E2G	1961b	2000/11/61/6335/OS
G 3 SE 43	C	1954pc	-	ac	6.54 4BooX3			E1G	1956b	5000/1/56 OS
M 3 SE 43	C/	1954mc	1959mc	ac	6.54 4BooX3			E2G	1961b	2000/10/61/6335/OS
G 3 SE 44	C	1954pc	-	ac	6.54 4BooX3			E1G	1956b	5000/1/56 OS
G 3 SE 45	B	1954pc	-	ac	6.54 4BooX3		1	E1G	1956b	5000/2/56 OS
G 2 SE 46	[B]	1948pc	1951	-	6.54 4BEoo3	RB	00	1EG	1952p	2500/3/53 ASE RCE
G 3 SE 46	B	1948pc	1951	-	1.51 4BooX3	RB	00	E1G	1952b	5000/9/55 OS
M 3 SE 47	B/	1952pc	1957mc	ac	6.52 4BooX3		11	E1G	1959b	2500/5/59/5900/OS
M 3 SE 47	B/	nd	-	ac	6.52 4BooX3		11	E1G	1959b	4000/9/61 ASE RCE
G 3 SE 50	D	1952pc	-	ad	6.52 4BooX3	R	11	E1G	1956b	5000/1/57
M 3 SE 50	D/	1952mc	1960mc	ad	6.52 4BooX3		11	E2G	1961b	2000/12/61/6335/OS
G 3 SE 51	B	1952pc	-	ac	6.52 4BooX3	R		E1G	1956b	5000/1/57
M 3 SE 51	B/	1952mc	1958mc	ac	6.52 4BooX3	R		E2G	1961b	2000/11/61/6335/OS
M 3 SE 52	D/:v	1954pc	1958mc	ac	6.54 4BooX3		11	E1G	1959b	2500/6/59/5900/OS

Erroneously classified M721 in the bottom margin.

M 3 SE 52	D//	1954mc	1961ch	ac	6.54 4BooX3		11	E2G	1961b	4000/11/61/6335/OS
G 2 SE 53	[A]	1947po	-	-	6.54 4BEoo3	RB	00	1EG	1952p	2500/1/53 ASE RCE
G 3 SE 53	B	1954pc	-	ac	6.54 4BooX4		11	2EG	1954b	5000/5/54 OS

1 2 3	4	5	6	7	8	9	10	11	12	13	14
G 3 SE 54	B	1954pc	-	ac	6.54	4BooX3		11	E1G	1956b	5000/12/55 OS
G 3 SE 55	B	1954pc	-	ac	6.54	4BooX4		13	E1G	1956b	5000/2/56 OS
M 3 SE 55	B/	1954pc	1957c	ac	6.54	4BooX4		11	E2G	1961b	2000/10/61/6335/OS
M 3 SE 56	C	1954pc	-	ac	6.54	4BooX3		1	E1G	1959b	2500/6/59/5900/OS
M 3 SE 56	C	nd	-	ac	6.54	4BooX3		1	E1G	1959b	3000/9/61 ASE RCE
M 3 SE 57	B/	1954mc	1959mc	ac	6.54	4BooX4			E1G	1960b	2500/1/60/5900/OS
M 3 SE 57	B/	nd	1959mc	ac	6.54	4BooX4			E1G	1960b	4000/9/61 ASE RCE
G 2 SE 60	A	1947pc	-	-	6.54	4Booo3	RB	0	1EG	1952p	2500/11/52 ASE RCE
G 3 SE 60	B	1952pc	-	ac	6.52	4BooX3	R	1	E2G	1955b	5000/10/55 OS
M 3 SE 61	C/	1952mc	1960mc	ac	6.52	4BooX3			E1G	1961b	2500/1/61/5900/OS
M 3 SE 62	B	1954pc	-	ac	6.54	4BooX4		11	E1G	1959b	2500/5/59/5900/OS
M 3 SE 62	B	nd	-	ac	6.54	4BooX4		11	E1G	1959b	3000/9/61 ASE RCE
M 3 SE 63	C	1954pc	-	ac	6.54	4BooX3		1	E1G	1958b	5000/5/58/5711/OS
M 3 SE 63	C/	1954mc	1959mc	ac	6.54	4BooX3		1	E2G	1961b	2000/11/61/6335/OS
M 3 SE 64	B	1954pc	-	ac	6.54	4BooX4		1	E1G	1958b	5000/5/58/5711/OS
M 3 SE 64	B/	1954mc	1960mc	ac	6.54	4BooX4		1	E2G	1961b	2000/11/61/6335/OS
M 3 SE 65	B	1954pc	-	ac	6.54	4BooX3			E1G	1958b	5000/5/58/5711/OS
M 3 SE 65	B/	1954mc	1960mc	ac	6.54	4BooX3			E2G	1962b	2000/1/62/6335/OS
M 3 SE 66	C	1954pc	-	ac	6.54	4BooX4		1	E1G	1958b	5000/6/58/5711/OS
M 3 SE 66	C	nd	-	ac	6.54	4BooX4		1	E1G	1958b	2000/9/61 ASE RCE
M 3 SE 67	C	1954pc	-	ac	6.54	4BooX3			E1G	1959b	2500/7/59/5900/OS
M 3 SE 67	C	nd	-	ac	6.54	4BooX3			E1G	1959b	4000/9/61 ASE RCE
M 3 SE 70	B	1953pc	-	ac	6.53	4BooX3		1	E1G	1959b	2500/6/59/5900/OS
M 3 SE 70	B	nd	-	ac	6.53	4BooX3		1	E1G	1959b	4000/9/61 ASE RCE
M 3 SE 71	B	1953pc	-	ac	1.53	4BooX3	R		E1G	1960b	2500/1/60/5900/OS
M 3 SE 71	B	nd	-	ac	1.53	4BooX3	R		E1G	1959b	4000/9/61 ASE RCE
G 3 SE 72	A	1947pc	-	-	1.48	4BooX3	RB		1EG	1952b	5000/12/52
M 3 SE 73	B/	1954pc	1956mc	ac	6.54	4BooX4		1	E1G	1958b	5000/9/58/5711/OS
M 3 SE 73	B/	nd	-	ac	6.54	4BooX4		1	E1G	1958b	2000/9/61 ASE RCE
M 3 SE 74	B	1953pc	-	ac	6.53	4BooX4		11	E1G	1958b	5000/9/58/5711/OS
M 3 SE 75	B	1954pc	-	ac	6.54	4BooX3		1	E1G	1958b	5000/7/58/5711/OS
M 3 SE 76	B	1954pc	-	ac	6.54	4BooX3			E1G	1958b	5000/7/58/5711/OS
M 3 SE 76	B/	1954mc	1960mc	ac	6.54	4BooX3			E2G	1961b	2000/11/61/6335/OS
M 3 SE 77	B/	1954mc	1958mc	ac	6.54	4BooX3			E1G	1959b	2500/4/59/5900/OS
M 3 SE 77	B/	nd	-	ac	6.54	4BooX3			E1G	1959b	3000/9/61 ASE RCE
G 3 SE 82	A	1946pc	-	-	1.47	4BooX3	RB		1EG	1953b	5000/1/53
M 3 SE 83	[B]	1953pc	-	ac	6.53	4BooX3		1	E1G	1959b	2500/8/59/5982/OS
M 3 SE 83	B	nd	-	ac	6.53	4BooX3		1	E1G	1959b	3000/9/61 ASE RCE
M 3 SE 84	B/	1953pc	1958mc	ac	6.53	4oooX3			E1G	1959b	2500/8/59/5982/OS
M 3 SE 84	B/	nd	-	ac	6.53	4oooX3			E1G	1959p	3000/11/61 ASE RCE
G 2 SE 85	[A]	1947po	-	-	6.54	4BEoo3	B		1EG	1952p	5000/11/52 ASE RCE
G 2 SE 86	[A]	1946po	-	-	6.54	4BEoo3	RB		1EG	1952p	5000/3/53 ASE RCE
G 2 SE 87	[A]	1946po	-	-	6.54	4BEoo3	RB		1EG	1952p	5000/11/52 ASE RCE
G 2 SE 88	[A]	1946po	-	-	6.54	4BEoo3	RB		1EG	1952p	2500/12/52 ASE RCE
G 3 SE 88	B	1954pc	-	ac	6.54	4BooX3			E2G	1955b	5000/8/55 OS
G 1 SE 89	[B]:r	1946po	-	-	1.47	4BEoo3	RB		WOE	1950p	4503/14 RE/Aug 50/5000
G 2 SE 89	[B]:r	1946po	-	-	6.52	4BEoo3	RB		2EG	1952p	9000/10/52 SPC RE
M 3 SE 89	C	nd	-	ac	6.54	4BooX3			E3G	nd:p	5000/6/62 ASE RCE
G 3 SE 91	C	1953pc	-	ac	1.53	4BooX3	R		E1G	1956b	5000/2/56 OS
G 3 SE 92	A	1947pc	-	-	6.47	4BooX3	R	0	1EG	1952b	5000/12/52
M 3 SE 93	B/	1953mc	1958mc	ac	6.53	4BooX3			E1G	1959b	2500/7/59/5982/OS
M 3 SE 93	B/	nd	-	ac	6.53	4BooX3			E1G	1959b	4000/9/61 ASE RCE
G 2 SE 94	[A]	1947po	-	-	6.54	4BEoo3	RB		1EG	1952p	2500/3/53 ASE RCE
G 3 SE 94	B	1953pc	-	ac	6.53	4BooX3			E2G	1955b	5000/8/55 OS
G 2 SE 95	B	1947pc	1952c	-	6.54	4Booo3	RB	0	1EG	1952p	7500/11/52 ASE RCE
G 2 SE 96	[A]	1946po	-	-	6.54	4BEoo3	RB	0	1EG	1952p	5000/12/52 ASE RCE
G 2 SE 97	A	1947pc	-	-	6.54	4Booo3	RB		1EG	1952p	5000/11/52 ASE RCE
G 3 SE 97	B	1954pc	-	ac	6.54	4BooX3			E2G	1955b	5000/8/55 OS
G 2 SE 98	A	1947pc	1951	-	6.54	4Booo3	RB		1EG	1952p	2500/11/52 ASE RCE
G 3 SE 98	B	1954pc	-	ac	6.54	4BooX3			E2G	1955b	5000/7/55 OS

1	2	3	4	5	6	7	8	9	10	11	12	13	14
G	1	SE 99+	[A]:r	1948po	-	-	1.48	4BEoo3	RB		WOE	1950p	4503/14 RE/Aug 50/5000
G	3	SE 99+	B	1951pc	-	ac	1.51	4BooX3	RB		2EG	1953b	5000/2/53 OS
M	3	SH 33	B	1956pc	-	ac	6.56	4oooX3		2	E1G	1959b	2500/4/59/5900/OS
M	3	SH 33	B/	1956mc	1960mc	ac	6.56	4oooX3		2	E2G	1961b	3000/11/61/6335/OS
M	3	SH 43	B	1956pc	-	ac	6.56	4oooX3			E1G	1959b	2500/6/59/5900/OS
M	3	SH 43	B/	1956mc	1960mc	ac	6.56	4oooX3			E2G	1962b	3000/1/62/6335/OS
G	3	SH 52	A	1953pc	-	ac	6.53	4BooX4		1	E1G	1954b	5000/12/54 OS
G	2	SH 53	A	1950pc	1952	-	6.54	4Booo3	R		1EG	1952p	5000/11/52 ASE RCE
G	3	SH 54	A	1953pc	-	ac	6.53	4BooX4			E1G	1954b	5000/12/54 OS
G	3	SH 55	A	1953pc	-	ac	6.53	4BooX4	R		E1G	1954b	5000/11/54 OS
G	2	SH 56	[A]	1950pc	-	-	6.54	4BEoo3	R		1EG	1952p	5000/10/52 ASE RCE
M	3	SH 60+	A/	1953mc	1960mc	ac	6.53	4BooX3			E1G	1961b	2500/2/61/6235/OS
M	3	SH 61+	A	1953pc	-	ac	6.53	4BooX3			E1G	1959b	2500/9/59/5900/OS
M	3	SH 61+	A/	1953mc	1960mc	ac	6.53	4BooX3			E2G	1961b	3000/10/61/6335/OS
G	3	SH 62	A	1953pc	-	ac	6.53	4BooX4			1EG	1953b	8000/11/53 OS
G	3	SH 63	A	1953pc	-	ac	6.53	4BooX4			1EG	1953b	8000/11/53 OS
G	3	SH 64	A	1953pc	-	ac	6.53	4BooX4			E1G	1954b	5000/12/54 OS
G	3	SH 65	A	1953pc	-	ac	6.53	4BooX4			E1G	1954b	5000/12/54 OS
G	3	SH 66	A	1953pc	-	ac	6.53	4BooX4			E1G	1954b	5000/12/54 OS
G	2	SH 67	[A]	nd:po	-	-	6.54	4BEoo3	R		1EG	1952p	5000/11/52 ASE RCE
G	3	SH 72	A	1953pc	-	ac	6.53	4BooX4			1EG	1953b	8000/11/53 OS
G	3	SH 73	A	1953pc	-	ac	6.53	4BooX4			1EG	1953b	8000/11/53 OS
G	3	SH 74	A	1953pc	-	ac	6.53	4BooX4			E1G	1954b	5000/12/54 OS
G	3	SH 75	A	1953pc	-	ac	6.53	4BooX4			E1G	1954b	5000/12/54 OS
G	3	SH 76	A	1953pc	-	ac	6.53	4BooX4			E1G	1954b	5000/12/54 OS
G	2	SH 77	[B]	nd:po	-	-	6.54	4BEoo3	R		1EG	1952p	5000 12 52 ASE RCE
G	3	SH 82	A	1953pc	-	ac	6.53	4BooX4			1EG	1953b	8000/11/53 OS
G	3	SH 83	A	1953pc	-	ac	6.53	4BooX4			1EG	1953b	8000/11/53 OS
M	3	SH 96	A	1951mc	-	ac	1.51	4BooX4			E1G	1960b	2500/2/61/6235/OS
M	3	SH 97+	A/	1950pc	1958mc	ac	1.50	4BooX3			E1G	1961b	2500/1/61/6235/OS
M	3	SJ 06	B	1956pc	-	ac	6.56	4oooX3			E1G	1960b	2500/5/61/6235/OS
M	3	SJ 07	B/	1950pc	1957mc	ac	6.50	3BooX2			E1G	1961b	2500/5/61/6235/OS
G	3	SJ 27	B	1953pc	-	ac	6.53	3BooX2			E1G	1956b	5000/2/56 OS
G	1	SJ 28	[A]:s	1946po	-	-	1.47	3BEoo1	R		1EG	1951p	6000/5/51
G	1	SJ 29+	[A]:s	1946po	-	-	1.51	3BEoo1	R		1EG	1951p	6000/2/51
G	3	SJ 37	B	1951pc	-	ac	1.52	3BooX2	R	10	1EG	1952b	5000/10/52
G	1	SJ 38	B:s	1951pc	-	ac	1.51	3BEoo1	RB		1EG	1951p	6000/8/51
M	3	SJ 38	B/	1951pc	1958c	ac	1.52	3BooX2	B		E1G	1962b	2000/5/62/6394/OS
G	1	SJ 39	[A]:s	1946po	-	-	6.46	3BEoo1	R		1EG	1951p	6000/5/51
M	3	SJ 45	C	1961mc	-	ac	6.61	4oooX3		1	E1G	1961b	2500/8/61/6235/OS
G	3	SJ 47	B	1952pc	-	ac	1.52	3BooX1	R		1EG	1952b	5000/10/52
G	1	SJ 48	B:s	1951pc	-	ac	1.51	3BEoo1	R	3	1EG	1951p	6000/8/51
M	3	SJ 48	B/	1951mc	1960mc	ac	1.51	3BooX1		3	E2G	1961b	2000/11/61/6335/OS
G	1	SJ 49	[A]:s	1946po	-	-	1.46	3BEoo1	R		1EG	1951p	6000/6/51
M	3	SJ 49	B/	nd	-	ac	6.53	3BooX2			E2G	nd:p	2000/6/62 ASE RCE
M	3	SJ 55	C	1959mc	-	ac	6.59	4oooX3		3	E1G	1961b	2500/2/61/6235/OS
G	3	SJ 57	A	1947pc	-	-	1.47	3BooX2	RB		1EG	1952b	5000/12/52
G	3	SJ 58	B	1950pc	-	ac	6.50	3BooX2	R	0	1EG	1952b	5000/11/52
G	3	SJ 59	B	1953pc	-	ac	6.53	3BooX2		1	1EG	1953b	5000/6/53 OS
G	3	SJ 67	B	1952pc	-	ac	6.52	3BooX2	R		1EG	1956b	5000/7/56
G	3	SJ 68	C	1952pc	-	ac	6.52	3BooX2	R	1	1EG	1953b	5000/1/53
G	3	SJ 69	B	1953pc	-	ac	1.53	4BooX4	R		1EG	1953b	5000/5/53 OS
M	3	SJ 69	B/	1953pc	1956mc	ac	1.53	4BooX4			E1G	1962b	2000/5/62/6394/OS
G	3	SJ 70	[B]	1951pc	-	ac	6.51	4BooX3	R	0	E1G	1956b	5000/1/57
G	3	SJ 71	B	1951pc	-	ac	6.51	4BooX3	R		E1G	1956b	5000/11/56
G	2	SJ 74	B	nd:po	-	-	6.54	4BEoo3	RB		1EG	1952p	2500/12/52 ASE RCE
G	3	SJ 74	[B]	1949pc	-	-	6.48	4BooX3	RB		E1G	1952b	5000/9/55 OS
G	2	SJ 75	[C]	1952pc	-	ac	6.54	4Booo3	R		1EG	1952p	2500/12/52 ASE RCE
M	3	SJ 75	C/	1952mc	1960mc	ac	6.52	4BooX3			E2G	1962b	4000/4/62/6355/OS
G	3	SJ 77	B	1952pc	-	ac	1.52	3BooX2	R	1	E1G	1956b	5000/1/57

1 2 3	4	5	6	7	8	9	10	11	12	13	14
G 3 SJ 78	B	1951pc	-	ac	1.51	3BooX2	R		1EG	1952b	5000/11/52
M 3 SJ 78	B//	1951mc	1960mc	ac	1.51	3BooX2			E2G	1961b	2000/12/61/6335/OS
G 1 SJ 79	B:r	nd:po	-	-	1.51	3BEoo1	R	0	1EG	1951p	6000/6/51
G 3 SJ 80	B	1952pc	-	ac	1.52	4BooX3	R	20	1EG	1953b	5000/1/53
G 3 SJ 81	A	1949pc	-	ac	1.49	4BooX3	RB	0	E1G	1956b	5000/1/57
G 3 SJ 83	C	1952pc	-	ac	1.52	4BooX3	R		E1G	1956b	5000/10/56
G 3 SJ 84	B	1951pc	-	ac	1.51	4BooX3	R		E1G	1956b	5000/8/56
G 3 SJ 85	B	1952pc	-	ac	6.52	4BooX3	R		E1G	1956b	5000/1/57
M 3 SJ 87	C	1957pc	-	ac	6.57	3oooX2			E1G	1958b	5000/4/58/5284/OS
G 3 SJ 88	C	1951pc	-	ac	1.51	3BooX2	R	00	1EG	1952b	5000/12/52
M 3 SJ 88	C//	1951mc	1960ch	ac	1.51	3BooX2		00	E2G	1961b	3000/11/61/6335/OS
G 1 SJ 89	B:r	nd:po	-	-	1.46	3BEoo1	RBC		1EG	1951p	6000/5/51

This sheet differs from the civilian edition B in that the corner extensions have been deleted.

G 3 SJ 90	B	1950pc	-	ac	6.52	4BooX3	R	2	1EG	1952b	5000/10/52
G 3 SJ 91	B	1952pc	-	ac	1.52	4BooX3	R		E1G	1956b	5000/8/56
G 3 SJ 92	B	1952pc	-	ac	6.52	4BooX3	R	1	E1G	1956b	5000/7/56
G 3 SJ 93	[B]	1952pc	-	ac	1.52	4BooX3	R		E1G	1956b	5000/1/57
G 3 SJ 94	B	1952pc	-	ac	6.52	4BooX3	R	3	E1G	1956b	5000/7/56
G 3 SJ 95	B	1950pc	-	ac	6.50	4BooX3	RB		E1G	1956b	5000/1/57
G 3 SJ 98	B	1952pc	-	ac	6.52	3BooX2	R	1	1EG	1953b	5000/1/53
G 2 SJ 99	B:s	1951pc	-	ac	1.52	3Booo2	RC		2EG	1952p	6000/5/52
M 3 SJ 99	B/	1951mc	1958c	ac	1.52	3BooX2	C		E3G	1962b	3000/1/62/6335/OS
G 3 SK 00	B	1951pc	-	ac	1.51	4BooX3	R		E1G	1956b	5000/3/56 OS
G 3 SK 01	B	1952pc	-	ac	6.52	4BooX3	R		E1G	1956b	5000/2/56 OS
G 3 SK 02	A	1949pc	-	-	1.49	4BooX3	RB	0	E1G	1956b	5000/2/56 OS
M 3 SK 05	B/	1951mc	1959mc	ac	6.51	4BooX3			E1G	1961b	2500/11/60/6235/OS
M 3 SK 06	B/	1950mc	1958mc	ac	6.50	4BooX3			E1G	1961b	2500/11/60/6235/OS
M 3 SK 06	B/	1950mc	1958mc	ac	6.50	4BooX3			E1G	1961b	2500/9/61 13 FSS RE
G 3 SK 08	D	1955pc	-	bd	6.55	4oooX3			E1G	1956b	5000/3/56 OS
G 3 SK 09	C	1955pc	-	ac	6.55	4oooX3			E1G	1956b	5000/3/56 OS
M 3 SK 09	C/	1955mc	1960mc	ac	6.55	4oooX3			E2G	1961b	2000/11/61/6335/OS
G 3 SK 10	B	1951pc	-	ac	1.51	4BooX3	RB		E1G	1956b	5000/4/56
G 3 SK 11	A	1948pc	-	-	6.48	4BooX3	RB	0	E1G	1956b	5000/2/56 OS
G 3 SK 12	A	1949pc	-	a	6.48	4BooX3	B	0	E1G	1956b	5000/2/56 OS
G 3 SK 19	B	1955pc	-	ac	6.55	4oooX3			E1G	1956b	5000/1/56 OS
M 3 SK 19	B/	1955mc	1958mc	ac	6.55	4oooX3			E2G	1961b	2000/12/61/6335/OS
G 3 SK 20	B	1952pc	-	ac	1.52	4BooX3	R		E1G	1956b	5000/5/56 OS
M 3 SK 20	B/	1952mc	1960mc	ac	1.52	4BooX3			E2G	1961b	2000/11/61/6335/OS
G 3 SK 21	B	1952pc	-	ac	6.52	4BooX3	R		E1G	1956b	5000/3/56 OS
M 3 SK 21	B/	1952pc	1957mc	ac	6.52	4BooX3			E2G	1962b	2000/4/62/6335/OS
G 3 SK 23	A	1948pc	-	-	1.52	4BooX3	RB	00	E1G	1956b	5000/3/56 OS
G 3 SK 24	A	1948pc	-	-	1.48	4BooX3	RB	0	E1G	1956b	5000/3/56 OS
G 3 SK 26	A	1949pc	-	-	1.49	4BooX3	RB		E1G	1956b	5000/5/56 OS
G 3 SK 27	[C]	1951pc	-	ac	1.51	4BooX3	RB		E1G	1956b	5000/6/56
G 3 SK 28	C	1951pc	-	ac	1.51	4BooX3	RB		E1G	1956b	5000/2/56 OS
M 3 SK 28	C/	1951mc	1960mc	ac	1.51	4BooX3	B		E2G	1962b	2000/2/62/6335/OS
G 3 SK 29	C	1955pc	-	ac	6.55	4oooX3			E1G	1956b	5000/2/56 OS
M 3 SK 29	C/	1955mc	1958mc	ac	6.55	4oooX3			E2G	1961b	2000/11/61/6335/OS
G 3 SK 30	B	1951pc	-	ac	1.51	4BooX3	RB		E1G	1956b	5000/4/56 OS
M 3 SK 30	B//	1951mc	1960mc	ac	1.51	4BooX3			E1G	1962b	2000/5/62/6394/OS
G 3 SK 32	B	1951pc	-	ac	1.51	4BooX3	RB		E1G	1956b	5000/5/56
M 3 SK 32	B/	1951pc	1958mc	ac	1.51	4BooX3	B		E2G	1961b	2000/11/61/6335/OS
G 1 SK 33	B:s	1947po	-	-	1.51	4BEoo3	RB		1EG	1951p	5000/5/51
M 3 SK 33	B/	nd		ac	6.52	4BooX3			E2G	nd:p	2000/6/62/ ASE RCE
G 3 SK 34	B	1951pc	-	ac	6.51	4BooX3	R		E1G	1956b	5000/2/56 OS
M 3 SK 34	B//	1951mc	1961ch	ac	6.51	4BooX3			E2G	1961b	2000/11/61/6335/OS
G 3 SK 36	B	1951pc	-	ac	6.51	4BooX3	R		E1G	1956b	5000/6/56 OS
M 3 SK 36	B/	1951mc	1960mc	ac	6.51	4BooX3			E2G	1961b	2000/11/61/6335/OS
G 3 SK 37	[C]	1952pc	-	ac	6.52	4BooX4	R		E1G	1956b	5000/6/56
G 2 SK 38	B:s	1951pc	-	ac	1.52	4Booo3	RB		2EG	1952p	6000/2/52

1	2	3	4	5	6	7	8	9	10	11	12	13	14
G	1	SK 39	B:r	nd:po	-	-	6.47	4BEoo3	RB		1EG	1951p	6000/6/51
G	3	SK 42	B	1952pc	-	ac	6.52	4BooX3	R	1	E1G	1956b	5000/6/56
M	3	SK 42	B//	1952mc	1961ch	acf	6.52	4BooX3		1	E2G	1961b	2000/11/61/6335/OS
G	3	SK 43	C	1952pc	-	ad	1.52	4BooX3	R		E1G	1956b	5000/6/56
G	3	SK 44	B	1951pc	-	ac	6.51	4BooX3	RB		E1G	1956b	5000/5/56
M	3	SK 44	B/	1951mc	1958mc	ac	6.51	4BooX3	B		E2G	1961b	2000/11/61/6335/OS
G	3	SK 45	[C]	1951pc	-	ac	6.51	4BooX3	R		E1G	1956b	5000/6/56
G	3	SK 46	B	1951pc	-	ac	1.51	4BooX3	RB		E1G	1956b	5000/5/56
M	3	SK 46	B/	1951mc	1960mc	ac	1.51	4BooX3	B		E2G	1961b	2000/11/61/6335/OS
G	3	SK 47	C	1953pc	-	bd	1.53	4BooX3	R		E1G	1956b	5000/6/56
M	3	SK 47	C/	nd	-	bd	1.53	4BooX3			E2G	nd:p	2000/6/62 ASE RCE
G	1	SK 48	A:s	1946po	-	-	6.46	4BEoo3	RB		1EG	1951p	6000/6/51
G	1	SK 49	A:s	1946po	-	-	6.46	4BEoo3	RB		1EG	1951p	6000/6/51
G	3	SK 50	B	1952pc	-	ac	6.52	4BooX3	RC		E1G	1956b	5000/4/56
M	3	SK 50	B//	1952mc	1962ch	acf	6.52	4BooX3	C		E1G	1962b	2000/7/62/6394/OS
G	3	SK 51	C	1951pc	-	ac	6.51	4BooX3	R		E1G	1956b	5000/5/56
M	3	SK 51	C/	1951pc	1958mc	ac	6.51	4BooX3			E2G	nd:b	2000/4/63 ASE RCE
G	3	SK 53	[C]	1952pc	-	ac	6.52	4BooX3	R		E1G	1956b	5000/5/56
G	3	SK 54	C	1951pc	-	ac	6.51	4BooX3	R	1	E1G	1956b	5000/5/56
M	3	SK 54	C	nd	-	ac	6.51	4BooX3	R	1	E2G	nd:p	2000/6/62 ASE RCE
G	3	SK 55	C	1951pc	-	ac	6.51	4BooX3	R		E1G	1956b	5000/6/56
M	3	SK 55	C	nd	-	ac	6.51	4BooX3			E2G	nd:p	2000/6/62 ASE RCE
M	3	SK 56	C:u	1952pc	-	ac	6.52	4Booo3	R		E1G	1957b	2500/10/57/333 2 SPC(Air)
M	3	SK 56	C/	1952pc	1958c	ac	6.52	4BooX3			E1G	1962b	4000/6/62/6394/OS
M	2	SK 57	B:t	1952pc	-	ac	6.52	4Booo3	R		E1G	1957b	2500/10/57/333 2 SPC(Air)
M	3	SK 57	B/	1952pc	1956mc	ac	6.52	4BooX3			E1G	1962b	4000/6/62/6394/OS
G	3	SK 58	[C]	1952pc	-	ac	6.52	4BooX3	R	0	E1G	1956b	5000/6/56
G	3	SK 59	B	1952pc	-	ac	6.52	4BooX3	R		E1G	1956b	5000/2/56 OS
M	3	SK 59	B	nd	-	ac	6.52	4BooX3	R		E2G	nd:p	2000/6/62 ASE RCE
G	2	SK 60	B	1951pc	-	ac	6.54	4Booo4	RB	0	1EG	1952p	2500/10/52 ASE RCE
G	3	SK 60	B	1951pc	-	ac	6.54	4Booo4	RB	0	1EG	1952b	5000/5/54 SPC RE
M	3	SK 61	B/	1951pc	1957mc	ac	1.51	4BooX3	B	0	E1G	1958b	2500/11/58/5756/OS
M	3	SK 61	B[/]	1951pc	1957mc	ac	1.51	4BooX3	B	0	E1G	nd:p	4000/6/62 ASE RCE
G	2	SK 62	B	nd:po	-	-	6.54	4BEoo3	RB		1EG	1952p	2500/11/52 ASE RCE
G	3	SK 62	B	nd:po	-	-	6.54	4BEoo3	RB		1EG	1952b	5000/5/54 SCE RE
G	3	SK 63	B/	1952pc	1956mc	ac	6.52	4BooX3		3	E1G	1956b	5000/6/56
G	3	SK 64	C	1951pc	-	ac	6.51	4BooX3	R	0	E1G	1956b	5000/6/56
M	3	SK 64	C/	nd	nd	ac	6.51	4BooX3		0	E2G	nd:p	2000/6/62 ASE RCE
G	3	SK 65	[A]	1948pc	-	-	6.48	4BooX3	RB		E1G	1956b	5000/6/56
M	3	SK 66	B	1951pc	-	ac	1.51	4BooX3	RB		E1G	1957b	2500/10/57/333 2 SPC(Air)
M	3	SK 66	B/	1951pc	1956mc	ac	1.51	4BooX3	B		E1G	1962b	4000/6/62/6394/OS
M	3	SK 67	C:u	1953pc	-	ac	6.53	4BooX4		1	E1G	1957b	2500/10/57/333 2 SPC(Air)

All the copies seen have colour tabs in the right hand margin.

M	3	SK 67	C	nd	-	ac	6.53	4BooX4		1	E1G	1957b	4000/9/61 ASE RCE
M	3	SK 68	D/	1952pc	1958mc	ac	6.52	4BooX4		1	E1G	1959b	2500/8/59/5982/OS

Worksop Aerodrome.

M	3	SK 68	D/	nd	-	ac	6.52	4BooX4		1	E1G	1959b	3000/9/61 ASE RCE
G	3	SK 69	B	1953pc	-	ac	6.53	4BooX4		11	E1G	1956b	5000/2/56 OS
M	3	SK 70	B	1951pc	-	ac	1.51	4BooX3	RB		E1G	1958b	2500/12/58/5756/OS
M	3	SK 70	B	nd	-	ac	1.51	4BooX3	RB		E1G	nd:p	4000/6/62 ASE RCE
M	3	SK 71	C/	1955pc	1957c	ac	6.55	4oooX3		3	E1G	1958b	2500/1/59/5756/OS
M	3	SK 71	C/	nd	-	ac	6.55	4oooX3		3	E1G	1958b	4000/9/61 ASE RCE
M	3	SK 72	C	1954pc	-	ac	6.54	4oooX3			E1G	1958b	2500/11/58/5756/OS
M	3	SK 72	C/	nd	-	ac	6.54	4oooX3			E2G	nd:p	4000/6/62 ASE RCE
M	3	SK 73	C	1954pc	-	ac	6.54	4oooX3		1	E1G	1958b	2500/11/58/5756/OS
M	3	SK 73	C/	nd	-	ac	6.54	4oooX3		1	E2G	nd:p	4000/6/62 ASE RCE
M	3	SK 74	C/	1955pc	1958mc	ac	6.55	4oooX3		1	E1G	1959b	2500/9/59/5982/OS

Syerston Airfield. There is also a blank area at 7841.

M	3	SK 74	[C/]	1955pc	1958mc	ac	6.55	4oooX3		1	E1G	1959b	4000/9/61 ASE RCE

1	2	3	4	5	6	7	8	9	10	11	12	13	14
M	3	SK 75	B/	1951pc	1956mc	ac	6.51	4BooX3	B		E1G	1959b	2500/9/59/5982/OS
M	3	SK 75	B/	nd	-	ac	6.51	4BooX3	B		E1G	1959b	4000/9/61 ASE RCE
M	3	SK 76	B	1959mc	-	ac	6.59	4oooX3		3	E1G	1959b	2500/12/59/5756/OS
M	3	SK 76	B	nd	-	ac	6.59	4oooX3		3	E1G	1959b	4000/9/61 ASE RCE
M	3	SK 77	C/	1953pc	1957mc	ad	1.53	4BooX3		1	E1G	1958b	2500/11/58/5756/OS
M	3	SK 77	C/	1953pc	1957mc	ad	1.53	4BooX3		1	E1G	1961b	4000/4/63 ASE RCE
M	3	SK 78	C	1953pc	-	ac	6.53	4BooX3			E1G	1959b	2500/9/59/5982/OS
M	3	SK 78	C	nd	-	ac	6.53	4BooX3			E1G	1959b	4000/9/61 ASE RCE
M	3	SK 79	B	1953pc	-	ac	1.53	4BooX3			E1G	1958b	2500/12/58/5756/OS
M	3	SK 79	B	1953pc	-	ac	1.53	4BooX3			E1G	nd:b	4000/4/63 ASE RCE
M	3	SK 83	B/	1955pc	1958mc	ac	6.55	4oooX3			E1G	1959b	2500/7/59/5982/OS
M	3	SK 83	B/	nd	-	ac	6.55	4oooX3			E1G	1959b	3000/9/61 ASE RCE
M	3	SK 84	B/	1955mc	1959mc	ac	6.55	4oooX3		11	E1G	1960b	2500/2/60/5982/OS
M	3	SK 84	B/	nd	-	ac	6.55	4oooX3		11	E1G	1960b	3000/9/61 ASE RCE
M	3	SK 85	B	1953pc	-	ac	6.53	4BooX3		111	E1G	1958b	5000/10/58/5711/OS
G	2	SK 90	B	1951pc	-	ac	6.54	4BEoo3	RB	01	1EG	1952p	2500/12/52 ASE RCE
G	3	SK 90	B	1951pc	-	ac	1.51	4BooX3	RB	01	1EG	1952b	5000/8/54 OS
M	3	SK 93	B/	1955mc	1958mc	ac	6.55	4oooX3		11	E1G	1959b	2500/9/59/5982/OS

Spitalgate Airfield.

1	2	3	4	5	6	7	8	9	10	11	12	13	14
M	3	SK 93	B/	nd	-	ac	6.55	4oooX3		11	E1G	1959b	3000/9/61 ASE RCE
M	3	SK 94	B/	1955mc	1959mc	ac	6.55	4oooX3		11	E1G	1960b	2500/1/60/5982/OS

Barkston Heath Airfield.

1	2	3	4	5	6	7	8	9	10	11	12	13	14
M	3	SK 94	B/	nd	-	ac	6.55	4oooX3		11	E1G	1960b	3000/9/61 ASE RCE
M	3	SK 95	A	1949pc	-	-	6.49	4BooX3	B	000	E1G	1958b	5000/6/58/5711/OS
G	2	SK 96	[A]	nd:po	-	-	6.54	4BEoo3	RB	10	1EG	1952p	2500/11/52 ASE RCE
G	3	SK 96	B	1953pc	-	ac	6.53	4BooX3		11	2EG	1954b	5000/5/54 OS
M	3	SK 97	B/	1953pc	1957mc	ac	6.53	4BooX4		110	E1G	1958b	5000/6/58/5711/OS
G	1	SM 80	B:r	1948po	-	-	1.48	4BEoo3		00	WOE	1950p	4000/11/50
M	3	SM 80	D/	1954pc	1958mc	ac	6.54	4BooX3		10	E2G	1958b	5000/9/58/5802/OS
M	3	SM 80	D/	nd	-	ac	6.54	4BooX3		10	E2G	1958p	5000/9/61/ASE RCE
G	2	SM 82	[A]	1951pc	-	ac	6.54	4BEoo4		0	1EG	1952p	2500/10/52 ASE RCE
G	1	SM 90	A:s	1948po	-	-	6.48	4BEoo3	R	1	WOE	1950p	6000/ /50
M	3	SM 90	B/	nd	-	ac	6.55	4BooX3		1	E2G	nd:p	5000/6/62/ASE RCE
G	1	SN 00	B:r	nd:po	-	-	6.48	4BEoo3	R	0	WOE	1950p	6000/ /50 (sic)
M	3	SN 00	B/	1950mc	1960mc	a	6.48	4BooX3		0	E1G	nd:b	4000/4/63 ASE RCE
M	3	SN 02	A	1951pc	-	ac	1.51	4BooX4			E1G	1959b	2500/9/59/5900/OS
M	3	SN 03	A	1951pc	-	ac	6.51	4BooX4	B		E1G	1959b	2500/4/59/5900/OS
M	3	SN 12	A	1950pc	-	ac	6.50	4BooX4	R		E1G	1959b	2500/9/59/5900/OS
M	3	SN 13	A	1950pc	-	ac	6.50	4BooX4	R		E1G	1959b	2500/8/59/5900/OS
G	2	SN 60	[A]	1947po	-	-	6.54	4BEoo3	R		1EG	1952p	5000/10/52 ASE RCE
G	2	SN 61	[A]	1950pc	-	ac	6.54	4BEoo3	R		1EG	1952p	5000/10/52 ASE RCE
G	2	SN 62	[A]	nd:po	-	-	6.54	4BEoo3	R		1EG	1952p	5000/11/52 ASE RCE
G	2	SN 63	A	1952pc	-	ac	6.54	4Booo4			1EG	1952p	5000/3/53 ASE RCE
G	2	SN 64	A	1952pc	-	ac	6.54	4Booo4			1EG	1952p	5000/10/52 ASE RCE
G	2	SN 65	A	1952pc	-	ac	6.54	4Booo4	R		1EG	1952p	5000/12/52 ASE RCE
G	2	SN 66	A	nd:po	-	-	6.54	4BEoo3	R		1EG	1952p	5000/10/52 ASE RCE
G	2	SN 67	[B]	1950pc	-	ac	6.54	4BEoo3	R		1EG	1952p	5000/11/52 ASE RCE
M	3	SN 69+	B//	1956mc	1960mc	ac	6.56	4oooX3			E1G	1961b	2500/2/61/6235/OS
G	2	SN 70	[A]	1948po	-	-	6.54	4BEoo3	RB		1EG	1952p	5000/11/52 ASE RCE
G	3	SN 71	A	1953pc	-	ac	6.53	4BooX4			E1G	1954b	5000/8/54 OS
G	2	SN 72	A	1952pc	-	ac	6.54	4Booo4	R		1EG	1952p	5000/3/53 ASE RCE
G	3	SN 73	A	1952pc	-	ac	6.52	4BooX4	R		E1G	1954b	5000/8/54 OS
G	3	SN 74	A	1953pc	-	ac	6.53	4BooX4			E1G	1954b	5000/8/54 OS
G	3	SN 75	A	1954pc	-	ac	1.54	4BooX4			E1G	1954b	5000/8/54 OS
G	3	SN 76	A	1953pc	-	ac	6.53	4BooX4			E1G	1954b	5000/8/54 OS
G	2	SN 77	A	nd:po	-	-	6.54	4BEoo3	R		1EG	1952p	5000/11/52 ASE RCE
G	2	SN 78	[B]	nd:po	-	-	6.54	4BEoo3			1EG	1952p	5000/11/52 ASE RCE
G	2	SN 80	[A]	nd:po	-	-	6.54	4BEoo3	R		1EG	1952p	5000/11/52 ASE RCE
G	3	SN 81	A	1953pc	-	ac	6.53	4BooX4			E1G	1954b	5000/8/54 OS
G	3	SN 82	A	1953pc	-	ac	6.53	4BooX4			E1G	1954b	5000/8/54 OS

1	2	3	4	5	6	7	8	9	10	11	12	13	14
G	3	SN 83	A	1953pc	-	ac	6.53	4BooX4			E1G	1954b	5000/8/54 OS
G	3	SN 84	A	1953pc	-	ac	6.53	4BooX4			E1G	1954b	5000/8/54 OS
G	3	SN 85	A	1953pc	-	ac	6.53	4BooX4			E1G	1954b	5000/11/54 OS
G	3	SN 86	A	1953pc	-	ac	6.53	4BooX4			E1G	1954b	5000/11/54 OS
G	3	SN 87	A	1953pc	-	ac	6.53	4BooX4			E1G	1954b	5000/8/54 OS
G	3	SN 88	A	1953pc	-	ac	6.53	4BooX4			E1G	1954b	5000/8/54 OS
G	3	SN 88	A	1953pc	-	ac	6.53	4BooX4			E1G	1954b	5000/11/54 OS
G	2	SN 90	[A]	1947po	-	-	6.54	4BEoo3	RB		1EG	1952p	5000/11/52 ASE RCE
G	2	SN 91	[A]	nd:po	-	-	6.54	4BEoo3			1EG	1952p	5000/11/52 ASE RCE
G	3	SN 92	A	1953pc	-	ac	6.53	4BooX4			E1G	1954b	5000/11/54 OS
G	3	SN 93	A	1953pc	-	ac	6.53	4BooX4			E1G	1954b	5000/11/54 OS
G	3	SN 94	A	1953pc	-	ac	6.53	4BooX4			E1G	1954b	5000/11/54 OS
G	3	SN 95	A	1953pc	-	ac	6.53	4BooX4			E1G	1954b	5000/11/54 OS
G	3	SN 96	A	1953pc	-	ac	1.53	4BooX4	R		E1G	1954b	5000/11/54 OS
G	3	SN 97	A	1952pc	-	ac	6.52	4BooX4	R		E1G	1954b	5000/11/54 OS
G	3	SN 98	A	1952pc	-	ac	6.52	4BooX4	R		E1G	1954b	5000/11/54 OS
G	2	SO 00	[A]	1947po	-	-	6.54	4BEoo3	RB		1EG	1952p	5000/12/52 ASE RCE
G	2	SO 01	[A]	1947po	-	-	6.54	4BEoo3	RB		1EG	1952p	5000/12/52 ASE RCE
G	3	SO 02	A	1951pc	-	ac	6.51	4BooX4	R		1EG	1952b	5000/11/52
G	3	SO 03	A	1951pc	-	ac	1.51	4BooX4	RB		1EG	1952b	5000/11/52
G	3	SO 04	A	1951pc	-	ac	1.51	4BooX4	RB		1EG	1952b	5000/12/52
M	3	SO 05	A/	1951pc	1957mc	ac	6.51	4BooX4	B		E1G	1959b	2500/7/59/5982/OS
M	3	SO 06	A	1951pc	-	ac	6.51	4BooX4	RB		E1G	1959b	2500/9/59/5900/OS
G	2	SO 10	A	1948pc	-	-	6.54	4Booo3	R		1EG	1952p	5000/10/52 ASE RCE
G	2	SO 11	[A]	1951pc	-	ac	6.54	4Booo4	R		1EG	1952p	5000/12/52 ASE RCE
G	2	SO 12	A	1951pc	-	ac	6.54	4Booo4	RB		1EG	1952p	5000/12/52 ASE RCE
M	3	SO 13	A/	1951pc	1956mc	ac	1.51	4BooX4	B		E1G	1959b	2500/4/59/5900/OS
M	3	SO 14	A	1951pc	-	ac	1.51	4BooX4	RB		E1G	1960b	2500/2/60/5900/OS
M	3	SO 15	A/	1951pc	1958mc	ac	1.51	4BooX4	B		E1G	1959b	2500/7/59/5982/OS
M	3	SO 16	A/	1951mc	1959mc	ac	6.51	4BooX4	B		E1G	1960b	2500/1/60/5982/OS
G	2	SO 20	[A]	nd:po	-	-	6.54	4BEoo3	RB		1EG	1952p	5000/12/52 ASE RCE
G	2	SO 21	[A]	nd:po	-	-	6.54	4BEoo3	RB		1EG	1952p	5000/12/52 ASE RCE
G	2	SO 22	A	1951pc	-	ace	6.54	4BEoo4	B		1EG	1952p	5000/12/52 ASE RCE
M	3	SO 23	A/	1952pc	1957mc	ac	6.52	4BooX4			E1G	1959b	2500/8/59/5982/OS
M	3	SO 24	A/	1951pc	1958c	ac	6.51	4BooX4	B		E1G	1959b	2500/6/59/5982/OS
M	3	SO 25	A/	1960^{c}mc	1960mc	ac	6.51	4oooX4			E1G	1961b	2500/4/61/5982/OS
G	2	SO 30	[A]	nd:po	-	-	6.54	4BEoo3	R		1EG	1952p	5000/12/52 ASE RCE
G	2	SO 31	A	nd:po	-	-	6.54	4BEoo3	R		1EG	1952p	5000/12/52 ASE RCE
G	2	SO 32	A	1951pc	-	ac	6.54	4BEoo4	RB		1EG	1952p	5000/12/52 ASE RCE
M	3	SO 40	B	1951mc	-	ac	1.51	4BooX3	R		E1G	1958b	2500/10/58/5756/OS
M	3	SO 41	B	1956pc	-	ac	6.56	4oooX3			E1G	1958b	2500/10/58/5756/OS
M	3	SO 41	B	nd	-	ac	6.56	4oooX3			E1G	1958b	3000/9/61 ASE RCE
M	3	SO 46	B	1956pc	-	ac	6.56	4oooX3		1	E1G	1958b	5000/6/58/5711/OS
M	3	SO 47	B	1957pc	-	ac	6.57	4oooX3			E1G	1958b	5000/5/58/5711/OS
M	3	SO 54	B	1957pc	-	ac	6.57	4oooX3	C		E1G	1959b	2500/9/59/5982/OS
M	3	SO 54	B	nd	-	ac	6.57	4oooX3	C		E1G	1959b	3000/9/61 ASE RCE
M	3	SO 56	B	1956pc	-	ac	6.56	4oooX3	C		E1G	1958b	5000/4/58/5711/OS
M	3	SO 57	B	1956pc	-	ac	6.56	4oooX3			E1G	1958b	5000/6/58/5711/OS
M	3	SO 65	B	1956pc	-	ac	6.56	4oooX3			E1G	1959b	2500/9/59/5982/OS

SO 67 appears as issued on the 1958 index, but has not been recorded.

1	2	3	4	5	6	7	8	9	10	11	12	13	14
M	3	SO 75	B	1961mc	-	ac	6.61	4oooX3			E1G	1961b	2500/8/61/5982/OS
G	3	SO 77	A	1949pc	-	-	1.50	4BooX3	RB		E1G	1956b	5000/6/56
G	3	SO 78	B	1951pc	-	ac	6.51	4BooX3	R		E1G	1956b	5000/5/56
G	3	SO 79	B	1952pc	-	ac	6.52	4BooX3	R		E1G	1956b	5000/5/56
G	2	SO 81	[A]	1947pc	1951c	-	6.54	4BEoo3	RB	1	1EG	1952p	2500/12/52 ASE RCE
G	3	SO 81	C	1952pc	-	ac	6.52	4BooX3	R	1	2EG	1954b	5000/6/54 OS
G	3	SO 85	B	1952pc	-	ac	6.52	4BooX3	R		E1G	1956b	5000/4/56 OS
G	3	SO 86	B	1951pc	-	ac	6.51	4BooX3	R		E1G	1956b	5000/4/56
G	3	SO 87	[B]	1951pc	-	ac	6.51	4BooX3	RB		E1G	1956b	5000/6/56
G	3	SO 88	B	1951pc	-	ac	6.51	4BooX4	BC		E1G	1956b	5000/5/56

1	2	3	4	5	6	7	8	9	10	11	12	13	14
G	3	SO 89	B	1952pc	-	ac	6.52	4BooX3	R	11	1EG	1953b	5000/1/53
G	2	SO 92	C	1952pc	-	ac	6.54	4Booo3	R	2	1EG	1952p	2500/10/52 ASE RCE
G	3	SO 92	C	1952pc	-	ac	6.54	4Booo3	R	2	1EG	1952b	5000/5/54 SPC RE
G	3	SO 95	A	1949pc	-	-	6.49	4BooX3	RB	0	E1G	1956b	5000/4/56 OS
G	3	SO 96	B	1951pc	-	ac	6.51	4BooX3	R		E1G	1956b	5000/4/56 OS
G	3	SO 97	B	1952pc	-	ac	1.52	4BooX3	R		E1G	1956b	5000/2/56 OS
G	1	SO 98	A:r	1947po	-	-	1.47	4BEoo3	RB		WOE	1950p	6000/2/51
G	2	SO 99	B:s	1951pc	-	ac	1.52	4Booo3	R		2EG	1952p	6000/1/52
M	3	SO 99	B/	1951mc	1960mc	ac	1.52	4BooX3			E2G	1962b	2000/6/62/6394/OS
M	3	SP 00	B	1959mc	-	ac	6.59	4oooX4			E1G	1959b	2500/12/59/5982/OS
M	3	SP 00	B	nd	-	ac	6.59	4oooX4			E1G	1959b	3000/9/61 ASE RCE
G	3	SP 05	B	1951pc	-	ac	1.51	4BooX3	RB		E1G	1956b	5000/6/56
G	3	SP 06	[B]	1952pc	-	ac	1.52	4BooX3	R		E1G	1956b	5000/4/56
G	3	SP 07	B	1952pc	-	ac	1.52	4BooX3	R		E1G	1956b	5000/3/56 OS
G	1	SP 08	[A]:s	1947po	-	-	1.47	4BEoo3	R		1EG	1951p	6000/5/51
G	2	SP 09	B:s	1951pc	-	ac	1.52	4Booo3	R		2EG	1952p	6000/3/52
M	3	SP 09	B/	1951mc	1958c	ac	6.51	4BooX3			E2G	1962b	2000/6/62/6394/OS
G	3	SP 15	[C]	1951pc	-	ac	6.51	4BooX3	R	0	E1G	1956b	5000/4/56
G	3	SP 16	B	1949pc	1953c	-	6.53	4BooX3	B	0	E1G	1956b	5000/1/56 OS
G	3	SP 17	B	1951pc	-	ac	1.51	4BooX3	R	1	E1G	1956b	5000/2/56 OS
G	2	SP 18	B:s	1951pc	-	ac	1.52	4Booo3	R	3	2EG	1952p	6000/4/52
M	3	SP 18	B/	1951mc	1960c	ac	6.56	4BooX3		3	E2G	1962b	3000/5/62/6394/OS
G	1	SP 19	A:s	1947po	-	-	1.47	4BEoo3	RB	1	1EG	1951p	6000/6/51
G	3	SP 25	B	1951pc	-	ac	6.51	4BooX3	RC	00	E1G	1956b	5000/2/56 OS
G	3	SP 26	D	1951pc	1953c	ac	6.53	4BooX3			E1G	1956b	5000/2/56 OS
G	2	SP 27	[B]	1951pc	-	ac	6.54	4BEoo3	RB	0	1EG	1952p	2500/1/53 ASE RCE
G	3	SP 27	B	1951pc	-	ac	6.51	4BooX3	RB	0	E1G	1952b	5000/10/55 OS
G	2	SP 28	[B]	1951pc	-	ac	6.54	4BEoo3	R		1EG	1952p	2500/12/52 ASE RCE
G	3	SP 28	B	1951pc	-	ac	1.51	4BooX3	R		E1G	1952b	5000/11/55 OS
G	3	SP 29	B	1951pc	-	ac	1.51	4BooX3	RB		E1G	1956b	5000/6/56 OS
G	2	SP 36	B	1951pc	-	ac	6.54	4Booo3	R		1EG	1952p	2500/11/52 ASE RCE
G	3	SP 36	B	1951pc	-	ac	6.51	4BooX3	R		E1G	1952b	5000/11/55 OS
G	1	SP 37	A:s	1947po	-	-	6.47	4BEoo3	RB	0	WOE	1950p	5000/2/51
G	1	SP 38	A:r	1947po	-	-	6.47	4BEoo3	RB	0	WOE	1950p	5000/2/51
G	3	SP 39	B	1952pc	-	ac	1.52	4BooX3	R	0	E1G	1956b	5000/1/57
M	3	SP 40	C/	1948pc	1957mc	ac	6.53	4BooX3	B	01	E1G	1958b	5000/5/58/5762/OS
G	1	SP 47	A:r	1948po	-	-	1.48	4BEoo3	RB	0	WOE	1950p	5000/1/51
G	3	SP 48	A	1948pc	-	-	6.48	4BooX3	RBC	00	E1G	1956b	5000/1/56 OS
G	3	SP 49	B	1952pc	-	ac	6.52	4BooX3	R		E1G	1956b	5000/2/56 OS
M	3	SP 50	B/	1952pc	1957mc	ace	6.52	4BooX4			E1G	1958b	5000/3/58/5737/OS
M	3	SP 55	B/	1951pc	1956mc	ac	6.51	4BooX3		0	E1G	1959b	2500/9/59/5982/OS
M	3	SP 55	B/	nd	-	ac	6.51	4BooX3		0	E1G	1959b	4000/9/61 ASE RCE
M	3	SP 56	B/	1951mc	1960mc	ac	6.51	4BooX3	B		E1G	1961b	2500/1/61/5982/OS
M	3	SP 56	B/	1951mc	1960mc	ac	6.51	4BooX3	B		E1G	1961b	no code 13 FSS RE
G	1	SP 57	A:r	nd:po	-	-	1.51	4BEoo3	RB		1EG	1951p	5000/2/51
M	3	SP 58	B//	1951mc	1958mc	ac	6.51	4BooX3		00	E1G	1959b	2500/6/59/5982/OS
M	3	SP 58	B//	nd	-	ac	6.51	4BooX3		00	E1G	1959b	4000/9/61 ASE RCE
M	3	SP 59	B/	1952pc	1958mc	ac	1.52	4BooX3			E1G	1959b	2500/9/59/5982/OS
M	3	SP 59	B/	nd	-	ac	1.52	4BooX3			E1G	1959b	4000/9/61 ASE RCE
G	3	SP 60	B	1948pc	1953c	ac	6.53	4BooX3	B	1	E1G	1956b	3000/4/56
M	3	SP 60	C	1960mc	-	ad	6.60	4oooX3		1	E2G	1962b	5000/1/62/6125/OS
M	3	SP 65	B	1955pc	-	ac	6.55	4oooX3			E1G	1959b	2500/9/59/5982/OS
M	3	SP 66	B/	1954pc	1958mc	ac	6.54	4oooX3			E1G	1959b	2500/8/59/5982/OS
M	3	SP 66	B/	nd	-	ac	6.54	4oooX3			E1G	1959b	4000/9/61 ASE RCE
M	3	SP 67	B/	1955pc	1958mc	ac	6.55	4oooX3			E1G	1958b	2500/12/58/5756/OS
M	3	SP 67	B/	nd	-	ac	6.55	4oooX3			E1G	1958b	4000/9/61 ASE RCE
M	3	SP 68	B/	1954pc	1958mc	ac	6.54	4oooX3		11	E1G	1959b	2500/6/59/5756/OS
M	3	SP 68	B/	nd	-	ac	6.54	4oooX3		11	E1G	1959b	4000/9/61 ASE RCE
M	3	SP 69	C/	1955pc	1958mc	ac	6.55	4oooX3			E1G	1958b	2500/11/58/5756/OS
M	2	SP 70	C	1948pc	1950c	ac	6.50	4BooX3	RB	0	E1G	1958b	5000/3/58/5737/OS

1	2	3	4	5	6	7	8	9	10	11	12	13	14
M	3	SP 75	[B]	1956pc	-	ac	6.56	4oooX4			E1G	1958b	2500/10/58/5756/OS
M	3	SP 76	B	1955pc	-	ac	6.55	4oooX3			E1G	1958b	2500/9/58/5756/OS
M	3	SP 76	B	nd	-	ac	6.55	4oooX3			E1G	1958b	2000/9/61 ASE RCE
M	3	SP 77	B	1954pc	-	ac	6.54	4oooX3	1		E1G	1958b	2500/10/58/5756/OS

Unusually wide roads cross the airfield, no doubt forming part of old runways.

1	2	3	4	5	6	7	8	9	10	11	12	13	14
M	3	SP 78	B/	1954mc	1958mc	ac	6.54	4oooX3	1		E1G	1959b	2500/1/59/5756/OS
M	3	SP 78	B/	nd	-	ac	6.54	4oooX3	1		E1G	1959b	4000/11/61 ASE RCE
M	3	SP 79	[B]	1954pc	-	ac	6.54	4oooX3			E1G	1958b	2500/10/58/5756/OS

An erroneous print code of 25000 was blocked out.

1	2	3	4	5	6	7	8	9	10	11	12	13	14
M	3	SP 85	[B]	1955pc	-	ac	6.55	4oooX4			E1G	1958b	2500/9/58/5756/OS
M	3	SP 86	C	1955pc	-	ac	6.55	4oooX3	13		E1G	1958b	2500/10/58/5756/OS
M	3	SP 86	C	nd	-	ac	6.55	4oooX3	13		E1G	1958b	4000/9/61 ASE RCE
M	3	SP 87	B/	1954pc	1958mc	ac	6.54	4oooX3			E1G	1959b	2500/2/59/5756/OS
M	3	SP 87	B/	nd	-	ac	6.54	4oooX3			E1G	1959b	4000/9/61 ASE RCE
M	3	SP 88	B/	1954mc	1958mc	ac	6.54	4oooX3	1		E1G	1959b	2500/2/59/5756/OS
M	3	SP 88	B/	nd	-	ac	6.54	4oooX3	1		E1G	1959b	4000/9/61 ASE RCE
G	2	SP 92	[A]	1947po	-	-	6.54	4BEoo3	RB		1EG	1952p	2500/12/52 ASE RCE
G	3	SP 92	B	1955pc	-	ac	6.55	4oooX3			2EG	1956b	5000/2/56 OS
M	3	SP 95	C	1956pc	-	ad	6.56	4oooX3	1		E1G	1959b	2500/9/59/5982/OS
M	3	SP 95	C/	1956mc	1960mc	ad	6.56	4oooX3	1		E1G	1965b	3000/9/65/6730/OS
M	3	SP 96	B	1955pc	-	ac	6.55	4oooX3	10		E1G	1959b	2500/9/59/5982/OS
M	3	SP 96	B/	nd	-	ac	6.55	4oooX3	10		E1G	1959b	4000/9/61 ASE RCE
M	3	SP 97	C/	1955pc	1958mc	ac	6.55	4oooX3			E1G	1959b	2500/5/59/5756/OS
M	3	SP 97	C/	nd	-	ac	6.55	4oooX3			E1G	1959b	4000/9/61 ASE RCE
M	3	SP 98	C	1955pc	-	ad	6.55	4oooX4	11		E1G	1958b	2500/11/58/5756/OS
M	3	SP 98	C/	1955mc	1960mc	ad	6.55	4oooX4	11		E1G	1962b	4000/6/62/6394/OS
G	2	SR 99+	nc:r	1948po	-	-	6.52	4Booo3			2EG	1952p	9000/10/52 SPC

Also used for GSGS Misc.1656 *Castlemartin Training Area* (500/6/54 OS).

1	2	3	4	5	6	7	8	9	10	11	12	13	14
M	3	SR 99+	B/	1956mc	1960mc	ac	6.56	4oooX3			E3G	1960b	5000/7/60/6125/OS
G	1	SS 09	B:r	nd:po	-	-	1.48	4BEoo3	R		WOE	1950p	6000/11/50
M	3	SS 09	B/	1949pc	1960mc	ac	1.48	4BooX3			E1G	1962b	4000/4/62/6394/OS
G	1	SS 22	A:s	1950pc	-	a	1.51	4BEoo3			1EG	1951p	5000/2/51
G	1	SS 32	A:r	1950pc	-	a	1.51	4BEoo3			1EG	1951p	5000/2/51
G	1	SS 42	A:s	1950pc	-	ac	1.51	4BEoo3	R		1EG	1951p	5000/2/51
G	1	SS 43	A:r	1950pc	-	ac	1.51	4BEoo3	R	0	1EG	1951p	5000/2/51
G	1	SS 44	A:r	nd:po	-	-	1.51	4BEoo3	R		1EG	1951p	5000/2/51
G	2	SS 53	[A]	nd:po	-	-	6.54	4BEoo3	R	0	1EG	1952p	2500/10/52 ASE RCE
G	3	SS 53	[A]	nd:po	-	-	6.54	4BEoo3	R	0	1EG	1952b	5000/5/54 SPC RE
G	2	SS 54	[A]	nd:po	-	-	6.54	4BEoo3	R		1EG	1952p	2500/10/52 ASE RCE
G	3	SS 54	[A]	nd:po	-	-	6.54	4BEoo3	R		1EG	1952b	5000/5/54 SPC RE
G	1	SS 58+	B:r	nd:po	-	-	1.48	4BEoo3	R		WOE	1950p	6000/11/50
G	1	SS 59	C:r	nd:po	-	-	6.48	4BEoo3	R	0	WOE	1950p	6000/12/50
G	2	SS 63	[A]	nd:po	-	-	6.54	4BEoo3	R		1EG	1952p	5000/11/52 ASE RCE
G	2	SS 64	[A]	nd:po	-	-	6.54	4BEoo3	B		1EG	1952p	5000/11/52 ASE RCE
G	2	SS 69	B:s	1951pc	-	ac	1.52	4Booo4	R		2EG	1952p	6000/4/52
G	3	SS 69	B	1951pc	-	ac	1.52	4BooX4	R		E2G	1952b	5000/8/55 OS
G	2	SS 73	A	nd:po	-	-	6.54	4BEoo3			1EG	1952p	5000/11/52 ASE RCE
G	2	SS 74+	A	1949pc	1951	-	6.54	4BEoo3	B		1EG	1952p	5000/11/52 ASE RCE
G	1	SS 78	A:r	nd:po	-	-	1.48	4BEoo3	R		WOE	1950p	6000/11/50
G	1	SS 79	B:r	1950pc	-	ace	6.50	4BEoo3	R		WOE	1950p	6000/11/50
M	3	SS 79	B/*	1950mc	1961r	ace	6.50	4BooX3			E2G	1962b	2000/5/62/6394/OS
G	2	SS 83	A	1949pc	-	-	6.54	4Booo3	B		1EG	1952p	5000/11/52 ASE RCE
G	2	SS 84	[A]	nd:po	-	-	6.54	4BEoo3	B		1EG	1952p	5000/10/52 ASE RCE
M	3	SS 87+	A/	1949mc	1959mc	a	6.48	4BooX3		2	E1G	1961b	2500/2/61/5982/OS
G	3	SS 88	[A]	1949pc	-	-	6.48	4BooX3	R	2	E1G	1956b	5000/6/56
G	3	SS 89	A	1949pc	1952	-	1.52	4BooX3	RB		E1G	1956b	5000/5/56
G	2	SS 93	[A]	nd:po	-	-	6.54	4BEoo3	B		1EG	1952p	5000/10/52 ASE RCE
G	2	SS 94	[A]	nd:po	-	-	6.54	4BEoo3	RB		1EG	1952p	5000/11/52 ASE RCE
M	3	ST 02	B	1961mc	-	acf	6.61	4oooX4			E1G	1968b	1000/12/67/7013/OS
G	3	ST 06	B	1951pc	-	ac	1.51	4BooX3	R	00	E1G	1956b	5000/3/56 OS

1	2	3	4	5	6	7	8	9	10	11	12	13	14
G	3	ST 07	B	1951pc	-		1.51	4BooX3	R		E1G	1956b	5000/6/56
G	3	ST 08	B	1950pc	-	-	1.48	4BooX3	R		E1G	1956b	5000/3/56 OS
G	1	ST 16	A:r	1948po	-	-	1.48	4BEoo3	R		WOE	1950p	6000/11/50
G	1	ST 17	A:s	1947po	-	-	1.51	4BEoo3	R		1EG	1951p	6000/1/51

The print code 6000/12/50 has been blocked out.

1	2	3	4	5	6	7	8	9	10	11	12	13	14
G	1	ST 18	B:r	nd:po	-	-	1.48	4BEoo3	R		1EG	1950p	6000/1/51
M	3	ST 18	B//	1947mc	1961ch	af	1.48	4BooX3			E1G	1962b	3000/5/62/6394/OS
G	1	ST 27	A:r	1947po	-	-	1.47	4BEoo3	R	0	WOE	1950p	5000/2/51
G	1	ST 28	B:r	1950pc	-	ac	6.50	4BEoo3	R		WOE	1950p	6000/1/51
M	3	ST 28	B/	1950mc	1961mc	ac	6.50	4BooX3			E1G	1962b	2000/5/62/6394/OS
G	1	ST 38	B:s	1947po	-	-	1.51	4BEoo3	R		1EG	1951p	5000/1/51
G	1	ST 47	[A]:s	1946po	-	-	1.51	4BEoo3	R		1EG	1951p	8000/1/51
G	1	ST 48	C:s	1950pc	-	ac	6.50	4BEoo3	R		WOE	1950p	8000/1/51
M	3	ST 48	C	1950pc	-	ac	6.50	4BooX3	R		E1G	1962b	3000/4/62/6394/OS
G	1	ST 56	A:s	1946po	-	-	1.51	4BEoo3	R	10	1EG	1951p	8000/2/51
G	3	ST 56	B	1951pc	-	ac	6.51	4BooX3	R	10	E2G	1955b	5000/9/55 OS
G	1	ST 57	A:r	1946po	-	-	1.47	4BEoo3	RB	0	WOE	1950p	8000/2/51
M	3	ST 57	B//	1951mc	1962ch	acf	1.51	4BooX3		1	E2G	1962b	5000/6/62/6394/OS
G	1	ST 58	C:s	1951pc	-	ac	1.51	4BEoo3	R	0	WOE	1950p	8000/1/50
M	3	ST 58	C/	1951mc	1959mc	ac	1.51	4BooX3		1	E1G	1962b	3000/6/62/6394/OS
G	1	ST 66	A:r	1946po	-	-	1.47	4BEoo3	R		WOE	1950p	10,000/2/51
G	1	ST 67	A:r	1946po	-	-	1.47	4BEoo3	R		WOE	1950p	10,000/2/51
G	1	ST 68	B:r	nd:po	-	-	1.48	4BEoo3	RB	1	WOE	1950p	10,000/2/51
G	1	ST 70	A:s	1949pc	1950	-	1.49	4BEoo3	B		1EG	1951p	1000/9/51
G	2	ST 70	[A]:s	1949pc	1950	-	6.52	4BEoo3			2EG	1952p	8000/9/52 SPC
M	3	ST 70	B	1958pc	-	ac	6.58	4oooX4			E3G	1965b	5000/12/64/6687/OS
M	3	ST 70	B	1958pc	-	ac	6.58	4oooX4			E3G	1965b	1/68/447/OA 13 FSS RE
G	1	ST 71	A:r	nd:po	-	-	6.48	4BEoo3	RB	0	1EG	1951p	6000/6/51
M	3	ST 71	B	1958pc	-	ac	6.58	4oooX3		3	E2G	1959b	6000/12/59/6157/OS
M	3	ST 71	B	1958pc	-	ac	6.58	4oooX3		3	E2G	1965b	4000/12/64/6687/OS
M	3	ST 71	B	1958pc	-	ac	6.58	4oooX3		3	E2G	1959b	1/68/448/OA 13 FSS RE
G	1	ST 72	A:r	nd:po	-	-	6.49	4BEoo3	RB	0	1EG	1951p	6000/6/51
G	1	ST 73	A:r	nd:po	-	-	1.49	4BE4X3	RB		WOE	1948p	5000/9/49
G	3	ST 73	A	1949pc	-	-	1.49	4BooX3	RB		2EG	1953p	6000/3/53 OS
G	1	ST 74	A	1949pc	-	-	1.50	4BEoo3	RB		1EG	1950p	not recorded
G	3	ST 74	A	1949pc	-	-	1.50	4BooX3	RB		1EG	1950b	5000/10/54 OS
G	1	ST 75	A:r	1947po	-	-	1.48	4BE2X3	RB		WOE	1947p	5000/9/47
G	3	ST 75	B	1947pc	1951c	ac	1.51	4BooX3	RB		2EG	1953b	5000/1/53
G	1	ST 76	A:r	1947po	-	-	6.47	4BE2X3	RB	0	WOE	1947p	5000/9/47
G	3	ST 76	B	1948pc	-	-	6.47	4BooX3	RB	0	2EG	1953b	5000/2/53 OS
G	2	ST 77	[B]	1951pc	-	-	6.54	4BEoo3	B	10	1EG	1952p	2500/12/52 ASE RCE
G	3	ST 77	[B]	1951pc	-	-	6.54	4BEoo3	B	10	1EG	1952b	5000/5/54 SPC RE
G	2	ST 78	A	1949pc	-	-	6.54	4Booo3	RB		1EG	1952p	2500/11/52 ASE RCE
G	3	ST 78	A	1949pc	-	-	6.54	4Booo3	RB		1EG	1952p	5000/5/54 SPC RE
G	1	ST 80	B:s	nd:po	-	-	6.48	4BEoo3	RB		1EG	1951p	1000/9/51
G	3	ST 80	B	1949pc	-	-	6.48	4BooX3	RB		2EG	1953b	6000/1/53
M	3	ST 80	C	1958mc	-	ac	6.58	4oooX3			E3G	1960b	6000/5/60/6157/OS
M	3	ST 80	C	1958mc	-	ac	6.58	4oooX3			E3G	1962b	5000/7/62/6394/OS
M	3	ST 80	C	1958mc	-	ac	6.58	4oooX3			E3G	1962b	1/68/449/OA 13 FSS RE
G	1	ST 81	A:r	nd:po	-	-	6.48	4BEoo3	RB		1EG	1951p	6000/6/51
M	3	ST 81	[A]	nd:po	-	-	6.60	4BEoo3	RB		E1G	1958b	5000/10/58/406/BP

Printed by No 2 SPC (Air). The letters "ST" have been added later, as have the magnetic variation values.

1	2	3	4	5	6	7	8	9	10	11	12	13	14
M	3	ST 81	B	1958pc	-	ac	6.58	4oooX3			E2G	1965b	3000/12/64/6687/OS
M	3	ST 81	B	1958pc	-	ac	6.58	4oooX3			E2G	1965b	1/68/450/OA 13 FSS RE
G	1	ST 82	A:r	nd:po	-	-	1.49	4BE4X3	RB		WOE	1948p	5000/9/48
G	1	ST 82	A:r	nd:po	-	-	1.49	4BEoo3	RB		1EG	1948p	3000/7/51
G	1	ST 82	A:r	nd:po	-	-	1.49	4BEoo3	RB		1EG	1948p	500/9/51

Overprinted for *Exercise Surprise Packet*.

1	2	3	4	5	6	7	8	9	10	11	12	13	14
G	2	ST 82	[?B]:s	nd:po	-	-	6.52	4BEoo3	RB		2EG	1952p	9000/11/52 SPC RE
M	3	ST 82	C/	1957mc	1960mc	ac	6.57	4oooX3			E3G	1965b	3000/12/64/6687/OS

1	2	3	4	5	6	7	8	9	10	11	12	13	14
G	1	ST 83	?A:r	nd:po	-	-	6.49	4BE??3	B		WOE	1949p	not recorded
G	1	ST 83	A:r	1949pc	1950	-	6.50	4BEoo3	B		1EG	1949p	3000/6/51
M	3	ST 83	B	1957pc	-	ac	6.57	4oooX3			E2G	1960b	5000/1/60/6176/OS
G	1	ST 84	A:r	nd:po	-	-	6.49	4BE??3	RB		WOE	1949p	not recorded
G	1	ST 84	B	1949pc	1950c	ace	6.50	4BEoo3	RB		1EG	1949p	not recorded
G	3	ST 84	B	1949pc	1950c	ace	6.50	4BooX3	RB		E1G	1949b	5000/11/57/5452/OS

An additional print code 5000/4/57 is blocked out.

M	3	ST 84	C	1958pc	-	ad	6.58	4oooX3			E2G	1962b	5000/10/62/6394/OS
G	1	ST 85	A:r	nd:po	-	-	1.49	4BE??3	RB		WOE	1949p	not recorded
G	1	ST 85	B:s	1949pc	1950c	ace	6.50	4BEoo3	RB		1EG	1949p	6000/6/51
G	1	ST 85	B:s	1949pc	1950c	ace	6.50	4BEoo3	RB		1EG	1949p	500/9/51

Overprinted for *Exercise Surprise Packet*.

G	3	ST 85	B	1949pc	1950c	ace	6.50	4BooX3	RB		2EG	1954b	5000/11/54 OS
G	1	ST 86	A:r	nd:po	-	-	6.48	4BE2X3	RB		WOE	1948p	5000/7/48
G	1	ST 86	A:r	nd:po	-	-	6.48	4BE2X3	RB		WOE	1948p	6000/12/50 SPC

GSGS Misc.495/1, 1950: Staff College Entrance Examination 1951, Tactics B.

G	1	ST 86	A:r	nd:po	-	-	6.48	4BE2X3	RB		WOE	1948p	500/9/51

Overprinted for *Exercise Surprise Packet*.

G	2	ST 86	B:s	1948pc	1951c	-	1.52	4Booo3	RB		2EG	1952p	9000/6/52
G	2	ST 87	[C]	1952pc	-	ac	6.54	4Booo3	R	12	1EG	1952p	2500/11/52 ASE RCE
G	2	ST 87	[C]	1952pc	-	ac	6.54	4Booo3	R	12	1EG	1952b	5000/4/58/364 2 SPC(Air)
G	2	ST 88	[B]	1951pc	-	ac	6.54	4BEoo3	RB	0	1EG	1952p	2500/12/52 ASE RCE
G	2	ST 88	[B]	1951pc	-	ac	6.54	4BEoo3	RB	0	1EG	1952b	5000/4/58/364 2 SPC(Air)
G	2	ST 89	B:t	1951pc	-	ac	6.51	4Booo4			1EG	1952p	10,000/9/52
G	1	ST 90	B:s	nd:po	-	-	1.48	4BEoo3	RB	0	1EG	1951p	1000/9/51
G	2	ST 90	B:s	nd:po	-	-	6.52	4BEoo3	RB	0	2EG	1952p	8000/9/52 SPC
G	1	ST 91	A:r	nd:po	-	-	6.48	4BE4X3	B		WOE	1948p	5000/4/49
G	1	ST 91	B:r	nd:po	-	-	6.48	4BEoo3	B		1EG	1948p	3000/7/51
G	1	ST 92	?A:r	nd:po	-	-	1.48	4BE??3	RB		WOE	1947p	not recorded
G	1	ST 92	B:s	1950pc	-	ace	6.50	4BEoo3	RB		1EG	1947p	3000/7/51
M	3	ST 92	C	1958pc	-	ad	6.58	4oooX3			E2G	1962b	5000/10/62/6394/OS
G	1	ST 93	B	1951pc	-	ace	1.51	4BEoo3	RB		1EG	1951p	not recorded
G	3	ST 93	B	1951pc	-	ace	1.51	4BooX3	RB		1EG	1951p	5000/6/54 OS
M	3	ST 93	C	1958pc	-	ad	6.58	4oooX3			E2G	1962b	5000/7/62/6394/OS
G	1	ST 94	A:r	nd:po	-	-	1.48	4BE2X3	RB		WOE	194?p	not recorded

This earlier version presumably existed for the next to have been printed with legends.

G	1	ST 94	A:r	nd:po	-	-	1.51	4BE2X3	RB		1EG	1950p	1000/6/51
G	3	ST 94	[A]	1947pc	1952	-	1.52	4BooX3	RB		1EG	1950b	5000/10/54 OS
G	1	ST 95	A:r	nd:po	-	-	1.48	4BE2X3	RB	0	WOE	1947p	5000/10/48
G	1	ST 95	A:r	nd:po	-	-	1.51	4BE2X3	RB	0	1EG	1950p	1000/6/51
G	1	ST 95	A:r	nd:po	-	-	1.51	4BE2X3	RB	0	1EG	1950p	500/9/51

Overprinted for *Exercise Surprise Packet*.

G	3	ST 95	B	1953pc	-	ae	6.53	4BooX3	RB	0	2EG	1954p	5000/9/54 OS
M	3	ST 95	C	1958pc	-	bc	6.58	4BooX3		3	E3G	1962b	5000/8/62/6394/OS
G	1	ST 96	A:r	nd:po	-	-	1.49	4BE??3	RB		WOE	1949p	not recorded
G	1	ST 96	A:r	nd:po	-	-	1.49	4BEoo3	RB		1EG	1949p	3000/7/51
G	1	ST 96	A:r	nd:po	-	-	1.49	4BEoo3	RB		1EG	1949p	500/9/51

Overprinted for *Exercise Surprise Packet*.

G	3	ST 96	B	1949pc	1953c	-	6.53	4BooX3	B		2EG	1954b	5000/9/54 OS
G	2	ST 97	A	1949pc	1952	-	6.54	4Booo3	RB	0	1EG	1952p	2500/11/52 ASE RCE
G	2	ST 97	A	1949pc	1952	-	6.54	4Booo3	RB	0	1EG	1952p	5000/5/54/573 RE
G	2	ST 98	A	nd:po	-	-	6.54	4BEoo3	RB	0	1EG	1952p	2500/12/52 ASE RCE
G	2	ST 98	A	nd:po	-	-	6.54	4BEoo3	RB	0	1EG	1952b	5000/4/58/364 2 SPC(Air)
G	1	SU 00	B:s	1947pc	1948c	-	6.47	4BEoo3	RB		1EG	1951p	1000/9/51
G	2	SU 00	B:s	1948pc	1948c	-	6.52	4BEoo3	RB		2EG	1952p	8000/9/52 SPC !
G	1	SU 01	A:r	1948po	-	-	6.47	4BE2X3	R		WOE	1947p	5000/1/48
G	1	SU 01	B:r	nd:po	-	-	6.47	4BEoo3	R		1EG	1947p	3000/7/51
G	1	SU 02	A:r	1948po	-	-	6.48	4BE2X3	B		WOE	1948p	5000/4/48
G	1	SU 02	B	nd:po	-	-	1.49	4BEoo3	B		1EG	1948p	3000/7/51
G	3	SU 02	B	1949pc	-	-	1.49	4BooX3	B		E1G	1948b	5000/10/55 OS

1 2	3	4	5	6	7	8	9	10	11	12	13	14
G 1	SU 03	B:r	nd:po	-	-	1.49	4BE4X3	RB		WOE	1948p	5000/4/49
G 1	SU 03	B:s	1951pc	-	ace	1.51	4BEoo3	RB		1EG	1950p	6000/7/51
G 3	SU 03	B	1951pc	-	ace	1.52	4BooX3	RB		1EG	1950b	5000/10/54 OS
G 1	SU 04	A:r	nd:po	-	-	1.49	4BE4X3	B		WOE	1948p	5000/1/49
G 1	SU 04	A:r	nd:po	-	-	1.51	4BE4X3	B		1EG	1950p	6000/1/51
G 3	SU 04	B	1948pc	1953c	-	6.53	4BooX3	B		2EG	1954b	5000/9/54 OS
G 3	SU 04	B	1948pc	1953c	-	6.53	4BooX3	B		2EG	1954b	5000/11/55 OS
G 1	SU 05	A:r	nd:po	-	-	1.48	4BE4X3	RB		WOE	1948p	5000/1/49
G 1	SU 05	A:r	nd:po	-	-	1.51	4BE4X3	RB		1EG	1950p	6000/1/51
G 1	SU 05	A:r	nd:po	-	-		4BEoo3	RB		1EG	1950p	nc

The five colour blocks are present in the upper right margin. Despite this, and the absence of some regular marginal detail, this is not a proof copy, since a batch of over one hundred copies has been recorded.

G 1	SU 05	A:r	nd:po	-	-	6.48	4BEoo3	RB		1EG	1950p	1000/7/51
G 1	SU 05	A:r	nd:po	-	-	6.48	4BEoo3	RB		1EG	1950p	500/9/51

Overprinted for *Exercise Surprise Packet*.

G 3	SU 05	A	1948pc	-	-	6.48	4BooX3	RB		1EG	1951b	5000/6/54 OS
G 3	SU 05	A	1948pc	-	-	6.48	4BooX3	RB		1EG	1951b	5000/1155 OS
M 3	SU 05	B	1958pc	-	ac	6.58	4oooX3			E2G	1962b	5000/9/62/6394/OS
G 1	SU 06	A:r	nd:po	-	-	1.49	4BE4X3	RB		WOE	1948p	5000/7/49
G 1	SU 06	A:r	nd:po	-	-	1.49	4BEoo3	RB		1EG	1948p	3000/7/51
G 1	SU 06	A:r	nd:po	-	-	1.49	4BEoo3	RB		1EG	1948p	500/9/51

Overprinted for *Exercise Surprise Packet*.

G 3	SU 06	[A]	nd:po	-	-	1.57	4BEoX3	RB		E2G	1955b	5000/3/55 SPC RE
G 2	SU 07	A	1948pc	1952	-	6.54	4Booo3	B	000	1EG	1952p	5000/3/53 ASE RCE
G 3	SU 07	A	1948pc	1952	-	6.54	4Booo3	B	000	1EG	1952b	5000/5/54 SPC RE
G 2	SU 08	[B]	1948pc	1951c	ace	6.54	4BEoo3	RBC		1EG	1952p	5000/12/52 ASE RCE
G 2	SU 09	[A]	1948pc	1951	-	6.54	4BEoo3	RB	00	1EG	1952p	5000/12/52 ASE RCE
M 3	SU 09	B	1959mc	-	ac	6.59	4oooX3		31	E2G	1962b	5000/7/62/6394/OS
G 1	SU 10	B:s	nd:po	-	-	6.47	4BEoo3	R	0	1EG	1951p	1000/9/51
G 2	SU 10	[B]:s	nd:po	-	-	6.52	4BEoo3	R	0	2EG	1952p	8000/9/52 SPC
G 1	SU 11	A:r	1948po	-	-	1.48	4BE2X3	RB		WOE	1947p	5000/1/48
G 1	SU 11	B:r	nd:po	-	-	1.48	4BEoo3	RB		1EG	1947p	3000/7/51
M 3	SU 11	C	1961mc	-	acf	6.61	4oooX3			E2G	1962b	5000/10/62/6424/OS
G 1	SU 12	A:r	1948po	-	-	6.48	4BE2X3	RB		WOE	1947p	5000/4/48
G 1	SU 12	B:r	nd:po	-	-	6.48	4BEoo3	RB		1EG	1947p	3000/7/51
M 3	SU 12	C	1958pc	-	ac	6.58	4oooX3			E2G	1960b	3000/5/60/6157/OS
M 3	SU 12	C	1958pc	-	ac	6.58	4oooX3			E2G	1962b	2000/9/62/6394/OS
G 1	SU 13	A:r	nd:po	-	-	1.48	4BE4X3	RB	10	WOE	1948p	5000/4/49
G 1	SU 13	B:r	nd:po	-	-	1.51	4BEoo3	RB	10	1EG	1950p	1000/7/51
G 1	SU 13	B:r	nd:po	-	-	1.51	4BEoo3	RB	10	1EG	1950p	6000/7/51
G 3	SU 13	B	1949pc	1950c	-	1.48	4BooX3	RB	10	E1G	1950b	5000/12/55 OS
G 1	SU 14	A:r	nd:po	-	-	1.48	4BE2X3	RB	10	WOE	1947p	5000/5/49
G 1	SU 14	A:r	nd:po	-	-	1.48	4BE2X3	RB	10	1EG	1947p	6000/1/51
M 3	SU 14	A	1948pc	-	-	1.48	4BooX3	RB	10	1EG	1947b	5000/6/54/OS
M 3	SU 14	A	1948pc	-	-	1.60	4BooX3	RB	10	E1G	1959b	5000/2/59/449 2 SPC(Air)
G 1	SU 15	A:r	nd:po	-	-	1.48	4BE4X3	RB	0	WOE	1948p	5000/4/49
G 1	SU 15	A:r	nd:po	-	-	1.51	4BE4X3	RB	0	1EG	1951p	6000/1/51
G 1	SU 15	[B]:s	1951pc	-	ace	1.51	4BEoo3	RB	0	1EG	1951p	1000/7/51
G 3	SU 15	B	1951pc	-	ace	1.51	4BooX3	RB	0	2EG	1954b	5000/9/54 OS
G 3	SU 15	B	1951pc	-	ace	1.51	4BooX3	RB	0	2EG	1954b	5000/11/55 OS
G 1	SU 16	A:r	nd:po	-	-	1.49	4BE4X3	RB		WOE	1949p	5000/6/49
G 1	SU 16	A	nd:po	-	-	1.49	4BEoo3	RB		1EG	1949p	5000/7/51
M 3	SU 16	A/	1949pc	1959mc	ac	1.49	4BooX3	B		E2G	1952b	5000/2/61/6157/OS
G 2	SU 17	[B]	1951pc	-	ace	6.54	4BEoo3	RB	0	1EG	1952p	5000/11/52 ASE RCE
G 3	SU 17	B	1951pc	1953rpb	ace	1.51	4BooX3	B	0	E1G	1952b	5000/2/57
G 2	SU 18	B	1951pc	-	ace	6.54	4BEoo3	RBC	0	1EG	1952p	7500/12/52 ASE RCE
G 3	SU 18	B	1951pc	-	ace	6.54	4BEoo3	RBC	0	1EG	1952b	5000/5/54 SPR RE
G 2	SU 19	[A]	nd:po	-	-	6.54	4BEoo3	RB	00	1EG	1952p	5000/11/52 ASE RCE
M 3	SU 19	C	1959mc	-	ac	6.59	4oooX3		13	E2G	1962b	5000/10/62/6394/OS

1	2	3	4	5	6	7	8	9	10	11	12	13	14
G	1	SU 20	A:s	1948po	-	-	1.48	4BEoo3	R		1EG	1951p	1000/9/51
G	2	SU 20	[A]:s	1948po	-	-	6.52	4BEoo3	R		2EG	1952p	8000/9/52 SPC
G	1	SU 21	A:r	1948po	-	-	1.48	4BE2X3		0	WOE	1947p	5000/1/48
G	3	SU 21	B	1948pc	-	-	1.48	4BooX3		0	2EG	1954b	5000/8/54 OS
G	3	SU 21	B	1948pc	-	-	1.48	4BooX3		0	2EG	1954b	5000/11/55 OS
G	1	SU 22	A:r	1948po	-	-	1.48	4BE2X3	RBC		WOE	1947p	5000/1/48
G	1	SU 22	C:r	nd:po	-	-	1.48	4BEoo3			1EG	1947p	3000/7/51
M	3	SU 22	D	1958pc	-	ac	6.58	4oooX3			E2G	1962b	5000/8/62/6394/OS
G	1	SU 23	A:r	1948po	-	-	6.48	4BE2X3	RB	0	WOE	1948p	5000/4/48
G	1	SU 23	B:r	nd:po	-	-	6.48	4BEoo3	RB	0	1EG	1948p	4000/7/51
G	3	SU 23	B	1949pc	-	a	6.48	4BooX3	RB	0	E1G	1948b	5000/1/57
G	1	SU 24	A:r	nd:po	-	-	1.48	4BE2X3	RB	0	WOE	1948p	5000/9/48
G	1	SU 24	A:r	nd:po	-	-	1.48	4BEoo3	RB	0	1EG	1948p	6000/7/51
G	3	SU 24	A	1948pc	-	-	1.48	4BooX3	RB	0	2EG	1954b	5000/8/54 OS
G	3	SU 24	A	1948pc	-	-	1.48	4BooX3	RB	0	2EG	1954b	5000/11/55 OS
M	3	SU 24	B	1958pc	-	ac	6.58	4oooX3		3	E3G	1959b	5000/7/59/5943/OS
M	3	SU 24	B	1958pc	-	ac	6.58	4oooX3		3	E3G	1965b	3000/7/65/6730/OS
G	1	SU 25	A:r	nd:po	-	-	1.48	4BE2X3	RB		WOE	1948p	5000/9/48
G	1	SU 25	B	nd:po	-	-	1.51	4BEoo3	RB		1EG	1951p	5000/2/51
G	3	SU 25	B	1948pc	1949c	-	1.51	4BooX3	RB		1EG	1951b	10,000/6/54 OS
M	3	SU 25	C	1958pc	-	ac	6.58	4oooX3			E2G	1959b	5000/7/59/5943/OS
M	3	SU 25	C	1958pc	-	ac	6.58	4oooX3			E2G	1965b	3000/8/65/6730/OS
G	1	SU 26	A:r	nd:po	-	-	6.49	4BE4X3	RB		WOE	1949p	5000/7/49
G	1	SU 26	A:r	nd:po	-	-	6.49	4BEoo3	RB		1EG	1949p	4000/7/51
G	3	SU 26	A	1949pc	-	-	6.49	4BooX3	RB		E1G	1949b	5000/12/55 OS
G	2	SU 27	[B]	1948pc	1950c	ace	6.54	4BEoo3	RB		1EG	1952p	7500/11/52 ASE RCE
G	2	SU 28	[B]	1950pc	-	ace	6.54	4BEoo3	RB		1EG	1952p	5000/3/53 ASE RCE
M	3	SU 28	C	1960mc	-	ad	6.60	4oooX4			E2G	1962b	5000/11/62/6394/OS
G	2	SU 29	B	1951pc	-	ace	6.54	4BEoo3	RB		1EG	1952p	5000/11/52 ASE RCE
G	1	SU 30	B:r	nd:po	-	-	6.47	4BEoo3	R	0	WOE	1950p	8000/7/50
G	3	SU 30	B	1948pc	-	-	6.47	4BooX3	R	0	E1G	1950b	5000/11/55 OS
G	1	SU 31	A:r	1946po	-	-	1.47	4BEoo3	R		WOE	1950p	8000/7/50
G	3	SU 31	[B]	1951pc	-	ace	1.51	4BooX3	R		2EG	1954b	5000/9/54 OS
G	1	SU 32	A:r	1947po	-	-	1.47	4BE2X3	RB		WOE	1947p	5000/10/48
G	3	SU 32	B	1949pc	-	-	1.47	4BooX3	RB		1EG	1952b	5000/10/52
M	3	SU 32	C	1958pc	-	ad	6.58	4oooX3			E2G	1958b	5000/1/59/5759/OS
G	1	SU 33	A:r	1948po	-	-	6.48	4BE4X3	RB	00	WOE	1948p	5000/7/48
G	1	SU 33	B:r	nd:po	-	-	1.51	4BEoo3	RB	00	1EG	1951p	5000/2/51

Coloured layers were added to some copies for use in *Exercise Hess*: see GSGS 4627A below.

1	2	3	4	5	6	7	8	9	10	11	12	13	14
G	1	SU 33	B:r	nd:po	-	-	1.51	4BEoo3	RB	00	1EG	1951p	1000/7/51
G	1	SU 33	B:r	nd:po	-	-	1.51	4BEoo3	RB	00	1EG	1951b	8000/6/54 OS
G	1	SU 33	B	-	-	-	1.69	4BEoo3	RB	10	1EG	1951b	500/5/67/380/OA 42 SER
G	1	SU 34	A:r	1948po	-	-	1.48	4BE2X3	RB	0	WOE	1948p	5000/7/48
G	1	SU 34	B:r	1950pc	-	ace	6.50	4BEoo3	RB	0	1EG	1948p	6000/7/51
G	2	SU 34	B:s	1950pc	-	ace	1.52	4Booo3	RB	0	2EG	1952p	5000/1/52
G	1	SU 35	A:r	nd:po	-	-	6.48	4BE4X3	B		WOE	1948p	5000/9/48
G	1	SU 35	A:r	nd:po	-	-	6.48	4BEoo3	B		1EG	1948p	1000/7/51
G	3	SU 35	A	1948pc	-	-	6.48	4BooX3	B		E1G	1954b	5000/12/54 OS
G	1	SU 36	A:r	nd:po	-	-	1.49	4BE4X3	RB		WOE	1948p	5000/4/49
G	1	SU 36	A:r	nd:po	-	-	1.49	4BEoo3	RB		1EG	1948p	5000/7/51
G	2	SU 36	A:s	1948pc	-	-	1.52	4Booo3	RB		2EG	1952p	10,000/5/52
G	2	SU 37	A	1949pc	1952	-	6.54	4Booo4	RB	0	1EG	1952p	5000/3/53 ASE RCE
M	3	SU 37	B	1960mc	-	ac	6.60	4oooX4		1	E2G	1960b	5000/1/61/6157/OS
G	2	SU 38	B	nd:po	-	-	6.54	4BEoo3	BC	0	1EG	1952p	5000/12/52 ASE RCE
G	2	SU 39	B	nd:po	-	-	6.54	4BEoo3	RBC	20	1EG	1952p	5000/12/52 ASE RCE
G	1	SU 40	A:r	1946po	-	-	1.47	4BEoo3	R	00	WOE	1950p	8000/7/50

1	2	3	4	5	6	7	8	9	10	11	12	13	14
G	1	SU 41	A:r	1947po	-	-	1.47	4BEoo3	R	1	WOE	1950p	8000/7/50
G	3	SU 41	A/	1947pc	1957c	ac	1.47	4BooX3		1	E2G	1957b	20,600/9/57
G	3	SU 41	A/(772)	1947pc	1957c	ac	1.47	4BooX3		1	E2G	1957b	20,600/9/57

Overprinted in red: *An overlay showing isodose lines as they might be plotted at a control centre from monitoring data corrected to one hour after detonation. At "A" an air burst nominal nuclear weapon has been assumed. At "B" a nominal nuclear weapon burst underwater in a shallow harbour has been assumed. Wind 10 miles per hour in both cases.* This is Figure 49 in an as yet unidentifed paper.

G	1	SU 42	A:r	1947po	-	-	6.47	4BE2X3	RB		WOE	1947p	5000/10/48
G	3	SU 42	B	1949pc	-	-	6.47	4BooX3	RB		2EG	1953b	6000/1/53
G	1	SU 43	A:r	1948po	-	-	1.48	4BE2X3	RB	1	WOE	1948p	5000/4/48
G	2	SU 43	B:s	1948pc	1949c	-	1.52	4Booo3	RB	1	2EG	1952p	10,000/1/52
M	3	SU 43	[C]	1967ᶜpc	-	ac	6.58	4oooX3		2	E3G	1967b	1500/12/67/405/OA 42SER
G	1	SU 44	A:r	nd:po	-	-	6.48	4BE4X3	R		WOE	1948p	5000/9/48
G	2	SU 44	A:s	1948pc	-	-	1.52	4Booo3	R		2EG	1952p	8000/5/52
M	3	SU 44	B	1958pc	-	ac	6.58	4oooX3			E3G	1965b	2000/6/65/6723/OS
M	3	SU 44	B	1958pc	-	ac	6.58	4oooX3			E3G	1965b	1000/1/67/1922/SPC
M	3	SU 44	B/*	1958mc	1966r	acf	6.58	4oooX3			E4G	1967b	1500/12/67/7009/OS
G	1	SU 45	A:r	nd:po	-	-	6.48	4BE4X3	R		WOE	1948p	5000/9/48
G	2	SU 45	A:s	1948pc	-	-	1.52	4Booo3	R		2EG	1952p	8000/6/52
G	1	SU 46	A:r	nd:po	-	-	6.49	4BE4X3	RB	0	WOE	1949p	7000/3/49
G	2	SU 46	B:s	1949pc	1950c	-	1.52	4Booo3	RB	0	2EG	1952p	9000/6/52
G	2	SU 47	[A]	nd:po	-	-	6.54	4BEoo3	RB	0	1EG	1952p	5000/12/52 ASE RCE
G	2	SU 47	[A]	nd:po	-	-	6.54	4BEoo3	RB	0	2EG	1952b	2500/3/59 13 FSS RE
M	3	SU 47	B	1960mc	-	ac	6.60	4oooX3		1	E3G	1965b	3000/9/65/6730/OS
G	1	SU 48	[B]	1949pc	1950c	ace	1.51	4BEoo3	RB	1	1EG	1950p	P11/244/1000/5/51 SMS
G	2	SU 48	[B]	1949pc	1950c	ace	6.54	4BEoo3	RB	1	1EG	1952p	5000/3/53 ASE RCE
M	3	SU 48	[B]	1949pc	-	ace	6.54	4BEoo3	RB	1	E1G	1960b	5000/6/60 13 FSS
M	3	SU 48	C	1960mc	-	ad	6.60	4oooX3		1	E2G	1966b	1000/12/67/406/OA 42SER
G	2	SU 49	[A]	1948pc	1951	-	6.54	4BEoo3	RB	12	1EG	1952p	5000/3/53 ASE RCE
M	3	SU 49	[C]	1959mc	-	ac	6.59	4oooX3		12	E2G	1966b	1500/2/66/6813/OS
M	3	SU 49	[C]	1959mc	-	ac	6.59	4oooX3		12	E2G	1966b	1000/12/67/407/OA 42SER
G	1	SU 50	A:r	nd:po	-	-	1.47	4BEoo3	R	33	WOE	1950p	5000/3/50
G	1	SU 50	A	nd:po	-	-	1.47	4BEoo3	R	33	WOE	1950p	8000/5/50
G	1	SU 51	B:r	nd:po	-	-	6.47	4BEoo3	RB		WOE	1950p	8000/8/50
G	3	SU 51	B	1949pc	-	-	6.47	4BooX3	RB		E1G	1950b	5000/11/55 OS
G	1	SU 52	B:r	nd:po	-	-	6.47	4BEoo3	B		WOE	1950p	5000/3/50
G	3	SU 52	B	1949pc	-	-	6.47	4BooX3	B		E1G	1950b	5000/1/55 OS
G	1	SU 53	[A]:r	nd:po	-	-	6.48	4BEoo3	RB		WOE	1950p	5000/3/50
G	3	SU 53	A	1948pc	-	-	6.48	4BooX3	RB		E1G	1953b	2000/2/53 OS
G	3	SU 53	A	1948pc	-	-	6.48	4BooX3	RB		E1G	1953b	5000/10/55 OS
M	3	SU 53	B	1967ᶜpc	-	ac	6.58	4oooX3			E2G	1967b	1000/12/67/408/OA 42SER
G	1	SU 54	A:r	nd:po	-	-	1.49	4BE4X3	R		WOE	1949p	7000/2/49

Also overdrawn for *Exercise Voice 2* (Restricted), Map P Serial 2.

G	2	SU 54	A:s	1948pc	-	-	1.52	4Booo3	R		2EG	1952p	9000/6/52
G	1	SU 55	A:r	nd:po	-	-	1.49	4BE4X3	RB		WOE	1949p	7000/1/49

With a layer bar at 0,400,500,600,700 feet bottom left beside the legend, for maps in GSGS 4627A.

G	2	SU 55	A:s	1948pc	-	-	1.52	4Booo3	RB		2EG	1952p	9000/5/52
G	1	SU 56	A:r	nd:po	-	-	1.49	4BE4X3	RB	03	WOE	1949p	7000/1/49

With a layer bar at 0,400 feet bottom left beside the legend, for maps in GSGS 4627A.

G	2	SU 56	B:s	1948pc	1950c	ace	1.52	4Booo3	RB	03	2EG	1952p	9000/5/52
M	3	SU 56	C	1961mc	-	ad	6.61	4oooX3		10	E3G	1968b	1000/12/67/7010/OS
G	1	SU 57	[B]	1948pc	1950c	ace	1.51	4BEoo3	RB	1	1EG	1950p	no code

A proof copy with colour blocks in the margin, received 16 November 1951.

G	2	SU 57	[B]	1948pc	1950c	ace	6.54	4BEoo3	RB	1	1EG	1952p	5000/3/53 ASE RCE
G	3	SU 57	[B]	1948pc	1950c	ace	6.54	4BEoX3	RB	1	1EG	1960b	5000/6/60 13 FSS
G	1	SU 58	[A]	1948po	-	-	1.51	4BEoo3	RB	1	1EG	1950p	P11/246/1000/5/51 SMS
G	2	SU 58	[A]	1948pc	1951	-	6.54	4BEoo3	RB	1	1EG	1952p	5000/11/52 ASE RCE
M	3	SU 58	[A]	1948pc	1951	-	6.60	4BEoX3	RB	1	2EG	1959b	5000/2/59/449/2SPC(Air)
M	3	SU 58	B	1960mc	-	ac	6.60	4oooX3		1	E3G	1966b	1000/2/66/6813/OS
M	3	SU 58	B	1960mc	-	ac	6.60	4oooX3		1	E3G	1966b	1000/12/67/409/OA 42SER

1 2	3	4	5	6	7	8	9	10	11	12	13	14
G 2	SU 59	A	1948pc	1952	-	6.54	4Booo3	RB	10	1EG	1952p	5000/11/52 ASE RCE
M 3	SU 59	A	1948pc	1952	-	6.54	4BooX3	RB	10	E1G	1960b	2500/6/60 13 FSS
M 3	SU 59	B	1960mc	-	ac	6.60	4oooX3		31	E2G	1968b	1000/12/67/7011/OS
G 1	SU 60	A:r	nd:po	-	-	1.47	4BEoo3	RB	1	WOE	1950p	11,000/3/50
G 2	SU 60	A:s	1946pc	-	-	1.52	4Booo3	RB	1	2EG	1952p	9000/5/52
G 1	SU 61	A:r	1947po	-	-	6.47	4BEoo3	RB		WOE	1950p	6000/8/50
G 3	SU 62	B	1948pc	1949c	-	1.49	4BooX3	RB		1EG	1952b	5000/12/52
G 3	SU 62	B	1948pc	1949c	-	1.49	4BooX3	RB		1EG	1952b	5000/11/55 OS
G 3	SU 63	B	1948pc	1949c	-	1.49	4BooX3	RB		1EG	1952b	5000/11/52
G 3	SU 63	B	1948pc	1949c	-	1.49	4BooX3	RB		1EG	1952b	1000/1/67/1923/SPC RE
M 3	SU 63	C	1967^{c}mc	-	adf	6.64	4oooX3			E2G	1967b	1500/12/67/410/OA 42SER
G 1	SU 64	A:r	nd:po	-	-	1.49	4BE4X3	B	0	WOE	1949p	7000/1/49
G 2	SU 64	A:s	1948pc	-	-	1.52	4Booo3	B	0	2EG	1952p	9000/5/52
G 2	SU 64	A	1948pc	-	-	1.52	4Booo3	B	0	2EG	1952b	250/10/66/1823/SPC
G 1	SU 65	A:r	nd:po	-	-	1.49	4BE4X3	RB		WOE	1949p	7000/1/49
G 2	SU 65	A:s	1948pc	-	-	1.52	4Booo3	RB		2EG	1952p	10,000/5/52
M 3	SU 65	C	1958pc	-	ac	6.58	4oooX3			E3G	1965b	5000/11/65/6783/OS
G 1	SU 66	A:r	1947po	-	-	1.47	4BE2X3	RB	0	WOE	1949p	7000/3/49
G 2	SU 66	B:s	1951pc	-	ace	1.52	4Booo3	R	3	2EG	1952p	10,000/6/52
M 3	SU 66	C	1961mc	-	ad	6.61	4oooX4		0	E3G	1968b	1000/12/67/7011/OS
G 1	SU 67	A:r	1947po	-	-	6.47	4BE1X3	RB		WOE	1949p	7000/3/49
G 2	SU 67	A:s	1947pc	1951	-	1.52	4Booo3	RB		2EG	1952p	10,000/6/52
G 1	SU 68	A:r	nd:po	-	-	1.49	4BE4X3	RB		WOE	1949p	7000/1/49
G 2	SU 68	B:s	1948pc	1949c	-	1.52	4Booo3	RB		2EG	1952p	10,000/6/52
M 3	SU 68	[C]	1960mc	-	ad	6.60	4oooX3			E3G	1967b	1000/12/67/411/OA 42SER
G 1	SU 69	A:r	nd:po	-	-	1.49	4BEoo3	B	00	WOE	1949p	7000/12/49
G 3	SU 69	A	1949pc	1953rpb	-	1.52	4BooX3	B	00	E1G	1949b	5000/9/54 OS
G 3	SU 69	A	1949pc	1953rpb	-	1.52	4BooX3	B	00	E1G	1949b	5000/11/55 OS
M 3	SU 69	B	1967^{c}mc	-	ac	6.59	4oooX3		11	E2G	1967b	1000/12/67/412/OA 42SER
G 1	SU 70	B:s	nd:po	-	-	1.51	4BEoo3	RB	1	1EG	1951p	6000/4/51 SPC RE
M 3	SU 70	C	1959mc	-	ac	6.59	4oooX4		1	E2G	1962b	5000/12/62/6394/OS
G 1	SU 71	C:s	nd:po	-	-	1.51	4BEoo3	RB		1EG	1951p	6000/4/51 SPC RE
M 3	SU 71	E	1958mc	-	ac	6.58	4oooX4			E2G	1962b	5000/10/62/6394/OS
G 1	SU 72	B:r	nd:po	-	-	6.48	4BEoo3	RB		WOE	1950p	5000/2/50
G 1	SU 72	B:s	nd:po	-	-	6.48	4BEoo3	RB		WOE	1950p	8000/6/54 OS
G 1	SU 73	B:r	nd:po	-	-	1.49	4BEoo3	RB		WOE	1950p	5000/2/50
G 1	SU 73	B:r	nd:po	-	-	1.49	4BEoo3	RB		WOE	1950p	8000/3/52 SPC RE
M 3	SU 73	[B]	nd:po	-	-	1.60	4BEoo3	RB		E1G	1959b	5000/6/59/472 2 SPC(AIR)

A non-standard printing derived from an earlier base edition, with sheet number letters inserted where numbers had been placed, before the dividing line - SU/73. The GSGS series information, grid north and print code are in black, and marginalia added in blue. Eastings are still drawn sideways.

M 3	SU 73	C	1964mc	-	adf	6.64	4oooX3			E2G	1965b	3000/10/65/6783/OS
G 1	SU 74	A:r	nd:po	-	-	1.49	4BE4X3	RB	0	WOE	1949p	7000/1/49
G 1	SU 74	A:r	nd:po	-	-	1.51	4BE4X3	RB	0	1EG	1950p	2000/5/50
G 1	SU 74	A:r	nd:po	-	-	1.51	4BE4X3	RB	0	1EG	1950p	5000/1/51
G 3	SU 74	A	1948pc	-	-	1.49	4BooX3	RB	0	E1G	1950b	5000/7/55 OS
G 3	SU 74	A	1948pc	-	-	-	4BooX3	RB	0	E1G	1950b	5000/8/58/380 2 SPC AIR
M 3	SU 74	B	1958pc	-	ac	6.58	4oooX3		1	E2G	1962b	5000/10/62/6424/OS
M 3	SU 74	B	1958pc	-	ac	6.58	4oooX3		1	E2G	1965b	1500/12/67/413/OA 42SER
M 3	SU 74	B/	1958mc	1963ch	acf	6.58	4oooX3		1	E2G	1965b	3000/11/65/6783/OS
G 1	SU 75	A:r	nd:po	-	-	6.48	4BE4X3	RB	0	WOE	1949p	7000/1/49
G 1	SU 75	A:r	nd:po	-	-	6.48	4BE4X3	RB	0	WOE	1949p	2000/5/50
G 2	SU 75	A	1948pc	1951	-	1.52	4Booo3	RB	0	2EG	1952p	10,000/6/52
G 3	SU 75	A	1948pc	1951	-	1.52	4BooX3	RB	0	E2G	1952b	5000/9/54 OS
M 3	SU 75	B	1958pc	-	ac	6.58	4oooX3		3	E3G	1964b	2500/8/64/6649/OS

6649 was the incorrect batch number, some copies of the sheets carrying it being altered by hand to 6640.

M 3	SU 75	B	1958pc	-	ac	6.58	4oooX3		3	E3G	1965b	5000/12/64/6686/OS
G 1	SU 76	A:r	1948po	-	-	1.48	4BE4X3			WOE	1949p	7000/2/49
G 1	SU 76	A:r	1948po	-	-	1.48	4BE4X3			WOE	1949p	2000/5/50

1	2	3	4	5	6	7	8	9	10	11	12	13	14
G	1	SU 76	B:s	1950pc	-		ac	1.51 4BEoo3			1EG	1950p	5000/1/51
G	3	SU 76	B	1950pc	-		ac	6.50 4BooX3			E1G	1950b	5000/5/55 OS
M	3	SU 76	C	1958mc	-		ad	6.58 4oooX3			E2G	1959b	5000/2/59/5707/OS
M	3	SU 76	C	1958c	-		ad	6.58 4oooX3			E2G	1959b	2000/9/61 ASE RCE
G	1	SU 77	A:r	nd:po	-	-	6.47 4BE2X3	R	1	WOE	1949p	7000/3/49
G	2	SU 77	B:s	1951pc	-		ace	1.52 4Booo3	R	1	2EG	1952p	10,000/7/52
G	1	SU 78	A:r	nd:po	-	-	6.48 4BE4X3	RB		WOE	1949p	500/7/49
G	1	SU 78	A:r	nd:po	-	-	6.48 4BE4X3	RB		WOE	1949p	5000/1/50
G	3	SU 78	A	1948pc	1952rpb	-	6.48 4BooX3	RB		2EG	1952b	6000/11/52
G	3	SU 78	A	1948pc	1952rpb	-	6.48 4BooX3	RB		2EG	1952b	5000/12/54 OS
G	1	SU 79	A:r	nd:po	-	-	1.49 4BEoo3	RB		WOE	1949p	7000/12/49
G	3	SU 79	A	1949pc	1952	-	1.49 4BooX3	RB		E1G	1949b	5000/11/54 OS
G	1	SU 80	B:s	nd:po	-	-	1.51 4BEoo3	RB	0	1EG	1951p	8000/2/51 SPC RE
G	3	SU 80	B	1948pc	1949c	-	6.48 4BooX3	RB	0	E1G	1951b	5000/11/55 OS
G	1	SU 81	B:s	1948pc	1951c	ae	1.51 4BEoo3	RB		1EG	1951p	8000/4/51 SPC RE
G	3	SU 81	B	1948pc	1951c	ae	1.51 4BooX3	RB		E1G	1951b	5000/2/56 OS
G	1	SU 82	B:r	nd:po	-	-	6.48 4BEoo3	R		WOE	1950p	5000/2/50
G	1	SU 82	B:s	nd:po	-	-	6.48 4BEoo3	R		WOE	1950p	8000/6/54 OS
G	1	SU 83	A:r	nd:po	-	-	6.48 4BE4X3	RB		WOE	1949p	5500/4/49
G	1	SU 83	B	nd:po	-	-	6.48 4BEoo3	RB		1EG	1948p	5000/6/51
G	3	SU 83	B	1948pc	1951c	ace	1.52 4BooX3	RB		E1G	1948b	5000/11/53 OS
G	3	SU 83	B	1948pc	1951c	ace	1.52 4BooX3	RB		E1G	1948b	5000/12/56
M	3	SU 83	B	1948pc	1951c	ace	1.60 4BooX3	RB		E1G	1959b	5000/2/59/449 2 SPC(AIR)
M	3	SU 83	C	1964mc	-		adf	6.64 4oooX3			E2G	1964b	2500/8/64/6649/OS

6649 was the incorrect batch number, some copies of the sheets carrying it being altered by hand to 6640.

M	3	SU 83	C	1964mc	-		adf	6.64 4oooX3			E2G	1965b	5000/12/64/6686/OS
G	1	SU 84	A:r	nd:po	-	-	1.49 4BE4X3	RB		WOE	1949p	7000/2/49
G	1	SU 84	B:r	nd:po	-	-	1.49 4BEoo3	RB		WOE	1950p	2000/5/50
G	2	SU 84	B:s	1948pc	1949c	-	1.52 4Booo3	RB		2EG	1952p	10,000/8/52
G	3	SU 84	B	1948pc	49c/52rpb	ac	1.52 4BooX3	RB		E2G	1952b	6000/12/56
M	3	SU 84	C	1958pc		bd	6.58 4oooX3			E3G	1958b	no code 13 FSS
M	3	SU 84	C	1958pc		bd	6.58 4oooX3			E3G	1958b	5000/9/58/5802 OS
M	3	SU 84	C	1958pc		bd	6.58 4oooX3			E3G	1963b	5000/4/63/6424/OS
M	3	SU 84	C	1958pc	-		bd	6.58 4oooX3			E3G	1958b	3000/10/65/6783/OS
M	3	SU 84	C/	1958mc	1965ch	bdf	6.58 4oooX3			E3G	1965b	1000/12/67/414/OA 42SER
G	1	SU 85	A:r	nd:po	-	-	1.49 4BE4X3	RB	00	WOE	1949p	7000/2/49
G	1	SU 85	A:r	nd:po	-	-	1.49 4BEoo3	RB	00	WOE	1950p	2000/5/50
G	1	SU 85	A:r	nd:po	-	-	1.49 4BEoo3	RB	00	1EG	1950p	5000/1/51

Also overprinted in purple and black for GSGS Misc.1538/1 *Exercise Try-back*, 1951.

G	2	SU 85	A:s	1948pc	1950	-	1.52 4Booo3	RB	00	2EG	1952p	8000/7/52
G	3	SU 85	A	1948pc	1950	-	1.57 4BooX3	RB	00	E2G	1955b	5000/3/55 SPC RE
G	3	SU 85	A	1948pc	1950	ac	1.52 4BooX3	RB	00	E2G	1952b	5000/1/57
G	3	SU 85	A	1948pc	1950	-	1.57 4Booo3	RB	00	E2G	1955b	5000/8/58/380 2 SPC(AIR)
M	3	SU 85	B/	1958mc	1961ch	ad	6.58 4oooX3		30	E3G	1961b	5000/2/61/6233/OS
-	3	SU 85	B	1958pc	-		ad	6.58 4oooX3		30	-	nd:b	8000/5/63/157/SPC

Printed in black with the overprinted information in blue.

M	3	SU 85	B	1958pc	-		ad	6.58 4oooX3		30	E3G	1961b	8000/5/63/157/SPC
M	3	SU 85	B/	1958mc	1961ch	ad	6.58 4oooX3		30	E3G	1961b	400/10/66/1823/SPC RE
G	1	SU 86	A:r	nd:po	-	-	1.49 4BE4X3	RB		WOE	1949p	7000/3/49
G	1	SU 86	A:r	nd:po	-	-	1.49 4BE4X3	RB		WOE	1949p	2000/5/50
G	1	SU 86	A:r	nd:po	-	-	1.49 4BEoo3	RB		1EG	1949p	5000/5/51

Also overprinted in purple and black for GSGS Misc.1538/2 *Exercise Try-back*, 1951.

G	2	SU 86	B:s	1951pc	-		ace	1.52 4Booo3	RB		2EG	1952p	9000/7/52
M	3	SU 86	C	1958pc	-		ad	6.58 4oooX3			E3G	1958b	5000/9/58/5802 OS
M	3	SU 86	C/	1958pc	-		adf	6.58 4oooX3			E3G	1962b	4000/5/62/6394/OS
M	3	SU 86	C/	1958pc	-		ad	6.58 4oooX3			E4G	1964b	2500/8/64/6649/OS

Another copy was altered to Edition 3-GSGS – Ref.SPC/M821 of 24 September 1964. 6649 was the incorrect batch number, some copies of the sheets carrying it being altered by hand to 6640.

M	3	SU 86	C/	1958pc	-		ad	6.58 4oooX3			E4G	1965b	5000/12/64/6686/OS

1	2	3	4	5	6	7	8	9	10	11	12	13	14
G	1	SU 87	A:r	nd:po	-	-	1.48	4BE4X3	RB	1	WOE	1949p	7000/2/49
G	3	SU 87	B	1948pc	1949c	-	1.48	4BooX3	RB	1	2EG	1953b	6000/1/53
G	3	SU 87	B	1949pc	-	-	1.48	4BooX3	RB	1	2EG	1953b	5000/12/55 OS
G	1	SU 88	B:r	nd:po	-	-	1.49	4BE4X3	RB		WOE	1949p	500/7/49
G	1	SU 88	B:r	nd:po	-	-	1.49	4BE4X3	RB		WOE	1949p	5000/1/50
G	3	SU 88	C	1948pc	1953c	-	1.49	4BooX3	RB		2EG	1953b	6000/4/53 OS
G	3	SU 88	C	1948pc	1953c	-	1.49	4BooX3	RB		2EG	1953b	5000/11/55 OS
G	1	SU 89	B:r	nd:po	-	-	6.49	4BE4X3	RB	0	WOE	1949p	500/7/49
G	1	SU 89	B:r	nd:po	-	-	6.49	4BE4X3	RB	0	WOE	1949p	5000/1/50
G	3	SU 89	C	1949pc	1950c	ace	6.50	4BooX3	RB	0	E2G	1953b	6000/2/53
G	3	SU 89	C	1949pc	1950c	ace	6.50	4BooX3	RB	0	E2G	1953b	5000/12/54 OS
G	1	SU 90+	A:s	nd:po	-	-	1.51	4BEoo3	RB	00	1EG	1951p	5000/1/51 A
M	3	SU 90+	C	1958mc	-	ad	6.58	4oooX3		11	E2G	1959b	5000/3/59/5759/OS
G	1	SU 91	B:s	1950pc	-	ace	1.51	4BEoo3	RB		1EG	1951p	5000/1/51 A
G	1	SU 91	[B]:s	1950pc	-	ace	1.51	4BEoo3	RB		1EG	1951b	5000/8/58/380 2 SPC(AIR)
G	1	SU 92	B:s	nd:po	-	-	1.51	4BEoo3	RB		1EG	1951p	5000/1/51 A
G	3	SU 92	[B]	nd:po	-	-	1.57	4BEoX3	RB		E2G	1955b	5000/1/55 SPC RE
G	3	SU 92	[B]	nd:po	-	-	1.57	4BEoX3	RB		E2G	1955b	5000/3/55 SPC RE
G	1	SU 93	A:r	nd:po	-	-	1.48	4BE4X3	RB		WOE	1949p	7000/2/49
G	1	SU 93	C:s	1948pc	1950c	ace	1.51	4BEoo3	RB		1EG	1950p	5000/1/51
G	3	SU 93	[C]	1948pc	1950c	ace	1.57	4BEoX3	RB		E2G	1955b	5000/1/55 SPC RE
G	3	SU 93	[C]	1948pc	1950c	ace	1.57	4BEoX3	RB		E2G	1955b	5000/3/55 +last SPC RE
M	3	SU 93	[C]	1948pc	1950c	ace	1.60	4BEoX3	RB		E2G	1959b	5000/2/59/449 2 SPC(AIR)
G	1	SU 94	A:r	nd:po	-	-	1.49	4BE4X3	RB		WOE	1949p	5500/4/49
G	1	SU 94	A:r	nd:po	-	-	1.49	4BE4X3	RB		WOE	1949p	2000/5/50
G	1	SU 94	B:s	1950pc	-	ace	6.50	4BEoo3	RB		1EG	1949p	5000/6/51
G	3	SU 94	B	1950pc	-	ace	1.50	4BooX3	RB		E1G	1949b	5000/8/53/380 2 SPC(AIR)
-	3	SU 94	B	1950pc	-	ace	6.50	4BooX3	RB		E1G	nd:b	no code

A proof copy, with incomplete marginalia.

G	3	SU 94	B	1950pc	-	ace	6.50	4BooX3	RB		E1G	1949b	5000/9/54 OS
G	3	SU 94	B	1950pc	-	ace	6.50	4BooX3	RB		E1G	1949b	5000/12/55 OS
G	3	SU 94	B	1950pc	-	ace	6.50	4BooX3	RB		E1G	1949b	5000/8/58/380/2SPC(Air)
M	3	SU 94	C	1961mc	-	ad	6.61	4oooX4			E2G	1964b	2500/8/64/6649/OS

6649 was the incorrect batch number, some copies of the sheets carrying it being altered by hand to 6640.

M	3	SU 94	C	1961mc	-	ad	6.61	4oooX4			E2G	1965b	5000/12/64/6686/OS
G	1	SU 95	A:r	nd:po	-	-	1.49	4BEoo3	RB		WOE	1949p	5000/8/49
G	1	SU 95	A:r	nd:po	-	-	1.49	4BEoo3	RB		WOE	1949p	5000/2/50
G	1	SU 95	A:r	nd:po	-	-	1.49	4BEoo3	RB		WOE	1949p	5000/2/50.6000/3/50

Also overprinted in purple and black for GSGS Misc.1538/3 *Exercise Try-back*, 1951.

G	2	SU 95	B:s	1948pc	1950c	ace	1.52	4Booo3	RB		2EG	1952p	8000/8/52
G	2	SU 95	B:s	1948pc	1950c	ace	1.52	4Booo3	RB		2EG	1952b	5000/3/56 OS
G	2	SU 95	B	1948pc	1950c	ace	1.52	4Booo3	RB		2EG	1952b	5000/4/58/364 2 SPC(AIR)
M	2	SU 95	[B]	1948pc	1950c	ace	1.52	4BooX3	RB		E2G	1960b	5000/6/60 13 FSS
M	3	SU 95	C	1961mc	-	ad	6.61	4oooX4			E3G	1964b	2500/8/64/6649/OS

6649 was the incorrect batch number, some copies of the sheets carrying it being altered by hand to 6640.

M	3	SU 95	C	1961mc	-	ad	6.61	4oooX4			E3G	1965b	5000/12/64/6686/OS
G	1	SU 96	A:r	nd:po	-	-	1.49	4BE4X3	RB		WOE	1949p	5500/4/49

Also used as GSGS Misc.424 [1950], with the south-west corner overprinted with concealment areas in solid green and yellow, with black detail.

G	1	SU 96	A:r	nd:po	-	-	1.49	4BE4X3	RB		WOE	1949p	2000/5/50

Also overprinted in purple and black for GSGS Misc.1538/4 *Exercise Try-back*, 1951.

G	2	SU 96	B:s	1951pc	-	ace	1.52	4Booo3	RB		2EG	1952p	10,000/9/52
M	3	SU 96	B/	1951pc	1956mc	ace	1.51	4BooX3	B		E2G	1958b	10,000/4/58/5707/OS
M	3	SU 96	C	1961mc	-	adf	6.61	4oooX3			E3G	1964b	2500/8/64/6649/OS

6649 was the incorrect batch number, some copies of the sheets carrying it being altered by hand to 6640.

M	3	SU 96	C	1961mc	-	adf	6.61	4oooX3			E3G	1965b	5000/12/64/6686/OS
G	1	SU 97	A:r	nd:po	-	-	6.48	4BE4X3	RB		WOE	1947p	5500/5/49
G	2	SU 97	B:s	1950pc	-	ace	1.52	4Booo3	R		2EG	1952p	10,000/6/52
M	3	SU 97	B///	1950mc	1961ch	acef	1.52	4BooX3			E2G	1952b	2000/6/62/6394/OS

1	2	3	4	5	6	7	8	9	10	11	12	13	14
G	2	SU 98	[B]	1950pc	-	ace	6.54	4BEoo3	RB		1EG	1952p	2500/3/53 ASE RCE
G	2	SU 98	[B]	1950pc	-	ace	6.54	4BEoo3	RB		1EG	1952b	5000/12/54 SPC RE
G	?	SU 98									2EG		not recorded
M	3	SU 98	C/*	1960mc	1962r	adf	6.60	4oooX3			E3G	1963b	5000/4/63/6394/OS
G	2	SU 99	B	1951pc	-	ace	6.54	4Booo3	RB		1EG	1952p	2500/11/52 ASE RCE
G	2	SU 99	B	1951pc	-	ace	1.52	4Booo3	RB		1EG	1952b	5000/1/55 OS
M	3	SU 99	C:x	1961mc	-	adf	6.61	4oooX3			E2G	1963b	5000/2/63/6424/OS
G	2	SW 54	[A]	1950pc	-	ac	6.54	4BEoo4			1EG	1952p	2500/11/52 ASE RCE
G	2	SW 61	[A]	1950pc	-	a	6.54	4BEoo4		0	1EG	1952p	2500/10/52 ASE RCE
M	3	SW 61	B	1960mc	-	ac	6.60	4oooX4		3	E2G	1962b	5000/9/62/6394/OS
G	2	SW 71+	A	1951pc	1951	a	6.54	4Booo3			1EG	1952p	2500/3/53 ASE RCE
G	3	SW 71+	A	1951pc	1951	a	6.54	4Booo3			1EG	1952b	5000/5/54 SPC RE
G	2	SW 72+	A	1951pc	1952	a	6.54	4Booo3			1EG	1952p	2500/1/53 ASE RCE
G	3	SW 72+	A	1951pc	1952	a	6.54	4Booo3			1EG	1952p	5000/12/54 SPC RE
G	2	SX 17	[A]	1950pc	-	a	6.54	4BEoo3			1EG	1952p	5000/10/52 ASE RCE
G	2	SX 18	[A]	1950pc	-	a	6.54	4BEoo3	R	0	1EG	1952p	5000/11/52 ASE RCE
G	2	SX 27	[A]	1950pc	-	a	6.54	4BEoo3			1EG	1952p	5000/11/52 ASE RCE
G	2	SX 28	[A]	1950pc	-	ac	6.54	4BEoo3	R		1EG	1952p	5000/10/52 ASE RCE
G	1	SX 35	A:r	1946po	-	-	6.46	4BEoo3	R		WOE	1950p	10,000/9/50
G	1	SX 36	A:r	1946po	-	-	6.46	4BEoo3	R		WOE	1950p	10,000/9/50
G	1	SX 45+	A	1947po	-	-	6.46	4BEoo3	R		WOE	1950p	no code
G	1	SX 46	A:s	1946po	-	-	6.46	4BEoo3	R	0	WOE	1950p	10,000/10/50
G	1	SX 54	A:r	1946po	-	-	1.47	4BEoo3	B		WOE	1950p	10,000/9/50
G	1	SX 55	A:r	1946po	-	-	6.46	4BEoo3	R		WOE	1950p	10,000/12/50
G	1	SX 56	A:r	1946po	-	-	6.46	4BEoo3	R	00	WOE	1950p	12,000/11/50
G	1	SX 57	B:r	nd:po	-	-	6.47	4BE4X3	R		WOE	1948p	5000/5/49
G	3	SX 57	[B]	1948pc	-	-	6.47	4BooX3	R		1EG	1948b	5000/10/54 OS
M	3	SX 57	B:w	1959mc	-	ac	1959	5oo6X5			E2G	1963b	5000/5/63/6394/OS
G	1	SX 58	A:r	nd:po	-	-	1.48	4BE4X3	R		WOE	1947p	5000/4/49
G	3	SX 58	B	1949pc	1950c	a	6.50	4BooX3	R		E2G	1955b	5000/8/55 OS
G	1	SX 59	A	nd:po	-	-	1.48	4BE??3	R		WOE	1947p	not recorded
G	3	SX 59	A	1949pc	-	-	1.48	4BooX3	R		1EG	1947b	5000/10/54 OS
M	3	SX 59	A/	1961c	-	ac	1.48	4BooX3			E2G	nd:p	5000/6/62 ASE RCE
G	2	SX 64+	[A]	1947po	-	-	6.54	4BEoo3			1EG	1952p	2500 12 52 ASE RCE
G	2	SX 64+	[A]	1947po	-	-	6.54	4BEoo3			1EG	1952b	5000/12/54 SPC RE
G	3	SX 65	B	1949pc	-	-	6.47	4BooX3	R		E1G	1956b	5000/4/56 OS
M	3	SX 65	B	1959mc	-	a	1959	5oo6X5			E2G	1962b	2000/11/62/6394/OS
G	3	SX 66	B	1949pc	-	-	1.47	4BooX3	R		E1G	1956b	5000/3/56 OS
M	3	SX 66	B	1959mc	-	ac	1959	5oo6X5			E2G	1962b	3000/11/62/6394/OS
G	1	SX 67	A:r	1947po	-	-	1.47	4BE2X3			WOE	1948p	5000/10/48
G	3	SX 67	[B]	1949pc	-	-	1.47	4BooX3			2EG	1954b	5000/11/54 OS
M	3	SX 67	B:w	1959mc	-	ac	1959	5oo6X5			E3G	1963b	5000/5/63/6394/OS
G	1	SX 68	A:r	1948po	-	-	1.48	4BE2X3			WOE	1947p	5000/1/48
G	3	SX 68	B	1948pc	-	-	1.48	4BooX3			E2G	1955b	5000/8/55 OS
M	3	SX 68	B/	nd	-	a	1.48	4BooX3			E3G	nd:b	5000/6/62 ASE RCE
G	1	SX 69	A:r	nd:po	-	-	6.48	4BE4X3	R		WOE	1948p	5000/5/49
G	3	SX 69	A	1949pc	-	-	6.48	4BooX3	R		E1G	1948b	5000/8/55 OS
G	2	SX 73+	[A]	1947po	-	-	6.54	4BEoo3			1EG	1952p	2500/3/53 ASE RCE
G	2	SX 73+	[A]	1947po	-	-	6.54	4BEoo3			1EG	1952b	5000/5/54 SPC RE
G	1	SX 77	A:r	1948po	-	-	6.48	4BE2X3			WOE	1947p	5000/3/48
G		SX 77									E2G		not recorded
M	3	SX 77	C	1959mc	-	-	6.59	4oooX3			E3G	1963b	5000/5/63/6394/OS
G	1	SX 78	A:r	nd:po	-	-	1.48	4BE2X3	R		WOE	1947p	5000/9/48
M	3	SX 78	B/	1951mc	1959mc	ac	1.51	4BooX4			E2G	1963b	5000/5/63/6424/OS
G	2	SX 83	[B]	nd:po	-	-	6.54	4BEoo3			1EG	1952p	2500/11/52 ASE RCE
G	3	SX 83	[B]	nd:po	-	-	6.54	4BEoo3			1EG	1952b	5000/5/54 SPC RE
G	3	SX 98	A	1948pc	-	-	6.48	4BooX3	R		E1G	1952p	2500/10/52 ASE RCE
G	3	SX 98	A	1948pc	-	-	6.48	4BooX3	R		E1G	1952b	5000/8/55 OS
M	3	SY 08+	A//	1948mc	1960mc	ac	6.60	4BooX3			E1G	1961b	3000/9/61 13 FSS RE
G	1	SY 58	A:r	nd:po	-	-	1.48	4BEoo3	RB		WOE	1950p	5000/8/50

1 2	3	4	5	6	7	8	9	10	11	12	13	14
G 1	SY 67+	A:r	1948po	-	-	6.48	4BEoo3	R		WOE	1950p	6000/9/50
G 3	SY 67+	A	1948pc	1953rpb	-	6.48	4BooX3	R		E1G	1950b	5000/12/55 OS
G 1	SY 68	B:r	1948pc	1950c	ac	6.50	4BEoo3	R		WOE	1950p	6000/9/50
M 3	SY 69	C	1958pc	-	ad	6.58	4oooX3			E1G	1959b	5000/3/59/5706/OS
M 3	SY 69	C	nd	-	ad	6.58	4oooX3			E1G	1959b	2000/9/61 ASE RCE
G 1	SY 78	B:r	nd:po	-	-	1.48	4BEoo3	RB		WOE	1950p	9000/9/50
G 1	SY 79	[B]:r	nd:po	-	-	1.50	4BEoo3	B		WOE	1950p	4503/14 RE/Sep 50/5000
G 3	SY 79	B	1949pc	-	-	1.48	4BooX3	B		2EG	1953b	6000/1/53
G 1	SY 88+	[A]:r	1948pc	-	-	1.50	4BEoo3	RB		WOE	1950p	4503/14 RE/Sep 50/5000
G 3	SY 88+	A	1948pc	-	-	1.48	4BooX3	RB		2EG	1953b	5000/2/53 OS
G 1	SY 89	[B]:r	nd:po	-	-	6.50	4BEoo3	B		WOE	1950p	4503/14 RE/Sep 50/5000
G 3	SY 89	B	1949pc	-	-	6.47	4BooX3	B		2EG	1953b	6000/2/53 OS
G 3	SY 97+	B	1948pc	-	-	6.54	4BooX3	RB		1EG	1952p	2500/11/52 ASE RCE
G 3	SY 97+	B	1948pc	-	-	6.54	4BooX3	RB		1EG	1952b	5000/5/54 SPC RE
G 1	SY 98	[A]:r	1948po	-	-	6.50	4BEoo3	RB		WOE	1950p	4503/14 RE/Sep 50/5000
G 3	SY 98	A	1948pc	1952	-	1.52	4BooX3	RB		2EG	1953b	6000/2/53 OS
G 1	SY 99	[B]:r	nd:po	-	-	1.50	4BEoo3	RB		WOE	1950p	4503/14 RE/Sep 50/5000
G 3	SY 99	B	1948pc	1952	-	1.48	4BooX3	RB		1EG	1950b	5000/10/54 OS
G 2	SZ 08	[B]	nd:po	-	-	6.54	4BEoo3	B		1EG	1952p	2500/12/52 ASE RCE
G 3	SZ 08	[B]	nd:po	-	-	6.54	4BEoo3	B		1EG	1952b	5000/5/54 SPC RE
G 2	SZ 09	[A]	1946po	-	-	6.54	4BEoo3	R		1EG	1952p	2500/12/52 ASE RCE
G 3	SZ 09	[A]	1946po	-	-	6.54	4BEoo3	R		1EG	1952b	5000/5/54 SPC RE
G 2	SZ 19	[B]	1950pc	-	ace	6.54	4BEoo3	R	31	1EG	1952p	7500/12/52 ASE RCE

The airfield layout at Hurn seems complete at the top, then blank from a quarter inch above the road.

G 2	SZ 29	B	nd:po	-	-	6.54	4BEoo3	RB	3	1EG	1952p	5000/11/52 ASE RCE
G 2	SZ 38+	[B]	nd:po	-	-	6.54	4BEoo3	R		1EG	1952p	2500/12 52 ASE RCE
G 3	SZ 38+	B	1949pc	-	-	6.47	4BooX3	R		1EG	1952b	5000/12/54 OS
G 3	SZ 39	A	1947pc	1951	-	1.52	4BooX3	R		E1G	1950b	5000/10/55 OS
G 2	SZ 47	B	nd:po	-	-	6.54	4BEoo3	B		1EG	1952p	2500/12/52 ASE RCE
G 3	SZ 47	B	nd:po	-	-	6.54	4BEoo3	B		1EG	1952b	5000/5/54 SPC RE
G 1	SZ 48	A:r	nd:po	-	-	1.48	4BEoo3	RB		WOE	1950p	5000/3/50
G 1	SZ 49	B:r	nd:po	-	-	6.47	4BEoo3	R	1	WOE	1950p	5000/3/50
G 1	SZ 49	B:r	nd:po	-	-	6.47	4BEoo3	R	1	WOE	1950p	8000/5/50
G 2	SZ 57	A	nd:po	-	-	6.54	4BEoo3	RB		1EG	1952p	2500/12/52 ASE RCE
G 3	SZ 57	A	nd:po	-	-	6.54	4BEoo3	RB		1EG	1952b	5000/5/54 SPC RE
G 2	SZ 58	[A]	nd:po	-	-	6.54	4BEoo3	RB	0	1EG	1952p	2500/12/52 ASE RCE
G 3	SZ 58	[A]	nd:po	-	-	6.54	4BEoo3	RB	0	1EG	1952b	5000/5/54 SPC RE
G 1	SZ 59	B:r	nd:po	-	-	6.47	4BEoo3	RB		WOE	1950p	5000/3/50
G 1	SZ 59	B:r	nd:po	-	-	6.47	4BEoo3	RB		WOE	1950p	8000/5/50
G 2	SZ 68	A	nd:po	-	-	6.54	4oEoo3	RB	0	1EG	1952p	2500/12/52 ASE RCE
G 3	SZ 68	A	nd:po	-	-	6.54	4oEoo3	RB	0	1EG	1952b	5000/5/54 SPC RE
G 1	SZ 69	B:r	nd:po	-	-	1.47	4oEoo3		0	WOE	1950p	5000/3/50
G 1	SZ 69	B:r	nd:po	-	-	1.47	4oEoo3		0	WOE	1950p	8000/5/50
G 1	SZ 79	B:s	nd:po	-	-	1.51	4BEoo3			1EG	1951p	8000/4/51 SPC RE
G 1	SZ 89+	A:s	nd:po	-	-	6.48	4BEoo3			1EG	1951p	8000/4/51 SPC RE
G 3	TA 01	B	1953pc	-	ac	6.53	4BooX3		11	E1G	1956b	5000/12/55 OS
G 1	TA 02	A:s	1946po	-	-	6.46	4BEoo3	R		1EG	1950p	5000/9/50 SPC RE
G 1	TA 03	A:s	1946po	-	-	6.46	4BEoo3	RB		1EG	1950p	5000/11/50 SPC RE
G 2	TA 04	[A]	1946po	-	-	6.54	4BEoo3	RB	0	1EG	1952p	2500/1/53 ASE RCE
G 3	TA 04	B	1953pc	-	ac	6.53	4BooX3		3	2EG	1954b	5000/5/54 OS

On some copies the print code is so feint as to be invisible.

G 2	TA 05	[A]	1947po	-	-	6.54	4BEoo3	RB	00	1EG	1952p	7500/11/52 ASE RCE
G 2	TA 06	[A]	1946po	-	-	6.54	4BEoo3	RB	0	1EG	1952p	5000/11/52 ASE RCE
G 2	TA 07	[A]	1947po	-	-	6.54	4BEoo3	RB		1EG	1952p	7500/3/53 ASE RCE
G 2	TA 08+	B	1951pc	-	ac	6.54	4BEoo3	RB		1EG	1952p	2500/3/53 ASE RCE
G 3	TA 08+	C	1954pc	-	bd	6.54	4BooX3			E2G	1954b	5000/1/55 OS
G 1	TA 11	B:s	1947po	-	-	6.47	4BEoo3	RB	00	1EG	1950p	5000/10/50 SPC RE
G 1	TA 12	A:s	1946po	-	-	6.46	4BEoo3	RB	0	1EG	1950p	5000/9/50 SPC RE
G 1	TA 13	A:s	1946po	-	-	6.46	4BEoo3	RB		1EG	1950p	5000/9/50 SPC RE

1	2	3	4	5	6	7	8	9	10	11	12	13	14
G	2	TA 14	B	1948pc	1949c	-	6.54	4Booo3	RB	0	1EG	1952p	2500/3/53 ASE RCE
G	3	TA 14	C	1953pc	-	ac	6.53	4BooX3		1	E2G	1954p	7000/12/54 OS
G	2	TA 15+	B	nd:po	-	-	6.54	4BEoo3	B	0	1EG	1952p	5000/3/53 ASE RCE
G	2	TA 16	A	1948pc	1952	-	6.54	4Booo3	RB	0	1EG	1952p	5000/3/53 ASE RCE
G	2	TA 17	A	1948pc	-	-	6.54	4Booo3	R		1EG	1952p	7500/1/53 ASE RCE
G	1	TA 20	A:r	1946po	-	-	6.46	4BEoo3	RB	0	1EG	1950p	5000/10/50 SPC RE
G	1	TA 21	A:s	1946po	-	-	6.46	4BEoo3	RB		1EG	1950p	5000/10/50 SPC RE
G	1	TA 22	A:s	1946po	-	-	6.47	4BEoo3	RB		1EG	1950p	5000/10/50 SPC RE
G	2	TA 23+	[A]	1946po	-	-	6.54	4BEoo3	B		1EG	1952p	2500/3/53 ASE RCE
G	3	TA 23+	B	1953pc	-	ac	6.53	4BooX3			E2G	1955b	5000/11/55 OS
G	2	TA 24	[A]	1947po	-	-	6.54	4BEoo3	RB		1EG	1952p	2500/12/52 ASE RCE
G	3	TA 24	B	1953pc	-	ac	6.53	4BooX3			2EG	1954b	5000/5/54 OS
G	2	TA 27+	[A]	nd:po	-	-	6.54	4BEoo3			1EG	1952p	5000/12/52 ASE RCE
G	1	TA 30+	A:r	1947po	-	-	1.47	4BEoo3	R	00	1EG	1950p	5000/10/50 SPC RE
G	1	TA 31+	A:r	1947po	-	-	6.47	4BEoo3	R		1EG	1950p	5000/10/50 SPC RE
G	2	TA 32+	[A]	1947po	-	-	6.54	4BEoo3	R		1EG	1952p	2500/3/53 ASE RCE
G	3	TA 32+	B	1953pc	-	ac	6.53	4BooX3			E2G	1954b	5000/12/54 OS
G	2	TF 11	A	1949pc	-	-	6.54	4Booo3	R		1EG	1952p	2500/11/52 ASE RCE
G	3	TF 11	A	1949pc	-	-	1.52	4BooX3	R		1EG	1952b	5000/6/54 OS
G	2	TF 28	[A]	nd:po	-	-	6.54	4BEoo3	RB	0	1EG	1952p	2500/3/53 ASE RCE
G	3	TF 28	B	1953pc	-	ac	6.53	4BooX3		1	E2G	1955b	5000/7/55 OS
G	1	TF 40	A:s	nd:po	-	-	1.51	4BEoo3	R		1EG	1951p	5000/3/51 A
G	1	TF 41	A:s	nd:po	-	-	1.51	4BEoo3	R	0	1EG	1951p	5000/2/51 A
G	2	TF 46	[A]	nd:po	-	-	6.54	4BEoo3	RB	0	1EG	1952p	2500/3/53 ASE RCE
G	3	TF 46	B	1953pc	-	ac	6.53	4BooX3		1	E2G	1955b	5000/7/55 OS
G	1	TF 50	A:s	nd:po	-	-	6.49	4BEoo3	R		1EG	1951p	5000/4/51 A
G	2	TF 51	B:s	1951pc	-	ac	1.52	4Booo3	R		2EG	1952p	5000/5/52 A
M	3	TF 51	B	1951pc	-	ac	1.52	4BooX3	R		E2G	1962b	2000/5/62/6394/OS
G	1	TF 52	A:s	nd:po	-	-	6.49	4BEoo3	R		1EG	1951p	5000/5/51 A
G	1	TF 60	A:s	nd:po	-	-	1.50	4BEoo3	R	0	1EG	1951p	5000/5/51 A
G	1	TF 61	A:s	nd:po	-	-	6.49	4BEoo3	RB		1EG	1951p	5000/5/51 A
G	1	TF 62	A:s	nd:po	-	-	1.50	4BEoo3	R		1EG	1951p	5000/5/51 A
G	1	TF 63	A:s	nd:po	-	-	6.49	4BEoo3	R		1EG	1951p	5000/7/51 A
G	1	TF 70	A:r	nd:po	-	-	1.48	4BEoo3	R	0	WOE	1950p	5000/4/50
G	3	TF 70	A	1949pc	-	-	1.48	4BooX3	R	0	2EG	1953b	6000/1/53
G	1	TF 71	A:s	nd:po	-	-	1.48	4BEoo3	RB	0	1EG	1951p	5000/10/51 A
G	1	TF 72	A:s	nd:po	-	-	1.48	4BEoo3	RB	0	1EG	1951p	5000/7/51 A
G	1	TF 73	A:s	nd:po	-	-	1.48	4BEoo3	RB	0	1EG	1951p	5000/10/51 A
G	1	TF 74+	A:r	nd:po	-	-	1.48	4BEoo3			1EG	1950p	5000/10/50 SPC RE
G	1	TF 80	A:r	nd:po	-	-	1.48	4BEoo3	R	00	WOE	1950p	5000/4/50
G	3	TF 80	A	1949pc	-	-	1.48	4BooX3	R	00	2EG	1953b	6000/3/53 OS
G	1	TF 81	A:s	nd:po	-	-	1.48	4BEoo3	R		1EG	1951p	5000/10/51 A
G	1	TF 82	A:s	nd:po	-	-	1.48	4BEoo3	R	000	1EG	1951p	5000/10/51 A
G	1	TF 83	A:s	nd:po	-	-	1.48	4BEoo3	R	00	1EG	1951p	5000/10/51 A
G	1	TF 84	A:r	nd:po	-	-	1.50	4BEoo3	R		WOE	1950p	4503/14 RE/Dec 50/5000
G	1	TF 90	A:r	nd:po	-	-	1.48	4BEoo3	RB	00	WOE	1950p	5000/4/50
G	3	TF 90	A	1949pc	-	-	1.48	4BooX3	RB	00	2EG	1953b	6000/3/53 OS
G	2	TF 91	A:s	1949pc	-	-	1.52	4Booo3	RB	00	2EG	1952p	5000/5/52 A
G	2	TF 92	A:s	1949pc	-	-	1.52	4Booo3	RB		2EG	1952p	5000/5/52 A
G	1	TF 93	A:s	nd:po	-	-	1.48	4BEoo3	RB	00	1EG	1951p	5000/10/51 A
G	1	TF 94	A:r	nd:po	-	-	1.50	4BEoo3	RB	0	WOE	1950p	4503/14 RE/Dec 50/5000
M	3	TF 94	B/	1955mc	1960mc	ac	6.55	4oooX3		1	E2G	1962b	5000/2/63/6394/OS
G	2	TG 00	C:s	1949pc	-	-	1.52	4Booo3	RB		2EG	1952p	5000/5/52 A
G	2	TG 01	B:s	1949pc	-	-	1.52	4Booo3	R	00	2EG	1952p	5000/5/52 A
G	2	TG 02	B:s	1948pc	1951c	a	1.52	4Booo3	R	0	2EG	1952p	5000/5/52 A
G	2	TG 03	A:s	1947pc	-	-	1.52	4Booo3	R		2EG	1952p	5000/5/52 A
G	2	TG 04	A:s	1949pc	-	-	1.52	4Booo3	RB	0	2EG	1952p	5000/5/52 A
G	2	TG 10	A:s	1946pc	-	-	1.52	4Booo3	R	0	2EG	1952p	5000/5/52 A
G	2	TG 11	A:s	1946pc	-	-	1.52	4Booo3	R	0	2EG	1952p	5000/3/52
G	2	TG 12	B:s	1950pc	-	a	1.52	4Booo3	R	00	2EG	1952p	5000/3/52

1 2	3	4	5	6	7	8	9	10	11	12	13	14
G 1	TG 13	A:r	nd:po	-	-	6.48	4BEoo3	RB	0	1EG	1950p	5000/9/50 SPC RE
G 2	TG 13	A:s	1949pc	-	-	1.52	4Booo3	RB	0	2EG	1952p	5000/4/52
G 1	TG 14+	A:r	nd:po	-	-	6.48	4BEoo3	R		1EG	1950p	5000/9/50 SPC RE
G 2	TG 20	A:s	1946pc	-	-	1.52	4Booo3	RC		2EG	1952p	5000/3/52
G 2	TG 21	A:s	1946pc	-	-	1.52	4Booo3	R	00	2EG	1952p	5000/2/52
G 2	TG 22	B:s	1950pc	-	ac	1.52	4Booo3	RB	0	2EG	1952p	5000/4/52
G 2	TG 23+	B:s	1949pc	1952c	-	1.52	4Booo3	R		2EG	1952p	5000/3/52
G 2	TG 30	A:s	1947pc	-	-	1.52	4Booo3	R		2EG	1952p	5000/4/52
G 2	TG 31	A:s	1947pc	-	-	1.52	4Booo3	R	0	2EG	1952p	5000/5/52
G 2	TG 32	A:s	1948pc	-	-	1.52	4Booo3	R	0	2EG	1952p	5000/3/52
G 2	TG 33	A:s	1949pc	-	-	1.52	4Booo3	R		2EG	1952p	5000/4/52
G 2	TG 40	A:s	1946pc	-	-	1.52	4Booo3	R		2EG	1952p	5000/4/52
G 1	TG 41+	A:r	1946po	-	-	1.47	4BEoo3	R	00	1EG	1950p	5000/10/50 SPC RE
G 2	TG 42	B:s	1948pc	1952c	-	1.52	4Booo3			2EG	1952p	5000/2/52
G 2	TG 50	A:s	1946pc	-	-	1.52	4Booo3	R		2EG	1952p	5000/5/52
G 2	TL 02	[A]	1946po	-	-	6.54	4BEoo3	RB		1EG	1952p	5000 12 52 ASE RCE
G 2	TL 03	[A]	1947po	-	-	6.54	4BEoo3	RB		1EG	1952p	5000/11/52 ASE RCE
M 3	TL 06	C	1955pc	-	ac	6.55	4oooX3		11	E1G	1959b	2500/9/59/5982/OS

Chelveston Airfield, Thurleigh Airfield.

M 3	TL 06	C	nd	-	ac	6.55	4oooX3		11	E1G	1959b	3000/9/61 ASE RCE
M 3	TL 07	B	1955pc	-	ac	6.55	4oooX3		11	E1G	1959b	2500/11/59/5982/OS

Molesworth Airfield.

M 3	TL 07	B	nd	-	ac	6.55	4oooX3		11	E1G	1960b	3000/9/61 ASE RCE
M 3	TL 08	B	1955pc	-	ac	6.55	4oooX3		1	E1G	1959b	2500/9/59/5982/OS
M 3	TL 08	B	nd	-	ac	6.55	4oooX3		1	E1G	1959b	3000/9/61 ASE RCE
G 1	TL 10	B:s	nd:po	-	-	1.51	4BEoo3	RB	10	1EG	1951p	5000/2/51 A
G 1	TL 20	B:s	nd:po	-	-	1.51	4BEoo3	RB	1	1EG	1951p	5000/2/51 A
G 2	TL 27	A	nd:po	-	-	6.54	4BEoo3	RB	000	1EG	1952p	2500/3/53 ASE RCE
G 3	TL 27	B	1955pc	-	ac	6.55	4oooX3		111	E2G	1955b	5000/11/55 OS
G 1	TL 30	B:s	nd:po	-	-	1.51	4BEoo3	RB		1EG	1951p	5000/2/51 A
G 2	TL 34	A	1949pc	1952	-	6.54	4Booo3	RB	00	1EG	1952p	2500/11/52 ASE RCE
G 3	TL 34	A	1949pc	1952	-	6.54	4Booo3	RB	00	1EG	1952b	5000/5/54 SPC RE
G 2	TL 35	B	1951pc	-	ad	6.54	4BEoo3	RB	0	1EG	1952p	2500/3/53 ASE RCE
M 3	TL 35	C	1956pc	-	be	6.56	4oooX3		1	E2G	1962b	5000/10/62/6394/OS
G 2	TL 40	B:s	1951pc	-	ace	1.51	4Booo3	RB	0	2EG	1952p	5000/3/52 A
G 1	TL 41	B:s	1951pc	-	ace	1.51	4BEoo3	RB	30	1EG	1951p	5000/10/51 A
G 1	TL 42	A:s	nd:po	-	-	1.51	4BEoo3	RB		1EG	1951p	5000/3/51 A
G 2	TL 43	B:s	1951pc	-	ac	1.52	4Booo3	B	0	2EG	1952p	5000/1/52 A
G 1	TL 44	A:s	nd:po	-	-	1.51	4BEoo3	R	00	1EG	1951p	5000/2/51 A
G 1	TL 45	B:s	nd:po	-	-	1.51	4BEoo3	RB	0	1EG	1951p	5000/3/51 A
G 2	TL 46	A	1947pc	1951	-	6.54	4Booo3	RB	00	1EG	1952p	2500/10/52 ASE RCE
G 3	TL 46	B	1955pc	-	ac	6.55	4oooX3		11	E1G	1956b	5000/2/56 OS

A copy is recorded with a manuscript note, top right: "This should be Edn 2. (Change on next reprint)".

G 1	TL 49	A:s	nd:po	-	-	1.51	4BEoo3	R		1EG	1951p	5000/3/51 A
G 1	TL 50	B:s	1949pc	1950c	ace	6.50	4BEoo3	RB	0	1EG	1951p	5000/4/51 A
G 1	TL 51	A:s	nd:po	-	-	1.49	4BEoo3	B	0	1EG	1951p	5000/4/51 A
G 1	TL 52	C:s	1951pc	-	ace	1.51	4BEoo3	RB	30	1EG	1951p	5000/10/51 A
G 1	TL 53	A:s	nd:po	-	-	1.49	4BEoo3	RB	0	1EG	1951p	5000/4/51 A
G 2	TL 54	B:t	1951pc	-	ac	1.52	4Booo3	RB	0	2EG	1952p	5000/6/52 A
G 1	TL 55	B:s	1949pc	1950c	ac	6.50	4BEoo3	R		1EG	1951p	5000/4/51 A
G 1	TL 56	B:s	1949pc	1950c	ac	6.50	4BEoo3	R		1EG	1951p	5000/4/51 A
G 2	TL 57	B:s	1951pc	-	ac	1.52	4Booo3	R	0	2EG	1952p	5000/3/52 A
G 2	TL 58	B:s	1951pc	-	ac	1.52	4Booo3	RB		2EG	1952p	5000/3/52 A
G 1	TL 59	B:s	1949pc	1950c	a	6.50	4BEoo3	R		1EG	1951p	5000/4/51 A
G 1	TL 60	B:r	nd:po	-	-	6.49	4BEoo3	RB		WOE	1950p	5000/10/50
G 1	TL 61	B:s	1950pc	-	ace	6.50	4BEoo3	B		1EG	1951p	5000/5/51 A
G 1	TL 62	A:s	nd:po	-	-	6.49	4BEoo3	RB	00	1EG	1951p	5000/6/51 A
G 1	TL 63	A:s	nd:po	-	-	6.49	4BEoo3	B	0	1EG	1951p	5000/5/51 A
G 1	TL 64	B:s	1949pc	1950c	ac	6.50	4BEoo3	RB	00	1EG	1951p	5000/6/51 A
G 2	TL 65	B:s	1951pc	-	ac	1.52	4Booo3	RB	0	2EG	1952p	5000/3/52 A

1 2	3	4	5	6	7	8	9	10	11	12	13	14
G 1	TL 66	A:s	nd:po	-	-	6.49	4BEoo3	RB	00	1EG	1951p	5000/6/51 A
G 1	TL 67	A:s	nd:po	-	-	6.49	4BEoo3	R	0	1EG	1951p	5000/6/51 A
G 1	TL 68	A:s	nd:po	-	-	6.49	4BEoo3	R	0	1EG	1951p	5000/6/51 A
G 1	TL 69	A:s	nd:po	-	-	6.49	4BEoo3	R		1EG	1951p	5000/7/51 A
G 1	TL 70	A:r	nd:po	-	-	1.51	4BEoo3	RB		1EG	1951p	5000/12/51
G 1	TL 71	A:s	nd:po	-	-	6.49	4BEoo3	RB	0	1EG	1951p	5000/7/51 A
G 1	TL 72	A:s	nd:po	-	-	1.49	4BEoo3	RB	0	1EG	1951p	5000/7/51 A
G 1	TL 73	A:s	nd:po	-	-	1.49	4BEoo3	RB	00	1EG	1951p	5000/7/51 A
G 1	TL 74	A:s	nd:po	-	-	6.49	4BEoo3	RB	0	1EG	1951p	5000/7/51 A
G 1	TL 75	A:s	nd:po	-	-	1.49	4BEoo3	B	00	1EG	1951p	5000/7/51 A
G 1	TL 76	A:s	nd:po	-	-	1.49	4BEoo3	RB		1EG	1951p	5000/10/51 A
G 3	TL 76	B	1949pc	1952c	-	1.52	4BooX3	RB		2EG	1953b	2000/3/53 13/RE
M 3	TL 76	C/	1955mc	1960c	ac	6.55	4oooX3			E3G	1962b	5000/10/62/6394/OS
G 2	TL 77	B:t	1952pc	-	ac	1.52	4Booo3		0	1EG	1952p	5000/7/52 A
M 3	TL 77	B	1952pc	-	ac	1.52	4BooX3		0	E1G	1962b	2000/6/62/6394/OS
G 1	TL 78	A:r	nd:po	-	-	1.49	4BEoo3	R	00	WOE	1950p	5000/4/50
G 2	TL 78	A:s	1949pc	-	-	1.52	4Booo3	R	00	2EG	1952p	9000/5/52
G 1	TL 79	A:r	nd:po	-	-	1.49	4BEoo3	R	00	WOE	1950p	5000/4/50
G 2	TL 79	A:s	1949pc	-		1.52	4Booo3	R	00	2EG	1952p	9000/9/52
G 2	TL 80	B:s	1948pc	1949c	-	1.52	4Booo3	RB		2EG	1952p	5000/1/52 A
G 2	TL 81	A:s	1949pc	-	-	1.52	4Booo3	RB		2EG	1952p	5000/5/52 A
G 2	TL 82	B:s	1949pc	1950c	ace	1.52	4Booo3	RB	00	2EG	1952p	5000/1/52 A
G 2	TL 83	A:s	1949pc	-	-	1.52	4Booo3	R		2EG	1952p	5000/1/52 A
G 1	TL 84	A:s	nd:po	-	-	6.49	4BEoo3	RB	0	1EG	1951p	5000/10/51 A
G 1	TL 85	A:s	nd:po	-	-	1.49	4BEoo3	RB	0	1EG	1951p	5000/10/51 A
G 1	TL 86	A:s	nd:po	-	-	6.49	4BEoo3	RB	0	1EG	1951p	5000/10/51 A
G 1	TL 87	A:s	nd:po	-	-	6.49	4BEoo3	RB	0	1EG	1951p	5000/10/51 A
G 1	TL 88	A:r	nd:po	-	-	1.49	4BEoo3	R	0	WOE	1950p	5000/3/50
G 1	TL 88	B:s	1949pc	1951c	ace	1.51	4BEoo3	R	0	1EG	1950p	5000/7/51
M 3	TL 88	C	1955pc	-	ad	6.55	4oooX3		2	E2G	1959b	5000/10/59/5943/OS
G 1	TL 89	A:r	nd:po	-	-	1.49	4BEoo3		00	WOE	1950p	5000/3/50
G 1	TL 89	B:s	1951pc	-	ace	1.51	4BEoo3		10	1EG	1951p	5000/9/51
G 3	TL 89	C	1955pc	-	ad	6.55	4oooX3		22	E2G	1956b	5000/6/56
G 2	TL 90	B:s	1949pc	-	-	1.52	4Booo3	RB		2EG	1952p	5000/3/52 A
G 1	TL 91	A:r	nd:po	-	-	1.50	4BEoo3	R	0	WOE	1950p	4503/14 RE/Dec 50/5000
G 1	TL 91	[A]	nd:po	-	-	1.50	4BEoo3	R	0	WOE	1950p	8000/6/54 OS
G 1	TL 92	A:r	nd:po	-	-	1.50	4BEoo3	RB	00	WOE	1950p	4503/14 RE/Dec 50/5000
G 3	TL 92	A	1949pc	1952rpb	-	1.49	4BooX3	RB	00	2EG	1953b	5000/4/53 OS
G 2	TL 93	A:s	1949pc	-	-	1.52	4Booo3	RB	0	2EG	1952p	5000/3/52 A
G 2	TL 94	A:s	1949pc	-	-	1.52	4Booo3	RB	0	2EG	1952p	5000/3/52 A
G 2	TL 95	A:s	1949pc	-	-	1.52	4Booo3	R	00	2EG	1952p	5000/3/52 A
G 2	TL 96	A:s	1949pc	-	-	1.52	4Booo3	R	0	2EG	1952p	5000/3/52 A
G 2	TL 97	A:s	1949pc	-	-	1.52	4Booo3	B	000	2EG	1952p	5000/3/52 A
G 1	TL 98	A:r	nd:po	-	-	6.49	4BEoo3	R	00	WOE	1950p	5000/3/50
G 2	TL 98	A:s	1949pc	-	-	1.52	4Booo3	R	00	2EG	1952p	9000/8/52
G 1	TL 99	A:r	nd:po	-	-	1.49	4BEoo3	R	00	WOE	1950p	5000/3/50
G 1	TL 99	A:r	nd:po	-	-	1.51	4BEoo3	R	00	1EG	1951p	5000/2/51
M 3	TL 99	[B]	1955pc	-	ac	6.55	4oooX3		12	E2G	1961b	2500/2/61/6233/OS
G 2	TM 00	B:t	1948pc	1950c	ace	1.52	4Booo3		0	2EG	1952p	5000/6/52 A
G 1	TM 01	B:r	nd:po	-	-	6.50	4BEoo3	R		WOE	1950p	4503/14 RE/Dec 50/5000
G 3	TM 01	B	1949pc	-	-	6.48	4BooX3	R		1EG	1953b	2000/1/53
G 3	TM 01	C	1955pc	-	ac	6.55	4oooX3			E2G	1956b	5000/2/56 OS
G 1	TM 02	[A]:r	1946po	-	-	6.50	3BEoo1	RB	0	WOE	1950p	4503/14 RE/Dec 50/5000
G 3	TM 02	A	1946pc	-	-	6.46	3BooX1	RB	0	2EG	1953b	5000/2/53 OS
G 1	TM 03	[A]:r	1948po	-	-	6.50	3BEoo1	R	00	WOE	1950p	4503/14 RE/Dec 50/5000
G 2	TM 03	C:s	1949pc	-	-	1.52	3Booo1	R	00	2EG	1952p	5000/5/52 A
G 2	TM 04	A:s	1947pc	-	-	1.52	3Booo2	RB		2EG	1952p	5000/5/52 A
G 2	TM 05	A:t	1947pc	-	-	1.52	3Booo2	RB	0	2EG	1952p	5000/6/52 A
G 2	TM 06	A:s	1946pc	-	-	1.52	3Booo2	RB	0	2EG	1952p	5000/5/52 A
G 2	TM 07	A:s	1947pc	-	-	1.52	3Booo2	R		2EG	1952p	5000/5/52 A

1	2	3	4	5	6	7	8	9	10	11	12	13	14
G	2	TM 08	A:s	1947pc	-	-	1.52	3Booo2	RB	00	2EG	1952p	5000/5/52 A
G	2	TM 09	A:s	1947pc	-	-	1.52	3Booo2	RB	000	2EG	1952p	5000/5/52 A
G	1	TM 11+	B:r	nd:po	-	-	1.49	4BEoo3	R		1EG	1950p	5000/10/50 SPC RE
G	2	TM 12	A:s	1946pc	-	-	1.52	3Booo2	R		2EG	1952p	5000/4/52 A
G	2	TM 13	A:s	1947pc	-	-	1.52	3Booo2	R		2EG	1952p	5000/3/52 A
G	2	TM 14	A:s	1947pc	-	-	1.52	3Booo2	RC	1	2EG	1952p	5000/3/52 A
G	2	TM 15	A:s	1947pc	-	-	1.52	3Booo2	RB		2EG	1952p	5000/3/52 A
G	2	TM 16	A:s	1946pc	-	-	1.52	3Booo2	RB	0	2EG	1952p	5000/3/52 A
G	2	TM 17	B:s	1947pc	1952c	-	1.52	3Booo2	R	00	2EG	1952p	5000/5/52 A
G	2	TM 18	A:s	1946pc	-	-	1.52	3Booo2	RB	000	2EG	1952p	5000/5/52 A
G	2	TM 19	A:s	1947pc	-	-	1.52	3Booo2	RB	0	2EG	1952p	5000/5/52 A
G	2	TM 22	A:s	1946pc	-	-	1.52	3Booo2	R		2EG	1952p	5000/1/52
G	2	TM 23	A:s	1946pc	-	-	1.52	3Booo2	RB	0	2EG	1952p	5000/3/52
G	2	TM 24	A:s	1947pc	1951	-	1.52	3Booo2	R	0	2EG	1952p	5000/3/52
G	2	TM 25	A:s	1947pc	-	-	1.52	3Booo2	R	0	2EG	1952p	5000/2/52
G	2	TM 26	B:s	1947pc	1952c	-	1.52	3Booo2	R		2EG	1952p	5000/1/52
G	2	TM 27	A:s	1946pc	-	-	1.52	3Booo2	R	0	2EG	1952p	5000/1/52
G	2	TM 28	A:s	1946pc	-	-	1.52	3Booo2	RB	01	2EG	1952p	5000/1/52
G	2	TM 29	A:s	1946pc	-	-	1.52	3Booo2	RB	1	2EG	1952p	5000/1/52
G	2	TM 33	A:s	1946pc	-	-	1.52	3Booo2			2EG	1952p	5000/4/52
G	2	TM 34	B:s	1949pc	-	-	1.52	3Booo2		0	2EG	1952p	5000/4/52
G	2	TM 35	A:s	1947pc	-	-	1.52	3Booo2	R	00	2EG	1952p	5000/1/52
G	2	TM 36	A:s	1947pc	-	-	1.52	3Booo2	RB	0	2EG	1952p	5000/1/52
G	3	TM 36	A	1947pc	-	-	1.52	3BooX2	RB	0	E2G	1952b	5000/1/55 OS
G	2	TM 37	A:s	1946pc	-	-	1.52	3Booo2	R	00	2EG	1952p	5000/2/52
G	2	TM 38	A:s	1946pc	-	-	1.52	3Booo2	RB	00	2EG	1952p	5000/2/52
G	2	TM 39	A:s	1946pc	-	-	1.52	3Booo2	RB	0	2EG	1952p	5000/2/52
G	2	TM 44	B:s	1949pc	-	-	1.52	3Booo2			2EG	1952p	5000/2/52
G	2	TM 45	A:s	1946pc	-	-	1.52	3Booo2	R		2EG	1952p	5000/4/52
G	2	TM 46	A:s	1946pc	-	-	1.52	3Booo2	R	1	2EG	1952p	5000/3/52
G	1	TM 47+	A:r	1946po	-	-	6.46	3BEoo1	R	0	1EG	1950p	5000/9/50 SPC RE
G	1	TM 48+	A:r	1948po	-	-	6.47	3BEoo1	RB	00	1EG	1950p	5000/10/50 SPC RE
G	2	TM 49	A:s	1946pc	-	-	1.52	3Booo2	R		2EG	1952p	5000/3/52
G	2	TM 59	B:s	1947pc	1951	ace	1.52	3Booo2	R		2EG	1952p	5000/3/52
G	1	TQ 00	B:s	nd:po	-	-	1.51	4BEoo3	RB	0	1EG	1951p	5000/1/51 A
G	1	TQ 01	B:r	nd:po	-	-	1.51	4BEoo3	RB		1EG	1951p	5000/1/51 A
G	1	TQ 02	A:s	nd:po	-	-	1.51	4BEoo3	RB		1EG	1951p	5000/1/51 A
G	3	TQ 02	B	1951pc	-	ace	1.51	4BooX3	RB		E2G	1955b	5000/11/55 OS
G	1	TQ 03	A:s	nd:po	-	-	1.51	4BEoo3	RB	0	1EG	1951p	5000/1/51 A
G	3	TQ 03	B	1948pc	1952c	-	6.52	4BooX3	RB	0	E2G	1955b	5000/11/55 OS
G	1	TQ 04	C:s	nd:po	-	-	1.51	4BEoo3	RB		1EG	1951p	5000/1/51 A
G	3	TQ 04	C	1948pc	1953c	-	6.48	4BooX3	RB		1EG	1951b	5000/5/54 OS
G	1	TQ 05	B:r	nd:po	-	-	6.48	4BEoo3	RB	0	WOE	1950p	5000/2/50
G	1	TQ 05	B:r	1950pc	-	-	1.51	4BEoo3	RB	0	1EG	1950p	14,000/1/51
G	1	TQ 06	B:r	nd:po	-	-	1.48	4BEoo3	R	00	WOE	1950p	5000/2/50
G	1	TQ 06	B:r	nd:po	-	-	1.51	4BEoo3	R	00	1EG	1950p	14,000/1/51
G	1	TQ 06	C:s	1951pc	-	ace	1.51	4BEoo3	R	00	WOE	1950p	5000/2/50
G	1	TQ 07	B:s	1951pc	-	ace	1.51	4BEoo3	R	31	1EG	1951p	1000/2/51
G	1	TQ 07	B:s	1951pc	-	ace	1.51	4BEoo3	R	31	1EG	1951p	9000/9/51
G	3	TQ 07	B	1951pc	-	ace	6.51	4BooX3	R	31	E1G	1951b	5000/10/55 OS
G	1	TQ 08	C:r	1950pc	-	acf	1.51	4BEoo3	RB	30	1EG	1951p	9000/1/51 A
G	1	TQ 09	B:s	nd:po	-	-	1.51	4BEoo3	RB	1	1EG	1951p	9000/1/51 A
G	1	TQ 10	A:s	1948po	-	-	1.51	4BEoo3	RB	0	1EG	1951p	5000/2/51 A
G	1	TQ 11	B:s	nd:po	-	-	1.51	4BEoo3	RB		1EG	1951p	5000/2/51 A
G	1	TQ 12	B:s	1950pc	-	ace	1.51	4BEoo3	RB		1EG	1951p	5000/2/51 A
G	1	TQ 13	C:s	1948pc	1950c	ace	1.51	4BEoo3	RB		1EG	1951p	5000/2/51 A
G	1	TQ 14	B:s	1949pc	1950	-	1.51	4BEoo3	RB		1EG	1951p	5000/2/51 A
G	1	TQ 15	C:s	1949pc	1950	-	1.51	4BEoo3	RB		1EG	1951p	5000/2/51 A
G	1	TQ 16	B:r	1947po	-	-	6.47	4BEoo3	RB		WOE	1950p	5000/7/50
M	3	TQ 16	D/	1956mc	1961ch	ad	6.56	4oooX4			E2G	1961b	5000/1/61/6239/OS

1 2 3	4	5	6	7	8	9	10	11	12	13	14
G 1 TQ 17	A:r	1947po	-	-	6.47	4BEoo3	R	11	WOE	1950p	5000/7/50
G 2 TQ 17	B:s	1950pc	-	acf	nd	4BEoo3	R	21	2EG	1952p	9000/11/52 SPC
G 1 TQ 18	B:r	nd:po	-	-	6.47	4BEoo3	R	3	WOE	1950p	5000/7/50
G 3 TQ 18	C	1951pc	-	ace	6.51	4oooX3	RC	3	E2G	1955b	5000/11/55 OS
G 1 TQ 19	B:r	nd:po	-	-	6.47	4BEoo3	RB	0	WOE	1950p	5000/8/50
G 3 TQ 19	B	1950pc	-	-	6.47	4BooX3	RB	0	E1G	1950b	5000/10/55 OS
G 1 TQ 20	A:s	1946po	-	-	1.51	4BEoo3	RB	0	1EG	1951p	5000/2/51 A
G 1 TQ 21	B:s	nd:po	-	-	1.51	4BEoo3	RB		1EG	1951p	5000/2/51 A
G 1 TQ 22	B:s	nd:po	-	-	1.51	4BEoo3	B		1EG	1951p	5000/2/51 A
M 3 TQ 22	B	1958mc	-	ac	6.58	4oooX3			E2G	1962b	2000/12/62/6394/OS
G 1 TQ 23	B:s	nd:po	-	-	1.51	4BEoo3	RB	0	1EG	1951p	5000/2/51 A
G 1 TQ 24	B:s	1950pc	-	ace	1.51	4BEoo3	RB	12	1EG	1951p	5000/2/51 A
G 1 TQ 25	C:s	1947pc	1950c	ace	1.51	4BEoo3	RB		1EG	1951p	5000/2/51 A
G 2 TQ 25	C	1947pc	1952c	ace	6.50	4BooX3	R		E1G	1951b	5000/10/55 OS
G 1 TQ 26	B:r	nd:po	-	-	6.47	4BEoo3	RB	0	WOE	1950p	5000/7/50
G 3 TQ 26	C	1950pc	1953c	-	6.53	4BooX3	B	0	2EG	1954b	5000/5/54 OS
G 1 TQ 27	B:r	nd:po	-	-	6.47	4BEoo3	RB		WOE	1950p	5000/7/50
G 3 TQ 27	C	1952pc	-	ace	1.52	4BooX3	R		E2G	1954b	5000/9/54 OS
G 1 TQ 28	B:r	nd:po	-	-	6.47	4oEoo3	RC	0	WOE	1950p	5000/7/50
G 3 TQ 28	B	1948pc	1952	-	1.55	4oooX3	RC	0	1EG	1950b	5000/5/54 OS
G 1 TQ 29	B:r	nd:po	-	-	6.47	4BEoo3	R	1	WOE	1950p	5000/7/50
G 1 TQ 30	B:s	nd:po	-	-	1.51	4BEoo3	RB		1EG	1951p	5000/2/51 A
G 1 TQ 31	B:s	nd:po	-	-	1.51	4BEoo3	RB		1EG	1951p	5000/2/51 A
G 1 TQ 32	B:r	nd:po	-	-	6.50	4BEoo3	RB		WOE	1950p	4503/14 RE/Nov 50/5000
G 3 TQ 32	B	1948pc	1949c	-	6.48	4BooX3	RB		1EG	1953b	2000/2/53 OS
M 3 TQ 32	C	1959mc	-	ac	6.59	4oooX3			E2G	1962b	5000/10/62/6394/OS
G 1 TQ 33	B:r	nd:po	-	-	1.50	4BEoo3	RB		WOE	1950p	4503/14 RE/Nov 50/5000
G 3 TQ 33	B	1948pc	49c/52rpb	-	1.48	4BooX3	RB		1EG	1953b	2000/2/55 OS
G 1 TQ 34	B:s	nd:po	-	-	1.51	4BEoo3	RB	2	1EG	1951p	5000/2/51 A
G 1 TQ 35	C:s	1951pc	-	ace	1.51	4BEoo3	B	1	1EG	1951p	5000/10/51 A
G 1 TQ 36	B:r	nd:po	-	-	6.47	4BEoo3	RB	1	WOE	1950p	5000/7/50
M 3 TQ 36	C	1957pc	-	ac	6.57	4oooX4		3	E2G	1962b	5000/4/63/6394/OS
G 1 TQ 37	C:r	nd:po	-	-	6.48	4oEoo3	R		WOE	1950p	5000/7/50
G 3 TQ 37	D	1952pc	-	ace	1.52	4oooX3	R		2EG	1954b	5000/5/54 OS
G 1 TQ 38	B:r	nd:po	-	-	1.48	4oEoo3	RC		WOE	1950p	5000/7/50
G 3 TQ 38	B	1948pc	1952	-	1.52	4oooX3	RC		E1G	1950b	5000/8/55 OS
G 1 TQ 39	B:r	nd:po	-	-	1.48	4BEoo3	RB		WOE	1950p	5000/5/50
G 1 TQ 40+	B:s	nd:po	-	-	1.51	4BEoo3	RB		1EG	1951p	5000/2/51 A
G 1 TQ 41	B:r	nd:po	-	ac	1.48	4BEoo3	RB		WOE	1950p	5000/7/50
G 1 TQ 42	B:r	nd:po	-	-	6.50	4BEoo3	RB		WOE	1950p	4503/14 RE/Nov 50/5000
G 1 TQ 43	B:r	nd:po	-	ac	6.50	4BEoo3	RB		WOE	1950p	4503/14 RE/Nov 50/5000
G 1 TQ 44	B:s	nd:po	-	-	1.51	4BEoo3	RB		1EG	1951p	5000/2/51 A
G 1 TQ 45	B:s	nd:po	-	-	1.51	4BEoo3	RB	0	1EG	1951p	5000/2/51 A
G 1 TQ 46	B:r	nd:po	-	-	6.47	4BEoo3	RB	0	WOE	1950p	5000/5/50
G 3 TQ 46	B	1949pc	1952	-	1.52	4BooX3	RB	0	E1G	1950b	5000/10/55 OS
G 1 TQ 47	B:r	nd:po	-	-	1.48	4BEoo3	R		WOE	1950p	5000/7/50
G 1 TQ 48	A:r	1948po	-	-	1.48	4BEoo3	RC	0	WOE	1950p	5000/7/50
G 1 TQ 49	B:r	nd:po	-	-	6.47	4BEoo3	RB	0	WOE	1950p	5000/7/50
G 1 TQ 50	B:s	nd:po	-	-	1.51	4BEoo3	RB		1EG	1951p	5000/3/51 A
G 2 TQ 51	B:s	1951pc	-	ace	1.52	4Booo3	RB		2EG	1952p	5000/1/52 A
G 1 TQ 52	[A]:r	1948po	-	-	6.50	4BEoo3	RB		WOE	1950p	4503/14 RE/Nov 50/5000
G 1 TQ 53	C:r	nd:po	-	-	6.50	4BEoo3	RB		WOE	1950p	4503/14 RE/Dec 50/3000
G 1 TQ 54	B:s	1948pc	1950c	ace	1.51	4BEoo3	RB		1EG	1951p	5000/3/51 A
G 2 TQ 55	D:t	1952pc	-	ace	6.52	4Booo3	R		1EG	1952p	5000/8/52 A
G 1 TQ 56	E:s	1950pc	-	ace	6.51	4BEoo3	R		1EG	1951p	5000/3/51 A
G 1 TQ 57	B:s	1950pc	-	ace	6.50	4BEoo3	R		1EG	1951p	5000/4/51 A
G 1 TQ 58	A:s	nd:po	-	-	6.48	4BEoo3	RB	1	1EG	1951p	5000/4/51 A
G 1 TQ 59	B:s	nd:po	-	-	6.48	4BEoo3	RB		1EG	1951p	5000/4/51 A
G 1 TQ 60	A:s	1947po	-	-	6.47	4BEoo3	RB		1EG	1951p	5000/5/51 A
G 1 TQ 61	B:s	nd:po	-	-	6.50	4BEoo3	B		1EG	1951p	5000/5/51 A

1	2	3	4	5	6	7	8	9	10	11	12	13	14
G	1	TQ 62	A:s	1948po	-	-	1.48	4BEoo3	RB		1EG	1951p	5000/5/51 A
G	1	TQ 63	B:s	nd:po	-	-	6.49	4BEoo3	RB		1EG	1951p	5000/5/51 A
G	1	TQ 64	B:s	nd:po	-	-	6.48	4BEoo3	R		1EG	1951p	5000/6/51 A
G	1	TQ 65	[B]:r	1949pc	1950	-	6.50	4BEoo3	RB	0	WOE	1950p	4503/14 RE/Dec 50/5000
G	1	TQ 66	[A]:r	1948po	-	-	1.50	4BEoo3	RB		WOE	1950p	4503/14 RE/Oct 50/5000
G	2	TQ 66	[A]:r	1948po	-	-	6.52	4BEoo3	RB		2EG	1952p	9000/11/52 SPC RE
G	1	TQ 67	B:r	nd:po	-	-	6.50	4BEoo3	RB	1	WOE	1950p	4503/14 RE/Sep 50/5000
G	3	TQ 67	B	1948pc	48c/52rpb	-	6.48	4BooX3	RB	1	1EG	1953b	2000/3/53 OS
G	3	TQ 67	B	1948pc	-	-	6.48	4BooX3	RB	1	E1G	1953b	5000/11/55 OS
G	1	TQ 68	B[2]:s	1951pc	-	ace	1.51	4BEoo3	R		1EG	1951p	5000/10/51 A
G	1	TQ 69	B:s	nd:po	-	-	6.48	4BEoo3	RB		1EG	1951p	5000/6/51 A
G	1	TQ 70	A:s	1947po	-	-	1.49	4BEoo3	R	0	1EG	1951p	5000/7/51 A
G	1	TQ 71	B:s	nd:po	-	-	6.47	4BEoo3	RB		1EG	1951p	5000/7/51 A
G	1	TQ 72	B:s	1951pc	-	ace	1.51	4BEoo3	RB		1EG	1951p	5000/10/51 A
G	1	TQ 73	B:s	nd:po	-	-	1.48	4BEoo3	RB		1EG	1951p	5000/6/51 A
G	1	TQ 74	A:s	1948po	-	-	6.48	4BEoo3	R		1EG	1951p	5000/6/51 A
G	1	TQ 75	A:r	nd:po	-	-	1.50	4BEoo3	RB		WOE	1950p	4503/14 RE/Sep 50/5000
G	3	TQ 75	A	1948pc	1952rpb	-	1.49	4BooX3	RB		2EG	1953b	6000/3/53 OS
G	1	TQ 76	[A]:r	1947po	-	-	6.50	4BEoo3	RB	1	WOE	1950p	4503/14 RE/Oct 50/5000
G	3	TQ 76	C	1951pc	-	ace	1.51	4BooX3	RB	1	1EG	1953b	2000/1/53
M	3	TQ 76	D/	1960mc	1962ch	bdf	6.60	4oooX4		1	E2G	1963b	5000/4/63/6394/OS
G	1	TQ 77	[A]:r	1947po	-	-	6.50	4BEoo3	RB		WOE	1950p	4503/14 RE/Oct 50/5000
G	3	TQ 77	B	1954pc	-	ace	6.54	4BooX3			E2G	1955b	5000/10/55 OS
G	2	TQ 78	B:s	1951pc	-	ace	1.52	4Booo3	R		2EG	1952p	5000/3/52 A
G	1	TQ 79	A:s	1947po	-	-	6.47	4BEoo3	RB		1EG	1951p	5000/6/51 A
G	1	TQ 81+	A:s	1947po	-	-	1.48	4BEoo3	RB		1EG	1951p	5000/10/51 A
G	2	TQ 82	A:s	1948pc	-	-	1.52	4Booo3	RB		2EG	1952p	5000/1/52 A
G	2	TQ 83	B:s	1948pc	1949c	-	1.52	4Booo3	RB		2EG	1952p	5000/1/52 A
G	2	TQ 84	A:s	1948pc	-	-	1.52	4Booo3	RB		2EG	1952p	5000/1/52 A
G	1	TQ 85	[A]:r	1947po	-	-	6.50	4BEoo3	RB	0	WOE	1950p	4503/14 RE/Sep 50/5000
G	3	TQ 85	B	1951pc	-	ac	6.51	4BooX3	RB	0	2EG	1953b	6000/2/53 OS
G	1	TQ 86	[A]:r	1947po	-	-	6.50	4BEoo3	RB		WOE	1950p	4503/14 RE/Oct 50/5000
G	3	TQ 86	B	1951pc	-	ace	1.51	4BooX3	RB		2EG	1953b	6000/2/53 OS
G	1	TQ 87	B:r	nd:po	-	-	6.50	4BEoo3	R		WOE	1950p	4503/14 RE/Oct 50/5000
G	3	TQ 87	B	1949pc	-	-	6.47	4BooX3	R		1EG	1950b	5000/6/54 OS
G	2	TQ 88	A:s	1946pc	-	-	1.52	4Booo3	R	1	2EG	1952p	5000/1/52 A
G	2	TQ 89	B:s	1950pc	-	ace	1.52	4Booo3	R		2EG	1952p	5000/1/52 A
G	1	TQ 91	[A]:r	nd:po	-	-	1.50	4BEoo3	R		WOE	1950p	4503/14 RE/Oct 50/5000
G	3	TQ 91	B	1949pc	-	-	1.48	4BooX3	R		2EG	1953b	6000/4/53 OS
G	1	TQ 92	B:r	nd:po	-	-	6.50	4BEoo3	RB		WOE	1950p	4503/14 RE/Dec 50/5000
G	2	TQ 93	B:s	1951pc	-	ace	1.52	4Booo3	RB		2EG	1952p	5000/5/52 A
G	1	TQ 94	A:s	nd:po	-	-	6.48	4BEoo3	RB		1EG	1951p	5000/10/51 A
G	1	TQ 95	B:s	nd:po	-	-	6.48	4BEoo3	RB		1EG	1951p	5000/10/51 A
G	1	TQ 96	B:s	nd:po	-	-	6.48	4BEoo3	RB		1EG	1951p	5000/10/51 A
G	2	TQ 97	B:s	1949pc	-	-	1.52	4Booo3	R		2EG	1952p	5000/3/52 A
G	2	TQ 98	A:s	1946pc	-	-	1.52	4Booo3	R		2EG	1952p	5000/3/52 A
G	2	TQ 99	A:s	1947pc	-	-	1.52	4Booo3	R		2EG	1952p	5000/3/52 A
G	1	TR 01	B:r	nd:po	-	-	1.50	4BEoo3	R		WOE	1950p	4503/14 RE/Dec 50/5000
G	1	TR 02+	A:r	nd:po	-	-	1.48	4BE2X3	R	0	WOE	1948p	5000/7/48
G	?	TR 02+									2EG		not recorded
M	3	TR 02+	C	1960mc	-	ad	6.60	4oooX3		3	E3G	1962b	5000/11/62/6394/OS
G	1	TR 03	A:r	nd:po	-	-	6.48	4BE4X3	RB		WOE	1948p	5000/9/48
G	1	TR 04	A:r	nd:po	-	-	6.48	4BE4X3	RB		WOE	1948p	5000/11/48
G	3	TR 04	A	1948pc	-	-	6.48	4BooX3	RB		1EG	1948b	5000/11/55 OS
M	3	TR 04	B/	1958mc	1961ch	ac	6.58	4oooX3			E2G	1963b	5000/1/63/6394/OS
G	2	TR 05	A:s	1948pc	-	-	1.52	4Booo3	RB		2EG	1952p	5000/5/52 A
G	2	TR 06	C:s	1949pc	1952c	ace	1.52	4Booo3	RB		2EG	1952p	5000/5/52 A
G	2	TR 07	B:s	1948pc	1951c	-	1.52	4Booo3	R		2EG	1952p	5000/5/52 A
G	2	TR 09+	B:s	1950pc	-	a	1.52	4Booo3			2EG	1952p	5000/2/52

1 2 3	4	5	6	7	8	9	10	11	12	13	14
G 1 TR 13	A:r	nd:po	-	-	6.48	4BE4X3	RB	10	WOE	1948p	5000/11/48
G 3 TR 13	C	1949pc	1953c	-	6.48	4oooX3	RB	10	E2G	1955b	5000/11/55 OS
G 1 TR 14	A:r	nd:po	-	-	1.48	4BE2X3	RB		WOE	1948p	5000/10/48
M 3 TR 14	B/	1949pc	1956mc	ac	6.48	4BooX3	B		E2G	1955b	5000/12/58/5759/OS
G 2 TR 15	B:s	1949pc	-	-	1.52	4Booo3	RB		2EG	1952p	5000/5/52 A
G 2 TR 16	A:s	1948pc	-	-	1.52	4Booo3	RB		2EG	1952p	5000/5/52 A
G 1 TR 23	[A]:r	1946po	-	-	1.50	4BEoo3	RB		WOE	1950p	4503/14 RE/Dec 50/5000
G 1 TR 24	B:r	nd:po	-	-	1.50	4BEoo3	RB		WOE	1950p	4503/14 RE/Oct 50/5000
G 3 TR 24	B	1950pc	-	-	1.47	4BooX3	RB		2EG	1953b	6000/3/53 OS
G 2 TR 25	B:s	1948pc	1949c	-	1.52	4Booo3	RB		2EG	1952p	5000/2/52
G 2 TR 26+	B:s	1949pc	-	-	1.52	4Booo3	R		2EG	1952p	5000/2/52
G 1 TR 34+	A	nd:po	-	-	6.46	4BEoo3	RB		WOE	1950p	5000/9/50
G 3 TR 34+	A	1946pc	-	-	6.46	4BooX3	RB		E1G	1950b	5000/2/56 OS
G 2 TR 35	A:s	1948pc	1951	-	1.52	4Booo3	RB		2EG	1952p	5000/1/52
G 1 TR 36+	A:r	1948po	-	-	1.48	4BEoo3	RB	01	1EG	1950p	5000/10/50 SPC RE
G 1 TV 59+	B:r	nd:po	-	-	1.48	4BEoo3	B		1EG	1950p	5000/9/50 SPC RE

1 2 3   4   5   6   7 8 9   10 11 12 13 14

GSGS 4627A coloured edition, with layers

See page 58 for an explanation of headings and layout. All but SU 33 have an "A" suffix to the GSGS number, in brown as part of the layer colour plate. The various composite coloured maps with layers, also classified GSGS 4627A, are not listed here. Copies of all these sheets are in the British Library.

G 1 SU 33 B:r nd:po - - 1.51 4BEoo3 RB 00 1EG 1951p 5000/2/51
    The recorded copy notes that "The voucher which accompanied these maps from the Production Centre states that they were produced for the Staff College, Camberley. Exercise Hess."
G 1 SU 45 A:r nd:po - - 6.48 4BE4X3 R   WOE 1948p 2000/3/49
G 1 SU 45 A:r nd:po - - 6.48 4BE4X3 R   WOE 1948p 2000/1/50 SPC
G 1 SU 46 A:r nd:po - - 6.49 4BE4X3 RB 0   WOE 1949p 2000/3/49
G 1 SU 46 A:r nd:po - - 6.49 4BE4X3 RB 0   WOE 1949p 2000/1/50 SPC
G 1 SU 55 A:r nd:po - - 1.49 4BE4X3 RB   WOE 1949p 2000/3/49
G 1 SU 55 A:r nd:po - - 1.49 4BE4X3 RB   WOE 1949p 2000/1/50 SPC
G 1 SU 56 A:r nd:po - - 1.49 4BE4X3 RB 03   WOE 1949p 2000/3/49
G 1 SU 56 A:r nd:po - - 1.49 4BE4X3 RB 03   WOE 1949p 2000/1/50 SPC
G 1 SU 84 A:r nd:po - - 1.49 4BE4X3 RB   WOE 1949p 2000/3/49
G 1 SU 84 A:r nd:po - - 1.49 4BE4X3 RB   WOE 1949p 2000/1/50 SPC
G 1 SU 85 A:r nd:po - - 1.49 4BE4X3 RB 00   WOE 1949p 2000/3/49
G 1 SU 85 A:r nd:po - - 1.49 4BE4X3 RB 00   WOE 1949p 2000/1/50 SPC

Cross references to other issues based on the 1:25,000 First Series

Sheets overprinted in green with rights of way information

SD 92, SH 52, SH 53, SH 61 & Part of SH 51, SH 62, SH 63, SH 64, SH 71, SH 72, SH 74, SH 83, SH 93, SN 69 & Part of SN 59, SO 49, SO 59.

Special Administrative Area Series Greater London

TQ 07, TQ 08, TQ 09, TQ 16, TQ 17, TQ 18, TQ 19, TQ 25, TQ 26, TQ 27, TQ 28, TQ 29 & Part of TL 20, TQ 35, TQ 36, TQ 37, TQ 38, TQ 39 & Part of TL 30, TQ 45, TQ 46 & Part of TQ 56, TQ 47, TQ 48, TQ 49, TQ 57, TQ 58 & Part of TQ 68, TQ 59.

North Pennines Rural Development Board Areas

NT 60, NT 70, NT 71, NT 81, NT 82, NT 83, NT 90, NT 91, NT 92, NT 93, NU 00, NU 01, NU 02, NY 37, NY 46, NY 47, NY 48, NY 50, NY 51, NY 52, NY 53, NY 54, NY 55, NY 56, NY 58, NY 59, NY 62, NY 69, NY 86, NY 87, NY 95, NY 96, NY 97, NY 98, NY 99, NZ 01, NZ 02, NZ 03, NZ 04, NZ 05, NZ 08, NZ 09, NZ 10, NZ 11, NZ 12, NZ 13, NZ 14, SD 57, SD 58, SD 59, SD 67, SD 76, SD 77, SD 85, SD 86, SD 95, SE 04, SE 05, SE 14, SE 15, SE 16, SE 18, SE 19, SE 25, SE 26, SE 27, SE 28.

Geological Survey

NB Only Geological Survey sheets on series sheet lines and based on First Series mapping are listed here.
NY 57, SH 75, SH 76, SK 06, SK 07, SK 15, SK 16, SK 17, SK 18, SK 26, SO 48, SO 49, SO 59, ST 45, ST 47, TL 81, TQ 81 & Part of TQ 80.

Soil Survey

NB Only Soil Survey sheets on series sheet lines and based on First Series mapping are listed here.
NY 53, NY 56, SD 58, SE 36, SE 39, SE 58, SE 60, SE 64, SE 65, SE 74, SE 76, SE 85, SJ 17, SJ 21, SJ 35, SJ 37, SJ 65, SJ 82, SK 05, SK 17, SK 66, SK 85, SK 99, SM 90, SM 91, SN 13, SN 24, SN 41, SN 62, SN 72, SO 09, SO 34, SO 52, SO 53, SO 61, SO 74, SO 82, SO 87, SP 05, SP 12, SP 30, SP 36, SP 47, SP 60, SP 66, SS 63, SS 74 & Part of SS 75, ST 10, SU 03, SU 88, SW 53, SX 18, SX 47, SX 65, TA 14, TF 04, TF 16, TF 28, TG 11, TG 31, TL 34, TL 54, TL 71, TL 99, TM 12, TM 28, TM 49, TQ 59, TQ 86, TQ 99, TR 04, TR 35.

NOTES

NOTES

NOTES

NOTES

NOTES

NOTES

NOTES

NOTES

PLATES

PLATES

Plate 2. Sheet 4009, [1944]. This is an undated printing, lettered largely in Caslon Old Roman, and with a green plate for woods (tree symbols) and parks (green tint). Minor details such as signal boxes and signal posts (S.B. and S.P.) on railways are shown. Road widths are exaggerated to fit in names.

Plate 1. Sheet 63/20, 1946. The city of Norwich. The sheet includes some hand-lettering, for example "Heigham Grove" and "Chapelfield Grove", and shows the River Yare in solid blue through the centre of the city.

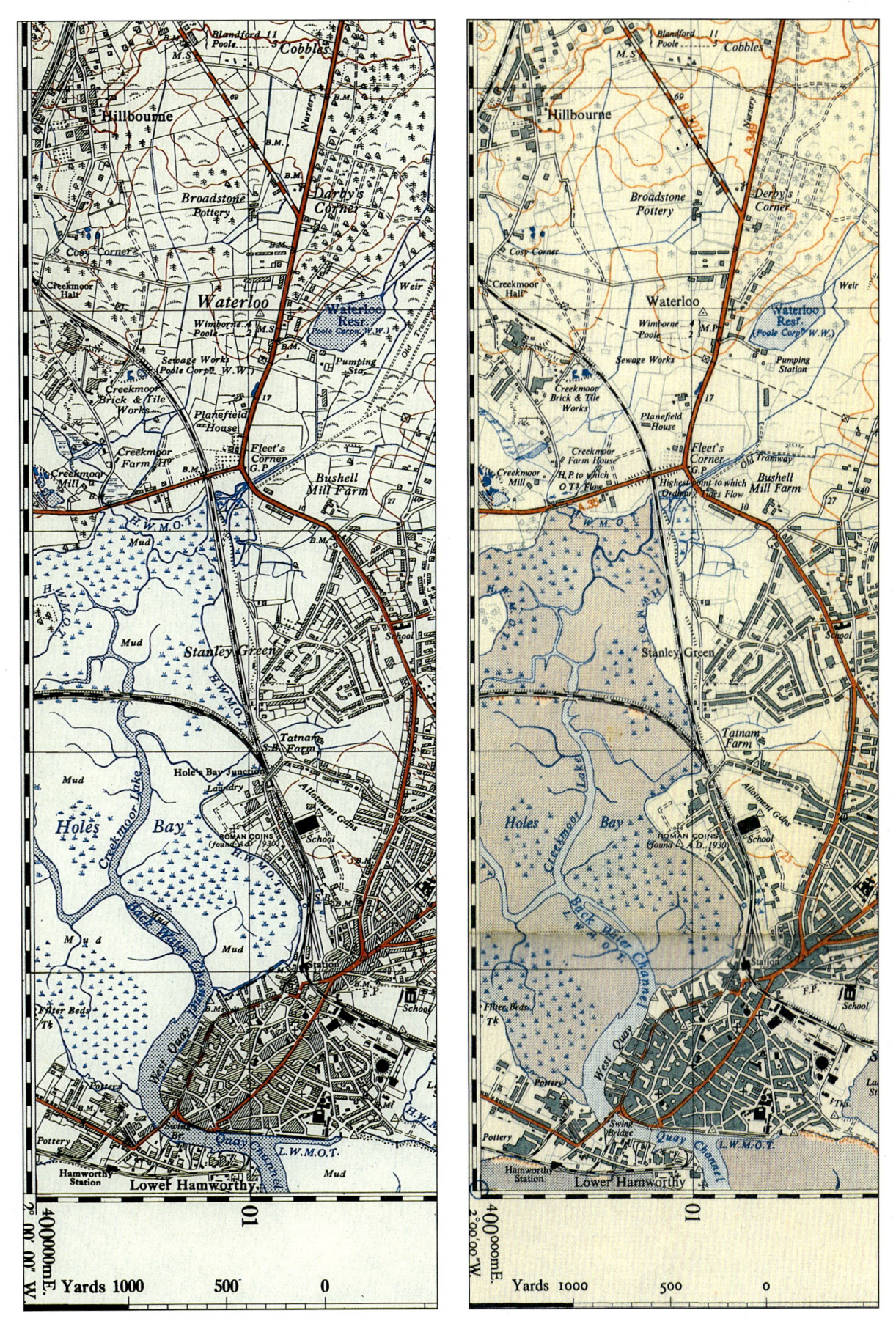

Plate 3. Left: sheet 40/09, 1944, an unpublished issue printed in three colours, with lettering mostly in Old Roman style no.5. Right: sheet 40/09, 1945, the first published sheet, printed in four colours, and lettered in Times Roman. This was the "definitive" style.

Plate 4. Regular Edition sheet SX 45 & part of SX 44, edition A, 1956, printed in five colours. The omission of detail in the naval dockyard at Devonport is apparent, perhaps less so are other security deletions [cp plate 5]. The function of HMS Raleigh is described, and the distances on milestones are given.

Plate 5. Regular Edition sheet SX 45 & part of SX 44, edition B/*/, 1967 [cp plate 4]. The naval dockyard at Devonport and detail elsewhere are now shown, but the description of HMS Raleigh and milestone distances have been deleted.

Plate 6. GSGS 4627A sheet 41/33, 1951, one of seven sheets embellished with layer colouring for use in the Staff College.

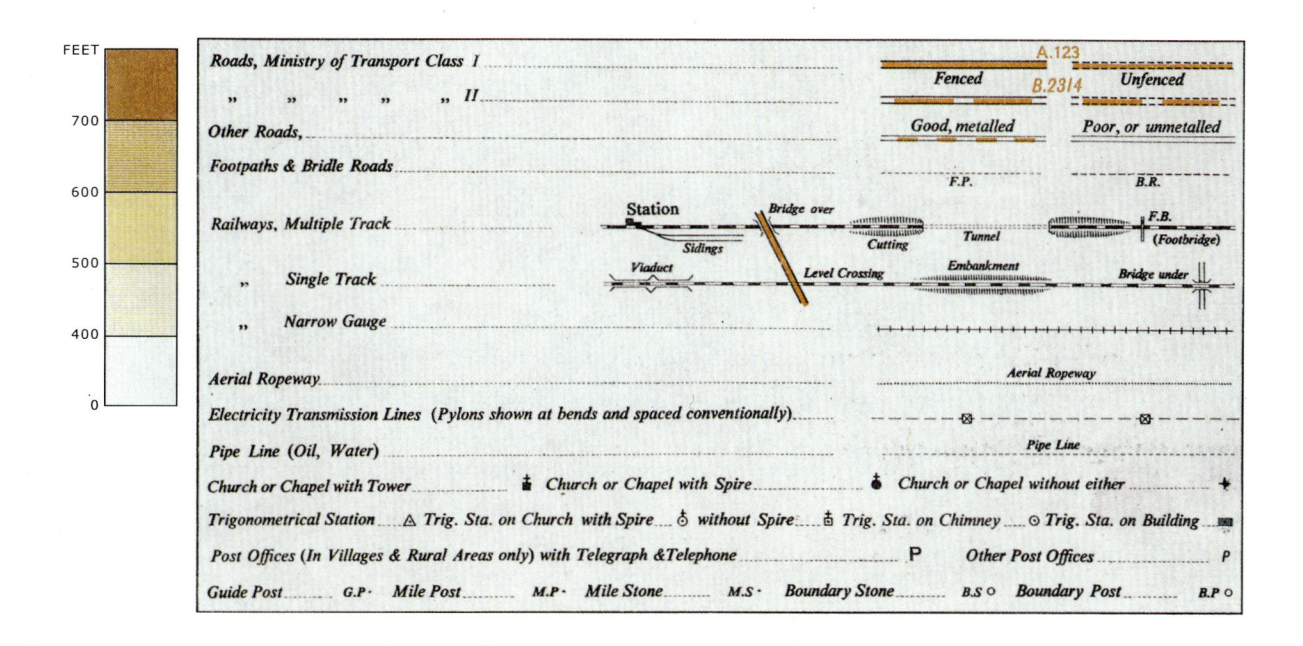

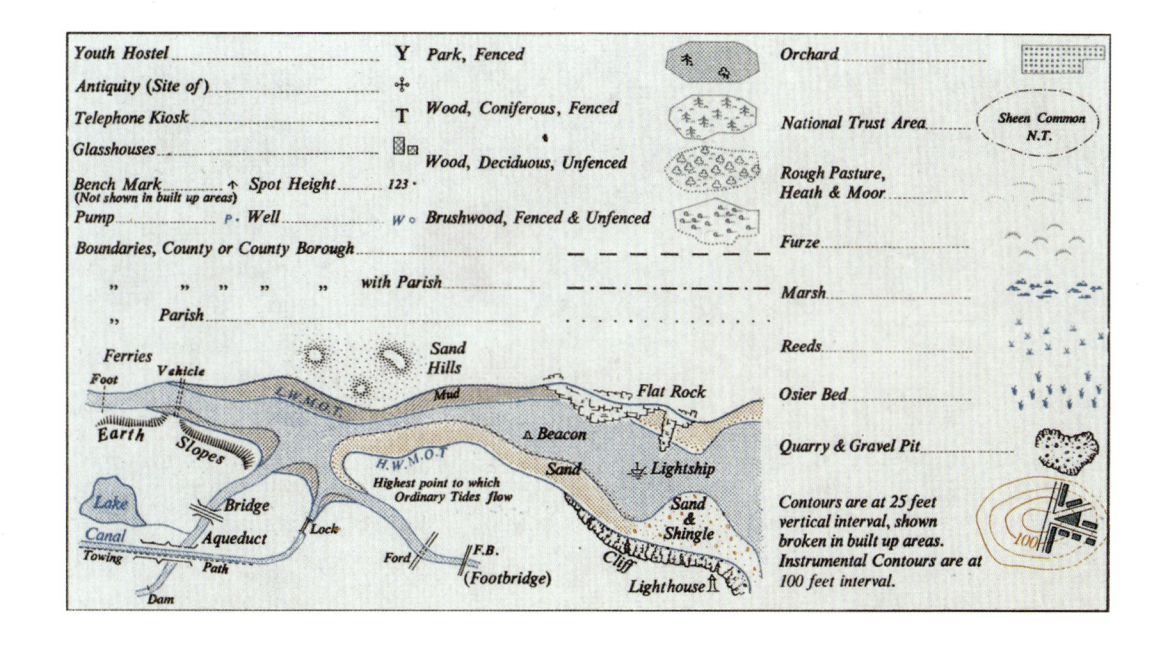

Plate 7. The layer and legend boxes present on GSGS 4627A sheet 41/55, 1949.

Plate 8. Sheet NY 30, edition C, 1959. This is in the so-called "original" three-colour style. Lettering is mostly in Old Roman style no.5, but replacement lettering (e.g. the milestone distances) is in Times Roman.